Encyclopedia of World Geography

MAN AND HIS WORLD TODAY

Foreword by
Professor Emrys Jones
London School of Economics

Encyclopedia of
World
Geography
MAN AND HIS WORLD TODAY
OCTOPUS
Octopus Books

First published 1974 by
Octopus Books Limited
59 Grosvenor Street, London W1

Planned and produced by
Elsevier International Projects Ltd, Oxford

ISBN 0 7064 0369 X

Printed in England by
Jarrold & Sons Ltd, Norwich

FOREWORD

The world, which was once thought to be too large for man's comprehension, can now be seen in a single photograph taken from space. Its form and its features are within the grasp of everyone. Not only can we see it as a whole, but modern technology enables us to experience any corner of it: Timbuktu is a centre for a package holiday, the stepped pyramids of Chichen Itza are within a day of London and Paris. Yet the seeming and diminishing size of the earth has done nothing to lessen its complexity and fascinations. Lands and climates, resources of every kind, and men and the way in which they use their habitat are all as varied and challenging as ever. The need to systematize and describe this variety is also as great as ever. At best, no one can personally experience more than a fraction of these bewildering riches. This encyclopedia brings the world to us in a new and exciting way.

First, it is an excellent reference book, thoroughly up to date and authoritative, with contributions by leading scholars with an intimate knowledge of each country, who are aware of political, social and economic problems. Every country is described individually, as are its people and economy. When every news bulletin makes us aware of yet another corner of the globe, both specialists and laymen need a source of quick reference whether for practical purposes or just for interest; famine in the developing world, a major discovery of fuel resources, an international incident in another continent – all need to be pinpointed and seen in their full context. For such purposes this book is ideal. Its succinct information is immediately intelligible to all readers.

Secondly, this book is delightful for the sake of its many superb illustrations. Colour photography brings each country to life, and makes the reader aware, as never before, of the setting in which men live and work in every part of the world.

Lastly, the book has a lengthy introduction which deals with fundamental aspects of the entire world. Maps, diagrams and photographs help the reader to understand why the continents are shaped as they are, why there are mountains and lowlands, why there are varied climates, what makes weather. Here we have a chance of seeing the world as a whole, the distribution of people, and of different races and religions, sources of energy and raw material, standards of living. By referring back to this introductory section the information for any one country can instantly be related to every other country. The introduction and the separate items for the countries supplement each other, and together give the reader a new insight both into the world as a home for man, and into man as a manipulator of his environment.

This book is a rare combination on which the publishers are to be congratulated, and I hope it will give pleasure to everyone who will use it to deepen his understanding of the modern world.

CONTENTS

9 ASPECTS OF GEOGRAPHY

41 EUROPE

135 ASIA

227 AFRICA

ASPECTS OF
GEOGRAPHY

THE DYNAMIC EARTH

6378.155 km 3963.205 mi

A

B

Man is born, lives, and dies on the surface of a small planet he calls Earth which orbits a minor star (the sun) some 30,000 light years from the centre of the galaxy he knows as the Milky Way. (A light year is nearly 6 billion miles.) The earth is only the fifth largest of the sun's satellites – Jupiter, Saturn, Uranus and Neptune are all many times larger – and it takes just one year to complete its elliptical journey round the parent star. The average radius of its orbit is 93 million miles (150 million km) and its orbital speed is a little under 70,000 mph (110,000 km/hr). The earth has one natural satellite of its own, the moon, which in turn orbits it every 28 days.

The earth is just the right distance from the sun for its surface temperature to have permitted the development of life. The next innermost planet, Venus, probably has far too hot a surface to support life and Mars, the earth's nearest neighbour farther out, has too variable and thin an atmosphere for life as we know it to have developed.

Three components in the earth's structure give rise to its ability to support plant, animal and human life – its solid surface, the waters which cover 71% of that surface, and its gaseous atmosphere, detectable perhaps as high as 5,000 mi (8,000 km) above the surface. Life exists in the comparatively narrow shell where these features come together and exchange materials and energy. Only at this interface can the atmosphere protect life from the cosmic radiation filling the solar system and the earth's crust shield it from the searing heat below.

Most of our present knowledge about the earth under our feet relates only to its crust – the outermost 20 mi (32 km) of the planet. Our understanding of the interior has been gathered only indirectly by the study of the shockwaves which travel across the globe from the centres of earthquakes through the various layers of its substance.

(A) The earth probably contains a solid core (1) having a radius of 800 mi (1,300 km). This is five times as dense as the earth's crust. Beyond this comes an outer liquid core (2), extending to a radius of 2,000 mi (3,200 km). The earth's solid mantle (3) is composed of iron and magnesium silicates. (B) An enlarged section of the outermost layers of the earth (4) to a depth of 187 mi (300 km). The upper part of the mantle (3) is fairly plastic and probably contains olivene, iron magnesium silicate. Above this occurs the earth's crust which is of two kinds. Basaltic material up to 5 mi (8 km) thick forms the oceanic crust (5) while granitic material underlies the continental land masses (6). Recent sediments (7) result from the erosion of the continental layer.

The geological time scale. Each period since the Cambrian 600 million years ago has been characterized by the appearance of new groups of animals. (All figures in millions of years).

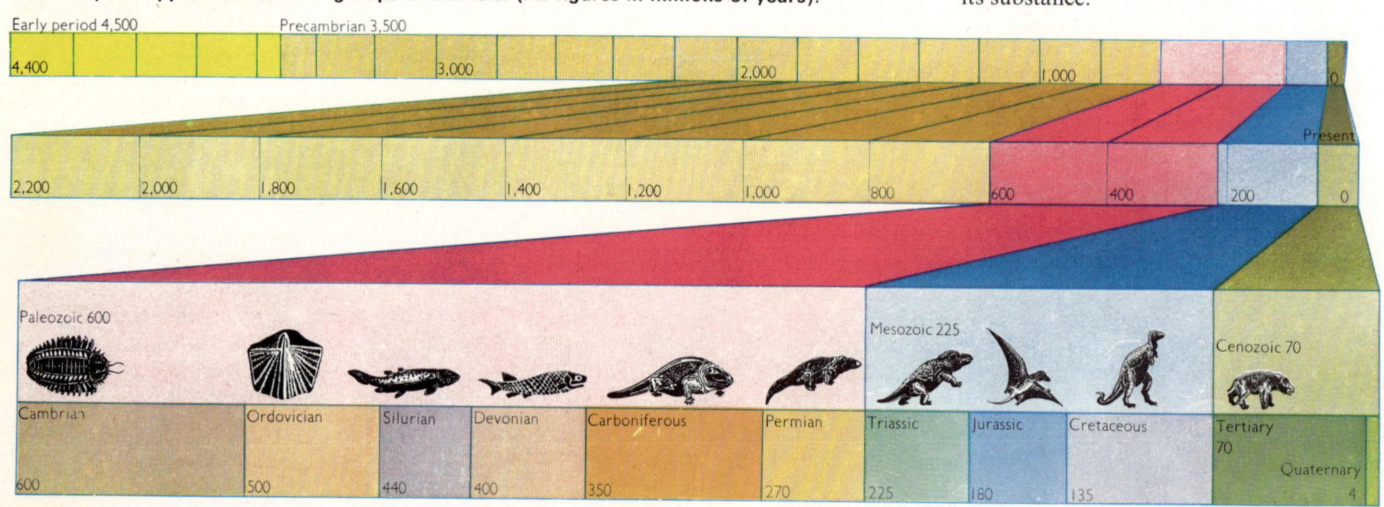

It is believed that at the heart of the planet is a solid core of very dense material at a temperature approaching 3,000°C. This is surrounded by a shell of liquid material, which gives way at a depth of about 1,900 mi (3,000 km) to the glassy substance of the mantle. On the surface of this floats the earth's crust, only a few miles thick. The white-hot lava ejected from volcanoes shows that parts even of the crust are often intensely hot.

The oldest portions of the earth's crust that we can identify today have been indicated by radiometric dating methods to be about 3,500 million years old though it is probable that the earth's crust first began to form some 4,550 million years ago.

The earliest known signs of life are fossilized bacteria discovered in rocks about 3,100 million years old. The rocks of succeeding ages contain fossils of increasingly complex organisms: primitive invertebrates appearing in the late Precambrian era (before 600 million years ago), fishes, land plants, and then reptiles in the Paleozoic (600 to 230 million years ago), with mammals and flowering plants first appearing in the Mesozoic era (between 225 and 70 million years ago).

Already the age of the giant dinosaurs (the Jurassic period, between 180 and 130 million years ago) had come and gone and the succeeding 70 million years (the Cenozoic era) would see the development of successive species of mammals and flowering and fruiting plants. Finally, the record of the rocks shows that the animal and plant species of the world began to resemble their present-day forms about 10 million years ago, with a particular line of development reaching out towards man. Although it is not possible to draw any sharp dividing line between modern man and the earlier hominids, the species *homo sapiens* is usually regarded as dating from some 200,000 years before the present.

In shape, the earth is not a true sphere but an 'oblate spheroid' – a sphere flattened somewhat at the poles. Observations on the precise orbits of artificial satellites have revealed the difference between the polar and the equatorial diameters of the earth to be 1 part in 298·26 – a mere 26·58 mi (42·77 km), taking the average equatorial diameter to be 7,926·41 mi (12,756·31 km).

The surface features of the earth only slightly modify its overall shape. The 12 mi (20 km) from the peak of Mount Everest (29,028 ft (8,848 m)) to the floor of the Marianas Trench in the Pacific (at 36,204 ft (11,055 m), the deepest ocean sounding recorded) would be no more apparent to an observer on the moon than the wrinkles on an orange.

Powerful forces have combined to create the face that the earth displays to man today. Huge circulatory currents carry large sections of the earth's crust in their slow wanderings across the surface of the planet. Similar though swifter-flowing currents agitate the waters of the oceans which fill the troughs between the continents. These in their turn are stirred by the winds, the fleetest currents of all, as they share out across the globe the solar energy falling on its surface. The gravitational pulls of sun and moon raise tides which daily scour the sea-shores while water fallen from the clouds in rain carves out deep valleys as in rivers it seeks again the sea.

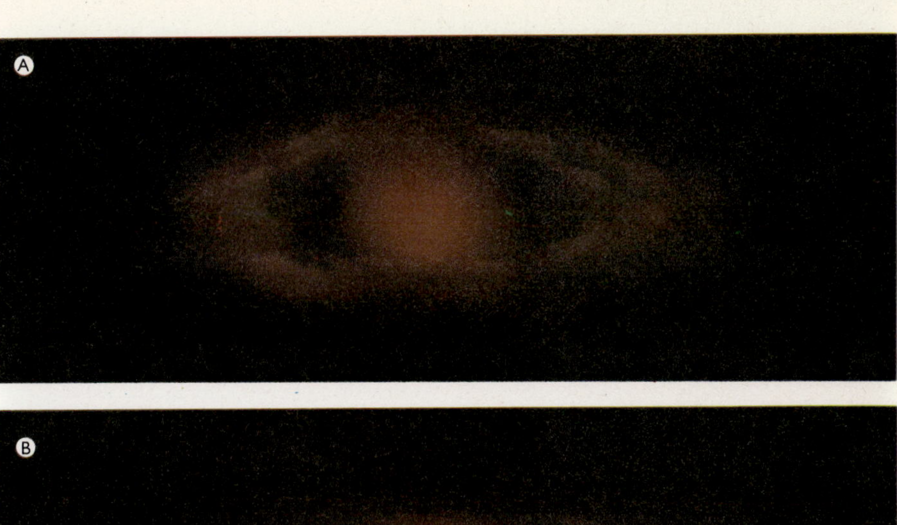

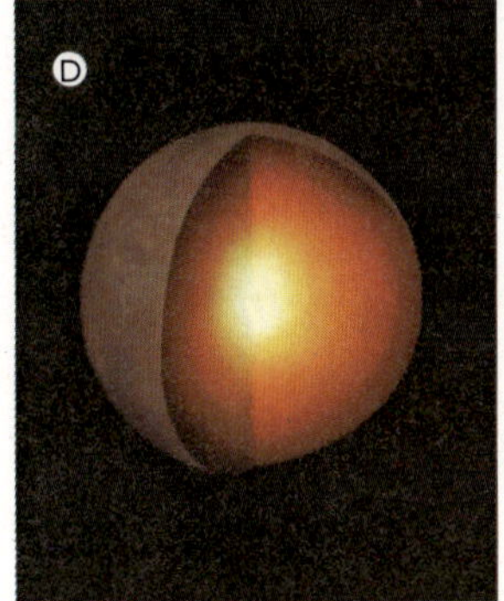

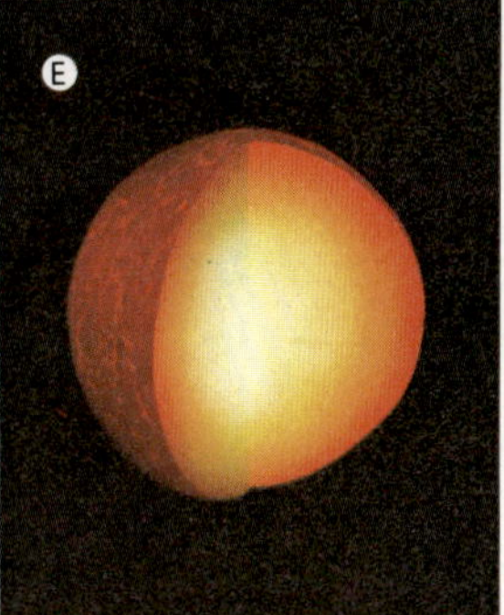

A probable history of the earth's first 1,500 million years. (A) A cloud of gas and dust spins around the infant sun. (B) Gradually, denser particles move towards the sun, while others form a flattened disc. As the sun shrinks, it begins to produce light and heat and the embryo planets lose their more volatile materials (C). (D) A more detailed section through the proto-earth reveals that the materials compressed near its centre have melted to produce a molten core. As the earth cools (E), its lighter elements float to the surface to form a crust. With more cooling (F), three layers separate out: a compact core; a semi-plastic mantle, and a crust.

In many respects the earth is still a mysterious planet, with probably many secrets yet to be discovered to human science. A uniform sphere could never have become the abode of man and the evolution of all forms of vegetable and animal life including the human species results in part from the very variety of the processes which go on near the surface of the dynamic earth.

WANDERING CONTINENTS

Men have speculated concerning the similarity between the shapes of the west coast of Africa and the eastern margin of South America since the 17th century. The hypothesis that the continents are continually wandering over the surface of the earth, although supported by much circumstantial evidence and particularly attractive to biologists seeking to trace the pathways of evolution, nevertheless failed to convince the majority of geoscientists before the discovery of a mass of new evidence in the 1960s. Alfred Wegner's famous theory of 'continental drift' (1915) always suffered from its inability to explain how the continental land masses could travel through the seemingly solid material which formed the earth's crust at the bottom of the oceans.

The solution to these difficulties came with W. J. Morgan's proposal in 1967 of the theory of 'plate tectonics'. According to this model, the surface of the earth comprises a mosaic of distinct units called plates which, although themselves rigid, can move relative to each other. The continents, including the 'continental shelf' – their submerged margins down to about 6,000 ft (2,000 m) – ride on the surface, of these plates, being composed of lighter materials than the rest of their substance. The continents cannot move relative to the plates in which they are situated and can only move with them. The motion of the plates has two components – one simply the relative motion of adjoining plates but the other involving the continual creation and destruction of the earth's crust at the active edges of the plates.

Running down the middle of most oceans is a double ridge where there is much volcanic activity and a high local heat flow. Semi-molten basaltic material rises to the surface at these ridges and solidifies, forming new oceanic crust which is added to the plates meeting at the ridge. These plates are pushed slowly apart as they grow, allowing more plastic basalt to rise to the ocean floor.

Perhaps the most striking evidence in favour of the theory of plate tectonics came from the measurement of the magnetic 'fingerprints' of the basalt rocks of the ocean bed. Most igneous rocks are magnetized in the direction of the earth's magnetic field prevailing at the time when they solidified from the plastic state. In the early 1960s it was discovered that the basalt of the Pacific floor was magnetized in alternating stripes of north-seeking and south-seeking polarity, arranged symmetrically on either side of the East Pacific Rise. It was known that the earth's magnetic field reverses its polarity roughly twice every million years and so the successive stripes were interpreted as crust formed at the ridge during the relevant periods. The symmetry of the pattern was seen to be strong evidence for the hypothesis of ocean-floor spreading.

Just as the oceanic plates are growing on

The world map *(top)* shows the active boundaries of the earth's crustal plates. At boundaries where semi-liquid material from the upper mantle rises to form new oceanic floor, adding to the plates and slowly pushing apart any continents carried on those plates, there is a mid-ocean ridge. Where a collision between two plates is resulting in the submersion of one beneath the other, there is a deep sea trench or mountain range. The section AB *(below)* shows how material rising at the mid-Atlantic ridge is causing the separation of the African and South American continents as they ride on their expanding plates. Where the South American plate collides with the Nazca (East Pacific) plate, the South American continental crust rides up on the oceanic crust of the Nazca plate, forming the Andes mountains and the Peru-Chile trench. As the Nazca plate plunges deep into the asthenosphere, its material melts and rises to fuel the volcanoes enlarging the continental crust in the Andes. Earthquakes occur on the line of friction where the descending oceanic plate rubs against the South American plate. The epicentres of these earthquakes define a Benioff zone plunging deep under the Andes. (1) oceanic trench, (2) Andes, (3) Atlantic ridge, (4) 100 km, (5) earthquake foci, (6) Benioff zone.

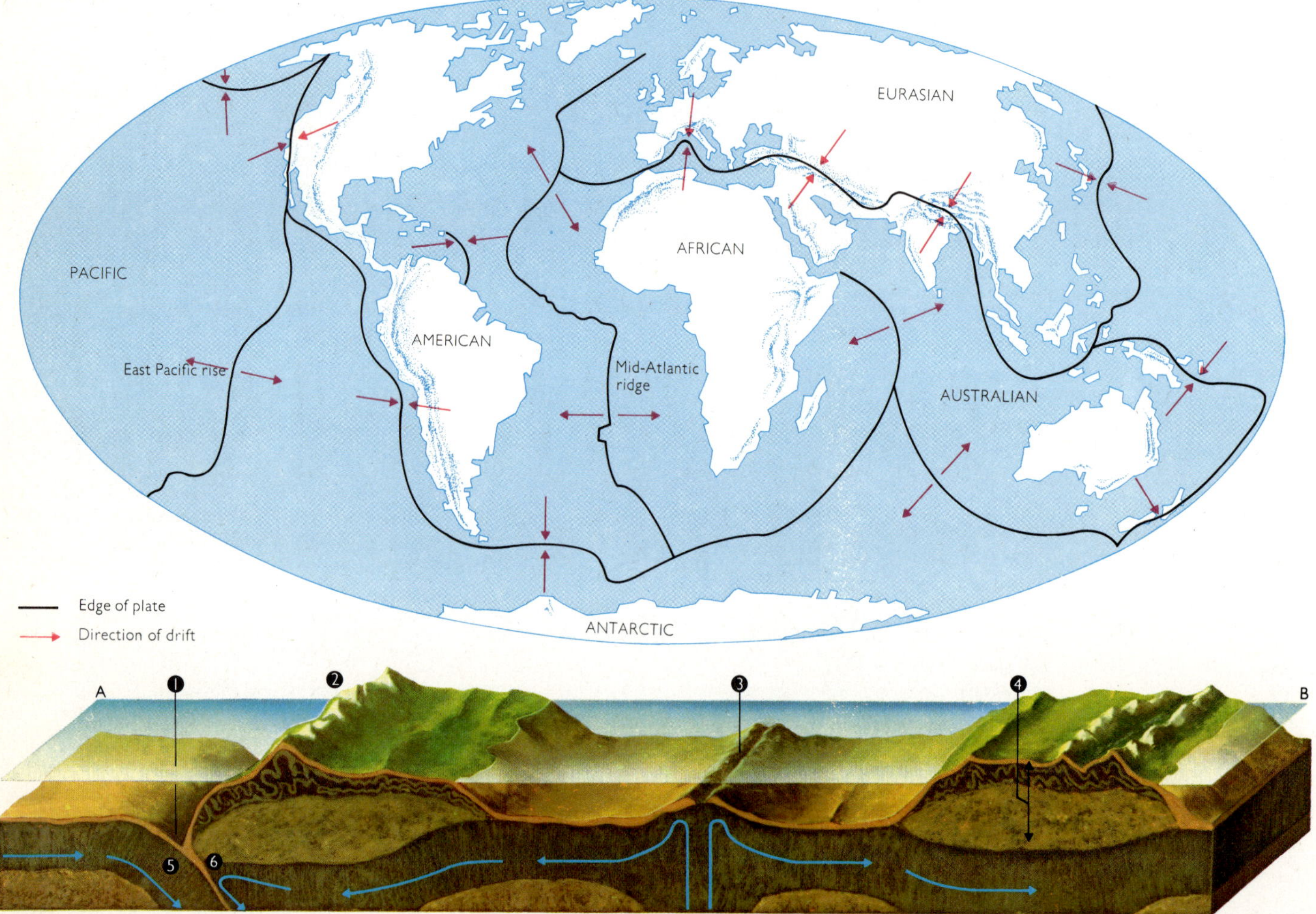

their mid-ocean margins, in other areas adjoining plates are converging with the resulting destruction of crustal material. Where two plates of oceanic material are converging, there is usually a deep ocean trench as one plate plunges beneath the other and remelts in the upper mantle. The descending portions of crust break up as they melt releasing elastic energy in the form of earthquake shockwaves. By tracing the centres from which these shockwaves originate, shallow near the surface trench, but successively deeper as they occur father back under the over-riding plate, it is possible to follow the melting crust on its journey down, often to depths of 450 mi (700 km). Material melted from the descending plate rises through the upper plate causing the arc of volcanoes which fringes the majority of mid-ocean trenches.

The rocks which form the continental masses are lighter than the basaltic material of the oceanic crust and thus ride on the surface of the basalt plates like parcels on a conveyor belt. When at length a continent is carried to the edge of a boundary where two plates are converging, it rides up on the descending plate forming high mountain ranges such as the South American Andes. Like the oceanic island arcs, such mountains are areas of great earthquake and volcanic activity.

By winding back the conveyor belts carrying the continents across the globe, it is possible to arrive at reconstructions of the face of the earth at successively earlier periods. This process must come to an end on reaching the world map of 200 million years ago because all the present continents are then found to be amalgamated in one large landmass known as Pangaea. But there are good reasons to suppose that plate-tectonic processes similar to those in action today operated for at least 400 million years before the formation of Pangaea and perhaps different plate processes began 1,400 million years before then. This takes us back to little short of the time when the shield regions at the heart of the major continents were formed. The Ural mountains were probably the result of the collision of two continents long before Pangaea came together, as was the Caledonian-Appalachian chain, now divided by the spreading of the North Atlantic ocean.

Some geologists have pointed to the implications of plate tectonics in the search for the earth's mineral resources. Conditions favourable to the trapping of oil and natural gas are expected to be found where there are beds of evaporites in continental-shelf regions close to the edges of expanding oceans, and much exploration is consequently planned in such regions. Similarly, metal ores are likely to be intruded into continental mountain blocks near zones of plate convergence, where molten minerals have risen from the disintegration of descending oceanic crust.

While the mechanism of the processes in the earth's mantle which power plate tectonics are still a matter for speculation among scientists, the pattern of developments on the surface of the earth can today be displayed in confident detail, a triumph for the defenders of a once-ridiculed idea.

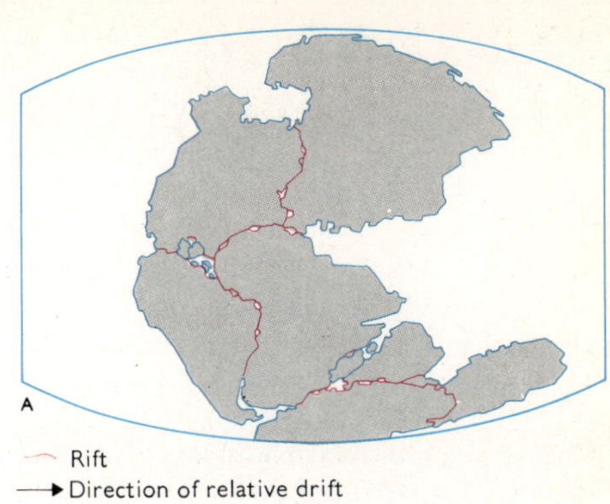

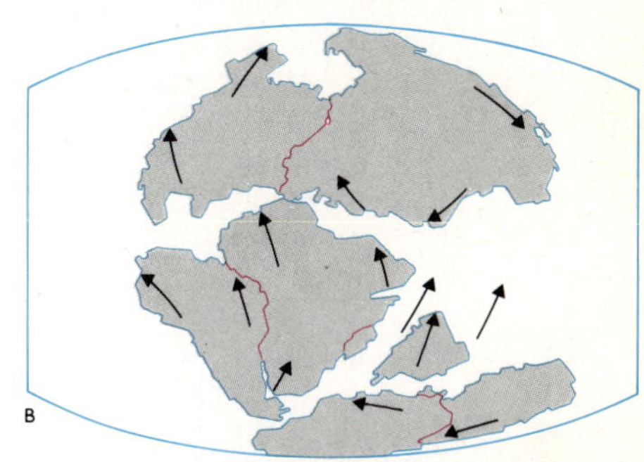

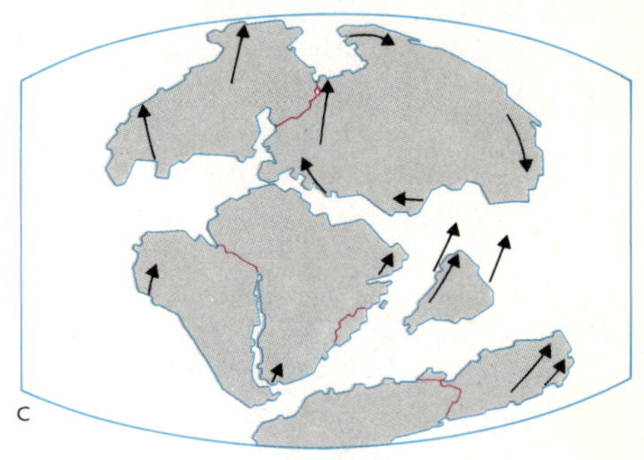

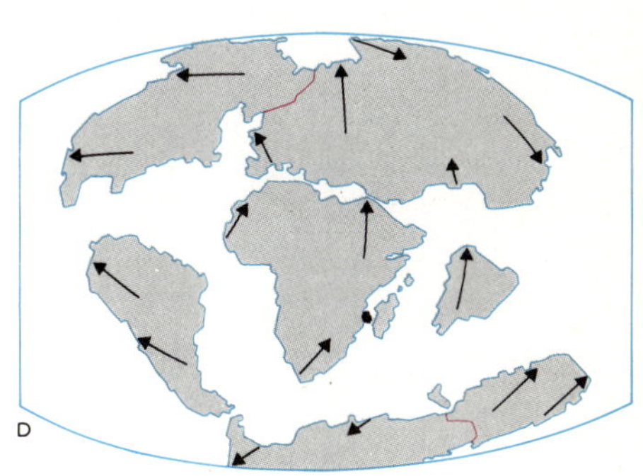

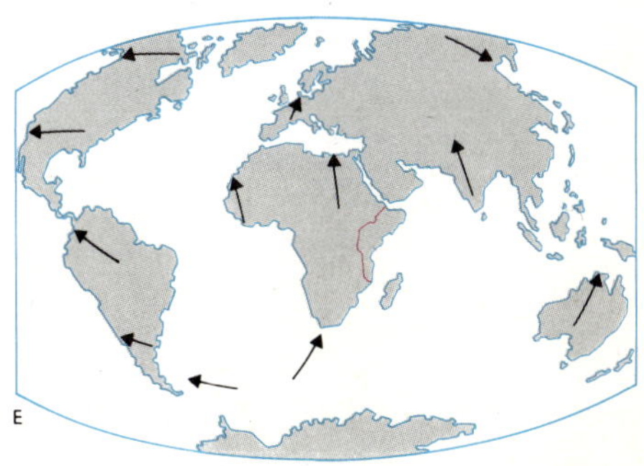

The past positions of the continents *(right)*. About 200 million years ago (A), earlier movement of crustal plates had assembled most of the present continental land masses into one large continent called Pangaea. Already a rift was appearing between Africa and Antarctica. By 180 million years ago (B), India had set off on its journey towards Asia. North America had begun to separate from the eastern section of Laurasia (the northern fragment of Pangaea) by about 130 million years before the present (C), followed by the opening up of the south Atlantic apparent 60 million years ago (D). More recently India has collided with Asia, piling up a particularly high system of mountains – the Himalayas – in the process, because of the reluctance of any section of continental crust to be returned down into the denser mantle. Today (E), the disposition of the continents is still changing with a new ocean opening up between Africa and Arabia (see illustration on page 9) and the Atlantic Ocean continuing to widen. Striking evidence of continuing activity along the mid-Atlantic ridge was provided by the appearance in November 1963 of a new volcanic island south of Iceland, soon given the name Surtsey *(below)*.

THE ROCKS OF THE EARTH

To the geologist all the materials which make up the earth's crust are rocks, whether they are hard crystalline granite boulders, combustible coal, soft clays, or loose fragments of sand or gravel. Most rocks have no definite chemical composition but primarily contain silicates – silicon and oxygen combined with a metal – though others, including chalk and limestone, are largely composed of one substance, usually calcite ($CaCO_3$). Rocks are usually classified as *igneous, sedimentary* or *metamorphic* according to how they were formed.

95% of the earth's crust is made up of igneous rocks. These have their origin in the molten mantle of the earth and the class is further subdivided according to the conditions under which they solidified from the magma, the rock-melt found at great depths. Plutonic rocks originally crystallized far below the surface but have often been exposed by subsequent erosion of the overlying material. Volcanic rocks on the other hand, have been formed on the surface from magma emitted through volcanoes and other fissures.

The texture of igneous rocks depends on the time they have taken to cool from the liquid state. The longer the period of cooling, the larger the crystals contained in the rock. Thus granite, a plutonic rock solidifying far below the surface, contains large crystals of feldspar, quartz and mica. Gabbro and peridotite, although formed in different circumstances, are similar in texture. Rocks such as basalt mainly contain small crystals because they have cooled relatively quickly. The volcanic rock, rhyolite, contains some large crystals embedded in a 'groundmass' of microscopic ones. The larger crystals were probably precipitated from the molten magma during the long slow cooling process before the material was expelled from the earth in a lava flow, while the subsequent rapid cooling deposited the mass of tiny crystals in the spaces in between.

Basaltic rocks are formed when magma rises to the surface at mid-ocean ridges and thus make up the major part of the oceanic crustal plates. Both the plutonic and volcanic rocks intruded into the overlying material in areas of volcanic activity probably have their origin in the remelting of basaltic rock from an over-ridden portion of oceanic plate.

Sedimentary rocks, although comprising only 5% of the earth's crust, cover about 75% of its land surface. The materials of which they are composed originated in the igneous rocks making up some ancient landscape. Mechanical and chemical weathering of these rocks followed by transportation by wind or water of the material thus eroded, has culminated in its deposition, usually on the bed of some ancient sea. There, overlain by great thicknesses of succeeding deposits, the sediment particles are gradually compressed into new rock. Coarse particles deposited by swift-flowing rivers give rough-grained sandstones, but the finer silt composing the deltas of slow meandering rivers will form softer clays and muds.

Other sedimentary rocks are the remains of once-living organisms. Chalk and limestones

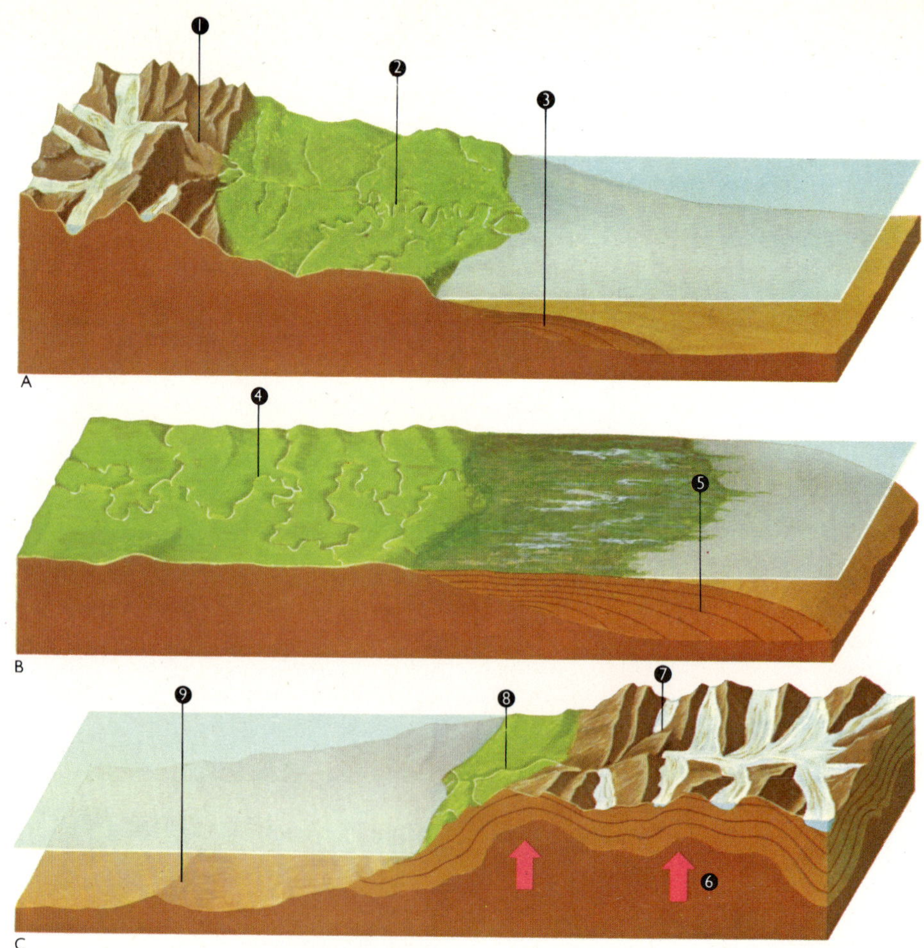

The geological cycle of erosion and deposition of rocks. In (A) erosion takes place at (1) and the resulting sediments, transported in rivers (2), are deposited in the sea (3). (B) After a long period of time, material has been transported from area (4) and deposited in area (5). The process is repeated (C) when uplift (6) creates high ground from the sediments deposited at stage (B). These are eroded (7), transported (8) and redeposited (9).

Folds in limestone. Under the conditions of high temperature and intense pressure encountered at even moderate depths, rocks which on the surface are hard and brittle become soft and pliable.

Thin sections of rock viewed under polarized light. (1) Sandstone, (2) Limestone containing fossils, (3) Gneiss, (4) Hornfels, (5) Peridotite, (6) Gabbro, (7) Granite, (8) Basalt, (9) Rhyolite.

originate in the calcerous shells of marine organisms and most types of coal are the compressed debris of plants which once occupied huge areas of swampy ground.

A third class of sedimentary rocks having considerable commercial importance are the 'evaporites', salts precipitated from the waters of confined or inland seas as their waters evaporate. In Texas evaporite deposits exceed 10,000 ft (3,000 m) in depth. Domes of evaporite materials often trap oil and natural gas in the porous rocks underlying them.

Many sedimentary rocks contain fossils – the chemically transformed remains of former living organisms. These once provided the principal means of dating rocks. Sedimentary rocks deposited in seas and lakes usually accumulate in horizontal layers called *strata*. Subsequent earth-movements can deform or fracture these strata so that today they can be found in almost any configuration.

Over large areas of the globe, the sequence of sedimentary rocks remains the same. While there may be gaps in the sequence and local variations in the thickness or composition of particular beds, the constancy of the 'succession of strata' often provides an important clue in the search for minerals and allows the identification of areas as having been formerly adjacent, which the process of plate drift has now separated by thousands of miles.

The metamorphic rocks result from the transformation *in situ* of both igneous and sedimentary rocks, caused by intense heat, great pressure or local changes in the chemical environment. Metamorphism is often associated with volcanic activity or the subsurface intrusion of molten rock or hot gases into pre-existing formations.

Marble results from the action of heat on limestone; slate from the compression of shales and mudstones, and hornfels from the intrusion of granite into shales. Where at depths of a few miles all these processes act together, large areas of quite general metamorphism containing gneiss may result.

The shape of the landscape is largely determined by the nature of the rocks underlying it. A line of hills often indicates the outcropping of a bed of rocks resistant to erosion; a waterfall may mark the point where a river flows from a hard to a soft bed. If the surface rock is soluble in rainwater, a river can disappear from the surface, only to reappear where its course takes it over impervious strata. Where rocks are of commercial value, man too contributes his quarries and spoil-heaps to the landscape, and whole cities grow up, dependent upon the wealth of the earth.

FORCES SHAPING THE LAND

The face of the land is the result of the operation of many interacting processes. Ultimately all depend on the cycle in which powerful plutonic forces raise rocks high above the surface of the sea, only for the processes of erosion to return their substance, in the form of fragmented sediments, to the floor of the ocean.

The initial breakdown of large beds of exposed rocks is called weathering, of which the principal agencies operate mechanically or chemically. Frost shattering is probably the most important mode of mechanical weathering. Water collects in fissures in the rock which are forcibly widened when the water freezes, eventually dislodging fragments from the rockface. Plant roots which grow into crevices in bedrock seeking underground water have a similarly destructive effect.

The end point in the breakdown of the rocks underlying a terrain is the formation of its soil. Soils contain organic materials – roots, animals and micro-organisms – as well as rock fragments and their structural properties and chemical composition are important in determining the type of vegetation-cover characteristic of the landscape.

Rivers are the most significant means of transporting rock debris from upland areas to the sea. In their youthful stages far upstream, their steep gradients give them considerable carrying power and quite large rock fragments can be transported. They cut deep gorges through bands of resistant rock, often following lines of natural weakness.

When a river leaves the mountains, it deposits most of its coarser sediments but retains the finer silt. It meanders sinuously across the plain before issuing into the sea by way of a broad estuary or, if it has much sediment to deposit, a fan-shaped delta.

The sea is rarely at peace with its shoreline. In one area it encroaches on the land, eating away at a line of cliffs; in another it deposits long bars of sand and shingle, enclosing a

Frost shattering on an exposed rock face. When water in rock crevices freezes, it expands, widening the crack and loosening fragments of rock. These eventually fall to the foot of the slope where they collect as scree.

A coastline showing both erosional and depositional features. Material eroded from cliffs during storms is often carried for several miles by longshore drift before being deposited on beaches.

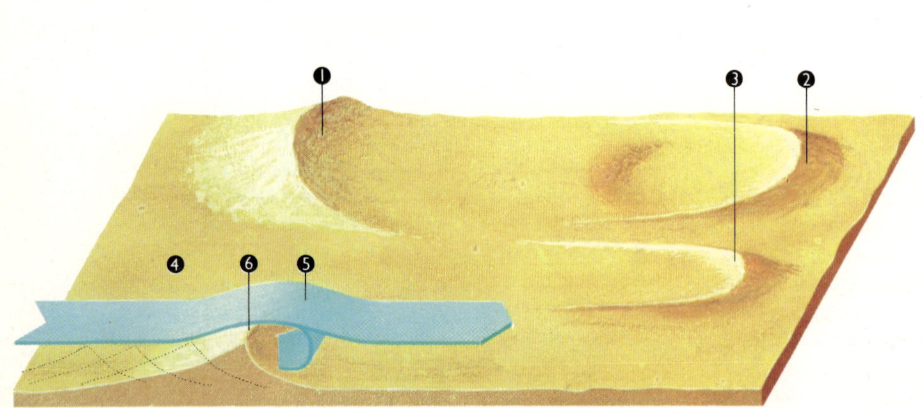

Wind is an important feature in shaping the landscape behind depositional coastlines and in sandy and rocky deserts. Sand dunes can be crescentic or barchan (1), parabolic (2), or hairpin (3). A crescentic dune (4) over which the wind usually blows in one particular direction (5) advances as its ridge (6) migrates forwards.

The river *(above)* has just emerged from a glacier as a melt-water stream. The highest portion of a stream's course, its youthful section, is characterized by swift currents and a steep overall gradient. Downstream *(below)* it enters the middle part of its course, its maturity, where it flows through broad valleys down gentler gradients. In its old age, it meanders sluggishly across a wide plain and empties into the sea through a large delta.

lagoon. Vegetation and wind-blown sand will gradually reclaim this – first as marshland, and later as pasture. Material plucked from retreating cliffs is often carried far along the coast by a process known as long-shore drift before it is deposited on some distant beach.

Wind and tide conspire to produce the sea's most destructive effects, as when tidewater, swept into the estuary of a river, effectively dams the stream and causes it to overflow its banks, spreading thick layers of silt over the floor of its flood-plain. Tropical cyclones too can sweep vast quantities of salt water over low-lying coastal areas.

Man the cultivator represents a powerful agency in transforming the face of the land. Imprudent farming practices have often speeded up natural erosion processes, despoiling the earth's precious resource of land.

The face of much of the landscape in Europe, North America and in high mountainous regions throughout the world is the result of the spread of large ice-sheets over these areas in the late Pleistocene (between 1·8 million and 11,000 years ago). Their upland areas accumulated vast depths of snow which, under the pressure of its own weight, was gradually converted into glacier-ice. A mixture of this ice and rock debris flowed down pre-existing valleys towards lower ground in glaciers which then flowed out onto the great sheets of ice and debris that spread across the low-lying land and frozen seas.

Remnants of the continental ice-caps still cover the land-surfaces of Greenland and Antarctica, but the ice has now retreated from the lower-latitude regions, leaving valley glaciers only in such areas as Scandinavia, Iceland, Alaska, the Alps, the Himalayas and the Andes.

The ice has left its mark on any upland areas which were once glaciated. River valleys formerly having a gentle V in cross-section have been deepened and their sides steepened to give the characteristic U-profile of a glaciated valley. Lakes fill hollows gouged out of the mountain shoulders (*cwms*, corries, or *cirques*). Knife-edged precipitous ridges (*arêtes*) run between sharp-pointed pyramidal peaks (called *horns* in the Alps). The valley-floors form the beds of ribbon lakes held back behind ridges of debris (moraines) deposited when the front of the retreating glacier remained stationary for a while. Pre-glacial rivers have been diverted through gorges cut by streams overspilling from lakes which were temporarily penned up by glaciers and ice-sheets. In a thousand ways the mountains show the marks of the ice.

The lowland areas overrun by the vast slow-moving ice sheets were modified by the ice no less than the highlands. As the ice retreated it gave up all the rock-debris it had carried with it from the ice-fields and deposited a thick layer of boulder clay across the landscape. Many more local glacial features are scattered across the boulder-clay plain –

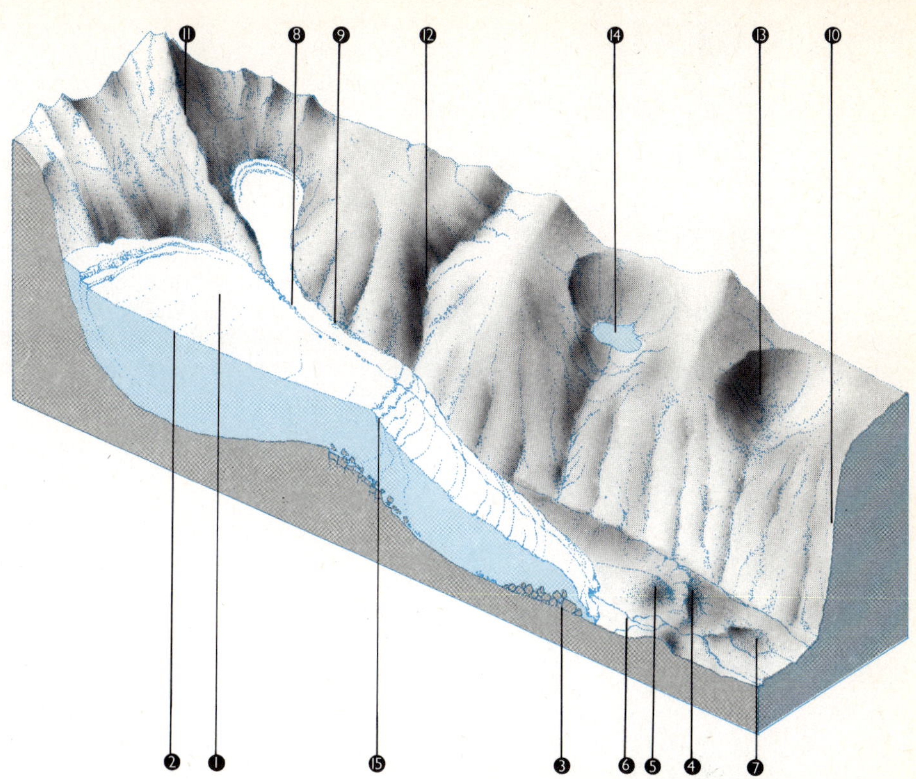

Features associated with glaciers and glaciated landforms. (1) head of glacier, (2) firn or névé, (3) region of ground moraine deposition, (4) terminal moraine, (5) drumlin, (6) braided stream, (7) kettle, (8) medial moraine, (9) lateral moraine, (10) U-shaped valley, (11) arête, (12) hanging valley, (13) cirque, (14) tarn and (15) ice fall.

moraines, where ice-sheets temporarily rested in their retreat, drumlin swarms giving the landscape a typically 'basket of eggs' topography and eskers, low ribbons of raised sediment meandering across the rolling plain marking the courses of debris-laden meltwater streams. Shallow depressions, kettles, recessed into the landscape indicate the places where large blocks of stagnant ice were deposited, melting away after the rest of the icesheet had abandoned its clay. Erratic boulders, isolated rocks carried perhaps hundreds of miles from the mountains from which they were plucked by the ice, are deposited just where the retreating ice-sheet left them. Vast areas of Finland and Canada boast a landscape of irregularly-shaped lakes alternating with forested strips of land, demonstrating the unevenness of the terrain left after the great ice-sheets had departed.

Changes in sea level have probably been the most dramatic result of the retreat of the ice. Although the melting ice added to the volume of water in the world's oceans and thus flooded the lower valleys of many rivers, great tracts of land which formerly had been depressed under the weight of overlying ice fields began to rise towards the end of the glaciation. The Baltic Sea is still shrinking as the lands on its northern coasts recover from the weight of their former burden of ice.

(Above) A glacier descending from an ice-field and issuing into a proglacial lake. _(Below)_ A U-shaped valley, the bed of a former glacier in Glacier National Park, Montana, USA.

THE OCEANS

Water in the liquid state covers about 71% of the surface of the earth, 74% if that frozen in the polar ice-caps is included. These surface waters, together with the groundwaters trapped under the land-surface and the vapour and ice held in the atmosphere, constitute the hydrosphere.

Over 97% of the hydrosphere is contained in the world's oceans; 2·16% is ice; 0·63% is groundwater; 0·03% is in rivers, inland seas and the soil, and less than 0·0001% of the total is in the atmosphere; the oceans are clearly man's most well stocked water resource.

The bed of the ocean displays a most varied topography. The continents are fringed by the relatively shallow waters of the continental shelf. This gives way at a depth of about 1,000 ft (300 m) to the abyssal depths of the deep ocean, the floor of which lies on average more than 12,000 ft (3,700 m) below the water's surface. The theory of plate tectonics accounts for the formation of the mid-ocean ridges, deep sea trenches, island arcs and individual seamounts which are scattered over the ocean floor.

The oceans probably constitute the earth's greatest potential source of valuable minerals. Although each cubic mile of sea-water contains about 165 tons of dissolved solids (35,000 parts per million) – mostly sodium

and chloride ions – only bromine, magnesium and salt are currently recovered commercially.

Other minerals of great potential significance occur on the sea bed, though there are at present considerable difficulties in extracting these economically. Gravels and sands are commonly dredged in shallow waters, and coal, iron and other valuable metal ores are sometimes mined in undersea workings pushed out from the land behind the coast. The most exciting prospects come from the discovery that in the deep ocean basins the ocean floor is strewn with metal nodules containing high concentrations of manganese.

The oceans find their greatest importance for mankind today as a source of food and a highway for trade. Rich fishing grounds occur where waters of different salinities and temperatures mix at the confluence of the various surface and deep-water currents. Such conditions are found in the North Sea, the Newfoundland Banks, off the west coast of the Americas, in parts of the South Atlantic and Indian Ocean and over the whole Antarctic Ocean.

The increasing frequency of international disputes as to fishing rights points to the fact that even the immense food potential of the world's oceans is not unlimited unless rational exploitation policies are agreed on.

The ocean floor has as varied a topography as has the land surface. (1) East Pacific Rise, (2) South-east Pacific Plateau, (3) Mid-Atlantic Ridge, (4) Mid-Indian Ridge, (5) Australian-Antarctic Rise, (6) Aleutian Trench, (7) Puerto Rico Trench, (8) Tonga Trench, (9) Kermadec Trench. (10) Peru-Chile Trench (11) Kuril Trench and (12) Ryukyu Trench.

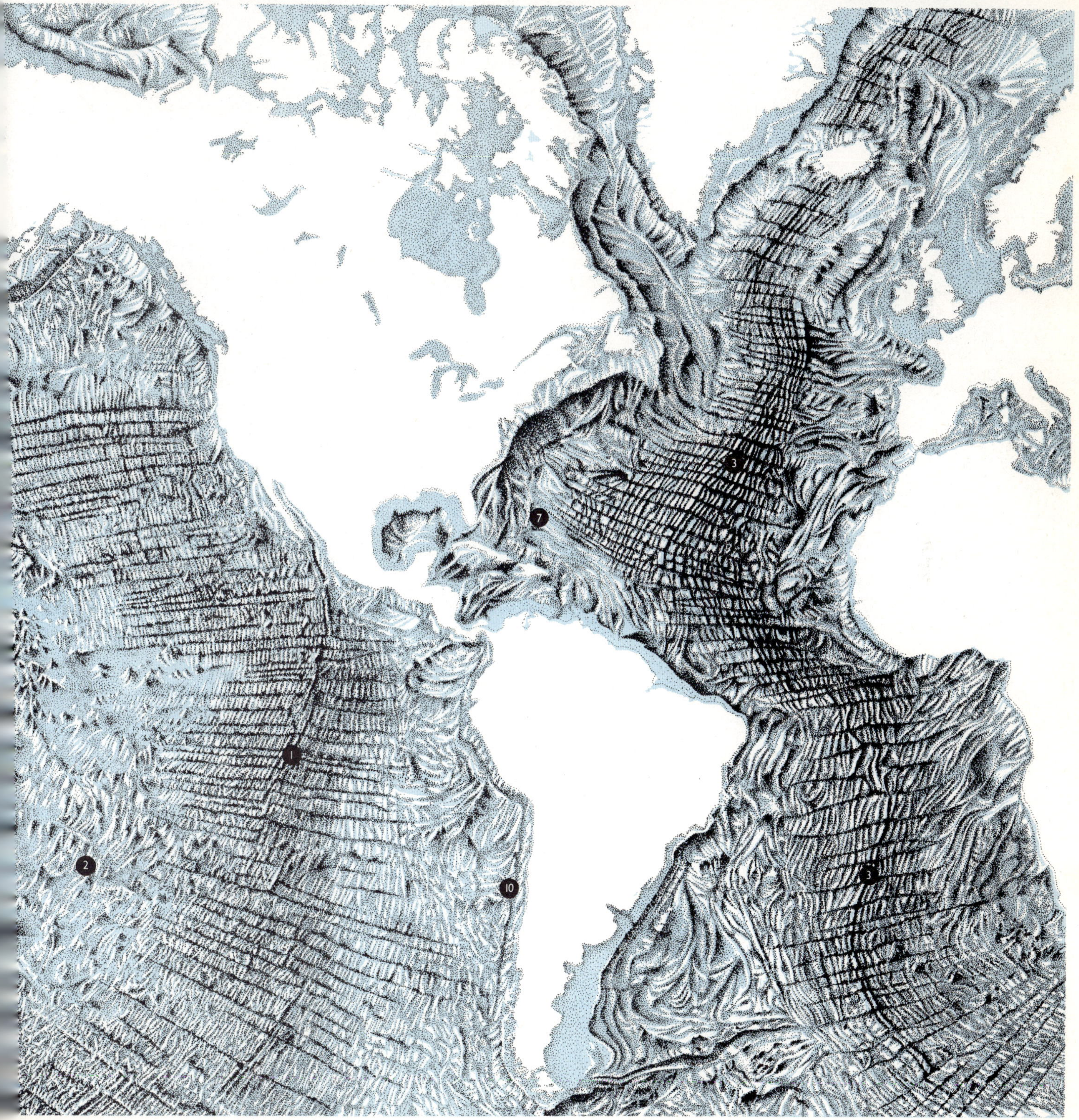

ENERGY AND WINDS

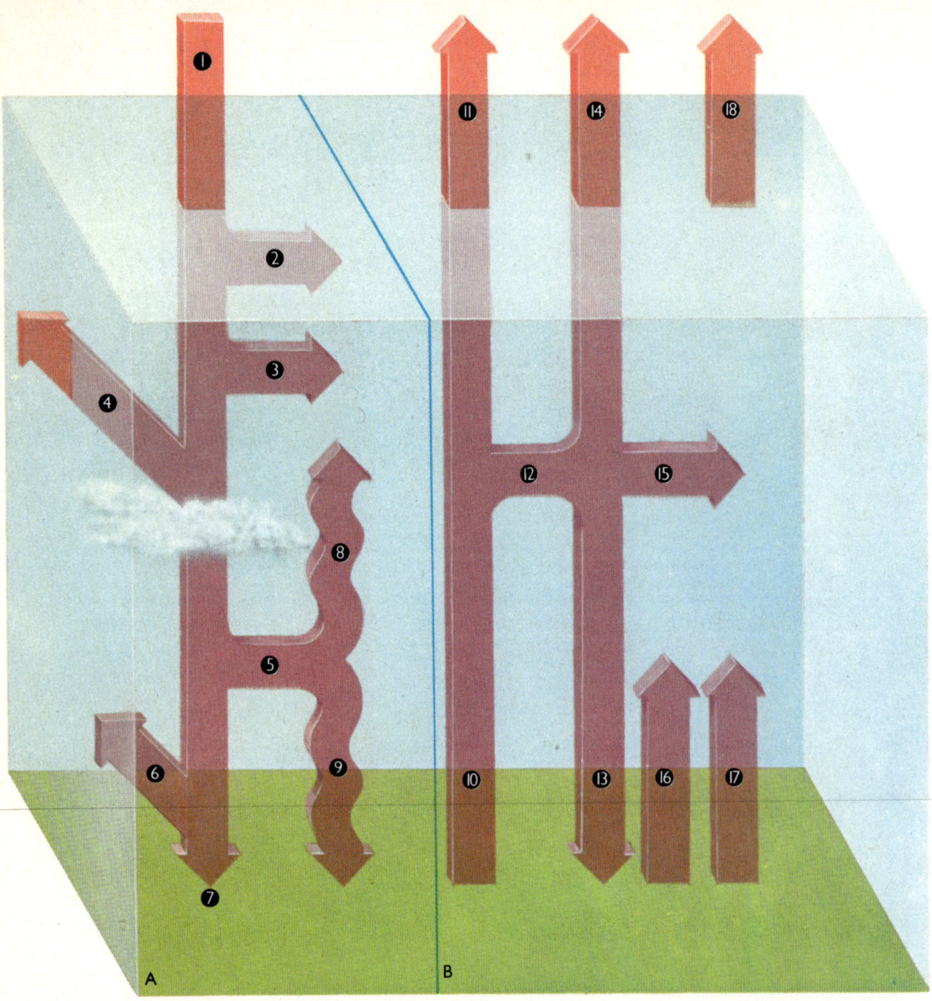

The earth's energy balance. Solar radiation reaching the earth is of shorter wavelength than that radiated from the much cooler earth. (A) Short-wave processes. Of every 100 units of solar energy arriving at the top of the earth's atmosphere (1), 2 are absorbed by the ozone in the upper atmosphere (2) and 15 by atmospheric water vapour, carbon dioxide and dust (3). 23 units are reflected back into space by clouds (4), 22 are scattered by atmospheric air and dust (5) and 38 reach the ground as direct solar radiation, 7 units of which are reflected back into space (6) and 31 are absorbed (7). Of the radiation scattered by the atmosphere, 6 units is lost into space (8) and 16 eventually reaches the ground as diffuse radiation and is absorbed (9). (B) Long-wave processes. Of 110 energy units radiated from the earth's surface in the far infra-red (10), 5 escape into space (11) and the remainder is absorbed by the atmosphere (12). At long wavelengths the atmosphere itself radiates 92 units back to the surface (13) and loses 57 into space (14), some of which energy has been supplied by the absorption of short-wave solar radiation (3) and the rest by non-radiative transfer – conduction and convection – from the surface: 5 units as sensible heat energy (16) and 24 units as latent heat in water vapour (17). The ozone in the upper atmosphere reradiates back into space at long wavelengths (18), the 2 units of energy it absorbs at short wavelengths (2).

The variation with latitude of the quantity of radiation arriving annually at the top of the earth's atmosphere (A) together with that for radiation absorbed and reflected at the earth's surface (B), by clouds (C) and by atmospheric air (D), all measured in kilocalories per cm².

Every day, energy equivalent to the output of 180 million electrical generating stations, each of 1,000 megawatt capacity, falls on the earth from the sun 93 million miles (150 million km) away at the centre of the solar system. What happens to this energy when it reaches the earth is displayed in the accompanying diagrams.

The diagram of the earth's energy balance shows that all the energy received by the earth is ultimately reradiated and lost into outer space. It can be seen that only a little over half the solar energy reaching the top of the earth's atmosphere reaches the surface, though somewhat more eventually finds its way into the atmosphere, mainly by reradiation from the surface at longer wavelengths.

The lower diagram shows that the majority of the earth's surface energy absorption occurs in a broad band on either side of the equator. At latitudes of less than 38°, the energy gain in short-wave radiation is greater than the loss at longer wavelengths – there is an energy surplus. A corresponding deficit is found if observations are made closer to the poles. Left to itself this would be an unstable situation with the tropics becoming hotter and hotter and the poles progressively colder. The redress of this energy imbalance is both the function of and the driving force behind the winds which redistribute hot and cold air over the surface of the globe.

The actual pattern of the earth's winds is both complex and constantly changing but a few broad principles can be picked out. Air-streams tend to flow in huge three-dimensional circulatory cells with the motion of warm low-level surface winds being counterbalanced by 'jet-stream' flows of colder air at higher levels (between 2 and 4 mi (3 to 6 km) up). Alternatively, the flow of warm high-level winds can balance low-level cold air flows.

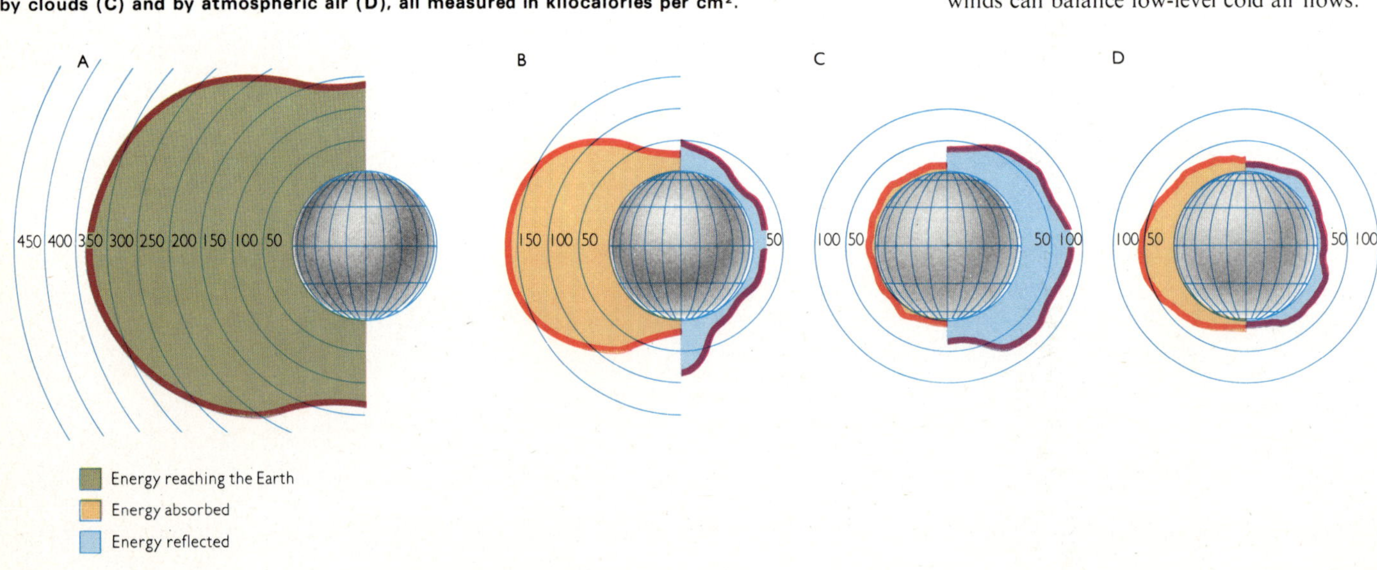

Ideally the earth's wind pattern might be expected to consist in air, heated by the equatorial energy surplus, rising and flowing in high-level air-streams to the poles, where it would give up its excess energy to finance the polar energy deficit. It would then return towards the equator in the form of cold low-level winds.

The mean atmospheric pressure diagrams show how this pattern is modified in fact by the effects of the uneven distribution of land and sea between the two hemispheres and the daily rotation of the earth. Only surface winds are shown as these most immediately influence local weather. In general winds flow from areas of high atmospheric pressure (highs) to areas of low pressure (lows), though because of the way circulations tend to be set up in moving air, actual surface winds often tend to spiral outwards from highs and inwards towards lows. In the northern hemisphere winds circulate around lows in an anti-clockwise direction and clockwise around highs, but the converse holds south of the equator.

Although in temperate regions these spiral motions are often much more apparent on charts of day-to-day weather than the general air-flows shown on diagrams averaging pressure and airflow over long periods, in many other parts of the world there is very little difference between daily weather patterns and the climatic averages.

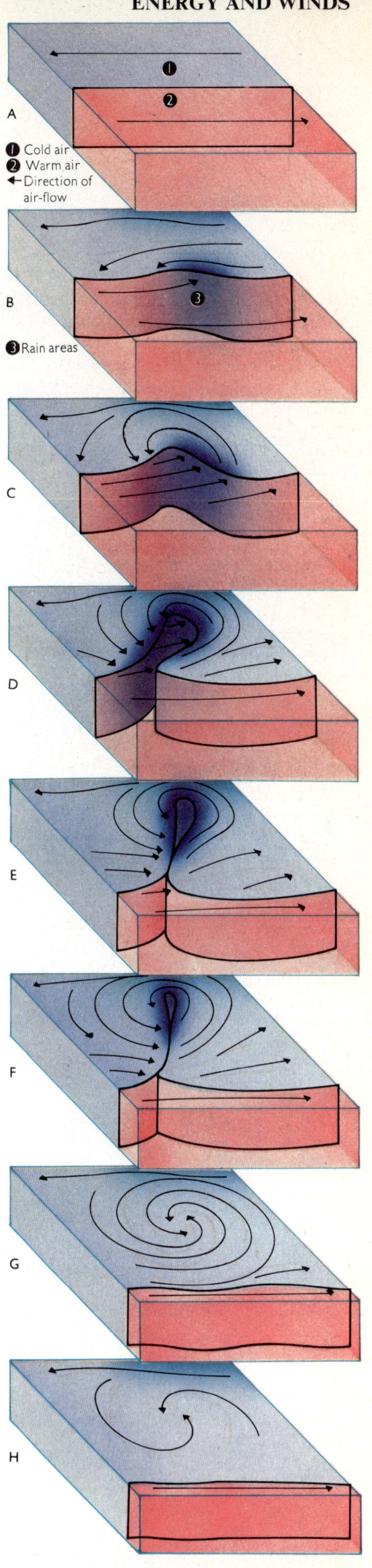

The birth and death of a depression. When adjacent warm and cold airstreams flow in opposite directions (A), a wave begins to form (B). A typical cyclonic system becomes apparent (C) with a broad band of rain at the warm front (where warm air is advancing against cold) and a narrow rainbelt at the cold front (where cold air is pushing into the warm). In (D) to (G) the cold front gradually chases and then catches the warm front, expelling the warm air to higher altitudes and producing an occluded front. The system, now a mere circulation of cold air (H), slowly dies away.

Weather patterns in the North Atlantic are characterized by sequences of secondary circulatory systems all travelling in the direction of the prevailing air flow. The accompanying diagram traces the life cycle of just one such secondary circulation – a depression or middle-latitude cyclone. While the diagrammatic sequence is in progress the depression should be thought of as being carried from east to west across the North Atlantic in the prevailing westerly air flow. The disturbance originates on the 'polar front' between the cold air mass covering Greenland and northern Canada and the warm moist air mass situated over the Gulf of Mexico and the middle Atlantic. It dies when the sector of warm air entrapped in a fold of cold air escapes upwards into the upper atmosphere, having precipitated the moisture it originally contained as rain or snow.

Fronts such as the polar front are very constant features of the earth's atmosphere, though they move in a regular pattern with the changing of the seasons.

The mean distribution of atmospheric pressure over the earth in (1) January and (2) July. The white lines are isobars, lines of equal pressure, and are labelled with the pressure in millibars at 5 millibar intervals. Air tends to move from areas of high pressure *(highs)* to areas of low pressure *(lows)* though the flow is modified by the earth's rotation. The directions of the prevailing surface winds are indicated by red arrows. The asymmetry of the January and July patterns is due to the preponderance of land in the northern hemisphere.

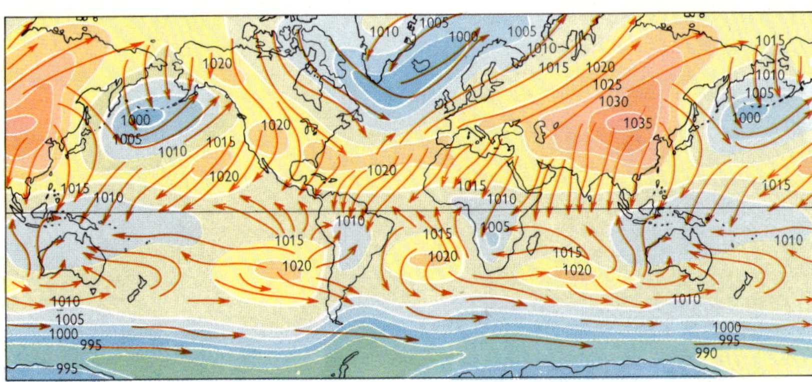

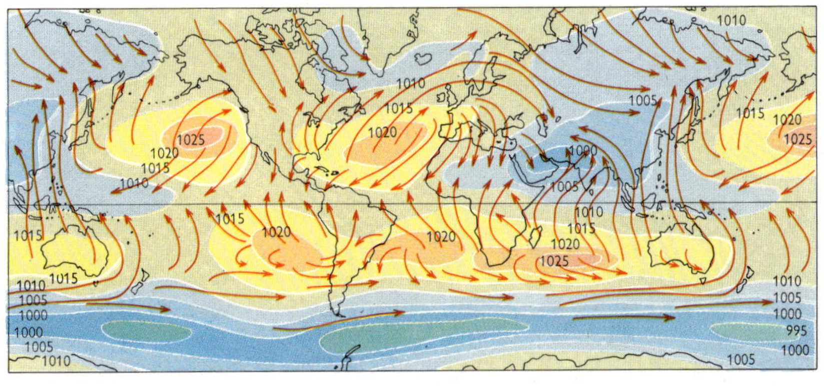

RAIN

Rain is the ultimate source of the water which fills all rivers, lakes, inland seas and reservoirs. Some rain is essential for the support of life, witness the expensive irrigation schemes which are built to bring water to areas having insufficient rainfall, but overmuch rain in too short a time brings with it the dreadful destruction associated with widespread flooding only too common across the inhabited world.

All rain has its origin in the world's seas and oceans as heat from the sun evaporates some of their water and raises it into the atmosphere as water vapour. This may drift with the prevailing air flow for thousands of miles before it is returned to the surface as some form of precipitation – rain, hail or snow.

Most precipitation occurs when bodies of warm moist air are lifted up to greater altitudes where their temperature drops until they become supersaturated with water vapour. If the air is not perfectly clean but contains say dust particles, liquid water droplets may begin to condense on the impurity nuclei and a complex process be set in train, the end result of which is the discharge from the cloud of its excess moisture in the form of rain.

There are three common mechanisms whereby moist air can be elevated until it becomes supersaturated. Convectional rainfall occurs when moist air, heated by some hot land surface, expands and rises rapidly to great heights at which it cools and precipitates rain. This process, common the year round in the humid tropics and in summer in more temperate climes, frequently gives rise to severe thunderstorms.

A warm front occurs when a body of moist warm air (1) advances against a mass of dry cold air (2). The warm air is pushed upwards and, as it becomes cooler, condensation takes place, resulting in a band of rainfall preceding the front (3). Although the front is in fact an inclined plane, it is represented on weather charts by the symbol shown at (4).

In orographic precipitation a range of hills or mountains causes the moist air to rise. Most of the rain falls on the windward slopes of the range leaving its leeward flanks in a drier 'rain shadow'.

In mid-latitude regions of the earth most rainfall is associated with atmospheric disturbances such as the fronts between adjacent warm and cold air masses. A typical depression has areas of rain associated with both its fronts – a broad band of light rain with the gently inclined warm front and a narrower belt of heavy rainfall with the much more steeply graded cold front.

Heavy precipitation is associated with several extreme weather phenomena. In thunderstorms, complex patterns of air movements conspire to produce electric charge separations between different parts of the cloud and between the cloud and the ground. The inequalities are discharged when intense electrical currents find a path through the ionized moist air causing a brief multiple flash accompanied by a loud report. The mechanism which gives rise to thunderstorms is the same as that which causes ordinary conventional rainfall.

Hurricanes, or tropical cyclones, although less frequent in occurrence than mid-latitude depressions, often cause considerable destruction of life and property. Originating only over warm seas, tropical cyclones are usually symmetrical disturbances containing neither cold nor warm fronts but having spiralling winds and a ring of torrential rain surrounding a calm and rainless centre or 'eye'.

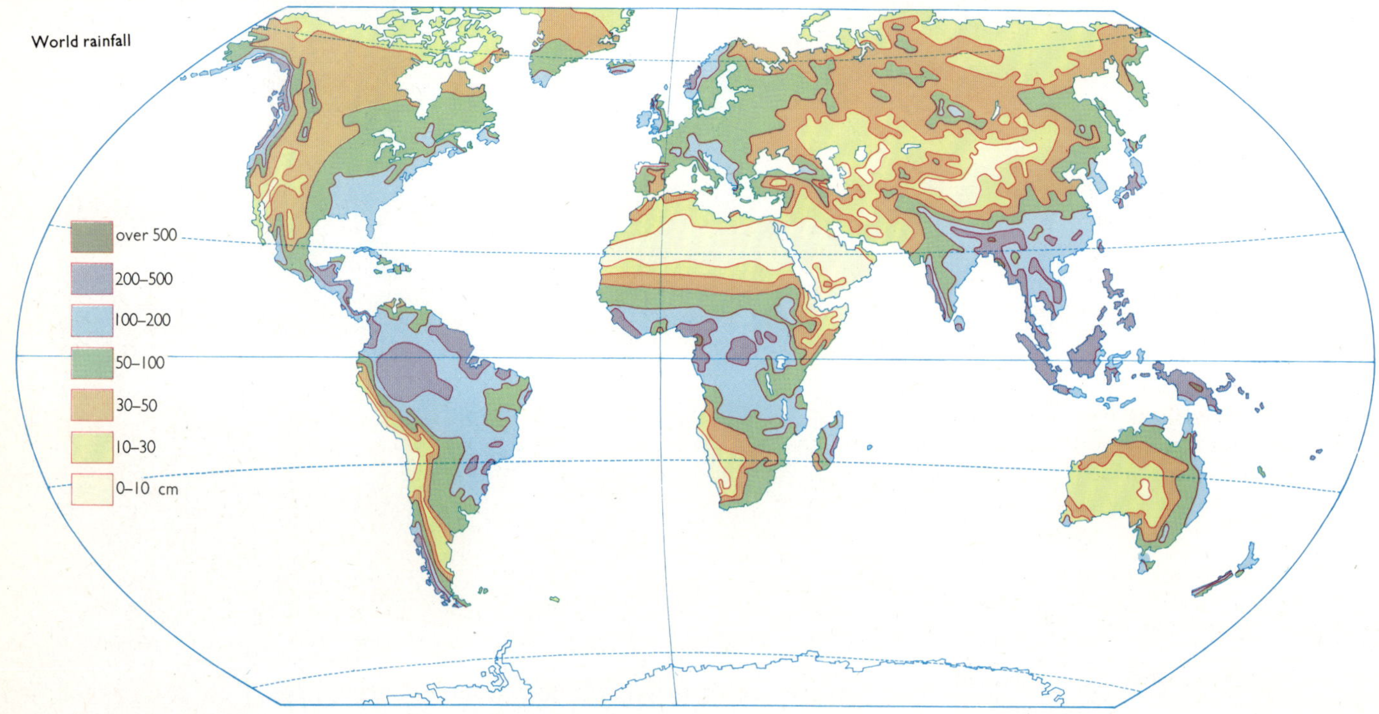

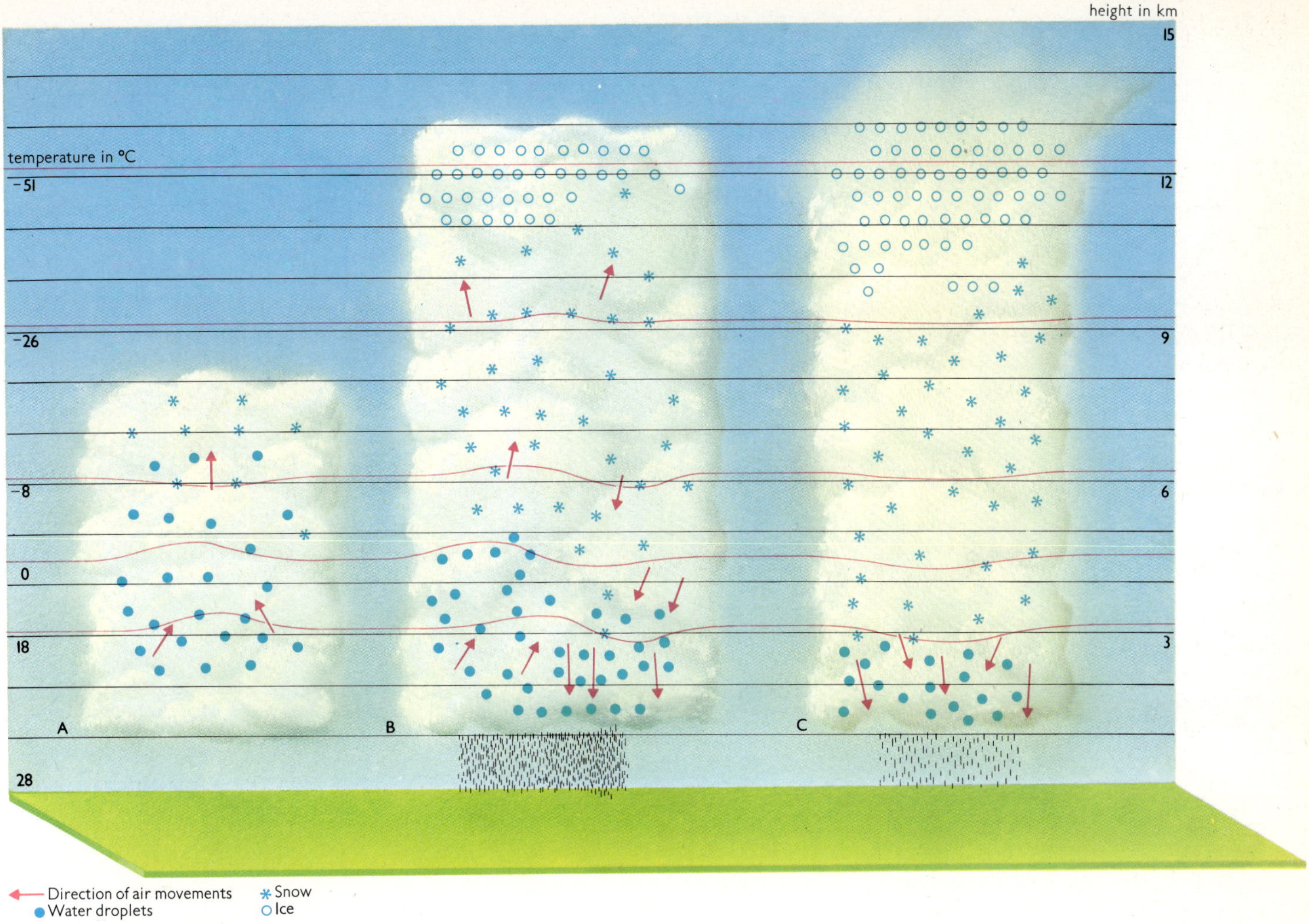

height in km
15
12
9
6
3
temperature in °C
−51
−26
−8
0
18
28
A
B
C

Direction of air movements
Water droplets
Snow
Ice

The development of a thunderstorm. (A) An updraught causes water droplets to be forced to heights of over 6 mi (10 km) resulting in the formation of ice and snow (B). Meanwhile compensating downdraughts force water droplets down to the surface as heavy rain. At a later stage (C), the cloud begins to dissipate as cool downdraughts produce only light rain.

Men have always longed for the ability to control the weather; rainmaking rituals are common in the ceremonial of the peoples of the earth's drier regions. Serious attempts at scientific weather modification date from the late 1940s and have included schemes for fog dispersal at airports and cloud clearing as well as for inducing precipitation.

The main line of approach to rainmaking has been the artificial introduction of impurity particles into clouds, thus providing nuclei for rain-drop formation. Progressing from laboratory experiments simulating the conditions inside rain-clouds, scientists held trials in which at first dry ice (solid carbon dioxide) and later silver iodide (having a crystal structure similar to that of ice) was dropped into clouds from aircraft. Although cloud-seeding experiments have been tried in many parts of the world, with even military applications in Vietnam, the results have been variable, some trials displaying evident success but others significantly failing.

The Taj Mahal in rain. In summer, an intense low pressure centre forms over northern India and draws in moist oceanic trade winds from south of the equator. As the South-West Monsoon, these bring heavy rains to much of the Indian sub-continent.

CLOUDS

Altocumulus clouds usually occur at heights between 10,000 and 15,000 ft (3,000 to 5,000 m).

Cumulonimbus clouds, forming quite near the surface often tower up to 20,000 ft (6,000 m). Cumulonimbus thunderheads often precede storms.

Clouds, a visible mass of tiny water droplets, ice crystals, or a mixture of both, are formed whenever atmospheric water vapour condenses. Their formation is thus an important stage in the production of rain.

Being usually confined to the troposphere, the lowest layer of the atmosphere (up to 50,000 ft deep (16 km) near the equator but only 30,000 ft (9 km) at the poles), clouds always exist in an environment where temperature decreases with altitude. This means that clouds are liable to be formed whenever moist air is uplifted and therefore cooled. Such uplift is commonly caused by the airstream carrying the moist air over mountains, in the atmospheric disturbances associated with warm and cold fronts and when great 'thermals' of warm air spiral upwards over local 'hot spots' on the ground.

The section below shows how different kinds of clouds are usually grouped in relation to the frontal system of a mid-latitude depression. The highest clouds, called cirrus by meteorologists, are entirely composed of ice crystals and may extend up to altitudes of 40,000 ft (12 km). Most rain falls from thick grey mid-altitude clouds like altostratus and the lower level nimbostratus.

Cumulus is an example of a widespread fair-weather cloud. Its base represents the condensation level and such clouds often build up to great heights throughout a warm day as more and more warm air rises into their base. They can often build up into the cumulonimbus clouds which are a clear sign of coming rain.

Satellite photographs of cloud formations provide meteorologists with important indications of daily atmospheric conditions over large areas of the earth's surface where there are few ground stations, particularly over oceans.

A section through a depression to show the cloud types associated with warm and cold fronts. (1) tropopause, (2) trailing cold sector air, (3) cold front, (4) warm sector air, (5) warm front, (6) leading cold sector air, (7) cumulus, (8) cirrus, (9) altocumulus, (10) altostratus, (11) cumulonimbus, (12) nimbostratus, (13) stratus, (14) stratocumulus, (15) cirrostratus.

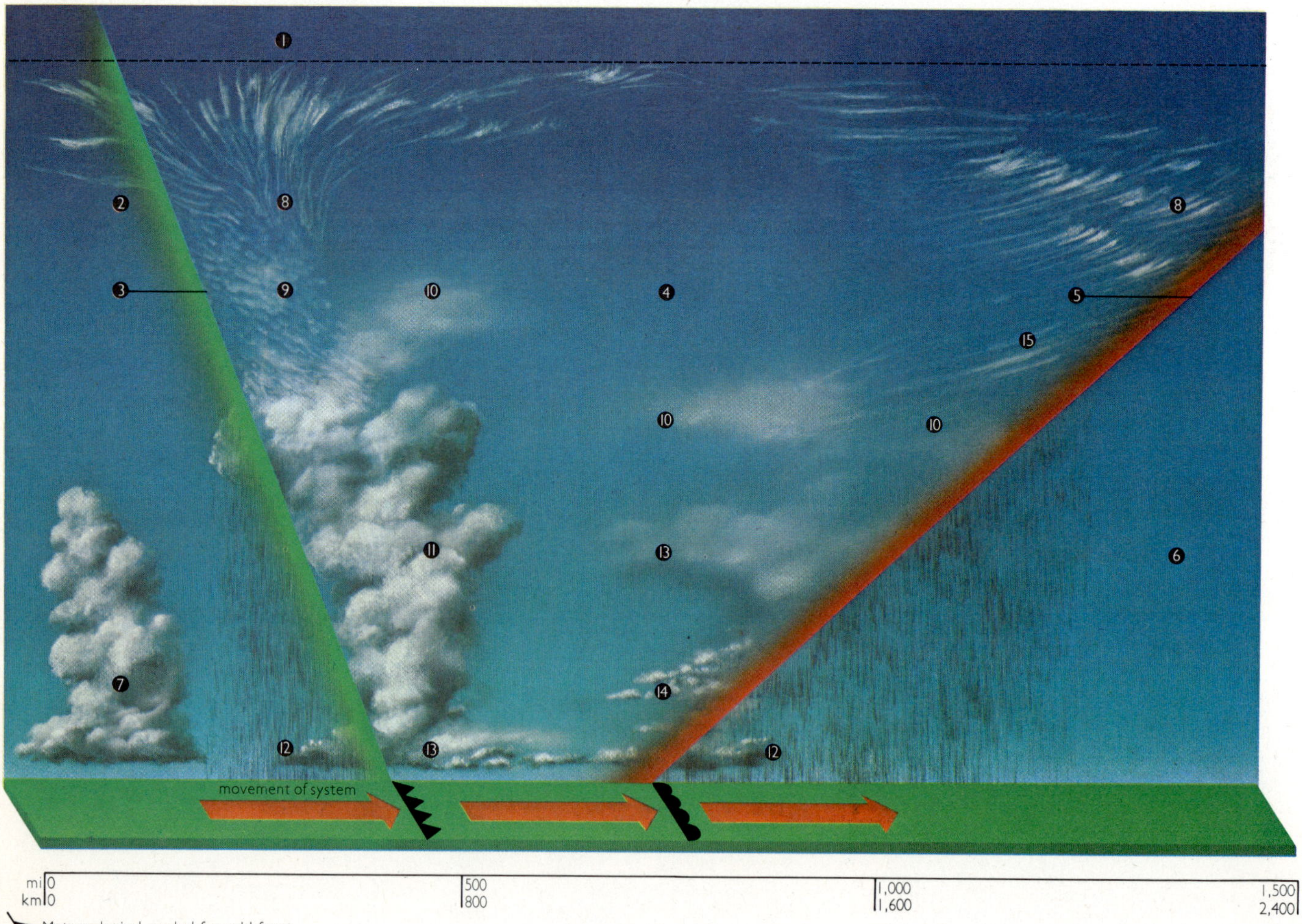

movement of system

mi 0 — 500 — 1,000 — 1,500
km 0 — 800 — 1,600 — 2,400

Meteorological symbol for cold front

Meteorological symbol for warm front

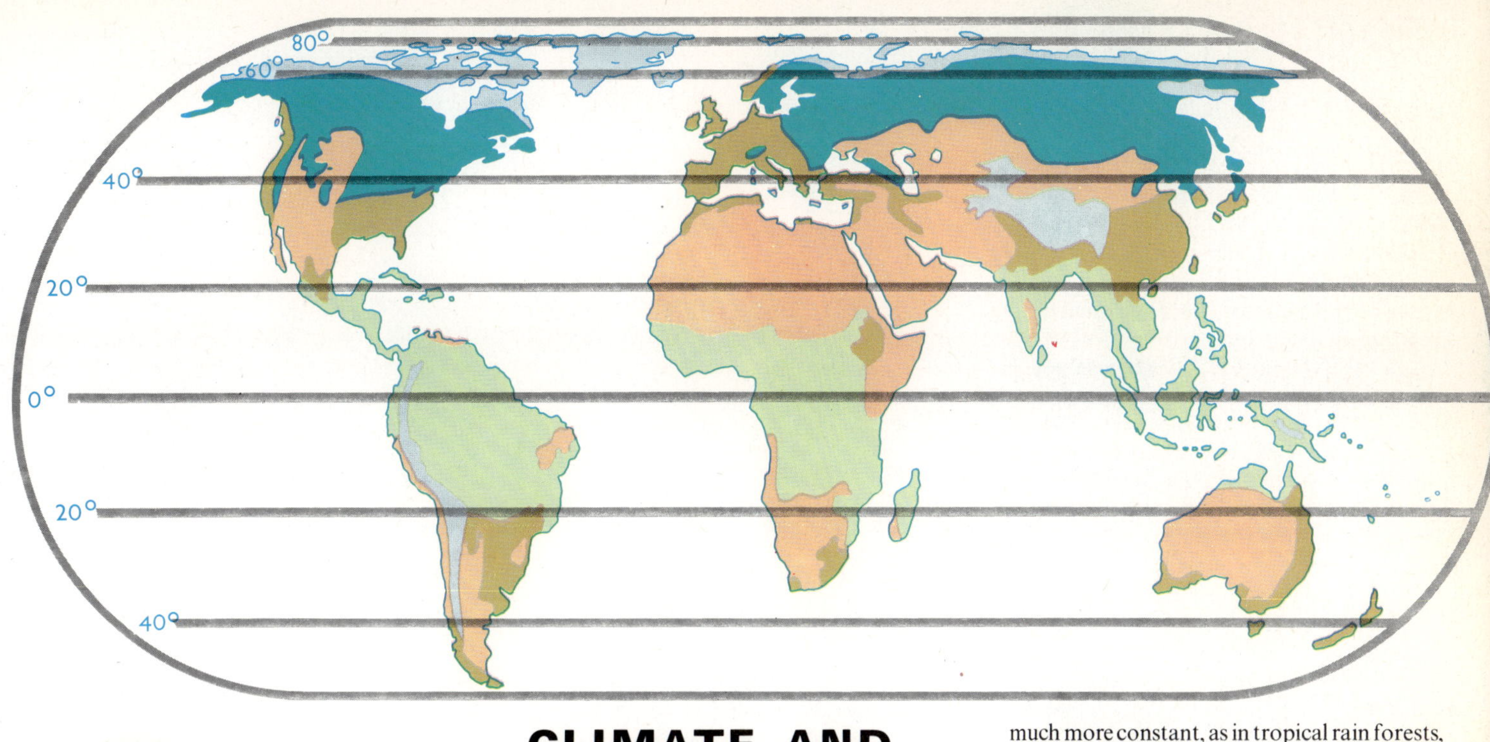

The climatic classification of Vladimir Köppen, still the most generally accepted, divides the climates of the world into five main types.

CLIMATE AND VEGETATION

Climate describes the characteristic weather conditions of an area. It takes account of temperature, rainfall, sunshine, wind, humidity and cloud-cover, usually averaging these over the year. In regions where the weather varies greatly from day to day, as in the UK, or there are great seasonal extremes of weather, as in Siberia, attempts at defining climates may seem to be of limited usefulness but in parts of the world where the weather is much more constant, as in tropical rain forests, the climate provides a good guide to the prevailing weather.

The generally accepted climatic division of the world proposed by Vladimir Köppen in 1900 is based on a correlation of annual temperatures and rainfall averages with vegetation types. The five broad categories of Köppen's classification are further subdivided according to the seasonal distribution of rainfall and temperatures.

The climate of a region depends largely on its latitude, relief and proximity to warm or cold ocean currents and the great continental and maritime air masses. Even vegetation can influence climate and there are several schemes for planting trees in deserts to encourage greater rainfall.

There is a close correlation between vegetation type and annual rainfall over most of the world.

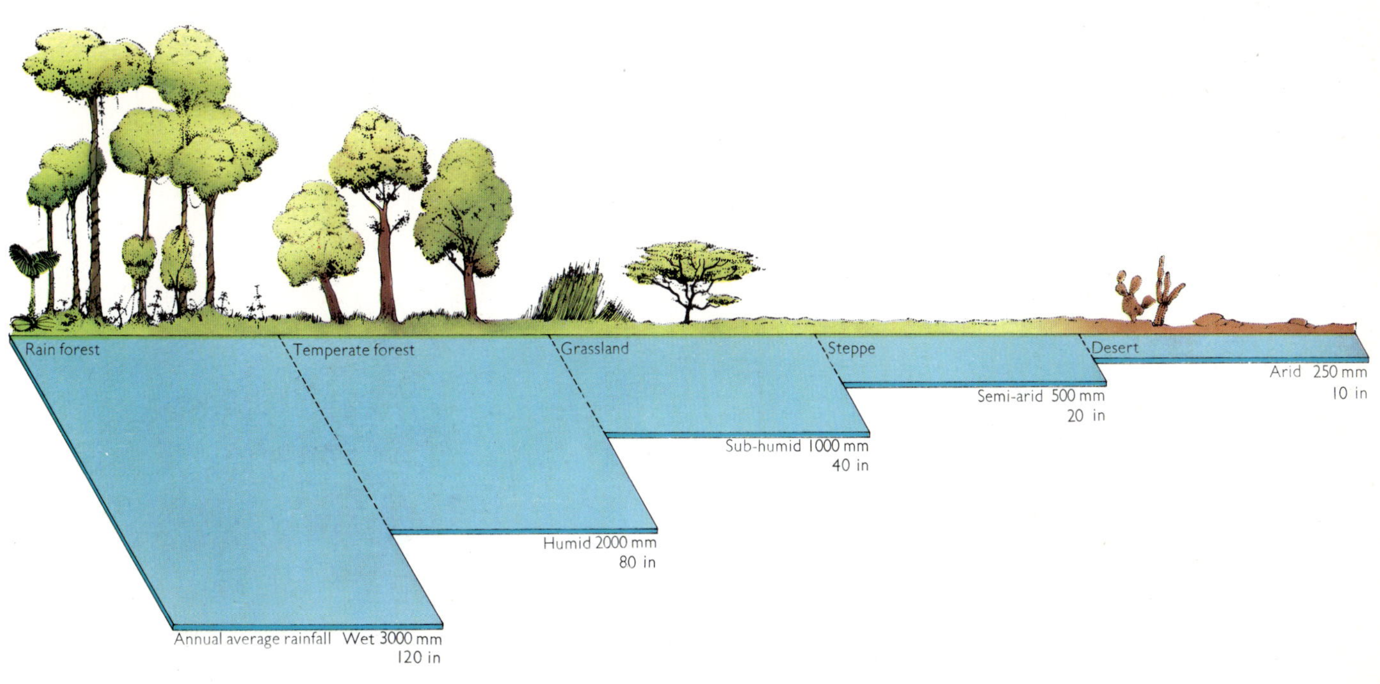

POPULATION

The UN has estimated that by 1975 the earth will seek to support 3,830 million people, more than twice the total in 1900. More significantly, if present trends continue, the world's population will reach 6,300 million by the end of the century.

In itself, this growth in the human population of the world can be regarded as a measure of the success of the human species. Man, a mere 5 million individuals when he first began to practise agriculture some 12,000 years ago, through it colonized vast tracts of the earth and carved from the landscape an environment favourable to himself. But in the latter part of the 20th century the achievement of mankind will be measured rather by his success in controlling the rate at which his population grows and in developing resources of food, energy and materials sufficient to maintain an increasing proportion of his numbers in a state resembling sufficiency.

Increases in population are not evenly distributed over the surface of the earth. Today, nations in general having the poorest food supply positions have the highest rates of population growth. This has not always been so; in the 19th century Europe led the way in population growth, expanding from 180 million in 1800 to 400 million at the end of the century.

Population increase results from the interplay of birth rates with mortalities. In early 19th-century Britain, although agricultural improvements and industrialization encouraged a high birth rate, the population increased only moderately. It was only with the development of surer techniques of childbirth and improved public hygiene and medical skill after 1850, that the death rate fell significantly.

In the industrialized nations of the world changing social attitudes greatly encouraged by economic pressures have tended to reduce the size of families in the 20th century. Low birth rates and low mortalities have resulted in a relatively slow rate of growth with all members of the population having a high life expectancy. While many demographers argue that any increase in population is too great in an overcrowded island like Britain, others point out that in a static population with high life expectancy, the proportion of the population of working age declines, with perhaps disastrous economic consequences for the support of the elderly majority. In Australia, the USA and Canada, although patterns of births and deaths are similar to those in the UK, the population pressure on resources is far less than that in western Europe.

In much of Africa, Asia and South America the produce of the land already falls short of providing a healthy diet for all the people. The annual rate of increase in the third-world nations averaged 2·1% in the late 1960s, nearly twice that in the developed nations (1·2%). After centuries of slow and erratic population growth similar to that experienced

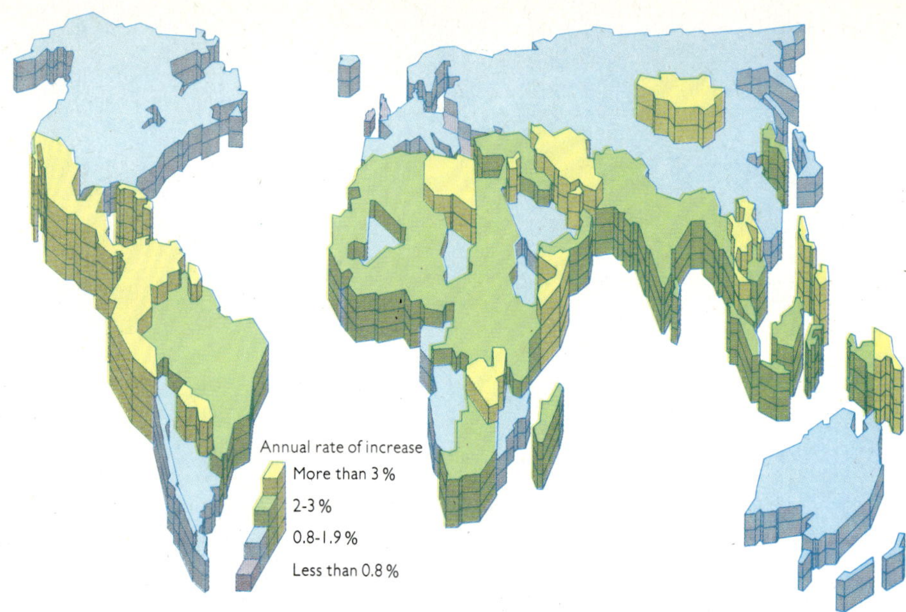

The global pattern of population growth

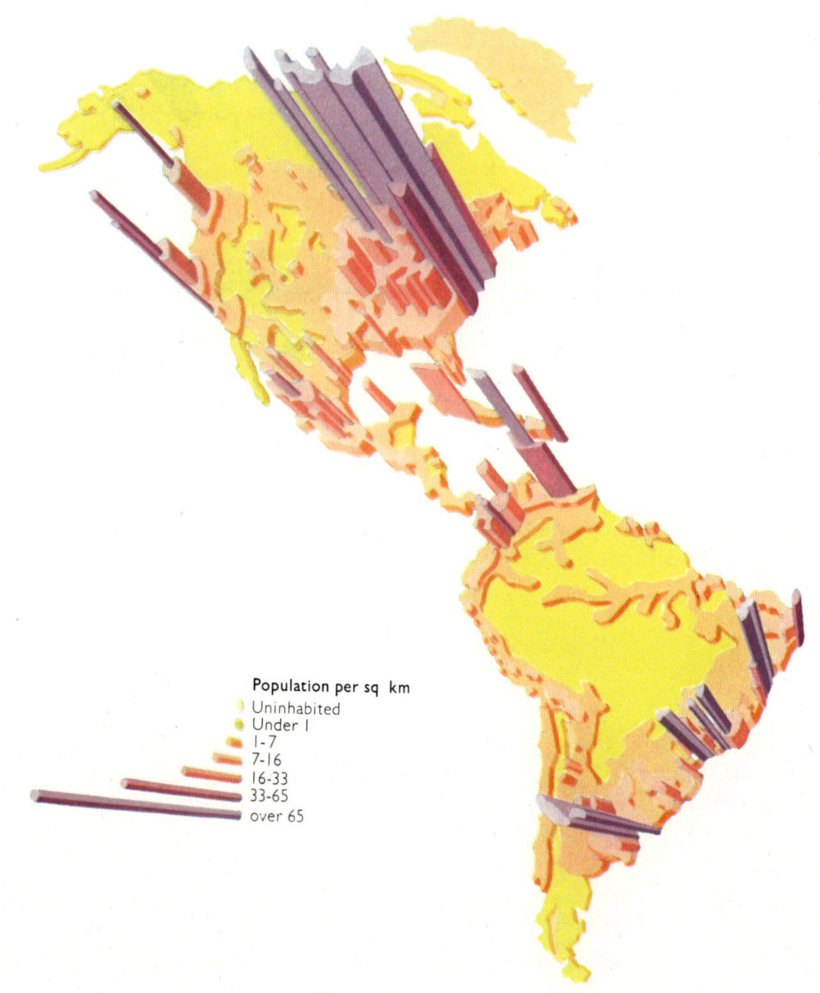

World population density.

in Europe before the 17th century, the impact of modern medicine, coming without any breathing-space in which to develop new social attitudes regarding the size of families, has been to produce a veritable explosion in the numbers of babies surviving into childhood at just the same time as adult life expectancies have begun to rise. With meagre resources of skilled manpower, capital and administrative experience, most developing nations cannot even keep pace with the new growth in their population; there is no possibility of their ever improving the quality of life of the mass of their people.

The population problem must be solved. Birth control, a popular answer, is no panacea and difficult to administer except in a tolerant social climate. Although the governments of many third-world nations are trying to promote family planning, as long as their economies are starved of technology and capital, the distress of the majority of their people will remain. The food supply position can be relieved both by developing more efficient techniques of agriculture and by taking under cultivation vast tracts of yet undeveloped land, but transportation and distribution systems will require parallel improvement. The best solution might combine all of these with large-scale migrations of people from over-populated regions like Bangladesh to under-exploited territories such as the USA.

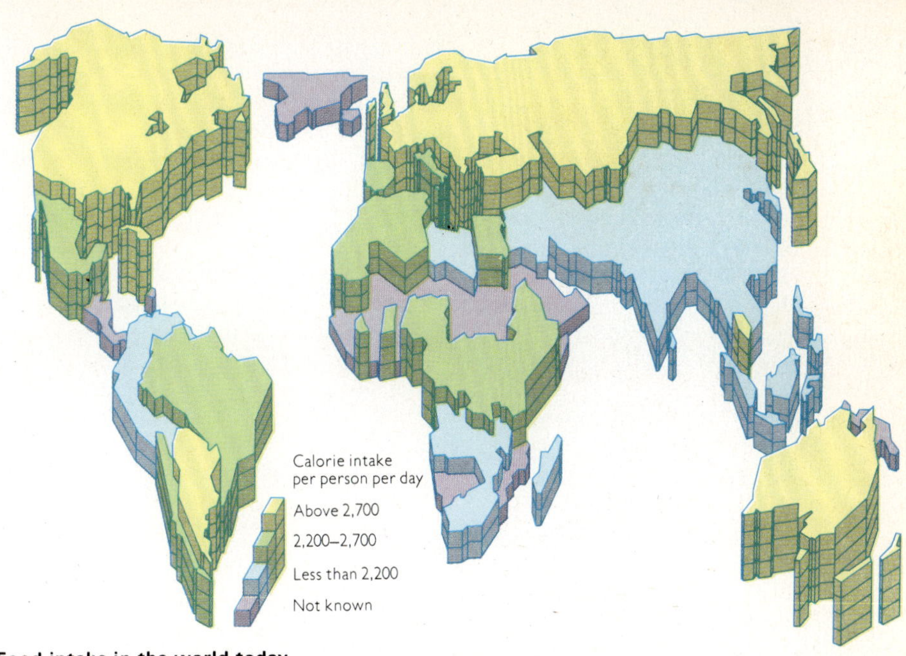

Food intake in the world today

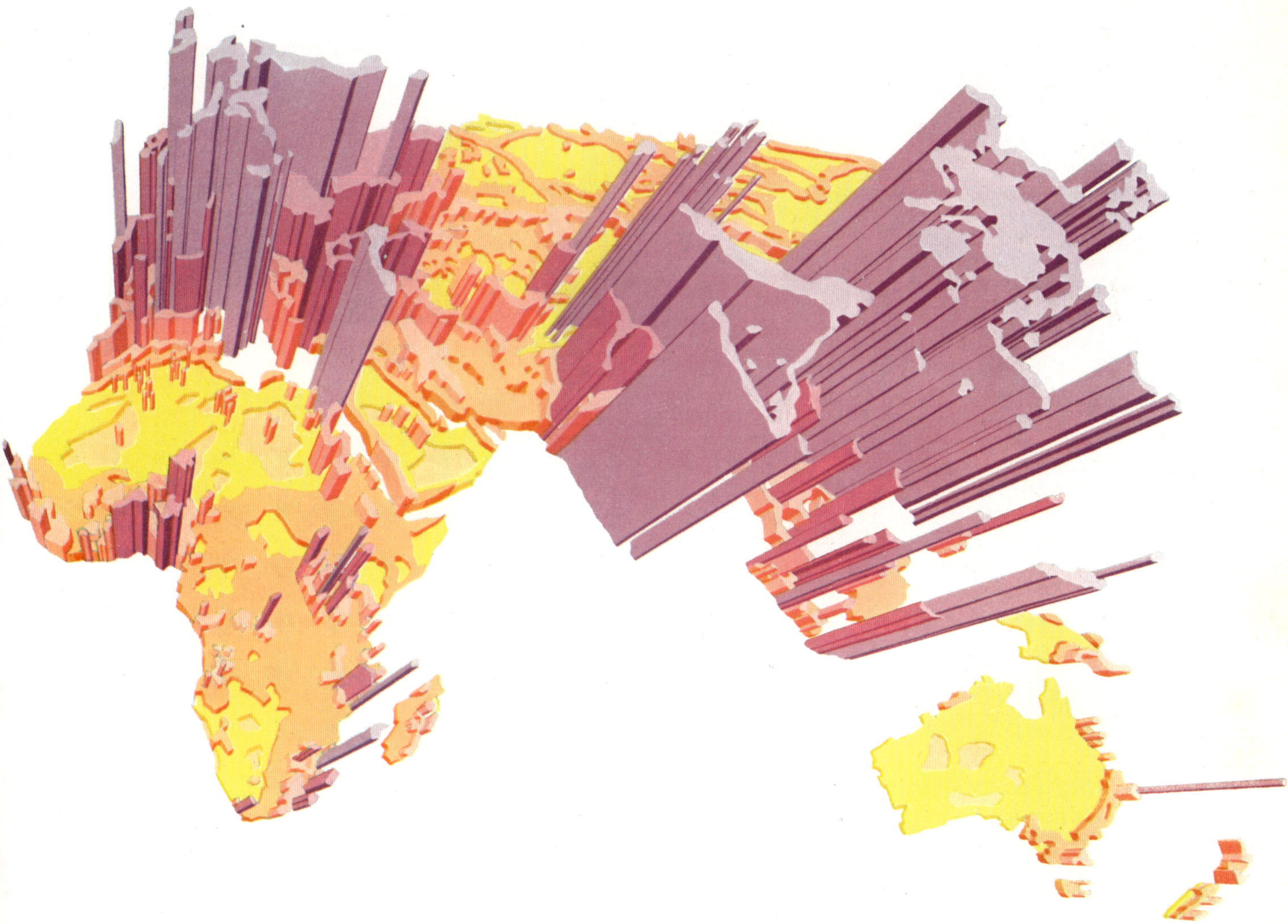

RACES AND CULTURE

Although mankind forms a single species which about 200,000 years ago became distinguishable from those other ancient species of proto-men which did not survive into the modern period, this species today exhibits several varieties which are known to the anthropologist as the races of men.

The physical anthropologist has classically distinguished three primary races of man – mongoloid, negroid and caucasoid – on the basis of the comparison of external features such as the general build of the body, the structure of the skull, face and jawbone, and the type of hair. More recently much attention has been directed to the study of the distribution of the various blood-groups among different peoples and to other fields of population genetics.

In addition to the primary races, there are several smaller groups known as composite races because they show features of more than one primary variety with which can be mixed some characteristics thought to belong to races no longer represented among men. The Bushmen of southern Africa, the aboriginal inhabitants of Australia and New Zealand and the peoples of the central Pacific are examples of composite races.

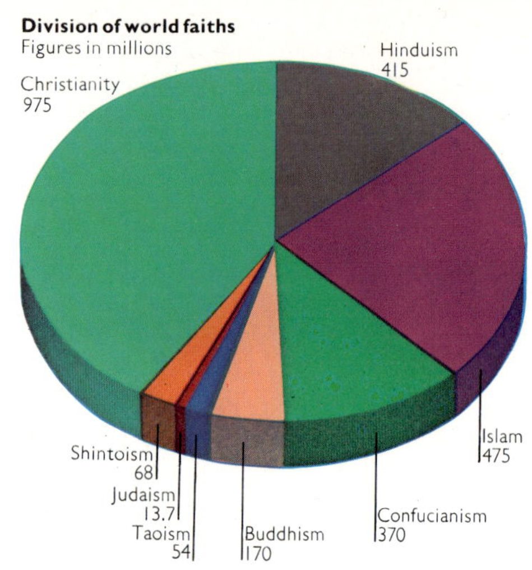

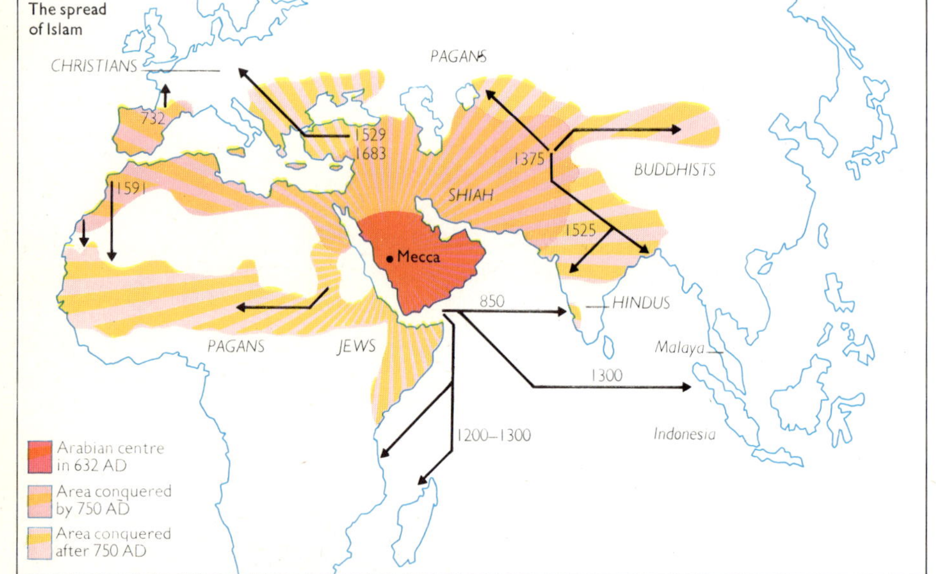

While the primary races of man themselves display many sub-varieties none of which is entirely typical, their general features are well characterized. Mongoloid peoples are round-headed and have broad jawbones, straight black hair, little body hair and yellow to brown skin. Typically they have an extra fold of skin (the epicanthic fold) in front of the eyelid which gives the impression of their having narrow, even slanting eyes. Negroid peoples are typically long-headed, woolly-haired and with a protrusive jaw. Their skin pigmentation ranges from light brown to nearly black. Caucasoids have straight or wavy blond to black hair, skin un-

Religions of the world today

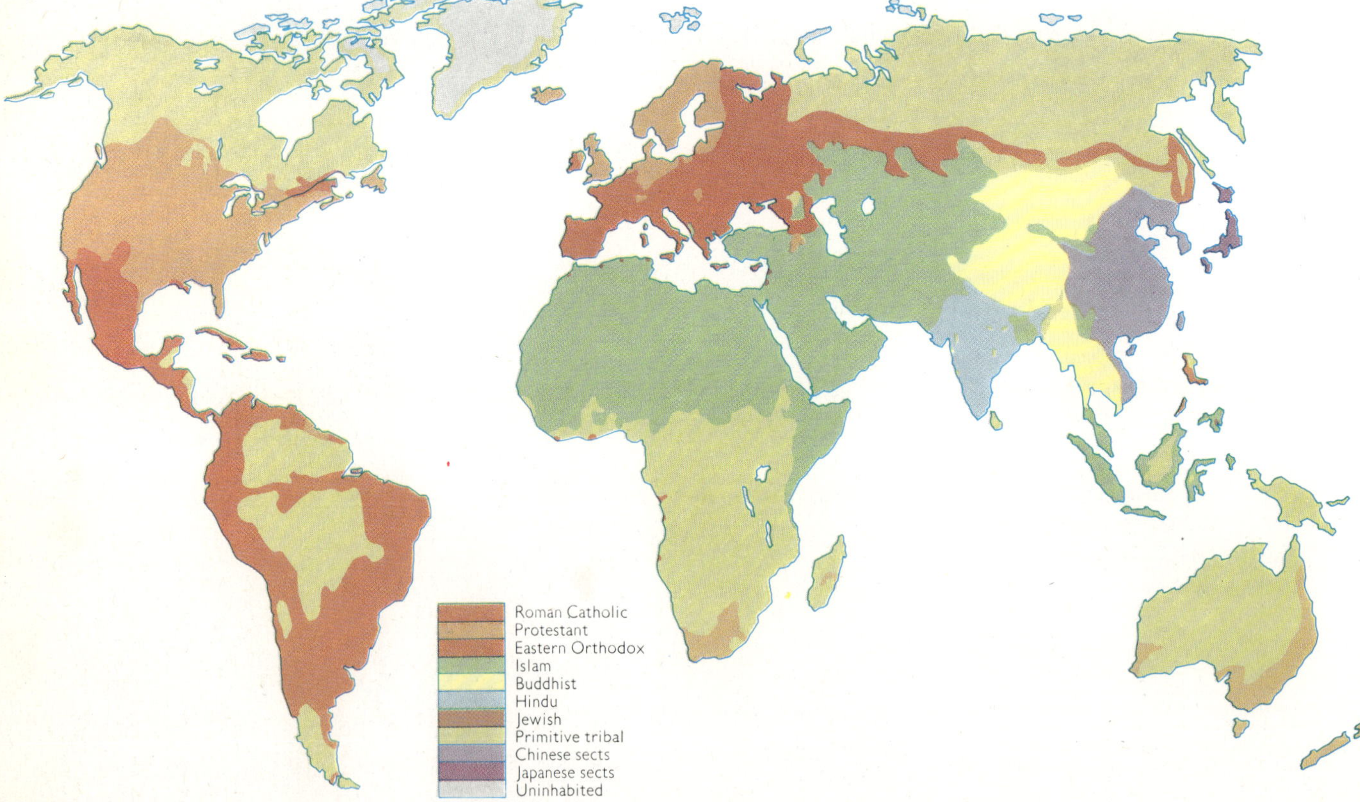

Roman Catholic
Protestant
Eastern Orthodox
Islam
Buddhist
Hindu
Jewish
Primitive tribal
Chinese sects
Japanese sects
Uninhabited

World races today

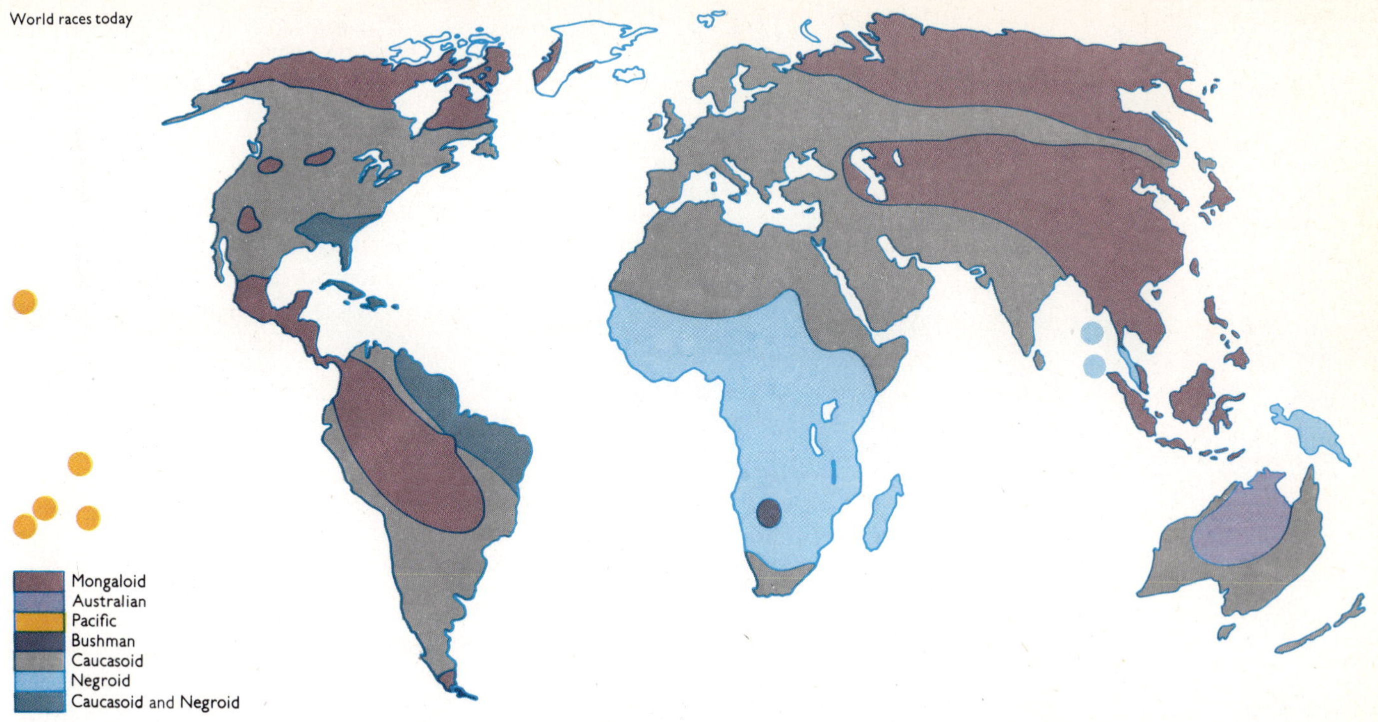

The present distribution of races throughout the world reflects the former migrations of whole nations of people, the colonization by some peoples of large tracts of territory in parts of the world formerly inhabited by other races and the forced movement of whole populations to work agricultural systems dependent on the institutions of slavery.

pigmented to dark brown and eyes blue, grey or brown.

Although the popular imagination probably first thinks of skin colour if the word 'race' is mentioned, it will be clear that this is a poor guide to ethnic classification. Indeed, because the colour of exposed skin depends so much on the accident of how much sunlight it has recently been exposed to, physical anthropologists in fact compare the colour of skin areas hidden on the inner part of the forearm.

Throughout the history and pre-history of the species, different groups of men have assembled storehouses of shared learning which have become characteristic of the group. The content of these cultures includes far more than mere information; shared customs, habits, laws, beliefs and art-forms are all vitally significant aspects of the culture of a society.

Most cultural anthropologists are concerned only with the study of comparatively isolated and 'primitive' societies but it should be recognized that all peoples share in some culture – not least those within the 'consumer culture' of modern western society.

Language is an important element in the development of culture and quite great degrees of cultural complexity can be achieved before the introduction of literacy, though as soon as information can be stored and transmitted through the written (or printed) word, a society's capacity for cultural innovation and diversification expands explosively.

The phenomenon of religion has been chosen from among many candidates to illustrate the geography of culture. Today's distribution results from the interplay of military conquest, active international proselytization, dispersal through nations and the migrations of whole peoples.

Anthropologists have long used the comparison of facial features as a guide to racial differentiation. The three primary races: mongoloid *(top)*; negroid *(middle, left and centre-left)* and caucasoid *(bottom)*, are each represented by some of their many variants. Composite races are represented by the Australian aborigine *(middle, centre-right)* and the Bushman *(middle, far right)*

THE CITY

Cities are as old as human civilization. Ur in Mesopotamia, founded about 3500 BC, contained all the functions that cities have provided through to the present, providing a trading, administrative, defensive, religious and cultural centre for the region surrounding it. It was in the city that man first began to practise that division of labour that so sets man apart from most other species. Specialized crafts and large-scale political and religious domination were possible only after men adopted permanent settlements, supported on the surplus products that farmers brought to exchange in the market place. Trade could only grow after the founding of cities offering fixed and specialized markets for merchants to travel between in days when their journeys might last for several years.

After the decline of the Roman Empire, urban life in Europe was virtually non-existent until the 11th century, although in the Orient and the Islamic world cities remained an important element in social organization. With the gradual break-up of the feudal system in the west, a drift of population from the countryside into the towns began, which, although hesitant at first, has gained momentum ever since. A wealthy and increasingly powerful merchant class began to develop and slowly wrested political power from the rural landholders. Urban architecture re-blossomed as a living art-form as the emergent cities vied with each other to display their commercial wealth in ever grander magnificence.

It was not until the 19th century that the industrial revolution in western Europe accelerated the process of urbanization. The city became as much the location of industry as it had ever been the centre for trading. Vast populations in the cities were crowded into dismally inadequate houses as the factories' appetite for labour grew insatiably and the countryside progressively became the mere larder of the towns.

The influx into the cities brought with it new problems of housing, sanitation, education, air pollution, traffic congestion, urban poverty and urban crime. Cities today retain many of these problems – and have discovered new ones – as the process of urbanization has spread like a contagion throughout the developed and developing world.

Recent urbanization has brought its gains as well as its losses. Were it not for the city, the world could probably not carry its present population as labour-intensive agriculture tends to be inefficient. Cities are necessary to provide both the pool of labour that industry requires to man its processes and the major consumer-markets of the world – conveniently close to the urban centres of manufacture.

Probably the most significant feature of the city in the 20th century is its ability to provide an increasing range and intensity of service

Negro Harlem
Spanish Harlem
Italian Harlem (top) and Little Italy (bottom)
Chinatown

The main areas of concentration of Negro, Spanish, Italian and Chinese elements in the population of New York. New immigrants to cities often settle in areas where there are communities of people from similar backgrounds.

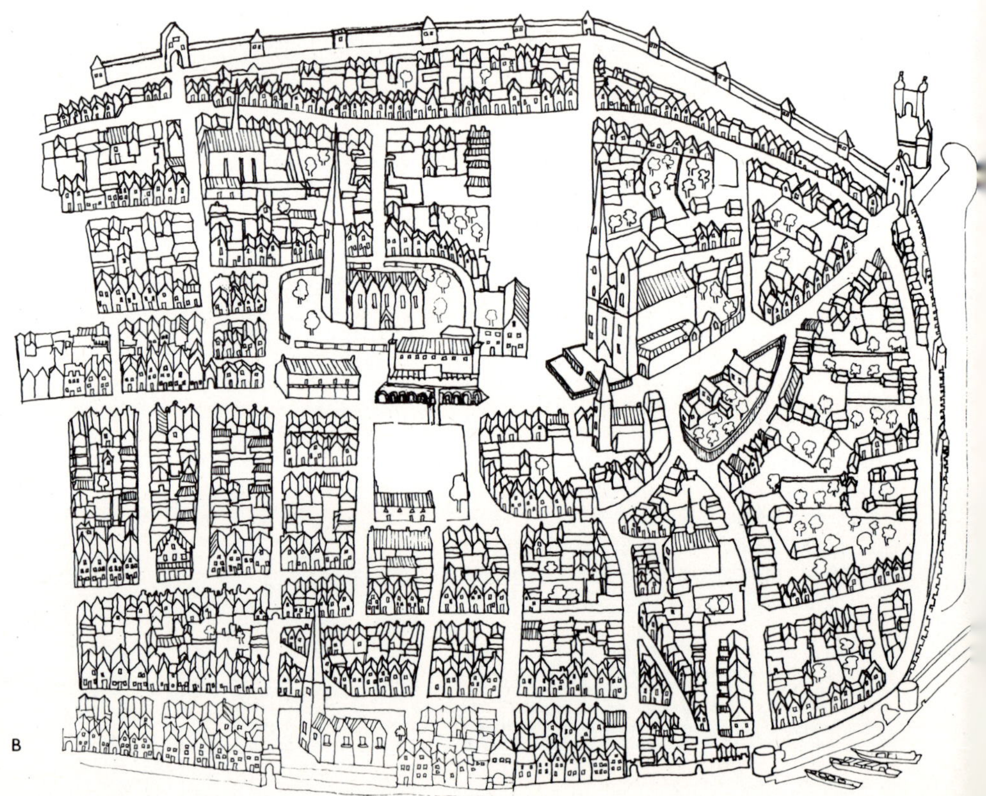

The centre of Sydney, Australia is built on a promontory extending into the sheltered waters of Port Jackson, while the suburbs range themselves around its shores. The celebrated Opera House (centre) symbolizes the cultural function of cities.

functions. It is not just the public utilities – gas, electricity, water, postal and telecommunications services – that are most efficiently provided within the urban context, but educational, medical, legal, financial and welfare services are also best provided where a population is not too widely scattered. Wholesale and retail distribution is a further sector where the comprehensiveness of the service can increase in direct proportion to the total population within economic range.

Many cities in recent years have experienced difficulty in persuading key workers – doctors, teachers, transport and local government employees – to live in the inner city. Attention is consequently being directed to making the urban environment an attractive place in which to live and work, chiefly by improving city transport systems and increasing recreational facilities.

Most established cities in the western world centre around a central business district which provides administrative, commercial and retail functions on sites of high value and consequently tall buildings. The retail core is traditionally surrounded by an area of wholesale facilities and inner-city industries with perhaps considerable space given over to railway goods yards. Residential areas range from inner-city slums through middle-class residential suburbs to exclusive outer suburbs. Usually housing densities fall with land values as the distance from the city centre increases. Other large uses of urban land include suburban factory estates, recreation space (parks and sportsfields) and horticulture.

Street patterns reflect the evolution and function of towns and cities *(below)*. Shops in central Mecca (A) have been built around the focal point of the city, the Great Mosque. Bremen in West Germany (B) has grown more haphazardly since its origins before the 9th century. The layout, with its large open square, is typical of many old market towns. In Versailles (C), a symmetrical arrangement of avenues is evidence of the more grandiose approach to planning that was current in France during the reign of Louis XIV. In Wolfsburg, Germany (D), there is a sharp distinction between the parts of the town that contain the Volkswagen works and those that function as residential areas. Similar clusters of dwellings around factories were built throughout Europe during the Industrial Revolution. In Cambridge, England (E), colleges have displaced commercial functions from their former situation near the River Cam. Neuf Brissac (F) lies on the frontier between France and Spain. It is a fortified town with limits sharply defined by the contours of the hill on which it stands and is typical of many built during the 16th century. Finally, the Spanish seaside resort of Sitges (G) has been developed as a strip along the coast that constitutes its attraction for tourists.

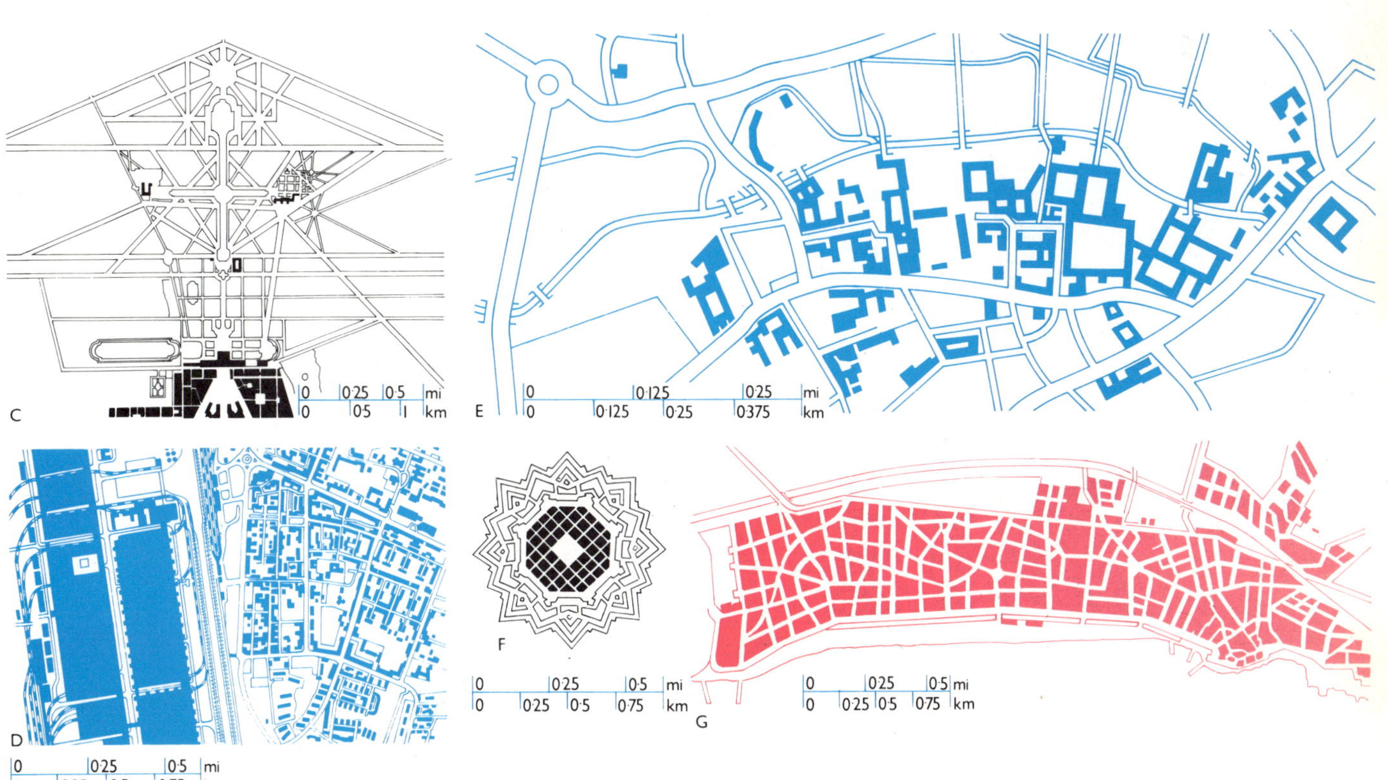

POLLUTION AND CONSERVATION

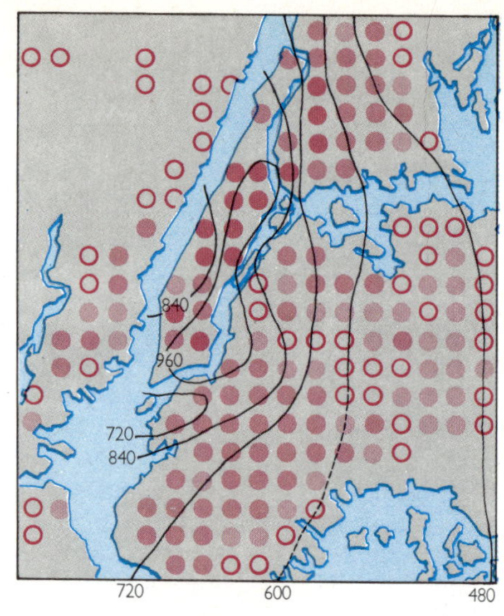

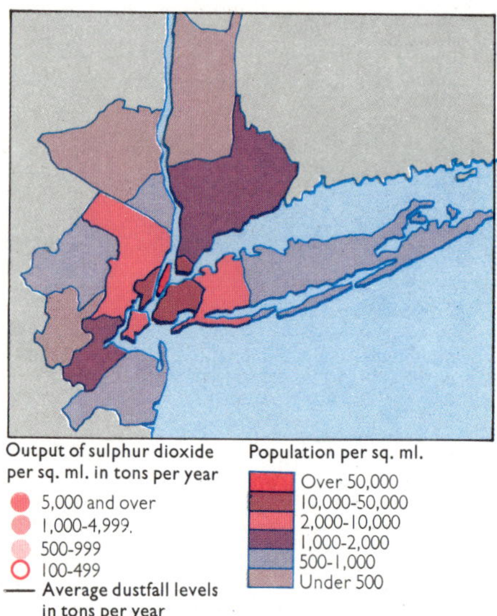

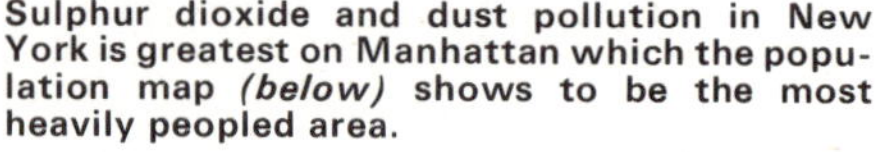

Sulphur dioxide and dust pollution in New York is greatest on Manhattan which the population map *(below)* shows to be the most heavily peopled area.

No keen observer of city, countryside or shore can have failed to notice that a direct consequence of man's increasingly reckless exploitation of the earth's natural resources is the contamination and destruction of the environment on which he depends to support his very existence.

Man is the dirtiest of animals. In the UK the average person produces about 1·5 lb (680 g) of junk every day and this is only a third of the US figure. All over the world, cities from Sydney to Milan, New York or Los Angeles suffer from acute problems of air pollution caused by the burning of domestic and industrial waste, coal and heavy oils for heating, and light hydrocarbon fuels by aircraft and motor vehicles.

The home probably comes top of the league of earth-polluters. Sewage and garbage – particularly the packaging materials in which all consumer products now seem to be double-wrapped – not only represent vast quantities of apparently useless waste materials but also a criminal misuse of the earth's precious and dwindling stock of natural resources. 'Planned obsolescence' in domestic products, if in the short run, commercially desirable and the

Chemical effluent, seen here polluting the River Calder in England, destroys any fish that formerly thrived in affected rivers and streams and renders their water unfit for drinking.

The city as a living body. The daily requirements of a typical US city of 1 million inhabitants *(left)* is contrasted with the waste it produces *(right)*.

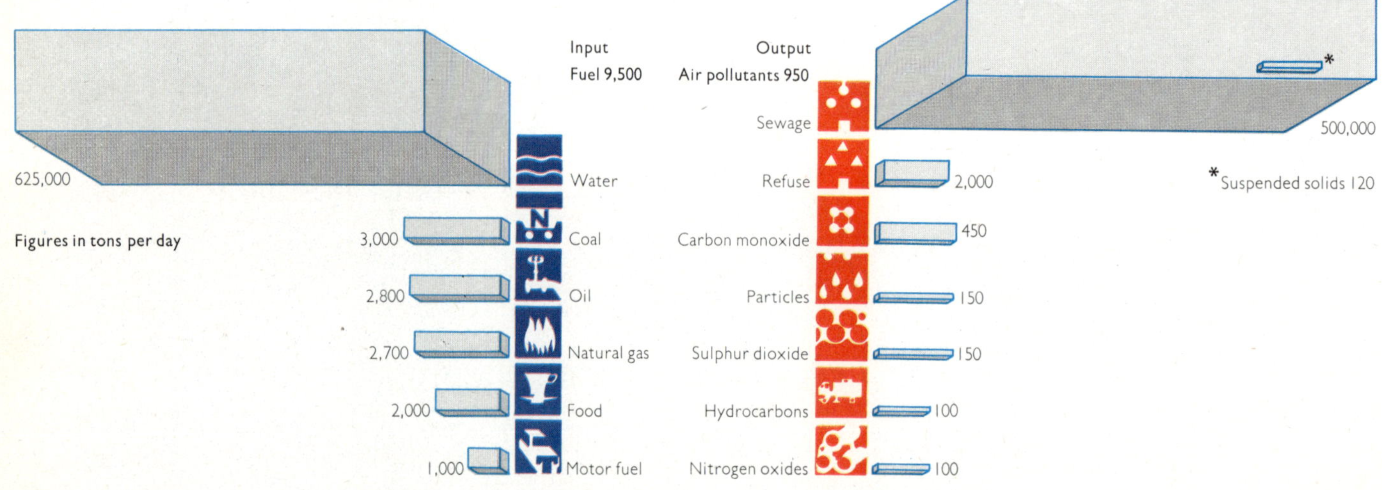

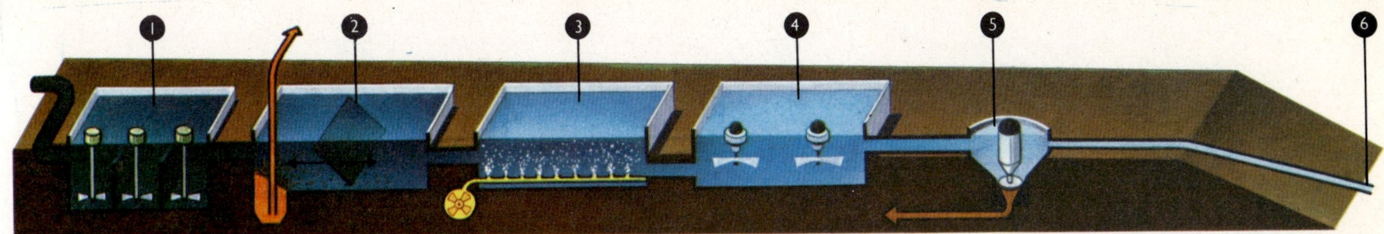

A process for the purification of industrial waste water, an essential step in cleaning up inland water bodies. Water from factories is neutralized with acid or alkali (1) and filtered to remove solids (2); the water is then reoxygenated (3) and mixed with bacteria (4) which are allowed to digest any remaining organic waste (5) before the water is pumped back into the river (6).

To protect threatened species of wildlife, it is sometimes necessary to catch some specimens (here, of Hunter's antelope, *Beatragus hunteri*) and remove them to a no-hunting area.

seemingly unavoidable consequence of technological progress, also results in vast accumulations of untreated junk littering both townscape and countryside. Particularly great problems are posed by synthetic plastics which are not 'bio-degradable', that is which cannot be naturally broken down and reused in biological cycles.

Industry produces its share of pollution – from the great spoil heaps associated with quarrying and mining to the air pollution issuing from chemical and metallurgical processes and from coal and oil-fired electricity generating stations.

A third more specialized type of pollution occurs when intensive agriculture employs herbicides, insecticides and inorganic fertilizers which break down to harmful residues. The natural concentration in the course of biological food-chains of the originally minute concentrations of these residues which enter rivers and thus the oceans, can result in sudden catastrophic drops in the fertility of the species at the ends of the chains.

The remedies for pollution are clear once its sources are identified. Air pollution results both from burning wastes from home and factory and fuels for heating and transport and from certain industrial processes which release dust and dangerous chemicals into the atmosphere. The development of more efficient means of burning fuels (through to carbon dioxide rather than the poisonous carbon monoxide) and processes for removing dust from factory flues (using electrostatic precipitators) offer partial solutions to air pollution.

The air is also the medium carrying noise pollution – from aircraft, traffic and industrial processes – which has so greatly increased during the 20th century. In modern cities complete silence is a most unusual experience. Solutions to noise pollution come from a combination of quietening the sources of noise and soundproofing homes and places of work.

Man uses more water than he does any other natural resource. Although supplies of water from precipitation are indefinitely renewable, in using them man contaminates the water with an increasing burden of dissolved and suspended refuse. As a result he not only drastically reduces the value of his water resources but eventually poisons the seas and oceans, destroying all marine life. The Mediterranean Sea already shows signs of becoming one vast basin of sewage and the many rivers which serve as the sewers of industry are lifeless, if watery, deserts. The solution to water pollution is clear; effluent water must be purified before it is discharged into rivers.

Conservation, the rational and well-managed exploitation of the earth's natural resources, is the other side of the pollution coin. The recycling of waste products, particularly of used paper and metallic products, not only reduces the problem of pollution but also conserves supplies of timber and metal ores. As the world runs short of energy-producing minerals and conservationists become increasingly concerned at the disturbing of the ecological balance in rivers into which water warm with the waste heat of industry is discharged, more efficient insulation of homes and the beneficial employment of thermal waste thus offer relief to problems both of thermal pollution and of energy conservation. With water itself becoming daily more precious, the elimination of pollution and the desirability of re-use come to be seen as identical causes.

Conservation of a second kind is concerned with the protection and preservation of as wide a range of living species as possible. Every animal and plant represents an irreplaceable reservoir of genetic material available to be used for the benefit of mankind; the extinction of any one of them would be a serious loss not only to science but also to the world at large. The very success of mankind has destroyed the natural habitats of many species and careful management of the local environment is necessary if such species are to be preserved.

A final division of the conservation movement, known as rural conservation, is concerned to prevent the continual encroachment by land-greedy cities upon the rural environment. The pleasure men feel when they visit or live in a relaxed country environment is at last recognized as being of as great value as benefits more easily evaluated financially.

Estimates of the dates by which the world's major mineral resources will have been exhausted.

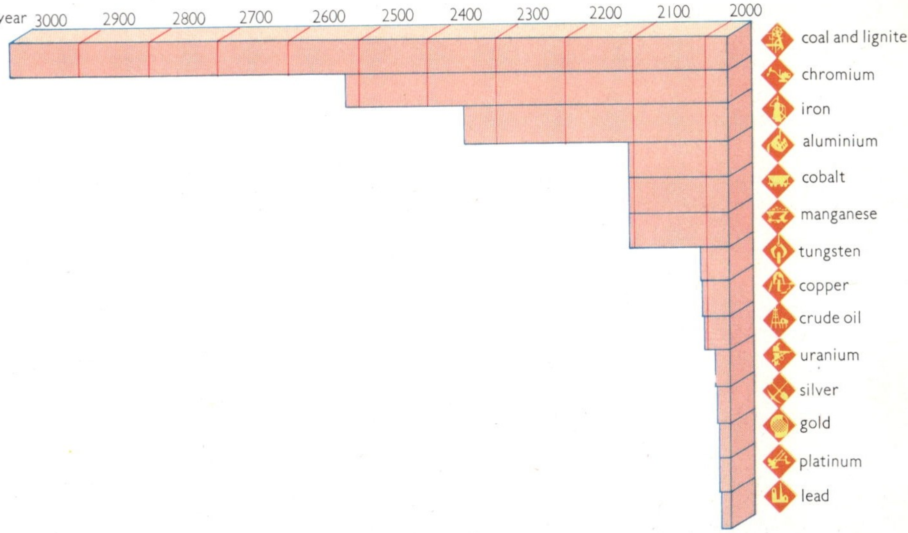

Major oil wells (red) and oil refineries (black) of the world

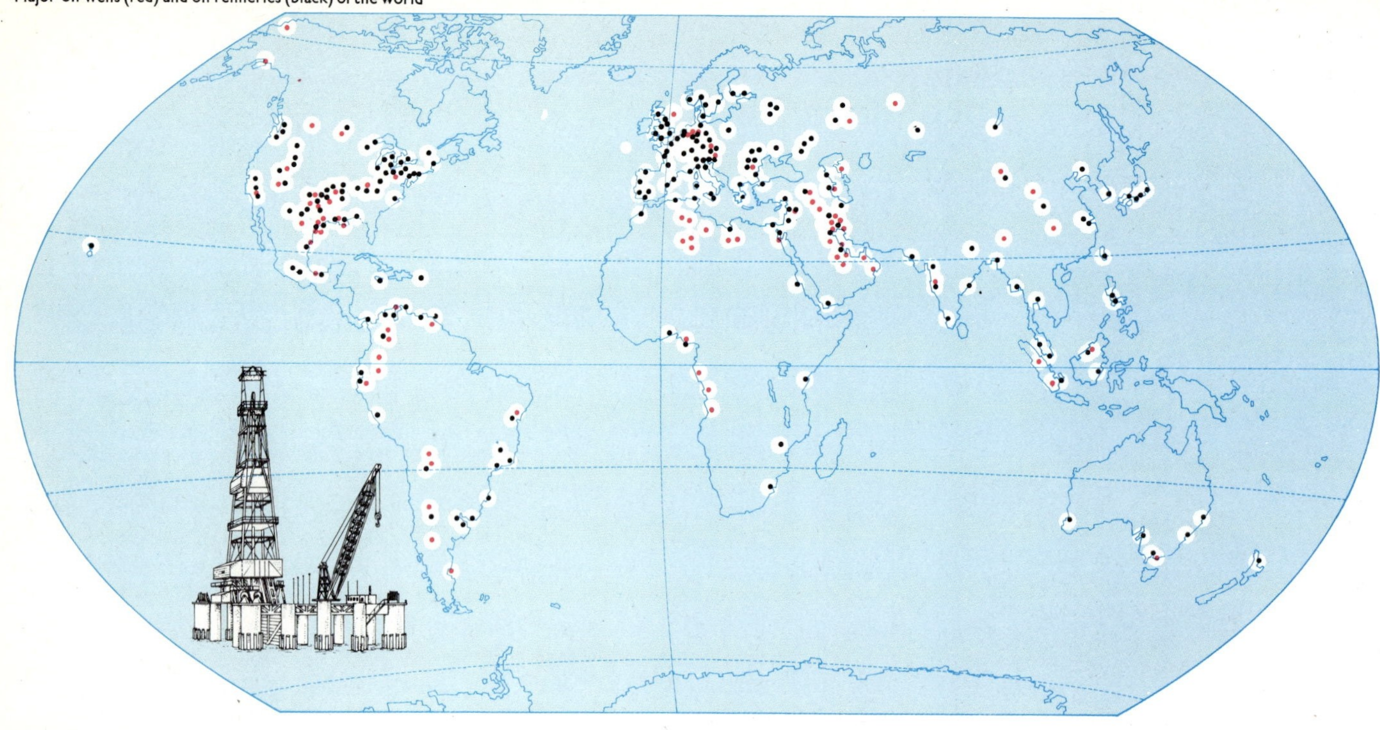

ENERGY AND MAN

Industrial man is nothing without energy. The story of his progressive mastery of his environment is the history of his discovery and development of new sources of useful energy – for cooking his food, to warm his shelter, draw his water and transport his person. The first breakthrough in energy technology came with the harnessing of steam power for pumping out mines in the 18th century. Within a century coal-raised steam was displacing all former energy sources – muscle, wood, wind and water – from their traditional employment. Steam both powered the factories and made mass transportation a reality, releasing men from the age-old constraint of locality.

The 20th century may have seen the passing of the age of steam but western man has become more than ever dependent on plentiful supplies of available energy. 'Available' is the key to the understanding of energy for although the total quantity of energy (or, more strictly, mass-energy) in the universe is constant – energy can neither be created nor destroyed – only a certain proportion of any given quantity can be converted into useful work.

A heat-engine is a machine which converts energy from one form to another. Typically it extracts chemical energy from a fuel (which is merely a convenient source or store of energy), puts some of it to work – in accelerating a car, turning a lathe or spinning a record turn-table – and expels the unusable energy in the form

Major coal (red) and lignite (blue) mines of the world

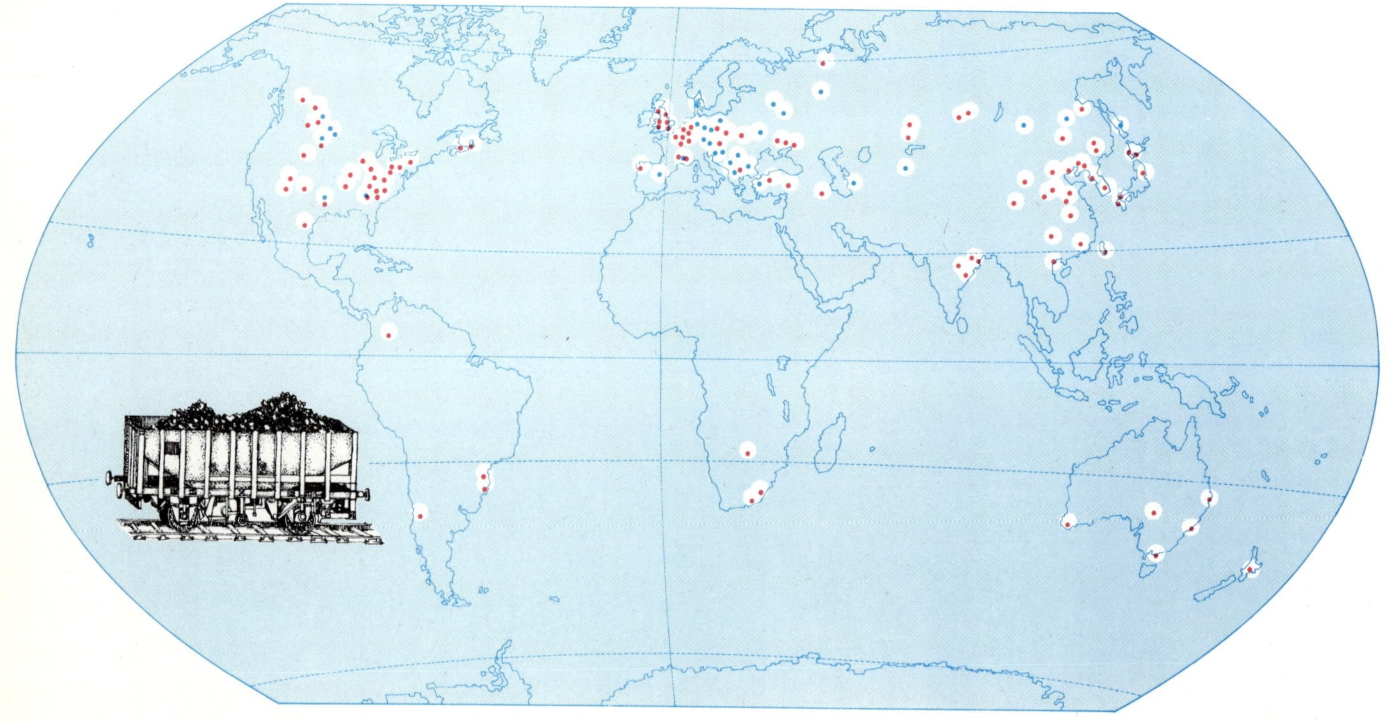

of waste (low-temperature) heat. But although a fuel may be used up its energy is never annihilated, for even the energy usefully converted into mechanical work is ultimately 'degraded' to unemployable heat. One solution to any energy shortage is to find ways of improving the efficiency of energy conversion – to discover how to get the maximum work out of the fuel.

Most of the world's energy resources ultimately originate in the solar radiation arriving at the top of the atmosphere. The sun is a massive nuclear furnace emitting energy in the form of heat and light radiation. Some of the energy reaching the earth warms the seas and land-surface, powering the winds and ocean-currents, while a minute proportion fuels the growth of plants.

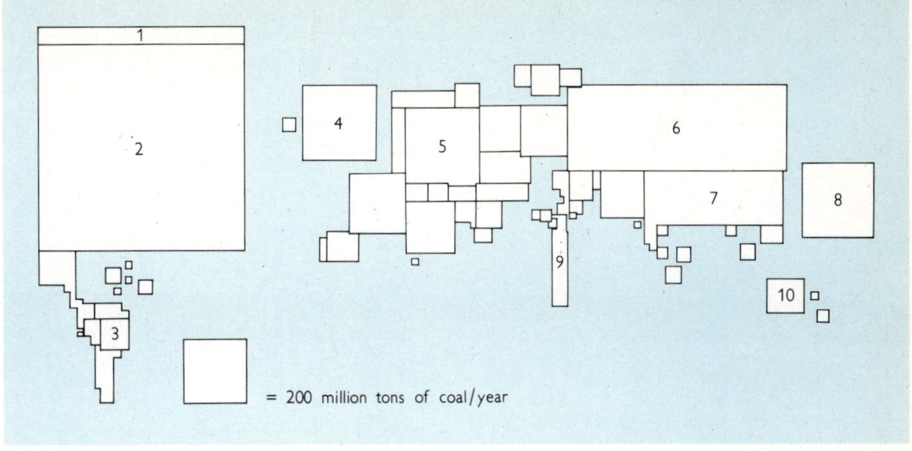

Consumption of Energy. The disproportionate share of the world's energy supplies consumed by the industrialized nations is highlighted on this diagram in which the size of each country is drawn in proportion to its energy consumption. (1) Canada, (2) the USA, (3) South America, (4) the UK, (5) Europe, (6) the USSR, (7) China, (8) Japan, (9) Africa, (10) Australia.

Although the potential for the direct conversion of solar energy is great, man has hitherto employed little ingenuity in harnessing this energy. Instead, while he still fuels his own body with vegetable foods and in many countries keeps himself warm by burning contemporary wood, 96% of his current energy consumption is supplied by burning 'fossil fuels' – the remains of plants which converted solar energy into carbohydrates many ages ago.

The principal fossil fuels are coal, petroleum oil and natural gas. Their importance to modern man is reflected both in the technological lengths to which he will go to get them out of the ground and in the political price he is willing to pay to secure access to adequate supplies. Although the stocks of any fossil fuel must ultimately become exhausted, if the capital is being eroded faster than it is replenished, recent experience has shown that known reserves tend to increase the more effort is put into looking for them. Further, the soaring price of any fossil fuel that appeared to be running out would force a switch to other sources of energy long before supplies were actually exhausted.

Although it is virtually impossible to store in useful quantities, the most versatile form of energy is electricity. The 'great blackout' which struck the north-eastern USA in November 1965 dramatically illustrates western man's dependence on electrical power. For 12 hours 30 million people were plunged into darkness and chaos resulted. While at present most electricity is generated by the conversion of fossil fuels, as these become more expensive there will probably be a switch to greater dependence on nuclear and water power. Most hydro-electric power is at present won by interrupting the hydrological cycle in which water evaporated from the oceans is precipitated on hills before flowing back down rivers into the oceans, but in future more electricity will be generated by trapping tidal energy behind barrages built across the mouths of large estuaries. In the long run nuclear fuels and the more efficient interception of solar energy offer fuller solutions to current energy shortages and the present fossil fuels will be released to serve as valuable raw materials for plastics and fibres manufacture.

Although potential energy supplies are not infinite (and there are serious problems regarding for instance nuclear waste disposal), if the necessary technology is developed, man need not fear that his energy requirements can be met throughout the forseeable future.

Major centres of electricity production in the world

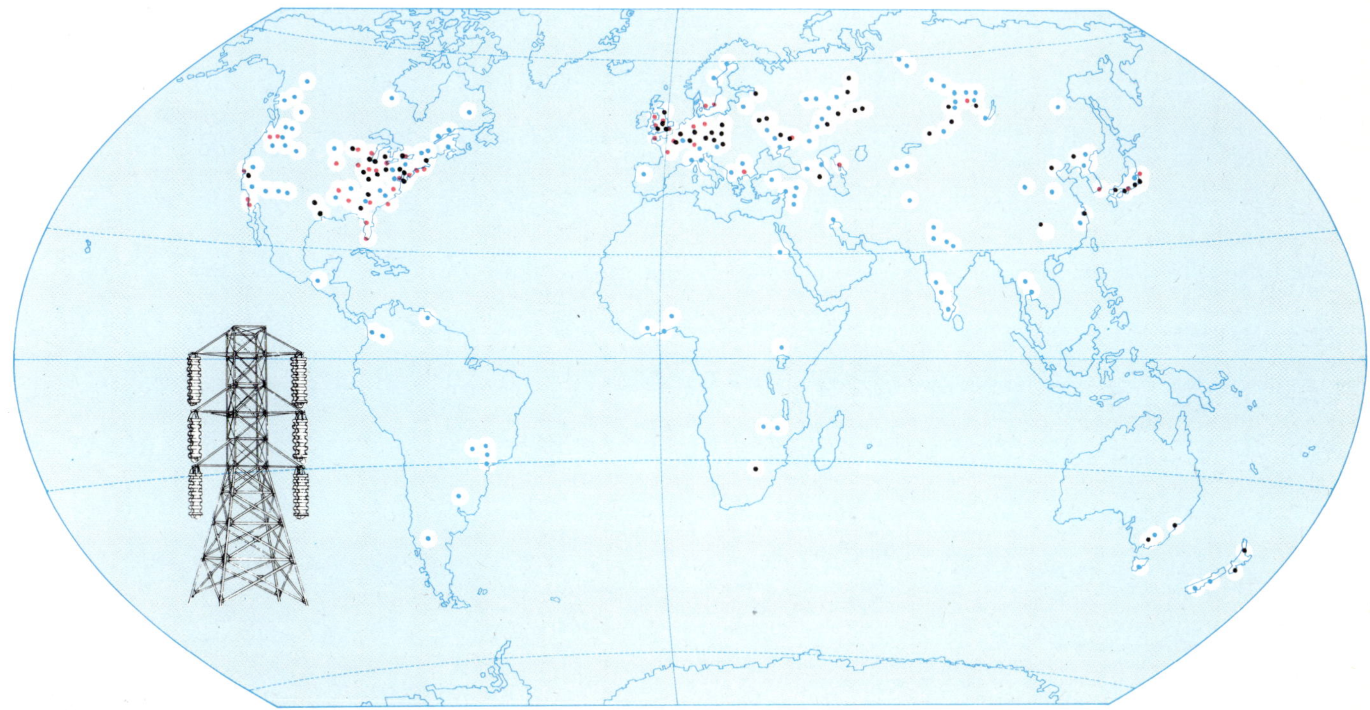

Major iron mines of the world

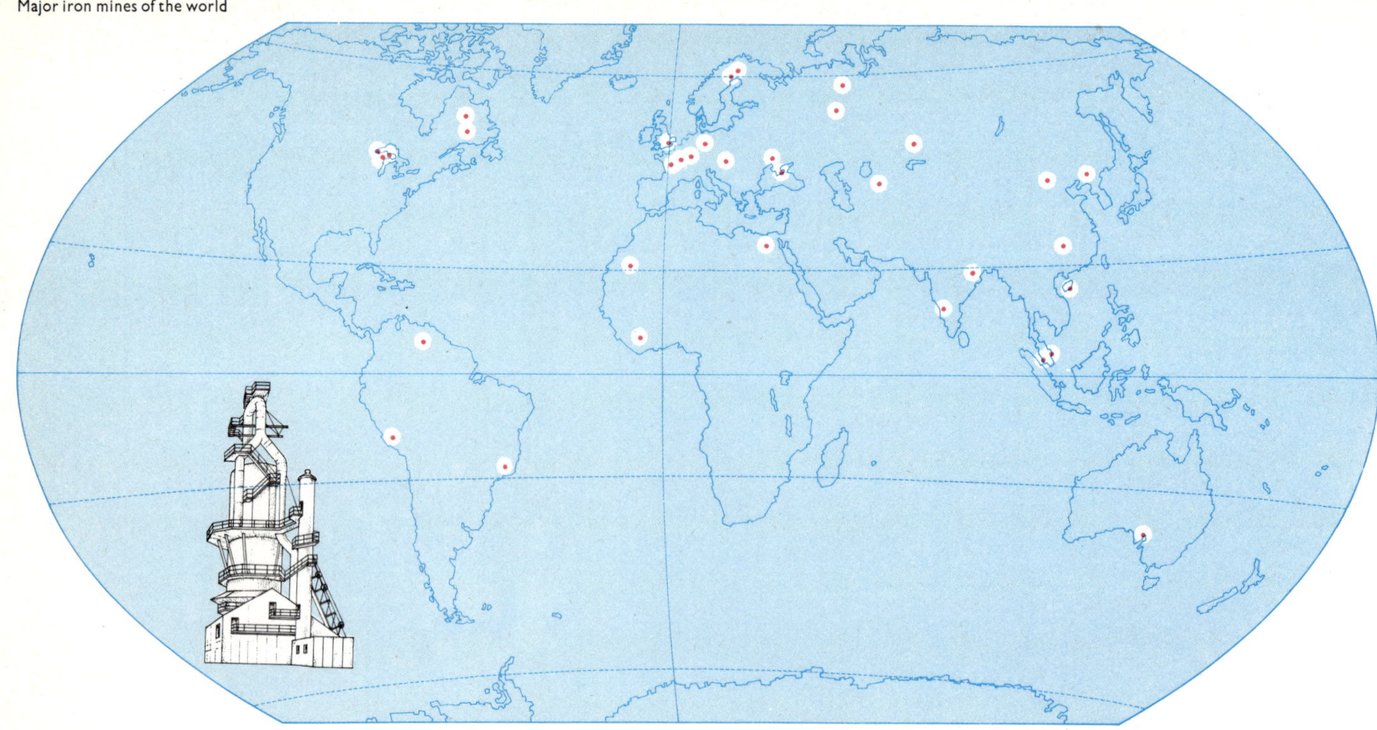

IRON AND STEEL

Steel forms the fabric of modern civilization. In 1970 world production of crude steel stood at 583 million long tons, 71% more than a mere decade earlier and 20 times the production in 1900. Most of the world's industry uses steel in the form of plate, wire, rail or tube as one of its raw materials and international steel production figures provide an accurate index to the health of the global economy.

The natural raw materials for steel production are the many ores of iron, coke (from coal) and limestone (used as a flux in smelting) but a considerable proportion of annual production represents the recycling of scrap steels. Iron makes up 5% of the weight of the earth's continental crust and although ores of progressively poorer quality will have to be worked, there is no danger that supplies of iron ore will run short before the year 2300.

After smelting in blast furnaces, most pig iron is converted into steel by the removal of some of the carbon that remains in the crude product. Increasingly this is being done by blowing pure oxygen gas across the surface of the liquid iron. The finished product is made either by casting shapes that can be directly machined or forged or by rolling solidified ingots into plate, girders, wire or tubes.

Because economies of scale are so significant in steelmaking, there is a continuing trend towards ever larger plants. In order to make 1 ton of finished steel 2·5 to 2·75 tons of raw materials must be collected on site so modern steel plants tend to be sited so as to minimize transport costs. This usually results in a coastal location being chosen, especially where raw materials have to be imported. At present nearly 80% of world steel production is concentrated in the USA, the USSR, Japan and the EEC. Other nations buy crude steel from the major producers and use it to manufacture their own finished products.

The reduction of iron ore. So that the raw materials used in iron smelting can be used as efficiently as possible, the blast furnace is surrounded by an extensive complex of auxiliary equipment.

A supply of iron ore, coke and flux (chiefly limestone) stored in nearby bunkers is brought to the furnace on conveyor belts or in loading cars (1). In amounts weighed out according to the existing programme, the materials are hauled in skip cars up a hoist to the top of the furnace (2). To minimize escape of gas, the receiving hopper is equipped with a gas lock consisting of a small and a large bell (3 and 4). The lock is operated by cylinders (3a and 4a) in the stock house (A) which also contains the electric winch (5) used to raise the skip cars.

The combustion process frees gas that can be used again for heating purposes. Before it can be reused, however, the solid constituents of the gas must be removed. This process is carried out by the dust catcher (B), the gas washers (C), and the electric precipitators (D).

The desired temperature in the blast furnace can be attained only by using preheated air. The blast-furnace gas is used for this purpose. It is taken to air-blast stoves (E), which are lined with firebrick. After the chocker work of firebrick has been raised to the proper temperature, the air blast passes through the stove and, mixed with cold air, reaches the furnace through the bustle pipe (F). The air temperature is regulated from the blowing house (G) to prevent fluctuations in the temperature within the furnace.

The smelted liquefied iron collects in the hearth at the bottom of the blast furnace, from which it is removed periodically through tapholes. The iron flows through a trough (H) to special hot-metal ladles (6) for transportation to the steel factories. The slag accompanying the iron is separated out and led to the slag ladles.

Slag from the blast furnace is tapped and led off through the slag trough and runner (J) to slag pits (K). In this process, slag that is unsuitable for making cement is separated out and placed in special cars (7) for later use in roadmaking.

2
3
4
C
D
B
E
E
E
G
F
F
H
3a
4a
5
1
1
6
J
J
J
K
K
A
7
1
2
D A S

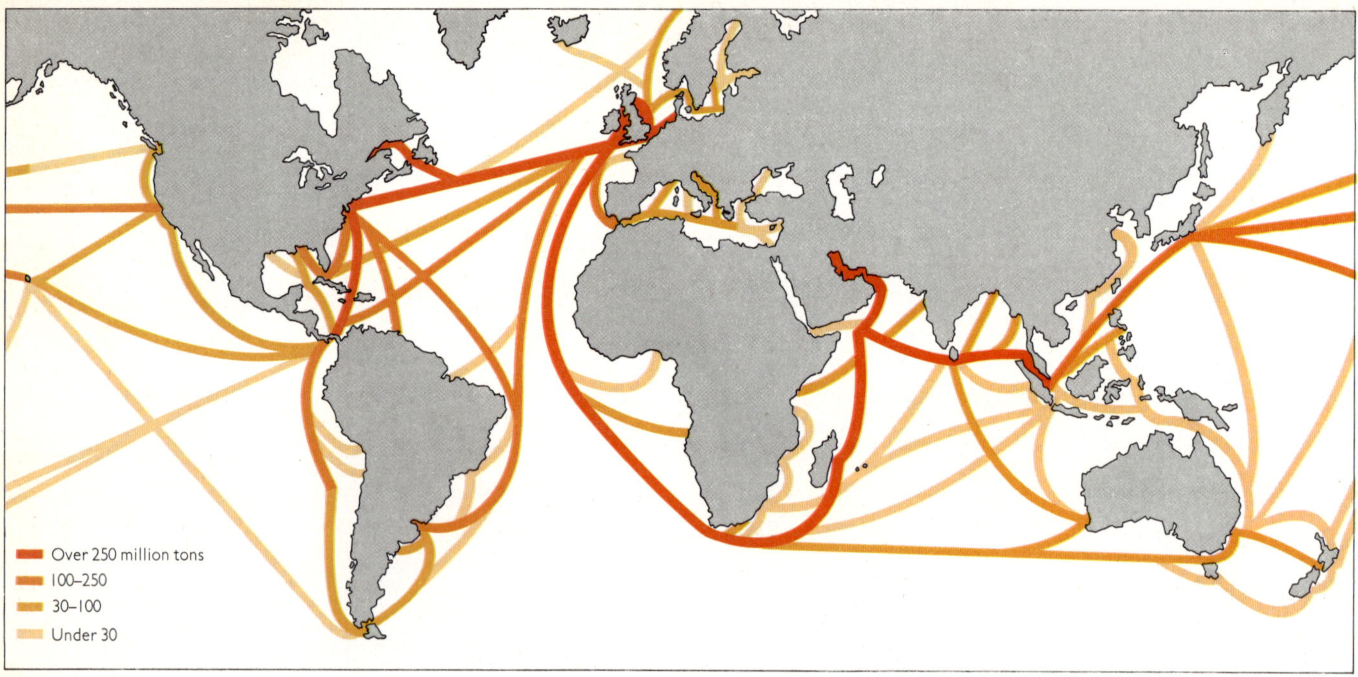

The major shipping routes of the world. By far the most important route links the USA with western Europe.

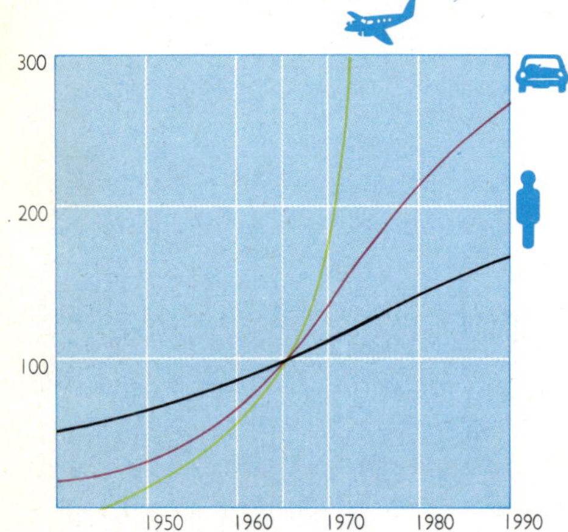

The actual and projected growth in the volume of air transportation (yellow) and motor vehicle use (red) compared with man's growing population (black). All the curves are drawn to an index of 100 for 1965. Air transport is unlikely to grow as quickly as the projection suggests because of the greatly increased cost and scarcity of fuel.

TRANSPORTATION AND INDUSTRY

The coming of the industrial revolution in the 18th century highlighted man's need for efficient means of transport. Until the time the first industrial canals were cut, overland transport remained an unreliable and dangerous enterprise and it was not until the development of the railway systems of Europe and North America that efficient large-scale manufacturing could become widespread away from coastal regions.

Across most of the world internal transport is dominated by the motor vehicle. In developed nations an increasing proportion of the people own their own cars and this has led to growing disuse of public transport systems. In most western countries road transport has also become dominant in the carrying of freight, though in several others, legislation aims to keep a high proportion of goods traffic on the railways. The motor vehicle, although very flexible in use, has a high social cost, particularly in overcrowded countries where road building swallows ever more of the precious land. As congested roads become increasingly inefficient, traffic is likely to return to the railways. A single-track railway can carry 48,000 passengers per hour at speeds far higher than the maximum of 12,000 which a single lane of motorway can accommodate. Railways have always remained the more attractive means for transportation of bulk freight and urban commuter traffic.

The growth of world air transport.

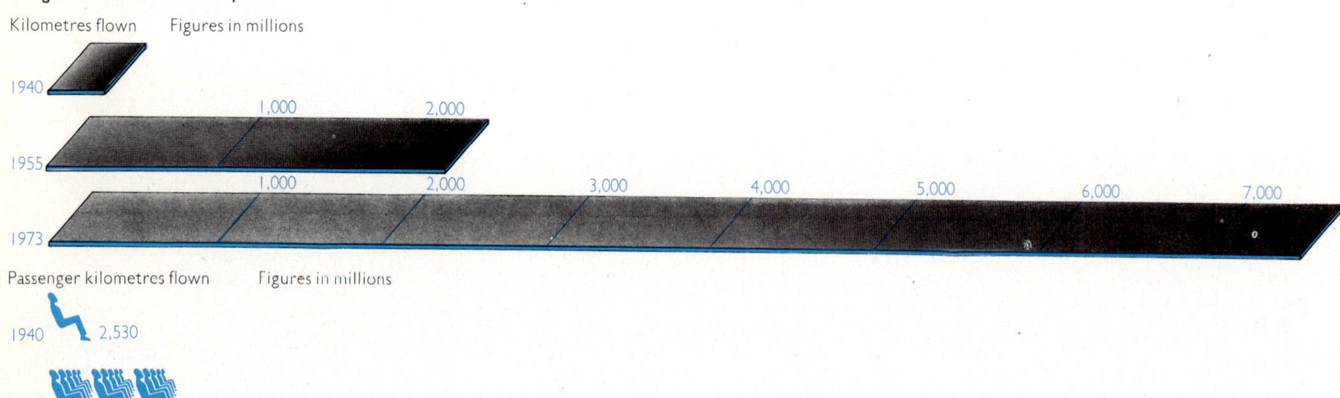

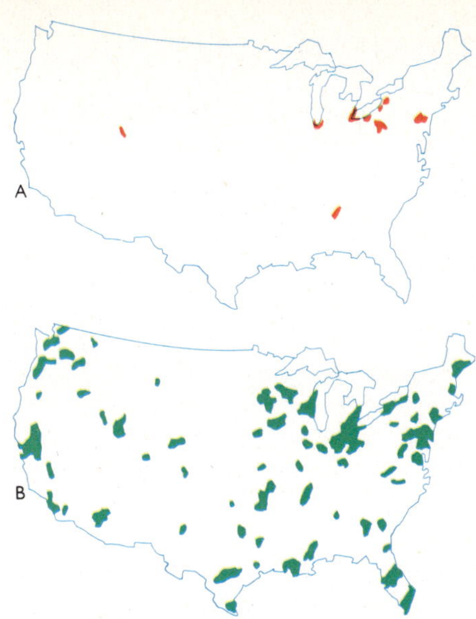

The location pattern of steel plants in the USA (A) reflects the high transport costs of raw materials associated with iron smelting. The food canning industry (B) is much less centralized because canning must occur in the agricultural source regions.

On the international level, ocean shipping remains the most efficient means of transporting freight although air-freighting of valuable and perishable light goods is likely to continue its growth. The largest oil tankers can deliver crude oil over 11,000 miles at rates of only £1 per ton and heavier-than-air flight, with its heavy fuel costs, is unlikely ever to challenge such rates. The growth in air passenger transport is likely to continue into the 1980s though perhaps at a slower rate than some observers have suggested largely because of increased fuel costs.

As with iron and steel making, the location of manufacturing industry is becoming increasingly dependent on transport costs, both for bringing in raw materials and for dispatching finished products to the market. Even as small a change as the opening of the St Lawrence Seaway in 1959 was sufficient to revitalize the Great Lakes steel industry in the 1960s.

The fate of industry depends on developments in transportation in a further important respect. The economies of many of the most highly industrialized nations are heavily dependent on the health of the industries manufacturing transportation equipment and motor vehicles. Three of the ten largest corporations in the USA manufacture automobiles and any setback in the sales of these companies in turn affects the profits of the thousands of firms which supply them with components. In more ways than one can transportation be seen to be the lifeblood of modern man.

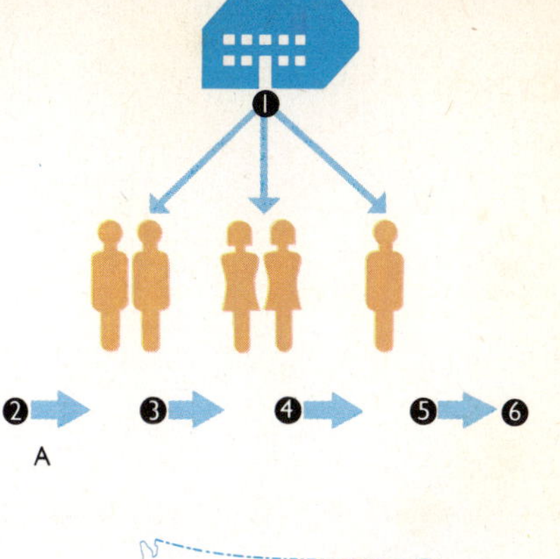

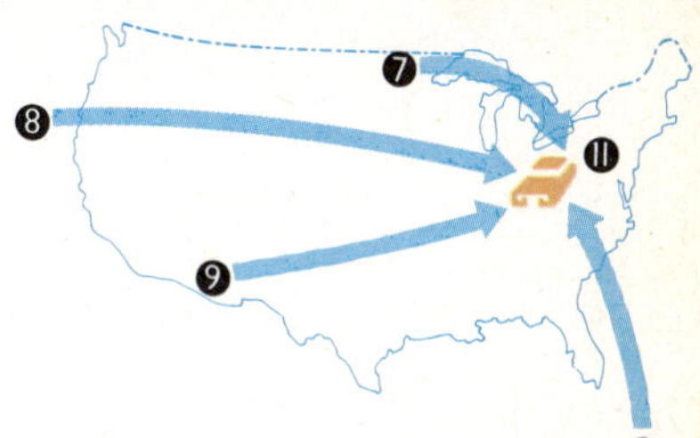

Contrasting patterns of manufacture. (A) Before the industrial revolution, the manufacture of woollen cloth in south-west England was organized on the 'putting out' system. A merchant 'put out' (1) the materials to farmers to process in their own homes, successive craftsmen carding (2), spinning (3), weaving (4) and finishing (5) the cloth, which was then sold (6) by the merchant. (B) Car manufacturing in the USA collects together steel (7), rubber (8), glass (9) and leather (10) from widely scattered sources; before assembling the product on a single site (11).

New cars coming off a Fiat assembly line in Italy. Each vehicle is carefully tested before dispatch from the factory.

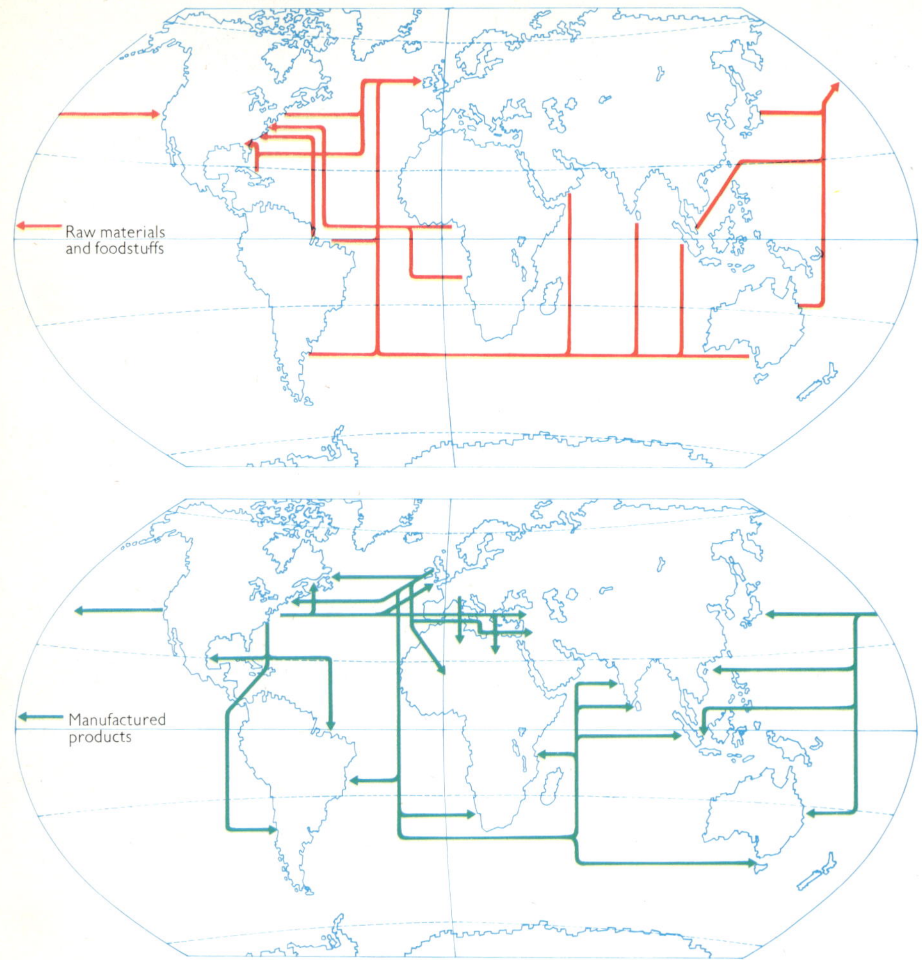

The general pattern of world trade involves a net flow of raw materials into the USA and Western Europe while the majority of manufactured products are exported to the rest of the world from these two regions. There is also a considerable trade in manufactured goods between the industrialized nations.

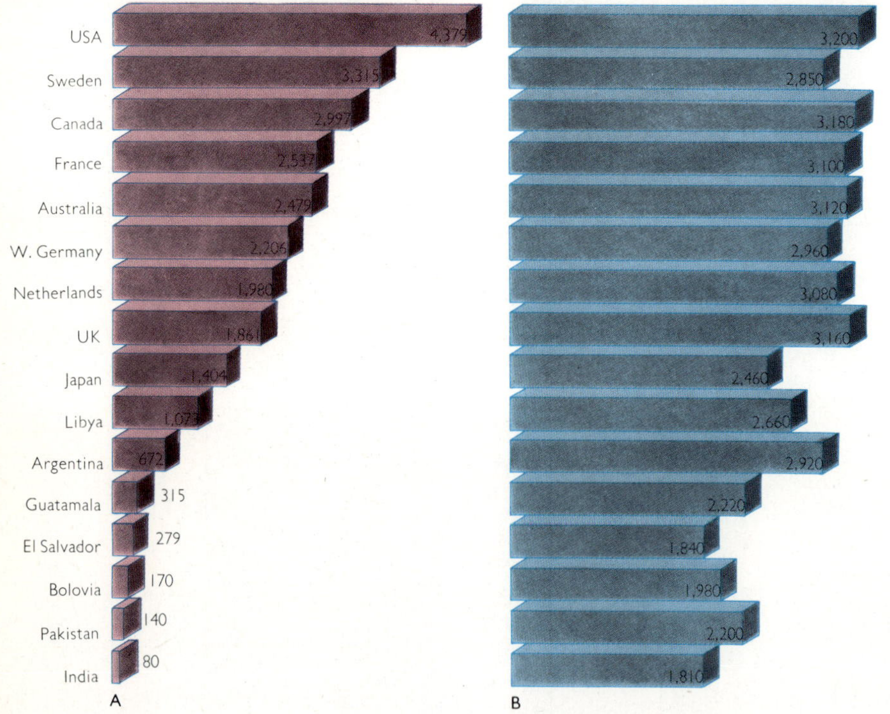

Standard of living can be measured in terms of (A), the average income per person per year in US dollars, or alternatively (B), the average number of calories consumed by each person per day.

TRADE AND STANDARDS OF LIVING

The most striking feature of the pattern of world trade is the flow of raw materials and foodstuffs from various parts of the developing world into the heavily industrialized nations of the West and the subsequent trade in manufactured goods back to the countries of the third world. Can it be an accident that the primary exporting countries (those exporting raw materials and importing manufactured goods) are also those with the world's lowest standards of living?

The only secure route to the improvement of a nation's standard of living is through economic growth. For a third world nation this involves the establishment of secondary (manufacturing) industries so that native raw materials can be processed internally thus saving on the high cost of import and perhaps producing a surplus for export.

The barriers to such development are great. Few third-world countries have sufficient capital or technological experience to develop manufacturing industries. They lack both pools of skilled labour and sufficient administrative expertise. Perhaps more serious is the fact that the economies of scale are so great in modern industry that the price of imported manufactures is often much lower than the minimum which could be charged for the home-produced product. One way round this problem is illustrated by the increasing number of developing nations in which there are plants for the final assembly of components bought in from the large-scale producers.

A serious problem is the supply of capital for such schemes. Many third-world nations, producing barely enough for the mere survival of their populations, have no surplus capital for investment. They are thus dependent on supplies of foreign aid and investment capital which are, however, seldom motivated by pure altruism. The true value of international aid is pitifully small; in 1971 the total flow of development aid, after allowing for the repayment of earlier aid-loans and the payment of interest charges, was only $8,000 million, less than 0·4% of the total gross national products of the USA, the USSR, the UK and the then six-member EEC.

Can standards of living in the developing nations be improved? Probably not while current patterns of world trade continue and the gap between rich and poor nations continues to widen. Nationalization of foreign investments tends to be counter-productive as it discourages inflows of new capital. Only when they can charge fairer prices for their raw-material exports, as has happened recently with oil, will the developing countries become able to take their rightful place among the nations of the world.

EUROPE

Europe's unique position in the world is the product of a rare and fortunate combination of physical and human assets. Although one of the smallest continents (only the island continent of Australia is smaller), Europe is among the most diverse. Well endowed with natural resources, it is blessed with a generally temperate climate that has favoured human settlement and endeavour. Add to these advantages the energy, enterprise, inventiveness and intellectual power of its peoples, and it is easy to see why this vast peninsula of the Eurasian land mass has been able to exert a world-wide influence.

Europe in History

Europe was the mother of Western civilization. The cradle was the Aegean shores where the ancient Greeks (building on Middle Eastern achievements) developed their distinctive culture. Under Rome, Europe south of the Danube and west of the Rhine became part of a Mediterranean world in which Christian ethics were presently added to Roman law and learning, but which eventually succumbed to the invasions and migrations of barbaric peoples. In the west the Christian Church preserved ancient learning and medieval Europeans gained new knowledge and skills from the Arabs, Indians and Chinese. Civilization in Europe flowered anew in the Renaissance, with its revived interest in the classical heritage and fresh advances in the arts and sciences, notably in the city-states of Italy. Even more remarkable was the Scientific Revolution which, beginning in the 17th century, changed the whole character of West European civilization. Preoccupation with the learning of the past gave way to study of the immediate and expanding world, and to application of scientific methods, in which the British, Dutch and French were leaders. Modern science originated in Europe.

Europe in the World. The great European voyages of discovery in the 15th century had set in motion processes of exploration and acquisition which profoundly altered the economic life of Europe, and which ushered in the age of overseas empire-building by Europe's maritime powers. To the largely backward lands that they exploited, the Europeans brought at first some of the institutions and material benefits of their own advanced civilization.

Most of such benefits were to flow from the Industrial Revolution – a distinctively European phenomenon which transformed the structure of society wherever it occurred, and which reached out in its effects across the world. The Industrial Revolution began in the later 18th century, when the British began developing a full-scale industrialized capitalist economy. By the mid-19th century Britain's example was being copied by other European countries. It is now that the continent's great mineral wealth, especially coal and iron, became of prime significance, as did the resources and markets available in the dependent territories of the imperial powers. Eventually technology and expertize, as well as manufactured goods, were to flow out from Europe. But many of these dependent countries had to await the great crumbling of empires after World War II before they were able to develop their own industrial potential to their own advantage. Some see the undeveloped Third World of today as a reminder that Europe has often taken too much and given too little.

About the same time as Britain was embarking upon her Industrial Revolution, Europe was stirring under the radical philosophies of the Enlightenment. Their challenge to benevolent despotism was unmistakable, and in France it culminated in revolution. French revolutionary ideas spread through Europe with the Napoleonic armies and they also found echoes in Latin America. After Napoleon's downfall revolutionary nationalism and the spirit of liberalism gathered force despite all attempts at repression by the old order, bringing independence to Greece and Belgium, and unity to Italy. Under Prussia's leadership imperial Germany was born.

By the end of the 19th century Europe with its still liberal civilization stood pre-eminent in world affairs. Its imperial powers and nation-states had an unprecedented stability. or so it seemed. Western Europe basked in its military strength, economic growth, scientific advances, artistic brilliance, glittering high society and prosperous middle classes, and there seemed no reason why this comfortable state of affairs should not continue indefinitely. But in 1914 rivalry between the great European powers culminated in the tragic conflict of World War I; Europe's economic and territorial stability was shattered. The post-war period was to see inflation and unemployment on a scale no one could remember, and this helped create the conditions which led to World War II.

Europe Today. World War II left Europe prostrate and divided, with two superpowers, the Soviet Union and the United States, in the ascendant. The political division between communist East and capitalist West Europe remains, but the chill climate of the Cold War years has given way to the warmer atmosphere of *détente*, marked by West Germany's treaty with East Germany, signed

Overleaf: **The Mont Blanc massif. Mt Blanc, at 15,781 ft, is the highest mountain of the Alps.**
Below: **A winter scene near Krasnoyarsk, capital of the Siberian territory of the same name.**

EUROPE
vegetation
snow and icecaps
tundras and high mountain flora
mountain grassland
mixed forest and northern coniferous forest
mediterranean vegetation
steppe grassland
cultivated areas, meadows and pastures
cultivated areas, arable land
cultivated areas, mediterranean agriculture
swamp
0 250 500 st. miles
scale 1 : 20,000,000
ARCTIC OCEAN
ATLANTIC OCEAN
GREENLAND
Arctic Circle
DENMARK STRAIT
Jan Mayen
ICELAND
Reykjavik
BARENTS SEA
Novaya Zemlya
North Cape
Hammerfest
Varangerfjord
KOLA PEN.
LAPLAND
WHITE SEA
TIMAN RIDGE
Kebnekaise 6962
LOFOTEN
VESTERAALEN
FAEROE ISLANDS
SHETLAND ISLANDS
ORKNEY ISLANDS
HEBRIDES
Ben Nevis 4406
GRAMPIANS
Edinburgh
NORTH SEA
IRELAND
Dublin
IRISH SEA
Snowdon 3560
PENNINES
St. George's Channel
Land's End
London
Thames
Strait of Dover
ENGLISH CHANNEL
Snøhetta 7500
Sognefjord
Hardangerfjord
Oslo
SKAGERRAK
KATTEGAT
Jutland
Copenhagen
Bornholm
Lake Väner
Lake Vätter
Stockholm
Gotland
Öland
BALTIC SEA
GULF OF BOTHNIA
FINNISH LAKE PLATEAU
Helsinki
Gulf of Finland
Leningrad
Lake Ladoga
Lake Onega
Hiiumaa
Saaremaa
Gulf of Riga
BALTIC HILLS
Rybinsk Reservoir
NORTH RUSSIAN RIDGE
VALDAY HILLS
Moscow
WEST RUSSIAN RIDGE
CENTRAL RUSSIAN UPLANDS
W. Dvina
Dnepr
Pripet Marshes
Kiev
Desna
Donets
Dnestr
Bessarabia
Odessa
Crimea
NORTH GERMAN PLAIN
Amsterdam
Brussels
Elbe
Weser
Rhine
Berlin
Oder
Vistula
Warsaw
Prague
Gerlachovka 8737
CARPATHIAN MTS.
Vienna
Budapest
Great Hungarian Plain
Tisza
Danube
Negoi 8346
Walachia
Brittany
Seine
Paris
Loire
Gironde
Garonne
BAY OF BISCAY
P. de Sancy 6188
MASSIF CENTRAL
Bern
Mt. Blanc 15781
Gross Glockner 12461
ALPS
Drava
Sava
Lombardy
Po
Belgrade
DINARIC ALPS
Dalmatia
Durmitor 8274
BALKAN MTS.
Macedonia
Olympus 9550
Istanbul
Bosporus
BLACK SEA
Ankara
ASIA MINOR
AEGEAN SEA
Athens
CYCLADES
SPORADES
Rhodes
PELOPONNESOS
Cape Matapan
IONIAN ISLANDS
IONIAN SEA
CRETE
CYPRUS
Cape Finisterre
CANTABRIAN MTS.
Douro
CASTILIAN PLATEAU
Madrid
Tagus
Lisbon
Guadiana
SIERRA MORENA
Guadalquivir
Andalusia
SIERRA NEVADA
Mulhacen 11424
Gibraltar
Str. of Gibraltar
Tangier
Cape Saint Vincent
Ebro
PYRENEES
Aneto Pk. 11168
Marseille
BALEARIC ISLANDS
Mallorca
Mt. Cinto 8891
CORSICA
Str. of Bonifacio
SARDINIA
APENNINES
Mt. Corno 9560
Rome
Vesuvius 3842
ADRIATIC SEA
TYRRHENIAN SEA
SICILY
Mt. Etna 10868
Strait of Messina
Tunis
MEDITERRANEAN SEA

Pollution of air and water around major industrial sites such as this one is a growing problem in many areas of Europe. Dutch scientists, for instance, claim to have developed film in the chemical-laden waters of the lower Rhine.

in December, 1972, and a European security conference at Helsinki in 1973.

In Western Europe, post-war reconstruction was accompanied by moves towards unity. Economic unity was largely accomplished by the creation and subsequent expansion of the European Economic Community (the Common Market), the Eastern counterpart of which is COMECON (the Council for Mutual Economic Aid), and monetary union by 1980 is also envisaged. Western *political* Union, however, is a much more complex matter involving national sovereignty and national pride. Despite the establishment of such bodies as the Council of Europe and the European Parliament, the creation of anything like a 'United States of Europe' seems far distant.

The Face of Europe

Location and Size. Europe covers some 4,063,000 sq mi (10,523,000 km²), about one-fourteenth of the earth's total land surface. It spans some 3,300 mi (5,300 km), from the Ural Mountains – its eastern boundary with Asia – to Cape St Vincent (Portugal); and from Norway's North Cape it extends southwards for some 2,450 mi (3,940 km) to Cape Matapan (Greece). Bounded by the Arctic and Atlantic oceans in the north and west respectively, and by the Mediterranean, Black and Caspian seas in the south, it has a long, irregular coastline, deeply penetrated by the sea, and offshore islands by the thousand, the largest group being the British Isles.

The Mountain Systems. Basic to the structure of Europe are its three mountain systems, known to geologists as the *Caledonian*, *Hercynian* and *Alpine*.

Some of the oldest rocks in the world form extensive areas of rather monotonous relief in Fenno-Scandia, the Baltic Shield of Finland and Sweden. It was against these ancient rocks, some 3,500 million years old, that the ranges of the Caledonian system were folded about 440 million years ago. This is the system which dominates Norway from the North Cape to the Skagerrak, and also extends through Scotland and Wales to Ireland.

The Hercynian system takes its name from the Harz Mountains of Germany and embraces the uplands, plateaux and mountains extending eastwards from Brittany and the Iberian peninsula to the Urals. It includes the rough upland region of Armorica (Brittany and western Normandy), the Central Massif of France, the Ardennes of Belgium and their continuation into the Rhenish uplands, the Vosges and Black Forest, the Thuringian Forest and the Harz Mountains, the Bohemian Plateau, the Lysa Gory, in southern Poland, the Ukraine Massif and the highly mineralized Urals.

Most of Spain and Portugal is formed of similar fold ranges, worn down to the *Meseta* (tableland) crossed by a number of sharp sierras. Aspromonte, in the 'toe' of Italy, and the block-like island of Sardinia are part of the same system. Ancient rocks occur again in the Balkan peninsula, but the greater part of eastern Europe is underlain by little disturbed primary rocks forming the 'Russian platform'.

Europe's most dramatic mountains, however, are the result of Alpine earth-movements which took place about 60 million years ago. Alpine Europe is the series of lofty mountain ranges extending across southern and south-central Europe from the Pyrenees through the Alps to the Carpathians and the Caucasus. The highest peaks occur in the Caucasus, where Mt Elbrus, Europe's highest mountain, reaches 18,481 ft (5,633 m). In western Europe the highest peaks are in the Alps, where Mont Blanc on the French-Italian border rises to 15,781 ft (4,810 m). In Switzerland the Pennine Alps attain 15,204 ft (4,634 m) in Monte Rosa. The central Pyrenees are permanently snow-capped, with Maladeta rising to 10,853 ft (3,308 m).

This fold system also appears in Spain, in the rugged Sierra Nevada, and in the Balearic Islands, Corsica and Sicily. In peninsular Italy the central Apennines have rugged limestone peaks, especially in the Abruzzi where Monte Corno, in the Gran Sasso d'Italia, reaches 9,560 ft (2,914 m). Yugoslavia has its Dinaric fold system, which continues in Greece as the Pindos Mountains and reappears as the mountain backbone of the island of Crete.

Europe's Volcanoes and Earthquakes. Vulcanism has helped to shape the relief of Europe for more than 250 million years. Its work is seen in the *puys* or volcanic cones in the Auvergne Mountains of south-central France such as the well-known Puy de Dôme; in the *Maare* (crater lakes) in the volcanic Eifel region of West Germany; and in the crater lakes around Rome. Volcanic rocks occur in many parts of Europe, especially Iceland.

Many volcanoes have been active in Europe in historical times. Italy has Vesuvius, Stromboli and Mt Etna, all still active volcanoes, and the island of Vulcano – inactive since 1890 – from which the word 'volcano' is derived. A strange and attractive feature of

EUROPE
political
CITY population more than 1,000,000
CITY population more than 500,000
CITY population more than 100,000
City population more than 50,000
City population less than 50,000
railways
roads
airport
scale 1 : 20,000,000
0
250
500 st. miles
ARCTIC OCEAN
ATLANTIC OCEAN
BARENTS SEA
WHITE SEA
NORTH SEA
BALTIC SEA
GULF OF BOTHNIA
Gulf of Finland
Gulf of Riga
ENGLISH CHANNEL
BAY OF BISCAY
MEDITERRANEAN SEA
TYRRHENIAN SEA
ADRIATIC SEA
IONIAN SEA
AEGEAN SEA
BLACK SEA
GREENLAND
Arctic Circle
Jan Mayen (Norway)
Bear Island (Norway)
Novaya Zemlya
North Cape
ICELAND
Reykjavik
FAEROE ISLANDS
SHETLAND ISLANDS
ORKNEY ISLANDS
HEBRIDES
NORWAY
SWEDEN
FINLAND
DENMARK
UNITED KINGDOM
IRELAND
NORTHERN IRELAND
FRANCE
SPAIN
PORTUGAL
ANDORRA
MONACO
SWITZERLAND
LIECHTENSTEIN
Luxembourg
BELGIUM
NETHERLANDS
GERMANY
EAST
POLAND
CZECHOSLOVAKIA
AUSTRIA
HUNGARY
ROMANIA
YUGOSLAVIA
BULGARIA
ALBANIA
GREECE
ITALY
SAN MARINO
CORSICA
SARDINIA
SICILY
Malta
CRETE
CYPRUS
Rhodes
Samos
UNION OF SOVIET SOCIALIST REPUBLICS
Hammerfest
Narvik
Kiruna
Kemi
Pechenga
MURMANSK
ARKHANGELSK
Kotlas
TRONDHEIM
BERGEN
Stavanger
OSLO
Sundsvall
STOCKHOLM
NORRKÖPING
GÖTEBORG
MALMÖ
Gotland
Öland
Bornholm (Dk.)
AARHUS
COPENHAGEN
TAMPERE
TURKU
HELSINKI
TALLINN
Vyborg
LENINGRAD
Lake Ladoga
Lake Onega
VOLOGDA
RYBINSK
YAROSLAVL
IVANOVO
GORKIY
KALININ
MOSCOW
RIGA
VITEBSK
SMOLENSK
TULA
OREL
VORONEZH
KAUNAS
VILNIUS
KALININGRAD
MINSK
KHARKOV
KIEV
DNEPROPETROVSK
LVOV
KISHINEV
ODESSA
Perekop
SEVASTOPOL
ABERDEEN
EDINBURGH
GLASGOW
BELFAST
DUBLIN
Cork
LIVERPOOL
LEEDS
MANCHESTER
BIRMINGHAM
CARDIFF
PLYMOUTH
SOUTHAMPTON
LONDON
A'DAM
R'DAM
BRUSSELS
LILLE
LE HAVRE
BREST
PARIS
NANTES
STRASBOURG
DIJON
BORDEAUX
LYON
TOULOUSE
MARSEILLE
TOULON
NICE
Monte Carlo
HAMBURG
BREMEN
HANNOVER
BERLIN
LEIPZIG
DRESDEN
COLOGNE
BONN
FRANKFURT
STUTTGART
MUNICH
BASEL
BERN
GDAŃSK
SZCZECIN
POZNAN
WARSAW
ŁODZ
WROCŁAW
KRAKOW
BIAŁYSTOK
PRAGUE
BRNO
KOŠICE
BRATISLAVA
VIENNA
LINZ
GRAZ
BUDAPEST
DEBRECEN
CLUJ
IAŞI
PLOESTI
BUCHAREST
ZAGREB
TRIESTE
BELGRADE
SARAJEVO
NIS
SKOPJE
SOFIA
VARNA
ISTANBUL
IZMIR
TIRANE
THESSALONIKI
ATHENS
MILAN
TURIN
GENOA
VENICE
FLORENCE
ROME
NAPLES
BARI
TARANTO
PALERMO
MESSINA
CAGLIARI
Ajaccio
LA CORUÑA
Cape Finisterre
SANTANDER
BILBAO
PORTO
VALLADOLID
ZARAGOZA
BARCELONA
MADRID
LISBON
VALENCIA
Mallorca
PALMA
BALEARIC ISLANDS
CORDOBA
SEVILLA
CADIZ
MALAGA
MURCIA
CARTAGENA
TANGIER
Gibraltar (Br.)
CASABLANCA
ALGERS
TUNIS
N. Dvina
W. Dvina
Volga
Oka
Don
Dnepr
Dnestr
Prut
Danube
Elbe
Oder
Vistula
Rhine
Rhône
Loire
Seine
Ebro
Douro
Tagus
Thames

The Lofoten Islands lie off the coast of Nordland in north-west Norway, just above the Arctic Circle. Their waters are rich cod and herring grounds and from January to April thousands of fishermen from all parts of Norway come to fish.

the Aegean is the partly submerged crater of the island of Santorini whose catastrophic volcanic explosion shattered ancient Greek cities more than 3,000 years ago. Some of vulcanism's most awsome current activity has occurred in Iceland. In 1963 the island of Surtsey rose dramatically from the sea off the south coast; in 1970 Hekla was again active; and on 23 January 1973, Helgafell roared into life, hurling ash and lava on the fishing town of Vestmannaeyjar on the island of Heimay, which had to be hurriedly evacuated.

Besides volcanic activity, earth tremors and earthquakes have occurred in the Mediterranean in historical times – particularly in the lower Tagus valley of Portugal, in southern Italy and Sicily, Malta, Greece, the Aegean islands and Yugoslavia.

Lowlands and Rivers. Europe's lowlands reach their greatest extent in the north-central part of the continent. Widening eastwards from the Low Countries across northern Germany, they extend into Poland and Russia, once described as 'an unending zone of plain', and continue northwards into Denmark and south-west Sweden. In northern France and southern and eastern England, scarplands dominate the landscape. Elsewhere are triangular-shaped lowlands associated with rivers like the Garonne, the axis of the Aquitaine basin, the Ebro in Catalonia, the Guadalquivir in Andalusia, and the Po in Lombardy. Other major river lowlands are those of the Rhine and its associated tributaries, the Main and the Neckar; the Danube, flowing eastwards to the Black Sea; and the Saône-Rhône corridor in south-east France. In the Mediterranean peninsulas, river basins and coastal plains are of limited extent.

Effects of the Ice Age. The surface features of widely-separated parts of Europe were altered by glaciation in Pleistocene times which

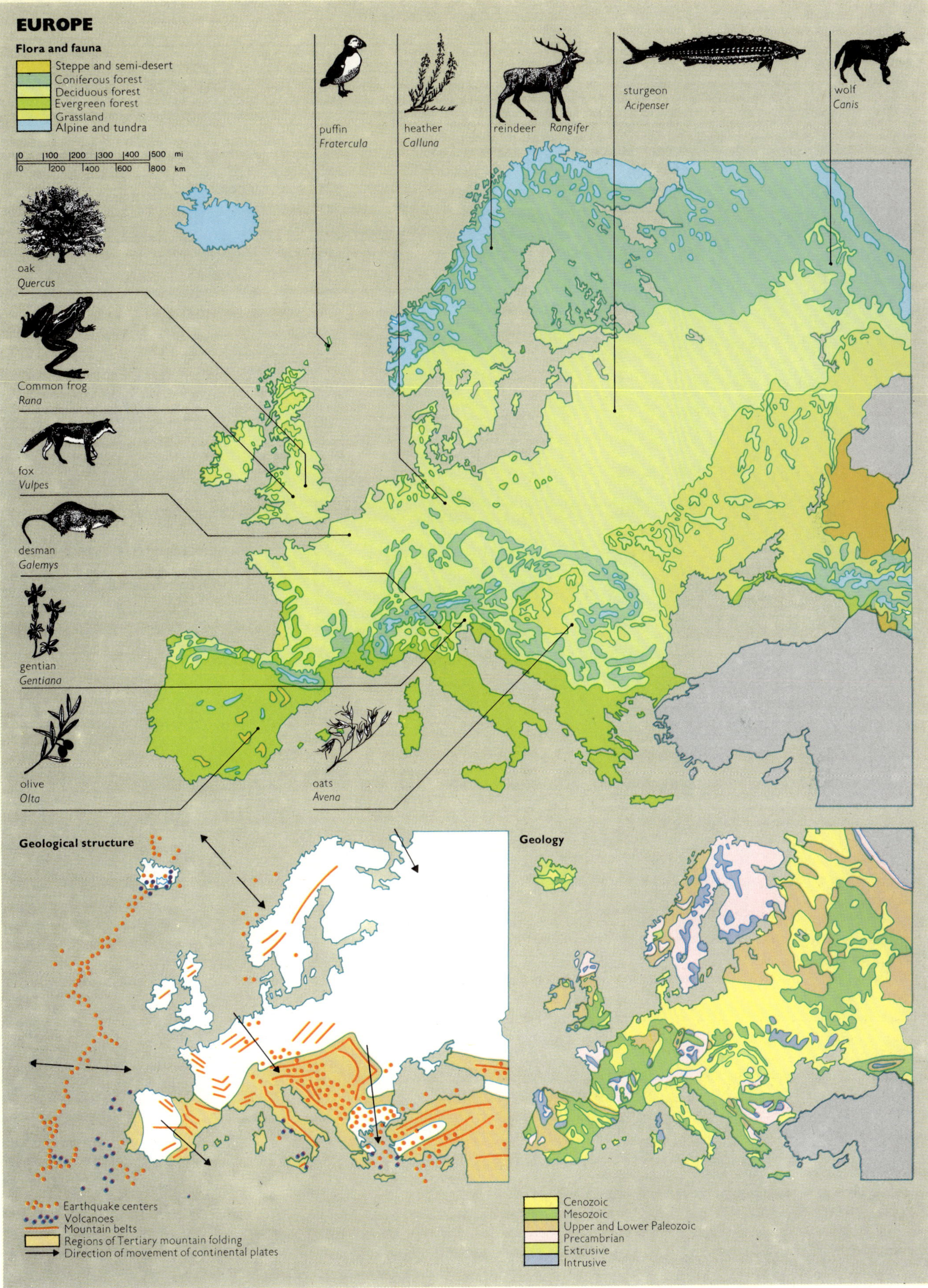
EUROPE
Flora and fauna
Steppe and semi-desert
Coniferous forest
Deciduous forest
Evergreen forest
Grassland
Alpine and tundra
0 100 200 300 400 500 mi
0 200 400 600 800 km
puffin
Fratercula
heather
Calluna
reindeer
Rangifer
sturgeon
Acipenser
wolf
Canis
oak
Quercus
Common frog
Rana
fox
Vulpes
desman
Galemys
gentian
Gentiana
olive
Olta
oats
Avena
Geological structure
Earthquake centers
Volcanoes
Mountain belts
Regions of Tertiary mountain folding
Direction of movement of continental plates
Geology
Cenozoic
Mesozoic
Upper and Lower Paleozoic
Precambrian
Extrusive
Intrusive

Amsterdam is an important port for grain. It is stored in harbourside silos, transferred onto barges, and from thence shipped to other parts of Holland and Europe.

began about 4 million years ago. Great ice sheets extended from Finland and Scandinavia across the Baltic and North seas and as far south as central Germany, and also across much of the Vistula basin in Poland and the Russian plains. Stages in the northward retreat of these ice sheets were marked by a variety of drift deposits – outwash sands, boulder clay, long terminal moraines, elongated eskers and hummocks called drumlins. The long river valleys of south-east Norway and Sweden are partly drift-filled and boulder clay is widespread in Denmark, southern Sweden and Finland.

Switzerland has the most spectacular examples of Alpine glaciation, notably in the Aar–St Gotthard region and in the Bernina Alps. Large-scale deposition of glacial material took place to the north of the Alps in the Swiss *Mittelland*, Bavaria (south of the Danube) and Upper Austria.

Associated with the Pleistocene period are the desposits of loess or *limon*. The distribution of this fine-grained sandy material is of major importance, for it weathers into a rich loam soil, as in the Paris basin, south-central Belgium, Dutch Limburg and other areas. The greatest loess areas are the Pannonic basin of Hungary, the lower Danube plains in Romania, and the steppes of the Ukraine.

Post-glacial features include accumulations of peat in the marshlands along the southern North Sea coastlands, interspersed through the forest zone of northern Europe, and becoming dominant in the permafrost zone of the Arctic tundra. Fine silt forms the deltas of the Rhine-Maas, Ebro, Rhône (Camargue), Po and many other rivers. Elsewhere coarse gravels have been deposited, as in the pebble delta of the Durance, east of the Rhône.

The westward sweep of the drift along the Mediterranean coast of France has created long shingle bars which partly cut off lagoons from the open sea along the Languedoc coast. The 'Silver Coast' of Gascony is no less famous for its sand dunes and lagoons fringing the *Landes*. Sand dunes also mark the North Sea coast from Dunkerque to the Skaw, at the northern tip of Jutland, and the *Haff* and *Nehrung* coast of the south-east Baltic.

Climate. Europe is fortunate in having a relatively mild climate. Except in the north and east, winters are not harsh; summers generally are warm, and rainfall is plentiful and well distributed. Along the Atlantic margin, the prevalent south-westerly winds bring moist, warm, tropical air into high latitudes. As a result Norway's west coast has the greatest temperature anomaly in the world. Although the waters of the fjords never freeze, the rapid increase in altitude behind the coast results in heavy winter snowfall. Summers are mild as a result of maritime influences, but there is often rain and low cloud. The small temperature range along the Atlantic coasts of north-west Europe means a long growing season, especially for grass (hence Ireland's nickname: the Emerald Isle).

Northern and eastern Europe have long, harsh winters, and snowfall is heavy when polar, maritime air moves in from the north-west. Otherwise skies are remarkably clear, winds comparatively light, but night frost severe. Such weather conditions also apply in Alpine Europe and other highland areas. Mountain regions develop their own type of climate. Temperature inversions are common, and the Alps have local winds such as the warm, dry southerly *Föhn* and the cold, northerly *Bise*. In the Mediterranean region, cold winds like the *Mistral* may funnel down the Rhône corridor, while the northerly *Bora* may bring bitterly cold spring weather to Venice and the Adriatic.

The Mediterranean coastlands have mild, wet winters and hot, dry summers. Local winds include the hot, dust-laden southerly *sirocco* which blows from north Africa across Malta and Sicily, and the onshore sea breeze (*le marin*) which may lower summer temperatures in coastal Languedoc. Seasonal contrasts are sharper in the eastern Mediterranean ranean than in the west, the summer heat being much more prolonged in Athens, for example, than in Marseille.

Europe's highest rainfall occurs on the mountain slopes behind the north-west Atlantic coasts. In southern Europe, the maximum rainfall occurs in the Dinaric Alps of Yugoslavia. Local variation in rainfall is much greater than in north-west Europe; some areas in the rainshadow of mountain ranges, such as parts of eastern Spain and Sicily, receive less than 20 in (508 mm) annually, and irrigation is essential for crop cultivation.

Vegetation. Little remains of Europe's original vegetation. Remote areas in high latitudes retain their tundra plants and bog-loving species such as cotton grass, while high mountain ranges have their 'alpine' flora. Coniferous forest spreads across northern Europe from south-east Norway, through Sweden and Finland, and into the Soviet Union, with outliers in north Germany and Poland. Its characteristic evergreen needle trees such as Norwegian spruce and pine, and some deciduous larch and birch, are also found in the Pyrenees, Alps, Carpathians and Caucasus. Fine stands of

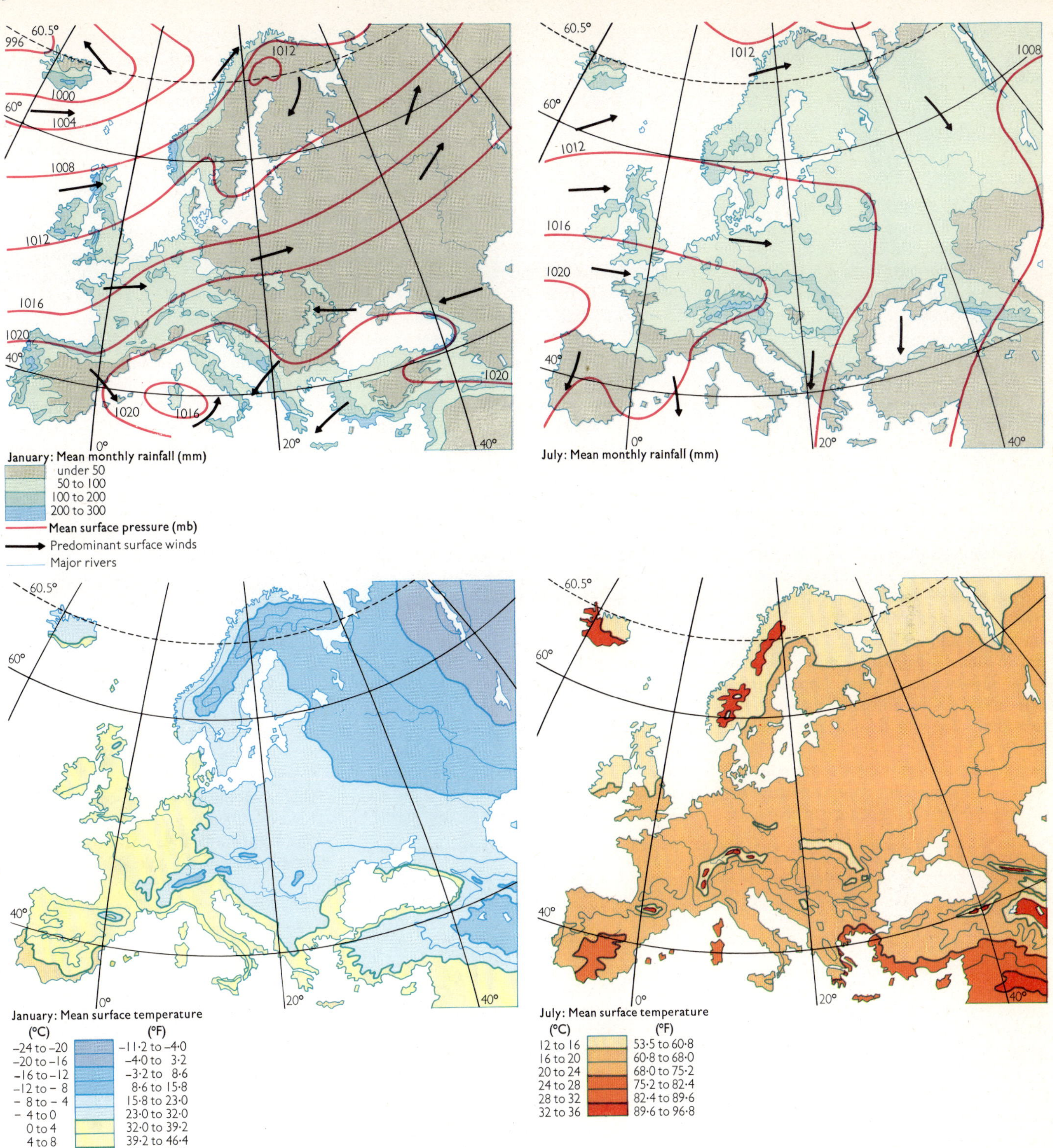

beech occur in Denmark and south-west Sweden. The rain-drenched slopes of the Savoy Alps in France have magnificent forests of spruce, silver fir, beech and sweet chestnut.

The vegetation of the Mediterranean lands is closely adapted to the climate. Where winter frost is uncommon, olive trees thrive; the cypress is widespread; the cork oak flourishes in Spain and Portugal, and the evergreen oak elsewhere. 'Aleppo' and other pines are common in the coastal areas.

Europe has few remaining natural heaths. In France the term *landes* is given to such areas, whether they be the windswept, gorse and broom covered ridges of southern Brittany or the sands of Gascony, where reclamation by the planting of pines has drastically changed the landscape.

Whereas western Europe is well-endowed with natural grassland, the south has little natural pasture, except where the climate is non-Mediterranean, as in Portugal and northern Italy. Burning, and grazing by sheep and goats, have reduced the original woodland to the dense scrub called *maquis* in French, *macchia* in Italian, and *matorral* in Portuguese and including thorny juniper, wild clematis, laurel and broom. The poorest form of plant life in the Mediterranean is *garrigue*, a creeping mat of thyme, dwarf lavender and other aromatic plants, found in Provence, Italy, Dalmatia, Greece and Malta. Irises, asphodels, lilies and other bulbous plants also flourish in the Mediterranean.

Wildlife has suffered greatly from man's activities, but bears, wolves, wild boar, reindeer and foxes still survive. The chamois, a peculiarly European 'goat-antelope', still roams the Pyrenees, the high Alps and the Caucasus. Bird life is rich and varied, but suffers heavy casualties from shooting and netting in those countries which have still to learn the importance of conservation.

The People

Europe is the home of more than 654 million people. It is the most densely-populated of the continents, and only Asia has a higher population. The distribution of people, and to some extent their nations, languages and creeds reflect both the diverse physical nature of the continent and over 2,000 years of history and economic evolution. High latitude countries such as Iceland, Norway and much of Sweden and Finland, with their rigorous environments, support densities of below 32 persons per sq mi (12 per km^2). Similarly parts of Spain's Meseta, the Pyrenees, Alps and Apennines are sparsely populated. Other regions of low density include the limestone regions of Yugoslavia and the French *Midi*; Arctic Russia, the Pripet Marshes and the salt steppes of the Caspian.

Rural population densities are highest in the lowlands, especially along the Atlantic, English Channel and North Sea coasts and in the river valleys of the Rhine, Rhône, Elbe and Po. Densities are also high in the Mediterranean coastlands, especially Catalonia, the Arno basin, the Naples region and coastal Sicily. Overpopulated and impoverished, these are areas of constant migration and emigration. The free flow of labour within the Common Market has brought large numbers of foreign workers to industrial centres in France and West Germany, for example, where the standards of living are much higher than in Mediterranean lands. Throughout Europe the higher wages offered by the manufacturing and service industries have helped to depopulate rural areas and overcrowd such cities as Paris, Munich and Zürich.

The greatest concentrations of population are in coal-field and industrial areas: the Ruhr conurbation of over 10 million people; the Sambre-Meuse coal-field; and the Nord-Pas de Calais coal-field of north-eastern France. In eastern Europe, the small Saxony coal-field in East Germany, the Upper Silesian coal-field in southwest Poland, and the coal-fields of western Czechoslovakia all contain a number of industrial towns.

Other outstanding areas of high density are associated with seaports such as Rotterdam, Antwerp, Hamburg, Göteborg, Leningrad, Barcelona and Naples; or with such large capital cities as London, Moscow, Madrid, Rome, Paris and Athens. In The Netherlands, the most densely populated major country in the world, the ring of urban population from Amsterdam via Utrecht to Rotterdam forms the Randstad, while the conurbation from Rotterdam along the New Waterway to Europoort is known as the Rijnmond.

In spite of urban overcrowding and local rural poverty, the European living standard as a whole is one of the highest in the world.

The Economy. Much of Europe is deficient in basic foods and essential minerals, and its economy is largely geared to importing raw materials from which its factories produce such sophisticated products as aircraft, motor vehicles, ships, electronic equipment, nuclear power equipment and plastics-products that are largely sold abroad. Most manufacturing is concentrated in Western Europe which alone possesses one quarter of the world's industrial capacity.

Majorca is the most popular of the Balearic islands, which lie about 100 miles off the east coast of Spain. Majorca's beaches attract tourists from all over Europe.

How Rural Land is Used. In spite of its industrial power, Europe relies heavily upon the products of the fields and forests that cover much of the continent. But there are great regional variations in land use. North of the Arctic Circle dwindling bands of nomadic Lapps herd reindeer from one grazing ground to another. Farther south, parts of Scandinavia and Finland are covered by (economically exploited) coniferous forest. The Vosges, Black Forest and Sudetes, with their steep slopes and poor soils, are similarly covered. Small mixed farms, however, have been established in cleared areas of south-eastern Norway, Sweden and Finland.

The lowlands of western Europe are now mainly improved for farming. Along the Atlantic fringe, where the climate is generally damp and the soils acidic, stockbreeding predominates, mainly on small farms which are uneconomic by modern standards. The largest and most prosperous farms producing grain as well as cattle are in Britain – in East Anglia, south-east England and eastern Scotland. Comparable crop yields and a wealth of meat and dairy produce come from the regions of *grandes cultures* in northern France and also from south-central Belgium, The Netherlands and Denmark. The only other comparable areas of large-scale crop-farming, combined with cattle-rearing, are in the *Börde* zone of Germany, in Hanover and Brunswick (in much of West Germany subdivision has made farms uneconomically small).

In eastern Europe, the steppe grasslands south of the mixed forest belt were long associated with pastoral nomadism. Only in the 19th century was it realized that these rich loess lands could be used for large-scale cultivation. Today the large state farms of the Ukraine grow mainly grain, sugar beet, fodder crops and occasionally cotton, as well as sunflowers for oil and seed. On the borders of Asia, the salt steppes of the lower Volga basin are useless for agriculture, except where irrigation has been introduced.

The basic crops of the Mediterranean lands are much as they were in Greek and Roman times – wheat, barley, olive oil and grapes. More intensive cultivation is possible in irrigated areas such as the *huertas* of eastern Spain, long noted for their citrus fruit. In northern Italy the land receives perennial supplies of water from the Cavour and other canals, and in the Po delta rice as well as

St Basil cathedral, built in the 16th century, dominates one end of Red Square in Moscow.

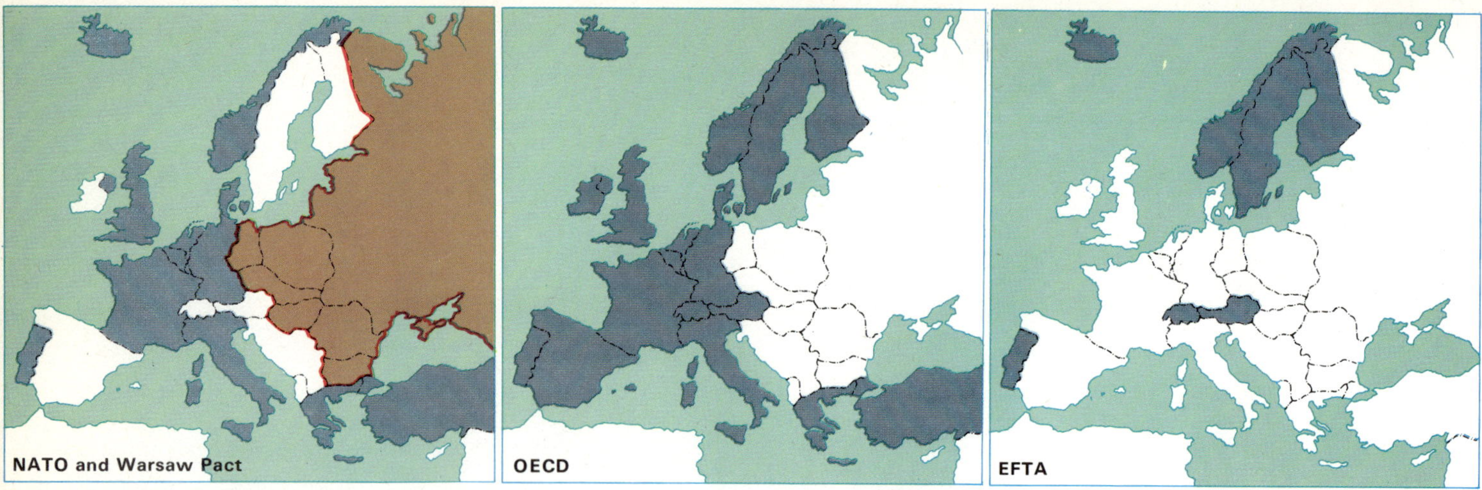

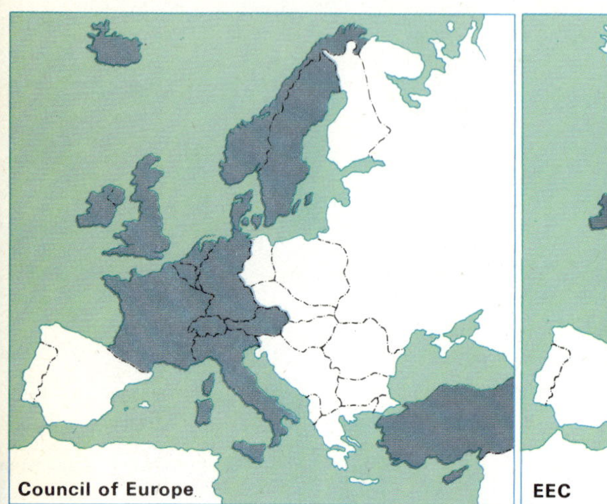

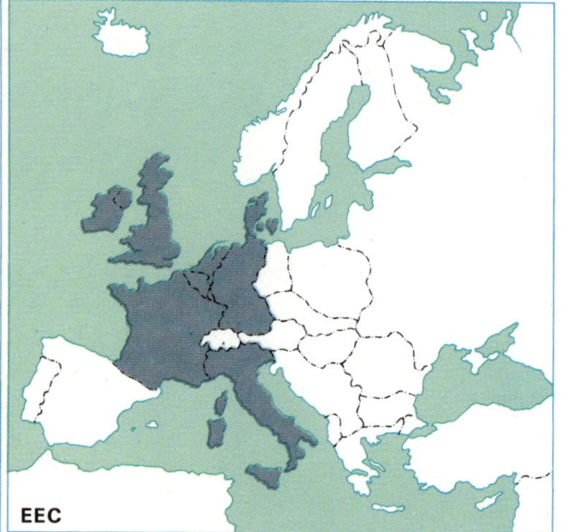

European International Organizations. NATO (the North Atlantic Treaty Organization) is a common defence group founded in 1949 and including the United States and Canada. The Warsaw Pact is a Soviet counterpart, including many Eastern European States, founded in 1955. The Organization for Economic Co-operation and Development (OECD) is a reconstituted (1961) body originally designed to direct the Marshall Plan. The European Economic Community (EEC), known as the Common Market, was founded in 1957 with the aim of moving towards a collective economic policy of its member states. The European Free Trade Association was organized in 1960 among countries not belonging to the EEC. The Council of Europe (1949) provides an opportunity for members to discuss matters of mutual interest.

sugar beet is grown. Some of the Rhône delta swamps of the Camargue, in southern France, are also sown with rice. Fruit orchards – peach, pear and apricot – cover the Mediterranean hillsides, and gnarled olives thrive on the lower slopes, often providing shade for peas, beans and other field crops, especially in Tuscany and Provence.

France is best known for specialized produce such as the cheeses and butter of Normandy, meat and fruit, and some of the world's finest wines. Pedigree dairy cattle like the white Charolais of the Massif Central are exported to build up herds abroad. Britain's Ayrshires, a dairy breed, and Hereford and Aberdeen Angus beef cattle are represented in most stockbreeding parts of the world.

The Netherlands, well-known for its Edam and Gouda cheeses, also produces winter vegetables, salad crops grown under glass, flowers and small fruit for export. Intensive production of field crops is also characteristic of Belgium, but the domestic market is a large and growing one.

Fisheries. The shallow North Sea waters overlying the Continental Shelf are some of the richest fishing grounds in the world. The chief fisheries are traditionally centered on the Dogger Bank, midway between Britain and south Norway; off the Norwegian coast; and in the North Atlantic off Iceland. The catch consists mainly of cod, haddock, plaice, halibut and herring. Iceland's determination to prevent the overfishing of her waters has led to disputes with Britain and West Germany.

Sardines and tunny are the main catch of Portuguese and Spanish fishing fleets, and are canned for export.

Iron and Steel. Western Europe was once supreme in world steel production, due to iron ore deposits conveniently near large reserves of bituminous coking coal, but the world leader today is the United States, with the Soviet Union and Japan in second and third places respectively. In the 19th century Europe's steel industry was located on the major coal-fields such as the Ruhr in northwest Germany. Today access to iron ore and scrap metal is far more important in determining location, and while, for example, the 'minette' ores of Luxembourg and Lorraine have made the middle Moselle valley a centre of heavy industry, the inland steel plants are now at a great disadvantage compared with the modern integrated plants on or near the coast where ore from Venezuela, West Africa, Labrador, Spain and Sweden can be brought to their doorsteps.

Sweden has large deposits of high grade iron ore in Swedish Lapland and also in the south, where it is the basis of the highly specialized Swedish steel industry. In eastern Europe there are small deposits of iron ore in Silesia and southern Poland. But East Germany, like Czechoslovakia, imports iron ore from Krivoi Rog in the Ukraine.

Coal is declining in relative importance due to increasing competition from oil, natural gas and (to a limited extent) nuclear power. In some countries, such as Sweden, Norway and Switzerland, hydro-electricity is the main source of energy.

In western Europe the chief source of coal is the North Rhine–Westphalian field. The major French coal-field, in the Nord and Pas de Calais, is a continuation of the Sambre-Meuse coal-field of Belgium. Of more importance to Belgium, however, are the deposits along the Dutch frontier in Kempenland. Spain mines coal in the Asturias, and Italy has small deposits in Sardinia. East Germany has the Saxony coal-field, and Poland most of the Upper Silesian deposits. The main Russian coal-field west of the Urals lies in the Don-Donetz basin, in the Ukraine, where it has given rise to a major industrial complex. Large reserves are said to exist in the Arctic Pechora basin. Lignite (brown coal), a low-grade fuel, is being used increasingly in central and eastern Europe.

Petroleum and Natural Gas. With the recent discovery of North Sea oil and natural gas, Europe has become a significant producer of these fuels, while still drawing supplies from Libya, the Arabia and Latin America (especially Venezuela). In mainland Europe oil has been struck at Bremerhaven and along the Ems valley; and modest quantities are produced in The Netherlands, West Germany and Austria. In eastern Europe oil production has declined in southern Poland and the Romanian Carpathians, although natural gas is of increasing importance. Poland, East Germany, Czechoslovakia and Hungary receive oil from the Soviet Union through the 'Druzhba' (Friendship) pipeline.

Natural gas is of increasing importance for both domestic and industrial use. Much of eastern and southern England now has piped North Sea gas. The main producing fields in western Europe are at Lacq, in southern Gascony, and at Slochteren in the northern Netherlands. This easily transported fuel is giving rise to new industrial complexes, and oil installations have encouraged the development of petro-chemical industries at such places as Strasbourg, Lavéra and Feysin (France), Karlsruhe (West Germany), and Porto Marghera (Italy). Thus a second industrial revolution is in progress whose new centres of industry are away from the traditional coal-fields.

Patterns of Transport and Trade. No continent can rival Europe's intensive road, rail, waterway and air networks. In western Europe, modern multi-lane highways now link many major cities and more are projected. Europeans own more cars than the peoples of any comparable region outside the USA, and freight travels partly by increasingly large fleets of 'juggernaut' lorries. But much goes by rail. Industrial north-western Europe is served by a dense network of railways, and railway tunnels bored through the Alps put northern Europe in direct touch with industrial northern Italy. Road transport systems tend to be weak in Eastern Europe. But only Europe's remotest mountains, tundra and steppes lack either railways or surfaced roads. Navigable canals linked with such big rivers as the Rhine, Danube and Volga provide valuable inland waterway systems permitting cheap transport of bulky freight and Europe's Maritime nations like Britain, France and West Germany handle most of the world's seagoing vessels. National and independent airlines serve all Europe's main cities and keep them in touch with those of the rest of the world.

Europe's political division into communist and noncommunist countries is reflected in trading patterns formalized through such trading groups as (Eastern) Comecon and the (Western) Common Market and EFTA. But Europe's nations also trade across tariff barriers with each other and abroad, largely for food and raw materials but also (as with Japan and the USA) for cars, electrical goods and machinery – sophisticated industrial products such as those that feature largely among Europe's own exports.

ICELAND

Area: 39,709 sq mi (102,846 km²)
Population: 214,000
Population Growth Rate: 1·8%
Capital: Reykjavik, pop 94,000
Language: Icelandic
Religion: Evangelical Lutheran Church
Currency: Krona = 100 aurar

Iceland lies in the North Atlantic Ocean, about 200 mi (325 km) east of Greenland and 650 mi (1040 km) west of Norway. It became a republic in 1944 and is a founder member of NATO. This cool northern volcanic island is the homeland for a few thousand energetic people remarkable for a long record of democracy and a high living standard based on the seas resources rather than upon largely barren land.

Fishing and fish products dominate the economy, and the republic has had several disputes with Britain over the rich fishing grounds, particularly in 1959, when Iceland extended its territorial limits to 12 mi (19 km) and more recently in 1971, when the limit was further extended to 50 mi (80 km). Britain and West Germany refused to recognize this limit, but neither negotiations nor the International Court at The Hague could budge the

Reykjavik is the capital and chief city of Iceland. Its name means 'city of smokes', after the steam of many hot springs in the area. Much of the city is heated with this natural steam.

Icelanders, and in the early 1970s there were repeated confrontations between vessels off the Icelandic coast.

Glaciers, Geysers and Volcanoes. Although there are grassy lowlands along the coast, most of Iceland consists of high, dissected mountains surrounding a rugged inland plateau. Nearly 80% of the island is wilderness covered only with mosses, lichens and occasional shrubs and trees; about 11% is covered by glaciers, including Europe's largest ice sheet, Vatnajökull, covering 3,140 sq mi (8,133 km²).

The glaciers have worn valleys and fjords deep into the mountains and the coasts. River valleys are often steep and gorge-like, but rivers draining the melting glaciers have built up large plains of sand and gravel.

There are many natural wonders: from the massive volcanic cones below Oraefajökull where Hvannadalshnúkur, Iceland's highest mountain, rises to 6,952 ft (2,119 m), to the long volcanic craters at Laki. Heflla, which erupted in 1947 and 1948, is probably Iceland's best known volcano. In 1963 an eruption off the south coast formed the new island of Surtsey and a decade later another eruption began smothering the major fishing port of Vestmannaeyjar. There are hot springs and intermittent hot-water ejections, called geysers after the Great Geysir, which once frequently spurted to 180 ft (55 m). Iceland's lava flow spreads over vast areas and accounts for one-third of all the world's lava outpourings over the last five centuries.

The climate is cool and temperate. Because Iceland lies in the path of the southwesterly air flow and the warm ocean current called the North Atlantic Drift, winters are milder than is usual in these latitudes. Reykjavík averages 30°F (−1·1°C) in January and 52°F (11°C) in July. The extreme north is slightly colder, but there the sun remains constantly above the horizon for some 18 days in June. Rainfall averages 34 in (864 mm) at Reykjavík, but nearly double this amount falls in the south-east. Weather changes are often so rapid that it is said 'if you don't like the Icelandic weather, just wait a minute'.

The People. Iceland has the lowest death rate and the third highest birth rate in Europe. Some 80% of the rapidly growing population lives in the towns, and more than half in or near Reykjavík, the capital. Most towns are around the coast, in the northern valleys or in the southern lowlands, and there are isolated farmsteads in the rural areas. Here, as in the towns, reinforced concrete buildings designed to withstand strong winds and earthquakes have largely replaced the earlier wood or corrugated iron structures.

The Icelanders are mostly fair-haired and blue-eyed like their Norse ancestors, and their language has changed little over the years. Education is of a high standard: schooling is compulsory from 7 to 15, and there is virtually no illiteracy. The University of Iceland at Reykjavík was founded in 1911. Most Icelanders belong to the state-endowed Evangelical Lutheran Church.

Fishing. The 6,000 Icelanders engaged in fishing, and those employed in the factories which process the catch, provide more than 90% of Iceland's exports. Cod, haddock, herring and other fish gather to spawn and feed in the shallow waters around the island where warm and cold ocean currents meet. The Vestmannaeyjar Islands off the south coast are an important centre for cod fishing, while Siglufjördhur in the north is the chief centre for herring. Many Icelandic boats are owned by the small coastal towns, but half the fishing fleet is concentrated at Reykjavík.

Farming and Industry. Grass is the dominant crop, and meat – beef and lamb – the basic agricultural product, although milk and milk products are important around Reykjavík. Where the soil is deep enough, potatoes and turnips are grown, and where hot springs can be used for heating and irrigation, tomatoes and cucumbers are grown under glass.

Although hot springs are a potential source of energy and hydro-electric power is abundant, industrial growth is restricted by limited natural resources and a small home market. Manufacturing is largely confined to processing Icelandic fish and milk, and producing fertilizers, clothing and food-stuffs made from inported materials. Industry is centred on Reykjavík and, to a lesser extent, on Akureyri.

Transport and Trade. Most of Iceland is covered by motorable roads and an extensive internal air network. There are international airports at Keflavík and Reykjavík, which is also the chief seaport.

Iceland has to import a wide variety of goods, including even fishing boats and nets. These imports are financed by fish and fish products which are exported frozen to Britain, the United States, Scandinavia and Germany, or dried to Mediterranean and North African countries. The republic's economy is helped by regular income from United States' bases in Iceland and by the tourist industry.

NORWAY

Area: 125,020 sq mi (323,802 km²)
Population: 3,917,773
Population Growth Rate: 0·8%
Capital: Oslo, pop 475,563
Language: Norwegian. Lappish and Finnish spoken in North
Religion: Evangelical Lutheran Church
Currency: Krone = 100 øre

Norway is a rugged, narrow northern nation whose people enjoy one of the highest standards of living in Europe. In 1949 it became a founder member of NATO on the understanding that neither foreign troops nor atomic weapons would be stationed on Norwegian soil. In the referendum of 1972 it rejected membership of the European Economic Community. The country has close trade and cultural links with its neighbours in Scandinavia, and cordial but somewhat more reserved relations with the Soviet Union, with which Norway has had a 122 mi (196 km) frontier since the Russians took the Pechenga area from Finland in 1944.

It was from Norway's many fiords that Viking sea raiders sailed long ago to settle in the Orkneys, Shetlands, Faeroes, Iceland and even Greenland, from which, in about 1000 AD, Leif Ericsson visited North America. The seafaring tradition has been unbroken; today Norway has the fourth largest merchant fleet in the world.

A Land of Coast and Mountains. Norway is a long and narrow country. Its coast, fringed by countless islands and skerries (rocky islets), is jagged with fjords, the longest of which, the beautiful branching Sognefjord, reaches inland for some 115 mi (185 km). The main outline of the coast is about 1,650 mi (2,655 km) long, but taking the many inlets into account increases this length to about 12,500 mi (20,117 km) – almost half the globe's circumference.

The coastal lowlands are major areas of settlement. The most extensive is around Oslofjord and embraces the valleys of the Glåma and other rivers, and long, narrow lakes like Mjösa. Because it is in the south and has fertile soils, this area, which includes Oslo, the capital, is economically the most important part of Norway. Another important lowland area has Trondheim as its focal point.

Most of Norway, however, is high plateau and mountain. From an economic viewpoint

The Feda Fjord in Norway. 'Fjord' is a Norwegian word referring to a narrow inlet to the sea. These were often originally river valleys which were later scoured and deepened by glaciers during the Ice Age. Norway is famous for the number and beauty of its fjords.

it is unfortunate that the main mountain block occupies much of the southern part of the country. Here, in Jotunheimen ('Home of the Giants') are Norway's highest peaks, Galdhöpiggen and Glittertind, both rising to more than 8,000 ft (2,438 m). Westwards lies the Jostedalsbreen, an icefield covering some 300 sq mi (777 km²), from which glaciers flow. Northwards the land is somewhat lower, with an irregular descent from the Swedish border to the discontinuous *strandflat* (narrow, coastal belt).

Climate and Vegetation. The mountain ranges and plateau divide the country into a western half characterized by cool summers, mild winters and high precipitation, and an eastern half with greater seasonal extremes of temperature and low precipitation. The irregular nature of the land surface often modifies this general pattern, and weather conditions can change from pleasant to severely cold over very short distances.

Some of the inland fjords are well protected by the mountains. This reduces rainfall, sometimes to such an extent that crops may have to be irrigated. A fair amount of the precipitation falls as snow or sleet. On the south-west coast the snow-cover lasts less than 30 days, but north of the Lofoten Islands winters are more severe and in Finnmark, even on the coast, snow covers the ground for more than six months.

Much of northern Norway is typically sub-Arctic, with scattered small trees such as birch, and tundra vegetation. Southwards from Narvik, coniferous forests, mainly of Scots pine, become increasingly prominent, though the high plateau of the south is largely treeless.

In the south and along the west coast where temperatures are normally higher, conifers increasingly give way to deciduous trees – yew, oak, ash, sycamore, beech and even the warmth-loving lime. Much of this woodland, however, has been cleared for agriculture.

People and Cities. Except for some 20,000 Lapps and 10,000 Finns living in the far north and speaking their own languages, the population is Norwegian. More than 40% of the population lives in small villages, and while there is a tendency for people to move into the towns, most of the major cities have little suitable land for building and expansion. Only in the Oslofjord area is a contiguous urban belt taking shape.

In general, the Norwegians have come to terms remarkably well with the problems of an urban and industrialized existence. Pollution is nowhere on the same scale as in some other parts of Europe and many city dwellers are able to spend a good part of their time in chalets in the countryside.

Oslo, the capital, known until 1925 as Christiana, stands in the most favoured part of Norway and has by far the largest concentration of industry. It is the seat of government, the largest port, and the home of Norway's oldest university, the Royal Frederiks University (founded 1811).

Of the other large centres, Bergen is a major port and the capital of fjordland; Trondheim, the gateway to the north, was the medieval capital and has a famous Norman Gothic cathedral; Stavanger, on the Jaeren Plain, has maritime industries and is a tourist centre.

Literacy, Language and Religion. In Norway compulsory education begins at 7 and continues until 14 or 16. There is virtually no illiteracy. The Norwegian language has two forms, Bokmål, the dialect of the towns, and Nynorsk, the dialect of the countryside, which are gradually merging into a single form, Samnorsk. Most Norwegians belong to the national, state-owned Evangelical Lutheran Church.

Farming, Forests and Fisheries. Norway is a prosperous country and has a remarkably low rate of unemployment. Yet in some ways the country is poorly endowed with resources. Because so much of the country is barren rock and mountain, less than 3% of the total land area is cultivated. The scattered meadows and pastures constitute about 40% of the total farmland. Pastoral products from countless small farms dominate Norwegian agriculture.

Østlandet is the major agricultural area, with a high concentration around Oslofjord. High-yield grain, root and fodder crops are intensively produced on its rich soils. The only other important farming areas are Tröndelag, where grain is also grown, and Jaeren, known for its livestock, root crops and vegetables. In the north, where conditions are less favourable, farming is normally combined with fishing.

Forests cover about 23% of Norway's land area and are a major natural asset. The number of people employed in forestry is relatively small, but many farmers own small woodlands which provide commercial timber. This fits well into the farming year, for felling takes place in autumn and winter, and transport from the forests continues as long as the snow cover lasts, usually into April.

The seas off Norway are rich in plankton and attract large numbers of fish, mainly herring off Vestlandet and cod in the north. Since much of the catch was landed at west and north coast ports, the fish was normally salted for export. Modern developments, however, have made frozen fillets and fish fingers a profitable export proposition. Mainland Norway has neither coal nor oil and it is the metalliferous ores which are important, although the situation may eventually be changed by the recent discoveries of oil in the Norwegian sector of the North Sea.

Iron ore leads, followed by pyrites; lead, zinc, titanium, molybdenum and niobium are also mined. Mines are scattered across the country, from the major iron ore mine at Sydvaranger near Kirkenes to the titanium mine at Sokndal near Stavanger.

Norway is a small but important industrial country. The abundance of cheap hydro-electric power has encouraged electro- chemical industries, especially the production of aluminium from imported alumina or bauxite. Apart from a few developments in the north, such as the steel plant at Mo-i-Rana, manufacturing is mainly in the south-east, especially in and around Oslo.

With spectacular mountain and fjord scenery, and excellent facilities for winter sports, Norways attracts tourists at all times of the year, especially from Sweden, the United States, Britain and Denmark. Their contribution to the national economy is important.

Exports, mainly chemicals, fish, metals, wood pulp and paper, pay only part of Norway's import bill, much of which is offset by shipping and tourist earnings. Imports include raw materials, machinery and foodstuffs. Norway's chief trading partners are West Germany, Britain, Sweden, the United States and Denmark.

Transport. Coastal shipping is still an important means of transport for some areas, but has to be subsidized in the face of competition from rapid overland transport. The rail network radiates from Oslo to Bodö, Andalsnes, Bergen, Stavanger and Sweden. Little more than a tenth of the roads are surfaced, but there are many good gravel roads. Aviation is of growing importance, although limited by the availability of suitable sites for airports.

Overseas Possessions. Norway has several outlying territories. Svalbard, a group of Arctic islands including Spitsbergen is important for its coal (mined by both Norwegians and Russians) and its radio and weather stations, and is the scene of oil and natural gas prospecting. Its capital is Longyearbyen. Several hundred miles north-east of Iceland is Jan Mayen Island, bleak, volcanic and named after a Dutch whaling captain. In the Antarctic, Norway has the uninhabited Bouvet and Peter I islands, and part of the Antarctic continent, Queen Maud Land.

SWEDEN

Area: 173,686 sq mi (449,847 km²)
Population: 8,127,396
Population Growth Rate: 0·8%
Capital: Stockholm, pop 1,307,000
Language: Swedish. Lappish spoken by tiny minority
Religion: Lutheran Church
Currency: Krona = 100 öre

Sweden, the fourth largest country in Europe, is the home of some of the world's most successful economic and social experiments. Sweden is basically a lake-strewn land with broad plains in the south and east and mountains and hills in the west and north. This division represents two areas of roughly similar size, but dissimilar character and economic development. In the north, Swedish Norrland has many of the characteristics of a developing country. It has only a small population, geared to the production of basic raw materials from timber and mineral resources using locally derived hydro-electric power. In contrast, southern and central Sweden contain some 80% of the population living in an agreeable physical environment with an advanced agricultural and physical economy.

From these two parts the Swedish people have created a nation internationally renowned for its very high standard of living, sophisticated culture, advanced democratic principles, and independent stand in world affairs. Sweden remained neutral in both World Wars, differed from most Western nations in strongly condemning American action in Vietnam, and decided to remain outside the European Economic Community.

The Land. The highest part of Sweden comprises the mountain range along the border with Norway which rises to a maximum height of 6,943 ft (2,117 m) at Kebnekaise in the northwest. Away from these mountains plateaux slope gently south-eastwards between 1,500 and 600 ft (500–200 m). The plateaux are in turn bounded in the east by the wide plains of the Bothnian coast and central Sweden. Lowland continues around the coast of southern Sweden, but inland the relief rises to 1,237 ft (337 m) to form the undulating plateau of the South Swedish Highlands.

The major rivers drain south-eastwards from the interior mountains to the Bothnian coast. Most of the rivers are interrupted by lakes. Lakes cover nearly one-tenth of the country and vary in character from small, irregular rock basins in the glacial drift-free areas of southern Sweden, to long ribbon

Modern office blocks in Stockholm, capital of Sweden. The city is located on several islands where Lake Malareu flows into the Baltic.

lakes in Norrland, and extensive depressions such as Lake Vänern (2,140 sq mi; 5,542 km²).

Climate and Plant Life. The Swedish climate is influenced by maritime air from the Atlantic, and continental air from Siberia. Owing to the westerly air masses which cause cool summers, mild winters, and relatively high precipitation (80 in; 2,000 mm in the high Sarek), the western mountains record Sweden's highest precipitation, and the coastlands of the south-west record the smallest mean annual temperature range (Göteborg 31°F; 17°C). Other parts of the country are more sheltered and the effects of the continental air masses are more pronounced. So the eastern parts of Sweden are characterized by a greater annual temperature range (maximum of 52°F; 29°C in Norrland), longer snow cover (up to seven months in the north), hotter summers, and smaller annual precipitation of which the maximum fall is recorded in July and August.

Over half the country is forested. Along the coastlands of south and south-west Sweden the natural vegetation is deciduous forest characterized by oak, beech, and hornbeam. Across the bulk of southern and central Sweden pine and spruce tend to crowd out the hornbeam and beech. The vast area of Norrland is covered by continuous tracts of pine and spruce, interrupted only by peat bog and pasture developed on the poorly drained lands. In the northern mountains the coniferous forest is replaced by subarctic vegetation dominated by birch and hardy dwarf shrubs. Only Swedish Lappland retains a pattern of vegetation unaffected by man.

The People. Sweden is one of Europe's most thinly-populated countries and its population of 8,127,396 enjoys one of the highest standards of living in the world. The majority of Swedes are tall and have fair or brown hair and blue eyes. About 10,000 of the short and stocky Lapps still live in the far north, now mainly as miners or lumberjacks. There have been well over half a million immigrants (mainly Finnish) since World War II.

Population densities vary greatly between central and southern Sweden, and Norrland where respectively 80% and 20% of the population live. As much as 78% of the Swedish population is considered urban, of which half live in the major urban concentrations around Stockholm, Göteborg, and south-west Skåne.

Sweden has been a constitutional monarchy since 1809 and has had effective parliamentary government since 1971. Uniformity and social awareness are major characteristics of Swedish life. Sweden's social security system is one of the most comprehensive and advanced in the world. Taxes are high but so are pensions, and medical service is largely free. However, the Swedish concern for forward planning and organization has other effects. There is little variety in the architecture of towns or farms. Also, in general, Swedish artists have been content to follow the lead of other Europeans.

Forestry and Forest Industries. Of Sweden's three great natural resources, timber, iron ore and water, timber is the most important. Over half the country is covered by forest, and timber is the basic raw material for the production of sawn wood, wood pulp, and manufactured timber products such as furniture, matches, prefabricated houses, and chemicals.

The distribution and productivity of the trees varies. In northern Sweden birch and pine predominate. Growth rates are slow, and wood quality high. In eastern Norrland and central Sweden spruce and pine are dominant and growth rates are three times as fast as further north. In the extreme south the growth rates are most rapid and beech is an increasingly important part of the forest.

Most of the more productive forest of southern and central Sweden and south Norrland is owned by the forest farmer and the timber companies, while the more marginal areas are owned by the state. Forest farming is well adapted to the natural environment as cutting is often done during winter months when farming is impossible and the ground is frozen. Floating the logs downstream to the mills is carried out in spring when the rivers are high due to snow melt. But such traditional seasonal patterns are now being superseded by more profitable practices involving year-round cutting, road transport, and a full-time labour force.

Saw mills in towns along the Bothnian coast produce sawn timber for direct export (50%), or for distribution to southern and central Sweden where it forms the basis of important local industries. Pulp production is located along the south coast of Norrland, around the northern part of Lake Vänern, and more recently, with new production methods capable of utilizing deciduous timber, in southern Sweden. Half the pulp is exported directly, while the remainder is used in the production of newsprint, wrapping paper, and rayon. Furthermore, with the byproducts of this industry, chemical plants have been established for the production of materials such as plastics, explosives, adhesives, and medicines.

Agriculture. Only 8% of the total area of Sweden is cultivated. Dairy farming and

The deeply indented rocky coast of the Bohuläu region north of Göteborg is popular with tourists. Its permanent inhabitants are mostly fishermen.

Sweden is more than half covered with forest; and forestry and wood products make up a quarter of her exports. Here logs are being rafted down a river to a sawmill in central Sweden.

livestock are the main sources of income. The length of the growing season varies considerably from 240 days in Skåne to less than 150 days in the north. And although long daylight hours do compensate for the short growing season, this factor combined with climate restricts the growth of sugar beet, oil-yielding plants, and wheat to the southern and central parts of the country. Only the hardy strains of barley, rye, potatoes, and grass survive in the north.

The proportion of cultivated land varies from 80% of the total area in Skåne to 20% in the South Swedish Highlands, and less than 1% in Norrland where nearly all the agricultural land is restricted to the lower parts of the valleys and the coastal plain. In Skåne large farms of 120–50 ac (50–20 ha) are geared to the production of arable crops such as sugar beet, wheat, and barley. Elsewhere the farms are smaller, the proportion of pasture is larger, and the farming economy is closely integrated with forestry. Livestock plays a vital part in Swedish agriculture. Livestock products such as milk, butter, cheese, and meat contribute, on average, some 70% of farm income.

Overall, 13% of the working population is engaged in agriculture, and in all except bad farming years Sweden is self sufficient in staple foods. The trend in Swedish agriculture is toward greater productivity from improved crop types, improved mechanization, and increased farm size. Mechanization in turn is causing rural depopulation and the abandonment of marginal land.

Minerals. Sweden is richly endowed in mineral resources, which have long made a significant contribution to her economic wealth. Iron ore is the country's second greatest natural resource, and iron ore from the Bergslagen area has been extracted since the 14th century. High grade iron ores from the same district and from the extensive ore fields at Kiruna, Gällivare, and Skappavaara in northern Norrland have been exploited only during the past century. In 1970 31·5 million tons of iron ore were produced.

Lead, zinc, tungsten and molybdenum are also extracted from the Bergslagen ore field. Around Skellefteå in Norrland concentrations of sulphide ores yield Sweden's largest output of copper, lead and arsenic. Other non-ferrous minerals of economic importance are limestone, used in the production of cement, and uranium-rich oil shales.

Industry. Industrial development in Sweden is of two types. In Norrland, timber and mineral resources have formed the basis of industries producing semi-finished products such as sawn timber, wood pulp and ore concentrates. Most important manufacturing developments such as the iron and steel plant at Luleå have been introduced by state intervention. The industrial development of southern and central Sweden is characterized by the manufacture of high quality, high value, finished products such as motor cars, ball bearings, and electrical equipment. Many of these products are known internationally.

Power for Swedish industry is provided by hydro-electricity in Norrland, wood burning power stations in the forest lands, and imported coal, oil, and nuclear-powered generating stations in central and southern Sweden. Large catchments, high discharges, and relatively steep gradients provide an estimated 20 million kilowatts of hydro-electric power. But as hydro-electric power sites become more expensive to develop nuclear power will play an increasingly important role in providing energy.

During the 19th century Sweden had to face the problem of industrial development without coal reserves. This was resolved by the production of high grade products with high sale value that could cover the costs of imported fuel. As a result, high grade, low phosphorous iron ore from the Bergslagen became the basis of high quality metal manufacture throughout central Sweden. Today Sweden produces one-fifth of the world total of ships. Motor vehicles mainly for export are made at the Volvo, Saab, and Scania-Vabis plants at Göteborg, Linköping, and Södertälje respectively. Other products range from electrical equipment such as refrigerators, calculators and telephone equipment, to specialized high quality steel products such as ball bearings, knives, scissors, machine tools and precision tools.

Paper manufacture and the manufacture of machinery for production of paper, pulp, and sawn wood is important throughout central Sweden. Of the domestic industries food production is concentrated around the large towns in Skåne. Cotton, linen and woollen textiles are manufactured in the Göteborg, Borås, and Norrköping regions and Stockholm in particular is important for the manufacture of finished textiles. Swedish furniture production and glass manufacture have developed styles that are now internationally famous and eagerly sought after on world markets.

Sweden has a well-balanced industrial base capable of adaptation to variations in world trade. But the largest problem is the high price of labour in a country with a very high cost and standard of living.

Transport and Communications. Internal transport in Sweden is largely by road and rail. Rail transport is particularly important for shifting freight in central and southern Sweden and minerals in Lappland between the inland ore fields and the Bothnian and Norwegian coasts.

International communications are handled mainly by air and sea. Roads and railways cross the international frontiers with Norway and Finland, but the change of gauge at Haparanda makes through traffic with Finland impossible.

External trade from Norrland is also inconvenienced by the freeze-up of the Gulf of Bothnia which closes ports such as Luleå for up to six months.

International Trade. The export and import of manufactured mechanical, electrical, and transport goods dominates Sweden's trade with the rest of the world. The value of this trade constitutes some 2% of total world trade. As the Swedish population is only 0·2% of the world population the value of trade per capita is very high.

Recent trends have been towards increased trade with other Scandinavian countries, and the expansion of the European Economic Community in 1973 must accelerate this trend unless Sweden is able to obtain from the Community some special commercial relationship that is independent of political attachment.

FINLAND

Area: 130,094 sq mi (336,943 km²)
Population: 4,684,000
Population Growth Rate: 0·5%
Capital: Helsinki, pop 804,000
Language: Finnish and Swedish (71%). Lappish spoken by minority
Religion: Evangelical Lutheran Church
Currency: Markka = 100 penni

Finland is a small Scandinavian republic, picturesquely endowed with lakes and forests. But it lives under the shadow of its immensely powerful neighbour, the Soviet Union, which has previously clashed with Finland and absorbed much Finnish territory. Maintaining friendly relations with the Soviet Union has been the keystone of Finnish policy. At the same time, Finland remains a Western democracy and, despite their official neutrality, the Finns see their country as a Western outpost. They have accordingly forged strong trade links with Britain and other Western countries. Finland is a member of the Nordic Council and is linked with the European Economic Community by an agreement concluded in mid-1972.

Plateau, Lake and Forest. Most of Finland is a low-lying, lake-strewn plateau. In the north, however, the hills and plateaux rise above 1,000 ft (305 m). In the north-west, Haltiatunturi rises to 4,344 ft (1,324 m); it is Finland's highest mountain.

This is a land of very ancient rocks, and its surface features bear the imprint of the ice sheet which covered northern Europe many thousands of years ago. When the ice melted, it left behind *eskers*, long and narrow ridges of sand and gravel, and the prominent Salpausselkä moraines running parallel to the south-west coast and marking the southern edge of the central lake plateau. Finland has about 60,000 lakes and lake-systems, the largest being Lake Saimaa covering about 680 sq mi (1,761 km²) and drained by the Vuoksi into Lake Ladoga. Many of the rivers have rocks and rapids along their courses, and some, such as the Kymi, have been harnessed for hydro-electric power.

A plain extends from the head of the Gulf of Bothnia right round the coast, which is low and irregular. Skirting the coast are some 30,000 small islands, including the Åland (Ahvenanmaa) group reaching towards Sweden.

About two-thirds of Finland is covered by forests of pine, spruce and birch. In relation to its size, Finland is the most heavily wooded country in Europe. Finnish Lapland has tundra vegetation.

Climate. Because of its position in the north-west of the immense Eurasian landmass, Finland has warm, moist summers and long, cold winters. Temperatures in Helsinki average 63°F (17·2°C) in July and 22°F (−5·5°C) in January, but the north is generally much cooler, although there the 'Midnight Sun' provides continuous daylight from about mid-May until the end of July. Rainfall, mainly in late summer, averages 27·6 in (701 mm) in the south, but much less in the north. The country is snow-covered from mid-November until April.

The People. The Finnish population is composed of three groups representing three successive waves of colonization: the Lapps, numbering about 2,500 and living in the north; the Finns, who form more than 92% of the population; and the Swedes, living mainly in the south and west. Each group retains its own language. This is reflected in such matters as place names; for example, Helsinki becomes Helsingfors in Swedish, while Turku is known as Åbo, and Tampere as Tammerfors.

The typical Finn is tall and fair-haired, with blue or grey eyes. Most probably he lives in a town, for only 25% or so of the population have farming or forestry as their livelihood. He is likely to be a member of the Lutheran National Church, or perhaps of the much smaller Greek Orthodox Church of Finland, and will be well educated. Schooling is free and compulsory from 7 to 15, and illiteracy is negligible. The oldest of the five universities is Helsinki, founded at Turku in 1640. There is an extensive social welfare system.

Towns and Cities. About a fifth of the population lives in and around Helsinki, the capital and chief port, seat of government and cultural centre. Turku, the former capital, is Finland's oldest city; standing on the banks of the River Aura, it is known for its 700-year-old cathedral and 13th-century castle, and for its timber exports. Tampere, Finland's second largest city, is a textile and manufacturing centre.

Inland towns such as Lahti are growing rapidly, due in part to the influx of Karelians but more to the development of wood industries – plywood, furniture and wood-

Helsinki is the capital and chief port of Finland. Its harbour is kept ice-free by using ice-breakers from January to May.

working machinery. Northern towns like Oulo (87,343), Kemi (30,000) and Rovaniemi (28,000), the capital of Finnish Lapland, are also largely dependent on the timber industry.

In rural Finland the large estates in the south-west contrast with the smaller farm-and-forest holdings scattered across the rest of the country. In the far north are the only true remnants of the old Finnish Lapp way of life, based on reindeer grazing and hunting. Development projects are depleting the grazing grounds and most of the reindeer are owned by settled Finnish farmers and not nomadic Lapps.

Forestry and the Economy. The prosperity of Finland is based mainly upon its forests which provide the raw materials for the rapidly expanding timber, pulp, paper, chemical and furniture industries. Most of the forests are owned by forest farmers, little more than a third being owned by the state and various corporations.

Integrated in the basic agricultural economy, forestry has traditionally been a seasonal activity. Each year the forest farmer contracts to cut a specific amount of timber for the local mill. The timber is felled in winter, when it is easier to transport the logs across the frozen ground to the waterways ready for floating downstream when the spring thaw comes. The system has its drawbacks, and forestry is now tending to become a full-time occupation, with timber being cut and moved to the mills throughout the year, and with greater use of road and rail.

Timber dominates manufacturing in Finland, accounting for about 28% of the production. Sawn timber, plywood, fibre board, mechanical and chemical pulps, paper and board are all produced at mills generally situated in river-mouth towns such as Kotka, Pori, Oulu and Kemi. These large and modern mills are the focus of an integrated system which includes a whole range of processes, from timber cutting to final manufacture and sale.

Farms and Fisheries. Latitude severely restricts agricultural development. Inevitably crops such as wheat, rye and sugar beet can be grown only in more southern areas, while crops in the far north are restricted by a growing season of only some 130 days in the year. On average only 9% of the country is cultivated, although the percentage varies from 30% in the south-west to less than 1% in parts of Lapland. Despite such limitations, the Finns have succeeded in developing new farmlands in the western, central and northern parts of the country.

Most farms are family enterprises of about 17 ac (7 ha) of cultivated land which are farmed in summer, the winter being devoted to forestry. Fodder crops like hay, oats and barley are grown and livestock raised. Only in the south-west is wheat grown for bread grain, along with sugar beet, oilseed and rape. Around the towns, and particularly in southern Ostrobothnia, market-gardening is locally important. Yields, usually low, are being increased by using special strains of crops, improved fertilizer and additional liming, and Finland is now self-sufficient in milk, butter, eggs, meat and potatoes.

Because Finland is remote from open seas, fishing has only a small part in the national economy. Most of the catch is Baltic herring.

Minerals, Power and Industry. Finland has a large variety of ore-bearing rocks, but because of their generally low ore content and the lack of capital, mining was relatively unimportant until recently. The country's largest mines are at Outokumpu in Karelia, where copper and sulphur are produced, and at Otanmäki, where iron ore is extracted. Nickel is mined at Kotalahti, zinc at Vihanti, apatite (for phosphate) near Kuopio, and chromite at Kemi.

Hydro-electric stations like those at Imatra on the River Vuoksi and Anjala on the Kymi provide most of Finland's power. Most of the hydro-electric potential has been developed, and imported fuels may therefore become more important. Two nuclear power stations are being built with the help of the Soviet Union, which will also be providing Finland with natural gas from 1974.

Finnish industry is still dominated by timber, but metal manufacturing, textiles, chemical and electrical industries are all expanding. At one time Finnish industry, except for forest products, was geared to domestic markets. Tampere provided textiles and Helsinki clothing, while food processing was important in the Turku area. These industries are still important but the Finns have now become known for specialized products such as machinery for forest industries and icebreakers and container ships. Television sets and electrical components are sold internationally.

Metal industries have advanced with the exploitation of local ores and their refining at coastal centres such as Pori (copper), Raaha (iron), Koverhar (iron), Harjavalta (nickel) and Kokkola (nickel and sulphur).

Some recent developments are intended to fill gaps in home production and reduce imports. Examples are the oil refinery near Turku, the fertilizer plants near Oulu, and the sulphate and sulphite plants near Harjavalta and Kokkola respectively. In conjunction with the Swedish firm Saab, a car manufacturing plant was established in 1969.

Transport. The integrated transport system comprises a well developed subsidized internal air service, a railway network linking major towns, ports and industries, and a short route bus service. Water transport along lakes, rivers and canals is now mainly used by tourists and the timber industry. Sea transport is restricted by the winter freeze-up of coastal waters. Northern ports may be closed for up to five months. Turku and Hango are kept open by the Finnish icebreaker fleet.

DENMARK

Area: 16,625 sq mi (43,065 km²)
Population: 4,978,106
Population Growth Rate: 0·7%
Capital: Copenhagen, pop 787,347
Language: Danish
Religion: Lutheran Church
Currency: Krone = 100 øre

The prosperous kingdom of Denmark, the smallest of the Scandinavian countries, lies between the Baltic Sea and the North Sea. Consisting of the northern two-thirds of the Jutland Peninsula and many islands, it is bounded entirely by sea, except in the south where it has a 42 mi (68 km) frontier with West Germany. Its eastern neighbour is Sweden; to the north, lies Norway.

The kingdom includes the Faeroe Islands (Faerøerne) and Greenland (Grønland). The Faeroes lie between Iceland and Scotland and have their own parliament and enjoy a large measure of autonomy.

Denmark is best known to the outside world for its high quality foodstuffs and interior design. But since 1945 there has been redevelopment in the economy, and manufacturing has replaced agriculture as the chief source of export income. In 1972 Denmark agreed to join the Common Market.

The Land. Denmark, a geologically youthful country, is a physical extension of the North European Plain. Apart from the easternmost island of Bornholm, a granite outlier of the Fennoscandian shield, the country is flat and low, consisting of a low-lying platform of younger Tertiary rocks on which various glacial materials were thickly laid by the retreating Quaternary ice sheet within the last million years. Glaciation affected the whole Jutland Peninsula, except the west.

Western, or peninsular, Denmark, extends northwards from the West German border to the Skaw, a great sandspit projecting between the Skagerrak and Kattegat. Behind the North Sea coast, with its sweeping beaches, its sand dunes and lagoons, are relatively infertile plains with leached sandy soils and peat bogs.

Eastern Jutland has hilly moraines, cut by deep inlets and valleys, and Denmark's greatest heights – Yding Skovhøj (568 ft; 173 m) and Ejer Bavnehøj (564 ft; 172 m) – rising above the upper valley of the Gudenaa, Denmark's longest river (98 mi; 158 km).

Eastern Denmark consists of an archipelago, of which the chief islands are Sjaelland, Fyn (Funen), Lolland and Falster. All have low morainic hill chains with surrounding loamy soils.

The Faeroe Islands, of which 17 are inhabited, have an area of 540 sq mi (1,399 km^2), and are very rugged. They are geologically comparable to Iceland. (The geography of Greenland is dealt with in a separate article.)

Climate and Vegetation. Denmark has a cool, temperate climate. Temperatures average 62·6°F (17°C) in July, the warmest month, and about 32°F (0°C) in February, the coldest month. Rainfall is low, averaging 23·6 in (599·4 mm).

Although most of the vegetation has been profoundly altered by man, there remain salt marshes and tidal mudflats behind the coastal islands of south-west Jutland, residual heathlands in the interior, and unreclaimed bogs. The steeper hill slopes of east Jutland and the islands support beech, oak and other deciduous timber trees. Conifers, widely planted for reclamation and windbreaks, are especially common in west and central Jutland.

Water fowl are abundant and storks, though less common than formerly, still migrate to parts of Denmark in summer.

The People. Denmark has an almost entirely Scandinavian population. A German minority of about 30,000 lives in south-western Jutland, and some 40,000 Danes live in German Schleswig.

Danes are mostly town-dwellers. The greatest concentration of population is in Sjaelland, where Copenhagen, the capital, and its suburbs have 1,460,000 inhabitants. The built-up area is continuous along the coast of the Sound (sometimes called 'the Danish Riviera') between the capital and Elsinore (Helsingør). The Danes and the Swedes anticipate a continuous conurbation around the Sound by 2000AD and are planning accordingly.

Copenhagen, the capital of Denmark is a city famous for its many canals. The name is from the Danish København, meaning 'merchants harbour'.

Peninsular Denmark has three main cities: Aarhus (117,7748; including suburbs, 187,342), generally regarded as Jutland's capital; Aalborg (85,632; including suburbs, 99,815), on Limfjorden; and Esbjerg (62,000), on the west coast. The chief city on Fyn, the central island, is Odense (107,531; including suburbs, 132,978), world-famous as the birthplace of Hans Christian Andersen.

Each major inlet has its port, which is usually a sizable manufacturing town as well. But hinterlands are everywhere restricted. The countryside is a pattern of brick-built villages gathered around medieval churches, with dispersed farmsteads.

Since 1958 regional development legislation has aimed at reducing the widening differences in growth between the various parts of Denmark. At the same time, the Danes have become increasingly aware of the extent to which fertile farmland is being covered by bricks and mortar, and of growing congestion along the coasts where summer homes and holiday camps are proliferating.

The Faeroe Islands have about 39,000 inhabitants, of whom more than 7,400 live in Thorshavn, the capital, on the island of Streymoy (Strømø).

The Danes are a cheerful, well organized people who have made their country widely known for its advanced social welfare and educational systems. Almost all Danes belong to the Evangelical Lutheran Church, the official church of the nation. The Danish language is closely akin to Norwegian and Swedish, and has richly varied dialects. Famous Danish cultural figures include the philosopher Søren Kierkegaard.

Industry. Manufacturing now provides more than 60% of Denmark's exports. In 1938 it only accounted for 30%. This is the extent of the redevelopment of Denmark's economy, and the decreasing importance of agriculture. Industry is highly diversified, but most factories are relatively small, the typical plant having about 100 employees. Traditional crafts have been the springboard for many successful industries, such as furniture, glass, porcelain, silverware, leather goods and clothing, in all of which a vital factor has been good design. The foodstuffs industries, ranging from brewing and distilling to canned and dairy products, and based principally on home-produced raw materials, command a world market.

Denmark's technical skills are displayed in engineering, especially in its marine branch

Just north of Copenhagen, in Elsinore, Frederick II built Kronborg castle. Designed in the Dutch renaissance style, it inspired the setting of Shakespeare's Hamlet.

and in the manufacture of diesel engines. Shipbuilding is centered mainly on Copenhagen. Expertise in bridges and harbour installations, developed over the years by domestic needs, now finds success in international markets. Agricultural engineering is also important.

Industry generally is dependent upon imported fuel for energy. Although coal is yielding partly to oil, solid fuel imports still provide both gas and electricity. Denmark's own major oil refineries lie on the coasts of the Sound, Little Belt and Great Belt, where the water is deep enough for large tankers. Oil refining has made possible a significant petrochemical industry.

Minerals are few. Some brown coal is produced in Jutland, but the thin coals of Bornholm are no longer worked. There is abundant chalk for cement, however, the largest quarries being along Limfjorden in north Jutland. Clay for brick-making and peat are plentiful, and Bornholm has china clay and granite. To what extent Denmark will benefit from North Sea oil and gas remains to be seen.

Intensive Farming. Agriculture was the historic backbone of the economy, with animal husbandry as the chief pursuit. In former times Denmark was renowned for its cattle and horses. Sheep, still basic to Faeroe Islands farming, were reared on the heathlands.

Farming methods changed radically in the last quarter of the 19th century. The predominantly owner-operated smallholdings, unable to compete with cereal exports from the New World, turned to dairying, pigs and poultry; and butter, cheese, bacon and eggs became the staple exports.

Today, with 60% of its surface area under the plough, Denmark has established a system of intensive production based on temporary grass, fodder grains, beet and potatoes. Farming owes much of its success to the co-operative organizations which had their origins in the 19th century. In addition to the processing and marketing farm produce, the co-operatives make bulk purchases of equipment and raw materials, and provide the farmers with financial help.

Danes are enthusiastic horticulturalists as well as farmers. Orchards are especially widespread in eastern Denmark, and greenhouse cultivation rivals that of the Netherlands in its perfection. Allotments ('colony gardens' as the Danes call them) provide rural retreats for the town dwellers.

Fisheries. Danish coastal waters contain a rich variety of fish. The North Sea harvest consists mainly of cod, herring and plaice, while Baltic waters, especially around Bornholm, provide herring. Most of the catch is exported. The best Danish oysters come from Limfjorden, and eels are also much esteemed. Trout nurseries are numbered in hundreds. Fishing is the chief activity of the Faeroese.

Transport. Denmark is a fragmented country and travel has always involved the sea. The ferry fleet is an essential of day-to-day transport. The island of Fyn is joined to Jutland by two suspension bridges, while Falster and Lolland are linked over Storstrøm (Great Stream) by a 2 mi (3·2 km) bridge, one of the longest in Europe.

Denmark was the first Scandinavian country to have railways, and the relatively dense network of 1,796 mi (2,890 km) is mostly state-owned. There are 33,523 mi (53,950 km) of surfaced roads. The bicycle remains a distinctive feature of the street scene.

The Danes strongly maintain their Viking seagoing tradition. Tankers are a major element in the Danish merchant fleet. The chief ports are Copenhagen, Aalborg, Aarhus, Odense and Esbjerg.

Jointly with Norway and Sweden, Denmark operates SAS (Scandinavian Airlines System), which pioneered trans-polar routes between Europe and Pacific America. Copenhagen's Kastrup Airport is the busiest in Scandinavia.

Trade. Because Denmark lacks a large domestic market and many essential raw materials, freedom to trade is all-important. In addition to its well-known dairy produce, Denmark exports an immense variety of high-quality consumer goods and, in relation to the size of its population, contributes a disproportionately large amount of technical and commercial talent. Tourism is an increasingly important source of foreign exchange. In 1970, more than 11 million tourists spent over $3 million. Denmark's chief imports are oil, solid fuels, fertilizers and feeding stuffs.

Denmark's principal trading partners have always been her neighbours. The North Sea port of Esbjerg was created in 1869 to handle exports to Great Britain, still her chief market. Since the establishment of the European Free Trade Association (1960), Sweden has become the chief market for Danish industrial goods.

The economic success of the Danes springs largely from their flexibility, capacity for innovation which has enabled them to adjust to changing economic circumstances more speedily and efficiently than many peoples in better endowed lands.

UNITED KINGDOM OF GREAT BRITAIN AND NORTHERN IRELAND

Area: 94,217 sq mi (244,022 km²)
Population: 55,348,364
Population Growth Rate: 0·6%
Capital: London, pop 7,379,014
Language: English. Welsh and Gaelic spoken by regional minorities
Religion: Church of England (60%); Roman Catholic (9%); Presbyterian, Methodist, Jewish
Currency: Pound = 100 new pence

The United Kingdom of Great Britain and Northern Ireland (popularly abbreviated to Great Britain, Britain or the United Kingdom) embraces most of the British Isles except the Republic of Ireland. The United Kingdom is a constitutional monarchy technically comprising four countries: the kingdoms of England and Scotland, the principality of Wales, and Northern Ireland. The pioneer of the Industrial Revolution in the 18th century, and once the world's greatest imperial power, Great Britain has lost its old industrial pre-eminence and renounced most of its overseas empire. After World War II, there were recurrent balance of payments crises: economic growth was slow and spasmodic and the rate of modernization not fast enough to keep abreast of the world's other trading nations. Meanwhile, new ideas of social justice, two world wars and colonial unrest had persuaded the British government to grant self-government to most of its overseas subjects by the 1960s. Britain's entry into the European Economic Community in 1973 thus reflected both continuing economic weakness and the need to find a new role in a world dominated by other powers.

The Land. The British Isles lie on the continental shelf of north-west Europe and are thus surrounded by shallow seas. Great Britain is the geographical term for the largest island separated from mainland Europe by the English Channel, which is only 20 mi (32 km) wide at the Strait of Dover. Broadly speaking the British Isles comprise two main natural regions: a rugged, cool, damp highland and island zone in the west and north; and an undulating, warmer, drier lowland zone in the east and south of Great Britain.

The island has a varied coastline of brooding cliffs, sand and shingle beaches, muddy estuaries and creeks, and (in western Scotland) deep and almost fjord-like inlets (sea lochs). The coast is so indented that no place on the island is more than 75 mi (121 km) from the sea. The many offshore islands include the Scilly Isles, Isle of Wight, Anglesey, and (off Scotland) the Hebrides, Orkneys and Shetlands.

England, covering 50,334 sq mi (130,365 km²), is the largest of the countries forming the United Kingdom. England's border with Scotland runs from the Solway Firth northeastwards to Berwick-upon-Tweed, and its border with Wales runs roughly southwards from the River Dee to the Severn Estuary.

England has as its 'backbone' the Pennines, a chain of hills running southwards from the Scottish border to Derbyshire in central England, and formed of massive limestone in the north and south, with shales and sandstones in the centre. The Pennines are highest in the north, where Cross Fell rises to 2,930 ft (893 m). At the southern end is the beautiful Peak District with its gorges and caves. The Pennines are rich in prehistoric remains and in coal and various minerals. Their Yorkshire dales are now a national park. Another scenic area lies west of the

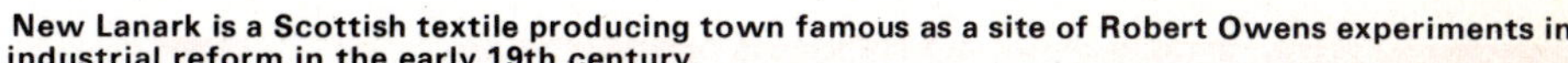

New Lanark is a Scottish textile producing town famous as a site of Robert Owens experiments in industrial reform in the early 19th century.

northern Pennines, where the rugged Cumbrian Mountains enfold the Lake District which contains England's highest peak, Scafell Pike (3,210 ft; 978 m), and lakes including Lake Windermere (England's largest lake). Lowlands occupy most of England east of a line from the River Tyne in the north-east to Exmouth in the south-west. South of the valley of the Tees, the Vale of York runs between the Pennines and the Yorkshire moors and wolds towards fertile, lowland Lincolnshire and the Fens bordering the Wash. The Fens, once marshes around the lower Ouse, Nene and Welland rivers, were gradually reclaimed for agriculture, and are known for their flower bulbs and potatoes. East of the Fens is another fertile area, the flat low plateau of East Anglia, noted for its grain and its sugar beet. West of the Fens is the Central or Midland Plain of England, linked by the Midland Gap with the Lancastrian Plain west of the Pennines.

The central feature of southern England is the London Basin, flanked by the Chiltern Hills and other chalk ridges and watered by the Thames, which flows from its sources in the Cotswolds to the city of London and on to the North Sea at the Nore. South of London is a region called The Weald, flanked on north and south respectively by the chalk ridges of the North Downs and the South Downs. The chalk country extends westwards to the rolling Salisbury Plain and the Isle of Wight.

South-Western (peninsular) England is largely a plateau of granite and sandstone which meets the sea in rugged cliffs and rocky coves. Its features include desolate Dartmoor, where granite crested hills called tors rise to more than 2,000 ft (610 m). Neighbouring Exmoor and Bodmin Moor are lower.

Scotland, with an area of 30,414 sq mi (78,772 km^2), has some of the finest mountain and coastal scenery in the British Isles. The English border country, marked by the Cheviot Hills, is part of Scotland's Southern Uplands, an attractive hill and moorland region which rarely rises to more than 2,000 ft (610 m). Road and rail routes through the uplands mostly follow the valleys of rivers like the Liddel, Annan and Nith, which in their upper reaches link with the valleys of tributaries of the Tweed and Clyde, thus providing natural approaches to the Central Lowlands of Scotland. These lowlands are deeply penetrated by the firths (estuaries) of Clyde, Forth and Tay. The Tay is Scotland's longest river (118 mi; 190 km), but by far the most important commercially is the Clyde leading to the port of Glasgow. The Firth of Forth is the gateway to Leith, the port of Edinburgh, the Scottish capital; it is spanned by two notable bridges, the Forth road bridge (suspension) and the Forth railway bridge (cantilever), the longest of their type in the United Kingdom.

Beyond the upland belt of the northern heights, where the Ochil Hills rise to 2,352 ft (717 m), and the red sandstone valley of Strathmore are the Scottish Highlands containing the highest mountains in the British Isles. Highest of all is Ben Nevis (4,406 ft; 1,343 m). Dissecting the plateau of the highlands are numerous glens (valleys) and straths (broader valleys), the most prominent being Glen More, the Great Glen, following the line of a great fault across Scotland from north-east to south-west and containing Loch Ness and other lochs which are linked by the Caledonian Canal. Scotland has many lochs; some, like Loch Ness and Loch Lomond, are lakes; others, like Loch Fyne, are inlets of the sea (sea lochs).

The Great Glen divides the North-West Highlands from the rest of Scotland. To its south-east are the Grampians, Cairngorms and the lowland plain of Buchan. To its north-west are bleak uplands with Carn Eige and Beinn Dearg rising to more than 3,500 ft (1,067 m). The west and north-west coast with its numerous bays and sea lochs, hundreds of islands and barren skerries, is often reminiscent of Norway. Among the largest islands are Lewis and Harris in the Outer Hebrides; Skye, Mull and Islay in the Inner Hebrides; and the Buteshire island of Arran. Off the north-east coast, separated from Duncansby Head by the Pentland Firth, are the mainly lowland Orkney Islands, a group of about 76 islands, the largest of which is Mainland. Agriculture is the chief activity, and Kirkwall the chief town. Still farther north are the Shetlands, a group of about 100 islands, most of which are uninhabited. Like the Orkneys they have some magnificent cliff scenery. Sheep farming is the main activity, and Lerwick on the island of Mainland the principal town.

Wales, covering 8,016 sq mi, (20,761 km^2), is largely a country of hill and mountain. More than 60% of its area is occupied by the Cambrian Mountains, which have their highest peaks in the north, where Snowdon rises to 3,560 ft (1,085 m) with lesser peaks like Carnedd Llywelyn nearby. Otherwise the Welsh massif lies mainly between 600 and 2,000 ft (183–610 m), falling away in the east to the border and the plain of Hereford, and in the south breaking into the famous narrow coalfield valleys such as the Rhondda. There are coastal plains in the north and south. The splendid natural harbour of Milford Haven is part of the rugged Pembrokeshire coast.

Many rivers radiate from the Welsh massif, among them the Conway, Clwyd, Dee, Severn, Usk, Wye, Taf, Towy, Teifi and Rheidol. The Severn, Britain's longest river (220 mi; 354 km), rises on the slopes of Plynlimmon and flows through Wales and England in a broad arc to the Bristol Channel. Welsh water resources serve several towns and cities in England.

Northern Ireland covers 5,452 sq mi (14,121 km^2) in the north-east of the island of Ireland. Its six counties (Antrim, Armagh, Down, Fermanagh, Londonderry and Tyrone) form the greater part of the old province of Ulster, and it is often called by this name. Its central feature is shallow Lough Neagh with its surrounding lowlands. The lough (a term meaning lake or sea loch) has only one outlet, the lower River Bann. Hill masses fringe the lowlands; to the west the Sperrin Mountains rise to 2,241 ft (683 m) and separate the Bann valley from the Foyle basin; to the east and north the volcanic Antrim Plateau fronts the sea in bold cliffs and headlands, and overlooks Belfast Lough and the Lagan valley. In the south-east the granite Mountains of Mourne rise to 2,795 ft (852 m) from the hummocky lowlands of Down and Armagh. Fermanagh, isolated in the far west, is lake country drained by the River Erne and rimmed by splendid limestone mountains.

Climate. The United Kingdom has a mainly mild and temperate climate which owes much to the North Atlantic Drift, the warm ocean current from equatorial regions which flows past the British Isles to the coasts of Norway warming the air above it. Thus the Orkneys, for example, although in almost the same northern latitude as Greenland, have the relatively high average annual temperature of 45°F (7°C).

Charles II (1660–1685) said the English summer was 'three fine days and a thunderstorm.' The major influence is the prevailing south-westerly winds from the Atlantic, and the corresponding succession of depressions which move north-eastwards over or near the British Isles. In winter the west of Britain is generally wetter, windier and warmer than the east, which may be affected by cold, dry winds from the Eurasian land mass. Average winter temperatures range from 39°F (4°C) in the Shetland Islands to 41°F (5°C) in the Isle of Wight. In summer, when the Atlantic depressions follow a more northerly course and the high-pressure belt of the Azores extends its influence northwards, average temperatures are 54°F (12°C) in the Shetlands and 61°F (16°C) in the Isle of Wight. In the south, summer temperatures sometimes exceed 80°F (27°C) and there is generally more sunshine than in the north.

The average annual rainfall is more than 40 in (1,016 mm), and is heaviest in areas exposed to the Atlantic and in the mountains. Thus western Britain has far more rain than the east or south. The Lake District and peaks like Snowdon and Ben Nevis may receive as much as 200 in (5,080 mm) annually. Patchy fogs are common in winter.

Vegetation. Over the centuries man has transformed the landscape. While considerable areas of moorland remain, the ancient pattern of mixed oak forest in the lowlands and birch and pine in the uplands has given way to arable land and pasture. The United Kingdom, however, does not lack woodland areas, and trees like oak, elm, ash and beech punctuate the hedgerows that criss cross the countryside. In Scotland the most common trees are pine and birch.

Wildlife. Encroachment of the town upon the countryside, the use of chemicals and the uprooting of hedgerows in the cause of more efficient agriculture have taken heavy tolls of Britain's wildlife, which is basically similar to that of other north-west European countries. Sporting interests help to ensure the survival of such animals as the red deer, fox and grouse, while trout hatcheries replenish the lakes and rivers. Additional conservation is at last helping badgers, otters and some other vulnerable species, and many nature reserves have been established. Some 200 species of birds breed in the British Isles, while many others come as regular migrants, and nearly all are now protected by law.

The People. The United Kingdom is one of the most densely populated countries in the world. England, with over 45,879,000 inhabitants, has a density of about 911 persons per sq mi (352 per km²). Scotland has over 5,223,000 people; Wales, over 2,724,000: and Northern Ireland, some 1,527,000. By the end of the present century the combined population is expected to exceed 63 million.

Immigration in recent years has affected both the size and ethnic composition of the population, and the United Kingdom now finds itself with a multi-racial society containing at least 1,500,000 mainly coloured migrants from 'New Commonwealth' countries like India, Pakistan and the West Indies. There was an unusually large influx in 1972 of some 27,000 Asians expelled from Uganda. Immigration is, however, normally regulated. Exploitation of immigrants led to the Race Relations Acts of 1965 and 1968 designed to prevent racial discrimination.

The United Kingdom has a predominantly urban and suburban population. There are now seven major conurbations, the largest being Greater London (7,393,000), followed by South-East Lancashire (2,387,000), the West Midlands (2,369,000), Central Clydeside (1,728,000), West Yorkshire (1,726,000), Merseyside (1,262,000) and Tyneside (804,000). The growth of the conurbations is such that a 'megapolitan' England is emerging from London to Manchester.

Government planning regulates new towns and land use, and helps to safeguard the best of the shrinking countryside. Since 1949, 10 national parks have been established in England and Wales (although the land is not nationalized nor reserved solely for recreational use) and many areas of outstanding beauty have been designated for protection.

The march of the Orange Men on July 12 in Belfast commemorates the victory, in 1691, of the Protestant forces of William of Orange over the Catholics, or Jacobites, under James II.

Beliefs and Culture. There is complete freedom of worship in the United Kingdom where, in addition to the major Churches, religious groups and societies of many kinds abound. There are two established Churches: the Church of England, with more than 27·5 million baptized members, is Protestant and episcopal; and the Church of Scotland, with more than 1·2 million members, which is Presbyterian. Church attendance has declined very markedly in the 20th century, however. The reigning sovereign must always be a member of the Church of England, and is its titular head. Other Protestant groups (Methodists, Congregationalists, Baptists and other 'free Churches') have about 2 million adherents, and there are also more than 5·3 million Roman Catholics. The Jewish community, numbering about 450,000, is centered mainly on Greater London.

Education is free and compulsory from 5 to 16, and most of the schools and centres of higher education are state maintained or aided. The independent (private) schools provide large numbers of public figures and executives. There are 44 universities, including Oxford and Cambridge, the oldest and best known, and a number of other institutions of equal status. A recent innovation is the Open University, which provides degree and other courses for part-time students by means of television, radio and correspondence courses.

Illiteracy is negligible, and is found mainly among the children of immigrants. In 1971 primary and secondary schools had more than 270,700 immigrant pupils. Thousands of students come from overseas, mainly from Commonwealth countries, to study in Britain.

The United Kingdom has been called a 'welfare state', and does indeed have comprehensive health and social services, now maintained largely from public funds but originally pioneered by voluntary bodies. Britain is also said to have a permissive society; this reflects changes in the traditional pattern of life which have occurred over the last 20 years, and in particular new and more legislation on such matters as divorce, abortion and homosexuality. Culturally Britain attracts world-wide interest through its drama, music, literature and other arts.

Government. Britain is a constitutional monarchy. The supreme legislative body is Parliament, consisting of the House of Commons with 630 members elected for a five-year term, and the House of Lords with about 1,000 members.

Northern Ireland, which is represented in the Commons by 12 members, had its own parliament from 1921 until 1972 when all powers were assumed by the central government at Westminster because of mounting terrorism and sectarian violence in Ulster.

The Economy. Manufacturing for export provides about one-third of the Gross National Product. Other important sectors of the economy include agriculture, which provides rather more than half Britain's food; and the 'invisibles', the shipping, banking, insurance and other international services which Britain provides. London has long been one of the world's leading financial centres.

Industry is mostly controlled by private enterprise, but some important industries are state-owned and operated. These include coal, atomic energy, steel, most of the aero-engine industry, electricity, gas and the railways. In manufacturing the range of products is enormous.

Britain stands fifth among the steel-producing nations of the world. The industry, which is now based on imported iron ore, is currently undergoing a £3,000 million modernization and expansion which involves the closure of obsolete open-hearth plants and general introduction of the efficient basic-oxygen method of production. Typical of the 10-year programme is the huge Anchor development at Scunthorpe, Lincolnshire, which began operating in 1972. Most of Britain's steel will eventually come from Scunthorpe and four other centres: Port Talbot and Llanwern in South Wales, Lackenby on Teesside and Ravenscraig in Scotland. Sheffield will remain the chief centre for stainless and high alloy steels.

Most of the steel produced is used in other British industries such as shipbuilding and vehicle manufacture. Ocean-going ships are built in four main localities: Clydeside in Scotland, Barrow and Merseyside in north-west England, Tyneside and Teesside in north-east England, and Belfast in Northern Ireland. The motor vehicle industry, once concentrated in the West Midlands (Birmingham and Coventry) and around London (Oxford, Luton, Dunstable, Dagenham) now also has plants on Merseyside and in Wales and Scotland.

Engineering centres include Birmingham, Wolverhampton, Rugby, Manchester, Leeds, Peterborough and Swindon. The aircraft and aero-engine industries are centred mainly on Bristol, Derby, Weybridge and Hatfield. Other products include industrial engines, farm machinery, mechanical handling equipment, machine tools, chemicals, synthetic rubber, plastics, electrical and electronic equipment, pottery, glass and paper.

Textiles are a major industry. Cotton goods are produced mainly in Lancashire, where Manchester is the commercial centre; woollens and worsteds chiefly in the West Riding of Yorkshire, where Bradford is the commercial centre, but also in the west of England and in Scotland. Artificial fibres are made in many areas, including North Wales and Northern Ireland (a region long known for its linen). Hosiery and knitwear are made in The East Midlands, north-west England and in Scotland. Dundee, in Scotland, has the world's oldest jute industry.

The Greater London area has a wide range of industries: notably printing, film production, clothing, furniture, food and drink, light engineering and consumer goods of many kinds.

Fuel and Power. Coal mining, a 300-year-old industry, has been in decline for some time. Less than 138 million tons were mined in 1971 compared with 289 million tons in 1913. About half of the present output comes from the Yorkshire, Derbyshire and Nottinghamshire coalfield. Other major coal-bearing areas are Durham and Northumberland, South Wales (the source of anthracite and other special coals), the Central Lowlands of Scotland, Lancashire and the West Midlands. With the increasing shortage and expense of other fuels the importance of native resources of coal and of the whole mining industry will have to be reviewed.

At present nearly all Britain's oil has to be imported and there are many tanker terminals around the coast. But the situation has changed with the discovery of oil and natural gas in the North Sea, where several major British fields are now in production. By 1980 the output may reach 75 million tons annually. Natural gas, mainly from the North Sea but also from small on-shore fields and, in liquid form, from the Sahara, now constitutes about 5% of Britain's total energy consumption. Electricity is generated mostly at coal-fired stations, although there are a number of hydro-electric stations in Scotland and Wales. Nuclear power stations at Windscale (Cumberland), Dounreay (Caithness), Winfrith (Dorset) Dungeness (Kent) and elsewhere also produce electricity for the national grid. About 20% of electricity is generated by oil-fired stations.

Agriculture in Britain is highly efficient and highly mechanized. It provides about 3% of the gross national product, employs 2·5% of the working population, uses 47 million ac (19 million ha) of land and has one of the world's heaviest tractor densities – one to every 35 ac (14·2 ha). Since World War II mixed farming has increasingly given way to intensive production of crops and livestock and to specialization, except in Northern Ireland. In England and Wales a broad division can be made between the pastoral west, where the greater rainfall promotes good pastures, and the arable east, drier and more favourable to grain crops and sugar beet. The east Scottish lowlands are also mainly arable, while the Fens and south Lancashire are noted for their potatoes and vegetables. Kent, Herefordshire and Worcestershire are known for their hops and orchards.

Dairying is widespread, with the Friesian breed predominant, except in Scotland where the Ayrshire is preferred. Many notable breeds of beef cattle, such as the Aberdeen Angus and Hereford, have been developed in Britain, which also has more than 30 breeds of sheep. Pig-breeding, mainly of Large White, Landrace and cross breeds, and poultry farming are practised in most areas.

Fisheries. The British are great fish eaters, and the fisheries industry is accordingly important. About 80% of the catch by weight is white fish such as cod, haddock and plaice; herrings account for some 15%, and shellfish for the remaining 5%. The main fishing grounds include distant waters like the west Atlantic and those off Iceland and Norway; middle waters (around the Faeroe Islands); and near waters (the North and Irish seas). The chief fishing ports are Hull, Grimsby, Fleetwood, Lowestoft, North Shields and Milford Haven. Fraserburgh, on the east coast of Scotland, is a major herring port.

Forestry. Nearly 7% of the total land area of Great Britain is classed as productive woodland. Of this 7% nearly half is in England and about two-fifths in Scotland. Nearly 60% of the total forest area is privately owned, the remainder being controlled by the Forestry Commission, which is particularly active in Scotland where large upland areas have been planted with conifers. Forestry is expanding, but at present Britain has to import about 92% of the timber needed.

Tourism earns substantial amounts of foreign currency; in 1971 Britain had more than 7 million overseas visitors, who spent an estimated £469 million.

Lloyds Insurance hall, London. The name 'Lloyds' is derived from a 17th century coffee house kept by Edward Lloyd, where shipowners and insurance men gathered.

Transport. Britain's road system has shown striking improvements, with the construction of more than 1,000 mi (1,609 km) of motorways by mid-1972. By the early 1980s there will be some 2,000 mi (3,219 km) of motorways as well as 1,500 mi (2,414 km) of other trunk routes.

Meanwhile the railway network contracts. More than 6,500 mi (10,461 km) of unprofitable routes have been closed since 1961, and further closures may be expected. Recent developments include electrification of certain profitable inter-city services, the introduction of containerized freight services, and the increased mechanization of bulk commodity services. An 'advanced passenger train' capable of speeds around 150 mph (241 kph) is being developed.

About 11% of the world's shipping is registered in Britain whose merchant fleet, the third largest (after Liberia and Japan), includes many oil tankers, bulk carriers and container vessels. London is still Britain's largest port, with Liverpool, the chief export outlet, second. Other major ports include Southampton (passenger traffic), Bristol (grain and animal feeding stuffs), Milford Haven (oil imports) and Manchester, an inland port linked with the Mersey by the Manchester Ship Canal. East coast ports like Newcastle, Hull and Felixstowe thrive on trade with north-west Europe.

British Airways, a merger of the two state-owned airlines (BOAC and BEA), currently flies scheduled services to some 84 countries. The independent airlines include British Caledonian Airways. Heathrow, London's chief airport, is supplemented by Gatwick, and a third London airport has been planned for the Maplin Sands at the mouth of the Thames. Other major airports include Luton, Manchester and Prestwick. There are extensive internal air services.

International Trade. The United Kingdom ranks among the world's leading trading nations and in 1971 handled more than 7% of international trade. Since 1962 trade links with Europe have been strengthened, while trade with Commonwealth countries has tended to decline. Major imports include crude oil, easily the biggest single item by value, and foodstuffs. Engineering products, especially electrical machinery, ships, aircraft, engines and cars, are the principal exports. Although more goods are usually imported (by value) than are exported, a balance is often achieved by the surplus on 'invisibles' – insurance, banking, shipping and tourism.

Entry, in 1973, into the European Economic Community offered new prospects of expanding trade.

Dependencies. The Channel Islands (off north-western France) and the Isle of Man (between England and Ireland) are not strictly parts of the United Kingdom, but dependencies of the Crown. The Channel Islands are all that remains of the old Duchy of Normandy; the Isle of Man (with ancient Norwegian associations) became subject to England in 1346. Both have their own legislatures. The Isle of Man's is the Tynwald, comprising the 24-member House of Keys and a Legislative Council. Those of the Channel Islands are the States of Jersey, the States of Deliberation (in Guernsey), the President and States of Alderney, and the quasi-feudal Chief of Pleas of Sark.

Britain still also controls remnants of its great overseas empire: associated states, colonies, protectorates and territories scattered over the globe.

The nearest such holding is Gibraltar in the Mediterranean, the remotest is British Antarctic Territory. Most holdings are in or near the West Indies, namely Bermuda, the Turks and Caicos Islands, Cayman Islands, Antigua, British Virgin Islands, Montserrat, St Kitts-Nevis-Anguilla, Dominica, Grenada, St Lucia, St Vincent and Belize (British Honduras). The Falkland Islands (disputed with Argentina), St Helena and other small islands stand in the South Atlantic Ocean. The Indian Ocean has the Seychelles and the British Indian Ocean Territory. In the Far East are Brunei and HongKong. The Pacific area contains the British Solomon Islands, Gilbert and Ellice Islands, Pitcairn Island, New Hebrides (governed with France) and Canton and Enderbury islands (shared with the USA). Some of these places are moving toward independence with British approval, but Rhodesia (still legally British) proclaimed its own independence in 1965. (Articles on most of these places appear in this book.)

EIRE

Area: 27,136 sq mi (70,282 km²)
Population: 2,971,230
Population Growth Rate: 0·5%
Capital: Dublin, pop 568,772
Language: Irish (Gaelic) and English
Religion: Roman Catholic (95%); Protestant (4%)
Currency: Pound = 100 new pence

Eire is a picturesque land of moors, mountains and soft green pastures that occupies all but the Protestant north-eastern sixth of the island of Ireland. Northern Ireland or Ulster, is part of the United Kingdom. Centuries of repressive rule from Protestant Britain spurred the south's underprivileged Roman Catholic majority to seek political freedom. By 1921 the British had agreed to the establishment of the Irish Free State as a British dominion, while the six Protestant Ulster counties opted to remain part of the United Kingdom. This division of Ireland and the continuing link with Britain was opposed by the Roman Catholic majority: in 1949 Eire declared itself a republic and withdrew from the commonwealth, reaffirming its claim to the Ulster counties. In pursuit of a united Ireland, the extremist IRA (Irish Republican Army) with a membership from both sides of the border, brought terrorism to the north and in the 1960s and 1970s, exploiting the grievances of the Catholic minority, plunged the whole of Ulster into sectarian strife. Understandably the Catholic cause found great sympathy in Eire, although the methods of the IRA had little public support. Late in 1973 the Irish Premier, in tripartite talks with Britain and representatives from Ulster, concluded an agreement at Sunningdale whereby a Council of Ireland was set up. Economically, the small, poor, sparsely-peopled nation of Ireland remains largely tied to its big British neighbour.

The Irish Landscape. Eire consists essentially of a broad central plain surrounded by a rim of higher land. The plain seldom rises to more than 400 ft (122 m), except where isolated ranges and plateaux project above its flat or gently rolling surface. The highland rim includes the scenic mountain ranges of west Connacht and Donegal which, like the eastern highlands, run north-east to south-west. The greatest expanse of upland is the Wicklow Mountains in the east, where Lugnaquillis rises to 3,039 ft (927 m). But Eire's most dramatic mountain scenery lies in Cork and Kerry; MacGillycuddy's Reeks, overlooking the beautiful lakes of Killarney, contain Carrantuohill (3,414 ft; 1,040 m), Ireland's highest mountain.

There are many lakes or *loughs*, mainly in the west, and many short, swiftly-flowing rivers descending to the coasts from the upland rim. The major river, the Shannon, is the longest in the British Isles and drains much of the central lowland; like the Erne, in the north-west, it is used for producing hydro-electric power.

Peat has been the main source of fuel in Ireland for centuries. Here an Aran Islander is loading peat which has been transported from the mainland.

The west coast has many bays and inlets with magnificent cliffs and attractive beaches, and offshore islands like Achill, the Aran group and Valentia Island. Long sea inlets are a feature of the south coast.

Climate and Vegetation. The climate is mild and wet. January temperatures range from 44°F (7°C) in the south-west to 41°F (5°C) in the midlands and north; July temperatures range from 60°F (16°C) in the south-east to 57°F (14°C) in the north-west. Much of the lowland has an annual rainfall of 30–45 in (762–1143 mm), but western mountain districts have more than 60 in (1542 mm). Skies are often overcast, with rain falling on two days out of three in the west and north. The east has lower and less frequent rainfall, more sunshine and fewer gales.

Remnants of the oak woodland which once covered the country survive only at Killarney and a few other places. The mountains and the west coast have heath vegetation. The extensive peat bogs lie mainly west of the River Shannon.

The People. The population today is well under half of what it was in 1841 due largely to persistent emigration, particularly from the rural areas, especially the poorer west where dependence on agriculture is very high. Nearly 40% of the population lives in the scattered farmsteads, houses and small market towns of the countryside. The larger towns are in the east and south, where land was better, access easier and foreign influence stronger than elsewhere. The largest city is Dublin, the capital, at the mouth of the River Liffey. Containing some 23% of the population, it is the major centre for government, transport, finance, manufacturing and education. Cork, the second most important centre, serves a large hinterland in the south, while Limerick is a notable regional capital.

More than 90% of the people are Roman Catholics. The Church of Ireland is the largest Protestant body. Education is free and compulsory from 6 to 14 and illiteracy is negligible. There are two universities: the National University and Dublin (Trinity College). Many Irish dramatists, poets and novelists have achieved world eminence.

A Small-Farm Economy. Although its role has declined, agriculture remains of prime importance, and its basis is the small owner-occupier farm of not more than 45ac (18 ha). Irish agriculture is predominantly mixed pastoral in type. The chief arable crops are barley, wheat, oats, potatoes, turnips and sugar beet; but cattle production is the leading enterprise, the eastern lowlands being the chief fattening area. Dairying dominates the south.

Ireland is not outstandingly rich in minerals. Coal is of limited extent and poor quality, and is mined only on a small scale. Peat, an important fuel, is cut by hand in the west, and on a much larger scale by mechanized techniques on the midland bogs. Most of the output is used to generate electricity. Other minerals include building stone, sand, dolomite, barytes and gypsum. During the 1960s large deposits of lead, zinc, silver and copper were discovered, and mines at Tynagh, Silvermines, Gortdrum and Avoca are now in production.

Since the 1950s, when emphasis was placed upon export-oriented industry, more than 300 factories have been established, chiefly by foreign firms attracted by government incentives. Many have been established in provincial towns, but the concentration is in and around Dublin.

Food and drink manufacture includes flour milling and sugar refining, vegetable and meat processing, preparing dairy products, distilling and brewing; it is more widely dispersed than other industries chiefly because the raw materials come from rural areas. About 25% of the total work force is engaged in textiles and clothing. Metal and engineering industries are on a smaller scale. Other industries involve tobacco, wood products, printing and publishing, chemicals, pottery and glassware.

Tourism and International Trade. Tourism is one of the republic's most important industries with earnings in 1970 exceeding £77 million. The chief attractions are the country's beautiful scenery and uncongested roads.

Imports are dominated by manufactured goods, but the purchase of raw materials, especially oil, is increasing as the economy develops. Agricultural products provide some 50% of exports, but there has been a substantial increase in the export of manufactured goods and metal ores. The strong commercial link with Britain accounts for more than half of her external trade. New opportunities appeared on January 1, 1973, when the republic became a member of the European Economic Community.

Transport. The closure of minor lines has greatly reduced the railway network, and transport is mainly by road. Dublin is the major seaport, handling two-thirds of the seaborne trade. Passenger services, chiefly car ferries, operate between several Irish and British ports. There are airports at Dublin, Shannon and Cork.

NETHERLANDS

Area: 13,018 sq mi (33,717 km²)
Population: 13,296,563
Population Growth Rate: 1·2%
Capital: Amsterdam, pop 1,035,999
Language: Dutch. Frisian in the North
Religion: Protestant (41%); Roman Catholic (40%)
Currency: Guilder = 100 cents

The Netherlands (popularly misnamed Holland) is a small but thriving nation in north-west Europe, bordering the North Sea. More than 40% of its predominantly flat, low-lying land was won from the sea; indeed the Dutch have been reclaiming land for more than a thousand years.

The Dutch recovered quickly from brutal German occupation during World War II. They became founder members of NATO and of the Common Market, and by granting independence to their Indonesian territory relinquished their status as a colonial power in order to be one of the cornerstones of a united Western Europe.

Land and Water. Nearly one fifth of the Netherlands is water, and over two-fifths of the land lies below sea level, while almost half has to be protected from sea and river floods by dikes.

Along the North Sea coast, from the West Frisian Islands in the north to the islands and river mouths in the south, is a narrow belt of sand dunes, about 20 ft (6 m) above sea level, reaching its lowest point in Koedood Polder (22 ft; 6·7 m below sea level). This flat region of rich clay soils contains the finest farmland and the largest cities of the Netherlands.

Farther inland are plains with low, sandy ridges and extensive pine forests. The course of the Lower Rhine, Lek, Waal and Maas (Meuse) rivers is marked by a broad tract of clay soils. As they approach the North Sea, these rivers, and the Western Scheldt River, create the complex pattern of islands and waterways that is now the scene of the Delta Plan. This scheme, begun in 1958, will seal all the sea entrances except the Western Scheldt River and the approach channel to the great port of Rotterdam. More land will be reclaimed and flooding prevented. The project is scheduled for completion in 1978. The other major reclamation scheme is in progress in the north, where the once-great bay of the Zuider Zee has been sealed off and converted into a comparatively small fresh-water lake, the IJsselmeer. Scheduled for completion in 1980, this project will provide 860 sq mi (2,227 km²) of new arable land.

The highest region of The Netherlands is in the south, where the uplands rise to Valser Berg (1,057 ft; 2,738 km) near Maastricht.

The Weather Pattern. Although The Netherlands has a mean annual temperature similar to that of Britain, the climate reflects its position in the lee of the British Isles: summers are marginally warmer and sunnier and precipitation is somewhat less. The effect of the European land mass is increasingly felt towards the north and east: there is a higher frequency of summer rainfall inland and the winters are perceptibly colder. Snow falls on 20–30 days each year and frosts are severe enough to freeze canals and other large bodies of water. The prevailing winds are

The Haringvliet dam, which spans a 3-mile channel, is part of a network of dams and dikes which protect south-western Holland from flooding.

The Netherlands is a world supplier of flower bulbs, especially tulips, which are grown mainly in the area between Leiden and Haarlem.

south-westerly or west; the North Sea coast has occasional violent storms.

Temperatures average 62°F (16·7°C) in summer and about 30°F (−1·1°C) in winter. Rain occurs throughout the year, but summer is the wettest season. Yearly precipitation averages vary from 22 in (558 mm) to 34 in (864 mm) in different parts of the country.

Vegetation and Wild Life. The Netherlands retains little of the original vegetation of blanket bog, marsh, fen and forest. Nor, since it is intensively cultivated, does it contain much wild animal life. Extensive bogs survive only in south-eastern Drente and the Peel (between North Brabant and Limburg). Heaths occupy the poorer sands, with marram and other coarse grasses on the coastal dunes, while salt-loving plants colonize the coastal mud-flats that are exposed at low tide. Numerous small mixed oak-birch woodlands occur in Overijssel and Gelderland, but while some are remnants of the former forest cover, most are the result of deliberate planting, as also are the extensive deciduous and coniferous woodlands of the Veluwe. In south Limburg beeches are common on the chalk.

The country's many open waters attract seagulls, waders and other water fowl; the heron is common beside the polder ditches. Red and roe deer and wild boar occur, notably in the Hooge Veluwe National Park. There, as elsewhere, their presence is largely the result of deliberate conservation. Fox, badger, and marten, absent in the west, become more numerous in the south and east.

The People. With a population of over 13 million (1970), The Netherlands is one of the world's most densely populated nations. Nearly half of the population lives in the provinces of North Holland and South Holland, which also contain the three largest cities of The Netherlands: Amsterdam, the capital, Rotterdam (670,060) and The Hague, the administrative capital (525,368). Some 11 other Dutch cities have populations of more than 100,000. The provinces of Friesland, Drente and Zeeland, with much lower densities, have little industrial employment and are areas of rural emigration. Attempts are being made to check this movement and relieve the overcrowded urban regions through the establishment of development areas and the fostering of industrial growth.

Dutch, a Germanic tongue that has many words surprisingly similar to English, is the official language of The Netherlands, but Frisian (also of Germanic origin), is spoken in the northern province of Friesland.

About 41% of the people of The Netherlands are Protestants, mostly members of the Dutch Reformed Church, and about 40% are Roman Catholics. These two religious groups have long held themselves apart from each other, creating a sharp division in Dutch life that is only now beginning to break down.

Education in The Netherlands is compulsory up to the age of 14. The government subsidizes all schools, including religious schools, which meet national educational standards. Universities are almost entirely financed by the government. The largest university is in Amsterdam, though the oldest (1575) is the State University of Leiden.

In recent years the Dutch have absorbed about 300,000 non-white immigrants from their former colonies or overseas territories, especially from Indonesia.

Agriculture and Forestry. About 70% of the land area of The Netherlands is devoted to agricultural use, but there is a considerable annual loss of farmland for urban expansion and roads which is uncompensated by new reclamation of land from the sea. Consequently, farming is both intensive and specialized and farms are small. Mechanization has contributed to the decline of the agricultural population since World War II, and agricultural workers now form barely 10% of the total labour force.

Livestock farming is the main branch of agriculture. In 1970 there were over 4 million cattle (72% of which were Friesians), 5·6 million pigs, 610,000 sheep and 56 million chickens. Besides dairy produce (notably butter, and Edam and Gouda cheese), eggs, meat products and bacon, the country exports fruits, vegetables, potatoes for seed and consumption, bulbs and flowers.

Groningen and Zeeland are the most highly arable provinces, although much of this land is given over to the livestock industry. Southern Limburg, Gelderland, Zeeland and North Brabant supply hard and soft fruits, and early vegetables and fruits come from the Westland district of Zuid Holland. Bulbs are produced in the *Bloembollenstreek* south of Haarlem, cut flowers and pot plants from the nurseries around Aalsmeer, and ornamental shrubs and trees from the specialized holdings around Boskoop.

Only 7% of The Netherlands is woodland, making it the least forested country of mainland Europe, and home-produced timber meets less than 10% of the country's needs.

Fisheries. In late medieval and early modern times the herring catch formed the basis of the country's prosperity. Today sea fisheries contribute only 0·3% to the national income. IJmuiden, the principal trawling centre, lands haddock, plaice, cod, sole and herring; Scheveningen and Vlaardingen are primarily herring ports. Closure of the Zuider Zee and implementation of the Delta Plan have caused the decline of several small fishing ports. The shrimp harvest and the output of the Zeeland oyster and mussel beds have been greatly reduced, but the IJsselmeer has developed freshwater fisheries.

Minerals. The post-war discovery of the world's largest natural gas field at Slochteren in Groningen province and reserves of oil in Drente province have added considerably to the country's known mineral resources. Coal in the south Limburg field has been worked since medieval times, but large-scale production began only in 1902. Today the coal fields are nearly worked out, and many mines are closed. However, some output continues until alternative employment is found for the miners.

Oil production, begun in 1943 during the German occupation, now stands at about 1·9 million tons annually. Nearly 32 million m^3 of natural gas is produced from the Slochteren field. Salt mining began on a large scale in the eastern provinces to meet shortages during World War I. Production of salt has further increased since the discovery of the Winschoten salt dome in 1958. Chalk, gravel and clay (for brickmaking) provide building materials; and peat, formerly important as a domestic and industrial fuel, is now chiefly dug for horticultural use.

Industry. The Netherlands has been famous since medieval times for its manufactures: woollen cloth from Leiden, Haarlem and Delft; linens from Haarlem; earthenware and china from Delft, and shipbuilding are traditional. However, the Industrial Revolution came relatively late to The Netherlands, and large-scale, modern industrial techniques were not widely adopted before about 1870. Since 1945, however, industrial expansion and diversification have been rapid, especially the growth of technologically advanced and science-based industries.

Petroleum and petrochemical industries have grown markedly since World War II. By 1970 there were five refineries in the Rotterdam-Europoort-Maasvaakte region which handled 100 million tons of crude oil a year, and a sixth refinery at Amsterdam.

Food processing industries include numerous dairy factories producing butter, cheeses, dried and condensed milk; and a variety of plants producing meat and bacon, biscuits, chocolate, canned fruits and vegetables and frozen fish.

Electro-technical industries have also expanded greatly since 1945. Products range from light bulbs and transistor radios to television transmitters and electron microscopes.

Shipbuilding in The Netherlands benefitted from the rebuilding of the country's shipyards after war-time destruction. Over half the industry is located in and around Rotterdam. Iron and steel production began on a large scale in 1918 at the Hoogovens plant at the mouth of the Noordzeekanaal. Today its output approaches 4 million tons annually. Other important industries include textiles and ready-made clothing, leather goods, printing and tourism.

Transport. With excellent inland waterways formed by 680 mi (1,098 km) of navigable river and 2,635 mi (4,245 km) of canals, all the provinces can be reached by vessels of up to 1,000 tons. The Noordzeekanaal and the Nieuwe Waterweg are of prime importance since they connect the two major cities, Amsterdam and Rotterdam, with the sea. Rotterdam, in terms of the amount of cargo handled yearly, is the largest seaport in Europe. The Amsterdam-Rijn Kanaal and the Waal and Lek arms of the Rhine provide routes to the Rhineland industrial areas.

Some 47,850 mi (77,000 km) of surfaced roads exist, of which 575 mi (927 km) have dual carriageways. Of the 1,955 mi (3,148 km) of railways, more than half are electrified. Utrecht is the administrative centre and main junction for the railways. KLM (Royal Dutch Airlines) operates international and domestic flights.

International Trade. Foreign trade is of great importance to The Netherlands and the government has therefore supported international efforts at economic cooperation, especially in its role as a founder member of the European Economic Community (EEC). Its chief trading partners are its fellow EEC members and the USA. Exports include meat, flower bulbs, petroleum products, chemicals, textiles, machinery and electrical equipment. Chief imports include grains, petroleum, chemicals, iron and steel products, textiles, machinery, and motor vehicles.

Dutch Dependencies. The Netherlands retains two relics of its colonial past in Surinam (Dutch Guiana) on the South American mainland, and The Netherlands Antilles, two island groups off Puerto Rico and Venezuela. The Netherlands, Surinam and The Netherlands Antilles officially form part of the Kingdom of the Netherlands. (Separate articles appear elsewhere on Surinam and the Netherlands Antilles.)

BELGIUM

Area: 11,781 sq mi (30,513 km²)
Population: 9,695,379
Population Growth Rate: 0·6%
Capital: Brussels, pop 1,074,726
Language: French and Flemish. German (2%)
Religion: Roman Catholic. Protestant minority.
Currency: Belgian franc = 100 centimes

Belgium is a small country that has prospered economically and preserved political unity despite limited natural resources, a chequered history and long-standing friction between its two peoples, the Flemings and the Walloons.

The Belgians have transformed their low-lying country bordering on the North Sea, creating a busy pattern of small farms and thriving industrial areas. Though Belgium was a battlefield in both world wars, they made a spectacular recovery after World War II as a Benelux country and a founder member of the European Economic Community. Like the Netherlands they gave up their status as a colonial power and asserted a European identity by granting independence to the Belgian Congo in 1960 and Ruanda-Urunda in 1962. They have also two major international organizations, NATO and the European Common Market, which both have their headquarters in Brussels.

Flanders and the Coast. Belgium's straight and shelving coast is marked by sandy beaches, and by a belt of dunes more than a mile (1·6 km) wide in places and safeguarded by marram grass, willows, conifers and other protective vegetation. The only major natural break in this 41 mi (66 km) coastline is near Nieuport (Nieuwpoort) where the River Yser (IJzer) enters the North Sea.

This coast is known for its sunshine which has resulted in a chain of popular seaside resorts such as Ostend, Blankenberge and Knokke. Some coastal towns like Ostend are also fishing ports, but it is the summer influx of British and other visitors that matters most.

Behind the dunes runs a strip of polders, land reclaimed from the sea and kept drained for agriculture. About 7 mi (11 km) wide on average, it merges with the Dutch polderlands along the River Scheldt in northern Belgium. Where crops now grow there were once shallow gulfs open to the sea. The silting up of such a gulf, the Zwijn, greatly contributed to the decline of Bruges as a port in the 1400s, and the city did not achieve even a partial recovery until the opening of the Zeebrugge ship canal in 1907.

The rest of Flanders is a plain rarely more than 150 ft (46 m) above sea level and gently undulating in places. Its flatness is relieved by rows of poplar trees that often chart the course of streams, canals and roads.

The Plateau Heartlands. Central Belgium, a broadening belt running from the Dutch frontier near Maastricht to the French frontier west of Tournai, and extending northwards to include Antwerp, is plateau country with altitudes ranging from 150 ft (46 m) to 1,000 ft (305 m). Large areas are covered by 'drift' rocks of sand and other fine materials deposited by the wind or by solifluction (a creeping movement of soil and rock fragments down slopes). *Limon*, a fertile, yellowish, windborne soil, thickly mantles the Brabant, Hainaut, Hesbaye and Herve plateaux. Here is some of Belgium's best arable land. In places there are remains of once extensive oak and beech forests such as the Forêt de Soignies on the southern outskirts of Brussels.

In addition to Brussels and Antwerp, this central region contains such historic centres as Malines (Mechelen), long famed for its lace and tapestry and its vast cathedral, and Louvain with its famous university (founded 1425), where Erasmus taught. In the north-east,

The Ardennes was a vast forest of ancient Gaul, which stretched from Koblenz to the Sambre river in Belgium. Though greatly reduced today, it is still a heavily forested region.

across the River Demer, is the Campine or Kempenland, a region of sand and gravel soils, heaths and bogs where coal-mining is important.

The Ardennes. The hills in southern Belgium are mostly 1,000–2,000 ft (300–600 m) above sea level, and are highest in the Hautes Fagnes ('High Heaths') in the north-east where Botrange, the highest point in Belgium, raises its gently rounded summit to 2,277 ft (694 m). Rivers such as the Meuse and its winding tributaries, the Semois and the Ourthe, cut deeply through the Ardennes, their wooded valleys and small scattered villages making this one of the most attractive parts of Belgium.

The Meuse, which receives the Sambre from the west at Namur, and the Scheldt (the French *Escaut* and the Dutch *Schelde*) together drain most of Belgium. Maritime Flanders is partly drained directly to the North Sea by the Yser, but interior Flanders and most of middle Belgium drain northwards to the Escaut systems (Lys, Dender, Senne and Dyle).

The Ardennes, still the home of the wild boar, has Belgium's most extensive woodlands containing beech, oak and hornbeam. There are marshes in flat and poorly drained parts of the Hautes Fagnes; but peat, once used as fuel, is now deep-ploughed and afforested, usually with conifers. Man's activities in the northern Ardennes have been largely influenced by coal-bearing rocks which outcropped in the Condroz area as well as around Liège and Charleroi. There are other coal-bearing areas westwards to the French frontier.

Climate. The Ardennes is noticably cooler and wetter than the rest of the country. Belgium generally is a country of cool winters and mild summers. The air masses coming from the west, and the sea, are paramount influences in its climate. Over much of the country rainfall is no more than 32 in (812 mm) annually, but in the Ardennes some 55 in (1,397 mm) falls annually in the higher parts of the Rocroi massif and the Hautes Fagnes. Snowfall in the Ardennes provides a limited skiing season based on the resort town of Spa, known for its health-giving mineral springs since the 1300s; hence the English term *spa*.

Flemings and Walloons. Linguistic and cultural differences between the two main groups of Belgians, the Flemish-speaking Flemings and the French-speaking Walloons, stem from the original settlement of the country. The Flemings represent the Germanic tribes which entered the north in the 4th century AD; the Walloons are descended from the Celts and Romans of the south. Despite considerable inter-mixing, the linguistic and cultural division remains and, indeed, was recognized by a constitutional amendment of 1970 establishing four linguistic regions and giving each autonomous cultural and economic powers; the Flemish north, with 55·7% of the population; the Walloon south (32·4%); the bilingual capital, Brussels (11·3%); and the German-speaking Malmedy region in east Belgium (0·6%).

Teaching in the schools is in the appropriate regional language. Many of the schools are controlled by the Roman Catholic Church, to which most Belgians belong. The illiteracy rate is nowhere more than 3%.

Towns and Cities. Today the tendency is for people to m ve from the rural areas to the larger towns. More than 25% of the people now live in the five major 'agglomerations' – Brussels, Antwerp, Ghent, Liège and Charleroi. Other major concentrations are the Hainaut industrial region extending from Mons to Charleroi, and the smaller urbanized areas around Verviers, Bruges, Malines and Ostend. The towns of the Sambre-Meuse coalfield were products of the 19th-century Industrial Revolution as much of their architecture shows. Brussels, the capital, stands in the valley of the River Senne. It is not only the seat of government, but an internationally important cultural, commercial and manufacturing centre. Because of its many ancient buildings, fine boulevards and fashionable shops it has been called 'Little Paris.' Antwerp, on the Scheldt, is Belgium's chief seaport with many industries. Ghent, or Gand, at the confluence of the Scheldt and Lys, is a leading textile centre and port, linked by ship canal to the West Scheldt at Terneuzen in The Netherlands. Liège or Luik, on the Meuse, is known for its iron and steel, engineering and armaments.

In the countryside, the settlement pattern varies. Scattered farmsteads are typical of maritime Flanders and the Famenne area in southern Belgium. Villages in Hainaut and much of the Ardennes are usually compact huddles of houses, while the Flemish north has 'street villages' straggling along its routeways.

Coal, Iron and Sand. Belgium is one of the primary manufacturing countries in Europe, but its factories depend mainly upon imported raw materials, for the country has few natural resources except coal.

The chief coal deposits lie beneath the younger rocks of the Campine and in a west-east series of basins between the French frontier and Liège. The Campine has the most productive coalfield. Total production has fallen dramatically since the closure of uneconomic mines.

Iron ore, long mined on the northern edge of the Ardennes and, since the late 1800s, in the south (where the Luxembourg and Lorraine deposits extend into Belgium), is also nearly worked out. But glass sand, quarried north of Charleroi and in the Campine, is the basis of an important industry whose products range from bottles and window glass to high-resistance glass for car windscreens and the beautiful Val-Saint-Lambert crystal.

Industry. Industry is centred mainly on the Sambre-Meuse and Campine coalfields, along the Brussels-Antwerp axis, and around Ghent. But other regions are industrialized to a greater or lesser extent.

The Liège area has been known for its metalworking since the Middle Ages; the establishment there of Belgium's first blast furnace (1825) encouraged the spread of the iron and steel and associated industries westwards to Charleroi, although development was less pronounced around coal-less Namur. The result is the almost unbroken line of industrial towns occupying the Sambre-Meuse valley.

Current development is concentrated around Charleroi and in the Meuse valley below Liège, and also on the outskirts of other established centres. Here are blast furnaces, rolling mills, foundries, heavy engineering plants, glass works, and zinc and copper refineries. The Borinage area, centred on Mons, has associated industries including cement and ceramics. Wire and nails are made near Charleroi.

But heavy industry is not confined to the coalfields. Clabecq, between Brussels and Charleroi, has important iron and steel works; the Sidmar iron and steel plant at Zelzate on the Ghent-Terneuzen ship canal, reflects the general shift of Europe's steel industry towards the ports receiving essential raw materials from abroad and exporting steel products.

Industries processing imported raw materials are prominent in and around Antwerp, and along the railways and canals radiating from this port. Oil refining and chemicals are important in the Antwerp area and around Ghent. Declining coal production and diversification in the Campine have led to new enterprises such as the Tessenderlo chemical works north of Diest and the Ford motor works near Hasselt.

Flanders has been making textiles since Roman times, and age-old skills, especially lace-making, survive in Bruges, Brussels, Ghent and other towns. The traditional Flemish linen industry is centred on Courtrai, where the mineral-free waters of the River Lys proved ideal for retting flax. Ghent produces cottons and man-made fibres, while the woollen industry thrives around Verviers.

Other industries include food-processing, based mainly on Brussels, Antwerp and Ghent; refining sugar from sugar beet, mainly at Tienen (Tirlemon); clothing and brewing.

Power for Industry. By-product gas is piped from various coking works to Antwerp, Brussels and the Sambre-Meuse belt, with offshoots to Ostend and Zeebrugge. A complex electricity grid also serves these and other industrial areas from coal-fired stations in the Campine and from thermal power stations usually sited beside the canals. Small hydroelectric plants associated with reservoirs in the Ourthe valley also supply the national grid. The Dutch national grid was extended into Belgium in 1965.

Farms, Forests and Fisheries. Just over half the total area of Belgium is used for agriculture. Much of the land is devoted to cereals and sugar beet, but most areas have pastures, and pastoral farming predominates in the Ardennes and in Belgian Lorraine.

Wheat, barley and oats are grown mainly in the loamy soils of central Belgium and in Belgian Lorraine, but some wheat and barley are grown on the polders and the Plain of Flanders, and in the Ardennes. Root crops, flax and hops are important on the Plain of Flanders, while central Belgium provides most of the sugar beet. Vegetables, glasshouse crops, soft fruits and orchards are common around the large cities.

Belgium is an intensively cultivated country of small farms. The typical farm is worked as a family unit, some members of the family regularly travelling to jobs in the nearby town. About 66% of the farms are rented.

The Town Hall in Leuven is a jewel of high Gothic architecture. Built by Mathew de Layens between 1448 and 1468, it justified the town's reputation for patronage of the arts.

Large fields are more common in the open countryside of Hainaut, Brabant and Hesbaye, where farms are larger than in other areas, usually averaging 125 ac (50 ha) compared with the 25 ac (10 ha) farms of the Flanders plain and the Campine.

Woodland covers about one-fifth of Belgium. Providing hardwoods for the furniture and building industries, and other timber for making paper and artificial fibres, is an important rural activity. Home production is insufficient, however, and softwoods are imported from Scandinavia for pulping.

By world standards the Belgian fishing fleet is small. Most coastal towns are harbours for vessels working the North Sea and Atlantic, but about four-fifths of the fishing fleet is based on Ostend and most of the remainder on Nieuport and Zeebrugge. The catch, mainly flatfish and herring, meets about a third of the demand, and fish is therefore imported.

Roads, Railways and Canals. Belgium has more than 7,300 mi (11,748 km) of metalled roads, and motorways link most of the chief centres. One of the historic natural routeways is being followed by a new motorway, the Walloon Highway, which will run from northern France, through Charleroi and Liège, to Aachen in West Germany. The main rail routes are electrified and local services carry commuter traffic to and from the large cities, notably Brussels.

The canal network, though less extensive than that of The Netherlands, plays an important part in the bulk transport of fuel, ores, timber and other heavy materials. Many sections, including the 'inclined plane' at Ronquières on the Brussels-Charleroi Canal, can now accommodate barges of up to 2,000 tons.

The Albert Canal is a valuable artery serving the hinterland of Antwerp. Plans exist for a new canal across Belgium to the German frontier and thence to Neuss near Düsseldorf; this will form a link between the Rhine and the Meuse, and thence, by the Albert Canal, with Antwerp. Links with The Netherlands are particularly important to Antwerp, and an improved waterway connecting the area with Rotterdam, Dordrecht and other centres will result from the Dutch Delta Plan, which also provides for an improved trunk road, the Benelux Motorway, between Antwerp and the Rotterdam-Europort area.

Belgium's Trading Partners. Belgium trades mostly with other countries of the European Economic Community, but also with the United States, Great Britain, the Scandinavian countries and former Belgian colonies, especially Zaïre. Fuel, raw materials and chemicals are the chief imports, whilst exports include manufactured goods, machinery, foodstuffs and manufactured chemicals. Antwerp, the chief seaport, attracts considerable trade from nearby regions such as northern France and parts of West Germany.

LUXEMBOURG

Area: 998 sq mi (2,586 km²)
Population: 345,000
Population Growth Rate: 0·7%
Capital: Luxembourg, pop 78,032
Language: French and German. Luxembourgeois in everyday use.
Religion: Roman Catholic (97%); Protestant and Jewish (3%).
Currency: Franc = 100 centimes

The Grand Duchy of Luxembourg is a remarkable example of the survival of a tiny landlocked country, surrounded by powerful neighbours. Luxembourg has not only maintained its independence, but has built up a role as one of the most highly industrialized countries in Western Europe and indeed in the world.

Grand Ducal History. Luxembourg began as Lucilinburhuc, the small castle of an obscure 10th century count. Later rulers played some part in the affairs of the Holy Roman Empire, and in 1354 their holding was made a duchy. Wars and treaties shuttled it successively to France, Spain and Austria, and back again to France. But by 1815 it had become a grand duchy linked by personal union with the Dutch Crown, a connection which continued until 1890. Meanwhile, in 1839, its western portion was annexed to newly-independent Belgium; this part is now the Belgian province of Luxembourg.

The Grand Duchy was occupied by Germany in both world wars. It was a founder member of NATO and, as a Benelux country, of the European Economic Community. Its economic identity has been progressively and completely merged with Belgium's since the two countries formed a 50-year economic union in 1921.

Uplands and Scarplands. Shaped like a pear, Luxembourg can be divided into two parts: the *Oesling* in the north, formed from parts of the forested Ardennes uplands and containing Luxembourg's highest point, Bourgplatz (1,834 ft; 599 m); and the *Bon Pays* or *Gutland* (Good Land), the hill and plateau country of the Lorraine scarplands covering the southern two-thirds of Luxembourg.

Luxembourg is largely the basin of two rivers, the Sûre and Alzette, both flowing to the Sauer, a tributary of the Moselle. Other streams also flow to the Moselle. The deep valleys cut by these rivers add much to the beauty of the countryside, which attracts many tourists to the Grand Duchy.

Luxembourg has long, snowy winters with an average temperature in January near freezing point. Summers are warm and wet, with an average July temperature of 65°F (18·3°C). Total rainfall is 35 in (889 mm).

The People. Culturally the people are closely linked with their neighbours, the Belgians, the French and the Germans. Education is compulsory from 6 to 15, and there is scarcely any illiteracy. Except for small Protestant and Jewish communities, the population is almost entirely Roman Catholic.

The standard of living is higher than that in most of Europe. But maintaining the industrial workforce depends upon immigration, especially from Italy. Luxembourg suffers from emigration, particularly of highly qualified people, such as graduates of foreign universities (Luxembourg has no university).

About one-fifth of the population lives in the capital, Luxembourg City, built around its ruined fortress on a series of plateaux split by the winding Alzette and Petrusse rivers. The old town and suburbs are linked by dozens of bridges and viaducts.

One-third of the population lives in the mining area, where Esch is the only true centre; the other small towns like Dudelange, Redange and Rodange being really industrial villages. The rest of the country has a few market towns and many attractive villages.

The Economy. Industry is the basis of Luxembourg's economic wealth and employs half of the working population. The chronic shortage of workers is offset by employing foreigners, who form 39% of the labour force. Luxembourg has a unique record in industrial relations; there has only been one strike, and that was against the Germans in 1942.

Industry is dominated by iron and steel, 90% of the production coming from the multi-national Arbed concern, the fourth largest iron and steel company in Europe. Production, at such centres as Esch, Dudelange and Differdang, is based on the iron ore deposits which extend into Luxembourg from French Lorraine. The output of pig-iron, steel, laminated products, phosphor slag, furnace clinker and gas generates more than 25% of the Gross National Product.

Other industry includes an older group (tanning, textiles, chemicals, food and drink

A general view of LaRoche-en-Ardenne in the valley of the Ourthe river in Luxembourg Province, the largest and the most southerly of the nine provinces of Belgium. The area characteristically has poor soil, many peat bogs and heavy upland forest. LaRoche-en-Ardenne is a tourist and market town.

– especially beer) and new ventures such as a tyre factory and chemical plant established as part of the government programme to diversify the economy and rejuvenate the rural areas.

In 1972, agriculture, which has long been uneconomic, supported about 10% of the population, but this percentage may well be halved in the next few years. Poor grains and root crops are grown in the Oesling, where forestry is also important. The warmer and more fertile *Gutland* produces better grains, fodder crops, fruits and grapes (for wine), and has pastures for meat and milk production.

Transport. The main roads radiate from the capital, and motorway links with neighbouring national networks have been planned. Luxembourg City is also the hub of the railway network, which has been modernized by diesel locomotives and electrification. The river port of Mertert on the Moselle, and its marshalling yard at Wesserbilig, serve the iron and steel region. The international airport is at Findel.

International Trade. The iron and steel industry provides 70% of the exports, which also include agricultural products, especially wine. By far the biggest earnings, however, come from 'invisible exports.' Liberal tax and company laws have attracted hundreds of company offices, in nominee form, to Luxembourg City, especially firms needing a Common Market headquarters. The economy is also stimulated by tourism and the revenues of Radio Luxembourg.

Imports fall into two groups: fuel and other materials for the steel industry, and consumer goods.

FEDERAL REPUBLIC OF GERMANY

Area: 95,980 sq mi (248,590 km²)
Population: 61,503,000
Population Growth Rate: 1%
Capital: Bonn, pop 278,800
Language: German
Religion: Protestant (51%); Roman Catholic (46%).
Currency: Deutsche Mark (DM) = 100 pfennige

The Federal Republic of Germany (West Germany) is best known for the so-called economic miracle whereby the West Germans mended and modernized industries shattered by World War II, and within a relatively few years of the massive German defeat at the hands of the Allies made their country the world's third most powerful trading nation. The West Germans' initial survival owed much to American aid, but their ensuing success owed more to the drive and organization which had brought their ancestors together as one nation in 1871. This post-war success was the more remarkable for the fact that West Germany was only a part (though the major part) of pre-war Germany. West Germany was formed in 1949 as a (capitalist) republic from German zones that had come under the control of the Western Allies during the war – the rest of the country had been under Russian control and came under separate (communist) rule as the German Democratic Republic (East Germany). Although West Germany became an independent sovereign state in 1955, the Western Allies had originally regarded its creation as a temporary expedient; as time passed the division of Germany remained apparently unalterable.

Under the leadership of Konrad Adenauer (1949–63) the country made an amazing economic recovery and took its place among the Western nations as a member of NATO and the European Economic Community. Its continuing prosperity is such that it has not only been able to absorb more than 9 million German refugees from the east, but has also found places for some 3 million foreign workers in its thriving industries. The economic prosperity of the Federal Republic is reflected today in the strength of the Deutsche Mark in international money markets.

In international relations, the most significant developments occurred after Willy

Brandt became chancellor in 1969. A Social Democrat, Brandt worked to improve relations with the German Democratic Republic and other Communist countries.

In building a new nation, West Germany's leaders have sought to bury the myth of German militarism fostered by previous leaders like Bismarck and Hitler. Thus the nation has paid reparations to Jewish survivors of Nazi persecution; forged close economic links with Germany's old enemies the British and French; and begun to make diplomatic and economic ties with the communist East.

The Land. West Germany shares three of Europe's major physical regions. The north is lowland, a part of the great European lowland which extends from Flanders and The Netherlands through the two Germanys and across Poland and into the Soviet Union. South of this lowland are the central uplands, which spread westwards into France and Belgium, and eastwards into Czechoslovakia and Poland. In the far south are the German Alps, a small northern part of the whole Alpine complex.

In West Germany, the North German lowland is about 100 mi (161 km) wide. Its North Sea coast, broken by the rivers Ems, Weser and Elbe, is shallow and silted, and fringed by the East Frisian Islands. Out to sea, off the mouths of the Elbe and Weser, is the sandstone island of Helgoland (Heligoland), once heavily fortified but now the haunt of holidaymakers, fishermen and bird-watchers. The Baltic coast has many bays such as the Keiler Bucht and Lübecker Bucht, and long, tortuous inlets (*Förden*).

In the west, behind the dyke-protected reclaimed coastal marshes, is a zone of low sandy hills (the *Geest*) and wet peat bogs (the *Moor*). The *Geest* is one of the most infertile areas in the whole of Germany; its low sand ridges rarely rise above 164 ft (50 m), and between the Ems and Weser rivers it is at most 50 ft (15 m) above sea level. The peat bogs between the ridges reach a depth of 50 ft (15 m) in some places. This *Moor* was once impenetrable, and while large areas have been reclaimed for farming over the last 100 years, parts remain unsettled and uncultivated. West of the Elbe the sandy Lüneburg Heath rises to almost 500 ft (152 m).

In the south, lowland embayments penetrate deeply into the central uplands, reaching to the heart of Westphalia and far into the Rhine basin. The southern border of the North German lowland is the wide loess belt which runs from the Cologne 'bay', a flat but highly fertile embayment of the North German lowland, along the Ruhr and the Weser hills (the *Börde*) and on into East Germany.

Some of the finest scenery in the Federal Republic is to be found in the central uplands, a complex region of worn-down mountains, high plateaux, lower hill country and deep river valleys which has its greatest heights and oldest rocks in the Bavarian Forest, the Harz Mountains, the Odenwald and the Black Forest, where Feldberg rises to 4,898 ft (1,492 m). The Eifel Mountains with their *Maare* (small crater lakes), the Vogelsberg, Kaiserstuhl, Rhön and Siebengebirge all betray their volcanic origins. Volcanic features are also found in the Swabian Alb.

Cutting through this region, and in the south forming part of the West German frontier with France, is the River Rhine. In its southern section, north of Basle, it flows through a broad rift valley, one of the more fertile and climatically favoured parts of Germany. North of Mainz, the Rhine leaves this wide alluvial plain to pierce the Rhine uplands through its famous gorge. This deeply-cut and narrow valley, known for its many castles and terraced vineyards, opens out around Bonn into the Cologne 'bay'.

Only a very small section of the Alps lies in West Germany, but the scenery is spectacular and makes this a favoured tourist area. In the western Bavarian Alps, the Zugspitze, Germany's highest mountain, rises to 9,725 ft (2,964 m).

Below 2,300 ft (700 m), the characteristic highly-glaciated Alpine landscape gives way to uneven, broken foothills. These mark the beginning of the extensive Alpine foreland which stretches as far as the Danube in the north. Swiftly-flowing rivers cross the foreland to the Danube, which here flows through a wide valley before breaking through the edge of the Bavarian Forest by a narrow gorge above Passau.

Climate. Broadly speaking, the West German climate is mild and wet on the coast but drier and colder inland. Yet the temperature variations between the extremes is relatively small. In the north-west, January temperatures average 34°F (1·1°C) on the coast and 35·5°F (2°C) inland; July averages are 62°F (16·7°C) on the coast and 64·4°F (18°C) inland. As one moves east and south, however, winters become slightly colder, although summer temperatures remain similar. Munich averages 28°F (−2·2°C) in January and 62·6°F (17°C) in July. Undoubtedly the most favoured area climatically is Württemberg and especially the Rhine rift valley. Here winters are mild with very few frosty days, and summers are very warm with average July temperatures approaching 68°F (20°C).

Regional rainfall varies more sharply than temperatures. Parts of the uplands, especially the Alps, the Black Forest and the Harz, have an annual precipitation of more than 79 in (2,007 mm). Most of the central uplands receive more than 29 in (737 mm), while parts of the north-west exceed this amount.

Vegetation and Wildlife. About 29% of the land area is forested. Heathland is now limited to a few comparatively small areas which are often protected as game or nature reserves, for example, Lüneburg Heath. Much of the *Geest* has been planted with conifers. The dense forests of the central uplands have been planted relatively recently, and in the process much of the natural beech has been replaced by conifers. The lower Alps also have extensive conifer forests, and there are smaller forests in the Alpine foreland, which also has many peat bogs, especially in the north. Alpine forest is supplanted at a low level (about 3,500 ft; 1,067 m) by pasture or more commonly by rough and rocky limestone terrain. Wildlife includes deer, wild boar and the chamois of the Alps.

The People. Rather more than half the population of West Germany is Protestant (mainly Lutheran), and about 45% are Roman Catholics. The once large Jewish community was almost wiped out by the Nazis and is now numbered in thousands.

Education is controlled by the governments of the various Länder and is compulsory from 6 to 18. Illiteracy is negligible. There are more than 50 universities and other institutes of higher education, the oldest being the University of Heidelberg, founded in 1386.

The Federal Republic has two main belts where the density of the population is high, one stretching along the Rhine valley from the Dutch border to Karlsruhe and including the Stuttgart area, and the other following the loess belt from the Ruhr to Hanover and beyond. There are also large concentrations of population in the North Sea ports of Hamburg and Bremen, in the Saar, and around Nürnberg and Munich. The *Geest* and many of the central upland areas are thinly populated.

West Germany is highly urbanized and the pattern of its towns and cities still bears a close resemblance to that of the 15th century. Most of the major towns of today were important then, and it is only on the coalfields and in one or two industrial areas that completely new towns have been built. Post-war reconstruction has given most of the cities modern centres, state housing schemes on the outskirts and private luxury housing beyond. Villages around the chief towns are often homes for wealthy commuters.

The villages vary greatly in character. Those in the north and in many parts of the central uplands and the Alps are generally small. On low-lying fertile lands settled early in the Middle Ages, large villages often grew up. The creation of these complex *Haufendörfer* was stimulated by partible inheritance in the south-west, where they are still characteristic.

The castle of Neuschwanstein, at the foot of the Bavarian Alps in West Germany, was commissioned in 1865 by Ludwig II of Bavaria.

Major Cities. West Berlin is regarded as a *Land* (state or province) of the Federal Republic (of which, however, it is not a constituent part; this is why its representatives in the Federal Diet are not permitted to vote). West Berlin lies almost 100 mi (160 km) within East Germany, and all routes to the city, apart from air routes, are under East German control. Access has improved since the four-power agreement on Berlin (1971) and the traffic treaty between the West and East German governments (1972). At the end of World War II, West Berlin had no obvious city centre, although typical city-centre functions had developed around the ruins of the Kaiser Wilhelm memorial church and the Kurfürstendamm, now West Berlin's busy main street. The Schöneberg town hall, south of this new centre, is the seat of the city administration. Architects of many nations have helped to rebuild West Berlin, notably its Hansa Quarter. Its industries have also been revived, especially the electrical, light chemical and engineering industries. Firms like Siemens and AEG Telefunken have re-established their works in West Berlin. But while the prosperity of West Berlin cannot be denied, there are many problems including an aging and decreasing population, shortage of labour and the continuing reliance on West German aid.

Hamburg is the most important West German port, although more German trade is handled by the Dutch port of Rotterdam. Hamburg, a proud independent Hansa city on the Elbe, has suffered greatly from the division of Germany, but remains one of the major industrial towns and also serves large areas of Schleswig-Holstein and Lower Saxony.

Heavily bombed in World War II, Berlin's Auguste-Viktoria Platz was rebuilt around the remains of Kaiser Wilhelm Church in 1968. The church has been preserved as it stood in 1945 as a permanent reminder of the evils of war.

Munich is the undisputed capital of the south. It is the cultural centre of Bavaria and has important industries, of which the most famous is brewing and the most characteristic, light engineering. Rebuilt Cologne, on the Rhine, is again the chief centre of the lower Rhinelands, although some of its supremacy has been lost to Düsseldorf, the capital of Nordrhein-Westfalen. Essen is the main shopping centre of the Ruhr, and an important business, administrative and industrial centre where the Krupp organization is still a major employer. Like Dortmund, the main centre of the eastern Ruhr and a hub of heavy industry, Essen has been completely rebuilt since World War II.

The financial capital of the Federal Republic is Frankfurt am Main. All the major West German banks and finance houses, together with many industrial concerns, have their headquarters here. The city is also a major industrial, commercial and exhibition centre. Stuttgart, the beautifully rebuilt capital of Baden-Württemberg, is the leading administrative centre of the south-west and a major industrial city. Bremen, like Hamburg an old Hansa port, is the smallest independent Land in the Federal Republic. With Bremerhaven it is West Germany's second port, with a wide range of port industries and a large modern iron and steel works. Hanover, the capital of Lower Saxony, is both a major industrial city and the centre of an important industrial region. It has a famous technical university.

The Economy. The Federal Republic is an industrial country – in terms of production the second largest in Europe (after the Soviet Union) and the fourth largest in the world (after the United States, the Soviet Union and Japan). Manufacturing provides 55% of the Gross National Product and employs 47% of the working population.

Industry. West Germany has one of the world's largest iron and steel industries. The industry consists of large units concentrated in the Ruhr, in the Saar, on the ore-field at Salzgitter and on the coast at Bremen. Firms like Thyssen and Krupp, well-known as giant concerns before World War II, have been extending their grip on the industry.

Large-scale organization is also found in the chemical industry, which has its main plants along the Rhine (Leverkusen, Hoechst and Ludwigshafen), with others in or near the Ruhr. Light chemicals are produced in most of the major industrial centres. Over the last few years oil refining has expanded rapidly, especially in market oriented locations in the south fed by pipeline from the Mediterranean. The main refineries are in the Ruhr, central Bavaria and the Karlsruhe area, and at the North Sea ports.

Heavy engineering is concentrated in the Ruhr, while shipbuilding is centred on Kiel and Hamburg. Machine construction and light engineering are widely distributed, the major concentrations being around old craft centres such as Stuttgart, Munich, Nürnberg, Frankfurt, Cologne, Hanover and Hamburg. The Federal Republic leads Europe in its electrical engineering, still represented in West Berlin but now centred mainly in the south and in the Rhine-Ruhr area. West Germany is second only to the United States in the manufacture of automobiles. The Wolfsburg-Braunschweig area of Lower Saxony is the home of Volkswagen, Mercedes-Benz have their main works near Stuttgart, and Opel are at Rüsselsheim and Bochum.

Textiles are produced at many centres including Nordhorn, Wuppertal, Augsburg, Bielefeld and Kassel. The optical industry, which in West Germany owes much to refugees from the east, is scattered throughout the country with major plants at Wetzlar, Oberkochen and Munich.

Minerals. The existence of abundant coal deposits was vital to the rise of Germany as an industrial power. It is no accident that the greatest concentration of people and industry in the Federal Republic occurs on the Ruhr coalfield. The Ruhr still produces some 100 million tons of bituminous coal annually, despite cut-backs in production and the closing of many pits since World War II. Most of today's output comes from large rationalized pits north of the Emscher River or west of the Rhine. The Ruhr valley itself has lost almost all its mining. Black coal is also mined in the Saar and near Aachen. Brown coal comes mainly from the Ville field west of Cologne, but also from Helmstedt (Lower Saxony), the Kassel area in Hesse, and from south of Munich in Bavaria.

Some 8 million tons of crude oil, about 12% of the domestic requirement, comes from wells on the northern lowland (especially Emsland, the *Geest* between Oldenburg and Osnabrück, and east of Hanover). Natural gas is also associated with the oil, and there are hopes for substantial natural gas finds in the North Sea.

Most industrial raw materials have to be imported. Domestic iron ores are of low quality. The largest deposits are in the Salzgitter-Peine region of Lower Saxony. Other minerals include potash and salt, and limited deposits of copper, tin, lead, zinc and silver.

Agriculture contributes only 4% of the Gross National Product and employs less than 9% of the permanent labour force, but it remains of major importance in political terms. Nearly 90% of the farms have less than 50 ac (20 ha), and many are so fragmented into strips that it is hard to work them efficiently. This situation is gradually being improved by government-supported consolidation at a rate of some 720,000 ac (291,374 ha) annually.

A toll castle against a background of vineyards at Kaub, on the Rhine between Mainz and Koblenz. This area, popular with tourists is famous for both its wines (Rheingau) and magnificent river scenery.

Pastoral farming is most in evidence in the north, where the coastal marshes and the low *Geest* form a belt of permanent pasture stretching from the Dutch border to western Schleswig-Hostein, and in the Alps and the Alpine foreland. Other upland areas have primarily pastoral farms, but these are of less significance.

About 45% of the agricultural land is devoted to grain and root crops. Grain is more important on the high Alb and in northern Bavaria, while root crops predominate around Nürnberg, Würzburg and in western Hesse around Frankfurt. In the north, rye is often the chief crop, whereas in the south, wheat is more common. Potatoes form a large part of the harvest on the Lüneburg Heath, while sugar beet is the major crop of the loess areas around Hanover and Braunschweig, and in the Cologne 'bay'. Vines are restricted to the Rhine rift valley and the upper parts of the Rhine gorge, the Mosel valley, the Heilbronn area north of Stuttgart and the middle Main valley. The Rhine rift valley has many orchards, which are also numerous in the lower Elbe valley below Hamburg and the upper areas of the Cologne 'bay'. North of Munich large areas are devoted to hops.

Forestry. The Federal Republic has considerable timber reserves and a deservedly high reputation for good forest management. Conifers predominate, and the beech, found on lower slopes throughout the country, is the commonest deciduous variety.

The amount of forest increases from north to south. In many of the southern uplands forestry is a major industry and employer of labour. Nevertheless home production cannot meet demand and timber must be imported.

Fisheries. In size of catch, the Federal Republic ranks third among the fishing countries of Europe. Bremerhaven is the chief centre of the industry, but Hamburg, Cuxhaven, Kiel and the small East Friesland harbours have some importance. Emden and the Weser ports are major herring drifter centres.

Transport. The Federal Republic has an excellent rail network, inland waterways of major importance, and easily the longest motorway system in Europe. Since the division of Germany, the chief rail routes run up the Rhine to Basle or across to Munich, and from Hamburg south to Hanover, Nürnberg and Munich.

The Rhine is the busiest inland waterway of

The Rhine river and Ruhr ironworks at night. The Ruhr is a West German industrial area containing 30 cities and occupying 1,700 square miles on the east bank of the Rhine. It has one of the world's largest coal deposits and produces coke, iron, steel and petroleum products.

Europe. It serves the Rhine-Ruhr, Rhine-Main, and Mannheim-Ludwigshafen industrial areas, and through its tributaries (such as the Mosel, Neckar and Main), Lorraine, Stuttgart and Bavaria respectively. The Weser, Elbe and Ems are also important transport routes. The most important canals are those serving internal transport in the Ruhr. The Dortmund-Ems Canal, linking the port of Emden with the Ruhr, suffers from both its small size and competition from the Rhine, while the Mittelland Canal linking the Ruhr with Berlin lost much of its traffic through the division of Germany. In the future the Main-Danube Canal and the projected Hamburg-Hanover canal may attract freight from road and rail. The Kiel Canal, running 61 mi (98 km) from Brunsbüttelkoog to Kiel, shortens the passage for ships plying between the North Sea and the Baltic.

West Germany's superb *Autobahn* (motorway) network is being extended, and the time will soon come when no place is more than 30 mi (48 km) from a motorway. The network is of great importance to the southern industrial centres, and its recent extension in the Ruhr is expected to attract new industry to this area. International and internal air services are operated by the Deutsche Lufthansa AG, in which the Federal Republic is the principal shareholder.

International Trade. West Germany is one of the leading trading countries of the non-communist world. Engineering products, electrical goods, cars and chemicals make up well over 50% of exports, while foodstuffs figure high on the list of imports. Trading partners include fellow-members of the European Economic Community and the USA.

GERMAN DEMOCRATIC REPUBLIC

Area: 41,768 sq mi (108,178 km²)
Population: 17,040,926
Population Growth Rate: 0·1%
Capital: East Berlin, pop 1,087,982
Language: German
Religion: Protestant (59%); Roman Catholic (8%); Other (6%)
Currency: Mark (M) = 100 pfennige

The German Democratic Republic (East Germany) emerged in 1949 from the eastern remnants of Nazi Germany. The new communist state, formed with Soviet Russian guidance, faced a seemingly bleak future. Shorn of lands handed to Poland and Russia, stripped of industrial wealth seized by Russia as war reparations, and partly depopulated by a westward exodus, East Germany for a long time seemed to be only a poor and estranged relation of thriving capitalist West Germany which was more than twice its size and had over three times its population.

Yet in the 1960s, East Germany entered an economic boom of its own. It built up the largest trading turnover in Eastern Europe, enjoyed that area's second-highest standard of living, had the world's fourth largest per-head energy supply and ranked ninth in the world in value of industrial output. Moreover, while East Germany remained Russia's main trading partner, by the early 1970s peace feelers from West Germany opened new prospects of increased trade with the West.

The Land. East Germany has arbitrary frontiers as a result of the post-war dissection of the German state. Its western frontier winds through the Thüringer Wald to the River Werra and then across the Harz Mountains to the Elbe, reaching the Baltic east of Lübeck. The eastern frontier is largely the line of the Rivers Oder and Neisse. In the south the border with Czechoslovakia follows the Erz Gebirge; while in the north, East Germany runs out to the Baltic. West Berlin stands as an island of Western capitalism in the Communist sea of East Germany.

The northern region of East Germany is part of the great lowland running eastwards from Belgium and The Netherlands through Germany and Poland, and into the Soviet Union. Its surface was formed by glaciation in recent geological times and is characterized by well defined end moraines and broad but shallow meltwater troughs (*Urstromtäler*) running roughly parallel from south-east to north-west, and by infertile sands and considerably more fertile boulder clay. The *Urstromtäler* form wet, marshy plains and their pattern of water courses has facilitated canalization in Brandenburg and elsewhere.

In the north behind the Baltic coast with its many small sandbars, are the forested ridges and innumerable lakes of the Mechlenburg lake plateau. Around Berlin the soil is more sandy and the relief less irregular. The southern limit of the lowlands is formed by a wide zone of loess which enters from West Germany to plunge down into the Leipzig embayment in the highlands of the south.

These highlands include the Harz Mountains, where Brocken, legendary scene of witches' revels on Walpurgis Night (30 April), rears its barren granite summit to 3,747 ft (1,142 m); the densely wooded massif of the Thuringian Forest; and the Erz Gebirge (Ore Mountains), where Keilberg, the Czech Klínovec, rises to 4,081 ft (1,244 m). The Harz and

Erz Gebirge have extensive plateau surfaces with deep river valleys leading down to the surrounding plains.

Climate. East Germany has a relatively mild climate, although its winters tend to be longer and more severe than those of West Germany. Average January temperatures are around 30°F (−1·1°C), while the July mean is around 64°F (17·8°C). Mid-winters are severe in the east, the Oder being frozen over for about 80 days a year and sea ice forming along the coast. Rainfall, like temperature, varies to some extent with local topography. Uplands like the Harz have areas where annual precipitation exceeds 79 in (2,006 mm), but most of the country receives less than 24 in (610 mm). While winter is a wet season, heavy rain occurs in summer, usually in the form of thundery showers.

Vegetation and Wildlife. Little survives of the once extensive heathland, but areas of marsh, moor and peat-bog remain. In many areas the sandy soils have been planted with conifers which play their part, together with grasses, junipers and willows, in anchoring the dunes along the Baltic coast. Upland areas like the Harz, the Thuringian Forest and the Erz Gebirge are densely forested with conifers. Wildlife includes deer and wild boar.

The People. The Communist government regards religion with unfriendly eyes, although the constitution proclaims complete freedom of worship. However, an estimated 59% of the population are Protestants, some 8% being Roman Catholics.

Education is under strict government control, and is free and compulsory from 6 to 15. Illiteracy is negligible. East Germany has some 54 universities and institutes of higher education, the oldest being Karl Marx University (formerly the University of Leipzig, founded in 1409).

The decline of about 6% in the population since 1950 is due mainly to the flight of more than 3 million East Germans to the Federal Republic (West Germany). Since the building of the Berlin Wall in 1961 and the strengthening of the western frontier, this movement has virtually ceased.

The towns and cities of the German Democratic Republic, like those of West Germany, suffered severe destruction in World War II. In their rebuilding, the state has consciously tried to achieve something different, an architecture and arrangement of streets and buildings which reflect the new political system. Common to most of the rebuilt cities are a large central square where demonstrations and mass rallies can be held, and the main processional route leading to it. Towns rebuilt in the early 1950s all have a quarter with grandiose 'Stalinistic' architecture, a style which was later abandoned as too expensive. Since then, full advantage has been taken of industrial building techniques, with monotony as the result.

Change has also come to the villages. Since the war, the large estates of the north have been broken up and the land redistributed. Present policy is to concentrate the rural population in large centres to ease the problem of servicing, and this may well cause many existing villages to decline.

The Largest Cities. East Berlin, the capital and and largest city of the Democratic Republic, has its new centre away from the border with prosperous, capitalist West Berlin, where post-war reconstruction was much more rapid. Early rebuilding centred on the Karl Marx Allee and the Alexander Platz, where the old Berlin town hall is the seat of the East Berlin administration. In the 1960s, Unter den Linden was carefully brought back to something of its former glory, but other famous streets such as the Wilhelmstrasse, where Hitler had his chancellery, are now relatively unimportant. A grassy mound marks the bunker where he committed suicide. The city's reconstructed industries include general, electrical and precision engineering, machine construction and the optical and light chemical industries.

Leipzig has regained its former supremacy as a great industrial fair centre, and is a major marketplace for East-West trade. It has not, however, regained its former pre-eminence as a centre of German publishing. Machine construction and electrical engineering are among its many industries.

Dresden was almost totally destroyed by air attack at the end of World War II. In its rebuilding, efforts were made to faithfully restore some of its most famous buildings along the Elbe. It is again a major industrial and tourist centre, producing chemicals, textiles, machinery and (at Meissen) the famous 'Dresden' china.

The Economy. The German Democratic Republic is now one of Europe's foremost industrial countries and the leading industrial country of eastern Europe, apart from the Soviet Union itself. Agriculture, which employs about 15% of the working population, is still regarded as important, however, on political as well as economic grounds.

Farms and Farming Systems. Nearly all agricultural land has been collectivized. Rather less than 7% of the land farmed is held by the state farms. Their average size is just over 1,600 ac (647 ha) and they are mostly run as trial stations and training establishments. Nearly all the remaining land is farmed by collectives. These average about 1,650 ac (668 ha) in size; but their number is falling all the time, while their size is increasing, and in the years to come farms of more than 5,000 ac (2,023 ha) will not be uncommon. What private land remains consists of plots where members of the collectives grow vegetables for sale in the local markets.

Almost 75% of the agricultural land is arable, slightly over 20% being classed as pasture and meadow. The emphasis placed on local specialization means that collectives with soils suitable for arable cultivation have little pasture or meadow, and only store-fed cattle. There are large areas where less than 15% of the agricultural land is pasture, and some areas where the percentage is even lower.

The Thüringer Wood in the south-west of the German Democratic Republic is an extensive area of woodland and pastureland.

Built in 1961, the Berlin Wall runs for 26 miles, effectively separating East and West Berlin.

Farming is at its most productive in the south, which is also the most densely populated industrialized area. Here, around the eastern Harz, is the loess belt, some 40 mi (64 km) wide. Its northern part, just south of Magdeburg, is perhaps the most fertile land in the whole of Germany. There is little pasture, and sugar beet is the most important crop. The loess thins out in a band across to Dresden, also an almost exclusively arable area. On the poor sandy soils of central East Germany, the proportion of arable is only slightly lower, though here the chief grain crop is not wheat but rye. On the better soils in the north, arable is again dominant. Most of the pastures lie in the *Urstromtäler* west and south of Berlin, and in the higher parts of Saxony in the south.

Forestry. About 27% of East Germany's land surface is woodland, the most extensive forests being in the Harz, the Thuringian Forest and the Erz Gebirge, and also on the poor sands of the Fläming and Lusatian Hills south of Berlin, and in the lakes region immediately north of the city. Fir and larch predominate in the lowland forests, spruce in the upland. The domestic timber industry is important, but timber is also imported.

Fisheries have been developed almost from nothing since World War II. Rostock and Sassnitz are the chief fishing ports, but despite good catches by the deep-sea fleet much fish has to be imported.

Minerals. Except for lignite (brown coal), East Germany is generally poor in mineral deposits. The output of bituminous coal from the Saxony field has declined sharply and may soon cease altogether, and large supplies have to be imported from Poland and the Soviet Union. Brown coal is an entirely different matter; East Germany is the world's largest producer and provides about 33% of the total world output. The two main fields are the West Elbe field between Halle and Leipzig, and the East Elbe (Lusatian) field between Cottbus and Hoyerswerda west of the River Neisse which is now the more productive of the two. Brown coal is used for electricity generation and, in briquette form, in home and industry. It is also used in producing metallurgical coke and as a raw material in the chemicals industry. The brown coal processing plant at Schwarze Pumpe near Cottbus is the largest in the country.

East Germany is a major world producer of potash, coming fourth to the United States, the Soviet Union and West Germany. The major deposits are on the upper Werra near the West German border. Salt is also mined in the same region. Copper has been mined in the Harz for centuries and is still produced near Sangerhausen and Mansfeld. Other mineral deposits are of little significance.

Industry, as in other Communist countries, is state-controlled. So far as value of production is concerned, the most important branch is machine construction, spread throughout the south, particularly in and around Magdeburg, Halle, Leipzig, Karl-Marx-Stadt and Dresden. Some of the plants are very large; the Ernst-Thalmann works at Magdeburg, for example, has a work force of 13,000 manufacturing heavy machinery. The machine tool industry is of great importance, especially in Karl-Marx-Stadt, Leipzig and Dresden. Vehicles are also a southern product, while a modest ship-building industry has been developed in the north at Rostock.

The chemicals industry, second in value of production only to machine construction, has its heart in the Halle area – at Leuna, Buna and Lutzendorf south of the city, and at Bitterfeld to the north. Products range from raw materials for the plastics and textile industries to films, dyestuffs and artificial rubber. There are other large plants at Guben on the Neisse, where artificial fibres are made, and at Schedt, which also has a refinery processing Russian oil.

The electrical, precision engineering and optical industries are mainly in the south, in the Erfurt (Jena), Leipzig, Dresden and Karl-Marz-Stadt districts. The electrical industry has also spread out from East Berlin into adjacent areas.

Transport. One of the greatest problems in the economic development of East Germany has been the labour shortage, and nowhere has this been more apparent than in the transport services. Partly for this reason, investment in transport has been low. Few new roads have been built, and surface quality is poor. Most goods traffic is handled by the railways, which also need modernization. Inland waters have only limited importance. The Baltic ports, especially Rostock, have been greatly improved. There are major airports at East Berlin and Leipzig, and international and domestic services are provided by Interflug.

International Trade. Exports are largely the manufactured products of the electrical, chemical and engineering industries. Imports consist chiefly of raw materials, foodstuffs and some manufactured goods from neighbouring Communist countries.

About 75% of East Germany's trade is with other Communist countries, the Soviet Union accounting for some 50% of trade by value. West Germany has now become the second most important trading partner by value.

POLAND

Area: 120,725 sq mi (312,677 km²)
Population: 33,202,000
Population Growth Rate: 1%
Capital: Warsaw, pop 1,326,200
Language: Polish
Religion: Roman Catholic (95%)
Currency: Zloty = 100 groszy

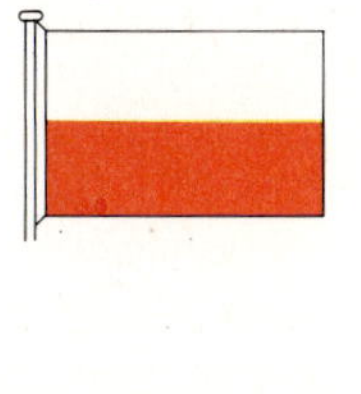

Poland's plains and rolling hills extend south from the Baltic Sea to Czechoslovakia. To the west lies East Germany, to the east is Russia – twin reminders that Poland's history has been one of bitter struggles for survival against its expansionist German, Russian and other neighbours. Today, Poland lies squarely in the Russian political and economic orbit. But the relative lack of collectivization and the survival of strong religious feeling in Poland are reminders that even under communist rule the Poles preserve some of their old individualism. Poland remains closely allied with the Soviet Union; she is a member of both COMECON and the Warsaw Pact organization. But her communism is essentially pragmatic and leavened by a surprising degree of liberalism. The Poles, for example, remain a devoutly Roman Catholic people. Poland is the only communist country to have an independent Roman Catholic university (at Lublin). Again, the Poles have persistently refused to collectivize agriculture, and their success in raising output and peasant standards of living is in striking contrast to the record of most other East European countries, including the Soviet Union. Also, while there is a censorship, some measure of discussion and criticism is permitted, even in the state-controlled press.

The Land. Poland is a country of the Central European Plain and most of its north and centre is lowland. The 310 mi (499 km) Baltic coastline curves gently and consists of broad sandy beaches, behind which is a belt of fixed and moving dunes rising in places to more than 100 ft (30 m). These dunes hold back rivers to form lagoons such as the Vistula and Szczecin lagoons and Lake Leba; some, like the long dune spits of Hel and Vistula, enclose bays such as the Bay of Gdańsk.

The plain proper begins south of the dunes, but is virtually squeezed out at Gdynia and Sopot by morainic hills which bound the plain to the south. These hills, often called the Baltic Heights and rising to some 400–1,000 ft (122–305 m), enclose thousands of lakes. This Polish lakeland or *pojezierze* is divided from north to south by the Vistula into western (Pomeranian) and eastern (Masurian) parts. The largest lakes, Śniardwy and Mamry, each cover more than 39 sq mi (100 km²).

South of the lakeland is the central plain dissected by shallow marshy valleys which the Poles call *pradoliny* ('ancient valleys'). The larger *pradoliny* provide vital east-west land and water routes: between the Warsaw area (Mazovia) and Berlin via sections of the rivers Bug, Bzura, Warta, Oder and Spree: and between the Toruń area (Kujawy) and Eberswalde in East Germany via the Vistula, Noteć, Warta and Oder systems. On either side of the *pradoliny*, the plain is low-lying and gently rolling, but highly diversified by sands, dunes, drift, morainic material and marsh.

In the south the plain ends where the Trzebnica Hills, Małopolska (Lesser Poland) upland and Lublin plateau rise to about 1,000 ft (305 m). Separated from the central plain by the Trzebnica Hills is the Silesian plain, a large lowland embayment of the upper Oder River covered by fluvial sands and loess. Similar materials cover the Sandomierz depression in the south-east, which is enclosed by the Małopolska and Lublin uplands. These uplands rarely exceed 1,000 ft (305 m) and are distinguished by their thick loess and an associated dryness, especially on the Lublin plateau, and by limestone areas in the Częstochowa-Kraków Jura. Only the ridges and upstanding crags of this Jura and the hard rock ridges of the Holy Cross Mountains (Góry Świętokrzyskie) are truly upland areas in south-central Poland, where they rise to more than 2,000 ft (610 m).

Poland's most impressive highlands lie along the Czech frontier, where the Sudety Mountains in the west and the Beskids (Carpathians) in the east are separated by the lowland Raciborz basin. This basin leads to the Moravian Gate and is occupied by both the Oder and Vistula rivers. The Sudety are the north-eastern extension of the Bohemian massif into Poland. Earth movements, glaciation and other geological processes have given them a diversified landscape of rounded mountains, narrow valleys connecting the intermontane basins, erosion surfaces and thermal springs. The highest peak is Sniezka (5,259 ft; 1,603 m). The Carpathians in Poland comprise primarily the flysch and limestone Beskids and their foothills, running west-east and rising to more than 3,300 ft (1,006 m). These forested uplands have many basins, the largest being the Nowy Targ (*Podhale*) depression. Extending southwards is the small but majestic Polish section of the glaciated Tatra Mountains. Zakopane, the chief tourist resort in this area, is near the geological divide which separates the jagged western limestone Tatras from the higher, crystalline eastern Tatras. The latter are at their most splendid at Morskie Oko ('Eye of the Sea'), a deep glacial lake at the foot of Poland's highest peak, Rysy (8,212 ft; 2,503 m).

Climate. Poland is predominantly a lowland country and relief, therefore, is climatically significant only in the extreme south. Climate varies only gradually from west to east and from north to south. The west is influenced by its proximity to the Atlantic. Warm air is brought from the Gulf Stream by the westerlies and winters tend to be mild compared with those in the east. Increasing distance from the Atlantic and growing proximity to

Silesia is a rich agricultural and industrial section of south-west Poland. The mountains in the background are the Sudety, which separate Polish Silesia from Czechoslovakian Sudetenland.

Siberian influences give the east a more 'continental' climate with colder winters and warmer summers. January temperatures average 28·4°F (−2°C) at Zielona Góra, but 24·4°F (−4·2°C) at Lublin, while July averages are 62·1°F (16·7°C) and 65·5°F (18·6°C) respectively. Increasing distance from the Baltic produces the same broad tendency from north to south.

Precipitation is low over most of Poland and is greater in summer than in winter. Szczecin has some 22·7 in (577 mm) annually; Gdańsk, due to its proximity to the morainic Kaszubian Heights, receives more (25 in; 636 mm). Averages are higher in most of the south, Sniežka and Zakopane receiving 53·1 in (1,349 mm) and 44·5 in (1,129 mm) respectively.

Vegetation and Wildlife. More than 25% of the country is forested, mainly by conifers, although beech and oak occur in the milder Baltic coastal areas and in the Beskid foothills. Forest management is good and has contributed, despite wartime depletions, to the preservation of substantial areas of primeval forest, including the *bory* (pine forests) of the lakelands and the Białowierża reserve. Marmot and chamois are conserved in the Tatra and other national parks, and a small number of European bison in the Białowierża forest.

Eighty five per cent of all Warsaw's buildings were destroyed in a terrible battle in World War II. Since 1945 much of the Old City *(below)* has been painstakingly restored.

The People. The population is entirely Polish except for small numbers of Ukrainians, Byelorussians, Germans and others. About 95% of the people are Roman Catholics, and there are also Polish Orthodox and Protestant (mainly Lutheran) communities. Education is free and compulsory from 7 to 15, and the illiteracy rate is about 5%. The many institutions of higher education include universities and theological academies.

During the Nazi occupation between 1939 and 1945 some 6 million Poles, about half of them Jews, were put to death in Oświecim (Auschwitz) and other German concentration camps in Poland: a further 2½ million were deported to Germany to work as slave labour. When peace came, the population of Poland was less than it had been in 1860. With the return of stability after World War II, the population grew rapidly again. Natural increase has remained high, although its rate began to slow in the late 1960s. In the north, where there is more forest and lakeland and less farming and industry, the population is thinly spread, except along the Vistula valley, around Trójmiasto (the tri-city of Gdańsk, Sopot and Gdynia) and Szczecin, and inland centres like Olsztyn and Białystok. In the south, higher density has been encouraged by good farming soils and exploitation of mineral and energy resources.

Industrialization since the war has brought many more people to the cities, although urban growth has varied. War destruction and the extermination of the Jews drastically reduced the population of many cities, notably that of Warsaw, which fell from 1,300,000 in 1938 to 160,000 in 1945 and did not completely recover until 1970. In the west and north, the size of many centres was restricted by the departure of large numbers of Germans and the arrival of considerably fewer Poles. But the slow or limited growth of such cities as Warsaw, Lódź and Wroclaw also owes much to the deliberate reduction of city housing densities and the creation of parks and open spaces near city centres to provide a healthier urban environment. Industry has actually been moved out of Lódź and Warsaw. Other major cities have grown rapidly as a result of planning, none more so than Kraków which was joined by the new steel city of Nowa Huta after 1950. The most urbanized region is Upper Silesia where some 15 towns now form a conurbation containing about 2 million people. The largest – Katowice, Zabrze and Bytom – each have more than 200,000 inhabitants.

The Economy. In 1939, Poland was second only to Czechoslovakia as the most industrialized country in east central Europe. Today more than 54% of Poland's Gross National Product is provided by industry, while a little more than 19% is provided by agriculture.

Agriculture has advanced more slowly than other sectors of the economy, but its output is nonetheless 60% greater than it was in 1938. Yields of wheat, potatoes, sugar beet, vegetables and, to a lesser extent rye, oats and barley, have greatly increased, as has pig production. Since 1970, peasant farmers have been encouraged by a 'new deal' which includes better rewards for farm produce, the stabilization of fodder prices, land tax reform and the extension of the free health service to rural areas. Agricultural imports, particularly of fodder grain, are still necessary. Some domestic market shortages are due to Poland's large food exports, especially of dairy products, meat and sugar. Agricultural exports account for 15–20% of all Polish exports by value.

Although 40% of Poland's land is state-owned, most of this is forest. Less than 14% of the farmland is socialized, mainly in the *Panstwowe Gospodarstwa Rolnicze* ('State Agricultural Enterprises') in the 'regained territories' of the north and west (Pomerania, East Prussia and Silesia) where the large *junker* estates were nationalized after the war. Otherwise the land is divided among more than 3,600,000 peasant farms, most of which have fewer than 25 ac (10 ha). Improvements on these farms have come mainly through the *Kólke Rolnicze* ('Agricultural Circles'), farming co-operatives which provide seed, fertilizers and machinery on credit.

Rye and potatoes are the chief crops of the poorer soils of northern and central Poland, with wheat of very secondary importance. There are large areas of meadow and forest, and pig-breeding and dairying are major activities, with flax and cattle prominent on the state farms of the north and north-east. The belt of better soils stretching from the Vistula delta to Upper Silesia has larger peasant farms, larger acreages under wheat and sugar beet, higher yields of rye, oats and potatoes, and important areas of fruit and vegetable production, especially west and south of Warsaw.

Lower Silesia stands apart as a region of intensive wheat production, sugar beet and dairy farming. The black-earth areas of Małopolska, the Lublin plateau and the Carpathian foothills provide wheat, barley, fodder, oilseed and potatoes. The Carpathian and Sudety areas are heavily forested and have a predominantly meadow and pasture, cattle and sheep orientation with potatoes and oats as the chief crops.

Minerals. Poland is a major coal-producer. Some 130 million tons of bituminous coal are mined annually, mainly in the Upper Silesian coalfield roughly between Kraków, Tarnowskie Góry and the Czech frontier near Cieszyn. Recently the demand for coking coal for the growing steel industry has necessitated the opening of the Rybnik basin in the south. Other coal resources include the small Lower Silesian field and the substantial brown coal reserves worked by open-cast methods at Turoszów and Konin.

Other minerals include copper, lead, zinc and sulphur, which are produced in sufficient quantities to provide an export surplus. Poland also has modest deposits of oil (between Bochnia and Jaslo in the Carpathian foothills), natural gas (Lubaczów), iron ore and salt.

Industry. World War II left Polish industry in ruins, but within a few years reconstruction was well advanced. Most industrial investment was channelled into rebuilding and expanding existing mines and factories in Upper Silesia and the engineering and chemical plants of Warsaw, Gdańsk, Poznań and other large centres. New steel mills were built at Nowa Huta (Kraków), Częstochowa and Warsaw. Nowa Huta alone now produces almost as much steel as the 12 Upper Silesian plants combined, and may be further expanded. The engineering and electrical industries have continued to expand and diversify. There has been some shift from metallurgy to chemicals, however, especially petrochemicals, fertilizers, plastics and artificial fibres. This reflects the increasing importance of Soviet oil and natural gas, and of sulphur, in the Polish economy. Consumer and other industries such as textiles, clothing, shoes and timber-processing have more or less maintained their former importance.

In the north, growth is associated with engineering, especially shipbuilding (Gdańsk, Szczecin) and timber-processing (Olsztyn). In central Poland the growth industries include engineering (Zielona Góra, Poznań, Warsaw, Lublin); chemicals, along the Vistula from Tarnobrzeg (sulphur), through Pulawy (fertilizers), Plock (oil-refining), Wloclawek, Toruń and Swiecko (using chemicals from Plock); and electricity and aluminium production in the Konin area.

Transport. Reconstruction of the war-devastated transport network severely strained Poland's scarce capital resources. Nor was the old network entirely suited to the new Poland. There has been considerable electrification of the trunk railway routes focussing on Warsaw and Katowice, and electrification of the former 'coal railway' (Katowice-Gdynia) is nearing completion. The railways still carry most of the freight and passenger traffic, for the development of long-distance road haulage has been inhibited by lack of sufficient vehicles. The road network is better in the west than in the east; poor roads are still common east of the Vistula and in the Kielce area.

Water transport has a minor role. Insufficient use is made of the Oder, and lack of money has delayed improvement of navigation on the Vistula and this river's integration with the Oder system. Land transport (rail, road and pipeline) dominates international commodity movements, especially to and from other COMECON countries. The major 'Friendship' pipeline crosses Poland en route from Kuybyshev in the Soviet Union to Schwedt in East Germany. Lot, the state airline, provides domestic and international services.

The Cloth Hall in the Old Market Square. Warsaw grew up in the 13th century around a castle on the banks of the Vistula. In 1536 it supplanted Kraków as the capital of Poland.

International Trade. Poland exports coal, sulphur, steel, chemicals, machinery, foodstuffs, metal products, textiles and shoes. Imports include basic fuels, ores, fertilizers and cereals. While 65% of Poland's trade is with other Communist countries, most food exports and shoes go to Western Europe or the United States.

CZECHOSLOVAKIA

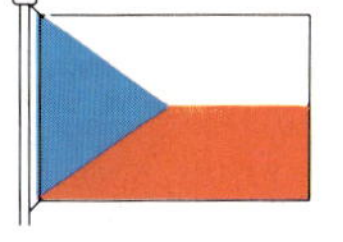

Area: 49,373 sq mi (127,876 km²)
Population: 14,406,772
Population Growth Rate: 0·5%
Capital: Prague, pop 1,081,608
Language: Czech and Slovak. Hungarian and German spoken by minorities.
Religion: Roman Catholic (77%); Czechoslovak Church (8%); Protestants (8%).
Currency: Koruna = 100 halérii

Czechoslovakia is a largely rugged and forested land; but it is also an industrially advanced communist state, sharing borders with communist and non-communist countries in central Europe. External pressures have helped to keep Czechoslovakia politically in line with other East European communist states, although some 30% of its trade is done with the non-communist West. In January 1968, Alexander Dubček became First Secretary (leader) of the Czech Communist Party and initiated a policy of liberalization –

Bohemia's tradition as a centre for fine glass-ware is kept alive today both by city factories and by industrious craftsmen dotted around the slopes of Czechoslovakia's Giant Mountains.

'socialism with a human face'.

This was abruptly ended in August, when hundreds of thousands of Russian and other Warsaw Pact troops invaded Czechoslovakia. By 1969 Dubček had been replaced by the more orthodox communist Gustav Husak and a purge of progressives was in full swing. Husak went so far as to publicly thank the Russians for invading his country (1971).

Land Features. Czechoslovakia is a long and narrow country, stretching some 450 mi (724 km) from east to west, and varying in width from only 60 mi (97 km) in the east to 175 mi (282 km) in the west. Geographically it can be divided into four distinct regions: the crystalline massif of Bohemia in the west; Moravia, a region drained by the river Morava; the mountain arc of the Carpathians, and the foothills and plains (part of the northern Danubian plain) which belong to Slovakia.

Bohemia is a diamond-shaped, broken plateau, sloping northwards and rimmed by mountains. In the north-west the Krůsné Hory (Erzgebirge, or Ore Mountains) rise to 4,080 ft (1,244 m) in Klínovec; in the north-east are the Giant and Sudety Mountains (Sněška, 5,259 ft; 1,603 m). The lower Moravian heights (the Českomoravská Vysočina) rise in the south-east. Along the West German border in the south-west are the Bohemian Forest Mountains.

In the south, the rolling uplands are pitted by the Plzeň (Pilsen) and České Budějovice depressions, and deeply cut by the meandering, northward-flowing River Vltava, which receives the Sázava and Berounka rivers before passing through Prague and joining the Elbe (Czech: Labe) on the northern Polabí-Ohřs lowlands, known for their rich black-earth soils. Such soils also occur in central Moravia (between Brno and Olomouc) and lowland Slovakia (between Bratislava and the River Hron). The Elbe leaves Bohemia through a long sandstone gorge in the Ore Mountains (the Děčín Walls).

Moravia, which lies in the centre of the country, is a region of rolling hill country and fertile loamy soils, drained chiefly by the Morava and its tributaries, but also in the north (Czech Silesia) by the Oder. The corridor formed by the two river valleys is a historic routeway between the Danube (and Vienna) and Polish Upper Silesia (and Kraków).

Slovakia is dominated by the parallel forested ridges of the Carpathians, running east-west, rising to some 3,000 ft (914 m) and separated by the Váh, Nitra and Hron rivers, all flowing to the Danube. Among the distinctive areas are the alp-like High Tatra, where Gorlachovka, the highest peak in Czechoslovakia, rises to 8,737 ft (2,663 m); the karst Low Tatra area and Slovenský Kras (Slovakian Karst); and the volcanic Štiavnica and Krupina Hills.

In the south-west, south-east and east are low-lying loess-covered extensions of the Danube-Tisa (or Pannonian) plain. A large alluvial island, Žitný Ostrov (Hye Island) lies in the Danube between Bratislava and Komarno.

Climate. Czechoslovakia has a central European climate with long, cold and rather dry winters, and hot, thundery summers. Winters are milder in the west, where Prague averages 31·1°F (−0·5°C) in January, than in the east, where Košice averages only 25·9°F (−3·4°C). Average July temperatures are 67·1°F (19·5°C) in Prague, and 66·4°F (19·1°C) at Košice. Variations of altitude aspect and exposure cause many regional and local variations.

Exposure to rain-bearing westerly winds from the Atlantic or Adriatic cyclonic systems brings heavy precipitation, exceeding 40 in (1,000 mm), to three highland areas: the western Bohemian rimlands, especially the Bohemian Forest; the Sudety Mountains, which are exposed to the passage of Atlantic air over the Ohře valley; and the entire Carpathians, from the Halé Karpath (Little Carpathians) near Bratislava to the Slovakian Karst near Košice, which are exposed to depressions moving north-east from the northern Adriatic. Each highland casts a rain shadow over the lowland to its east. Thus much of plateau and lowland Bohemia, and Moravia and eastern Slovakia, have less than 25·6 in (650 mm) annual precipitation.

Vegetation and Wildlife. Nearly 30% of Czechoslovakia is forested. Conifers predominate in the Bohemian highlands and the higher, northern Carpathians; beech on the lower hills of Bohemia and Moravia, and especially on the southern Carpathians; oak in the lowlands of the Elbe, Moravia and Slovakia.

Wildlife is varied and hunting is a popular sport. Wolves still range the Carpathians, but the brown bear, the wild boar and the wild cat are becoming rarer. Endangered species like the chamois of the High Tatra are strictly protected. Roe deer are common in the woodlands, while the lowlands are the home of hares and badgers.

The People. The Czechs form about 65% of the population and the Slovaks about 30%, the remaining 5% being formed by Hungarian, German, Polish and Ukrainian minorities.

The natural rate of increase is very much higher among the Slovaks (about 10 per 1,000) than among the Czechs (about 3 per 1,000) – one of the factors which has aggravated labour shortages, particularly in Bohemia, and made it more difficult to develop Slovakia and raise living standards there. Language barriers have tended to inhibit the movement of Slovaks to farms and factories in Moravia and Bohemia.

The population is unevenly distributed. The most populous regions are north and central Bohemia, the Plzeň and České Budějovice basins, Moravian Silesia, central Moravia and south-western Slovakia. The plateaux and highlands are thinly populated, except for the Carpathian valleys and the basins in the Bohemian rimlands.

Towns and Cities. Some cities emerged as major transport, trade and administrative centres. Typical of these are Prague, the only city with more than a million habitants; Brno (339,000), an important manufacturing city, dominated by the grim Špilberk Castle and known for its Augustinian monastery where Gregor Mendel discovered the principles of genetics; and Bratislava (291,000), now the capital of Slovakia and once the capital of Hungary (1536–1683).

Plzeň (147,000), Ostrava (280,000) and Most (55,000) developed as industrial centres during the 1800s, while Gottwaldov (65,000) – formerly Zlín – was built in the 1930s by the famous Bata footwear company. Examples of communist town-planning include Havířov (82,000) near Ostrava, and the new section of Košice (142,000). Czechoslovakia has many small, solidly-built medieval or baroque towns.

The Economy. Czechoslovakia continues to be what it was before World War II, a highly industrialized country, but with this difference: industrial production has been increased six-fold, and employment in mining and manufacturing has risen by 930,000. Industry provides more than 40% of the Gross National Product, compared with about 12% from agriculture.

Industry has been given high investment priority for various reasons. Czechoslovakia has the skills and the plant to meet the manufacturing needs of less developed members of the Council for Mutual Economic Aid (COMECON), and is well suited to specialization in growth industries, especially chemicals, coke, metallurgy, engineering, armaments, precision instruments and glass. While there has been some expansion of the large pre-war shoe and textile industries, growth has been mainly in heavy industries, and engineering and chemicals now lead.

Another factor was the need to develop Slovakia, a very backward and overpopulated rural region which, however, had the natural resources, labour force and transport facilities, in relation to the Soviet Union, that Bohemia lacked. Since 1946 Slovakia has been given priority, and the rate of industrial growth there has consequently been much faster.

Industry in Slovakia is based on local natural resources (hydro-electricity, metal ores and timber) and labour, and upon transport links with the Soviet Union (which supplies iron ore and coking coal for the Košice steel complex, and oil and gas for the petrochemical complexes of Bratislava and south-western Slovakia), and with Hungary (which supplies bauxite to the Žiar aluminium combine in the Hron valley).

Industry in Bohemia and Moravia is widely dispersed. Among the most important areas are Ostrava (mining, coke and metallurgy); Plzeň-Kladne (steel and heavy engineering); the Sudety area (glass and textiles); the Bohemian Ore Mountains (brown coal, chemicals and electricity); and Prague (engineering, pharmaceuticals and consumer goods). Bratislava is known for its electrical, food and chemical products; Brno for its engineering; Gottwaldov for its shoes; and Košice for its steel, engineering and chemicals.

Minerals. Czechoslovakia is not self-sufficient in minerals and has to import iron ore and bituminous coal. Brown coal, mined in increasing quantities in the Cheb-Chemutov-Most fields in the Bohemian Ore Mountains, is used by the chemical-processing plants and for generating electricity. Slovakia has brown coal in the Nitra valley, but eastern Slovakia relies partly on coal imported from the Ukraine.

Small amounts of oil and natural gas are produced south of Hodonín in the Morava valley, but national needs are met mainly by imports from the Soviet Union. Oil from Kuybyshev (the Volga oilfield) is conveyed by the 'Friendship' pipeline to a new chemical complex near Nitra.

Scattered small-scale mining still continues in the Bohemian and Slovakian Ore Mountains, long the source of various ores. Lead-zinc ores are still extracted at Pribram and Banská Štiavnica. Iron ore deposits between Plzeň and Prague are still worked, while the pyrites at Chvaltice in the Elbe valley are mined as basic chemical material. Uranium is mined in Bohemia and Moravia.

Agriculture. In 1970, Czechoslovakia had 17,527,183 ac (7,093,000 ha) of farmland, of which about 90% was 'socialized' in co-operation and state farms. Farming has a regional pattern. On the southern fringing plains from Znojmo in south-western Moravia to the Soviet border near Ushgorod (Ukraine), wheat, maize and sugar beet are grown along with some tobacco and hemp.

The Ohre-Polabi lowlands and central Moravia provide wheat, sugar beet and fodder crops, but also have vineyards, orchards and forests. The foothills and plateaux of southern and central Bohemia, upper Moravia and Slovakia have large forests and provide crops of rye, oats and potatoes. Hops are important around Plzeň and České Budějovice. The Bohemian rimlands, the Hruby Nizky (northern Moravia) and the Carpathians have forests, meadows and upland pastures, and grow potatoes, rye, oats and flax.

But Czechoslovak agriculture has had the slowest rate of post-war growth in production in all Europe. After World War II, labour shortage was especially critical in Bohemia, where the flight or expulsion of the Sudeten Germans reduced the agricultural work force by some 600,000. Farms in Bohemia and Moravia (officially 'the Czech lands') still produce 5% less than in 1936. By contrast, Slovak farms now produce 55% more than they did before the war.

Transport. Until the 1950s the country suffered from an obsolete transport network focussed on two former imperial capitals beyond its frontiers: Vienna and Budapest. The key to improvement has been railway electrification, first on the west-east line from Most in north-western Bohemia to Cierna on the Soviet frontier, and then progressively on the more

Road transport has expanded rapidly and heavily-used lines radiating from Prague.

General view of the city of Prague. St Vitus' Cathedral, in the background, sits in the centre of Hradcany Palace, once the residence of the ancient Kings of Bohemia. Prague, one of the oldest and most historic cities of Europe, is justly famous for its architecture.

now handles more freight and passengers than the railways. River transport, mainly on the Danube between Bratislava and Komarno, and on the Vltava and Elbe, handles increased cargo but no longer carries significant numbers of passengers. Czechoslovakia operates a few ocean-going ships from the Polish port of Szczecin. Prague has a busy international airport.

Trade. Czechoslovakia trades mainly with other communist countries, notably the Soviet Union and other members of COMECON. Its imports are principally raw materials (especially iron ore), fuels (coal, oil and natural gas) and foodstuffs (especially cereals and livestock products). Manufactured goods dominate among exports, followed by raw materials and fuels.

HUNGARY

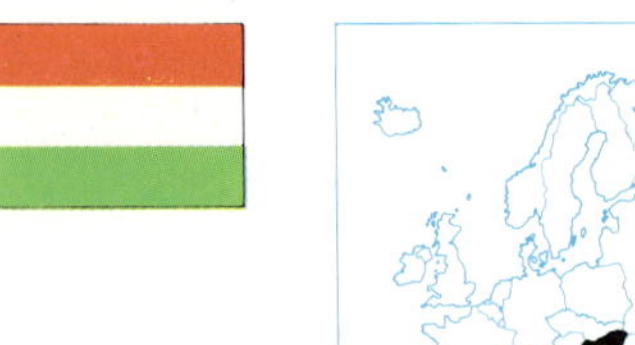

Area: 35,920 sq mi (93,031 km²)
Population: 10,395,000
Population Growth Rate: 0·3%
Capital: Budapest, pop 2,023,200
Language: Hungarian
Religion: Roman Catholic (65%); Protestant (25%); Orthodox (3%); Jewish (1·5%).
Currency: Forint = 100 fillér

Hungary is a landlocked central European country with a communist government. Most of the land is a low level plain whose soil is the nation's main natural asset.

For centuries, Hungary has been either a minor part of an empire or deeply influenced by powerful neighbours. Thus, as recently as 1956, Russian tanks helped to crush a revolt against unpopular communist rule and some 160,000 Hungarians fled the country. However, by the 1970s, Hungary's communist rulers were adopting relatively independent policies which permitted a welcome increase in once-scarce consumer goods, a limited amount of private enterprise, and closer trading links with the West.

The Plains and the Danube. Hungary is a predominantly lowland country. In the northwest, an area of about 1,900 sq mi (4,921 km²) is occupied by the Kisalföld (Little Plain), composed mainly of alluvial material from the Danube and its tributaries, with much reclaimed marshland in the north. Central and eastern Hungary consist mainly of the Nagyalföld (Great Plain), which stretches eastwards from the Danube bluffs and across the River Tisza to the Romanian border. Some 140 mi (225 km) by 110 mi (177 km) at its greatest extent, it is one of the flattest and most monotonous plains in Europe, but contains several well-defined regions bearing specific names – Cumania (Kiskun), the low loess plateau between the Danube and Tisza; and, east of the Tisza, the sandy Nyírség plateau, the marshy flood plains of the Bodrog and Nyírség rivers, the Hajduság loess area near Debrecen, and the marsh and salty arid steppe of the extensive Hortobágy.

In contrast to these plains are the Transdanubian Mezöföld (Middle Plain) and hill country, a more rolling area dissected by the Kapos, Zala and other tributaries of the Danube, and the middle and northern mountains. The southern Transdanubian hills include the isolated Mecsek massif, thickly forested and rising to 2,240 ft (741 m). Separating the Transdanubian hills from the middle mountains is Lake Balaton, the largest natural lake in Europe, covering some 230 sq mi (596 km²) and with popular summer resorts along its shores. Its fisheries are important.

The Danube divides the mountains into two sections: the western 'Transdanubian' ranges which include the Bakony Forest, the Vértes, Gerecse and Pilis Hills, built mainly of limestone; and the eastern 'Carpathian' foothills (the Börzsöny, Cserhát and Mátra mountains) and the Bükk, Zemplén and Tokaj mountains. Hungary's highest mountain is Kékestetö (Mt Kékes), which rises to 3,330 ft (1,015 m) in the wooded Mátra.

Climate. Hungary has a continental climate with cold winters and hot, dry summers, but the country lacks the great extremes of temperature that might be expected from its position far from the tempering influence of an open ocean. This fact is due to a slight penetration of Mediterranean influences from the Adriatic: the infiltration of Atlantic air into the Danube valley: and the shelter provided by the Carpathians, which restrict the penetration of icy winter Siberian winds from the east.

Winters are relatively mild, and precipitation relatively high in western Hungary and Budapest. January temperatures in Budapest average 30·6°F (−0·8°C); July temperatures, 71·4°F (21·9°C). Winters are colder and precipitation lower on the Great Plain, where summer droughts sometimes cause crop failure. Flooding also occurs with the spring thaw on the plains and in the mountains, which often coincides with the heavy early summer rains. The floods are at their worst when downriver ice on the Danube holds back the flood water in the Tisza and upstream Danube.

The government has spent large sums of money on straightening and embanking the rivers, and on the construction of the Eastern and other drainage canals in the Great Plain to combat flood, provide irrigation water and reclaim saline soil areas like those in the Hortobágy.

Vegetation and Wildlife. Cultivation has destroyed most of the natural vegetation such as the deciduous forests of Transdanubia and the wooded steppe of the Great Plain, but 17% of the country is still forested, due in part to post-war reforestation, mainly with beech and oak, in the mountains. Willows line the river banks on the plains, and acacia has been planted to stabilize sand dunes in Cumania. Poplars on the flood plains of the Tisza valley provide wood for the cellulose industry.

Wildlife is like that in some other parts of central Europe. It includes wolves, jackals, wild boars and such spectacular birds as the black stork and great bustard – a turkey-sized bird that roams the immense plains, but in dwindling numbers.

The People. About 95% of the population consists of Magyars, the descendants of the Finno-Ugrian and Asiatic Turkish peoples who settled the area long ago. There are small German, Slovak, Romanian and other minorities, although many non-Hungarians returned to their homelands after World War II. Population growth is slow, due to a relatively low birth rate and relatively high death rate. Urbanization has also been comparatively slow, due to a housing shortage and restrictions on migration to overcrowded Budapest, where a high proportion of the urban population lives.

No other city compares in size or importance with Budapest. Straddling the Danube, this city is not only the capital of Hungary, but its cultural, commercial and industrial hub. Of the major provincial towns, Miskolc is an industrial centre; Debrecen, a historic cultural centre and seat of Kossuth's short-lived government (1848–49), also is a manufacturing town and a market centre; Pécs, an industrial and university town; Szeged, on the Tisza, a commercial and cultural centre for the southern Great Plain. Outside Budapest, the population is distributed relatively evenly in market towns and villages, except in the middle mountains. The Great Plain, however, has a thinner spread of people due to depopulation during the Turkish occupation; its large villages and small towns are separated by uninhabited tracts or areas with only a few scattered farms. Most Hungarians are Roman Catholics, but there are Calvinist, Lutheran, Orthodox and Jewish communities. All churches are under state control and receive state subsidies, although the communist government has tended to discourage religion, particularly where it divides the people's loyalties. The Hungarians are a cultured people with a rich literature, and its folk melodies inspired Liszt, Kodály, Bartók and other composers. Illiteracy has been virtually eradicated. Education is free and compulsory from 6 to 16, and there are major universities at Budapest, Pécs, Szeged and Debrecen, and 14 specialized universities.

The Economy. Industry has been greatly expanded and now contributes more than 38% of the Gross National Product. Engineering is highly localized in Budapest, which has a considerable reputation for its transport vehicles (especially Ikarus buses and trucks), agricultural machinery and machine tools. Several smaller cities like Miskolc and Györ have engineering traditions, and the policy of decentralizing industry has established engineering in southern cities.

Growth has been outstanding in the electrical and chemical industries. The former developed from pre-war workshops in Budapest and has increased the industrial dominance of the capital. The latter, however, has been developed, as in Poland, at smaller centres some distance from the capital: coal-based chemicals at Várpolata, gas-based chemicals at Szeged, Szolnok, Tiszapalkonya, and at the new town of Kazincbarcika north of Miskolc.

The region around Lake Balaton in Hungary produces fine wines made principally from the Italian Riesling grape. Lake Balaton (48 miles long) is the largest lake in Central Europe.

The major new industrial complex of the post-war era is Dunáujvaros, a new city on the Danube about 35 mi (60 km) south of Budapest. In addition to Hungary's largest steel combine (based on Soviet ore and coal), it has inter-linked coke-chemicals, cement and straw-cellulose plants.

Although now of lesser importance, the food-processing and textile industries have also advanced, with a shift in emphasis from primary processes such as milling and spinning to manufacturing (sugar, jams, canned fruits and vegetables and meat, and clothing). Industry in general remains predominantly north of a line joining Keszthély and Miskolc, but the shares of Budapest and this area have decreased with new development on the Great Plain and around Pécs.

Minerals. The extent and type of industrialization has been severely limited by the relatively poor mineral resources. Energy resources are varied but inadequate, although strenuous efforts have been made to increase production. Stretching diagonally across Hungary from Balaton to the Sajo valley is the 'coal axis.' Brown coal comes mainly from the small fields at Ajka, Tatabánya, Dorog, Salgótarján and Ózd-Miskolc, while bituminous coal is mined at Komló near Pécs and in the Mecsek Mountains. Imports from the Soviet Union, Poland and Czechoslovakia make good any deficiencies.

Oil is tapped in the Zala county south-west of Balaton, where it is refined; natural gas near Hortobágy and Békéscsaba in the east, whence it is piped to Budapest and to metallurgical industries in the Miskolc and Szeged areas. Even so, large quantities of crude oil have to be imported from the Volga fields in the Soviet Union via the Friendship II pipeline for refining near Budapest, while gas is imported from Romania.

Iron ores mined near Miskolc support a local metallurgical industry, but some 75% of Hungary's iron ore needs are provided by the Soviet Union.

The only major mineral resource which is exported in large quantities is bauxite, mined chiefly in the Bakony Forest and the Vértes Mountains, and north-east of Lake Balaton. Exports go mainly to Poland, Czechoslovakia and the Soviet Union in exchange for fuels or electric power.

Agriculture. Agriculture provides nearly 19% of the Gross National Product and employs about 26% of the labour force. Its progress has been comparatively modest, production now being only a third more than it was in the 1930s. This is due mainly to insufficient investment in agriculture through the priority given to industrial development, and to the rapidly ageing and declining farm labour force. Such increase as has occurred is the result of collectivization, changes in land use, and improvements in means and methods.

Collectives and state farms have largely replaced the small peasant holdings and the many scattered and inefficient farms of the Great Plain. Larger acreages have now been given over to high-yield hybrid maize, irrigated green fodder and silage maize, and tree crops. Maize has now replaced wheat as the leading crop, for Hungary has deliberately become the European equivalent of America's Mid-West 'corn-hog belt.' Such mixed farming has improved farm efficiency. Areas devoted to oats, rye, potatoes, turnips and pasture have declined in favour of the more intensive forms of land use. Irrigation and drainage in the eastern Great Plain have brought fertile virgin areas under cultivation. Fertilizers have raised yields, and greater mechanization has facilitated faster harvesting.

Farmers of the Kisalföld grow cereals, flax and potatoes and raise dairy cattle. On the Great Plain east of the Tisza, maize, wheat, sugar beet and sunflowers are important crops, and pigs, sheep and cattle are reared. Some areas specialize in rice (the Tisza valley), paprika (around Szeged), hemp (the south-east) and tobacco (the north-west). Cumania is a region of rye, potatoes, silage maize, fruit and vines. Kecskemét is noted for its apricots and the potent brandy made from them, while wine from the large state vineyards goes to make vermouth and the German *Sekt*. Hungary's most famous wines, however, come from vineyards elsewhere – at Eger and Tokaj in the north-east, and around Pécs and on the northern shores of Lake Balaton.

Tobacco, rye, potatoes, apples and sunflowers are the chief crops of the sandy Nyírség region. Potatoes are also grown, along with cereals and sugar beet in the Mezőföld-Transdanubian hill area, and in the hill and mountain country along with rye and oats, although the highlands also have much pasture and forest.

Transport. Although Hungary straddles two important rivers, the Danube and the Tisza, with its capital and leading industrial city on the former, surprisingly little use is made of water transport. Both freight and passengers travel mainly by rail and road, and the reasons are not far to seek. Road and rail networks are relatively good; distances are short; and economically, train and truck have advantages over barges, even for bulky fertilizers, cement and building materials. Commercial use of the Danube has been restricted by political difficulties (it is an international river) and by the physical hindrances of spring flood and summer low water. Much of the immediate hinterland of the Danube and Tisza is agricultural plain whereas the railways and roads have access to the mineral centres in the mountains.

The radial pattern of roads and railways, centring on Budapest, has been a major factor in the supremacy of the capital in industry and services. Post-war peripheral development involving such centres as Györ, Pécs, Szeged, Debrecen and Miskolc, has been hindered to some extent by the lack of good routes by-passing Budapest. This has served to confirm the historical role of the Danube and Tisza as barriers to, rather than carriers of, traffic. Outside Budapest there are only two road and rail crossings of the Danube: at Komarno and Baja.

International Trade. Fuels, electricity, raw materials and semi-manufactured goods account for near 30% of Hungary's exports; the main items being rolled steel, bauxite, alumina, and the thermal electricity supplied to neighbouring regions of Austria and Czechoslovakia. Other exports include transport equipment, pharmaceuticals and shoes. The chief imports are raw materials, machinery and vehicles, and foodstuffs.

About 70% of Hungary's foreign trade is with other Communist countries, the Soviet Union alone accounting for 36%, and the rest being taken mainly by East Germany, Czechoslovakia and Poland, with West Germany and Italy each having a modest share.

AUSTRIA

Area: 32,374 sq mi (83,850 km²)
Population: 7,456,403
Population Growth Rate: 0·5%
Capital: Vienna, pop 1,643,100
Language: German. Slovenian, Croatian and Hungarian spoken by minorities.
Religion: Roman Catholic
Currency: Schilling = 100 groschen

Austria, a small republic in the Alpine heart of Europe, is all that now remains of the sprawling Austro-Hungarian Empire. Today, Austria is best known as a Mecca for tourists who come to admire its majestic mountains and gem-like lakes, and to visit Vienna, one of Europe's great cultural cities.

Austria was forcibly incorporated into Hitler's Third Reich in 1938 and then occupied by the Allies for the decade after the end of the war. In 1955 it was re-established as an independent democratic republic and made a declaration of permanent neutrality. Nevertheless Austria's natural sympathies with the West have been reflected in its membership of the Council of Europe (1956) and moves towards a special relationship with the European Economic Community.

The Land. Lying mainly in the eastern Alps, Austria is one of the most mountainous countries in Europe. The Austrian Alps extend across the country from west to east in three main series of ranges: the central crystalline Alps, flanked by the northern and southern calcareous Alps.

The Austrian Alps are wider than the Swiss Alps, but structurally younger and not so high. Some of the oldest rocks occur in the Hohe Tauern range and in the Semmering Pass area southwest of Vienna. The Hohe Tauern also contains Austria's highest peak, Gross Glockner, rising above the Pasterze Glacier to 12,461 ft (3,798 m), and the permanent ice and snowfields of the Gross Venediger massif (12,008 ft; 3,660 m).

The rocks forming these ranges vary. Metamorphic schists and slates occur in the ranges dominating the valley of the River Inn south of Innsbruck. Marble and similar rocks underlie the colourful Austrian Dolomites on the border with Italy. White limestone is a feature of the Dolomites in East Tyrol, and provides a striking contrast with the turquoise blue of the lakes in the Salzkammergut region of Salzburg province.

The outer (northern) Alpine zone, which rarely exceeds 6,000 ft (1,828 m), dies out in the wooded slopes of the Wienerwald (Vienna Forest) northwest of Vienna. Its rounded foothills contrast sharply with the frost-shattered, needle-like peaks of the calcareous Alps.

Within the Alps there are several broad enclosed basins. Typical of these is the Klagenfurt basin in Carinthia containing the long and narrow Wörthersee, an appealing lake that has been called 'the Austrian Riviera'.

The Danube, more grey than blue, traverses northern Austria from west to east for some 217 mi (349 km). From Passau (Bavaria), where it is joined by the Inn, it surges into turbulent rapids, cutting between the Austrian Alps and the crystalline massif of the Bohemian Forest through a long gorge. This is the Danube at its most beautiful.

Below Linz the hills recede and the acid soil *Waldviertel* ('forest quarter') gives way, north of Vienna, to the *Weinviertel* ('wine quarter'), noted for its vineyards. The open lowland area to the north is drained by the March (Morava).

The Vienna basin, south of Vienna and between the Alps and the River March, is separated from the Burgenland by the limestone Leitha Hills. Their steep southeastern slopes dominate the reed-girdled Neusiedler See, the lake home of many rare birds. Burgenland, so named from its many castles, is a northern extension of the Hungarian plains.

Wild Plants and Animals. Forests cover more than 36% of the total land area. Carinthia is especially well-forested: Klagenfurt, its capital, is the scene of an important annual European timber fair.

Throughout the Alps there is a marked altitudinal zoning of vegetation. In most areas, the timber-line rises to at least 7,000 ft (2,134 m). Above this is a zone of Alpine plants and summer pastures. Higher still, edelweiss and other rare plants are found growing among the bare rocks.

Wild animals include chamois, deer, hares, marmots and Alpine species of birds.

Climate. Austria has a Central European climate with moderately severe winters and warm summers. In winter, the weather is bright and clear, although temperatures often fall below freezing point and snow falls, especially in the Alps. In most winters the Danube has to be kept open by icebreakers.

In summer, high pressure gives rise to high temperatures. Vienna averages 68°F (20°C) in July, but midday temperatures are much higher. Thunderstorms are liable to occur, especially in the Aips. Austria is remote from westerly air streams and consequently has a rather low total precipitation. Maximum rainfall usually occurs in July.

The People. The Austrians are a German-speaking and mainly Roman Catholic people. In eight of the nine provinces the old Bavarian dialects of German are spoken, but in Vorarlberg people speak with the Alemannic accents of neighbouring Switzerland.

The population of about 7,460,000 (1971 estimate) includes small Slovene, Croat and Magyar minorities. The population map is dominated by the city-province of Vienna which, with 1,643,100 inhabitants, contains about 22% of the total population. The city itself retains much of the cosmopolitan character and distinctive culture of the glittering imperial capital. Graz (250,300), the second city of Austria, stands in the Alpine foothills where the Mur valley opens on to the plain of the Drau. The capital of Styria, it has a cathedral and a university, and various light industries. Styria is the home of the *Steirergewand*, the green-lapelled grey suit with wide green stripes on the trousers that has become widely popular throughout Austria.

Austria's third city is Linz (203,983), capital of Upper Austria, a Danube port and steel centre. Salzburg (115,720), the birthplace of Mozart and capital of Salzburg province, derives its name from ancient salt mines in the area that are still worked. Innsbruck (110,209), capital of the Tyrol, is a medieval city and tourist centre commanding the route from Munich (Bavaria) to the Brenner Pass.

The Austrians, especially the Viennese, are cultured people with long traditions in music, literature, the theatre and other arts. The country has maintained an astonishing record of musical achievement from the late 18th century (with Haydn and Mozart), through the 19th century (with Schubert, the Strauss family, Brahms and Bruckner) to the 20th century (with Mahler, Schönberg and Berg).

The educational system, distinguished by its high standards, had its roots in the medieval monastic schools of the later 11th century. Vienna still has the historic 'Schottengymnasium', a school founded in 1155. Vienna university founded in 1365, is the oldest university in the German-speaking countries.

Farm and Forest. The Austrian countryside has many distinctive regional features – the wood and stone chalets of Tyrol, the solid foursquare *Vierkanthof* farmhouses of Upper Austria, the Montafon cattle of Vorarlberg, the sturdy draught horses of the Enns valley in Styria and the open-strip cultivation of the Burgenland. All these reflect a traditional form of agriculture in which the basic unit is generally the small peasant farm, where mechanization is slowly making progress.

Little more than 21% of the total land area is under cultivation. The Burgenland, Danube valley, and the March and Vienna basins account for most of the arable land. Crops include wheat, barley, maize (corn) and sugar beet, as well as clover, hay and other fodder crops. In some provinces dairying is important.

A large share of the forests is owned by small farmers. Softwood timber, a traditional building material, is also used in making matches, boxes and other products, and for fuel. The forests also provide wood pulp, volatile oils and charcoal. They are of immense importance in checking avalanches and preventing landslides and soil erosion, and have a considerable influence on climate.

The modern gothic Rathaus, or Town Hall, in the city of Vienna. In the background is the Votive Church, built in 1879 to commemorate the death of the Emperor Maximilian in Mexico.

Austria's Tourists. Austria is one of the leading tourist countries in Europe. In 1970, it was visited by nearly nine million foreigners, more than the total population of the country. The Tyrol, especially Kitzbühel, has an international reputation for winter sports. Salzburg, beautifully set between the Alpine foothills and dominated by its medieval castle and baroque cathedral, is famed for its annual musical festival. Vienna, with its rich imperial traditions, is renowned for its great music,

Ski lifts in the Austrian Alps, the goal of many winter holidaymakers. Tourism is one of Austria's major sources of income with more than 9 million visitors each year.

under state control. Iron and steel production is centred on Bruck-Leoben in Styria, and Linz, on the Danube. Austria also has one of the world's largest aluminium plants at Ranshofen in Upper Austria, and large chemical plants at Linz (fertilizers and pharmaceutical products) and Lenzing (man-made fibres). Steyr is known for its machinery and automotive products. Graz (Styria) for its engineering and textiles; and Bregenz (Vorarlberg) for its textiles, especially lace and embroidery.

But the main industrial centre is Vienna; its products range from rolling stock and electrical goods to fashion clothing, leather goods, jewellery and fine porcelain. The city has long been renowned for its printing and publishing (especially music).

Foreign Trade. Exports include machinery, iron and steel, textiles, chemicals, timber, paper and electric power. Imports include raw materials, fuels and foodstuffs. In 1969, West Germany, Italy, Switzerland and Great Britain accounted for 62·1% of Austria's imports and 49·5% of its exports.

Transport. Its position in central Europe makes Austria, and in particular Vienna, a focal point of international rail routes. The Semmering Tunnel route is the oldest Alpine railway in Europe. Much profitable transit freight, mainly from Germany, Italy and Yugoslavia is carried by Austrian railways.

The well-developed road system includes the Vienna-Salzburg and Carinthia motorway by way of the Gross Glockner Pass (8,212 ft; 2,503 m). A Vienna-Tarvis motorway is under construction.

baroque architecture, coffee houses, whipped cream and *heuriges* (wine shops).

Mineral Resources and Industry. Iron ore has been worked in Austria since the days of the Celts. Today about 75% of the output comes from the Erzberg in Styria, and the balance from Huttenberg in Carinthia. Austria also produces magnesite and is one of the world's chief sources of graphite. The major coal deposits are in Styria, Carinthia and Burgenland.

Austria is one of the leading oil-producing countries of Europe (excluding the Soviet Union), ranking fourth, after Romania, West Germany and Yugoslavia. Oil and natural gas are produced in the Zistersdorf area, near Vienna and Gösting in Lower Austria. The main refineries are outside Vienna.

Austria has hydro-electric power in abundance. Huge dams have been built in remote valleys to harness the rivers and the great Hohe Tauern-Kaprun complex in Salzburg province, built with American aid, is one of the largest hydro-electric plants in Europe.

Since 1946, basic industries such as mining, electric power, iron and steel have been

LIECHTENSTEIN

Area: 62 sq mi (160 km²)
Population: 21,530
Population Growth Rate: 1·4%
Capital: Vaduz, pop 3,921
Language: German
Religion: Roman Catholic
Currency: Swiss Franc = 100 rappen

Liechtenstein is a tiny principality which commands attention by the beauty of its scenery, the contrasts of old and new in its way of life, and the surprising variety of its products. Once part of the Holy Roman Empire and associated with Austria until the end of World War I, Liechtenstein subsequently failed to gain admission to the Swiss Confederation, but has been linked with neighbouring Switzerland in a customs and postal union since 1924. Its army was disbanded in 1868 and, like Switzerland, the principality has maintained neutrality in time of war.

The Landscape. Liechtenstein consists of a flat plain along the right bank of the Upper Rhine and a region of foothills and mountains, the Pre-Alps, along the Austrian border. At one time the plain was marshy and liable to spring flooding. Today, however, most of the land is drained and used for mixed farming, with apple, cherry, pear and plum trees lining many of the roads and scattered through the fields. Farms are mostly small and crops are grown in traditional long parallel strips.

Most of the villages and farms lie above the former limit of flooding. Vaduz, the capital, dominated by its 13th-century castle, stands on the lower slopes of the Alpine foothills. Vineyards occur on the terraced slopes, but above Vaduz there are forests of fir and larch.

The People and their Work. The Liechtensteiners are mainly of German descent and are mostly Roman Catholics. A number of foreign workers, mainly Germans, Austrians and Italians, also live in the principality. Vaduz is the only town, and most of the people live in the villages scattered over the Rhine plain, all connected by modern roads.

The constitution provides for a 15-member Diet elected for four years by proportional representation. Women do not have the vote.

The unspoilt landscape attracts a growing number of tourists, some coming for winter sports. Many people work in the tourist service industries, and others in light industry. Liechtenstein is now increasingly industrialized and produces textiles, ceramics, precision instruments, knitting machinery, artificial teeth and coated lenses. In 1930, some 70% of the population was occupied in farming; today the figure is 6%.

SWITZERLAND

Area: 15,941 sq mi (41,287 km²)
Population: 6,311,000
Population Growth Rate: 1·2%
Capital: Berne, pop 259,000
Language: German, French, Italian. Romance spoken by minority.
Religion: Protestant (53%); Roman Catholic (45%).
Currency: Swiss Franc = 100 rappen (German) or centimes (French)

Switzerland is a small, landlocked republic renowned for its striking mountain scenery; dairy products, watches and precision engineering; and for a long tradition of neutrality and independence which has made it the home of many international organizations. The nation consists in a confederation of 22 cantons. The efficiency of Swiss agriculture and industry, together with a sophisticated banking system and an important tourist trade, have enabled the Swiss to overcome their lack of natural resources and achieve a high and enviable standard of living.

The Landscape. Switzerland covers some 15,900 sq mi (41,270 sq km). More than half the country lies in the Alps, and 11% in the Jura Mountains. Between these mountain chains, a rolling plateau region strewn with lakes – such as lakes Constance, Zürich, Lucerne and Geneva – contains the bulk of the population as well as the largest cities, the richest agricultural land and most of the manufacturing industry.

The rugged Swiss Alps form one of the principal watersheds of Europe. The Rhine and the Rhône rise close to one another near the centre of the chain; the Rhine flowing into the North Sea, the Rhône into the Mediterranean. The Ticino flows into the Po River and the Adriatic Sea, and the Inn into the Danube and the Black Sea. The upper Rhine and Rhône, running roughly east and west, split the Alps into northern and southern ranges. In the south are such peaks as the famous Matterhorn (14,717 ft; 4,482 m) and Monte Rosa (14,691 ft; 4,478 m) and, in the north, are Jungfrau, Mönch and Eiger, dominating the Bernese Oberland. Pilatus and the Rigi overlook the Lake of Lucerne and, in eastern Switzerland, the Säntis rises to 8,215 ft (2,504 m). The Alps are crossed by several high passes, including the St Gotthard, Simplon and St Bernard, which carry an increasing amount of international traffic, but are blocked by winter snows. The railway tunnels beneath the St Gotthard and Simplon passes carry the heaviest amount of traffic, both freight and passenger. The recent opening of the St Bernard road tunnel has made possible year-round traffic between western Switzerland and Italy.

The Jura Mountains, a series of narrow ridges and valleys rising to about 6,000 ft (1,830 m), run in a crescent along the French border to the west, roughly parallel with the Alpine ranges but separated from them by the plateau region.

Climate. Switzerland has a central European climate modified by the extremes of altitude. About one fifth of the Alps lies under permanent ice and snow and snowfall is considerable throughout the country. Fog blankets, caused by cold air from the mountains, often cover the low-lying areas. Summer, however, can be warm and temperatures on the plateau may exceed 80°F (26·6°C). The warm southerly *foehn* wind may bring sudden clear skies, while the northerly *bise* brings bitter cold in winter.

Vegetation and Wildlife. After centuries of overgrazing, little of the original alpine vegetation remains. Natural woodland is found on only about one-fourth of the land area. Below the zone of Alpine flora, fir forests, with larch and pine, mantle the slopes, often scarred by avalanches. In the more temperate regions, beech, walnut, oak and sweet chestnut occur, while luxuriant Mediterranean types of vegetation flourish on the sheltered shores of the Swiss-Italian lakes. On the high Alpine slopes, chamois, marmot, hare and game birds are still seen.

The People. Switzerland's population of over 6 million is concentrated mostly in the plateau region and in the largest cities. Of these Zürich, with over half a million people, is the largest and fastest growing. It is the chief centre of banking and commerce. Basle, at the head of Rhine navigation, is important as a rail terminal, port and centre of industry. Geneva is famous as a world conference centre. Berne, which retains much of its medieval charm, is one of the smallest European capitals.

There are three official languages. German (or the local variant called *Schwyzzerdütsch*), is spoken by almost three-quarters of the people. The French speaking cantons centre around Geneva in the west. About 10% speak Italian in the south. A fourth national language, Romansh, is spoken by a small minority in Graubünden.

Swiss education is thorough. Children from 6 to 14 must attend school. There are seven major universities, the oldest being the University of Basle (1460).

A little over half the people are Protestant (Swiss Reformed Church), the rest mostly Roman Catholic. Because of a shortage of labour, about half a million foreign workers are employed in Switzerland, chiefly for the less skilled tasks, a situation which creates many social problems.

The Swiss people are extremely independent, and there are still marked differences in customs and traditions, as well as language, from valley to valley and even village to village. Having fought for their independence, they still maintain a citizen army, all males from 20 to 50 being liable for service and undergoing regular training.

Agriculture. The major farming region is the central plateau, the source of most commercial crops and dairy produce. Dairy farming for cheese and chocolate production is carried on in the more humid west. In the mountain valleys winter feeding of livestock depends largely on the hay crop, whether derived from

Zurich lies at the northern end of Lake Zurich where the Limmat river issues from the lake; a site which has been occupied in the past by neolithic lake dwellers, Helvetii and Romans.

lush valley meadows or high Alpine pastures. Flocks and herds are driven from the valleys to the high mountain meadows above the tree line for the spring and summer. Throughout Switzerland farm holdings are small, and though they are intensively worked, mechanization has been slow, and about two-fifths of the country's food supplies must be imported.

Industry and Trade. Because of Switzerland's lack of resources and its high level of industrialization, many of the basic materials for manufacturing – coke, coal, iron ore, petroleum – have to be imported. Even some electricity must be brought in, though Switzerland generates more power per person than most European countries. Specializing in high quality products for export, the factories are mostly small and scattered so as to utilize the sources of hydro-electric power.

The Jura region specializes in the making of clocks and watches, jewellery, photographic and other precision equipment and sewing machines. St Gallen is traditionally the centre of the textile industry, including lace and embroidery, now made on complex machines. Heavy engineering is located in Zürich, Winterthur, Baden, Brugg and Biel, while Basle combines the manufacture of machinery with textiles and dyestuffs; several firms at Basle specialize in pharmaceutical products for a world market.

Switzerland's economy is kept in balance by adding to these extensive exports the earnings from tourism and banking, enabling the country to import expensive raw materials and power. Swiss international banking benefits from the security of the country's traditional neutrality, as well as its own reputation for caution and secrecy. Tourism, which is highly organized, brings in nearly 7 million visitors a year for skiing, sailing, mountain-climbing and sight-seeing. Altogether, tourism swells Switzerland's annual income by about £140 million, while banking services add another £60 million.

Transport and Communications. Switzerland's command of the major Alpine passes, the St Gotthard, Simplon and St Bernard, gives it great importance in international transport. Here traffic from western and central Europe converges en route to Milan and northern Italy. In addition some 1,800 mi (2,900 km) of railways, including a network of Trans-Europa expresses, and 11,200 mi (18,000 km) of roads criss-cross the country.

A system of modern arterial highways is rather belatedly under construction. Apart from steamer services on the larger lakes and the Rhine between Basle and the North Sea, water transport is negligible. There are international airports at Kloten, near Zürich, Geneva, Basle and Berne. The privately-owned international airline, Swissair, serves around 40 countries.

Nearly 500 newspapers, published in German, French and Italian, serve the country. Over half of the daily papers are published in German, *Der Blick* of Zürich having the largest circulation. The efficient postal, telegraph and telephone services in Switzerland are operated by the government, as are the radio and television stations. The latter broadcast in the three official languages, French, German and Italian.

FRANCE

Area: 211,208 sq mi (547,029 km²)
Population: 51,590,000
Population Growth Rate: 0·9%
Capital: Paris, pop 8,196,746
Language: French. Breton, Spanish, Flemish and Italian spoken by minorities
Religion: Roman Catholic. Protestant (2%) Jewish (1%)
Currency: French franc = 100 centimes

France is the largest nation in western Europe and, historically, one of the world's major cultural and political powers. Despite the appalling human and material devastation of two world wars, during which France was one of the focal points of conflict, the nation has never enjoyed such great economic strength as it has today. Vast tracts of fertile farmland have made France the granary of western Europe, and massive industrial growth has helped to underpin France's position as one of the world's main trading powers.

This high economic status has given weight to the remarkably independent foreign policies followed by France since World War II, especially under Charles de Gaulle. Determined to raise France's international standing after the bitter humiliations of World War II when most of defeated France was occupied by Germany, de Gaulle helped develop France's independent nuclear defences, pulled France out of NATO, pursued trade with communist Russia and worked to give France a leading voice in the European Economic Community. Under de Gaulle, and his successors the French have enjoyed an unprecedented rise in the standard of living. In the 10 years to 1973, the country's economic growth rate of 5·8% per annum was second only to that of Japan. National income tripled and per capita income increased by 150%.

France is a presidential democracy, with a president elected by direct universal suffrage for a seven-year term. The president chooses the prime minister.

Highland France. Most of the surface of France lies less than 1,000 ft (300 m) above sea level but there are several distinctive upland areas. The most imposing highlands are the young fold mountains which mark the frontiers in the south-west and south-east: the Pyrenees, the Alps and the Jura. All these were heavily glaciated, but only the Alps are high enough to have permanent icefields. The Pyrenees rise sharply from the plains of the Aquitaine basin as a formidable but narrow barrier between France and the Iberian peninsula. In the west and centre, their French flank is well wooded, and the central sector has hydro-electric stations, mineral spas and winter sports centres. Wedged between France and Spain is the ancient co-principality of Andorra, north-west of which is Pic de Montcalm (10,105 ft; 3,080 m).

The Alps and Jura dominate France's borders with Italy and Switzerland; here, too, hydro-electric power and winter sports are important. Mont Blanc, in the Alps, rises to 15,781 ft (4,810 m); it is the highest peak in western Europe. Separated from the Jura by the Belfort gap, are the Vosges, the link between the Rhine and Rhône basins. The Vosges are faulted highlands which rise to some 4,400 ft (1,341 m); their lower slopes are forested and their grassy, rounded summits are known as *ballons*. Part of the Ardennes massif also comes within the French border.

About one-sixth of France is covered by the Massif Central, a plateau-like upland block whose faulted south-eastern edge forms the Cévennes mountains, the watershed between rivers flowing to the Atlantic and those flowing to the Mediterranean. Bounded on the east by the rivers Saône and Rhône, the Massif Central dips north-westwards. Since it is also faulted internally, it can be divided into distinct smaller regions such as the faulted troughs of the rivers Allier and Loire, and the areas of geologically recent activity marked by the denuded volcanoes of Cantal and Mont Dore, and by the ash and lava cones (*puys*) such as Puy de Dôme, which rises to 4,872 ft (1,485 m).

A smaller and much lower massif, Armorica, occupies the two north-western peninsulas, Brittany and the Cotentin, and parts of Normandy and the Vendée. On average some 300 ft (90 m) above sea level, it is a low plateau with hilly ridges rising to a maximum of only 1,118 ft (341 m).

The magnificent twin towers of Notre Dame (1163–1300), the gothic spire of Sainte Chapelle (1248) and, to the left, the Palais de Justice dominate this view of Ile de la Cité in Paris.

Lowland France, the basins between the massifs and the mountains, lies mainly in the north and west of the country. Gaps or sills (*seuils*) in the rims of the basins provide obvious transport routes. For example, the Paris Basin has several *seuils* all with major transport routes: the chalk plateau of Artois to the north, the Lorraine escarpments to the east, the Langres plateau to the south-east and the 'gate' of Poitou to the south-west.

The Paris Basin, covering about 29,000 sq mi (75,110 km^2), is the largest of the basins and, as its name suggests, is roughly centred on the city of Paris. It extends from the English Channel in the north to the Massif Central in the south, and from Armorica in the west to the Ardennes and Vosges in the east. Within these confines, the scarps and plains, valleys and rivers create a mosaic of *pays* (sub-regions). The contrasting landscapes include the flat and fertile plain of Beauce, the fertile loess-covered chalklands of Picardy, the woods and marshes of the sandy Sologne, and the rolling plains of the chalky Champagne country. Its central feature is the Île-de-France, a limestone plateau with fertile plains rich in *limon* (loess) and bisected by the River Seine, the historic heart of France.

In the south-west is another major basin, the triangular Aquitaine lowland bounded by the Atlantic, the Pyrenees and the Massif Central, and covering some 24,000 sq mi (62,160 km^2). It is drained by the River Garonne and its tributaries. The sandy Landes district forms its western edge, and its Pyrenean border is marked by valleys radiating from the Lannemezan plateau. Access to the basin is limited, only the Naurouze Gap providing an easy route.

There are many lesser lowlands. In the north, part of the Flanders basin comes within France. In the east, Alsace (the French portion of the Rhine rift valley) is isolated by the Vosges from the Paris Basin, but opens south-westwards via the Belfort Gap to the Rhône-Saône corridor. This corridor, really a series of basins between the Alps and the Massif Central, is a link between northern Europe and the Mediterranean coastlands, where there is a similar series of basins (Rousillon, Hérault and the Rhône delta).

The Major Rivers. Most of France is drained by five river systems: the Seine, Loire, Garonne-Gironde, Rhône-Saône and Rhine. The Seine, flowing from its source on the Langres plateau to the English Channel near Le Havre, and fed by the Aube, Yonne, Marne, Oise and other tributaries, is the most important river, historically and commercially, in France. Well-suited to navigation and with numerous canal links, it serves Paris and has Rouen as its port for ocean-going ships. By contrast the Loire, because of its sand and gravel banks and seasonal fluctuations, is the least used of the major rivers. The beauty of its valley and the many historic chateaux along its banks, however, have made the Loire an internationally famous tourist attraction. The Garonne, too, is ill-suited to navigation; its left-bank tributaries are short and steep, the right-bank tributaries flow through gorges in the limestone Causses, and no large channel develops before Bordeaux and the Gironde estuary formed by the confluence of the Garonne and Dordogne at Bec d'Ambès.

The Rhône-Saône system is a different story. The Saône, flowing from the southern Vosges to join the Rhône at Lyon and canalized along its lower course, forms with the Rhône a major corridor between northern Europe and the Mediterranean. The Rhône, long notorious as the wildest of French rivers, has now largely been tamed and its course is marked by an imposing succession of dams, hydro-electric stations, locks and river ports. This great project, which has already taken several decades, is expected to reach its full development in 1976, when the Rhône will be completely canalized from Lyon to the Mediterranean. It is not limited to navigation

The Chateau of Chambord, a royal hunting lodge rebuilt in the Renaissance style in the 16th century, is one of the most famous of the chateaux of the Loire valley.

and hydro-electric power, but includes irrigation schemes and the development of agriculture and industry along the river. It has already brought far-reaching changes and benefits, and more are to come.

The French portion of the Rhine system includes a section of the Rhine itself, the upper and longer course of the Moselle, which has been partly canalized, and the upper reaches of the Meuse and Scheldt (Escaut). These rivers and their associated canals link eastern France with the Rhine world.

The Coasts of France, about 2,300 mi (3,701 km) in total extent, are very varied. In the north-east, the sea is bordered by dunes. From Cap Gris Nez to the Seine estuary, chalk cliffs tower over the Channel and are broken by only a few inlets such as Dieppe. Westwards the long, broad beaches of Normandy stretch to the Cotentin, beyond which are the much indented coasts of Brittany. The 'Rade de Brest' (Brest roadstead) is the deepest and largest of the many rias (long, narrow inlets) all round the peninsula. Reefs and high cliffs are characteristic, and the tidal range reaches 50 ft (15 m) in the Bay of Mont St Michel.

South of the Vendée and the sandbanks of the Gironde is the sand dune coast of the Landes, broken only by the Bassin d'Arcachon, one of the chain of lagoons extending southwards. At Biarritz, the famous resort of the Côte d'Argent, the dunes give way to the rocky Pyrenean coast.

The Mediterranean coast, west of the Rhône delta, consists of long sand bars enclosing isolated marshy lagoons noted for their oyster beds. The Rhône delta itself is the largest example and consists of two sections: the western marshes, the rice and cattle country of the Camargue; and the stony eastern section called the Crau. Of the delta's marshy lagoons, the most important is the Étang de Berre, connected by canal with the sea and serving as an annexe to the port of Marseilles. The adjoining Golfe de Fos is now the scene of a major industrial development, at Fos-sur-Mer.

Eastwards, from Berre to the Italian frontier near Menton, the coast was formed by the drowning of the limestone and granite ridges of Provence. The mountains reach the sea in the Massif de Maures, the Monts d'Estérel and the Alpes Maritimes. Many bays have formed, such as those of Marseille, Toulon and most of the glittering resorts of the Côte d'Azur, the world famous French Riviera, which has as an enclave the small independent principality of Monaco.

Corsica is a mountainous and rocky-coasted island which in some things is still a law unto itself. Covering 3,367 sq mi (8,722 km²), it is known for its rugged scenery, fragrant *maquis* and ancient towns, the basis of a flourishing tourist industry. Ajaccio, its capital, is the birthplace of Napoleon Bonaparte.

Climate. While France in general has a temperate oceanic climate, with oceanic influences penetrating almost every region, there are many regional climatic variations. Nearly all the country receives between 20 and 50 in (508–1,270 mm) of rain annually; the average annual lowland temperatures vary from 48°F (8·9°C) at Lille to 59°F (15°C) at Nice, and the average annual range of temperature varies from 18°F (10°C) at Brest to 35°F (19°C) at Strasbourg. About 60% of the country has fewer than 80 days with frost per annum.

The north lies within the westerly wind belt, and with the west and south-west generally has cool summers and mild winters with rain mainly in autumn. The climate of the Massif Central and eastern France is more continental, with cold winters and warm and rainy summers. The south coast is Mediterranean, with hot, dry summers and warm winters with occasional rain. A well-known local feature is the *mistral*, a cold dry wind most prevalent in winter, blowing strongly from the north down the Rhône valley.

Vegetation. The main distinction is between the Mediterranean south and the rest of France except for Aquitaine, which has a mixed vegetation. Beech and oak are typical trees of north and central France. The Mediterranean south with its reddish soils (especially *terra rossa* formed on limestone) is distinguished by drought-resistant scrub (*maquis*) with myrtle and wild olive; by *garrigue*, a poorer and more scattered form of scrub containing thyme, lavender and other aromatic plants; and by the olives of the southern coastlands. Trees include evergreen oak, cork oak and various species of pine. Evergreen oak and other southern species are occasionally found in competition with typically northern trees in Aquitaine. Larch is widespread in the Briançon and Haut Champsaur region in the Alps, and there are extensive conifer plantations in the Landes.

The People. Since 1946 the population has increased by more than 10 million and is currently increasing at the rate of about 500,000 a year. About two-thirds of this growth is due to the excess of births over deaths, and the remaining third to immigration. Immigrants have been pouring into France at a rate of about 120,000 per year, and an increasing number are becoming French citizens. In all they now number about 3 million, the largest groups being the Spaniards, the Italians and the Algerians. Others include Portuguese, Moroccans, Belgians, Tunisians and Yugoslavs. For the most part they take the poorer paid jobs in building, industry and agriculture.

More than 70% of the population are town-dwellers, by far the largest conurbation being Paris, which contains more than one-sixth of the population. Three other conurbations, Lyon, Marseille and Lille-Roubaix-Tourcoing, have populations around the million mark; and Bordeaux, Toulouse, Nantes, Nice, Rouen, Toulon, Strasbourg, Grenoble and St-Étienne all have more than 300,000 inhabitants. Movement from country to town is accelerating.

The dominance of Paris reflects not only its importance as the seat of government, but its educational, cultural and transport facilities and the number and quality of jobs it offers. Attempts made since 1945 to transfer industry from the Paris region to the provinces have not been entirely successful, and in 1965 an overall blueprint (the *Schéma Directeur*) was published, providing for large new suburbs and the decongestion and reconstruction of the city centre. At the same time, a counter-balance to the influence of Paris has come with the development of regional capitals such as Lille, Nantes, Strasbourg, Nancy-Metz, Lyon, Marseille, Toulouse and Bordeaux.

Beliefs and Culture. France is a predominantly Roman Catholic country. Protestants form about 2% of the population, Jews about 1%, and there are also some 600,000 Moslem immigrants from North Africa. France has a highly centralized educational system under the control of the Ministry of National Education.

Agriculture, although it now provides only some 8% of the Gross National Product and employs only some 13% of the working population (compared with 20% in 1962), remains a highly important activity which provides about one-fifth of all French exports and contributes significantly to the French balance of payments. In the European Economic Community, France is the leading producer of grain, sugar beet, beef, milk and wine.

French farms are of two kinds: the traditional small family-owned farm which strives to be entirely self-supporting, and the large, mechanized and highly efficient farm operated as a commercial enterprise. The total number of farms has been decreasing considerably (from 2,286,000 in 1955 to 1,580,000 in 1970), and the government has in fact been encouraging farmers to leave the land so that the smallest farms can be amalgamated into larger and more viable units. Farmers have been leaving the land at the rate of about 150,000 per year. Unfortunately, those that remain are usually elderly, and the government is now offering grants to young farmers in the hope of achieving a balanced agriculture, in which the family farm survives as a desirable social and economic unit.

More than 40% of the land suitable for crops and grass is occupied by medium-sized farms of 49–124 ac (20–50 ha). Farms of less than 49 ac (20 ha) are found mainly in areas where farming is a secondary activity or where alternative work is scarce and there is no pressure to consolidate holdings. Many small farms in the south specialize in market-gardening, vines or flowers. Some farms are organized in co-operative societies. Farms of more than 124 ac (50 ha) are mostly found in the grain-growing areas of the Paris basin.

Crops and Livestock. Wheat is grown in every department and there is normally a large surplus for export. About 40% of the crop comes from high yield farms in the Paris Basin and Flanders. This area also provides about 50% of the French barley crop. Maize, once grown only in Aquitaine and the Saône plain, is now also grown as far north as the Loire as a result of the introduction of American hybrid varieties. Sufficient rice for domestic needs is grown in the Carmargue.

Sugar beet and other root crops are cultivated mainly in the north from Brittany to Flanders. Potatoes and forage crops are grown everywhere; early crops, together with vegetables, coming from the coastal plateaux of Brittany. In Flanders and Brittany, the forage crops sustain dairy cows as part of intensive mixed farm systems. Fruit-growing is widespread. Brittany and Normandy are known for their apples and other temperate fruits, Aquitaine for its plums, and the Rhône valley for its peaches and apricots. Provence provides an abundance of fruit and vegetables, mainly from irrigated areas along the lower Durance. Apricots and early vegetables are grown in Roussillon.

The massif central is a highland in south central France. Except in the valleys, the soil of this region is thin and makes poor farmland.

Milk and dairy products are the speciality of the north-west coastlands, especially Normandy, noted for its butter, Camembert and other cheeses. Dairying is also important in Savoy, in mountain areas such as the Jura and the Vosges, and on the Massif Central. In the limestone Causse de Larzac is the town of Roquefort, famous for its cheese made from ewe's milk. Meat (beef, lamb, pork and veal) is produced mainly in central France.

Wine. France is known throughout the world for its age-old skills and traditions in wine-making. France is the world's second largest producer by volume, and the largest producer of high quality wines whose standards are ensured by carefully controlled cultivation and production. The areas particularly associated with high quality wines include Champagne, Bordeaux and Burgundy. Beaujolais and Alsace are other noted areas. By contrast southern France, especially Languedoc, produces an abundance of low quality wine. In most years there is a surplus which is difficult to sell, and in some areas plans have been made to convert vineyards to irrigated vegetable production or to intensive mixed farming.

Forestry. Woods and forests occupy one-fifth of the country and provide enough timber for domestic needs and export. The chief wooded regions are Alsace, Franche-Comté and Burgundy (with the Vosges and Jura mountains) in the east, and Provence (with the Alps) and Aquitaine (with the Landes) in the south.

Fisheries. Some 500,000 tons of cod, herring, tunny, lobsters, oysters and other fish and shellfish are caught annually. Fishing centres on the many ports along the rocky coast of Armorica, although the leading port by volume of landings is Boulogne, outside the region. Both Boulogne and Lorient are fully equipped with canning, refrigeration and repair services. Arcachon, on the south-west coast, and Cancale in northern Brittany have large commercial oyster farms.

Mineral and Power Resources. The chief coalfields, the Nord-Pas de Calais and Lorraine, are extensions of fields across the north-eastern frontier. About 50% of the output comes from the Nord field where production, however, is declining, and some 30% from Lorraine, where reserves are much greater and production has been expanding. Despite ostensibly favourable geological structures, France produces relatively little petroleum, mainly at Parentis in the Landes. The Lacq field near the Pyrenees is the main source of natural gas, which is piped to Paris and other centres. Hydro-electric power is generated in the mountainous areas of the south and east: on the short, steep rivers of the Pyrenees, and along the Rhine, Rhône and Durance. Nuclear power stations, using uranium from Brittany and the Massif Central, include those at Marcoule on the Rhône and Chinon and St Laurent on the middle Loire. The Rance estuarty on the Golfe de St Malo in Brittany has the world's first major tidal power station.

Metallic and Other Ores. France is the world's third largest producer of iron ore, after the Soviet Union and the United States. About 90% of the output comes from Lorraine, with small quantities of high-grade ore from Normandy. France is also the chief producer of bauxite (named after Les Baux and mined in Provence and Languedoc) in the European Economic Community. Non-metallic minerals include potash (southern Alsace) and by-product minerals such as the sulphur of Lacq and phosphates from the Lorraine iron ores.

Industry. Expansion, mergers and government intervention have transformed the structure of French industry which was once, like farming, a matter of small-scale family enterprise. Many large concerns have been created, some with government encouragement or participation. Renault, the biggest car manufacturing firm in France, is now government-owned. Paris remains the chief manufacturing centre with a wide range of products from cars and aircraft equipment to the exquisite products of the *haute couture* houses.

The iron and steel industry is mainly concentrated in the north and east. An important new unit is the steel complex at Dunkerque, where iron ore and fuel can be readily imported. A second large coastal steel plant has been built at Fos-sur-Mer on the Mediterranean west of Marseille. Fos, with its port, oil storage, petrochemical, steel and other industries, is destined to be one of the world's largest industrial complexes; only in Japan is there anything comparable. It has road, rail and water links with most of industrial Europe.

Oil refining and petrochemicals are currently concentrated mainly at Dunkerque and along the lower Seine in the north, and at the Étang de Berre in the south. Crude oil is piped to inland refineries near Paris, Strasbourg, Lyon and Rennes. There are small port refineries at Nantes, Bordeaux and Sète.

The automobile and aircraft industries are centred mainly on Paris. Toulouse is also involved in aircraft manufacture; and Valenciennes, Rennes, Strasbourg and Sochaux (Doubs) are concerned in automobile manufacture. Great advances have been made in the telecommunications industry. Textiles remain important at Roubaix (woollens), Lille and Tourcoing (cottons) and especially Lyon, long noted for its silk and now also for

Grape picking in the Sauternes district. Sauternes wine is considered the leading white wine of Bordeaux. It is a sweet wine made from picking grapes when they are over-ripe and thus contain a higher proportion of sugar to water than most wine-making grapes.

synthetic fibres. Other textile areas include Paris, Castres, the Vosges and the Rhône valley.

Among the many other industries are electrical engineering (Paris and Alpine France), the rubber industry (Paris, Lyon, Clermont-Ferrand and Montluçon), leather, ceramics, jewellery, watch-making, and the famous perfume industry.

Transport. France has a dense road network whose major routes extend radially from Paris. *Autoroutes* (motorways) include the Lille-Marseille motorway, completed in 1970, and sections of what will eventually be a national network linked with other European motorway systems.

Large sections of the state-owned railway system have been electrified or converted to diesel traction, and uneconomic lines eliminated. The most profitable routes are those converging on Paris, and the Bordeaux-Marseille and Metz-Valenciennes lines. In 1970 the 110 mph (180 kph) gas turbine turbotrain came into service on the Caen-Paris line.

France has a heavily used network of inland waterways, including the Seine, the Rhône, the Rhine, the Moselle (up to Nancy), the Scheldt, the Dunkerque-Valenciennes canal and the Canal du Nord. Rouen, Paris and Strasbourg, by virtue of their inland water traffic, are major ports in terms of tonnage handled. The chief seaports include Marseille, Le Havre and Dunkerque. International air services are provided by Air France and UTA; domestic services by Air Inter.

International Trade. Exports include cars, textiles, arms and military aircraft, wine, wheat, dairy products and chemicals. The chief imports are oil and other fuels, raw materials, and manufactured goods covering weak sectors in domestic production. About 45% of French trade is with countries of the European Economic Community, and about 12% with those of the Franc Zone.

Overseas Departments and Territories. The French republic now consists of metropolitan France, administratively organized into 95 departments and four overseas departments (Martinique, Guadeloupe, Réunion and Guiana). France also has seven overseas territories: French Polynesia, New Caledonia, French Territory of the Afars and the Issas, the Comoro Archipelago, Saint-Pierre and Miquelon, Wallis and Fortuna Islands, and the Southern and Antarctic Territories. The New Hebrides is an Anglo-French condominium.

The overseas departments and territories are remnants of the once considerable French empire which were either too small or unwilling to accept independence in 1958 as member states of the French Community, a form of association proposed by the Constitution of the Fifth Republic. While some states such as the Malagasy Republic and Senegal are still formally members of the Community, many other former French states in Africa withdrew in 1960 in favour of complete independence.

MONACO

Area: 0·7 sq mi (1·8 km²)
Population: 23,152
Population Growth Rate: 1·3%
Capital Monte Carlo, pop 9,948
Language: French and Monégasque
Religion: Roman Catholic
Currency: French franc = 100 centimes

Monaco, a tiny Mediterranean principality, is a fashionable pleasure resort visited by more than 600,000 tourists annually. It is renowned for its gambling casino at Monte Carlo; its international motor sports (the Monte Carlo Rally and the Monaco Grand Prix); its oceanographical museum; and its reigning prince, Rainier III, and his wife, the former American actress Grace Kelly.

The continuance of the princely line is important to the Monégasques, who do not pay income tax.

Were it to lapse, Monaco would fall under French protection and they would become subject to increased taxation. Meanwhile Monaco has customs union and other special relationships with France.

Towns and People. Monaco is a series of connected towns: Monaco-Ville on a rocky headland overlooking the port; La Condamine, the commercial centre; Fontvieille, a small industrial zone where land for homes and office buildings has been reclaimed from the sea; and, across the port, Monte Carlo with its casino, opera house, grand hotels, shops and villas.

The population includes about 3,000 Monégasques, more than 12,000 French and many Americans, British, Italians and Germans. The principality has an 18-member National Council and a government headed by a Minister of State, who must be a French citizen nominated by the French government.

Tourism is the chief money-earner. Tourists were first attracted to Monaco by its mild winter climate. The principality is well protected from the *mistral*, a cold northerly winter wind, and has an average January temperature of 50°F (10°C). The casino and many of the best hotels are owned by the Société des Bains de Mer, once controlled by the Greek shipping magnate Aristotle Onassis, but owned by the Monégasque government since 1967. Among the other sources of income are frequent issues of attractive postage stamps.

Attracted by tax concessions, many foreign companies have registered in Monaco. But in 1963 French pressure compelled Monaco to tax business profits.

ANDORRA

Area: 179·5 sq mi (465 km²)
Population: 20,000
Population Growth Rate: 4·1%
Capital: La Vella, pop 7,500
Language: Catalan. French and Spanish also spoken
Religion: Roman Catholic
Currency: French franc. Spanish peseta

Andorra, one of Europe's micro-states, is an ancient co-principality on the borders of France and Spain. It may date from the 8th century, and by tradition it is associated with the Emperor Charlemagne. Its two princes are the French President and the Bishop of Urgel in north-eastern Spain.

Andorra is a group of high valleys on the southern flanks of the Pyrenees. No part of the state is below 3,000 ft (900 m). Being little more than a syndicate of villages, it looks much like neighbouring Pyrenean areas. The climate is severe in winter, but mild in spring and summer.

There are road links with both France and Spain, but because the Spanish route is much flatter and lower (and therefore less frequently blocked by snow), and because it leads directly to the sizeable town of Seo de Urgel, Andorra has always been more Catalan than French in character. France joins with Spain in providing primary education, post and telegraph services, and television.

A Pastoral Country. There is a well-developed pastoral life, the high summer pastures being traditionally held communally. Cattle winter in the villages, but sheep are largely moved to lower land over the border. Animals and man have destroyed much of the forests of pine, fir, oak and box. The most fertile soil is on the terraced slopes. The yields of potatoes and cereals are rather poor, and the main crop is tobacco, a basis of the flourishing smuggling industry.

There are two types of villages: the permanent villages on south-facing slopes and at lower elevations; and the temporary 'summer' villages, called *cortals*, at 5,000–6,000 ft (1,500–1,800 m) which are occupied only when livestock graze the alpine meadows.

The population has increased rapidly, from 5,660 in 1954 to about 21,500 today, due to an influx of immigrants attracted by Andorra's tourist potential and tax advantages. Only a third of the population is now native Andorran. A popular skiing centre, and a place for bargains (owing to absence of customs duties), Andorra now bulges with hotels, shops and expensive villas, and is visited by some 2·5 million tourists every year.

Lavella, capital of Andorra. High in the Pyrenees, Andorra attained autonomous status under Charlemagne, and since 1278 has been a co-principality of France and Spain. The church in the background is romanesque.

SPAIN

Area: 194,884 sq mi (504,750 km²)
Population: 34,320,000
Population Growth Rate: 1%
Capital: Madrid, pop 3,121,000
Language: Spanish. Catalan, Galician and Basque spoken by regional minorities
Religion: Roman Catholic
Currency: Peseta = 100 centimos

Spain occupies four-fifths of the Iberian Peninsula at the south-west corner of Europe. The lofty Pyrenees, straddling the country's only land connection with the rest of Europe, have, in the past, isolated Spain from the rest of the continent. The victory of the right wing elements of society, backed by the army and led by General Franco, in the civil war of 1936, was greeted with anger and despair by idealists of many nations who thought the cause of the left wing Republican faction was just. Only recently, with some degree of liberalization in General Franco's authoritarian regime, have foreign aid and the effects of tourism brought Spain into the mainstream of European economic development.

The Land. The chief feature of Spain is a vast plateau, the Meseta Central, at an average altitude of 2,000 ft (600 m) and tilting slightly downwards towards Portugal. It is bounded in the north by the Cantabrian Mountains, stemming from the Pyrenees; in the east by the Iberian Mountains; and in the south by the Sierra Morena and the Betic Cordilleras, where Mulhacén, Spain's highest peak, rises to 11,411 ft (3,478 m) in the Sierra Nevada massif south-east of Granada. In the north are the fertile valleys of Galicia, Asturias and the Basque country. In the north-east, the long triangular basin of the River Ebro divides the Meseta from the Pyrenees. In the south, between the Sierra Morena and the Betic chains, is the fertile plain of Andalusia, watered by the Guadalquivir, the only major navigable river in Spain.

The Meseta itself – high, largely barren and remote – is divided by the forbidding central mountain chain, the sierras de Gata, de Gredos and de Guadarrama. South of the Sierra de Guadarrama, at almost the exact centre of the country, lies Madrid, the capital.

Three large rivers, the Douro, Tagus and Guadiana, drain the Meseta into the Atlantic, all flowing partly through Portugal. Spain has only rather short rivers flowing into the Mediterranean, except for the Ebro which flows some 480 mi (772 km) from the Cantabrian Mountains and at length breaks through the Mediterranean coastal ranges in a deep gorge to reach the sea. Most Spanish rivers are aptly summarized in the saying 'a long name, a narrow channel, and a little water'.

The coasts of Spain are mostly rocky, with occasional narrow plains. Sandy beaches are a feature of the Mediterranean coast, and are now a considerable source of income. Majorca, Minorca, Ibiza and some of the other Balearic islands, lying off the Mediterranean coast of Spain, have equally benefited from the tourist influx.

Climate, Vegetation, and Wildlife. Due to its shape, size and position, Spain has perhaps the widest range of climates of any country in Europe. The most striking contrast is between the humid north and north-west, which receive up to 80 in (2,032 mm) of rain, coming in all seasons, and the arid rest of the country. The interior has hot summers and cold winters with snow lying on the Meseta for some 20 days a year. Temperatures in Madrid average 76°F (25°C) in August and 40°F (5°C) in January. The Mediterranean coast is generally warm and dry, with much heat in summer and rainfall mainly in winter. Valencia has an average January temperature of 50°F (10°C).

Spain has great areas of barrenness, but about half the land is covered with spontaneous vegetation – trees (mainly conifers), scrub and poor pasture. Grazing, burning and cultivation have restricted forests to the mountain areas. One-fifth of the land is covered by *matorral*, a Mediterranean scrub like the French *maquis*. In the south-east, where the rainfall is less than 15 in (380 mm), the *mattoral* merges into esparto grass steppe.

A few bears still roam the Cantabrian Mountains, while chamois survive there and in the Pyrenees. Both the northern mountains and the Meseta have wolves, foxes, hares, rabbits, and squirrels. The Spanish lynx, mountain goat and Andalusian wildcat are found in the Betic Cordilleras and in Andalusia. A number of migratory birds are seasonal visitors to the country.

The People. The Spaniards are a highly composite people deriving from a great variety of ethnic groups that settled in this West European cul-de-sac at different times over several thousands of years. Thus although Castilian

Spanish is the national language, the Basques of northern Spain speak their own unique tongue; the Catalans of north-eastern Spain speak Catalan, a language, akin to southern French Provençal; and the Galicians of the north-west speak Galician, a Portuguese dialect.

There are many other regional differences. For instance some Andalusian women still veil their faces in much the same way as the Moorish women who lived there five centuries and more ago. Houses vary from the sombre stone buildings with pitched roofs found in the Pyrenees to the dazzlingly whitewashed, flat-roofed cubes of the south.

Spain's formal cultural life has been somewhat restricted under Franco's authoritarian regime, and artists including the painter Pablo Picasso and the 'cellist Pablo Casals have preferred working abroad. Spain's basic system of education has long been poor and still leaves much to be desired, although illiteracy fell from 29% to 9% between 1935 and 1965. Schooling is now compulsory from 6 to 14, but less than a third of those receiving primary education are enrolled in the secondary schools, and only about 10% of these proceed to one of Spain's 13 universities. A growing number, however, are now receiving higher technical education. Most of the population are devout Roman Catholics, and the Roman Catholic Church – the State Church – plays a powerful role in Spanish life.

The Pattern of Population. Almost half the population of Spain is urban. Madrid, the administrative, financial and principal cultural centre, is the largest city, with over 3 million inhabitants. Madrid features broad tree-lined boulevards and the Prado – one of Europe's major museums of art. Unlike the national capital (which stands in the centre of Spain), most other major cities stand on or near the coast. They include Barcelona, the capital of Catalonia and Spain's leading port and manufacturing city; Bilbao, the leading Basque city and also a large industrial centre; and Valencia, famous for its exports of fruit. Seville is a leading cultural centre and inland port, and like Zaragoza and many other major cities, serves a mainly agricultural area.

In rural areas, settlement patterns include the small hamlets of Galicia, many descended from the ancient Celtic hill-top *castros*; the agglomerated villages of the northern Meseta, many of them still dominated by the lord's castle, commemorating their foundation during the Reconquest; and the much larger villages of the south, some of which are towns in size but not function.

The Economy. Spain is still poor compared with most West European nations but Spanish economic progress has been dramatic since 1960. Much of the growth has been stimulated by foreign aid and investment, and by the tourist boom (24 million visitors in 1970 earned the country over £700 million in foreign currency).

Economic expansion has been largely controlled by the government. In the 1960s Spain launched its first so-called General Economic Development Plan (1964–67) and various regional plans designed largely to irrigate dry land and modernize transport systems. The overall plan (intended to create a million more jobs and sustain a 6% annual industrial growth rate) proved the first of a series that promises to turn Spain into an industrially developed country, much as Italy has already been transformed since World War II.

Industrial growth has already been rapid, especially in Catalonia (noted for textiles and chemicals), the Basque provinces (producing iron and steel and machinery) and Asturias (with iron ore and other mines). Madrid now has metallurgical and chemical industries,

Alhambra is a 13th century citadel and palace built by the Moors above the city of Granada. After the expulsion of the Moors in 1492 it seriously deteriorated. However, restoration began in 1828 and it remains one of the most beautiful examples of Moorish architecture in Spain.

Jerez de la Frontera is famed for its sherries. The word 'sherry' is said to derive from 'Jerez' and 'Jerez' from the moorish 'Shiraz'.

and shares the manufacture of motor vehicles with Barcelona.

But the government policy of establishing industry in depressed areas that lack both raw materials and technical skills seem so far to have absorbed much capital with few results or prospects. Manufacturing remains concentrated in the traditionally industrial areas: the north, north-east and south-west.

Spain's industrial growth is partly helped be its rich endowment with minerals. Fossil fuels help to provide industrial power. The chief of such fuels is coal, ranging from anthracite to low-grade lignite, and mined largely in the northern region of the Asturias. Oil was discovered near Burgos in 1964, but the field there provides less than 1% of national consumption and hopes now centre on the off-shore fields near the Ebro delta, which may be able to supply 10% of the home demand. However, more than half of Spain's electricity supply is hydro-electrically generated from over 1,000 power stations harnessed to mountain streams.

Spain remains the world's largest producer of mercury. The Almadén deposits are unequalled. Near Huelva are large deposits of copper pyrites and sulphur and the famous Río Tinto mines. Iron ore has long been mined and exported in the north, but many smaller mines have now closed. Other minerals include lead, tin and potash. Uranium mining has recently begun in Jaén and Lérida provinces, and Spain's first nuclear power station, at Zorita on the Tagus, began supplying Madrid with electricity in 1968.

Agriculture. Farming is still the mainstay of the economy but now accounts for less than half of the value of Spain's exports. Agriculture, indeed, has not kept pace with industrial development, and remains largely unmechanized – much of the field work still being done by teams of poorly paid peasants. In some areas farm output has, in fact, declined, as peasants are attracted to the prosperous towns and cities.

In the rainy north-west, farmers produce cattle, rye, maize, potatoes and fruits such as pears and apples. More than half of Spain's cattle are reared in this region, while Galicia grows much of Spain's maize. Wheat is the main crop of the *secano*, the type of poor, dry agriculture practised over most of Spain and especially on the Meseta. The extensive, but sparse, pastures of the interior are suitable only for sheep, and large numbers are reared.

One-tenth of Spain's cultivated land is occupied by olive trees. Spain leads the world in olive oil production and is also the world's third largest wine producer. The main vine areas are Valencia, Catalonia and La Rioja (the upper Ebro plain). Jerez, near Cádiz, is world-famous for its sherry. Oranges are concentrated around Seville and along the coast at Valencia. Malaga and Almería are noted for table grapes and raisins.

The *huertas* (irrigated lands) of the south and east coasts owe much to Arab influence, and traditional methods of water-raising and distribution are still in use. Elsewhere irrigation is more recent, associated with rice cultivation in the Ebro delta and with large dams on the major rivers. The Badajoz Plan on the Guadiana has already been completed, and a similar scheme is under construction on the Tagus.

The farms of north and central Spain are mostly small, scattered holdings. Large estates, the *latifundia*, cover the southern plains. Agrarian reform has thus proceeded on two fronts: the consolidation of the small farms, and the establishment of village communities to work compact, irrigated farms on the large estates.

Forestry and Fisheries. Timber resources are insufficient to meet domestic demands, and supplies have to be imported, mainly from Scandinavia. Spain, however, is second only to Portugal as a producer of cork, mainly from the cork forests of Extremadura.

Fishing is an important industry, especially in the north-west where the coast has many fine harbours such as Vigo and Corunna. The catch is mainly sardines, anchovies, tunny, cod and hake. In the late 1960s, Spain's total catch ranked as the world's sixth largest.

Transport. Surfaced road mileage is still inadequate by modern standards, although major highway improvements are in hand. Spain's buses and coaches carry more people than the trains and Spain has now improved on the 1967 figure of only 41 cars per 1,000 inhabitants – far fewer than that for any other West European nation except Portugal. Donkeys, mules and horses still greatly outnumber cars, vans and lorries in much of rural Spain. A number of new airports have been built to cope with tourist traffic, and the state-owned airline, Iberia, operates scheduled international services. Domestic services are few, except between the major centres.

The railways, also a state enterprise, have been improved and extended under a 10-year modernization programme beginning in 1964. As a maritime nation, Spain still has a number of shipping lines, and coastal traffic has survived the challenge of road transport, although it is now greatly diminished.

International Trade. Trade expanded sharply during the 1960s, although the value of exports still remained substantially below that of imports. Exports include citrus and other fruits, olive oil, non-ferrous metals, textiles and ships; imports are mainly of heavy machinery, transport equipment, raw materials, foodstuffs, chemicals and fuel oil.

Spain's chief trading partners are the United States, West Germany, France, the Netherlands and Great Britain. Latin America is a

Bullfighting is a popular entertainment in Spain. It was probably introduced into Spain by the Moors, but contests between men and bulls took place in ancient Greece, Crete and Rome.

valuable outlet for the Spanish publishing industry and surplus industrial products. The adverse balance of trade (largely caused by the demand for mineral imports sparked off by rapid industrialization) is considerably redressed by earnings from tourism and remittances from Spaniards working abroad.

Overseas Holdings. The largely volcanic Canary Islands comprise two provinces of metropolitan Spain set in the Atlantic Ocean off North-West Africa. The Province of Spanish Sahara on the coast of West Africa is a region rich in largely unexploited phosphates. Spain has also kept a colonial foothold in North Africa where Morocco was once partly a Spanish colony. There, Ceuta, Melilla, Chafarinas, Alhucemas and Peñón de Vélez are still under Spain's sovereignty.

GIBRALTAR

Area: 2·25 sq mi (5·8 km²)
Population: 28,694
Population Growth Rate: 1·2%
Capital: Gibraltar, pop as above
Language: Spanish and English
Religion: Roman Catholic
Currency: Pound = 100 pence

Gibraltar has lost much of its old importance as a British base guarding the western entrance to the Mediterranean. This remarkable wedge-shaped limestone mass, rising almost sheer to 1,396 ft (426 m) and honeycombed with caves, tunnels and fortifications, was likened by the English novelist Thackeray to 'an enormous lion crouched between the Atlantic and the Mediterranean'. A narrow isthmus, almost at sea level, links it with Spain, the whole area being a mere 2·5 sq mi (6·5 km²).

Gibraltar was ceded to Britain by the Treaty of Utrecht (1713). Since 1964, Spain has persistently sought to recover this lost possession, negotiating, harassing and eventually, in 1969, closing the frontier (and thus depriving nearly 5,000 Spaniards of their daily work in Gibraltar). Earlier the Gibraltarians had overwhelmingly affirmed their desire to stay with Britain.

The Rock and its People. The town of Gibraltar is built on the north-west slopes; and the dock area, like the airport runway, is built almost entirely on reclaimed land. Water comes from wells, catchment areas and a desalination plant, and is also imported. Except for olives and stone pines, there is little vegetation, although many gardens have fruit trees. The famous Barbary apes still roam around the Rock.

The population of 26,833 is of very mixed descent – Spanish, Genoese, Maltese, Portuguese and Jewish. Spanish and English are spoken, and most of the people are Roman Catholics. Gibraltar has a partly elected Council and an elected chief minister and House of Assembly.

Gibraltar has no agriculture and only a little light industry. Most people are employed in government service and the port, or in trading, construction and the growing tourist industry, which also attracts labour from Morocco. The economy is heavily dependent upon British support.

PORTUGAL

Area: 35,510 sq mi (91,971 km²)
Population: 9,668,000
Population Growth Rate: 0·9%
Capital: Libson, pop 831,000
Language: Portuguese
Religion: Roman Catholic
Currency: Escudo = 100 centavos

Portugal rivals Albania as Europe's poorest nation. Yet this small republic in the extreme south-west of Europe receives income from overseas holdings over 20 times larger than itself; is a major exporter of cork, sardines and port wine; and has an expanding tourist industry based on its golden, sun-soaked beaches. Old-fashioned inefficient farming methods and lack of industry are largely to blame for Portugal's poverty. Beginning in 1953, the government launched a series of national plans to get the economy moving. But 20 years later Portugal still lagged far behind most European countries.

During more than three decades of personal but discreet dictatorship by the Prime Minister Dr Antonio de Oliveira Salasar, who was incapacitated by a stroke in 1968, the country became a semi-fascist state with a rigid apparatus of government. By the early 1970s, Portugal was seeking entry into the European Economic Community and therefore its leaders were striving to dispel the unfavourable image created by an authoritarian government, and by the country's long refusal to unshackle its territories in Africa.

The Land. Structurally, Portugal consists of the western edge of the Spanish *Meseta* (tableland), which here slopes down westwards to broad coastal plains. Most of Portugal north of the central River Tagus is hilly and mountainous, with inland plateaux. The greatest heights occur in the north-central ridges, where the Serra da Estrela reach 6,532 ft (1,991 m). Most of Portugal south of the Tagus is lowland. The Tagus valley itself is a flat alluvial region rising gradually southwards to the plains of Alentejo, which are separated from the Algarve coast in the far south by the Serra do Caldeirao. Portugal's 500 mi (805 km) Atlantic coast is mostly flat and sandy, and often fringed by dunes impounding lagoons. Rocky outcrops occur near Lisbon and at Cape St Vincent.

Portugal's largest rivers – the Minho, Douro, Tagus and Guadiana – all have their origins in Spain. The Mondego and Sado are the chief rivers flowing wholly within the country. The drowned estuary of the Tagus provides Lisbon with a fine harbour; Setubal has an analogous position on the Sado.

Climate, Vegetation and Wildlife. Despite influences from the humid Atlantic and the continental *Meseta*, a Mediterranean type of climate predominates. Mild rainy winters contrast with hot, dry summers; summer heat and aridity being greater in the south. There is a marked rain-shadow effect in the north where the windward slopes of mountains receive 100 in (2,500 mm) of rain annually, while the eastern Trás-os-Montes receive less than 20 in (500 mm). All of Portugal south of the Tagus receives less than 32 in (800 mm) per year, and less than half this amount falls in the eastern Algarve. Winds are generally westerly and sea fogs are common along the Minho coast. The whole Portuguese coast is invariably warmer than the interior of the country.

Plant and animal life are very similar to those of Spain, except that the humid climate of Northern Portugal makes this region the most densely forested in the Iberian peninsula. Because climatic elements from the Atlantic and Mediterranean intermingle, plants from northern Europe flourish side by side with Mediterranean varieties making Portugal a botanist's paradise. Distinctive regions include the Algarve, with its figs, carobs and almonds. Wolves and wild boars are found in the more remote areas. Bird life includes many migrant species en route between north-western Europe and Africa.

The People. The Portuguese are a typically Mediterranean people, short or of medium stature, with brown eyes and dark hair, though taller, lighter-eyed types are found in the north-east. About 70% of the population lives in the coastal north, but the pattern of distribution has been changed by the growth of Lisbon and Oporto, the only cities with more than 50,000 inhabitants, and by the development of the Algarve coast as a tourist and retirement area.

The population has been increasing rapidly since the 19th century; its growth rate is one of the fastest in Europe. Emigration has thus been almost always a necessity and was for long directed towards Brazil, Angola and Mozambique. Today's movement is chiefly to France.

There are many regional variations in the Portuguese countryside. Scattered dwellings are a feature of the Minho and the irrigated Algarve, which has distinctive colour-washed cottages with strangely patterned chimneys. Small villages occur south and east of the Minho, becoming larger and more widely-spaced in Alentejo, where the large farms of the cereal wheatlands can often be traced back to Roman villas.

Portuguese towns are of ancient and varied foundation. The Celtic character of the north gave rise to thousands of *castros*, fortified hilltop villages, from which many towns and villages are descended. Braga, Beja, Évora and Coimbra were important Roman towns, and their original gridiron street patterns are still discernible. In the Algarve, narrow winding streets, high walls and enclosed courtyards speak of Moorish influences.

Most Portuguese are Roman Catholics. Education is compulsory from 7 to 12, and the illiteracy rate is about 30%. Coimbra University was founded in 1920, and there are also universities at Lisbon and Oporto. Portugal has a literary tradition dating from the 12th century and is also known for its folk music.

Agriculture. Although there has been industrial development in recent decades, Portugal remains a largely agricultural country. Agriculture, characterized by small, intensively-cultivated family holdings, is based largely on traditional peasant techniques. Despite schemes on the Mondego, Sado and Sorraia rivers, irrigation is developing only slowly. The water wheel and periodic floods still irrigate larger areas than dams or canals. Efforts at modernization, however, are evident in the cereal estates of Alentejo. Rice cultivation is a new development in coastal areas.

Wine was made in Portugal at the time of the Greeks, and the vine is now widely cultivated. The most famous vineyards are on the terraced slopes of the middle Douro valley, which provide port wine. Britain, which assured port of a market by the Methuen Treaty (1703), still takes most of the output. But changing social habits have caused a decline in consumption so that many of the Douro terraces have been replanted with olives. In Portugal generally, vines are rarely a monoculture.

Beans, potatoes and the poorer cereals – rye, oats and maize – are grown in the north; wheat on the vast, treeless fields of the large estates of Alentejo. About 1 million ac (420,000 ha), mainly in the drier regions, are devoted to olives.

The basic contrast between northern, humid Atlantic Portugal and the more Mediterranean south is reflected in livestock distribution. Cattle predominate in the north, especially in Minho, and sheep and goats in the south. Pigs are ubiquitous. Animal power – draught cattle and mules for transport – still dominates much of Portuguese farming, although there has been some mechanization, especially on the larger farms of Alentejo.

Forestry. In western Alentejo, some 500,000 ac (200,000 ha) of cork oak constitute a third of the world total by area and nearly half by output. Cork and evergreen oaks are scattered over vast areas, used occasionally for cereals but normally left fallow – the *montado* system. The trees provide cork (the bark), wood, and pannage for pigs. The extensive pine forests are mainly north of the Tagus and provide valuable softwood and resins.

Fisheries. The Portuguese cod fleet still fishes annually off the Grand Banks of Newfoundland as it has done for some 300 years. But the sardine catch, brought into Setúbal, Matozinhos, Portimão and Olhão, is of far greater importance, and is mostly canned in home-produced olive oil for export. Some 50,000 Portuguese earn their living by fishing.

Industry and Trade. Portugal has still to fully exploit its varied mineral resources. Small quantities of iron ore, manganese, tin and copper are mined, but the only minerals of export importance are wolframite, of which Portugal is a significant world producer, and cupriferous pyrites. Fuel resources are very limited but there are small coal deposits near Oporto. Some of the rivers, notably the Douro and the Zezere, provide hydro-electric power, the main source of the nation's electricity.

Lisbon and Oporto are the chief industrial centres. Lisbon has such recently-established enterprises as oil-refining, shipbuilding, and steel production (at nearby Seixal), while Oporto has a textile industry of longer standing, based on home-grown wool, and cotton from the overseas territories. Otherwise manufacturing is mostly associated with agriculture and fishing.

An important part of the population of Portugal still derives its income from fishing.

Wine is one of Portugal's important exports and is often still made in traditional ways. The fortified wine, port, is particularly famous.

A thriving tourist industry has been developed and plays a significant part in the economy. Well over 4 million tourists visit Portugal.

Portugal is the leading world producer and exporter of cork, sardines and port wine, which together make up two-thirds of her exports. Wolfram, timber, resins, textiles and fruit are also exported. About 25% of exports go to the overseas territories in return for products including coffee, sugar and raw cotton. Lack of industrial development compels Portugal to import manufactured goods (especially vehicles), petroleum, coal, and industrial raw materials like iron and steel, mainly from the United States, Britain and West Germany. A constantly adverse balance of trade is redeemed by tourist earnings, emigrant remittances and merchant and shipping services.

Transport. The main axis of Portugal's road and rail networks is Oporto-Coimbra-Lisbon. East-west routes are less important, except for the rail links with Spain and the rest of Europe via the Tagus and Douro valleys. The country's most important commercial links are by sea and air. Lisbon is one of the world's great ports and has a commanding position on sea routes from Europe to the Mediterranean, Africa, South America and the Far East. Oporto is also a major trade centre, but its function is more regional.

Overseas Holdings. The Portuguese lost their centuries-old footholds in India in 1961, when Indian forces seized Goa, Damão and Diu. Portugal still retains the following overseas provinces: Portuguese Guinea, Angola and Mozambique in Africa; Cape Verde Islands, and São Tomé Principe in the Atlantic Ocean; and Macao and Portuguese Timor in the Far East. Madeira and the Azores in the Atlantic politically rank as parts of Portugal.

ITALY

Area: 116,303 sq mi (301,225 km²)
Population: 54,023,211
Population Growth Rate: 0·8%
Capital: Rome, pop 2,800,849
Language: Italian. German and French spoken by minorities.
Religion: Roman Catholic
Currency: Lire = 100 centesimi (nominal)

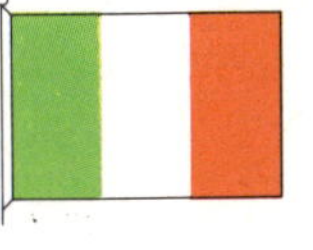

Italy's long, boot-shaped peninsula is rich in historic Roman remains, Renaissance art cities, and the natural beauty of Alpine peaks, with tranquil lakes and a countryside bathed in the brilliant light of the Mediterranean sun. It is small wonder that the Republic of Italy is the greatest national tourist attraction in Europe. Indeed, before World War II when Mussolini's grandiose schemes to revive the imperial power of Ancient Rome involved Italy disastrously on the side of the Nazis, this economically poor country seemed little more than a Mecca for tourists despite public works projects begun by the Fascist dictator. But after the war, spectacular economic growth with American aid made northern Italy one of Europe's most affluent industrial zones. The rural south, however, remained stubbornly poor. Raising the south's living standards was one of the main problems facing Italy's government in the 1960s and early 1970s.

Beehive huts of an ancient Mediterranean type in Alberobello, near Bari, on the Adriatic.

Alpine Italy. The outstanding features in Italy's relief are the Alps, which swing across northern Italy in a massive west-east arc from Liguria to Carnia, and the Apennines, which ridge peninsular Italy from Liguria in the north-west to Calabria in the south-west. Both ranges belong to the young fold mountain (Alpine) system of Europe, dating from the Tertiary period about 60 million years ago.

Except where the Swiss canton of Ticino intrudes almost to the Lombardy Plain, the Italian frontier coincides with the Alpine watershed for most of its length. The highest sectors are the Graian Alps, the Mont Blanc massif, and the Pennine and Rhaetic Alps, all of which have peaks of more than 13,000 ft (3,962 m). The characteristic Alpine landforms are the product of glacial erosion, largely during the Ice Age. Glaciers overdeepened the valleys, some of whose mouths were choked in places with debris that acted as natural dams; lakes Maggiore, Como, Iseo and Garda were formed in this way.

Fortunately for Italy's trans-Alpine communications, the mountain rampart is penetrated by a number of deep valleys. The Dora Riparia, leads up to the Mont Cenis pass; the Dora Baltea to the Great and Little St Bernard passes and the Mont Blanc tunnel; the Ticino valley to the St Gotthard pass, and the Adige valley to the Brenner pass. Most of these passes are closed in winter, but all-weather road tunnels have been built, for example under Mont Blanc and the Great St Bernard and San Bernardino passes. Railway traffic is concentrated on the Mont Cenis, Simplon, St Gotthard and Brenner routes.

The North Italian Plain, more than 200 mi (322 km) long and up to 120 mi (193 km) wide, occupies a vast depression filled in by material eroded from the Alps and the Apennines. The Po's alpine tributaries have cut the upper plain into a number of separate platforms where the permeability of the gravels reduces the effectiveness of irrigation. In the lower plain the major problem is the maintenance of efficient drainage. The reclamation of the marshy zone along the Po, which widens downstream into the deltas, lagoons and sandspits fringing the Adriatic, is mainly the work of the last hundred years. Even now the Po delta is occasionally flooded, a problem aggravated by the slow subsidence of the area. The agricultural prosperity of the plain is attributable less to any natural fertility than to the success of many generations in controlling and using its waters. Irrigation from the rivers is reinforced by thousands of reliable springs (*fontanili*) welling up along the junction of the upper and lower plains.

Peninsular Italy. The main Apennine ridge swings across from Liguria to the Adriatic before continuing down through the Abruzzi and Campania to Calabria. It reaches 9,560 ft (2,914 m) in the Gran Sasso d'Italia, but generally varies between 2,500 and 6,000 ft (762–1,828 m). Sharp peaks are unusual; in the north undulating ridges mark the watershed, while in the centre and south massive blocks and intermontane basins are characteristic. All the many trans-Apennine routes are long and tortuous.

The outer flank, the Adriatic Sub-Apennines, is fretted by scores of rivers into a comb-like succession of ridges and valleys. The coast is straight; the coastal plain, very narrow. In Apulia the Tavoliere plain and the River Bradano separate the Apennines from the limestone platforms of Gargano and Le Murge. West of the Apennines, in Tuscany, Umbria, Lazio and Campania, lies the hill and plateau country of the Anti-Apennines, penetrated by the Arno, Tiber, Liri, Volturno and other rivers. Some areas are mountainous, such as the Apuan Alps, source of the famous Carrara marble. The uplands between Monte Amiata and the Alban Hills are of volcanic origin. Volcanic, too, are the Phlegraean Fields, the islands of Ischia, Procida and Ponza, and the Lipari Islands, and Italy's most famous volcanos, Vesuvius and Etna. Lakes Bolsena, Vico and Bracciano occupy volcanic craters.

The Ligurian coast is steep and rocky, but has some good natural harbours such as Genoa and La Spezia. Farther south, rocky headlands alternate with shallow bays behind which alluvial plains have developed. For centuries these plains were waterlogged in winter and malarial in summer. The Pontine Marshes and Tiber delta were reclaimed in the 1930s, and since 1950 water control has enabled the Maremma, between Pisa and Civitavecchia, and the lower Volturno and Sele plains to be resettled. Similar projects have been carried out in the Tavoliere (Apulia) and along the Metaponto coast (Basilicata) and lower Crati valley (Calabria).

Sicily is mostly hill and mountain country, and the development of its largest plain (the plain of Catania) has been hindered by unruly rivers. The Apennines continue along the northern coast and volcanic activity is spectacularly represented by Mt Etna (10,705 ft; 3,263 m). The dreary hills and plateaux of the interior give little reward to the cultivator.

Sardinia is an island apart. Its surface is largely composed of crystalline rocks formed during a much earlier mountain-building phase than that which raised the rest of Italy.

Although not impressively high (Gennargentu, 6,018 ft; 1,834 m), Sardinia has some of Europe's most rugged country. The outstanding problem is river control; the taming of the Tirso and Flumendosa for power and irrigation of the Campidano lowland has gone far towards solving it.

Climate. In climate there is a marked distinction between continental and peninsular (and insular) Italy, the divide occurring along the Ligurian and Tosco-Emilian Apennines.

The Apennine screen reduces the influence of the sea on continental Italy, giving the North Italian Plain long warm summers and short, but raw and often foggy winters. Temperatures in Milan average 35°F (1·7°C) in January and 77°F (25°C) in July. The long, warm growing period, combined with adequate rainfall (30–45 in; 762–1,143 mm), favours a wide variety of crops. Except for a few very mild districts such as the shores of Lake Garda, winter temperatures are too low for delicate crops like olives and citrus fruit. In the Alps the summers are predictably shorter and cooler, and the winters cold and snowy. The deeper valleys can be stiflingly hot in summer but extremely cold in winter when temperature inversion occurs. Occasionally a warm, dry wind – the *Föhn* – blows down the valleys.

The climate of peninsular and insular Italy might be called variations on a Mediterranean theme. Summers, if sunnier, are only marginally hotter than those of the northern plain; the chief contrast is in the mildness of the winters. Summer drought is increasingly prolonged towards the south. The problem this presents to the farmer is aggravated by intense evaporation, the unreliability of the rainfall and its incidence in brief and often destructive showers. Precipitation and relief are correlated; the higher Apennines receive over 60 in (1,524 mm), while the total is below 20 in (508 mm) in Apulia. In winter deep depressions over the Ligurian Sea generate strong winds. Sardinia is occasionally swept by the northerly *maestrale*, and the Ligurian and Tuscan coasts by the westerly *libeccio*. In spring the *sirocco*, blowing from North Africa, brings hazy, stifling conditions to southern Italy.

Vegetation. Italy was originally clothed in forests, but these have mostly fallen victim to man's depredations. In the Alps, where forestry is still an important activity, oak, sweet chestnut and beech are common up to about 3,000 ft (914 m), and thereafter spruce, fir, pine and larch. Much depends on aspect, but above 7,500 ft (2,286 m) trees – mainly larch, juniper, alder and mountain pine – survive only in stunted form, finally giving way to humble alpine plants.

Extensive oak and chestnut woodlands have survived on the infertile morainic hills along the Alpine margin of the Northern Plain, and poplars are a valuable 'crop' along the rivers.

In Mediterranean Italy, drought-resistant evergreen oaks survive in parts of the Anti-Apennines and more extensively in Sardinia where cork oak is also commercially important. Where woodland has been cleared and the land then abandoned, a secondary growth of laurel, broom and other woody shrubs – the *macchia* – frequently takes over. Some particularly stony deforested areas support nothing but *garriga*, a sparse heathy cover composed of lavender, thyme, rosemary and other fragrant herbs.

Wildlife suffers heavily from the national passion for shooting and hunting. The larger mammals that survive include the chamois of the Alps, wolves, wild boars and deer. Edible fishes are plentiful in offshore waters.

The Fiat car factory at Mirafiori, a former municipal park outside Turin. The mountains in the background are the Alps.

The People. Italians range in physical type from the fair-haired blond individuals found in the Alps to the swarthy, dark-haired peoples of the peninsula and islands. Such differences largely reflect prehistoric settlement patterns although they take into account more recent population changes such as the influx of Germanic Lombards into northern Italy and the Arabs who occupied Sicily during the Dark Ages.

The population of Italy has been increasing in recent years at the rate of about 500,000 per annum. Both the birth rate and the death rate are declining, and with some 25% of the population under 15 years of age, a sizeable annual increase may be expected for some time. Birth rates and infant mortality rates are predictably highest in the south.

Since 1950 there has been a radical redistribution of the population. There is a general drift from the countryside into the larger cities. Southerners have swarmed into the major industrial cities of the north, and there has been a smaller but locally important drift into Rome from the central provinces. Many Italians, mainly from the south, have migrated, either temporarily or permanently, to West Germany and Switzerland, or have gone overseas to the United States, Australia or Canada.

In the countryside, where farms are small and fragmented, or where large landowners once predominated, as in the south, the peasantry is concentrated in large villages. Many of these, from their sites, witness to the importance of defence in the past. Where holdings have been consolidated, as in Marche or Tuscany, homes and villages are more dispersed. Dispersal is also typical of recently colonized areas.

Italy's urban tradition can be traced back through the Middle Ages to the Romans, Etruscans and Greeks. Many of the major cities originated as route centres or ports. In the northern plain, for example, the Alpine fringe is studded with towns (Turin, Ivrea, Bergamo, Brescia, Verona), most of which control important trans-Alpine routes. A similar string (Piacenza, Parma, Modena, Bologna, Forlì) skirts the Apennine fringe along the ancient Via Aemilia. Cremona and Piacenza, both Roman foundations like most cities of the Plain, grew up as bridging points on the Po.

Turin, for long the capital of Piedmont and the Kingdom of Sardinia, and briefly the capital of Italy, is now one of the world's great centres of automobile engineering and a major focus of southern immigration. Its rectangular layout seems to reflect the Piedmontese respect for order and discipline. Milan, ringed with a cluster of satellite factory towns, is the industrial, commercial and financial capital of Italy. Venice and Trieste have contrasting sites, the one almost afloat among its lagoons and mudflats, the other on a peninsula between the two bays where the alluvial coast of the Plain gives place to the rocky coast of Istria. Both cities have their problems: Venice is threatened by subsidence and old age, while Trieste, originally an outlet for Austria-Hungary, is over-dependent on its extra-Italian hinterland and has had to struggle to find a new role as an industrial centre. The Ligurian ports are firmly linked to the northern industrial hinterland. Genoa, with its steel, oil refining, shipbuilding and engineering, is a major industrial centre. Savona serves Turin, while La Spezia is mainly a naval base.

One of the main attractions of central Italy is its provincial towns, nearly all of them Roman (if not Etruscan) foundations. Such are Lucca, Pisa, Arezzo, Orvieto, Viterbo, Spoleto and Perugia, each with its special niche in history and all of them heirs of a rich cultural legacy which is nowadays turned to good account by the tourist industry. Rome and Florence also rely heavily on their illustrious past, but Rome owes even more to its twin roles of capital and focus of Roman Catholicism. Its industries are typical of a capital – transport servicing, printing, consumer goods (especially high fashion clothing) and film production. Foremost of the southern cities is Naples, Italy's second port and the most important industrial centre outside the north.

Culture and Beliefs. Italy is one of the world's most cultured countries, known especially for its long tradition of learning and for its contributions to the arts. From Italy, Ancient Rome deployed its legions and culture across much of Europe and the Mediterranean world: after the Roman Empire collapsed, Rome emerged as the fount of Western Christianity: and later, in the 15th century, the Italian city-states saw the beginnings of the European Renaissance. Italy was the birthplace of opera, which remains the great love of many Italians. Among its greatest literary figures is Dante Alighieri, who not only wrote the *Divine Comedy* but was the decisive force in establishing the Tuscan form of Italian as the common language of the country.

Few younger generation Italians are illiterate. Education is compulsory from 6 to 14, and there is a profusion of universities and institutes of higher education, the oldest being the University of Bologna which dates from the 11th century.

Italy is a Roman Catholic country. Within Rome is the independent Vatican City state, the spiritual and administrative hub of the universal Roman Catholic Church. The influence of the Church is far-reaching, and at times Italians have been faced with grave problems of conscience, most recently over divorce and contraception.

The Economy. Immediately before World War II, Italy was still largely a peasant country. Today 39% of the Gross National Product is

Above: The Ponte Vecchio once joined the ducal residence of Palazzo Vecchio with the government offices in Palazzo degli Uffizi. Today it is lined with shops. The Ponte Vecchio is the only Florentine bridge to survive World War II. *Right:* The Italian coast near Genoa has many attractive fishing and tourist towns. Camogli is typical and is remarkable for the height of its harbour houses.

derived from industry and only 11% from agriculture. Since the mid-1950s Italy's industrial expansion has provided jobs for millions of peasants, and today less than 20% of the working population is engaged in farming, although the percentage is much higher in the south. If this trend continues, there will be more opportunity to rectify the basic structural weaknesses in Italian agriculture. This will involve the reversion of large tracts of hill and mountain country to forest; consolidation of farm units (more than 50% of Italy's farms have less than 5 ac, or 2 ha); less concentration on cereals and more on meat and dairy produce for an increasingly affluent home market; increased specialization on crops such as fruit, for which Italy has a natural export advantage; abandoning share-cropping and other archaic forms of operation; and developing co-operation and modern methods of processing and marketing. Despite shortcomings in its agriculture, Italy is self-sufficient in wheat, maize, rice, olive oil, wine, sugar (beet), fruit and vegetables.

Alpine agriculture is mainly concerned with cattle and the hay and fodder needed to support them. In some areas, cattle are still taken to summer pasture in the mountains to graze (transhumance). In the Adige valley, farmers specialize in vines and fruit, especially apples. The Northern Plain is Italy's most favoured agricultural area. Wheat, maize, lucerne and other fodder crops are widely grown; sugar beet is confined to the damp soils, particularly of the lower Emilian plain. In lower Lombardy, the *fontanili* (springs) enable successive hay crops to be grown, often eight or nine a year, in support of the dairy industry. Apples, pears and peaches are grown for export near Verona and Ferrara, and between Bologna and Forlì. Vines are grown almost everywhere, but the high quality wines come from the Monferrato plateau and the Alpine foothills of Veneto.

In central Italy farming practice must adjust to the summer drought against which irrigation offers protection in only a few relatively small areas. Much of the hill country is suitable only for wheat, but where the soils are better, wheat, lucerne, beans, tomatoes and vegetables are commonly grown between rows of olives, almonds, vines and fruit trees. In the Chianti and Alban hills wine production is more specialized.

The continental south (Molise, Apulia, Campania, Calabria and Basilicata) has all the physical disadvantages of central Italy writ large. The summer drought is longer and many areas, such as the erodible clay hills of Basilicata, are suitable only for low-grade wheat production and sheep-rearing. The most flourishing area is the Campanian lowland, which has naturally fertile volcanic soils and abundant water from rivers and wells. This makes possible the cultivation of a wide range of field and tree crops including fruit and vegetables for export. Citrus fruits are the speciality of the irrigated areas along the Calabrian coast, while Le Murge produces more wine, olive oil and almonds than any other southern region.

Part of the Forum Romanum, centre of the Roman world. To the right is the arch of Septimus Severus, in the centre the two columns which remain of the Temple of Vespasian, and to the left the colonade of the Temple of Saturn. The buildings behind belong to modern Rome.

To alleviate poverty in the south a fund (the *Cassa del Mezzogiorno*) was set up for rehabilitating vast tracts of land by flood control, drainage, and soil conservation, and for improving the infrastructure (roads, power and water supplies etc.) and financing land reform. Today the Cassa's task in the agricultural sphere has been largely accomplished, and its attention is now directed to industry.

Sicily has irrigable zones along the northern and eastern coasts from Palermo to Catania which provide crops of many kinds. Oranges, lemons and early vegetables are the main exports. Most of the hilly interior is devoted to wheat and grazing. Dry tree crops (olives, almonds, figs) are of local importance, and the Marsala district specializes in wine.

Forestry and Fisheries. About one-fifth of Italy's surface is classed as forest. The most extensive softwood timber is in the Alps, especially in Trentino-Alto Adige. Cork oaks and resinous pinewoods are locally important along the coastal fringes, but Italy is a major importer of timber.

Except on the continental shelf at the head of the Adriatic, the seas around Italy are not rich in fish. Sardines, anchovies, octopus and mullet form most of the catch. The annual migration of tunny into the Tyrrhenian Sea is exploited by the Sardinians and Sicilians. The industry as a whole is shared among thousands of fishermen operating in small boats from scores of ports, including Chioggia, Leghorn and Naples.

Minerals. Italy has a wide variety of minerals but the individual deposits are rarely large or easily worked. Small amounts of inferior coal are mined at Carbonia (Sardinia) and in the Alps; lignite, mainly in Tuscany. The only notable oilfield is in Sicily (Gela, Ragusa); its poor quality output supplies only a tiny fraction of the country's needs. The main natural gas field in the Po valley, and lesser deposits in the south, provide about 10% of Italy's energy requirements. Italy's poverty in fuels obliges the country to import huge quantities of oil as well as appreciable amounts of American coal. It is planned to import Russian natural gas by pipeline from Austria.

Other minerals include salt and pyrites (mainly in Tuscany); sulphur and potash

(Sicily); small deposits of iron ore (Elba); lead-zinc ores (Sardinia); manganese (Piedmont); mercury (Tuscany) and bauxite (Abruzzi).

Hydro-electric Power comes mainly from the Alps, with modest contributions from the central Apennines and Calabria. There are few opportunities for furthe expansion and although some nuclear power is now produced, imported oil will remain Italy's chief energy source for some decades.

Industry. Italy has a dual economy. Electricity generation and the railways, for example, are nationalized and the state, through its majority holdings, has indirect control of steel, shipbuilding, engineering and other concerns. Private enterprise, which claims such famous names as Fiat, Pirelli, Olivetti and Montedison, is nonetheless strongly represented. State control has been useful in directing new development to the south, where basic industries such as steel and oil-refining have been established as 'poles of development' in the Naples area and the Bari-Taranto-Brindisi triangle. Salerno, Foggia, Catania, Palermo, Augusta, Gela and Cagliari are also growing industrial centres.

At one time most of Italy's steel was made in the north (Milan, Turin and Bergamo). Today the new plant at Taranto in the south, one of the most modern in the world, is the major producer. Another new plant has been built at Genoa, while the Bagnoli and Piombino plants have been modernized.

Engineering and metal-working are the most important sector of Italian industry. Heavy industry is concentrated in the Genoa–Turin–Milan triangle. Genoa, Monfalcone, Trieste and Castellammare are the leading shipbuilding centres. The automobile industry is dominated by Fiat, Europe's second biggest manufacturer. Assembly is mainly concentrated at Turin, but Brescia, Varese and Bolzano are also involved. Milan, the home of Alfa Romeo and Innocenti, takes second place. Both Fiat and Alfa Romeo have built new plants near Naples.

The rapidly expanding chemicals industry is concentrated in the major ports (Savona, Genoa, Leghorn, Venice and Naples) and along the Turin–Vercelli–Novara–Milan axis, with lesser plants in the Alps, Tuscany and the south. The spectacular expansion in petrochemicals is a mainly southern development typified by the synthetic fibre plants at Foggia, Caserta and Pisticci. But the north still has the lion's share of production, notably at Ravenna (with Europe's biggest synthetic rubber plant), Mantua, Ferrara, Porto Marghera and above all Milan. With the largest oil refinery capacity in Western Europe Italy is able to export large quantities of refined products.

The textile industry, with its long-established reputation for quality and design, is centred mainly in the north, with the cotton mills of Naples and Salerno, and the woollen mills of Prato, as notable exceptions. Lombardy is outstanding for its cottons and rayons; Biella and Bergamo province for woollens; Veneto for its knitwear; and Como for its natural silk. Food processing is widely distributed, with notable concentrations in Milan, Bologna and other northern cities as well as in Campania. Mass production has not yet destroyed Italy's traditional handicraft industries, and most towns have their specialities. Venice, for example, is noted for its glass, silver, lace and knitwear; Florence for its leatherwork and jewellery.

The Vatican church of St. Peter's. Its facade is by Bernini and its dome by Michelangelo.

Tourism. Few countries have more to offer the tourist than Italy. Some areas have become so popular and so developed that their natural attractions are threatened, but the rewards in foreign currency and employment are rich.

Transport. Except for some minor lines the Italian railway system is owned and run by the state. About half of the network is electrified. Italy has good conventional roads and a network of *autostrade* (motorways) that is surpassed in Europe only by the West German *autobahnen*. The rivers are ill-suited to navigation, but the Italian lakes carry a heavy tourist traffic. Alitalia, the state airline, operates to most parts of the world and provides internal services centred on Rome. Italy ranks among the world's major shipping nations, and its large merchant fleet contributes to the movement of goods through Italy's ports. The most important of these are Genoa, Leghorn, Naples, Savona, Trieste and Venice.

International Trade. Over the last 15 years Italy has become one of the world's leading trading nations. Oil, coal, ores, chemical raw materials, timber and fibres figure largely among its imports, and on balance Italy is a major food importer. Vehicles, machinery, textiles and refined oil products are notable exports.

About half of Italy's trade is with its partners in the European Economic Community, of which Italy is a founder member. The United States, Britain, Switzerland and the oil-producing Arab states are also important trading partners, and efforts are being made to increase trade with the Communist countries. Emigrant remittances and tourist earnings make large contributions to the balance of payments.

VATICAN CITY

Area: 0·15 sq mi (0·4 km²)
Population: 1,000
Language: Italian. Latin (administrative and legislative)
Religion: Roman Catholic
Currency: Lira = 100 centesimi (nominal)

Vatican City is the smallest independent state in the world. Lying in north-western Rome, near the west bank of the River Tiber, it has been the seat of the Pope, the Supreme Pontiff, and the spiritual and administrative headquarters of the universal Roman Catholic Church since 1929. The administration of the Vatican City is directed by a commission appointed by the Pope.

Extending westwards from the Castel Sant' Angelo, the Vatican City occupies a roughly triangular area whose perimeter, for most of its length, is clearly defined by massive Renaissance walls. It is normally entered from the east through Bernini's famous colonnaded *piazza* which provides an impressive approach to the Basilica of St Peter, the largest Christian church in the world. Other famous buildings include the 1,400-room Vatican Palace containing the Sistine Chapel with Michelangelo's celebrated paintings and many art treasures; and various libraries and museums.

The Vatican maintains diplomatic relations with many other states. It issues its own coins, postage stamps and car licence plates, and has its own bank, railway station, gaol, police force, printing works, daily newspaper (*L' Observatore Romano*) and radio station.

SAN MARINO

Area: 23·5 sq mi (61 km²)
Population: 19,000
Population Growth Rate: 1·8%
Capital: San Marino, pop 4,500
Language: Italian
Religion: Roman Catholic
Currency: Lira = 100 centesimi (nominal)

San Marino is a small, landlocked democratic republic on an eastern spur of the Apennines in northern Italy. Entirely surrounded by Italian territory, it consists for the most part of the craggy, triple-peaked Monte Titano (2,480 ft; 756 m), on which stands the city of San Marino, the capital (pop. 4,500), overlooked by its ancient fortress. Triple walls encircle the mountain peaks, and there are many historic buildings. Below the city is its chief suburb, Borgo Maggiore. Elsewhere, except for the small town of Serravalle, there are only tiny villages.

Virtually all San Marino's citizens are Roman Catholics; and more of them live abroad than in the republic. San Marino lives mainly by tourism, the sale of postage stamps and agriculture. Over 2,340,000 tourists visited it in 1970. Wine, textiles, ceramics, furniture and building stone are exported.

MALTA

Area: 121 sq mi (313 km²)
Population: 322,070
Population Growth Rate: 0·1%
Capital: Valetta, pop 15,401
Language: Maltese and English. Italian widely spoken
Religion: Roman Catholic
Currency: Pound (M) = 100 cents

Malta commands the narrows between Sicily and North Africa, and thus the main passage between the eastern and western Mediterranean. The tiny Maltese islands paid for their strategic importance by centuries of foreign rule.

From 1814, Malta was a British colony and in 1964, the islands became the independent State of Malta within the Commonwealth of Nations. However, after the Malta Labour Party won the 1971 election its neutralist leader, Dom Mintoff, broke with NATO, negotiated a new defence pact with Britain (to run until 1979), established closer relations with Libya and welcomed Chinese economic aid. He also aspired to make Malta economically independent – a difficult task for a state whose only major resources are limestone and cheap labour. Meanwhile tourism, a major money-earner, has slumped, and the new tax regulations for settlers are less attractive.

The Land. Malta consists of three inhabited islands – Malta (the largest), Gozo and Comino – and several islets. All consist of fractured limestone strata rising from the sea. The island of Malta is crossed by ridges and valleys in the north; the east coast, where most Maltese live, has many inlets such as the Grand Harbour at Valletta and St Paul's Bay. There is much worthless land, but the precious shallow soil in the tiny stone-walled fields of the terraced valleys is intensively cultivated.

The Maltese. Malta is extremely densely populated and the people are very mixed. Some regard the Phoenicians as their distant ancestors, and the difficult Maltese language (Malti) has a Punic/Arabic base and elements from several other languages. Whatever their origins, they are devout Roman Catholics almost to a man.

The farmers, some of whom work part-time in other industries, grow wheat, potatoes, onions and grapes (for wine), and raise pigs, sheep and the ubiquitous goat. Irrigation is important, water coming from underground sources and from desalination. Cut flowers and vegetables are grown under glass for export.

The former naval dockyard, a major employer of labour, is now a civil ship-repair establishment and older industries such as food processing, brewing, textiles and gloves, have been reinforced by light engineering, metal working, car assembly, plastics, clothing and rubber products.

The Maltese islands have a pleasant Mediterranean climate whose only tiresome feature is the enervating *sirocco*, a sand-laden southerly wind from North Africa. The islands have much to offer the visitor, and many modern hotels have been built.

Valletta, the capital and chief sea-port, is girdled by the massive fortifications and packed with the Baroque palaces and churches of the Knights of St John.

Valletta, the capital of Malta, was built in 1565 and given to the military order of the Knights of St John of Jerusalem after they had been expelled from Rhodes by the Turks.

YUGOSLAVIA

Area: 98,650 sq mi (255,504 km²)
Population: 20,738,000
Population Growth Rate: 1·1%
Capital: Belgrade, pop 1,209,360
Language: Serbo-Croatian, Slovenian and Macedonian
Religion: Orthodox (42%) Roman Catholic (25%) Moslem (10%)
Currency: Dinar = 100 para

Yugoslavia is a largely mountainous land flanking the eastern shore of the Adriatic Sea. It has gained a reputation as the most individualistic and independent country in the Communist world. Under the leadership of its President Tosip Broz Tito, Yugoslavia has maintained friendly relationships with western Europe and has avoided the dependence on Soviet Russia which is characteristic of much of eastern Europe.

When Yugoslavia was invaded by the Nazis in World War II, the communist Partisans led by Tito received Allied support in resisting, and eventually by 1944, driving out, the Germans; in 1945 the monarchy was abolished and since then Yugoslavia has been a communist federation of six republics.

The Land. Uplands and mountains cover much of Yugoslavia. The most extensive lowlands are in the north, in fertile Slavonia, west of the river Danube, and in Vojvodina, sometimes called the larder of Yugoslavia. The monotony of the plains is broken by imposing rounded hills in Slavonia, and by bluffs and terraces in Vojvodina.

Highland Yugoslavia is varied. In northwest Slovenia, alpine ridges alternate with deep glaciated valleys filled by lakes or with depressions such as the Ljubljana basin. Here, in the Julian Alps, is Triglav (9,393 ft; 2,863 m), Yugoslavia's highest peak. Stretching south-eastwards as far as Lake Shkoder on the Albanian frontier are the Dinaric Mountains, a series of parallel ridges and longitudinal furrows broadening to a width of some 175 mi (282 km) in the south and approaching 8,000 ft (2,438 m) in height near the coast. Their descent to the island-fringed Dalmatian coast is everywhere abrupt, except where the River Neretva cuts its gorge through to the Adriatic Sea.

The northern Dinarics are forested; the southern are pure limestone and have extensive karst areas – plateaux pitted, broken and dissected by hollows, collapsed caverns and canyons, and drained by underground streams. The dry, barren appearance of the karst belies the heavy average annual rainfall it receives and the large underground water resources it contains.

Eastern Yugoslavia consists of the lower, rolling country of the once-forested Šumadija area of Serbia, and the Vardar and Morava river systems. Linking broad valleys and depressions by gorges cut through the mountains, these two rivers form one of the historic routes from central Europe to the Aegean (at Salonika) and the Mediterranean. The Morava and its tributary the Nišava provide a good route to Sofia.

The whole of highland Yugoslavia is earthquake-prone, the most vulnerable area being centred on Skopje, which was devastated by an earthquake in 1963. Structural lines of weaknesses in the earth's crust are many and tremors are frequent, no fewer than 2,266 having been recorded between 1945 and 1965.

Climate and Vegetation. Yugoslavia has a varied climate. Along the Adriatic coast the pattern is truly Mediterranean, with warm winters, and hot summers tempered by the sea, and adequate winter rainfall. The Dinaric 'wall' causes heavy precipitation in the highlands, never less than 80 in (2,000 mm) but reaching 210 in (5,300 mm) on Mount Lovcen above Kotor in Montenegro. The 'wall' shuts out warm air, and interior basins only a short distance from the Adriatic thus have January temperatures below freezing point.

The north and east experience continental conditions. Summers on the northern plains tend to be hot and humid, the winters cold. But in winter, depressions pass regularly from the Adriatic to Pannonia. Contact between the warm, moist air and the cold, dry air from the Russian anticyclones induces strong local winds like the raging *Bora* in Dalmatia.

More than a third of the country is covered by forests of beech, oak and fir. The north-east was once largely steppe, and thin scrub occurs on the poor limestone mountains.

The People. No other European country, except the Soviet Union has such a diverse population. South Slavs form more than 80% of the population, but they comprise five nationalities – Serbs (the most numerous), Croats, Slovenes, Macedonians and Montenegrins. They practise three religions (Greek Orthodox, Roman Catholic and Moslem). They use two alphabets, Latin (Slovenes and Croats) and Cyrillic (Serbs, Montenegrins and Macedonians), and have Serbo-Croat as a *lingua franca*. There are Shiptar (Albanian), Magyar (Hungarian), Turkish, Bulgarian as well as some 13 other minorities.

The population is distributed unevenly. Some areas still reflect the depopulation that occurred in Ottoman times when people took to the mountains for refuge. About two-thirds of the population lives in villages and most of the remainder in the largest cities, especially Belgrade, the federal capital, and the capitals of the republics. Large regional centres include Niš (Serbia), Maribor (Slovenia), the ports of Rijeka and Split (Croatia), Banja Luka (Bosnia) and Subotica (Vojvodina).

Despite modern expansion, many towns and cities retain examples of their historic architectural styles, 'Alpine' in the north-west, 'Hungarian' in the north, 'Turkish' in the centre and south, and 'Venetian' along the coast. Some places, like Ljubljana, the capital of Slovenia, have suffered repeated destruction by earthquake. Belgrade has often been fought over and destroyed due to its strategic position at the confluence of the Danube and Sava.

The many national groups and languages make education a problem. Education is compulsory from 6 to 15 and all the national minorities have their own schools, but illiteracy has not yet been completely eradicated. Each major group has its culture, and traditional costumes and folk art are widespread.

Since 1971 a collective ruling body, called the Presidency, has had overall control under Tito, who is President of the Republic for life. The national government features a Federal Assembly with five chambers of 120 members each, except the Chamber of Nationalities, which has 140 members. All members are elected by popular vote for a 4-year term and candidates who are not members of the Communist Party are allowed to stand for election. There is also a Federal Executive Council headed by the President. Each republic and autonomous region has its own assembly and executive council. These have become much more important since decentralization became the policy of the federal government. 'Self-management' has become the function of the republics. With the federal government concentrating on overall matters such as foreign policy, defence and economic planning, but ready to step in, as in Croatia, should any republic show signs of excessive independence.

The Economy. Industrialization has been the basis of economic growth since World War II. Since 1945 a large percentage of Yugoslav capital investment has gone into industry, and industry now contributes some 48% of the Gross National Product.

Agriculture, while its importance has relatively declined, still employs half the labour force and contributes 27% of the Gross National Product. Peasant farmers still own about 84% of the cultivated land and 91% of the livestock. Their farms are very small, averaging little more than 10 ac (4·6 ha), fragmented into tiny strips worked by draft animals and few tools. Yields are low. The remaining 16% of cultivated land is worked by state (experimental) farms, collectives and co-operatives. These spread new techniques among the peasants and usually grow industrial and fodder crops.

Peasant crops are primarily for subsistence. The expansion of socialized farming, the wider use of machinery and fertilizers, the increasing substitution of vegetables, fodder and tree crops for cereals and industrial crops, and improvement in livestock yields have doubled farm output since 1956. The north-east is the granary of Yugoslavia. The foothills bordering Pannonia produce fruits, vines and livestock.

Pastoral farming predominates in the higher Dinaric region, and is combined with horti-

Montenegro is a Balkan republic, now part of Yugoslavia which takes its name from the 'Black Mountain',Mt Lovcen, the historical centre of the state. Montenegro consists mostly of poor agricultural land, but it is famous for the bellicosity of its citizens.

Right: A bridge in Mostar, the capital of Hercegovina in Yugoslavia. This bridge which is 20 metres high dates from the 16th century when the Turks controlled Yugoslavia. *Left:* Part of the chain of the seven Blitvice lakes which are joined by a series of waterfalls.

culture along the Adriatic coast and its islands, and with fruit and tobacco in Macedonia.

Minerals are plentiful. Coal, mainly low quality brown coal and lignite, is mined in several republics, Bosnia and Serbia being the major producers. Oil and gas, discovered mainly since 1950, are produced in increasing quantities in Slavonia and Vojvodina. Adequate iron ore is available in central Bosnia and western Macedonia; copper, near the Iron Gates; and bauxite in the karst zone from Istria to Montenegro.

Yugoslavia is self sufficient in non-ferrous metals and a leading European exporter of antimony, bauxite, chrome, copper, lead, molybdenum, mercury and zinc.

Industry. Industrial growth has been most rapid in the electrical industries, engineering, chemicals, paper and steel. Industry was formerly concentrated in the north. Since 1945, efforts have been made to promote industrial growth in the centre and south, far richer in natural resources. The largest industrial centres are, however, Zagreb, Belgrade, Ljubljana, Maribor, Sarajevo and Niš, in that order.

Industries which have been developed near the raw materials they use include steel (central Bosnia and Skopje), aluminium (Mostar and Titograd), non-ferrous metals (Serbia) and wood-processing throughout the Dinaric zone.

The generation of electricity, once confined to the north, is now spread throughout central Yugoslavia in association with lignite mining and with hydro-electric plants on the major Dinaric rivers and the Danube. The joint Romanian-Yugoslav Iron Gates hydroelectric power scheme, was opened in 1972.

Tourism, is an important and growing source of foreign currency. In 1971, 5,238,000 foreign tourists visited Yugoslavia, compared with only 500,000 in 1957.

Transport. The north has the best road, rail and river facilities, and its routeways carry more than 66% of all the freight and passenger traffic. Much has been done since 1945 to expand and improve transport. Major projects recently completed include the Adriatic Highway from Rijeka to Skopje via Dubrovnik; the Belgrade-Bar railway; several new ports, including the iron ore-petrochemicals port at Bakar near Rijeka; and new airports at the major cities and at Dubrovnik, services by the national airline, Yugoslav Airlines.

International Trade. Rapid economic growth has increased demand for capital goods from abroad. Imports have therefore increased, but exports have scarcely doubled since 1939, although the character of both has changed.

Machinery, metal products, ships, electrical appliances, leather and tobacco products have replaced livestock, fresh food, sawn timber and ores as the main exports. Textiles and clothing, once major imports, have been overtaken by engineering components, steel, food, fodder and fertilizers.

Until 1949 Yugoslavia traded mainly with the Soviet bloc, and when this trade ceased, with Western Europe. Since 1955, her trade with the Third World has increased and, more importantly, trade with the Soviet bloc has been resumed. Today, Western Europe and the Soviet countries are of equal importance, each sharing about 35–40% of Yugoslav imports and exports, the rest being shared by North America and the Third World.

ALBANIA

Area: 11,099 sq mi (28,748 km²)
Population: 2,226,000
Population Growth Rate: 3%
Capital: Tirana, pop 169,000
Language: Official language based on Tosk (South), secondary on Gheg (North)
Currency: Lek = 100 qintars

Albania is a mountainous Balkan republic on the south-east shore of the Adriatic Sea, flanked by Greece and Yugoslavia. Communist Albania is the smallest nation in both area and population in eastern Europe, and one of Europe's poorest countries. Agriculture is primitive, and industry underdeveloped. Independence from Turkey in 1912 did not bring prosperity. Badly devastated by the Germans in World War II, Albania became a

Communist state in 1945. The Albanians had been early converts to Christianity, but under Turkish rule mostly embraced Islam, which remained the predominant religion until 1967 when mosques and churches were closed, and Albania was proclaimed the first atheist state. In foreign relations, Albania has established close relations with China and broken her ties with the Soviet Union.

With the highest birthrate in Europe, Albania has doubled her meagre population since 1945. Nearly 75% of the people live in the coastal lowlands, where most of the agriculture, transport facilities and centres of new industry are located. Cities with more than 50,000 people include the capital, Tirana, Durrës, Vlorë and Elbasan in this region. Elsewhere, only Korçë reaches this figure. Most Albanians live in villages on the plains or in smaller villages or mining centres in the hills.

The Land. Albania is mostly mountainous, but low-lying plains form a narrow strip along the north and central coasts, extending in an embayment (the Myzeqe) from Durrës and Elbasan in the north to Vlorë and Berat in the south. Here the steeply-descending rivers have deposited fertile alluvium and built up small deltas into the Adriatic. Along the rivers and behind the sandbars much reclamation has taken place.

Two-thirds of Albania is highland, the east and the south entirely so. Parallel mountain ranges follow the northwest-southeast Dinaric trend, rising eastward to 2,000 m (6,500 ft). These ranges are separated by the longitudinal valleys of the rivers Drin, Mat, Shkumbin Seman and Vijosë.

The Alp-like Prokletije (the 'accursed mountains'), bordering Yugoslavia on the north, are succeeded southwards by the serpentine Mirdite, limestone Krrabe and southern ranges. The two deep fault troughs forming the eastern frontier with Yugoslavia are occupied respectively by the River Drin, draining Lake Ohrid, and by Lake Prespa. The area is prone to earthquakes.

Climate. Coastal Albania has a 'Mediterranean-maritime' climate with average temperatures of 46·6°F (8·1°C) in January and 76·5°F (24·7°C) in July. In the hills, a 'transitional' climate gives average January and July temperatures of 40·6°F (4·8°C) and 75·9°F (24·4°C). In the east and north, there is a 'Mediterranean-continental' climate with January averaging 29·5°F (−1·2°C) and July 70·5°F (21·4°C).

Precipitation increases from low in the west to high on mountains which cast the eastern troughs into a rainshadow, and is heaviest between September and May.

Agriculture. Much of the land surface is too rugged or otherwise unsuitable for agriculture but farming employs about 75% of the population. The most fertile areas are the Adriatic coastal plains and the large Korçë basin. The small fragmented farms once worked by poor, illiterate and tradition-bound peasants have been supplanted by large-scale collectives similar to those of the Soviet Union. State farms serve as experimental stations to develop and diffuse new agricultural techniques. It is hoped that rice and cotton crops will thrive on the marshy plains. Tobacco is a major export.

Industry. Faced with building on a negligible foundation, the Albanians have expanded heavy industries (power, metallurgy, chemicals and timber-processing) more rapidly than light industries like textiles, shoes and food, even though the latter still predominate. Expansion is based upon the local processing of domestic raw materials – coal, oil, gas, metal ores, food and industrial crops, and timber. Most new industry is in the Shkodër-Elbasan-Vlorë triangle. The largest manufacturing centre is Tirana.

Improving Transport and Communication. The post-war Albanian regime has channelled more investment into transport than any other economic sector except industry, seeking to build a basic modern inter-regional road network; adequate rail links between the main port, Durrës, and its hinterland centres of Tirana and Elbasan; and good port facilities at Durrës and Vlorna.

International Trade. Most exports derive from expanding primary activities such as farming, mining and forestry. Imports are mainly manufactured goods or scarce minerals needed by a society that is being industrialized. This industrialization is also beginning to reduce some imports of manufactures – for example textiles and electrical equipment.

The Albanian town of Pulajt. Located in Shkodër, in the northern part of the country where the foothills of the Albanian Alps meet the marshy coastal plain, this area has served from ancient times as a market centre for the remote villages of the northern mountains.

The geographic pattern of trade has been strongly influenced by Albania's political alignment first with the Soviet Union, and since 1961 with China, but her trade with eastern Europe generally has not declined, and in 1970 improved trade relations were established with Yugoslavia, followed in 1972 with Italy.

Albania's future is uncertain. Her neighbouring states are by no means friendly, industrial progress depends on Chinese aid, and the country still lacks skilled labour and technicians.

Albania's rocky, mountainous land is a great hindrance to transport and communication and roads are few. Despite the land, and the changes that have come with the socialist regime, Albania is still principally agricultural.

GREECE

Area: 50,943 sq mi (131,944 km²)
Population: 8,957,000
Population Growth Rate: 0·7%
Capital: Athens, pop 2,540,241
Language: Greek. Turkish, Albanian and Armenian by minorities
Religion: Greek Orthodox, Moslem (1·3%)
Currency: Drachma = 100 leptae

Greece is a small European country with an illustrious past. It is situated in the eastern Mediterranean at the lower end of the Balkan Peninsula. About one-fifth of the land area is made up of islands, including the large islands of Rhodes, Crete and Euboea. By European standards Greece is poor and underdeveloped. Less than a third of the land is arable, yet nearly half the people are farmers, and agriculture is the country's leading industry. The Greeks, however, with their deeply indented coastline and many islands, have always been seafarers, and today their merchant fleet is the seventh largest in the world.

Greek exports (traditionally of farm produce) increasingly feature manufactured products while imports are largely desperately-needed raw materials and manufactured goods. The resulting trading imbalance is partly offset by considerable earnings from shipping, remittances by Greeks living abroad, and tourism based upon the country's blue skies, warm seas and historic classical ruins.

Ancient Greece was the birthplace of democracy, yet the history of modern Greece is one of political instability and repressive governments. In 1967 a group of army colonels established a military dictatorship and exiled the king. Since then Greece has been ruled by the 'colonels', now civilians, who have denied individual rights and the freedom of the press. On 1 June 1973 the monarchy was abolished and replaced by a republic.

The Land. Greece, being the most southerly part of the Balkan Peninsula, forms an extension of the peninsula's basic features. However, extensive subsidence of the land has resulted in a partial drowning of the coastlands and much interpenetration of sea and land, as well as island arcs across the Aegean Sea. The mountainous island of Crete and Rhodes' central range are extensions of the mountain systems of the mainland, mostly limestone ranges which run generally from north-west to south-east. From their slopes flow rushing streams which are reduced to dry gravel beds during the summer drought. The Pindus Mountains in central Greece rise above 8,000 ft (2,400 m); the highest mountain in Greece, Mt Olympus in Thessaly, rises to 9,570 ft (2,888 m). The whole region is one of structural instability, earth tremors being common. Active volcanism has taken place in historic times in the Aegean region.

The uplands of Greece have poor, stony soil, usually deforested and bare or covered with shrubs. The lower hill country provides olives, wine, fruits and pasturage. The plains and valleys, about a third of the land area, have rich soils in which rice, tobacco and cereals are grown. Northern Greece consists of Thrace, Macedonia and Epirus. The marshy coastal plains of Thrace are backed by rugged mountains and fertile valleys. Macedonia continues the same features westward towards Epirus, which flanks the Ionian Sea as far south as the Gulf of Patras. The Macedonian plain of Salonika (Thessalonike) with its port constitutes one of the country's richest agricultural regions and the second industrial area after Athens. Thessaly is another rich plain, edged by very high mountains. Separating Thessaly from Epirus is the inhospitable Pindus range. East of the Pindus lies the heartland of ancient Hellas, extending from Corinth to the plains of southern Thessaly. Its outlook, then as now, is seawards towards the Aegean. Inland the relief is rugged and broken, the lowlands and basins being small and separated. The narrow plain of Attica extends to Cape Sounion on the Saronic Gulf. The southern coast on the Gulf of Corinth is mountainous, and between the city of Corinth and the Saronic Gulf the Corinth Canal, cut in 1893, extends 4 mi (6·4 km). It is now little used.

Below the isthmus with its canal is the Peloponnesus, virtually a large peninsula, mountainous and with little arable land. Long, parallel mountain ranges end in the south in rocky promontories divided by deep inlets. There are small, isolated basins such as that of Sparta, and in the south-east several compact plains opening to the sea. There is a narrow but almost continuous coastal plain.

Over 430 islands, some very small, account for 9,000 sq mi (24,000 km²) of the nation's land area, or about 20%. The larger Ionian Islands off the west coast include Corfu to the north, quite densely populated, and Leukas, Cephalonia and Zante. Islands close to the east coast mainland include Euboea, which runs parallel to the whole of Attica. Off Thrace are Thasos, Samothrace and Lemnos, sunken

Greece is largely mountainous, and herding is an occupation which survives from the once predominant nomadic pasturalism of the area. However over thousands of years, overgrazing has significantly altered the ecology of parts of Greece which was once more fertile than it is now.

The Corinth canal was cut through the four mile limestone neck joining the Peloponnesus to mainland Greece in 1893. In ancient times the Corinthians prospered when they built a stone causeway over the isthmus and in AD 67 the Emperor Nero began, but never completed, a canal.

remnants of the Thracian massif. To the south-west lie the Northern Sporades (Skiros is the largest), and in the centre of the Aegean Sea the rocky Cyclades, including the islands of Paros, Naxos and Mykonos, as well as spectacular Santorin, which has seen many eruptions in historic times. The islands off Turkey include Lesbos, Chios and Samos. To the south-east the Dodecanese, a string of small islands, end in the large and famous island of Rhodes. Finally Crete, the largest Greek island, lies 60 mi (97 km) off the Peloponnesus and is 160 mi (260 km) long. Its ports face the Aegean.

Climate. Most of Greece has a typical Mediterranean climate with mild winters and long dry summers, but Thessaly, Macedonia and Thrace in the north have a continental climate, with hard winters and hot summers. Generally temperatures average about 40°F (4·4°C) in winter and 75°F (23·9°C) in summer. There are wide ranges in temperature however, as well as in rainfall, which occurs everywhere in the winter months. The western coast is much wetter than eastern Greece, precipitation diminishing from north-west to south-east.

Vegetation. Greece was once well-endowed with natural woodland, but because of centuries of overgrazing much of the original woodland has been reduced to scrub or a ground cover of aromatic herbs. Only the remote mountain slopes, notably in the Pindus range, still support forest trees. Lower down, oak, sweet chestnut and Aleppo pine occur, or have been replaced by olive groves and vines. On the lowlands and the formerly marshy coastal areas, crops such as wheat, rice, maize and tobacco are grown.

The People. The total population of Greece is about 9,000,000, roughly a quarter of whom live in Athens with its port, Piraeus. Greater Athens now totals about 2,500,000; Salonika about 340,000. Crete has a population of under 400,000, Heraklion, its capital, having about 64,000. Since the war, there has been a marked exodus from the once over-populated hill villages to the cities and plains, or overseas. The formerly high Greek population figures have been affected by renewed massive emigration since the war to West Germany, the USA, Canada, Australia and other countries. The highest densities of population are now found on the narrow plain of Attica, the west coastlands of the Peloponnesus, the Ionian Islands, the plains of Thrace and the large islands off Asia Minor.

Greeks are racially very mixed, but display a strong homogeneity owing to a common language and the tenacious culture it represents. The family (which includes an extended series of relatives on both sides) is all-important in Greek society and commands an intense loyalty and devotion, though urban and industrial life, as everywhere, is tending to break down these older traditions. At least 97% of the people belong to the Greek Orthodox faith, which is also the official religion. Greek Orthodox festivals, of which the chief one is Easter, play an important role in the lives of the people, every village or town having its own patron saint and its own feast day.

Education is much prized in Greece. All public education is at present free, while children are required to go to primary school from the age of 6 to 12. Above that there is a system of gymnasiums and lyceums. There are some 20 universities and colleges, headed by the University of Athens and the University of Salonika (the largest, with about 20,000 students). Illiteracy in Greece has been

Acropolis, 'high place' in Greek, refers to a fortress or citadel built on a hill which the rest of the city may lie beneath. The most famous example is the Acropolis in Athens, surmounted by the Parthenon, built between 447 and 438 BC and dedicated to Athena, patroness of the city.

reduced in recent years to less than one-fifth of the population.

The Economy. Enemy occupation during World War II and the long civil war between monarchists and communists that followed nearly ruined Greece's economy. Since 1948, however, there has been a remarkable recovery, with an average and continuing growth rate of over 7%. Agriculture is still the principal industry and manufacturing is little developed. Tourism however has grown strikingly, bringing in 50% more revenue in 1971 than the year before, and it continues to develop.

Farming and Fishing. Farming methods are quite primitive, and holdings are generally small – less than 25 ac (10 ha). Lack of capital and often difficult terrain limits mechanization. There is subsistence farming in the remote areas, and sheep and goats are ubiquitous, often migrating in the summer to the cooler hill pastures, with consequent overgrazing. The mule is a common draught animal, cattle and horses being less common except in Thrace and the north. In the remote mountain regions the Vlachs, a traditionally pastoral people, migrate seasonally with their flocks and herds, wintering in tents or reed huts.

The most profitable cash crops of the lowlands are rice, wheat and cotton. Vines, grown mainly for currants, show fluctuating returns. The Greek resinous wines have a generally local market. Greece is still one of the leading producers of lemons, olives and tobacco, the latter having been cultivated in Thessaly and Thrace since the times of the Turks. Agriculture has greatly expanded since World War II, partly through irrigation, draining of marshy areas, better seed and the use of fertilizers.

Fishing for local consumption is important to supplement the insufficient food crops. The eastern Mediterranean is not rich in fish, except for mullet, tunny and sardines, though dried octopus is a delicacy.

Mining and Manufacturing. Minerals are few and dispersed. Lignite, or brown coal, used to generate electric power, is the chief mining product. Limited quantities of bauxite and chromite, as well as barite, iron ore, lead, limestone and marble are also mined. The famous white marble of Greece has been quarried from the hills of Attica and from many of the Aegean islands since Classical times.

Since the Greek economy still depends primarily upon agriculture, trade and shipping, manufacturing industry is very little developed, and is in general still largely a matter of small-scale family businesses. Greek manufacturers also produce almost exclusively for domestic consumption. Yet the annual growth rate of manufacturing has been as high as an average of about 10% for many years. With the enormous migration of population into the Athens–Piraeus area in recent years, industry has expanded there and along the Saronic Gulf towards Corinth. Shipbuilding and repair yards have long been important in the harbour area of Piraeus and at Eleusis. The manufacture of cheap cotton goods and the processing of food-stuffs are now being supplemented by cement, chemical and light engineering industries, including the assembly of trucks and motor vehicles. Large-scale industry is represented by such new enterprises as Salonika's Esso-Pappas group of chemical, petroleum and metallurgical factories, and a new aluminium refinery on the Corinthian Gulf. The nation's entry into the European Economic Community in 1962 as an associate member is stimulating considerable modernization of industry for competitive reasons.

The power sources of Greece remain inadequate. However, American and West European capital has been invested in schemes to build hydro-electric and thermal power plants, and to ensure an adequate water supply. The country's public power system is almost entirely state owned.

Transport and Communications. Maritime transportation has always been important to Greece as a seafaring nation. Yields from its commercial fleet constitute one of the major sources of invisible earnings. Of all Greek shipping, Piraeus handles over 60%, while Salonika is the next major port, with Patras third in tonnage. While the railway network remains fixed at around 1,500 mi (2,400 km), Greek roads have become far more important since the war and have been greatly improved, while the building of secondary roads have brought many isolated villages into touch with the world. Olympic Airways, a private company, maintains both domestic and international air services.

The government controls the Greek national telephone, telegraph and radio services. Greece does not yet have a nationwide television network.

International Trade. Agricultural products such as dried fruit, cotton, currants, oranges and tobacco still make up much of the nation's exports, though manufactured goods were poised to outstrip them in value soon after 1970. Increased earnings from exports, tourism and shipping are outweighed by heavy imports of raw materials and industrial equipment required in modernization, so that Greece is perennially plagued by an adverse trade balance. Most Greek trade is now with the Common Market countries of Western Europe, although the USA still takes a large share. Greek exports to the communist countries, though small, were increasing.

BULGARIA

Area: 42,823 sq mi (110,911 km²)
Population: 8,540,000
Population Growth Rate: 0·7%
Capital: Sofia, pop 910,242
Language: Bulgarian. Turkish, Greek and Armenian by minorities
Religion: Bulgarian Orthodox (70%) Moslem (9%)
Currency: Lev = 100 stotinki

Bulgaria occupies a roughly rectangular area in the north-east of the Balkan Peninsula. It is bounded on the north by Romania, on the east by the Black Sea, on the south by Turkey and Greece, and on the west by Yugoslavia. This communist country's picturesque peasant costumes, folk dances and creaking bullock carts appeal to the growing trickle of Western tourists whose cash Bulgaria is trying hard to attract. But this old-fashioned charm is also evidence of archaic economy. By the end of the 1960s no other European country had a higher percentage of its population tied to the poorly-paid task of farm labouring, and none had fewer cars per 100 inhabitants. By the early 1970s, however, industrialization was changing this pattern.

The Origins of Modern Bulgaria. In ancient times Bulgaria was part of Thrace and Moesia. The Thracians were overcome by the Slavs, who were themselves conquered by the Turkic-speaking Bulgars. These 'Old Bulgarians' assimilated Slav language and culture, but their empire was finally extinguished by the Byzantine emperor Basil II.

A second empire, centred on Türnovo, was established about 1185 whose ruler, Ivan II Asen (1218–1241) led a great cultural and religious revival, but by 1396 Bulgaria had fallen to the Ottoman Turks.

In the 1870s, an autonomous Bulgarian state was created with Russian aid, and in 1908 the independent kingdom of Bulgaria was proclaimed. Bulgaria sided with Germany in both world wars and was occupied by the USSR (1944). The child-king Simeon II was deposed (1946) and Bulgaria became a people's republic.

Mountain, Valley and Plain. Bulgaria is a mountainous country. Its ranges, running east-west across the country, create a series of distinctive physical zones. The south-west is dominated by the high Macedonian-Thracian massif of the Rhodope Mountains, scenically most impressive in the rugged Pirin and Rila Dagh Mountains, where Musala Peak, the highest point in Bulgaria, rises to 9,595 ft (2,925 m). The rivers Struma and Mesta cut deep trenches through the mountains on their southward courses to the Aegean.

North of the Rhodope Mountains is the broad Thracian Plain, which extends eastwards as the Burgas lowland, dominating south-eastern Bulgaria. It is drained by the River Maritsa and its tributaries, including the Tundzha. Flanking the plain on the north-east is a broken and detached mountain range, the Sredna Gora (Middle forest).

From Sofia in the west, through Kazanlük to Sliven in the east, runs a series of depressions surrounded and separated by forested hills. North of this are the steep scarps of the Balkan Mountains, the backbone of Bulgaria, difficult to penetrate from the south, except through the Shipka Pass, between Gabrovo and Kazanlük, and the Iskür gorge, between Sofia and Vratsa. Known as the Stara Planina (Old mountain range), the Balkan Mountains rise to 7,791 ft (2,375 m) at Mt Botev. They are rounded mountains with limestone and crystalline plateaux descending gently to the low Danubian platform of northern Bulgaria. This is rarely more than 400 ft (122 m) above sea level and ends in steep cliffs overlooking the Danube.

Climate. Bulgaria has a predominantly continental climate with hot summers and cold winters. The low annual precipitation reaches its maximum in summer. The build of the country, however, permits air from the Mediterranean to penetrate some regions, and there are thus significant local variations.

The climate is most typically continental on the Danube platform, which is open to the north and east winds bringing cold in winter and heat in summer. Rainfall in this region occurs mainly as summer thunderstorms, averaging 23 in (584 mm) annually.

The Balkan Mountains restrict the southward passage of continental air, and the Thracian Plain thus has a warmer climate with more winter rain brought by the Mediterranean air that penetrates the Maritsa valley. Even stronger Mediterranean influences occur farther south.

Climate in the mountains varies with

The rose gardens of central Bulgaria are the largest in the world, and Bulgarian attar of roses, a base for perfume, is world famous.

The church of Shipka is a 19th century building commemorating the heroism of a Bulgarian contingent in the Russian army which held off the Turks at Shipka Pass in 1877.

altitude, being typically alpine near high summits like Musala Peak. The mountains also give the Sofia and other intermontane basins their specific climatic features: cold winters resulting from temperature inversions and shelter from the wind throughout the year. It is this shelter that enables damask roses to be grown in Bulgaria's famous valley of roses, extending from Karlovo through Kazanlük, where the buds are distilled into attar of roses, used in making perfume. Bulgaria is the world's chief supplier of rose oil.

Vegetation and Wildlife. The natural vegetation ranges from the steppe-like grasslands of the Danube platform to the forests of the mountain slopes and the alpine plants, including edelweiss, of the high Balkan, Rila and Pirin ranges. Oak and beech forests clothe the limestone slopes of the Stara Planina, walnut and chestnut are typical of the Maritsa valley, while the higher slopes of the Rhodope Mountains are thickly forested with conifers.

Wildlife in the forests and mountains includes bears, wolves, jackals, wild boars, eagles and falcons.

The People. About 88% of the population is Bulgarian. The Turkish-speaking minority accounts for 8·6% and there are small groups of Armenians, Romanians, Jews and Gypsies. More than half of the population lives in the towns and cities.

Traditionally, Bulgaria was a land of villages and regional craft with market centres like Pleven, Pazardzhik and Blagoevgrad. With postwar development, the rural population has declined and villages have become fewer and larger as a result of the collectivization of agriculture. This trend seems likely to continue. In 1971 a start was made in grouping co-operative and State farms into 'agricultural-industrial complexes', the first two being centred on the towns of Vratsa and Plovdiv. The largest cities are the capital, Sofia (910,242); Plovdiv (247,500), Varna (219,000) and Ruse (149,600).

The Bulgarians are a people of mainly peasant stock, hardworking, patient, and inclined to be taciturn. They have a reputation for living to a great age; some centenarians attribute their longevity to regular consumption of yoghourt, a popular item in Bulgaria. Bulgaria has a rich folk art (ceramics, wood-carving, metalwork and weaving) and attractive regional costumes and dances. But the standard of living is lower than that in nearly all other European countries.

The Bulgarian language, belonging to the south Slavonic group, is closely related to Russian and is written in the Cyrillic script.

Industrial Developments. Before World War II Bulgaria was a backward country with a predominantly peasant agricultural economy. Under Communist rule not only has agriculture been transformed, but the emphasis has been switched from farm to factory.

Heavy industry, producing energy, metals, machinery, chemicals, building materials and capital goods, takes the leading place. Consumer goods industries are of secondary importance even though many former scattered handicraft workshops have been replaced by modern plants.

Sofia is the hub of the largest industrial concentration. Industry is also prominent in the Varna, Plovdiv and Stara Zagora districts, but is otherwise widely dispersed for social and strategic reasons. Many growing industries (electricity generation, metallurgy, chemicals, building materials, timber-processing, food and textiles) are tied to the sources of their raw materials, which are widely spread throughout the country. Engineering works have been developed in many centres along the four parallel east-west routes: the River Danube, and the road and rail links between Varna and Vratsa, Burgas and Sofia, and Svilengrad and Sofia. Recent major projects include the large Kremikovsti iron and steel combine at Sofia, the Burgas petro-chemicals complex and the Pleven oil refinery.

Mineral Resources and Power. Bulgaria has sufficient metal ores for both home needs and export. There are three main mining areas: the western Stara Planina between Pirdop and Belogradchik (lead, zinc, copper and iron ore); the eastern Rhodope Mountains, the main source of lead and zinc; and the Burgas-Varna area (copper and manganese). Iron ore is mined near Belogradchik (Martinovo), Kremikovsti (Sofia) and Krumovo (Marbas).

Oil is extracted near Tulenovo, in the Balchik district on the Black Sea; from off-shore wells north of Varna; and since 1962 at Dolni Dubnik near Pleven. But Bulgaria lacks good quality energy minerals, which must be imported, chiefly from the USSR. Brown coal and lignite, mined by opencast methods at Pernik near Sofia and at Dimitrovgrad in the 'Marbas' (Maritsa basin) are used for making gas and generating electricity. An atomic power station being built, with Russian aid, at Kozlodni on the Danube, was due to be completed in 1974.

Agriculture. Bulgaria has fertile soils. Cereals are grown throughout the lowlands, but the Danube region is the chief source of wheat, barley and maize, and also of sugar beet. Thrace specializes in cotton, rice, vegetables and fruits; the Black Sea coast in vines, vegetables and fruit; the south-western valleys (Arda, Struma and Mesta) in apples, tobacco and 'southern' fruits such as apricots and peaches; and the foothills of the Stara Planina in plums. Large numbers of beef and dairy cattle, sheep, pigs and poultry are raised.

Since World War II, agricultural output has doubled, while the number of farm workers has been halved. Increased productivity has been achieved by more and better mechanization, extensive use of fertilizers and, above all, by increased irrigation.

Intensively-cultivated crops (including fruit, vegetables, vines, and root crops) are now more important than cereals, and farming has become more 'mixed', as in western Europe.

Tourism. A chain of attractive tourist resorts has been developed along the Black Sea coast, including Zlatny Pyassazy (Golden Sands) in the Varna area, and Slantshev Bryag (Sunny Beach) and Nassebur north-east of Burgas. Bulgaria receives more than a million foreign visitors every year.

Transport. Bulgaria has 18,680 mi (30,062 km) of roads, including 1,446 mi (2,327 km) of motorways, and 3,736 mi (6,013 km) of railways, of which 442 mi (712 km) are electrified. Road transport has supplanted the railways as the chief carrier; and has brought a decline in river-borne traffic, although the Danube ports of Ruse, Lom, Vidin and Svishtov remain active. All Black Sea traffic is handled by Varna and Burgas. Balkan, the national airline, provides domestic and international services.

Trade. Farm produce is still the major export, but high-value fruits and vegetables have supplanted cereals in importance, and manufactured and canned foods now occupy pride of place. Exports now include refined and fabricated metals, chemicals, engineering and electrical products, vehicles and clothing. Imports, including fuels, metal products, factory machinery, transport equipment and farm machinery, are such that Bulgaria usually has a trade deficit. The balance is restored by credits and technical assistance mainly from the USSR.

Some 80% of Bulgaria's trade is with her partners in COMECON (Council for Mutual Economic Aid), the Soviet bloc's equivalent to the European Common Market.

ROMANIA

Area: 91,698 sq mi (237,500 km²)
Population: 20,469,658
Population Growth Rate: 1·1%
Capital: Bucharest, pop 1,591,784
Language: Romanian. Hungarian, German, Ukrainian and Yiddish by minorities
Religion: Romanian Orthodox (80%) Greek Catholic (10%) Roman Catholic (9%)
Currency: Leu = 100 bani

Romania is a communist country with a difference. Although associated with its powerful neighbour, the Soviet Union, by membership of the Warsaw Pact and the Council for Mutual Economic Assistance (COMECON), the Socialist Republic of Romania has maintained a surprising and often daring independence of Russia in its economic and foreign policies. Romania has consistently opposed the integration of its economy with other Comecon countries, and has forged trade and other links with the West. In foreign affairs, Romania has found friends in both East and West, and sometimes deviated from the Moscow line.

Thus Romania refused to follow Moscow in condemning Israel during the Six-Day War (1967) and did not cooperate in the invasion of Czechoslovakia by Warsaw Pact forces (1968). While strictly neutral in the Moscow-Peking ideological conflict, Romania maintained friendly relations with China and as a go-between helped to create a favourable climate for President Nixon's historic visit to China in 1972. Nixon's earlier visit to Romania (1969) was the first made by a United States President to an Iron Curtain country in 24 years.

Relatively poor and underdeveloped, Romania has nonetheless made much progress in developing agriculture and industry, especially since 1967 when Nicolae Ceauşescu became President. But future economic development depends largely upon Romania's continued adroit political handling of the Soviet Union.

Mountains and Plains. The heart of Romania, the Transylvanian basin, is enclosed by the Carpathian Mountains and Apuseni or 'Western' Mountains which, in turn, are fringed on all sides, except the north and the extreme south-west, by plains: the Tisza plain to the west, the Moldavian plain to the east, and the Wallachian plain to the south.

The Transylvanian basin is rolling hill country through which the River Mureş and its Tîrnava tributaries flow, creating a small undulating plain between Cluj and Tîrgu Mureş and dividing the northern Someş hill country from the higher Tîrnava platform in the south. This platform is separated from the southern Carpathians by the Fagaraş and Sibiu depressions, providing a good east-west route from Braşov towards the corridor formed by the River Mureş.

The thickly forested Carpathians enter Romania in the north to sweep, scythe-like, around the Transylvanian basin. They are highest in their southern section, the Transylvanian Alps, where Mt Negoiul, Romania's loftiest mountain, rises to 8,347 ft (2,544 m). Several depressions lie within the mountains, such as Braşov, Giurgiu and Dorna in the eastern Carpathians, and Petroşeni and Hunedoara in the southern Carpathians. Linked by relatively low passes with the outer plains and valleys, and with the inner Transylvanian basin, these depressions provide good natural routeways across the Carpathians, as do the valleys and gorges cut by the several major Danubian tributaries that rise in Transylvania or in the intermontane depressions. The most important are the Olt, which cuts the 'Red Tower' gorge; the Timiş, which provides the 'Porta Orientalis' linking Timişoara and the Iron Gates; and the Mureş and Someş corridors.

Romania's lowest, most extensive and flattest plain lies in Wallachia, spreading southwards from the loess-covered slopes where the rivers emerge on to the plain from the Carpathians, to the sand dunes of the Baragan district and the Danube marshlands. The Tisza plains are similarly flat and loess-covered but less extensive within Romania. By contrast, Moldavia is more of a platform of horizontally-bedded rocks broken by the rivers Siret and Prut, and by the Jijia depression in the north. The low-lying Dobruja lies between the Danube and the Black Sea.

Climate. Romania has long winters, with heavy snow and temperatures below freezing everywhere. In winter, the western and southern Carpathians prevent milder westerly air

Women mending a road in the Transylvanian Alps near Sinaia. Although an unusual sight to many north-west Europeans and North Americans, women do hard physical labour in most countries.

from the Adriatic or Atlantic from entering western Romania, except for some favoured areas such as Timişoara and Turnu Severin, where January temperatures are generally higher (though still below freezing). The Carpathians, with the Ukrainian Podolian uplands, also restrict the penetration of much colder continental Russian air. This fact, and proximity to the Black Sea, give even eastern Moldavia less severe winters than might be expected. Winter is mildest (31·5°F; −0·3°C) along the Black Sea coast.

In summer, July temperatures average 70°F (21°C). Rain and snow are heaviest in the Carpathians, where Virful Omul averages 53 in (1,346 mm) annually. Western Romania averages more than 24·8 in (630 mm) annually, compared with less than 15 in (381 mm) in Dobruja.

Vegetation and Wildlife. Forests cover more than a quarter of the land surface, with conifers predominant in the eastern Carpathians, and deciduous hardwoods elsewhere in the mountains. The Danube is bordered by willows and poplars, except in its broad delta, where reeds are widespread and wildlife, especially water birds including rails, ducks and pelicans abound. The Danube delta is one of Europe's richest nature reserves – one of the few remaining undrained, large areas of marshland habitat. The remote mountains of Romania harbour chamois and eagles; and hamsters and other small rodents occur on the open steppes.

The People. Romania has one of the faster growing populations of Europe. Despite the death toll caused by World War II fighting, the extermination of most Romanian Jews and the emigration of nearly 400,000 Transylvanian Germans, the population has increased by nearly 40% since 1930 due to a high (though now declining) birth rate and a more rapidly declining mortality rate associated with much improved medical services.

Relatively few people live in the Carpathians and the marshy Danube delta, or in the Dobruja and the Baragan plain in the south-east. The fertile plains of Moldavia, Wallachia and Pannonia have rather more people, due in part to recent industrial development. Still more densely populated is the Carpathian foothill zone running from Suceava in the north-east to Turnu Severin in the south-west and including such important centres as Bacău, Ploieşti and Piteşti. The main distinguishing feature of the Transylvanian basin is ethnic; it is the home of nearly one million Hungarians, including the related Szeklers, and 400,000 Germans. The former inhabit the north, centre and west, Cluj being their cultural centre and Tîrgu Mureş the capital of the Autonomous Magyar Region; the Germans dominate the south. Romania also has small Ukrainian, Serbo-Croat, Greek and Turkish minorities.

Urbanization has been increased by industrialization, and about 40% of the population now lives in the towns. By far and away the largest city is the capital Bucharest (1,591,784). Standing on the River Dambovița, it is the chief cultural and commercial centre, a focal point of roads and railways, and Romania's largest manufacturing city. It is largely modern and has imposing public buildings and pleasant parks.

Celts, Latins, Slavs, Magyars, Germans, Turks and Byzantine Greeks (Phanariots) have all played a part in Romania's ethnic history, but it is the Latin strain that now predominates. The Romanians have a literary tradition dating from the 1500s. Today education is free and compulsory from 6 to 15 and there is little illiteracy. There are five universities and more than 50 other institutes of higher education. The largest of the 15 religious groups is the Romanian Orthodox Church.

Exploiting Mineral Resources. Romania has a range of energy sources, metallic ores and non-metallic minerals which few countries in Europe can rival, although quantities are in some cases limited.

The most important mineral resources are petroleum and natural gas. In Europe, Romania is second only to the Soviet Union as a producer and exporter of hydrocarbon fuels. The Ploieşti oilfield is no longer the chief source; most of the production comes from around Tîrgu-Jiu, Craiova, Piteşti and Bacău. Natural gas is produced in all oilfields, but the chief source is the Cluj-Tîrgu Mureş area of the Transylvanian basin. Gas is piped to all the large cities and to Hungary. Salt for chemical industries is mined near Turda and Ploieşti.

Coking coal and anthracite are mined near Petroseni and Reşița in the south-west Carpathians, but high-grade coal has to be imported from the Soviet Union to meet growing industrial demands. Lignite, used mainly to generate electricity, is mined near Oradea and other places on the Carpathian margins. The output of electric power has been appreciably increased by the opening, in 1972, of the hydro-electric project at the Iron Gates gorge on the Danube. A joint Yugoslav-Romanian undertaking, it has helped improve navigation on the Danube. Romania is also building atomic power stations which will use uranium from deposits found in the Apuseni Mountains.

All metal ores are available in sufficient quantities except iron ore, and the expansion of the Romanian steel industry is now based mainly on supplies from the Ukraine. Precious metals are mined and refined in the Apuseni Mountains; copper, lead and zinc near Baia-Mare. There are open-cast bauxite mines near Oradea.

Industrial Growth. Since World War II Romania has maintained one of the highest rates of industrialization (and hence economic growth) in the world. Many factors have contributed to Romania's success in developing industry: among them, improvements in the transport and electricity networks, the construction of more pipelines, the introduction (from the West) of advanced technologies, the ready availability of labour, and the skilful siting of new industrial plants. Thus the coal, iron ore and hydro-electric power of the Reşița-Hunedoara-Iron Gates area sustain steel and engineering plants. Chemicals, metallurgy and engineering, based on oil, natural gas, lignite, salt and hydro-electric power from the River Argeş, are important in the Târgu-Jiu-Craiova-Buzau triangle; chemicals, timber-processing and engineering, in the Moldavian Carpathians and Siret basin; chemicals and engineering, in the Transylvanian basin; and metallurgy and chemicals in the Maramureş-Bihor region.

Food processing, textiles and shoe industries are dispersed on the plains and developed in major cities, all of which also have engineering plants. Braşov is known for its vehicles, tractors and roller-bearings; Craiova for its electrical equipment; and Tîrgovişte for its oil-drilling machinery. Industries based on imported raw materials are being developed along the lower Danube, the most important being the Galați steelworks, which uses Soviet ore and coking coal, and which is now the largest steel producer for the Romanian market. Bucharest, the largest manufacturing city, has leather, engineering, textile and chemical industries among many others.

Agriculture. Since 1949 nearly all Romania's farmland has been collectivized and agricultural production has increased by about 70%. The creation of some 5,400 state and collective farms has made it easier to re-locate crops and carry out land reclamation schemes. For example, terraced orchards have replaced grain fields in the Carpathian foothills, an area exceptionally prone to soil erosion and gullying; waterlogged areas in Wallachia and along the Danube have been reclaimed for intensive cultivation of vegetables, lucerne and silage maize. Parched areas of the plains have been made fertile by irrigation, and the contraction of the grain area has been offset by the use of fertilizers which has more than doubled the yield in some areas.

Grain crops dominate the plains, and are grown in association with sugar beet, oilseeds and, locally, tobacco (eastern Wallachia) or hemp (northern Tisza plain). Around large centres such as Bucharest, Brăila-Galați, Timişoara and Iaşi, market gardens provide vegetables and fruit, and dairy farming is important. An orchard and vineyard belt extends from Suceava to the Iron Gates,

Peles castle in Sinaia, a resort in the Transylvanian Alps, was built in the 16th century and is a museum of Romanian art.

terraced tree crops being an integral part of the scientific management of the Carpathian foothill lands. Mixed farming (grain, fodder and tree crops, and livestock) is general in the Transylvanian basin and Mureş corridor. Cattle and other livestock are reared on pastures in the Carpathians and in the Danube marshlands.

Transport. The network of good roads remains thin. Road transport is concerned mainly with short hauls and is a long way from threatening the railway monopoly of long-distance traffic. Rather does it feed the railways, whose network has been extended and modernized.

Water transport on the Danube is hampered by winter ice, and by silted channels in the delta. Ferry services across the Danube are important, because there is only one bridge, the Giurgiu-Ruse Friendship Bridge, linking Romania and Bulgaria. The major river ports are Brăila and Galaţi, which handle foods, minerals and manufactured goods. All Romanian ocean-going traffic is handled by Constanţa.

International Trade. Fresh and processed foods are still important exports, but minerals and petroleum have been surpassed as exports by engineering products and manufactured goods. The industrialization which made this possible has necessitated, in turn, a large increase in imports of machinery, coal and ore.

While the Soviet Union is still the major trading partner, West European countries (Italy, West Germany and Great Britain) are seriously challenging Czechoslovakia and East Germany for second, third and fourth place.

UNION OF SOVIET SOCIALIST REPUBLICS

Area: 8,649,412 sq mi (22,402,000 km²)
Population: 245,066,000
Population Growth Rate: 1·1%
Capital: Moscow, pop 7,172,000
Language: Russian. Numerous regional languages
Religion: Officially discouraged. Russian Orthodox, Armenian, Georgian, Baptist and others
Currency: Rouble = 100 kopeks

The Union of Soviet Socialist Republics (popularly called the USSR, the Soviet Union, or Russia), is by far the world's largest political unit. In the course of its history, it has grown from a small principality to a world super-power covering approximately one-sixth of the earth's land surface. Russia became the first communist state when the Bolsheviks (communists), headed by Lenin, assumed power in 1917. Today, the Soviet Union is more than twice the size of Communist China, nearly three times as big as the United States or Australia and more than 90 times the size of the United Kingdom.

From its western limits, where it has boundaries with six European states (Norway, Finland, Poland, Czechoslovakia, Hungary and Romania), the country sweeps eastwards for nearly 6,000 mi (10,000 km) to the shores of the Pacific Ocean and the Bering Strait, which separates the USSR from North America. Maximum distances from north to south vary from 1,800 to 3,000 mi (2,900–4,800 km). In the north there is a lengthy coastline on the Arctic Ocean, and in the south long land boundaries with Turkey, Iran, Afghanistan, China and Mongolia. The great bulk of this vast territory is in high latitudes, at least three-quarters of the area lying north of the 50th parallel.

The population of this enormous area is now over 246 million: only Communist China and India have greater populations. The USSR today is a federation of 15 Soviet Republics, each representing the territory inhabited mainly, though not exclusively, by one of the country's major ethnic groups. In theory every constituent republic has the right to secede from the USSR, though in fact all are inextricably bound by the pyramidal communist system of government headed by the Kremlin in Moscow. Externally, the Soviet Union exerts immense pressure on the lesser states of Eastern Europe whose communist governments were mostly installed subsequent to Russian occupation in and after World War II. There are also numerous smaller political units – Autonomous Republics, Autonomous *Oblasts* (regions) and National *Okrugs* (areas) – representing the homelands of smaller minority groups.

The largest and most important of the republics is the Russian Soviet Federated Socialist Republic (RSFSR), which covers more than three-quarters of the whole country and has a population of 130,700,000. Its capital, Moscow (7,172,000), is also the capital of the USSR and the world's fifth-largest city.

The other 14 republics can be divided into four main groups: 1) the western republics of the Ukraine, Belorussia and Moldavia; 2) the southern Caucasian republics of Armenia, Azerbaijan and Georgia; 3) the Central Asian republics of Kazakhstan, Kirgizia, Tadzhikistan, Turkmenistan and Uzbekistan; 4) the Baltic republics of Estonia, Latvia and Lithuania (whose incorporation into the USSR in 1940 has still to be officially recognized by the United States and some other countries).

With abundant and immensely varied natural resources, the USSR is virtually self-sufficient. In agriculture and industry, its output has become second only to that of the United States. However, the USSR still has a long way to go to match the living standards of affluent Western societies. Prolonged concentration on the development of heavy industry has resulted in the relative neglect of consumer industries. The USSR has nonetheless made remarkable economic, educational and technological progress. Russian achievements in science and technology continue to win worldwide acclaim, and despite economic shortcomings Russians enjoy a standard of living far higher than before.

The Land

In terms of surface relief, the USSR may be divided into two roughly equal parts along the line of the Yenisei River. In the western half of the country, lowlands predominate, dissected by low plateaux and hill ranges and bounded in the south by a series of high mountain chains. East of the Yenisei, though there are extensive lowlands, notably along the Arctic coast, mountains and plateaux are predominant. The USSR has over 100,000 rivers and 270,000 lakes. The lakes are unevenly distributed, most of them being in the north-western and south-eastern parts of European Russia, in western Siberia and in the northern areas of eastern Siberia.

The USSR can be divided into several principal geographical regions.

The East European Plain occupies most of the territory west of the Ural Mountains. It is bounded by the Carpathian Mountains and the Central European Plain in the west, the Baltic, White and Barents seas in the north, and the Black and Caspian seas in the south.

Practically the whole of this large region is less than 1,000 ft (300 m) above sea level, with the exception of certain hill masses – such as the Valdai Hills, the Central Russian Uplands, the Volga Heights and the Ufa Plateau. In the extreme north, Mount Khibiny on the Arctic Kola Peninsula reaches 3,907 ft (1,191 m). The region is drained to the Baltic

The Church of the Resurrection in Leningrad, the second largest city in the Soviet Union.

by the Western Dvina, to the Arctic Ocean by the Northern Dvina and the Pechora, to the Caspian by the Volga (at 2,290 mi; 3,690 km – Europe's longest river) and to the Black Sea by the Dnieper and the Don. In the north-west are lakes Ladoga and Onega.

About 75% of the Soviet Union's population live in the East European Plain. The region has highly fertile soils and many of the country's principal industrial centres.

The Caucasus Mountains extend for 745 mi (1,192 km) north-west to south-east across the isthmus between the Black and Caspian seas. In the north, the main Caucasian range runs from the Kerch Strait to the Apsheron Peninsula, separating the very distinctive Transcaucasian region from the rest of the USSR. Noted for its wild and majestic alpine scenery, most of the range rises above 10,000 ft (3,000 m), culminating in Mount Elbrus (18,510 ft; 5,633 m), Europe's highest peak. The main range is separated by the lowland regions of Rioni and Kura from the Lesser Caucasus, which comprises a series of mountain ranges and volcanic plateaux along the borders with Turkey and Iran.

The Ural Mountains are traditionally the boundary between Europe and Asia, though they present a relatively minor barrier to movement. Stretching nearly 1,500 mi (2,400 km) from north to south, they consist of a series of geologically old hill ranges worn down by erosion to heights of between 1,500 and 2,250 ft (500–785 m). There are only occasional ridges and peaks above that level, reaching their highest point with Mount Narodnaya (6,214 ft; 1,894 m) in the north. The Urals are rich in minerals and have many important industrial centres.

The West Siberian Plain is the largest area of continuous lowland in the USSR if not in the world. It stretches east from the Urals to the Yenisei River and south from the Arctic Ocean to the Kazakh Hills, covering about 1 million sq mi (2·6 million km²). The whole plain is less than 600 ft (200 m) above sea level, and more than half the area is below 300 ft (100 m). It slopes gently northward and is drained by the large, slow-flowing Ob and Irtysh rivers (which together form the world's fifth-longest river, 3,460 mi; 5,540 km). The region contains vast expanses of marshy, ill-drained land, particularly in the north.

The Aral-Caspian Lowland lies south of the West Siberian Plain, extending westwards between the Urals and the Caspian Sea to the Stavropol uplands bordering on the northern Caucasus and southwards to the mountain fringes of Central Asia. The Caspian and Aral seas and Lake Balkhash are in this region, and the principal rivers are the Volga, Ural, Amu Darya, Syr Darya and Chu.

There are grassy plains and plateaux, but much of the region is semi-arid. Two great sandy deserts, the Kara Kum and the Kyzyl Kum, lie in the south. Though most of the region's rivers flow towards the Aral Sea, only the Amu Darya and Syr Darya actually reach it, many of the smaller rivers disappearing into the desert sands. There are large salt-steppes around the northern Caspian – where the Karagiye Depression on the Mangyshlak Peninsula forms the lowest point in the USSR, 433 ft (132 m) below sea level.

The Kazakh Uplands separate the West Siberian Plain from the lowlands of Soviet Central Asia, extending southwards to Lake Balkhash. They comprise a series of plateaux and hill ranges, with isolated heights and steep, rugged valleys. Heights range mainly between 1,500 and 3,000 ft (500–1,000 m).

The Central Asian Mountains close in the southern fringes of the Aral-Caspian Lowland, separating Soviet Central Asia from neighbouring Iran, Afghanistan and China. This is a region of high plateaux and ranges, with several major ranges over 10,000 ft (3,000 m) protruding westwards into the lowlands, and between them deep valleys and basins. In the extreme south, the lofty Pamir ranges include Peak Communism (24,590 ft; 7,495 m), the Soviet Union's highest mountain.

The Central Siberian Plateau occupies most of the area between the Yenisei and Lena rivers: it has an average height of 2,000 ft (600 m), but slopes northwards to the coastal plains of the Arctic Ocean. It contains valuable mineral deposits. The highest mountain ranges are the Sayan (over 9,800 ft; 3,000 m) and Baikal (over 6,500 ft; 2,000 m) ranges in the south. The latter fringes Lake Baikal, the world's largest and deepest freshwater lake – over 5,700 ft (1,740 m) deep and 11,780 sq mi (30,500 km²) in area.

The East Siberian Uplands stretch from the Lena River in the west to the Pacific Ocean in the east and from the Arctic Ocean in the north to the Amur River in the south. The area is predominantly mountainous, with lowland regions along the middle reaches of the Lenask, along the Indigika and Kolyma rivers in the north, and along the Amur and Ussuri rivers on the border with China.

The north-east has high mountain ranges rising to about 10,000 ft (3,000 m) and highland plateaux, curving off from the coast of the Sea of Okhotsk and extending into the Kamchatka Peninsula. The Kamchatka Peninsula has 22 active volcanoes and many more dormant ones.

Climate. The USSR has, on the whole, a continental climate, with sharp contrasts between winter and summer temperatures. Winters are long and severe. More than half the country is snow-covered for six months of the year. The Arctic coast is icebound for most of the year, and the rivers are frozen over for between two and eight months. Summers, however, are generally warm.

In eastern regions of the USSR, the climate is relatively severe, while the northern shores along the Arctic Ocean have a cold polar climate. The climate in the south is sub-tropical.

With the exception of the southernmost regions, January mean temperatures are below freezing over most of the country. Winter cold intensifies towards the interior of Siberia: while most of the European areas have January means between 3°F and 25°F

The Bolshoi Theatre, Moscow, one of the most celebrated of Russian cultural institutions.

(−16°; —4°C), they range from −10°F to 3°F (−24°; −16°C) in western Siberia and fall below −40°F (−40°C) in north-eastern Siberia. On the north-eastern edge of Siberia, the temperature can drop as low as −97°F (−72°C). The temperature rises farther east, as one approaches the Pacific.

In the summer months, sea-level temperatures are closely related to latitude. Mean July temperatures range from 36°F (2°C) in the far north to 89°F (32°C) in the hottest part of the Central Asian desert.

Precipitation over most of the country is moderate or light. The wettest areas are in western Transcaucasia, where annual rainfall exceeds 90 in (2,286 mm), and in the far east, where the summer monsoon brings annual totals of 30–40 in (762–1,016 mm) along the coast. Apart from mountain areas, the rest of the country averages less than 26 in (660 mm) annually. Over most of the East European and West Siberian plains annual totals are 16–20 in (406–508 mm), but central and eastern Siberia receive only 10–16 in (254–406 mm), while most of Soviet Central Asia has less than 8 in (203 mm).

Vegetation. Over most of the country, vegetation is arranged in a series of fairly well-defined latitudinal belts which generally reflect climatic variations from the Arctic north to the deserts of Central Asia.

The *tundra*, occupying a northern coastal belt which varies in width from 100 to 250 mi (160–400 km), has a cold and damp climate. Here, the subsoil is permanently frozen, only the top few feet thawing out in summer, and drainage is extremely poor. The highly acid, waterlogged soils support only a scanty vegetation of mosses, lichens and, in more favoured districts, coarse grasses, dwarf trees and shrubs. Tundra conditions also occur in high mountain areas in lower latitudes.

The *taiga*, or coniferous forest zone, stretches right across the USSR in a belt 600–700 mi (960–1,120 km) wide from the Baltic to the Pacific, and is the largest of the natural vegetation zones, commonly subdivided into a western section occupying much of the East European and West Siberian plains, and an eastern section covering most of Siberia and the Far East. In the west, the principal components of the forest are pine and fir, with spruce, larch and birch; while the east beyond the Yenisei is dominated by the larch. Zones in which birch is the main species occur along the southern fringes of the taiga in western Siberia and in Kamchatka. Forest is not continuous throughout the taiga zone. In poorly drained areas, particularly in the western taiga, there are large stretches of peat bog, while the drier parts of the eastern taiga, notably in the Lena valley, carry a coarse grassland vegetation.

The *mixed forest* zone lies between the taiga and the steppe and is most extensive in the European part of the country, though mixed forest also occurs in the Amur lowlands of the Far East. As climatic conditions ameliorate towards the south, the proportion of deciduous trees increases until they become dominant. In contrast to the taiga, the great bulk of which is still in its natural state, the mixed forest zone has been used for agriculture since prehistoric times and much of the natural vegetation has been cleared.

The *wooded steppe* is a narrow transitional zone between the mixed forest and the steppe proper. Soils are more fertile than those to the north.

The *steppes* of the USSR, with their rich *chernozem*, or black earth soils which are the most productive in the country, run in a continuous belt from the western frontier to the Altai mountains. Steppe vegetation and soils also occur farther east, in basins along the southern fringes of Siberia. Relatively little of the natural grassland remains, particularly in the European sector, having been replaced over large areas by arable farmland.

The *semi-desert* region emerges as conditions become hotter and drier towards the south, and both soils and vegetation become poorer. *Chernozems* are replaced by alkaline chestnut soils with scanty grass cover.

The *desert*, with its still drier conditions, has even poorer soils and vegetation, but there is a considerable variety within the desert zone. Clay deserts, like those of the Ust Urt plateau between the Aral and Caspian seas, are the poorest of all, with virtually no vegetation cover. Sand desert, interspersed with stony areas, occupies most of the southern half of this region and supports a scanty natural vegetation of poor grasses and saxaul, though dense thickets of poplar and tamarisk may occur along water courses.

The *humid sub-tropical lowlands* of Transcaucasia, including the small Talysh lowland on the Caspian and the large Rion lowland on the Black Sea, have special conditions. Hot summers, mild winters and heavy rainfall give luxuriant natural vegetation, now largely replaced by intensive cultivation.

Mountain areas show a great diversity of soil and vegetation conditions. In the Caucasus, various types of forest, mainly deciduous, occur at intermediate levels, with extensive mountain grasslands above 7,000 ft (2,130 m). In the mountains of Soviet Central Asia, where conditions are much drier, there is relatively little forest and grassland types of vegetation are dominant. The mountain regions of Siberia are mainly under coniferous forest, with tundra at higher levels, particularly in the north and north-east.

Wildlife. Apart from numerous species of birds in summer, the most conspicuous wildlife in the tundra comprises lemmings, Arctic foxes and a small number of reindeer. Common animals of the taiga are the wood lemming, wolverine, red-backed vole, sable, brown bear and elk. The mixed forest regions have wildcats, deer, pine martens and polecats, with leopards, tigers and black bears in the Far East. The steppes have many reptiles, and mammals that include susliks, marmots, gazelles and jerboas. Caracal lynxes, gerbils, shrews, wild asses, wildcats, wild boars, jackals, Persian gazelles and numerous birds and reptiles are found in the desert areas. Animals peculiar to mountain regions include snow leopards, wild goats, grouse and ibexes.

The People

The USSR is a multi-national state with 126 different nationalities and national groups, speaking as many different languages. Some national groups, like the Yukagirs of the Yenisei or the Itel'mens of Kamchatka, number only a few hundred persons. Among the ethnic minorities there are some 2 million Jews, over 1·8 million Germans and over 1·1 million Poles.

Nearly three-quarters of the total population belong to the three main Slavic groups. Russians constitute the largest and most widespread group (over 53% of the total), and predominate in the government of the country. The Ukrainians constitute just under 17% of the population and the Belorussians, also in the western part of the USSR, represent nearly 4%.

The Uzbeks (3·8%) are the most numerous of the Turkic peoples. They live in Soviet Central Asia, along with other Turkic groups such as the Kazakhs, Kirgiz, Turkmens and Karakalpaks. The Tatars, Baskirs and Chuvashes live along the lower Volga River valley, while Siberian Turkic groups include the Yakuts, Siberian Tatars and Altai.

Siberia also has Mongol-speaking Buryats, Tungusic-speaking peoples like the Evenki, and smaller groups such as the Chukchi, Koryaks and Eskimos. Among the peoples of the Caucasus are the Armenians, the Georgians and the Turkic-speaking Azerbaidzhanis.

Baltic peoples include the Latvians and Lithuanians (who speak related Indo-European languages), and the Estonians, a Finno-Ugric people like the Finns, Karelians, Mari and Udmurts. Other peoples of the Soviet Union include the Moldavians (who are related to the Romanians) and the Tadzhiks, and Iranic-speaking people of Soviet Central Asia.

Population distribution and growth. The Soviet Union's population is very unevenly distributed. About 65% live on the East European Plain, where there is a zone of continuous settlement extending from the western border to the Urals. Throughout this area, densities are above 65 persons per sq mi (25 per km^2), reaching much higher levels around the main urban-industrial agglomerations and in the

Khiva, in Uzbekistan, was a Moslem city ruled over by the Mongol Golden Horde and later Tamerlane. It remained a Khanate until 1920.

Wheat fields in the Ukraine. The USSR is the world's leading producer of wheat.

most productive agricultural regions. One-fifth of the entire population of the USSR lives within a 300 mi (500 km) area around the capital, Moscow. The Moscow Region itself has a population density of about 647 per sq mi (250 per km²).

In contrast, population densities fall off very rapidly to the north and south-eastwards towards the Central Asian desert. The Asian part of the USSR has an average of less than 10 persons per sq mi (3·8 per km²).

In addition to the main settled area of the East European Plain, there are two outlying areas of population concentration. More than 12 million people live in the Transcaucasian republics, mainly in high densities in the agricultural lowlands, while the bulk of Soviet Central Asia's 20 million inhabitants is to be found in the narrow piedmont zone between the mountains and the desert.

The situation in Siberia and the Far East is particularly striking: occupying well over half the territory of the USSR, these areas have only 25 million inhabitants – little more than 10% of the total population. Most of the population of these regions is to be found in the south, along the line of the Trans-Siberian Railway, leaving vast areas to the north so thinly populated as to be virtually uninhabited.

Between 1959 and 1970, the population of the USSR as a whole rose by 15·8%, though there was a marked reduction in the annual growth rate during the 1960s, due to a decline in the birth rate and a marginal increase in the death rate. There are, however, pronounced regional variations in the birth rate. Birth rate decline has been most marked among the Slav populations of the RSFSR, the Ukraine and Belorussia, and in the Baltic republics. On the other hand, non-Slav ethnic groups, particularly those of Central Asia and to a lesser extent the Transcaucasians, continue to show very high birth rates. The addition of net gains due to migration has meant that Kazakhstan, the Central Asian republics, Azerbaijan and Armenia have had population growth rates well above the national average.

Way of Life. Pre-Revolutionary Russia was mainly agrarian, but the Soviet Union's economic progress led to a radical change in the ratio between rural and urban populations. Today, some 56% of the population live in urban centres: living conditions in rural areas are considerably harder.

Most cities are overcrowded and the majority of urban families live in small apartments, often shared with other families. Television sets and cars are still luxuries, and some foodstuffs, notably meat, are expensive and not always available. Some everyday expenses, however, are notably lower than in the West. Rents, services (such as electricity and gas) and public transport are cheap. Health services and higher education are free. Cultural recreations and, particularly, sports are well catered for.

A feature of life in the USSR is the part played by women. They form more than half the country's labour force and are to be found in all kinds of occupations, from street-cleaning and construction work to science and technology. Most of the Soviet Union's doctors and teachers are women. There are thousands of government nursery schools to care for young children.

Education. The Soviet Union places great emphasis on high standards of education, particularly in science and technology. Educational programmes are supervised by the Communist Party and administered by the state: Marxist-Leninist doctrines are emphasized.

Education is free and compulsory from 7 to 15 or 16 years of age. There are more than 218,400 primary and secondary schools with a combined enrolment of over 58,300,000 pupils. There are over 4,000 professional, technical and trade schools, and also many schools for specially gifted children. The USSR has more than 700 institutes of higher education and some 40 universities, attended by about 4·5 million students. On completing their studies, students are obliged to spend three years working in their specialization in jobs assigned by the state, often in underdeveloped regions. Moscow University, founded in 1755, is the oldest and largest university, with more than 30,000 students.

There are three organizations aimed at bringing children and young people into the Communist Party and providing a core of dedicated party leadership. Children aged 7 to 9 join the 'Little Octobrists', going on from there to the 'Young Pioneers', and finally joining the Komsomol, or Young Communist League, which caters for those aged 15 to 28.

Religion. The Soviet Constitution promises 'freedom of religious worship' but in fact religious observance of any kind is discouraged. Nevertheless, many religious groups are still active. The largest group belongs to the Russian Orthodox Church which claims about 30 million adherents. Moslems constitute the second-largest religious group. There are substantial numbers of Armenian and Georgian Christians, Protestants, Roman Catholics, Jews and Buddhists.

The Economy

The estimated Gross National Product of the USSR in 1971 was about $536,000,000,000, of which 77·5% came from industry and 16·5% from agriculture. Only the United States produces more than the Soviet Union.

The USSR has a planned, heavily centralized, socialist economy. The government controls all the means of production and distribution, fixes wages and the prices of nearly all products, and directs all domestic and foreign trade. The chief controlling authority is Gosplan, the State Planning Commission.

Since 1928, by a series of five-year plans, the USSR has been transformed from an essentially agricultural country into the world's second most powerful industrial nation. Heavy industry has long been given priority but production of consumer goods has been increasing and further increases were promised in the 1971–1975 plan. Since 1965, there has been some cautious relaxation of the over-centralized system of industrial management. Industrial plant managers have been given greater freedom to operate on a profit basis related to consumer needs, with the plants themselves benefiting from the achievement of sales and profit targets.

Agriculture. Only about 10% of the land of the USSR is under crops and just over 25% is used for agriculture of any sort, including grazing. These low figures reflect the existence of vast areas where farming is difficult or impossible owing to the nature of the physical environment. Such areas are most extensive in the Arctic and Sub-Arctic zones of Siberia and the Far East and in the arid interior of Soviet Central Asia. As a result of these factors, approximately 68% of the sown area of the USSR is found in the European section of the country, 18% in Kazakhstan, the Central Asian republics and Transcaucasia, and only 14% in the whole of Siberia and the Far East. Whereas about 25% of the land of the European USSR is under crops, the proportion falls to about 8% in Kazakhstan and Central Asia and is little more than 2% in Siberia and the Far East.

Cereals, of which wheat and barley are by far the most widespread, account for nearly 60% of the sown area and fodder crops (including sown grasses) for nearly 30% more. The remainder is devoted to industrial crops, potatoes and vegetables. The USSR is the world's leading producer of wheat, barley, rye, flax, potatoes and sugar beet. Recently, however, drought has led to serious shortages of grain, and large quantities have had to be imported.

Livestock numbers rose rapidly over the past decade, after a long period during which they had increased only slowly. In 1971, the USSR had 99·2 million cattle, 137·9 million sheep and 67·5 million pigs.

Soviet citizens throng through a square in Moscow on May Day, a traditional socialist holiday around the world.

The tundra and much of the taiga are little used for agriculture. In the northern part of the European USSR and over all but the southern fringes of Siberia and the Far East, the density of population is very low and its distribution is related mainly to non-agricultural activities such as lumbering, mining, reindeer herding, hunting and fishing. A few small, scattered patches of agricultural land in this zone are devoted to cereals and cattle rearing and there is some dairying and vegetable production around the larger towns, but the agricultural potential and productivity of this vast territory are low. Along the valleys of the Lena and its tributaries more favourable conditions permit the rearing of cattle on natural pastures, mainly for beef, and the cultivation of spring-sown crops of barley, rye and wheat.

A zone of mixed farming covers large parts of the European USSR, a smaller area of western Siberia and the lowlands of the Amur and Ussuri valleys in the Far Eastern region. These regions are characterized by relatively intensive forms of mixed farming, producing large quantities of a wide range of crops and livestock products. The greatest diversity is found in the European section where, on the arable side, the emphasis is on cereals (wheat, oats, barley and rye), potatoes and fodder. Flax and sugar beet are also important. Livestock farming is also well developed, particularly the dairying and pig-rearing sides. In most districts, these varied activities are closely integrated in modern mixed farming systems which support a dense and numerous population. In the west Siberian section of this zone, agriculture is less intensive, population densities are much lower and farmland occurs in scattered blocks separated by stretches of forest, grassland and marsh. Grain, beef and dairy goods are the main products here. In the Far Eastern mixed farming zone, cereals and cattle are again dominant in the Amur and Ussuri lowlands. The main crops include wheat, oats, rice, maize, soya bean, sorghum and sugar beet.

A number of characteristic features serve to distinguish the agriculture of the steppelands from that of the forest zones to the north. The advantages of the rich *chernozem* soils and high summer temperatures are to some extent offset by the problem of drought, which becomes progressively more severe towards the east and south. Thus there is a tendency for the yields of most crops to be lower, particularly in the east. While the mixed forest zone has been the scene of settled farming since the early part of the Christian era at least, the steppelands remained a zone of nomadic pastoralism until modern times. Conversion to arable farming in the Ukraine took place mainly in the 18th and early 19th centuries, and was even more recent in areas farther east. In the initial stages of this conversion, the steppe was devoted predominantly to the cultivation of wheat. During the present century, however, other crops have been introduced on a large scale and the livestock side has become progressively more important. Thus this zone, too, is now a zone of mixed farming, but one in which cereals remain dominant to a greater degree than elsewhere in the Soviet Union.

Effective precipitation, population density, the intensity of farming and the range of agricultural products all decrease from west to east, so that the steppelands may be further sub-divided on this basis. The most intensively farmed and most productive section is that to the west of the Volga, covering much of Moldavia, the Ukraine and the North Caucasus region. Here, the main crops are wheat and barley, but large areas are also devoted to other cereals (including maize, grown mainly in Moldavia, the western Ukraine and the Kuban lowlands), sugar beet, sunflowers, hemp and potatoes. Large numbers of cattle, both beef and dairy, are reared, mainly on the basis of fodder crops, sown grasses and the residues from industrial crops.

In the steppelands of the Volga region, cereals and cattle are again the dominant items, but there are larger areas of natural pastures and the area devoted to crops other than cereal diminishes. This transition is taken still further in the broad belt of steppeland which stretches across western Siberia and northern Kazakhstan from the Urals to the Altai. Development of this region dates from the end of the 19th century, but the most rapid expansion of arable farming took place as recently as 1954–60, when huge areas were ploughed up under the Virgin Lands Scheme. There is a greater degree of specialization in cereal farming than in any other part of the country, despite the low and variable rainfall which results in great fluctuations in output from year to year. During the 1960s, however, the emphasis on cereals was significantly reduced as a result of expansion of the livestock side, based on an increasing area of fodder crops as well as on the extensive natural pastures which remain. Lastly, beyond the River Ob, pockets of steppe and wooded steppe in southern Siberia are devoted to a similar kind of grain and livestock farming.

Muscovites enjoying chess at tables set up in the open-air in Gorki Park. Chess is perhaps the 'national sport'.

The arid zones of Kazakhstan and Soviet Central Asia are devoted entirely to livestock farming with an overall emphasis on sheep, though cattle are also important in the semi-desert. Nomadic pastoralism has been replaced, during the Soviet period, by a system in which the stock depend to a greater extent on hay and fodder crops produced in the more favoured areas, rather than on the scanty natural pastures. Livestock densities are in any case low compared with those of the forest and steppe zones.

In a number of southern regions, particularly favourable physical conditions support agricultural systems in which cereals and livestock play a relatively minor role and the emphasis is on various forms of high-value crop production. The most important areas in this category are the irrigated lands of Soviet Central Asia and, on a smaller scale, eastern Transcaucasia. In these districts, the dominant crop on irrigated land is cotton, followed by lucerne, maize, rice and jute. The lower hill slopes support dry-farmed cereals and vines and there are also extensive areas of orchards and vegetable gardens. The tradition of irrigated farming extends over millennia and these areas support the highest rural population densities in the whole of the USSR.

In the trans-Carpathian Ukraine, in central Moldavia and in the Crimea, high summer temperatures and a long frost-free period permit the intensive cultivation of fruit, vines and tobacco; while in the humid lowlands of western Transcaucasia conditions are sub-tropical and the main crops are tea, citrus fruit, tobacco, maize, and vegetables. The lower part of the Volga valley is devoted to vegetables and melons.

In the Caucasus and the mountains of Soviet Central Asia and southern Siberia, the dominant activity is the raising of cattle and sheep, which often involves seasonal movement from high-level summer pastures to winter grazing in the adjacent valleys and basins.

The USSR has a unique system of agricultural organization. Following the Revolution, the old system of large estates and small peasant holdings was completely replaced by two new types of organization, the Collective Farm (*kolkhoz*) and the State Farm (*sovkhoz*). Under the collective system, a large block of land, allocated during the collectivization process to the peasants of a particular district, is worked as a single farm, management being in the hands of a committee elected by the members of the collective. These collective farms sell much of their produce to the state but a proportion of the output, together with a share of the cash received from sales, is allocated to each member of the collective according to the amount of work he has put in on the collective's lands. On the State Farm, however, the labour force consists of paid employees.

In both cases, individual workers have their own 'private plots', separate from the collective or state farm lands. On these private plots they work as individuals, consuming what they produce or selling it on the free market. The existence of this 'private sector' in agriculture has often been critized on the grounds that it tempts the individual to put more effort into his private plot than into his work on the collective or state farm. Private plots undoubtedly play a major role in some branches of Soviet agriculture: although they cover only 3·5% of the total sown area, they include nearly half the land devoted to vegetables and fruit and support 40% of the Soviet Union's dairy cattle, 29% of its pigs and 20% of its sheep. Thus the Soviet population depends quite heavily on the private sector for its supplies of vegetables, fruit, meat and dairy products.

Both collective and state farms are very large units. There are now some 33,600 collectives with an average sown area of 7,290 ac (2,950 ha) each. State farms are even larger: there are 14,994 of these, with an average sown area of 15,113 ac (6,116 ha).

Another striking feature is the great size of the rural and agricultural populations, which are much larger than in other developed countries. In 1970, 105·7 million people – 44% of the Soviet population – lived in rural areas and the farm labour force exceeded 45 million, more than a third of the working population. This reflected the low productivity of Soviet agriculture and the major role which farming still plays in the life of the people.

Forestry and Fisheries. The Soviet Union's vast forests make it world leader in lumbering. The largest areas are in the Asian regions, along the northern coasts, in the Urals and in the north-western part of the country. However, transportation problems hamper the exploitation of the more remote forest regions, and most cut timber comes from areas west of the Urals – where there is about a quarter of the country's forests.

Soviet fishing fleets range far afield in the Atlantic and Pacific oceans. The Baltic, Black and Caspian seas are now less important, though sturgeon caught in the Caspian still provide the highest grade of caviar. Today, most of the annual catch – which exceeds 6,000,000 tons – comes from the northern and Far Eastern seas. The USSR has large whaling fleets, and ranks second only to Japan in the size of its whale catch.

Minerals. The Soviet Union is extremely well endowed with industrial raw materials and energy resources, and there are few items in which it is not self-sufficient. The USSR rivals the USA as the world's leading coal producer and had a 1970 output of 464 million tons of hard coal (including 165 million tons of coking coal) and 160 million tons of lignite. The bulk of this production comes from a small number of major fields. Over 60% of the total is derived from four fields – Donbass, Kuzbass, Karaganda and Pechora – considered to be of 'all-Union' significance, in the sense that they produce quantities of coal well in excess of the requirements of local industry and thus have large surpluses for movement to other regions.

Prior to the Revolution, the only coalfield developed on a large scale was the Donbass in the eastern Ukraine. Although its share of total Soviet output has progressively declined and now constitutes little more than one-third, the actual volume of production from this field has continued to increase and now exceeds 200 million tons a year. Donbass coal, in addition to supplying the major industrial complexes of the Ukraine, is sent to most other regions of the European USSR.

The second producer is the Kuzbass, in western Siberia, which now produces nearly 100 million tons a year. First developed on a large scale as a result of the establishment of the Ural-Kuznetsk industrial combine in the 1930s, this field has continued to increase its output ever since and has large reserves to support further expansion. The Kuzbass supplies coking coal to the Urals as well as to its own iron and steel industries, and sends smaller quantities to eastern Siberia and Soviet Central Asia.

The Karaganda field (30 million tons) was also first developed in response to demand from the Urals, and continues to supply coal to that region as well as to the Central Asian republics. The output from northern Kazakhstan has risen rapidly since 1950 with the development of a second major source at Ekibastuz in the Irtysh valley. Output from the Pechora field is less than 20 million tons, but the field provides an important additional supply to the northern regions of the European USSR.

The other coalfields are of local rather than all-Union significance. In terms of volume of production, the several fields of the Urals take a high place with a combined annual output in excess of 60 million tons. However, a large proportion of this is low-grade coal, including much lignite, and there is a marked shortage of coking coal. The Moscow basin's annual output of more than 40 million tons is entirely in the form of lignite, used locally for electricity generation.

The output of the east Siberian and Far Eastern regions together exceeds 75 million tons: this comes from a number of widely separated fields of which the most important are the Cheremkhovo, Bukachacha, Bureya, Sakhalin and Kansk-Achinsk fields. Vast reserves of coal exist in the more northerly parts of these two regions, particularly in the Tunguska and Lena basins, but as yet these reserves are virtually untouched. Finally, mention should be made of the small-scale but locally important production of the various fields in the Central Asian and Transcaucasian regions, and of the Lvov-Volynsk and Dnieper valley fields in the Ukraine.

A subtropical health resort at Gagry where the Caucasus rise from the Black sea in Georgian SSR.

A striking feature of the last two decades has been the rapid growth of the Soviet oil and natural gas industries. In the case of oil, output has more than doubled since 1960, reaching a total of 353 million tons in 1970. More than 70% of the present production comes from the Volga-Ural field, which has been almost entirely developed since 1940. The bulk of the remainder comes from the Baku and North Caucasus fields which, in the inter-war years, accounted for some 80% of a much smaller Soviet total. While the Volga-Ural field is expected to remain dominant in the Soviet oil industry during the 1970s, rapid expansion may also be expected in a number of new fields which have come into production in recent years, notably the Tyumen field in the West Siberian Lowland and the Mangyshlak field in western Kazakhstan. Thus production is, and seems likely to remain, heavily concentrated in the western half of the country. At the moment, the only significant out-

Odessa, a Ukrainian port on the Black sea was the scene of the famous 1905 revolt of the sailors of the battleship 'Potemkin'.

put east of the Yenisei is that from the small Sakhalin field, though sizable reserves are also believed to exist in the Lena valley.

The production of natural gas has expanded even more rapidly than that of oil. In 1955, total Soviet output of natural gas was only 317,800 million cu ft (9,000 million m^3): it rose to 4,520,000 million cu ft (128,000 million m^3) in 1965 and 7,063,000 million cu ft (200,000 million m^3) in 1970. At present, about half the total comes from regions which are also major oil producers, for instance the Baku-North Caucasus and Volga-Ural zones. Rapid expansion is now under way in areas where there is relatively little oil, such as at Dashava and Shebelinka in the Ukraine and in the Central Asian republics of Uzbekistan and Turkmenistan. Large deposits have also been found in the northern part of the West Siberian Plain and are now being developed.

Oil and natural gas are now the leading sources of energy in the USSR, accounting for 58% of Soviet energy supplies as against 37% derived from coal. Their exploitation has involved the construction of a complex system of oil and gas pipelines linking the producing areas to the consuming centres. These pipelines cross the western frontiers of the USSR carrying Soviet oil and gas to its East European neighbours.

Certain broad zones are of particular importance in the production of ores. They include the Urals, the southern part of the European USSR, the Kola Peninsula, the Caucasus, Kazakhstan and the Central Asian republics, and a number of districts in southern Siberia. The following account deals first with ferrous and then with non-ferrous ores.

Production of iron ore reached 196 million tons in 1970. Over most of the Soviet period, the iron and steel industry has been based mainly on high-grade iron ores from Krivoi Rog in the Ukraine and from several sites in the Urals, notably near Nizhni Tagil and Magnitogorsk. In 1955, the Ukraine and Urals together accounted for well over 80% of all Soviet iron ore. Important supplementary sources included the deposits at Kerch in the Crimea, Atasu near Karaganda, Olenegorsk in the Kola Peninsula and Temir Tau and Tashtagol in the mountains to the south of the Kuzbass. Since 1955, however, these high-grade ores have been unable to support further expansion of the iron and steel industry and increasing use has been made of lower-grade ores which are often worked opencast. These include the ores of the Kursk Magnetic Anomaly, Kachkanar in the northern Urals, Rudny in north-western Kazakhstan and (since 1965) Zheleznogorsk in eastern Siberia.

The USSR is among the world's leading producers of several other metals used in the iron and steel industry, and production of most of these is concentrated at a small number of sites. Manganese, which is exported to several European countries, is produced almost entirely at Nikopol in the Ukraine and Chiatura in Georgia; chrome ore comes mainly from Khromtau in northern Kazakhstan; nickel is mined in the southern Urals, at Nikel in the Kola Peninsula and at Norilsk near the mouth of the Yenisei.

The production of non-ferrous metals is more widely distributed. Soviet output of copper is second only to that of the USA and until the 1930s took place mainly in the Urals. Since that time, while the Urals have remained important, production has expanded rapidly in other areas, notably at Dzhezkazgan and Balkhash in Kazakhstan, at Alaverdi in the Transcaucasus and, during the 1960s, at Almalyk in Uzbekistan. Copper is also produced along with nickel in the Kola Peninsula and at Norilsk.

Lead-zinc ores occur mainly in the south of the country, and these metals are produced near Ordzhonikidze in the Caucasus; at Achisay, Almalyk and other sites in Central Asia; at Salair in the Kuzbass and at Tetyukhe in the Far East. The most important single region is the upper basin of the Irtysh around Ust-Kamenogorsk.

Valuable deposits of bauxite are worked at Boksitogorsk in the north-western region, at several places in the Urals and at Arkalyk in northern Kazakhstan. However, these have proved insufficient to support the growing demand for aluminium and there is an increasing reliance on other sources of the metal, such as nephelite from the Kola Peninsula, alunite from the Transcaucasus and even kaolin from Uzbekistan. The large new aluminium works recently constructed at hydro-electric power sites in Siberia are expected to rely mainly on nephelite from Belogorsk and elsewhere, although at present their raw material comes mainly from the Urals and from Arkalyk.

Mention should also be made of tin, one of the few metals in which the Soviet Union is not yet self-sufficient. Several tin mines have been opened up in recent years, all of them in the east Siberian and Far Eastern regions. The remote northern parts of these two regions are believed to have large resources of many minerals. As yet, apart from nickel and copper mining at Norilsk, the only items to be developed in these inhospitable high-latitude areas are tin and gold.

Industry. The Soviet Union has developed an advanced industrial economy which is second only to that of the United States in overall production. This development has, however, been somewhat unbalanced in the sense that, until quite recently, effort has been concentrated on the heavier, more basic branches of industry (fuel and power, iron and steel, heavy engineering, chemicals, transport), while lighter industries producing goods for consumption by the population at large, together with agriculture, have been relatively neglected. The balance is now being redressed, and consumer goods industries are expanding rapidly, thus affording a higher living standard to the Soviet people.

Before the Revolution, such industrial development as did occur took place almost entirely in the European section of the country and this concentration of activity in the west remains a dominant feature, though the Soviet period has witnessed important industrial developments in the Asian regions. More than 70% of Soviet industrial production comes from a zone stretching from the Urals to the western frontier, and a number of industrial concentrations may be distinguished within this zone. These include the two metallurgical bases of the eastern Ukraine and the Urals, which between them produce some 70 million tons of steel out of a Soviet total of 116 million tons. These two areas are also dominant in the heavier branches of engineering, producing large quantities of equipment for use in other industries.

A third industrial concentration of a different type occurs in the Centre, the region focussed on Moscow. Despite a lack of industrial resources and a heavy reliance on other regions for its raw materials, the Centre plays a leading role in the more complex forms of engineering (machinery, machine tools, motor vehicles, electrical equipment), and in textiles, chemicals and a wide range of consumer goods industries. In addition to these three main concentrations, a variety of manufacturing industries are to be found in the larger cities of the European USSR, notably in Leningrad, the towns along the Volga and major regional centres such as Kiev, Kharkov, Minsk, Riga, Voronezh and many others.

Total industrial capacity outside the European sector is small, but has increased rapidly during the Soviet period. The most important single element is the iron and steel and heavy engineering complex of the Kuzbass in western Siberia. The only other steel producing districts of any significance are at Karaganda (in Kazakhstan) and in the Transcaucasus. The central Asian and Transcaucasian republics are of growing importance in non-ferrous metallurgy, chemicals, textiles and light engineering. Eastern Siberia and the Far Eastern region have vast industrial resources but have been little developed so far.

Electricity. Much attention has been paid by Soviet planners to the development of the country's electricity supplies, and the output of electric power has more than doubled since 1960, reaching a total of 740,400 million kilowatt-hours in 1970. This still represents a production figure of about 3,000 kwh per head of the Soviet population, a per capita output well below that of the USA and several West European countries.

Some 83% of Soviet electric power is derived from thermal generating stations based on coal, oil, natural gas, lignite or peat. Hydro-electric stations tend to fall into two main categories: a large number of relatively small plants have been built in hilly or mountainous areas such as the Caucasus, the Urals, and Central Asia, but the biggest stations are found along the major rivers of the East European Plain and Siberia. More than a dozen large barrages have been built across the Dnieper and Volga, converting those rivers into a series of man-made lakes, while in Siberia massive plants have been established on the Irtysh, Ob, Yenisei and Angara. Several large stations have also been built in the mountains of Soviet Central Asia. The largest untapped hydro-electric potential is now that of the major Siberian rivers, but the wisdom of further large-scale capital investment in remote, thinly populated parts of the country has recently been questioned.

International Trade. The USSR trades mainly with other communist countries in Europe, though its trade with the West is on the increase. All foreign trade is organized by the state and is administered by the Ministry of Foreign Trade and 29 import and export organizations. Policies and distribution of trade are largely determined by the Council for Mutual Economic Aid (COMECON). This communist economic planning organization was established in 1949, and its members are the USSR, Bulgaria, Czechoslovakia, Hungary, Poland, Romania, East Germany and Mongolia. Cuba, North Korea, North Vietnam and Yugoslavia are represented by observers.

The Soviet Union imports chiefly machinery, foodstuffs and various consumer goods. Imports of machinery and equipment represent some 35% of the total. The country's main exports are crude oil, coal, iron ore, iron and steel, paper, textiles, vegetable oil, motor vehicles, foodstuffs, lumber and clocks and watches. The main trading partners are East Germany, Poland, Czechoslovakia, Bulgaria and Hungary.

Transport. The USSR depends on its railways to a far greater extent than many other industrialized countries. The railways – which have about 82,830 mi (132,530 km) of track – are responsible for about two-thirds of all freight movement and half the passenger traffic, even though recent rapid growth in other forms of transport has led to a relative decline in their importance. Road transport, the main rival, is used mainly over short distances or where railways do no exist. Waterways and pipelines (for goods) and airways (for passenger transport) are used extensively for long distances.

The densest part of the railway network is in the European section of the country. Moscow, the transportation hub of the USSR, is a focal point for 11 main railways, including the famous Trans-Siberian route to Vladivostok on the Pacific coast. Outlying regions are served by a small number of main trunk routes with numerous short branches.

Road transport still accounts for only 5% of total freight movement, mainly over short distances to the nearest railhead. Over the last decade there has been rapid growth in passenger movement by road. This consists mainly of commuting journeys within the major urban agglomerations and medium-distance journeys by bus between the cities of the European USSR. There are about 300,000 mi (480,000 km) of surfaced roads.

Most of the traffic on the country's 89,974 mi (143,958 km) of inland waterways is on the waterways of the East European Plain. A network of linking canals permits traffic from the Baltic to the Black Sea, or from the Arctic to the Caspian. The Volga is the most heavily used river. The major rivers of Siberia provide links between north and south, but carry only a small amount of traffic.

The national airline, Aeroflot, has provided the most rapidly expanding form of passenger transport in recent years, with an international network and an extensive internal network.

A huge dam at Bratsk on the Angara river in south-central Siberia. Russia's potential for hydro-electric power is enormous.

THE ARCTIC

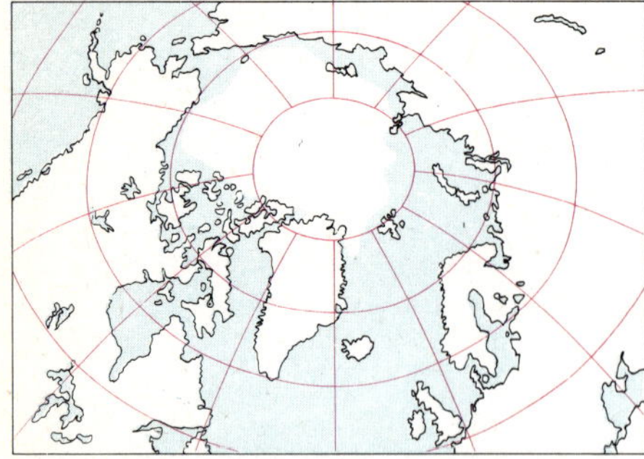

The Arctic is the vast northern polar expanse of land and sea embracing the 5·5 million sq mi (14·2 million km²) of the Arctic Ocean, innumerable islands, and northern parts of Asia, Europe and North America. Its name is derived from the Greek *Arktos* (the Great Bear, a constellation in northern skies).

Until the last century, apart from such hardy peoples as the Eskimos, only explorers, whalers and fur traders entered this grimly inhospitable region. Today the Arctic has an unexpectedly important place in man's affairs. The Soviet Union, Canada and the United States all have bases and defence posts there. In civil aviation, the north polar route greatly shortens flights between Europe and the Pacific and is used by various airlines. Information from Arctic weather stations aids meteorologists in forecasting storms approaching North America or Europe. The Arctic has rich fishing grounds and considerable mineral resources. Large industrial towns have been built within the Arctic Circle. Specially strengthened ships now sail the icy waters.

The Limits of the Arctic. Astronomers place the southern limit of the Arctic along the Arctic Circle (66°30′N), the latitude above which the sun does not rise during the winter solstice and does not set during the summer solstice. But most geographers prefer the tree line, the northern limit of forest growth, as the southern boundary of the Arctic.

The tree line roughly coincides with the July isotherm of 50°F (10°C), which embraces all places which have an average summer temperature of 50°F. This line thus brings within the Arctic broad belts in Asia, Europe and North America, the whole of Greenland and most of Iceland. South of this boundary is the Sub-Arctic, a region of cold winters and warm summers.

The Face of the Arctic. The hub of the Arctic is its great ocean, most of which is covered with pack ice throughout the year. The Arctic lands in and around the ocean are partly mountainous (Alaska for example), and parts are liable to earthquakes and tremors. The small Norwegian island of Jan Mayen has the world's northernmost volcano, Beerenberg (7,470 ft; 2,277 m), which erupted in 1970.

The typical landscape, however, is tundra – monotonously flat and treeless land with myriads of small lakes and bogs and perpetually frozen subsoil (permafrost). Temperatures for most of the year are below freezing point; winters are long and dark, and are intensified by the piercing cold winds which rage across the snow-covered ground. January temperatures average less than −40°F (−40°C) in north-eastern Siberia, and temperatures of −90°F (−68°C) and lower have been known at Verkhoyansk. In the two or three months of summer, when temperatures may be as high as 50°F (10°C), the tundra comes to life with mosses, lichens, low shrubs, cushion grasses and small flowering plants such as saxifrage and lupin. Their blaze of colour is as transient as it is vivid.

Long rivers flow across the tundra to the Arctic Ocean, among them the Mackenzie and Back rivers in Canada; the Colville River in Alaska; and the Ob, Yenisei and Lena in the Soviet Union.

Wildlife. Many birds and mammals found in the Arctic during summer are visitors that return south as winter sets in. But the polar bear, Arctic fox, musk-ox, snowy owl, raven and ptarmigan are among year-round denizens. The wide range of visitors includes caribou, grizzly bear, wolf, fox, wolverine, ermine and many migrant birds.

The peoples of the Arctic depend heavily on the musk-ox, caribou, and sea mammals (seals, walrus and whales) for food, clothing, tents, and so forth. Nowadays they hunt with rifles, and the increased slaughter effected by these modern weapons has seriously depleted the animal population; protective legislation has been introduced in most Arctic lands.

Peoples of the Arctic. Arctic North America and Greenland have only one indigenous racial group, the Eskimos, but the Eurasian Arctic has a number of distinct peoples, divided linguistically into three broad groups: the Uralians in the west, the Altaic groups farther east, and the Palaeo-asiatics.

Arctic man is typified by the Eskimos, whose forbears were living in the Arctic at least 12,000 years ago. Today they face all the inevitable difficulties of transition from an isolated primitive culture to a modern one of great mineral wealth featuring very different values and skills. In Alaska, the finds of potentially vast oil deposits in the Prudhoe Bay area attracted huge investments from international oil interests and seems likely to revolutionize the economy. Conservation interests have so far blocked pipeline construction however, and the feasibility of transporting oil by a year-round sea route remains to be proved.

The Soviet Arctic has a better developed general economy than other parts of the Arctic. It has two towns above the Arctic Circle, each with over 100,000 inhabitants, and 12 more with over 10,000. These do not face typically severe Arctic conditions; the port of Murmansk is ice-free throughout the year owing to warm ocean currents. Rich and varied mineral deposits have been proved at numerous Soviet sites. Norilsk is the centre of a vast complex producing nickel, copper, cobalt, platinum and coal. Natural gas is extracted at Tazovski, also above the Arctic Circle. The lumber industry, though based on Sub-Arctic regions, exports via the tundra and the North-East Passage.

Canadian eskimos building an igloo at Pond Inlet. A low tunnel at the entrance, below the level of the sleeping platform, prevents cold air from entering. A hole in the roof lets out smoke.

ASIA

Asia is the largest of the continents, occupying about one-third of the land surface of the earth and containing well over half of the world's population, much of it living in poverty yet expanding at an astonishing rate.

The term Asia is a Western invention originally describing only the plains of Ephesus in Asia Minor, and it imposes an artificial unity upon a vast area of the earth. Asia stretches from its mountain and sea boundaries with Europe and Africa in the west, to the shores of the Pacific Ocean in the east (an extreme distance of 6,000 mi; 9,650 km) and from the Siberian Arctic coast in the north to the tropical lands bordered by the Indian Ocean in the south (over 5,000 mi; 8,050 km).

This great region contains the world's highest peak – Mt Everest at 29,028 ft (8,848 m) – and the world's lowest exposed land surface – the shores of the Dead Sea at 1,294 ft (394 m) below sea level. While intense cold grips dry Arctic Asia, the south-east has sweltering heat and daily downpours. Vegetation ranges from lichens and mosses in nearly barren northern Siberia to the dense canopy of tropical rain forest found from India to Indonesia. Wildlife varies from polar bears and reindeer in the far north to tigers and crocodiles in the south. Asian people include both technologically advanced Russians and the primitive Negrito hunters of the Andaman Islands.

Geographers divide Asia into four regions. Soviet Asia (the north of the continent from the Urals east to the Bering Sea and south to Iran, Afghanistan, Mongolia and China); South-West Asia (the mainly dry lands from Turkey and Arabia east to Afghanistan); Monsoon Asia (lands with rainfall patterns largely geared to seasonal wind reversals and including all south, south-eastern and east Asian countries); and High Asia (the high dry plateau heart of Asia from Tibet to Mongolia).

Asia in History

Man's Earliest Civilizations. Mesopotamia, the low-lying land in south-west Asia between the Tigris and Euphrates rivers, was the homeland of the inventive Sumerians, more than 5,000 years ago. Here, also, the Babylonians and Assyrians later flourished, while farther north the Hittites built their cities. The Indus valley, in what is now Pakistan, was the site of another highly developed civilization more than 4,000 years ago at such centres as Harappa and Mohenjo-Daro. China's remarkable civilization began in the Hwang Ho (Yellow River) valley at about the same time. The plough, the wheel, writing, printing, navigation, silk and most domesticated animals including chickens, horses, cows, cats, and dogs – all these came from Asia, as did all the world's great religions.

Today, in terms of industry and the general standard of living, the West has far outstripped the East. But the great scientific, technological, and economic triumphs of the West came only after 1500. Before that date the contributions of Eastern civilizations to the advancement of learning and living standards were probably as great if not greater than those of the West.

From the Steppes of Central Asia. An outstanding feature on the map of Asia is the great rampart of the Himalayas, containing many of the world's highest mountains. This enormous natural barrier has had a permanent influence on the distribution and migration of Asian peoples, effectively impeding north-south movement between Central Asia and India. No such barrier exists between Central Asia, Europe and China, and throughout history movement east and west from Central Asia has brought some degree of cultural continuity to these areas.

Over the last 2,000 years nomadic peoples surged from the steppes of Central Asia to conquer and pillage. Some pushed east into China, and Huns, Tatars and Turks migrated westwards. During the 13th and 14th centuries the Mongols carved out a vast but short-lived empire, but were never able seriously to threaten India across the Himalayas or subdue the densely forested lands of South-East Asia.

Centuries earlier, Asia had made its mark in western Europe where Moslem armies conquered most of Spain in the early 700s, and Spain's last Moslem ruler was only ousted in 1492. In Eastern Europe, Istanbul remains today as a Moslem reminder of the Ottoman Turkish empire that ruled the Balkan area from the late 1300s to the early 1900s.

Europe and Asia. First-hand European contact with China dates only from the Middle Ages. European interest in East Asia quickened after 1295 when Marco Polo returned from his remarkable travels in China, and described the wonders to be found there – including coal, asbestos and 200,000 river craft laden with spices.

The most important travel was by sea. Maritime traders from the Western nations arrived on Asia's southern shores in the 16th century, following old-established Arab trade routes. European colonizers followed from the 17th century onward, competing with one another for the rich profits of Asian trade. Britain gradually gained control of India, France seized Indochina, and the Dutch Republic took the East Indies, while Russia began to push into Siberia, Central Asia, Mongolia and Manchuria. When World War II broke out in 1939, European political and economic domination of most of Asia was at its height.

The war, and the rise of nationalism in Asia, largely demolished the Europeans' imperial power. Today Asia outside Russia is ruled by its own peoples, whose governments range from capitalist to communist. Much of the continent remains remote and economically unexploited.

The Land

The most significant and immediately obvious fact about Asia is its vastness. Asia's roughly 16·9 million sq mi (43·7 million km^2) makes it more than four times the size of Europe, a third as large again as Africa and as big as North and South America combined.

Although Europe and Asia may be regarded as a single continent, Eurasia, it is usual to divide the two along the line of the Black Sea, the Caucasus Mountains, the Caspian Sea, the Ural River and the Ural Mountains. South-West Asia extends westwards to the Mediterranean and is separated from Africa only by the Red Sea and the Suez Canal. Great oceans wash Asian shores: the Arctic Ocean in the north, the Pacific Ocean in the east, and the Indian Ocean in the south.

Large parts of coastal Asia take the form of great peninsulas, for instance, Asia Minor, Arabia, the Indian sub-continent, mainland South-East Asia, Korea and Kamchatka. Japan, Indonesia, the Philippines and Ceylon comprise major offshore islands. But interior Asia forms one of the world's largest unbroken land masses.

The Structure of Asia. The continent has three main structural features: older blocks of the earth's crust, younger blocks, and Tertiary folds from 10 to 70 million years old. The older blocks are the foundation of much of Siberia, China, Arabia and peninsular India. The younger blocks stretch like a warped S from the Urals and the Kirghiz steppes in the west, across the Gobi Desert, and south to Malaysia and Borneo. The Tertiary folds occupy areas where mountain-building has occurred most recently, and thus cross Asia from Turkey to include the Tibetan Plateau and the Himalayas before turning south to embrace part of Sumatra.

The high mountains associated with these folds are Asia's most impressive surface features. In Turkey they reach 12,000 ft (3,700 m) in the Taurus and Pontic ranges separated by a plateau some 3,000 ft (900 m) above sea level. In the Caucasus several peaks exceed 15,000 ft (4,600 m). Farther east the Elburz and Zagros ranges partly enclose the plateau of central Iran. In the Pamirs, where the Hindu Kush and other ranges converge, heights exceed 20,000 ft (6,100 m) and from here several ranges fan out eastwards: the Karakoram with the peak of K2 at 28,250 ft (8,611 m); the Tien Shan and Kunlun Shan, which almost enclose the high plateau of the Tarim river basin; and the multiple ranges of the Himalayas forming the southern boundary of the great plateau of Tibet, which lies on average 15,000 ft (4,600 m) above sea level and from where several ranges poke eastwards into China.

Of all these ranges the Himalayas are the

most striking, partly because they are the highest, and partly because they rise so abruptly from the plains of India: you can stand at a mere 800 ft (244 m) above sea level in the steaming jungle of northern West Bengal and contemplate the majestic ice slopes of the Kanchenjunga group rising to 28,169 ft (8,585 m) little more than 40 mi (60 km) away.

East of India the ranges continue southwards to form the Arakan Mountains of Burma and to reappear in the Andaman and Nicobar Islands, and in Sumatra, Java and smaller islands eastwards. Another line of ranges extends from eastern Siberia through the Japanese islands and the Philippines to Indonesia.

Along the Tertiary fold belt the earth's crust is relatively weak. Adjustments are taking place all the time, and these are apparent in earthquakes and volcanic activity. Slight earth tremors occur daily. There have been disastrous earthquakes in Turkey and Iran in recent years, and also in Pakistan and Japan. Active volcanoes are prominent in Sumatra, Java, the Philippines and Japan.

High relief is not confined to the Tertiary fold belt. Along the west coast of peninsular India, for example, a raised plateau edge of older rocks runs close to the sea; from the sea the resultant steep slope appears as a dramatic mountain range. The relief of the older and younger blocks is generally more subdued and is typified by the plateaux of central Siberia, and by the vast lowland of western Siberia where large areas of hard rocks lie buried beneath glacial deposits and alluvium. Plateau country is also characteristic of central Arabia.

Asia's Great Rivers. The continent's big rivers are on the grand scale in both size and volume of discharge. Many have deposited immense spreads of alluvium along their middle and lower courses. In Siberia the Irtysh and Yenisei flowing to the Arctic Ocean have built up an alluvial plain extending 1,500 mi (2,400 km) from east to west. Other rivers constructing plains on a similar scale include the Indus in Pakistan, the Ganges and Brahmaputra in India, and the Hwang Ho and Yangtze in China.

The Yenisei, flowing some 3,690 mi (5,940 km), is Asia's longest river. Both the Ob-Irtysh system and the Yangtze flow for more than 3,400 mi (5,470 km), while the Indus and Brahmaputra approach 2,000 mi (3,200 km). The Salween, unbridged throughout its entire length of 1,750 mi (2,820 km), except for recent structures across its headwaters on the Tibetan border, flows in a narrow gorge to the Indian Ocean at Moulmein in Burma. The sacred Ganges (1,560 mi; 2,510 km) shares its huge delta on the Bay of Bengal with the Brahmaputra. Dams to store water for irrigation and to provide hydro-electric power are helping to add immensely to the potential of Asia's great rivers for advancing agriculture and industry.

Most Known Climates can be found in Asia, from polar north of the Arctic Circle to equatorial in islands astride the equator. Apart from latitude, Asia's climatic patterns are largely influenced by the distribution of land and sea: particularly the great expanse of the interior land mass and the seas that fringe the south and east.

Remote from any ocean and bordered on the south by a massive natural barrier of mountains, interior Asia is almost out of reach of the mild influence of ocean winds and the rain they bring. Thus Central Asia has great extremes of temperature between winter and summer, and it is also very dry.

South and South-East Asia, on the other hand, are largely open to moist sea breezes. This factor and nearness to the equator gives much of this part of Asia abundant rain and temperatures that remain high all year round.

The seasonal changes in rainfall that occur in much of Asia are largely influenced by changes in the atmospheric pressure over Central Asia. In summer, when the sun is directly overhead at noon as far north as central India and southern China, the thermal equator moves north too and the land mass of Central Asia gets very hot, warming the air above it which thus becomes less dense and rises, creating a huge low pressure area. Winds then tend to flow clockwise around and into this low pressure area from the cooler seas that edge the continent (except where blocked by mountains), bringing often torrential rain to much of India, South-East Asia, China and Japan.

In winter the situation is different. The thermal equator moves south, and the central land mass cools quickly to freezing point and below. The resulting contraction and sinking of the cold air above the land produces an atmospheric high pressure area from which cold dry winds flow outwards with a broadly anti-clockwise circulation. Thus the northerly winter winds that affect most of India bring no rain, and the winds affecting China from the interior ensure a cool or cold dry winter (but after passing over the sea, bring rain or snow to the west coast of Japan).

Atmospheric pressure changes in Central Asia therefore produce reversing wind systems or *monsoons*, that help to bring seasonal rains alternating with drought to many parts of southern and eastern Asia.

Extremes of Temperature. The greatest extremes occur in north-east Asia where temperatures range between −58°F (−50°C) in winter and 60°F (15°C) in summer. The north also has the longest, harshest winters. The seas along the entire Arctic coast are frozen except for a few weeks in summer. The rivers are frozen for at least five months throughout Siberia, and north of the 70th parallel for as much as eight months. The Aral Sea and the northern part of the Caspian also freeze.

Extreme temperatures are also characteristic of the high plateau of Central Asia and as far east as Peking there are sub-freezing winter temperatures but a July average as high as 80°F (27°C). However the high mountain core of Asia screens areas to the south from the cold winter air of the north and centre, and only light frosts normally occur in winter in northern India and the highland areas of Burma.

The Mediterranean coastlands generally have warm winters and hot summers. The highest temperatures of all occur in the hot, largely desert countries of Saudi Arabia and Iraq. At Baghdad, noon shade temperatures in summer may exceed 120°F (49°C) although the January average is only 49°F (9°C). In equatorial Asia, temperatures vary relatively little: in Singapore, for instance, they hover around 80°F (27°C) all year round.

Rainfall and the Farmer. Asia's average annual rainfall varies from practically nil in the deserts of the south-west and centre to 120 in (3,048 mm) and more in the south-east. Sumatra, Java, Borneo and the coastal slopes of Burma all receive over 120 in (3,048 mm). At Cherrapunji on the edge of the Assam plateau 4,300 ft (1,290 m) above sea level, the average is 457 in (11,420 mm), and 900 in (22,500 mm) and more have been recorded in exceptional years.

In the many islands of South-East Asia rainfall is well distributed throughout the year; near the Mediterranean rain tends to fall only in the winter; but winter drought, with rainfall concentrated in the summer months (June–October) is typical of much of Asia including most of India and Burma.

The distinction between areas which receive rain throughout the year and those in which rain is confined to one season has profound implications for the farmer. Where there is no long dry season, grain crop can be grown throughout the year. Two successive crops a year can be raised on the same field; if the first crop fails because of rain shortage, there is the chance that enough rain will fall for the second crop to flourish. Thus serious local food shortage may be avoided. Where there is a long dry season, grain can be grown under natural conditions only in the wet season. If rainfall is then low and the crops fail, the farmer must wait a full year for the next wet season for a new sowing of his fields. This failure of rainfall accounts for the devastating famines which have swept India and China in the past.

Deserts, Steppe and Forest. Asia's great belts of natural vegetation include all the major types on earth. The northernmost zone is the tundra which stretches along the harsh, cold Arctic coast; here subsoil is permanently frozen and no large trees can grow, although

ATLANTIC OCEAN
ARCTIC
Spitsbergen
Franz Josef Land
Barents Sea
NOVAYA ZEMLYA
Kara Sea
Arctic Circle
North Sea
Baltic Sea
White Sea
North Cape
Lisbon
Madrid
London
Paris
Amsterdam
Berlin
Oslo
Stockholm
Helsinki
Leningrad
Arkhangelsk
Bern
Prague
Vienna
Warsaw
Budapest
Rome
Corsica
Sardinia
Algiers
Tunis
Sicily
Athens
Istanbul
Kiev
Moscow
EUROPE
EAST EUROPEAN LOWLANDS
Narodnaya 6184
URAL MOUNTAINS
WEST SIBERIAN LOWLAND
Yamal Pen.
Sverdlovsk
Kuybyshev
Volga
Kama
Ural
Dnepr
Danube
MEDITERRANEAN SEA
BLACK SEA
CASPIAN SEA
Crete
Cyprus
Ankara
CAUCASUS MTS
Elbrus
Ararat 16,945
Damascus
Jerusalem
Syrian Desert
Mesopotamia
Baghdad
Tigris
Euphrates
Tehran
Demavend 18934
ELBURZ MTS
ZAGROS MTS
PLATEAU OF IRAN
Ust-Urt Plateau
Aral Sea
KIRGHIZ STEPPE
Turan
Kara Kum
Syr Darya
Amu Darya
Golodnaya Steppe
Lake Balkhash
Dzungaria
ALTAI MTS
GREATER ALTAI MT
Belukha 14783
Omsk
Tomsk
Novosibirsk
Baraba steppe
Tashkent
Urumchi
TIEN SHAN
TURKESTAN
Tarim Basin
Takla Makan
ALTIN
KUNLUN
Lenin Peak 23405
Soviet Union Peak
PAMIRS
HINDU KUSH
Kabul
Kashmir
Godwin Austin 28250
TRANSHIMALAYA
Lake Helmand
Indus
Sind
Thar
Delhi
Ganges
Varanasi
Mt. Everest 29028
BENGAL
Calcutta
Karachi
MAKRAN
Gulf of Oman
Persian Gulf
BAHRAIN ISLANDS
AFRICA
Libyan Desert
Tropic of Cancer
Nile
Cairo
An Nafud
ARABIA
Hejaz
Riyadh
Mecca
RED SEA
Khartoum
Rub al Khali
Yemen
Aden
Hadramaut
Bab el Mandeb
Perim
Gulf of Aden
KURIA MURIA ISLANDS
Socotra
Cape Guardafui
Addis Ababa
Equator
ARABIAN SEA
Bombay
VINDHYA MTS.
INDIA
THE DECCAN
WESTERN GHATS
EASTERN GHATS
Madras
Malabar Coast
Coromandel Coast
Nilgiri 8640
LACCADIVE ISLANDS
Palk Strait
Cape Comorin
Gulf of Mannar
CEYLON
Adams Peak 7352
Colombo
MALDIVE ISLANDS
BAY OF BENGAL
Lake Victoria
Zanzibar
Dar es Salaam
INDIAN OCEAN

ASIA
vegetation
snow-and icecaps
tundras and high mountain flora
mixed forest and northern coniferous forest, in Russia and Siberia, taiga
tropical rain forest
monsoon forest and thorn scrub
steppe- and mountain grassland
desert and semi-desert
cultivated areas
irrigated areas
oases
swamp
scale 1: 40.000.000
0
500
1000 st. miles
PACIFIC OCEAN
BERING SEA
SEA OF OKHOTSK
SEA OF JAPAN
YELLOW SEA
EAST CHINA SEA
SOUTH CHINA SEA
SULU SEA
CELEBES SEA
PHILIPPINES
MONGOLIA
CHINA
MANCHURIA
INDOCHINA
Tokyo
Peking
Tientsin
Seoul
Shanghai
Nanking
Canton
HONG KONG
Taipei
FORMOSA
Manila
Bangkok
Rangoon
Singapore
Brunei
BORNEO
NEW GUINEA
Tropic of Cancer
Equator

ASIA
Flora and fauna
Alpine and tundra
Coniferous forest
Deciduous forest
Evergreen forest
Grassland
Steppe and semi-desert
Desert
0 200 400 600 800 mi
0 400 800 1200 km
Wild horse
Equus
Wild sheep
Ovis
Arctic skua
Stercorarius
pheasant
Phasianus
Snow goose
Anser
beaver
Castor
millet
Setaria
Giant salamander
Megalobatrachus
goat
Capra
pangolin
Manis
Giant panda
Ailuropoda
yak
Bos
cryptomeria
Cryptomeria
dunnock
Prunella
tea
Camellia
camel
Camelus
hyaena
Hyaena
bamboo
Bambusa
gavial
Gavialis
breadfruit
Artocarpus
peacock
Pavo
Flying lemur
Cynocephalus
rice
Oryza
cotton
Gossypium
sal
Shorea
Indian elephant
Elephas
rhinoceros
Rhinoceros
Siamese fighting fish
Betta
coconut
Cocos
Spectral tarsier
Tarsius
Monkey eating eagle
Pithecophaga

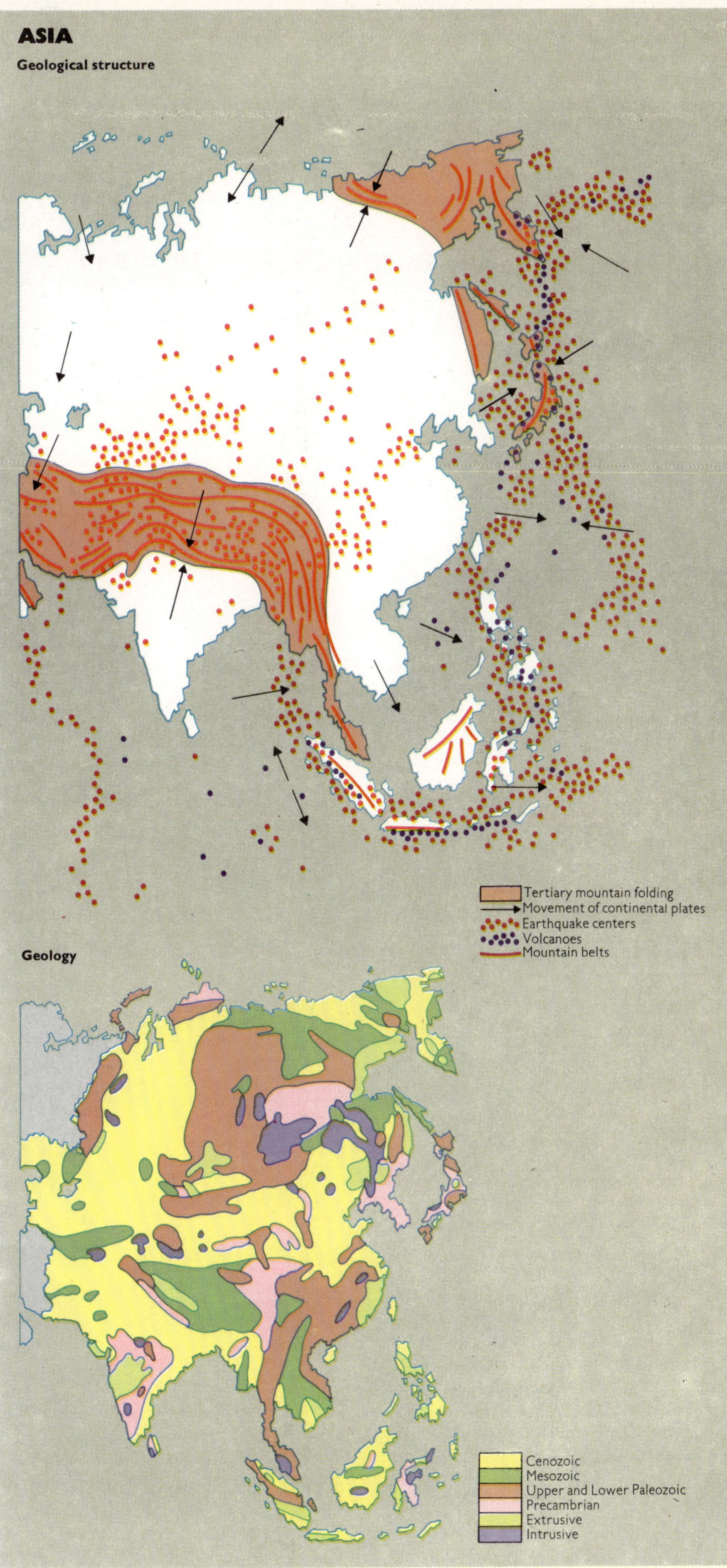

mosses, lichens and small flowering plants appear everywhere during the brief summer. South of the tundra is the *taiga*, the bleak, seemingly endless Siberian forest belt ranging across the continent, its sombre conifers (pine, fir, larch and spruce) covering an area larger than the United States. South of the taiga are the broad grassy treeless steppes and the semi-deserts of Central Asia; and south of these, a zone of deserts and salt seas runs from Arabia to Mongolia supporting date palms in the hotter oases and sparse scrub and tamarisks elsewhere. High barren plateaux and mountains with low-growing alpine plants stretch from Afghanistan to China. To the south, south-east and east of this mountain core lie the partly forested land of Monsoon Asia including the Indian subcontinent, South-East Asia, China and Japan. Wet evergreen rain forests containing teak and other valuable hardwoods dominate much of the south and south-east. China and Japan include mixed deciduous and coniferous forests.

Several kinds of tree are associated with particular parts of Asia. The banyan, which in maturity resembles an entire grove in itself, is common in India and Ceylon. The beautiful deodar, a form of cedar, is important in the western Himalayas. Other types of cedar, including the majestic Cedar of Lebanon, are found in Mediterranean Asia. Date palms thrive in the oases of the south-western deserts, and coconut palms and mangroves line low shores in south and south-east Asia.

Wildlife. The Arctic Coast is the home of seals, walruses, polar bears and many seabirds. On the tundra, lemmings feed on mosses; and reindeer, wolves and Arctic foxes are summer visitors. The wolves and brown bears of the Siberian forests have been relatively unmolested by man. Other forest animals include lynxes, reindeer, elk and squirrels.

Until recently the steppes were roamed by herds of gazelles, saiga antelopes, wild horses and camels, but their numbers are now much reduced. Among the burrowing rodents of the steppes are the jerboa and the marmot. Bustards, quails, sand grouse, hoopoes, sand martins and bee-eaters are common birds.

Wildlife is much scarcer in the deserts of Central Asia. There are lizards and mammals including the Dzeren antelope, the now rare wild ass, the cheetah and gerbil.

Animal life in India, South-East Asia and China is rich and varied, although some species, such as the tiger, are becoming rare. The giant panda of the highlands of China is one of the world's rarest animals, and the wild yak is now almost extinct. The jackal, civet and mongoose are common throughout the area, and monkeys abound in the tropical forests of southern Asia, Assam and the southern tip of India.

Wild boars, deer, antelopes and the Indian

ASIA
political
CITY population more than 1,000,000
CITY population more than 500,000
CITY population more than 100,000
City population less than 100,000
INDEPENDENT AFTER 1945
former British territory
former French territory
former Dutch territory
former Japanese territory
former American territory
railways
roads
airport
scale 1: 40,000,000
0
500
1000 st. miles
BERING SEA
East Cape
Arctic Circle
Anadyr
Vrangel Island
ALEUTIAN ISLANDS (U.S.)
NORTH LAND ISLANDS
NEW SIBERIAN ISLANDS
OCEAN
SOCIALIST REPUBLICS
Khatanga
Verkhoyansk
Yakutsk
Lena
Okhotsk
Magadan
Ust Kamchatsk
SEA OF OKHOTSK
SAKHALIN
Sovetskaya Gavan
KOMSOMOLSK
KHABAROVSK
Amur
KURIL ISLANDS
Lake Baykal
IRKUTSK
ULAN-UDE
Yenisey
ULAN BATOR
MONGOLIA
CHICHIHAR
HARBIN
VLADIVOSTOK
SAPPORO
HOKKAIDO
SEA OF JAPAN
HONSHU
SENDAI
TOKYO
YOKOHAMA
KANAZAWA
NAGOYA
KYOTO
OSAKA
KOBE
SHIKOKU
KYUSHU
KITAKYUSHU
NAGASAKI
KAGOSHIMA
JAPAN
SHENYANG
NORTH KOREA
PYONGYANG
WONSAN
SEOUL
SOUTH KOREA
PUSAN
PEKING
TIENTSIN
LUTA
YELLOW SEA
TSINAN
TSINGTAO
TAIYUAN
Yellow River
LANCHOW
SIAN
NANKING
SHANGHAI
CHINA
WUHAN
Yangtze
CHENGTU
CHUNGKING
CHANGSHA
FOOCHOW
TAIPEI
FORMOSA
RYUKYU ISLANDS
Salween
Mekong
Lhasa
BHUTAN
Brahmaputra
KUNMING
CANTON
MACAO (Port)
HONG KONG (BR.)
VICTORIA
HAINAN
NORTH VIETNAM
HANOI
MANDALAY
BURMA
Akyab
RANGOON
VIENTIANE
LAOS
THAILAND
BANGKOK
KHMER (CAMBODJA)
PHNOMPENH
SOUTH VIETNAM
SAIGON-CHOLON
ANDAMAN ISLANDS (INDIA)
ANDAMAN SEA
NICOBAR ISLANDS (INDIA)
SOUTH CHINA SEA
LUZON
PHILIPPINES
MANILA
QUEZON CITY
Mindoro
ILOILO
CEBU
Palawan
DAVAO
MINDANAO
SULU ARCHIPEL
TALAUD ISLANDS
BONIN ISLANDS
VOLCANO ISLANDS
MARIANA ISLANDS
Guam (U.S.)
PACIFIC OCEAN
CAROLINE ISLANDS
Tropic of Cancer
Equator
BRUNEI
Brunei
MALAYSIA
SUMATRA
MEDAN
KUALA LUMPUR
SINGAPORE
PONTIANAK
BORNEO
CELEBES
MANADO
Morotai
Halmahera
SULA ISLANDS
Ceram
Buru
Amboina
KAI ISLANDS
ARU ISLANDS
SCHOUTEN ISLANDS
Djajapura
IRIAN BARAT
INDONESIA
TERR. OF NEW GUINEA
PAPUA
80°
70°
60°
50°
180°
170°
160°
150°
140°
130°
100°
20°
10°
0°

The Ganges River is sacred to the Hindus, and thousands of pilgrims come each year to cleanse and purify themselves; others come to cure ailments, and, in some cases to die in the river, believing that they will then be carried away to paradise. Varanasi (Benares), above, is a famous centre for pilgrimages and for Sanskrit scholarship. Steps, known as Ghats, line the river bank and are surmounted by many temples.

rhinoceros occur in India, and Indian elephants and some other large mammal species range from India to Sumatra. India has many beautiful birds, including eagles, peacocks, parrots, pheasants, kingfishers, cranes and herons. The many reptiles include crocodiles and cobras. The Komodo dragons of remote Indonesian islands are giant lizards, weighing up to 350 lb (159 kg) and measuring up to 10 ft (3 m), which have survived from prehistoric times.

Wildlife preservation is less well supported in Asia than in Africa. Only some 300 Asiatic lions are left, and wild asses are in danger of dying out. Arabia's splendid herds of gazelle have been ruthlessly pursued by truck and machine-gun, and the magnificent white oryx is nearly extinct.

The Peoples of Asia

Distribution and Main Divisions. Asia's population of over 2,600,000 is nearly 60% of the world total. As much as 55% of the world total is concentrated in the lowlands of Monsoon Asia, where densities exceed 1,000 per sq mi (385 per km^2) over parts of the Ganges plains and plains of China and reach 3,000 per sq mi (1,155 per km^2) in coastal areas of Java. Half of the world's 50 largest cities lie in Monsoon Asia, among them Tokyo and Shanghai which perhaps rank as the two largest cities on earth. No other nations have populations remotely approaching those of China and India. Yet Monsoon Asia also contains large thinly peopled mountain areas.

By contrast, South-West Asia holds a mere 105 million people, of which about one-third live in Turkey and one-quarter in Iran. Soviet Asia is more thinly peopled still, with a recorded 57 millions inhabiting this colossal tract of land – mainly concentrated in Central Asian oases, settlements strung out along the Trans-Siberian Railway, Pacific ports and new Arctic and sub-Arctic cities. But the region with the fewest people is High Asia where scattered oases support most of the settled communities.

Food and Famine. In many areas the pressure of dense population upon productive land has long presented grave problems of malnutrition. Today they are particularly acute because of the current population explosion.

Every 24 hours Asia's population increases by more than 100,000. Even to maintain the present situation in the face of this growth, food production must increase annually by 2%; in fact, the increase ran at 2·7% by the early 1970s, little more than enough to keep pace. The development of improved medical services, has thus proved a mixed blessing. Every fall in infant mortality further increases the population, and many Asian governments are only just beginning to promote birth control. The introduction of strains of crops which are two or three times as productive (the 'green revolution') is only partly helping the food problem: Asia's traditional areas of high rural population cannot do enough to add to their cultivable land. India, for example, is already cropping most of its potential arable land, and desert areas are still spreading in both India and Pakistan.

Racial and other Differences. All three main racial groups are represented in Asia: the Negroid, in parts of the Philippines and in mainland South-East Asia; the Mongoloid (including the Mongols, Chinese, Japanese and Koreans), in Central Asia and the Far East; and the Caucasoid (including the Arabs, Afghans, Jews, Iranians, Pakistanis and most of India's millions), in South-West and South Asia. The peoples of South-East Asia – the Burmese, Thais, Malays, Laotians, the Khmers of Cambodia and the Vietnamese – show a mixture of races. The Malays are the largest group, accounting for most of the people of Indonesia and the Philippines.

Cultures vary greatly. Some peoples, like the Chukchi and Eskimos of north-eastern Siberia, are semi-nomadic hunters or primitive fishermen. Central Asia has nomadic pastoral peoples, but most Asians are now settled farmers. However as many as one in every three Asians is a city-dweller.

Asia's Many Tongues. The languages and dialects spoken by Asians can be numbered in thousands. They derive from more than 20 families, of which the most important are the Indo-European, Altaic, Semitic and Sino-Tibetan. The Indo-European family includes most of the languages of the Indian subcontinent and some of those spoken in the parts of Asia nearest to Europe, but Turkish belongs to the Altaic family. In Asia the Semitic languages, Hebrew and Arabic, are confined to Israel and other countries of the south-west. Chinese and many South-East Asian languages belong to the Sino-Tibetan family, but Malay in its various forms belongs to the Indonesian group. Some languages such as Japanese and Korean belong to no classified family.

In countries like India, which has 179 languages and 544 dialects, the profusion of tongues is a major hindrance to trade, communication and cultural development. In China, the home of widely divergent dialects, the government is encouraging the use of the official 'National Language'.

The importance of education is recognized by most Asian countries, but the sheer number of schools and teachers needed and the lack of money for them are serious obstacles to progress. Less than half of Asia's children attend school, and the adult illiteracy rate exceeds 50%.

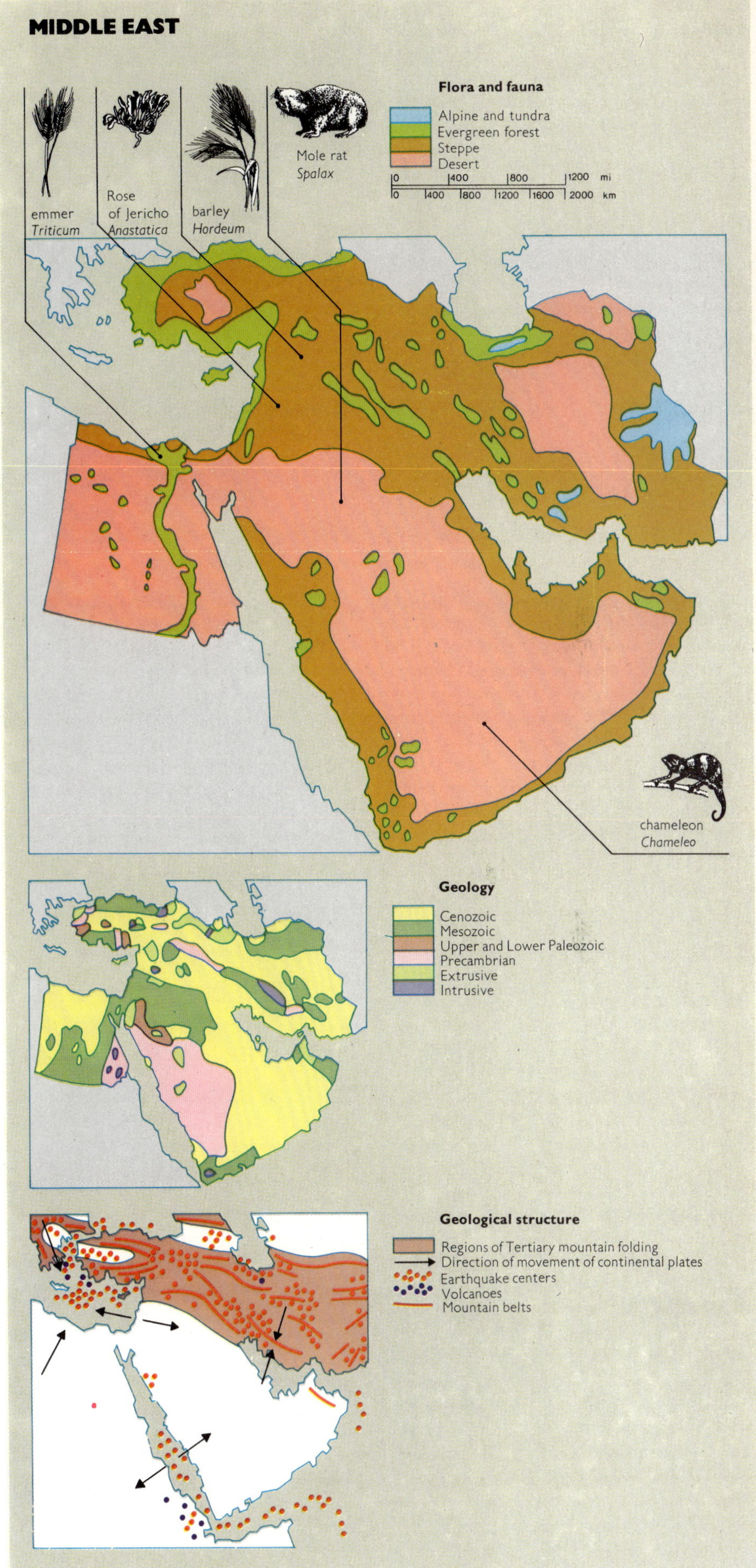

Tea picking on a plantation in India. Tea was introduced into Europe in 1610.

The Home of Great Religions. Asia is the home of the world's great religions. Hinduism, the ancient religion of India based on a rigid caste system, claims the greatest following – about 20 out of every 100 Asians. Buddhism has fewer followers (about 8%) but is more widespread, having been carried from India to Ceylon, South-East Asia, Tibet, China and Japan. Islam retains its hold on South-West Asia, Pakistan and Bangladesh, and also has many followers in Burma, Indonesia and West Central Asia. About 19% of Asians are Moslems.

Christianity was founded in South-West Asia and is now represented mainly by the Russian Orthodox Church and to a lesser extent by the Armenian Church. About half the population of Lebanon is Christian, as is more than half the population of the Philippines. There are also Christian communities in India, China, Japan and other countries, but only 4% of Asians are Christians.

Two ancient philosophical religions, Confucianism and Taoism, survive in China but there, as in other communist countries, religious belief is discouraged. Shinto ('the Way of the Gods') has almost as many adherents as Buddhism in Japan.

The Economy

Agriculture. Roughly two out of every three Asians work on the land, largely with little help from modern farm tools and know-how. Yet most of Asia is too cold, dry or mountainous for farming. Despite these drawbacks, careful husbandry and rich alluvial or volcanic soils in the east, south-east and south yield several crops a year, locally supporting more people per acre than perhaps any other land on earth.

There are big contrasts in the patterns of agriculture in Soviet, South-West, High and Monsoon Asia.

Much of Soviet Asia remains undeveloped, but a productive farming belt runs east-west along the Trans-Siberian Railway, with wheat, rye, oats and cattle as the major products, largely flowing from vast collective farms at least partly exploiting modern farming methods and machines. The steppes are now a leading source of Soviet grain. Irrigation projects in the arid south-west, including the Tashkent-Samarkand oasis zone, have enabled that area to provide 90% of the Soviet Union's cotton. Central Asia and Transcaucasia also yield sunflower seeds, grapes and tea. Sheep and goats thrive on the drier steppe, and reindeer herding persists in the far north.

In South-West Asia, farming still occupies most of the people (nearly 90% in Afghanistan), but aridity makes farming impossible over large areas and in some countries less than 2% of the land is cultivated. Nonetheless, farm production has generally increased at an annual rate of around 4%. This is due to the extension of irrigation (especially in the great river valleys) and of dry farming on the semi-arid steppe lands. The money for flood control and water distribution and storage has come largely from national oil royalties. Major crops include grains (grown largely in Turkey, Iraq and Iran), and vegetables and citrus fruits produced by the small nations facing the eastern Mediterranean. Goats and sheep range the steppes and mountain pastures from Turkey, and Arabia to Afghanistan.

Similarly, nomadic herdsmen tend flocks of sheep and herds of goats and horses on the high, dry and largely uncultivable steppes of High Asia, particularly north of the desert belt in Mongolia where extensive grasslands form the traditional grazing grounds of Mongol tribes. There are also prosperous farms in the Kashgar, Khotan, Turfan and Yarkand oases watered by snow-fed streams flowing from such ranges as the Altyn Tagh and Tien Shan. Scattered farmers win a living from the valleys of south-eastern Tibet.

Monsoon Asia, with over half the world's population, depends largely on the traditional and inefficient production of food grains from countless small farms, many of them occupying less than 1 ac (0·4 ha). Apart from using cattle for ploughing, the peasant farmers cultivate entirely by hand and lack modern pesticides and fertilizers. There are exceptions to this pattern, notably Japan where artificial fertilizers, light machinery and improved strains of seed have raised per acre rice yields dramatically, to treble those of India. A general lack of capital suggests that Monsoon Asia as a whole will remain agriculturally underproductive.

Nevertheless, this region produces most of Asia's rice output which itself forms 90% of the world total. Much of this comes from great river valleys in Bangladesh, India, Burma, Thailand, Cambodia, North and South Vietnam and China (the world's largest rice grower), but crops vary locally. Drier areas of India and northern China depend on wheat, millets or soyabeans. Sugar cane, sugar beet and vegetables (in India and China), and manioc and bananas (in South-East Asia) are other valuable foods.

Cash crops include natural rubber and copra from South-East Asia (where countries including Malaysia and Indonesia provide most of the world supply); cotton from India and China; tea from China, India and Ceylon; and jute from Bangladesh and India.

Food taboos and a lack of pastures keep Monsoon Asia generally poorly supplied with meat and milk; in China, pigs and poultry are the main food animals.

Fisheries. In the late 1960s, Asia (excluding Soviet Asia) produced about 37% of the world fish catch, with Japan, China, India and Indonesia ranked among the world's main fishing nations. Fisheries provide Japan's main source of protein (with seaweed as a supplementary seafood). Japan, China and Indochinese countries extensively farm freshwater fish in ponds. Soviet Asia is famous for its sturgeons, and its Pacific ports are springboards for fleets that fish the North Pacific Ocean. Indonesian and Indian offshore waters offer scope for huge expansion of these nations' fishing industries.

Forestry. Excluding Soviet Asia, the continent has only some 13% of the world's forest area. But Soviet Asia's northern belt of softwood conifers is the greatest forest area on earth – much of it untapped except near the Trans-Siberian Railway. Siberia is supplying an increasing proportion of Soviet timber.

Monsoon Asia's forests have been largely cleared for cultivation. But in India valuable

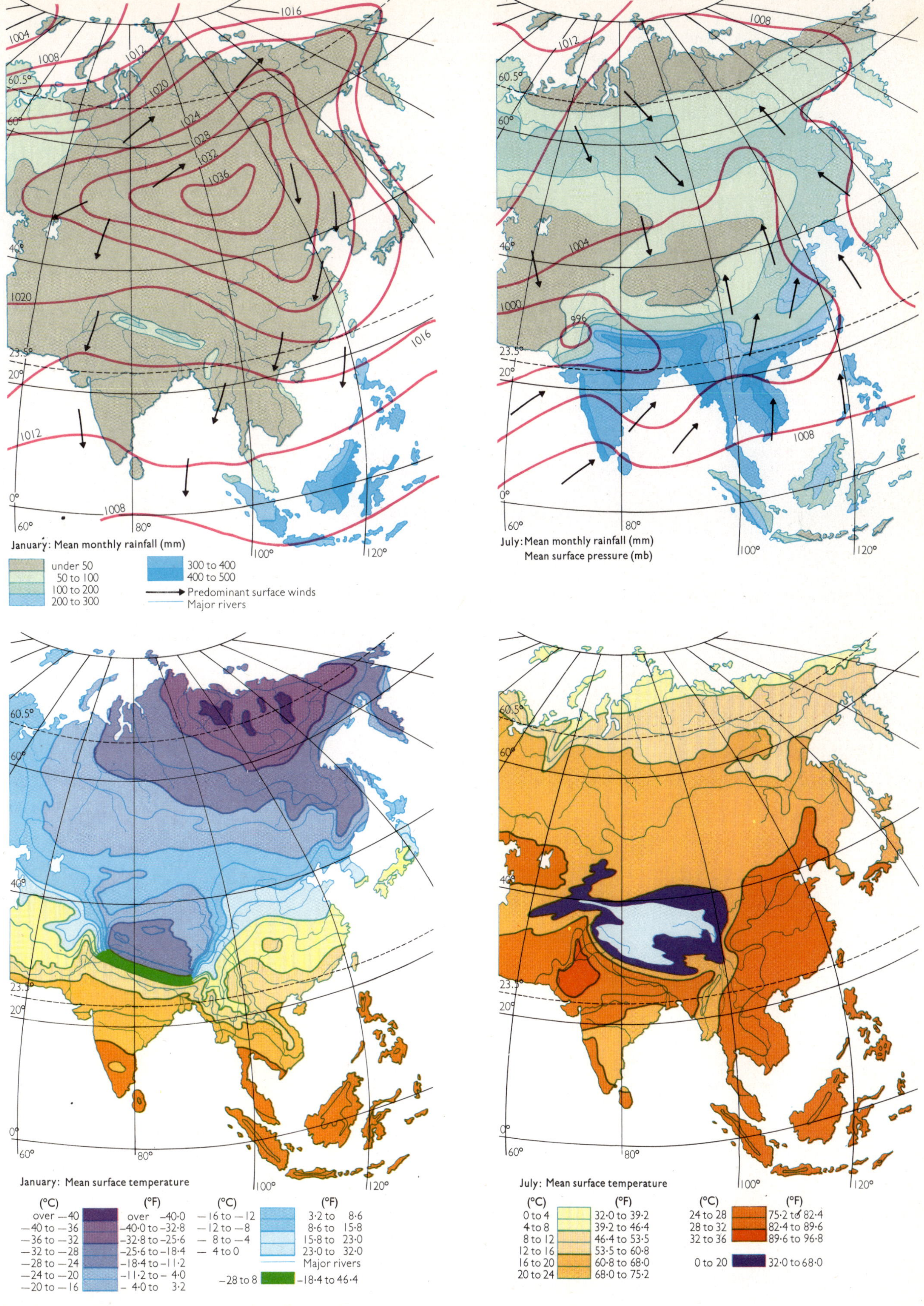
January: Mean monthly rainfall (mm)
under 50
50 to 100
100 to 200
200 to 300
300 to 400
400 to 500
Predominant surface winds
Major rivers
July: Mean monthly rainfall (mm)
Mean surface pressure (mb)
1004
1008
1012
1016
1020
1024
1028
1032
1036
1000
996
60.5°
60°
40°
23.5°
20°
0°
60°
80°
100°
120°
January: Mean surface temperature
(°C) (°F)
over −40 over −40·0
−40 to −36 −40·0 to −32·8
−36 to −32 −32·8 to −25·6
−32 to −28 −25·6 to −18·4
−28 to −24 −18·4 to −11·2
−24 to −20 −11·2 to − 4·0
−20 to −16 − 4·0 to 3·2
−16 to −12 3·2 to 8·6
−12 to −8 8·6 to 15·8
− 8 to −4 15·8 to 23·0
− 4 to 0 23·0 to 32·0
Major rivers
−28 to 8 −18·4 to 46·4
July: Mean surface temperature
(°C) (°F)
0 to 4 32·0 to 39·2
4 to 8 39·2 to 46·4
8 to 12 46·4 to 53·5
12 to 16 53·5 to 60·8
16 to 20 60·8 to 68·0
20 to 24 68·0 to 75·2
24 to 28 75·2 to 82·4
28 to 32 82·4 to 89·6
32 to 36 89·6 to 96·8
0 to 20 32·0 to 68·0

pines and deodars clothe the lower Himalayan slopes. Burmese and Thai teak forests have been long exploited, but the equatorial hardwoods of Malaysia and Indonesia remain relatively untouched. Mountain forests in Manchuria contain China's main accessible forest resources, and Chinese reafforestation projects are attempting to make good losses due to centuries of forest clearance.

Mineral Resources. Asia is rich in almost all major minerals, and much of its mineral resources remain unexploited – even unassessed.

As a whole, the continent contains abundant mineral sources of energy, largely in the form of fossil fuels. In the late 1960s South-West Asia alone held 60% of the world's computed oil reserves, concentrated around the Persian Gulf. Before 1945 only Iran and Iraq were significant oil producers in the area, and both remain important. But in recent years, big reserves have been located beneath Saudi Arabia and the Gulf sheikdoms. Saudi Arabia and Kuwait together contribute over 40% of the oil output of South-West Asia, which itself amounts to nearly one-third of the world total. Qatar, Abu Dhabi, Dubai and Oman, latecomers in the field of oil production, are likely to play a progressively larger part as new reserves are discovered there and world oil shortages intensify. Smaller oil reserves lie in Soviet Asia, Assam, Burma, Sumatra, China and elsewhere.

Russian geologists claim that eastern Siberia has half the world's entire coal reserves. Much of this Russian coal is in seams over 20 ft (6 m) thick and can be worked so easily by open-cast methods that thermal stations using it can produce electricity more cheaply than hydro-electric stations. The open-cast mines of Kazakhstan have seams 40 ft (12 m) thick; and in the Kansk-Achinsk field, rich reserves extend for 400 mi (640 km) along the Trans-Siberian Railway. China, too, has at least adequate supplies of coal and India and Japan produce substantial quantities.

Rubber is produced from a gum called latex which lies just under the surface of the bark of the rubber tree. The common rubber tree originated in the Amazon basin and trees were tapped in their wild state. However, now they are grown on plantations in many parts of the world.

Russia is rich in natural gas supplies and pipes gas from Krasnoyarsk on the Trans-Siberian Railway 1,000 mi (1,600 km) north to Norilsk, a new city in the Arctic Circle.

Asia's sources of nuclear fuels include Central Asia (the Soviet Union's major source of uranium) and south-west India where there are valuable monazite deposits.

Many Asian countries possess considerable hydro-electric potential, among them India, Pakistan, Bangladesh, the Indochinese countries, China, Japan and Soviet Asia. Lack of capital delays development in some poorer countries, but oil royalties are helping to finance projects in South-West Asia, and Japan and India already have substantial outputs of hydro-electric current. Some of the biggest projects undertaken have been those in Soviet Asia. At Bratsk, on the Angara River, the 4,500 Mw hydro-electric scheme was the world's largest when it was opened in 1961. Even larger is the 6,000 Mw Krasnoyarsk plant (opened in 1967), itself destined to be overtaken by the 6,300 Mw Sayano-Shushensk plant on the Yenisei and a massive 20,000 Mw construction on the lower Lena.

Besides its mineral power resources, Asia is rich in minerals valuable for manufacturing. The continent produces nearly 75% of the world's tin (mainly from Malaysia, Burma and Thailand). China is a leading source of tungsten and antimony; Soviet Asia contributes strongly to Russia's role as the world's top producer of iron ore, manganese, chromium and lead. Additionally, much of the world's supply of manganese, mica and chromite comes from India, Turkey or the Philippines.

Manufacturing Industry. Much of Asia lags industrially far behind Europe and North America. Some notion of Asia's generally low level of industrialization may be gained from its energy and steel consumption which in the later 1960s stood at around 12·5% of the world total (if we exclude Soviet Asia). Lack of capital, of mineral resources and of an affluent home market are all factors limiting industrial growth in much of South-West and Monsoon Asia.

Soviet Asia is the continent's most strongly industrialized area, with heavy industry entrenched largely in the Kuznetsk Basin; engineering located in cities strung out along the Trans-Siberian Railway; textiles concentrated in Central Asia; and new industrial centres springing up in the far north, as at Norilsk (based largely on locally mined non-ferrous metals) and Bratsk (where cheap hydro-electricity underpins aluminium production from Russian bauxite).

Japan, in Monsoon Asia, ranks second in industrial development, its energetic peoples compensating for a lack of many raw materials by developing a wide range of sophisticated manufacturing skills. Japan's major modern industries involve heavy engineering, shipbuilding, vehicles, electronics, chemicals and textiles – manufacturers being largely concentrated in major cities on southern Honshū and northern Kyūshū islands.

Similarly, Hong Kong, Singapore and Formosa have exploited special geographical or political positional advantages to develop important manufacturing concerns.

Elsewhere in Monsoon Asia, China is now fast building up its industries, with heavy industry concentrated in the mineral-rich north-east and light (notably textile) industry centred on such ports as Canton, Shanghai and Tsingtao.

India, too, has important industrial zones: heavy industry dominates eastern India between Jamshedpur and Calcutta; and textile manufacture occupies a belt in western India from Ahmadabad south to Bombay. The main world jute processing centres are Calcutta and nearby towns in Bangladesh.

But for the most part, industry in Monsoon Asia is confined to small-scale concerns supplying local needs – for instance, textile factories in Burma, Indonesia and Thailand.

Similarly, South-West Asia and High Asia generally lack major manufacturing enterprises, although in South-West Asia such cities as Tel Aviv, Jaffa, Ankara and Baghdad are centres of limited industrial production (largely of textiles and processed foods). By 1970 Turkey was South-West Asia's sole steel producer, and the region's engineering industry was negligible.

Trade. Colonial rule in Asia encouraged a trading pattern whereby Asian colonies supplied Europe with tropical farm products and other raw materials and relied increasingly upon Europe's new factories for manufactured products. To some extent, this pattern still prevails. Thus India exports large quantities of tea, jute, iron ore and cotton; Malaysia ships out tin and rubber; Arabian states sell vast quantities of (largely unprocessed) oil; and agricultural produce and minerals are major Chinese exports.

Most such nations rely for sophisticated machinery, vehicles and electronic products upon imports, largely from the Western world.

Exceptions to this trading pattern include Japan, Hong Kong, Singapore and Formosa which all import raw materials (Japan, for instance taking vast quantities of Australian iron ore) and export sophisticated manufactures. Soviet Asia and China trade internationally but from positions of intended self sufficiency. By the late 1960s, Asian nations as a group accounted for a mere 14% of all world exports.

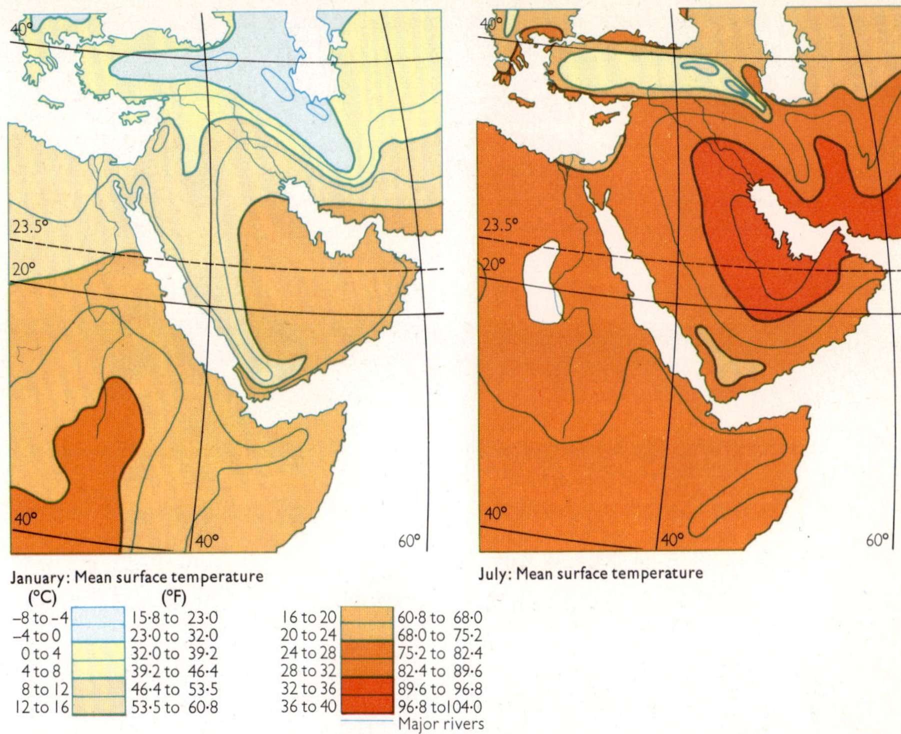

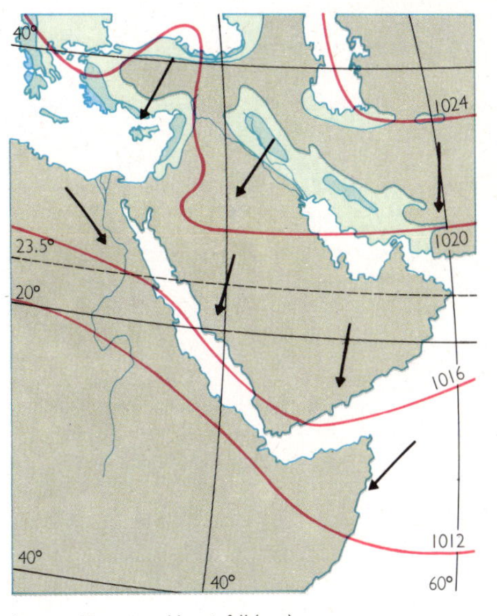

January: Mean monthly rainfall (mm)

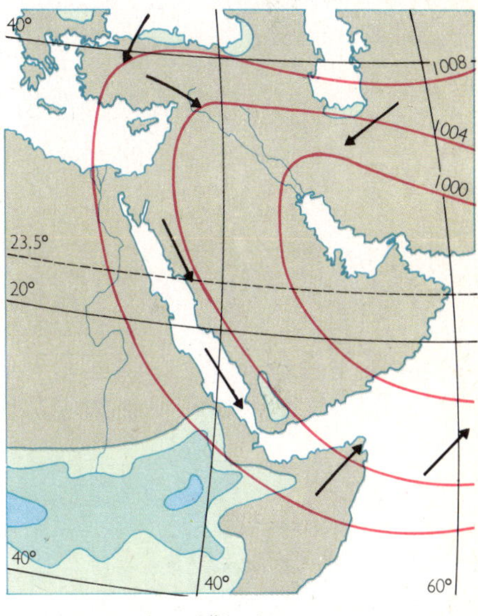

July: Mean monthly rainfall (mm)

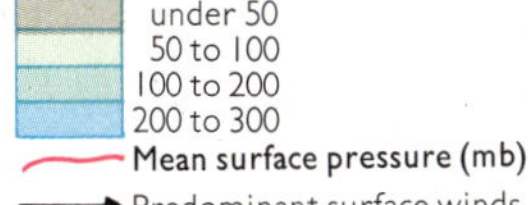

Transport. The great Himalayan and other mountain barriers have always hampered overland travel between northern and southern Asia. The Trans-Siberian Railway alone crosses the entire continent and no railway penetrates the mountains to link north and south. India, China and Japan all have relatively well developed railway systems.

In ancient times the trackway known as the Silk Road crossed Central Asia to link China with Europe, but Asian road systems today remain largely primitive and the roads unsurfaced. Slow bullock carts, pack animals and bicycles are common means of transport. Motor vehicles are generally few: in 1967 the number of cars in China was too small to be analysed; in India it was 1 per 1,000 people. (In the USA about 400 per 1,000.)

Great rivers like the Yangtze, Irrawaddy and Ganges, and canals (especially in China) provide much of Monsoon Asia with valuable water highways bearing heavy barge and other traffic. Sea trade is well served by such big ports as Bombay, Calcutta, Colombo, Rangoon, Singapore, Hong Kong, Canton, Shanghai, and Yokohama.

TURKEY

Area: 300,868 sq mi (779,248 km²)
Population: 35,666,549
Population Growth Rate: 2·5%
Capital: Ankara, pop 1,208,791
Language: Turkish; Kurdish and Arabic spoken by minorities
Religion: Moslem (98%) Jewish and Christian
Currency: Turkish pound (or lira) = 100 kurus (piastres)

Turkey straddles two continents – Asia and Europe – and covers more land than any nation in Europe except Russia. Most of the republic comprises the mountains and plateau of oblong Asia Minor which is separated by a newly-bridged strait from European Turkey – a small area on which stands Turkey's largest city, İstanbul. Turkey was once the core of the East Roman (Byzantine) Empire, and later of a great Turkish empire which controlled much of the Middle East and south-eastern Europe. This was finally dismembered in World War I and a secular republic proclaimed which attempted to adopt western laws and customs. Modern Turkey remains strategically important because of its position guarding Russia's narrow outlet from the Black Sea into the Mediterranean. Internationally, Turkey has remained aligned to the west and joined organizations such as NATO as protection against the Soviet Union. Internally also communism has been regarded as a threat and the army has repeatedly assumed government control to ensure the dominance of right-wing elements. By the early 1970s Turkey stood economically below most of the poorest European countries and suffered nagging balance of payments problems. Yet this mineral-rich country had a gross national product that (per head of population) outranked most Asian nations, and it seemed poised for impressive industrial growth.

Turkey was the centre of two great empires, first the Byzantine Roman Empire and then the Ottoman Turkish Empire, which stretched from Hungary to the Caspian. After the Ottoman empire's defeat and dismemberment in World War I Turkey's survival seemed precarious. Then, the wartime hero Mustafa Kemal (later surnamed Atatürk: 'Father of the Turks') rallied his countrymen, routed a Greek army in Anatolia, deposed the hereditary Sultan, and proclaimed a republic (1923).

The Land. Trakya (European Turkey), is mostly rolling plain, with low mountains in the north-east and highland in the south-west. Asian Turkey (by far the larger part) consists almost entirely of high plateaux and mountain ranges, and has only limited areas of plain. Much the most extensive and fertile coastal plain is the Cilician Plain around Adana in the south. The broad lower valleys of the Gediz and Menderes rivers in Aegean Turkey are also alluvial plains.

The Hagia Sofia, in Istanbul, was built in the 6th century A.D. its name means 'Holy Wisdom.'

The heart of Anatolia is an immense lozenge-shaped plateau averaging about 3,000 ft (900 m) above sea level, its corners roughly marked by the cities of Ankara, Kayseri, Konya and Afyon, and its centre occupied by the salty Lake Tuz. Low hills or chains of volcanoes divide the plateau into a number of separate *ovas* or basins. The underlying rock is largely limestone; domes, swallow-holes and other karstic features are common. To north and south respectively, the plateau is bounded by the fold ranges of the Pontus and Taurus Mountains. Of these two systems, the Tauric is the broader. Since it is also composed of more porous rocks and receives less rainfall it is much less eroded and dissected by surface drainage than the Pontic system, where complex patterns have been cut by the rivers Sakaria, Kızıl Irmak (the ancient Halys) and Yesil Irmak.

In south-central Turkey, the Taurus and Nur ranges converge to form the Anti-Taurus, which curve eastwards towards Lake Van whose salty waters cover more than 1,400 sq mi (3,626 km²). In this extreme eastern region, prolonged volcanic activity has blocked lakes, levelled the deeper valleys and produced such high peaks as Ararat (*Ağrı Dağı*), which at 16,945 ft (5,165 m) is Turkey's highest mountain. Within these folds are the headstreams of the Euphrates (known in Turkey as the Firat), while the upper Tigris (the Turkish Dicle Nehri), flowing through south-eastern Turkey, drains the Kurdistan plateau.

Climate. The south and west coasts enjoy a typically Mediterranean climate without extremes of heat or cold and with a moderate rainfall, mainly in winter. The north coast and its immediately adjacent ranges, however, receive considerable rainfall in summer from the steady north winds which have gathered moisture while crossing the Black Sea. The central plateau remains dry; its average annual rainfall of about 12 in (305 mm) is always unreliable and mostly comes in convectional storms. Summers on the plateau are very hot; winters, harsh and cold.

Mountainous far eastern Turkey is rather better watered than the plateau. Here, the snow lies for four months, and the region's seasonal range of temperature is more extreme than that of the west.

Vegetation and Wildlife. The central plateau is the threshold of the steppes of Central Asia and, except for its dry heart, is monotonously grass-covered. Occasional clumps of poplars grow on irrigated land near the villages, and where high volcanoes like Hasan Dağ and Erciyas Dağı break the surface there are modest but richly varied woods. Similar steppe conditions prevail in the high intermontane basins in the east.

The mountain forests have suffered greatly from centuries of felling and grazing. On the western coastal ranges, for example, this has produced *maquis*, dominated by low ilex and myrtle. In more remote parts of the Taurus, however, pine woods survive down to sea level, and parts of the Anti-Taurus still have cedars. Above the conifers is a zone of deciduous oaks which in turn gives way to the rich summer pastures along the broad crests of the ranges.

Turkey's finest forests are on the Pontic folds flanking the Black Sea, especially on their seaward slopes. Here oaks and hazels predominate from sea level up to the rhododendron scrub just below the snows.

Wildlife includes brown bears, wolves, wild boars, deer, leopards and more than 500 species of birds.

The People. When the Turks first came to Asia Minor they entered an area that was already ethnologically complex. Assimilation of the people they found there, and later mingling following their advance into Europe, have made the Turks far more Balkan than Asian in appearance. They are a proud, tough people with a long military tradition which is seen today in the special status of the army in national affairs.

The population, which has doubled since 1940, is predominantly Turkish, but includes over 2 million Kurds, some 366,000 Arabs and numbers of Circassians, Greeks, Armenians, Georgians, Lazes and Jews. European Turkey and the Straits have the main concentration of people; but the Ankara district, the north coast, the central western valleys, and the plains of Cilicia and the Hatay are also well populated. Most of the people still live in the countless small villages.

Beliefs and Culture. Nearly 99% of the people are Moslems, mainly of the Sunni sect, and despite secularization, Islam remains a powerful influence. There are small but significant Christian (Orthodox, Gregorian, Roman Catholic and Protestant) and Jewish communities.

Although education is compulsory from 7 to 15, lack of schools and teachers keeps the illiteracy rate above 33%. Istanbul and Ankara each have three universities, and there are others at Izmir, Erzurum and Trabzon.

The Economy. By the 1970s Turkey was refashioning its old image as a land almost totally dependent upon farming and handicrafts. The rapidly expanding economy was now entering a boom period, with considerable growth in the industrial sector which featured such major projects as the İzmir oil refinery, the Seydişehir aluminium plant and the huge Keban dam across the Euphrates. Turkey is still heavily dependent upon foreign, especially American, financing, but is aiming to pay an increasing proportion of the cost of economic expansion from its own improving resources.

Agriculture still dominates the economy. It provides a little more than 28% of the gross national product but employs more than 60% of the working population. The staple crops are wheat and barley, which are sown in the autumn to ripen in the following summer. Other crops – vegetables, fruits, nuts – are available in wide variety, but it is the size of the grain harvest that matters most. In years of surplus, wheat and barley may be exported, and there is a steady export of dried fruits from the western valleys through the port of İzmir. The main commercial crops are tobacco, grown for export in the western districts, and cotton, grown mainly for domestic use on the wide deltaic plains around Adana and also in irrigated districts of the north and west. Animals, the chief kinds being sheep and goats, are fitted into the farming pattern in most parts of the country.

Irrigation, mechanization, financial aid and the establishment of co-operatives have vastly improved Turkish agriculture over the last two decades. The government has also checked the decline of the forests, and promoted conservation and improvement.

Minerals. Turkey has considerable mineral resources including copper, chrome, coal and iron. Copper has long been mined at Ergani; and chrome, of which Turkey is the leading producer in Europe and western Asia, comes mainly from the Maraş and Fethiye areas. Turkey's main coal-field, the only large field in South-West Asia, is near Ereğli on the Black Sea coast. Iron ore is mined at Divriği. The small Garzan-Raman oil field in the upper Tigris basin produces about 10% of Turkish requirements.

Power and Industry. Turkey naturally benefits from its proximity to the oil producing and exporting countries of the Levant, and has large refineries at Mersin and İzmit. The country has invested heavily in the generation of electricity, which is produced partly by burning lignite but mostly from hydro-electric plant geared to high dams.

The iron and steel industry is concentrated on the Black Sea coast around Karabük near the Ereğli coal-field, with Kirşehir, south-east of Ankara, as a minor centre. İstanbul's commercial and transport facilities have

Wheat and other grains are grown on the Anatolian plateau. Despite the import of farm machinery from the United States, Canada and other countries, much of the crop is still harvested by sickle. Four-fifths of Turkey's agricultural produce are cereal crops.

attracted many industries, including tobacco, chemicals and food-processing. Government policy, however, is to disperse industry so that all areas may benefit. Thus Eskişehir has become the main textile centre; Uşak, the largest centre for refining sugar; Kars, a processing centre for dried milk. Manufacturing is still largely concerned with satisfying the domestic market, although some craft products like the *kilims* (woven rugs) of Gaziantep, the Kütahya pottery, Beykoz glass and Eskişehir meerschaum pipes find their way abroad.

The tourist industry is of growing importance. Turkey's historic sites and dependable summers attract increasing numbers of visitors to its Aegean and Mediterranean coasts.

Transport. Turkey has good road and rail networks. Main seaports, such as Istanbul and İzmir, have become congested, but İzmir, İskenderun and the Black Sea ports of Samsun, Sinop and Trabzon, have all been modernized, and İstanbul is to have a relief port and new industrial zone on the Marmara coast. There are international airports at İstanbul and Ankara, and international and domestic airlines services.

International Trade. Turkey exports chrome ore, tobacco, cotton, wool, carpets, olive oil, dried fruits and (in some years) wheat and barley. Imports include crude oil and a wide range of manufactured goods. Turkey's associate membership of the Common Market was negotiated in 1971.

CYPRUS

Area: 3,572 sq mi (9,251 km²)
Population: 639,000
Population Growth Rate: 1%
Capital: Nicosia, pop 115,000
Language: Greek (80%). Turkish
Religion: Eastern Orthodox (80%), Moslem (20%)
Currency: Cyprus pound = 1,000 mils

Cyprus, the third largest island in the Mediterranean, is a divided country. Independence from Britain came in 1960, but Britain retains military bases on the island. Greek Cypriots, forming about 78% of the population, stand bitterly opposed to the far less prosperous Turkish minority, a little over 18% of the population. The Turkish Cypriots want the island to be partioned. Intercommunal talks, often interrupted by civil strife, had not found a solution to the problem by the early 1970s, and the United Nations peacekeeping force, first sent to the island in 1964, continued its task of preventing a civil war in which Greece and Turkey would also inevitably be involved. Meanwhile the Turkish Cypriots had established their own virtually autonomous areas in the countryside and in the larger cities.

A view of Kyrenia, an ancient Cypriot port, and in the background, the Kyrenian mountains. The island of Cyprus has long been contested between Asian and European powers.

Physical Geography. In the north, the coast is flanked by the rugged peaks of the limestone Kyrenia Mountains (called the Karpas in the north-east). The chief mountains, the Troodos massif, lie in the south-west, where Mount Olympus, the island's highest peak, rises to 6,408 ft (1,953 m). Between these two mountain systems is the broad and fertile Mesaoria lowland stretching eastwards to the Bay of Famagusta.

The climate is Mediterranean; winters are mild and wet, summers hot and dry. Annual rainfall is relatively high in the mountains (40 in; 1,016 mm), but not more than 20 in (508 mm) in the lowlands. The Troodos Mountains are snow-covered for some 10 weeks and the rivers, dry in summer, are raging torrents during the short rainy season. Forests of pine, cypress and juniper clothe the mountains; elsewhere there is scrub and rock bared by erosion.

The People. In addition to the Greek and Turkish communities, Cyprus has small numbers of Armenians, Maronites and British. Most of the people live in the coastal lowlands and the towns. Education is free and compulsory for children aged 6–12 and there is an extensive social insurance scheme.

Cyprus is a republic and a member of the Commonwealth of Nations, established under its first president, Archbishop Makarios. The constitution provides for a Turkish vice-president, but since 1963 the Turkish Cypriots have taken no part in the government.

Half the population lives by agriculture, growing wheat, barley, vegetables, vines, citrus fruits, almonds and tobacco. Sheep, goats and donkeys are reared, and stall-fed cattle kept.

The mountain forests satisfy local timber requirements. Cupreous pyrites, iron pyrites, asbestos and chromite, chiefly from the Troodos, are the leading minerals. Wine-making and the processing of farm products are the chief industries. The island is popular with tourists and is well served by air and shipping lines. Cyprus exports minerals, fruits and wine, and imports manufactured goods and food, trading mainly with Britain.

SYRIA

Area: 71,479 sq mi (185,131 km²)
Population: 6,794,998
Population Growth Rate: 3%
Capital: Damascus, pop 835,000
Language: Arabic. Armenian, Turkish and Kurdish spoken by minorities
Religion: Moslem (87%) Christian (13%)
Currency: 1 Syrian pound = 100 piastres

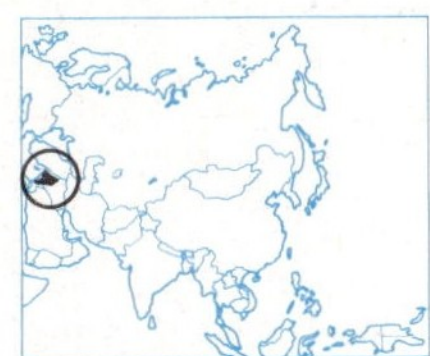

Syria (officially The Syrian Arab Republic) is a politically unstable land where military rule has alternated with socialist civilian government.

In 1971 Syria joined with Egypt and Libya in forming the Federation of Arab Republics, and has loudly urged the total destruction of Israel, but lacked the military power to accomplish it.

The Land. The Mediterranean coastal plain is everywhere narrow except in the extreme south where the low-lying Tripoli-Homs gap separates Mt Lebanon from the Jebel Ansariya range, which stretches northwards to the Turkish border near Antakya, the ancient Antioch. East of Jebel Ansariya is the Ghab Depression, part of the rift valley running southwards to the Red Sea and continuing in East Africa. From its Lebanese sources the Orontes River flows through the Ghab on its way to Turkey; its marshes are now being drained and reclaimed. East of the river is the Jebel ez Zawiye, a part of the elevated rim of the Syrian desert and steppe plateau. Desert and steppe stretch eastwards to the Euphrates, which enters northern-central Syria from eastern Turkey and flows south-eastwards to Iraq. Its only large perennial tributary in Syria is the Khabur. The desert, mostly gravel and pebble, has a close network of shallow *wadis* (seasonal watercourses).

Syria's south-western border with Lebanon follows the crest of the Anti-Lebanon and Mount Hermon ranges at an average altitude of 7,000 ft (2,134 m), Mount Hermon itself rising to 9,232 ft (2,814 m). Streams descending from the mountains irrigate valleys and oases on the desert verge, the largest of which is the Ghouta, the Damascus oasis, watered by the Barada and Awaj.

The monotony of the Syrian desert and steppe is broken by several low ranges extending northwards to the ancient oasis city of Palmyra with its imposing ruins and on to the Euphrates. In the south, volcanic hills (the Jebel ed Druz) rise to 5,689 ft (1,734 m); between them and Mount Hermon is the Hauran, a stony but fertile plain that is the Bashan of the Bible.

The citadel of Aleppo was once the capital of the Hittites. The present structure dates from the 12th century after which it was destroyed by the Mongols and restored in the 16th century.

Climate, Vegetation and Wildlife. About a third of Syria receives less than 5 in (127 mm) of rain annually; but much of the west and north receive 10–15 in (254–381 mm), while the higher mountains and coastal Syria receive more than 30 in (762 mm) during the rainy season (September–May).

Winters are mild on the coast but much colder inland, and temperatures drop below freezing in the mountains, which have heavy snowfalls. Summer is a period of drought, with temperatures inland often reaching 104°F (40°C).

The Syrian Desert is not entirely devoid of plant life. In spring it is bright with wild flowers, and it provides scanty but locally adequate pasture for sheep, goats and camels. Western Syria has low scrub with thin forests of evergreen oak, pine, cedar, wild olive and cypress.

Gazelles are still found in the desert as are hares, foxes and rodents. The remoter parts of the mountains are the haunt of wild cats, wolves, wild boars and even bears.

The People. Most Syrians are Arabs but there are Armenian communities in Aleppo and Damascus, and Kurds concentrated in Damascus and near the Turkish border. About 60% of the people live in villages, mainly in the west of the country, but the towns are growing rapidly, especially Damascus. More than half the population is illiterate.

Syrians are predominantly Moslems of the Sunni (orthodox) sect. Christians, about 10% of the population, include Maronites and members of all the Eastern Orthodox and Uniate churches. Two important Moslem sects, whose beliefs are by no means those of orthodox Islam, dominate two areas: the Jebel Ansariya, which is inhabited mainly by Alawites, and the Jebel ed Druz, the land of the Druzes. The influence of Alawites and Druzes in national politics has been considerable.

Agriculture. Syria has always been a rich country agriculturally. The most fertile areas are the broad steppe zone in the west and north, where wheat and barley are grown without irrigation, and the numerous small islands of intensively irrigated land, such as the Damascus oasis and favoured areas along the Euphrates, which provide a wide range of crops. Since 1945 the area under dry farmed crops has been doubled, while that under irrigation has nearly trebled. Much of the increased irrigated cultivation depends on the Euphrates. A large new dam constructed on the river at Tabqa in northern-central Syria will again double the irrigated area and provide hydro-electric power.

Besides cereals, Syrian farmers grow high-quality grapes, fruit and vegetables. Export tobacco is grown in the Latakia coastal district, and rice in the Euphrates valley. Large numbers of goats, sheep and cattle are reared.

Minerals and Industry. Although Damascus steel was famous in the Middle Ages, Syria has few minerals except oil, produced commercially at Süveydiye, Karachuk and Rumailan in the extreme north-east and piped to Homs for refining, and to the port of Tartus for export. Oil pipelines from Iraq cross Syria to Baniyas and Tripoli.

Handicrafts, a feature of the bazaars of Aleppo and Damascus, still provide work for many craftsmen. Modern factory industry – textiles, processed foods, glass, cement – is centred on Damascus, Aleppo and Homs.

Transport. Syria has railway links with Turkey, Iraq and Lebanon. Two new routes have been opened recently, one from Akari to Tartus, the other from Aleppo to Tabqa. A network of good roads links the main cities with Tripoli and Beirut in Lebanon, and Mafraq in Jordan.

International Trade. Syria's chief exports are agricultural products. Italy and France are major suppliers of manufactured goods, but much of Syria's trade is with the Soviet Union, China and East European countries. Syria also has a considerable transit trade with Lebanon and Iraq. The visible trade deficit is largely made up by pipeline royalties, earnings from tourism, and remittances from Syrians working abroad.

LEBANON

Area: 4,015 sq mi (10,399 km²)
Population: 2,700,000
Population Growth Rate: 2·9%
Capital: Beirut, pop 700,000
Language: Arabic. English and French widely spoken
Religion: Moslem (50%) Varied Christian (50%)
Currency: Lebanese pound = 100 piastres

Lebanon in some ways seems a country apart from much of the Arab world to which this tiny eastern Mediterranean republic belongs. Lebanon stands alone in its record of relative political stability, in its culturally mixed yet largely harmonious society.

Lebanese independence from France was proclaimed in 1941. Democratic Lebanon has weathered government crises and grown rich as a middleman trading between the West and Middle Eastern states. The main threat to Lebanon's stability in the early 1970s was its Arab standpoint in the prolonged Arab-Israeli confrontation. This attitude helped to attract about 150,000 Palestinian refugees whose anti-Israeli guerrilla activities have provoked reprisals and brought embarrassment to Lebanon. Tension between the Lebanese government and refugees has led to brief waves of bloodshed.

A general view of Beirut, the capital city and chief port of Lebanon. Beirut is of great antiquity and was probably first associated with the Phoenicians.

The Land. Mountainous Lebanon is undoubtedly the most beautiful country in the eastern Mediterranean. The well-wooded slopes of the Lebanon Mountains, the higher parts of which are snow-covered from December until June, stand out as an island of fertility and greenness by comparison with the parched, brown Syrian desert and steppe. The coastal plain and mountain valleys owe their luxuriance to the many small mountain streams which ensure an abundance of water throughout the year.

The Lebanon Mountains lie entirely within the country; they are highest in the north near Hermil and Bsharri, where the summit of Qurnet as Sauda reaches 10,125 ft (3,086 m).

Lebanon's eastern border runs along the higher crests of the Anti-Lebanon range to Mount Hermon (9,232 ft; 2,814 m), turning westwards to form the southern border with Israel which reaches the Mediterranean at En Naqura.

Between the Lebanon and Anti-Lebanon ranges lies the Bika Valley, a fertile plain about 10 mi (16 km) wide. Lebanon's two major rivers, the Litani and the Orontes, rise in the Bika near Baalbek, the ancient Heliopolis, noted for its Roman remains. The Litani is a wholly Lebanese river and is now being harnessed for irrigation and power. The Orontes flows mainly in Syria.

Climate, Vegetation and Wildlife. Lebanon has a Mediterranean climate with very mild winters on the coast and hot summers. Precipitation (concentrated from October to April) varies from 15 in (381 mm) in the north to more than 60 in (1,524 mm) in the mountain heights, which have heavy snows. The coast is humid in summer. By comparison the mountains are fresh and sunny, and there are numerous resorts, some of which provide winter skiing. This and the attractive scenery have made Lebanon popular with tourists from the hot arid lands of the Persian Gulf as well as from Europe.

Lebanon has some large forests of evergreen oak and Aleppo pine, but only six small groves of the famous cedars of Lebanon survive. Many migrant birds pass through Lebanon, and a few wolves and bears survive in remoter areas.

The People. About half the population is Christian, predominantly Maronite. The Moslem population is about equally divided between the Sunni and Shi'a sects. The Druzes, numbering at least 100,000 and living mainly in the south, represent a schismatic offshoot of Islam and are politically influential.

Most Lebanese are of Arab stock. There is a small but commercially important Armenian community, and many European and American businessmen temporarily reside in Lebanon, making Beirut (about 700,000) the most cosmopolitan city in the Arab world.

Beirut and its suburbs contain nearly a quarter of the population, and there are many large villages and small towns in the mountains within close range of the capital. About half the population still live in villages. Many of the Palestinian Arab refugees are housed in camps at Saida (Sidon), Beirut and Baalbek.

The country's education system is highly developed, but organized welfare is lacking.

Agriculture. Much of the cultivable land, about a quarter of the total area of the country, is used as rough grazing for sheep and goats, or for unirrigated crops of wheat and barley. Parts of the coastal plain, the middle western slopes of the Lebanon Mountains and the central Bika Valley are the most productive areas.

The main export crops are bananas, citrus fruit and apples grown under irrigation. Potatoes, onions, tomatoes, vines and olives are also important, and some tobacco is grown. The once flourishing silk industry, based on mulberry tree cultivation in the mountains, has now died out.

Most farms are less than 25 ac (10 ha), but the Bika has a few large farms. Farming has become a part-time occupation in many of the mountain villages near Beirut, which now serve as dormitory suburbs or resorts.

Minerals and Industry. Lebanon has few minerals except building stone, although small deposits of iron ore, lignite and other minerals exist. Oil from Saudi Arabia and Iraq is piped to terminals at Saida and Tripoli respectively, where there are small refineries serving the local market. Much of the very varied manufacturing industry is in or near Beirut and Tripoli, but wine is made near Zahle in the Bika Valley. Products range from cement to processed foods and light metal goods.

Transport and Trade. Lebanon has an excellent road system and rail links, now used only for heavy goods traffic, with Homs and Damascus in Syria. Beirut, the chief commercial port, also has the busiest international airport in the Middle East.

Although Lebanon exports oil products, fruit and vegetables, there is always a substantial trade deficit. This is made good, however, by the large transit trade with Syria, Iraq, Kuwait and Jordan through the free ports of Beirut and Tripoli; royalties on the oil pipelines; the profits of Lebanese overseas banking, insurance and business; remittances from the many Lebanese who live abroad; tourism and the unknown (but perhaps substantial) profits from illicit trade in cannabis and other drugs.

ISRAEL

Area: 8,017 sq mi (20,764 km²)
Population: 2,998,400
Population Occupied Territories: Approx 982,700 (1970)
Population Growth Rate: 2·8%
Capital: Jerusalem, pop 291,700
Language: Hebrew and Arabic. Also English
Religion: Judaism, Moslem, Christian minorities
Currency: Israeli Pound = 100 agorot

Israel is a tiny Jewish republic occupying a large part of the historic land of Palestine, and surrounded by hostile Arab states: Lebanon to the north, Syria and Jordan to the east and Egypt to the south. After nearly 2,000 years of foreign rule the Jews of Palestine proclaimed the State of Israel (*Medinat Yisrael*) on 14 May 1948, the day before a British mandate over Palestine ended. Since that date – through the War of Independence (1948–49), the Sinai Campaign (October 1956) and the Six-Day War (June 1967) and further conflict in October 1973 – Israel's policies have been dogged by the need to build a self-sufficient modern state to face the continued hostility of its Arab neighbours and their Palestinian Arab refugees.

Various factors influence political and social realities in Israel: a pioneering ideology dating from the end of the 19th century; the need to build a self-sufficient modern state; the necessity of integrating Jewish immigrants from vastly different cultural backgrounds; and the central position held by religious laws and customs in an otherwise secular democracy. These and other factors have created a dynamic society with contradictions and tensions peculiar to modern Israel.

During the Six-Day War, Israeli forces occupied the Arab territories of the Gaza Strip, Sinai, the West Bank of the Jordan and the Golan Heights (Syria). This article deals only with the areas which were part of Israel before June 1967.

The Land. For a small country, Israel has a surprising variety of physical relief, soil and climate. This is due partly to its geological structure and partly to its position on the borderland between a typical Mediterranean climate and the desert conditions of Arabia. Essentially, the country consists of three contrasting regions extending from north to south.

The first region, the coastal plain, is broken into two portions by the rocky spur of Mount Carmel, the only promontory on the otherwise smooth, dune-fringed coastline. The plain is a few miles wide around Haifa Bay but very narrow from Acre (Akko) northwards. South of Haifa, as the Sharon Plain, it gradually widens. Farther south, near Gaza, the plain is up to 25 mi (40 km) wide.

The central hills are highest in the north, in Galilee, where Mount Meron reaches 3,962 ft (1,208 m). Upper Galilee consists predominantly of limestones. In Lower Galilee the central range is broken by several fertile valleys, the most important being the Plain of Esdraelon, which provides a historically significant route from the Mediterranean to the Jordan Valley. The hills of Samaria and Judea average 2,000 ft in height, and form a much dissected plateau sloping eastwards to the Jordan Valley. South of Hebron the Judean hills are separated from the central Negev mountain ranges by the predominantly dry valleys of Wadi Beersheba and Wadi Salt.

The third region consists of the Jordan Valley and its southward extension, the Arava, which are part of the Great Rift Valley extending to Mozambique in Africa. The Jordan tributaries rising near Dan and in southern Lebanon flow southwards to Lake Tiberias. From there the Jordan flows to the Dead Sea – the lowest point on earth, 1,294 ft (394 m) below sea level.

More than half the country to the south is the Negev Desert; the northern part of which is an extension of the coastal plain, and is semi-desert, while the rest is typical hot desert.

Climate and Natural Vegetation. Summers in Israel are warm or hot with virtually no rain between early May and late September. Most of the annual rainfall comes in December, January and February, and ranges from 15 to 25 in (400 to 625 mm) on the plain and from 20 to 35 in (500 to 900 mm) on the hills. At the farthest point south, Eilat on the Gulf of Aqaba, rainfall drops to a mere 1 in (25 mm) per year. Mean mid-day temperatures range between 86°F (30°C) and 95°F (35°C) in summer, and between 50°F (10°C) and 68°F (20°C) in winter. Cold spells are marked by occasional snow, and hot *khamsin* winds from the south and east may bring extreme heat and drought in early summer and autumn, with temperatures rising to 113°F (45°C).

The higher and more rugged areas of Galilee and the Carmel range carry thin forests of Aleppo pine and oak, but the most characteristic vegetation is Mediterranean *maquis* and *garrigue* scrub. Farther south, there has been substantial afforestation since the establishment of the state, with pine, eucalyptus and acacia predominating.

Bird species are numerous. Wild boar and gazelles are among the few remaining large mammals.

At Masada, the Israelites had captured Herod's natural rock fortress in the Judean desert. Here they made their last stand against the Romans in 73 A.D. Because of the dry climate articles not destroyed in the siege are well preserved and include several Old Testament scrolls.

The People. Israel's population increased rapidly after 1948, when it numbered about 879,000. This was largely due to the immigration of Jews from Europe, America, Asia and Africa. Between 1948 and 1970 there were more than 1,350,000 immigrants. The vast number of immigrants has been the major factor in the country's development. The early pioneering settlers were reinforced by large numbers of highly skilled professionals and businessmen. These were balanced by a large influx of Jews from Asia and Africa, who for the most part lacked education or industrial skills.

Israel has built up a highly skilled labour force which, backed by contributions from Jews abroad amounting to more than $200 million annually, has led to the growth of a well advanced economy and a high standard of living. Four out of every five Israelis live in urban areas. About half the population lives in and around the three major cities of Tel Aviv-Jaffa, Jerusalem and Haifa. Population is densest in the central part of the coastal plain. The Arab community numbers some 400,000 (with almost one million more inhabitants in the areas occupied since 1967).

Cultural and artistic life is unmistakably Western and secular. Only a third of the people consider themselves practising Judaists. Defence is an important feature of life in Israel: all males between the ages of 18 and 29 are obliged to serve in the armed forces for three years, unmarried women under 26 must serve for 20 months. The Israeli armed forces are renowned for their fanatical courage in the face of superior numbers and devotion to the national ideal.

A group of Kibbutzniks at En Yahav in the Negev Desert tend an irrigated field.

Israel has a republican form of government. The *Knesset* (parliament) is a one-chamber house of representatives elected every four years by universal suffrage and secret ballot. The president is elected for a five-year term by secret ballot of the members of the Knesset. A system of proportional representation has encouraged the proliferation of political parties. Most Israeli cabinets have been coalitions of several parties in which *Mapai*, the Labour Party, has been the dominant influence.

Agriculture. Since 1948, Israel has made remarkable progress in agriculture, though in fact only some 12% of the total employed population work in agriculture, forestry and fishing. Farming is highly capitalized, market-oriented and efficient. The country is largely self-sufficient in foodstuffs except for bread grains and certain kinds of meat. There is a substantial surplus of fruit for export, particularly oranges and grapefruit. The country also grows enough cotton for its needs, and sugar beet and peanuts have been successfully introduced.

The basis of Israeli farming is irrigation – almost half the total area under crops is irrigated. Water supply and distribution are the responsibility of a quasi-governmental body, *Tahal* (Water Planning for Israel), and most settlements and towns draw their supply from a national piped water system. The backbone of this system is the National Water Carrier, a major canal and pipeline drawing water from the Jordan at Lake Tiberias, and with which all other sources of water are linked. In the Galilee and Judean hills, dry hill farming still predominates, with the emphasis on tree crops, vegetables and tobacco. There is little cultivation in the Negev, owing to lack of water, but some pioneer collective settlements and research stations have irrigated their lands with water piped from the Jordan and Yarkon rivers.

Poultry and dairy cattle are a very important element in the mixed farming of Israel where maize and lucerne (alfalfa) for fodder are grown under irrigation. Beef cattle are less important. Sheep are largely confined to the Galilee hills and the Negev – where most of Israel's 38,000 Bedouin Arab population live. About half the roughly 25,000 tons of fish caught or landed in Israel annually are from freshwater sources. Carp and trout are cultivated in artificial ponds. Most Israeli saltwater fishing is done in the distant waters of the Atlantic and Indian oceans.

Industry and Mineral Resources. Almost 30% of the employed population are engaged in industry, and industrial exports are now greater in value than agricultural exports. The large investments in the country by private and institutional Zionist funds since the creation of the state, and reparation payments from the West German government after 1952, stimulated industry and the development of the necessary infrastructure of a modern state.

Among the more important industries are textiles, chemicals, pharmaceutical products, cut and polished diamonds, and processed foods. A very wide range of consumer goods are manufactured or assembled from cars and trucks, radio and electrical goods, to fashions and cosmetics. The two main industrial areas are around Tel Aviv and Haifa. Most of the heavy industry, oil refining, cement manufacture, super-phosphates, steel making and engineering are located around Haifa Bay, near the country's major port. Car and truck assembly and component manufacture is carried on near Haifa and at Ashqelon in the south. The small size of the country has encouraged the wide distribution of industry, particularly since the opening of the port of Ashdod in 1965.

The *Histadrut* (General Federation of Labour) is a powerful trade union organization which is actively involved in the ownership of a workers' health service, insurance, retail trade and banking. The economy of Israel is thus a peculiar blend of private enterprise, state ownership and planning, in which the trade unions and various cooperative organizations play an important role.

Israel is relatively poor in economically exploitable minerals. Some oil and natural gas are produced at Helez, Bror and Kokhav in the southern Coastal Plain; and natural gas near Arad in the northern Negev. Less than 10% of the country's petroleum requirements are supplied from these sources. The most valuable mineral deposits are potash, bromine and other dissolved salts from the Dead Sea. These are extracted at a plant at Sedom. Potash and superphosphates are exported and the glass sand, kaolin and copper of the Negev are used in local industry. There are abundant supplies of good building stone, and these are the basis of an important cement industry.

The Town Hall of Bat Yam, a suburb of Tel Aviv, typifies the spirit of modern architecture in Israel. Tel Aviv, the first capital of Israel, is a modern city, built largely since World War II.

International Trade. In the early years of the state, Israel had a massive excess of imports over exports. The situation has improved considerably, but exports still amount to only slightly more than 50% of the value of imports. The most important exports are cut and polished diamonds, citrus fruits, fertilizers and chemical products, textiles and a wide range of manufactured metal and plastic goods. Imports are much more varied; but food, particularly cereals and meat, amounts to some 15% of total imports by value. Tourism provides foreign currency earnings second only to the earnings provided by diamond exports, and is rapidly increasing.

Transport and Communication. Israel has an excellent system of major and minor roads which link all settlements. Public bus services are widely distributed, frequent and cheap. Apart from the fast direct railway from Haifa to Tel Aviv, most passenger transport is by road. Internal air services are operated by the domestic Arkia airline, and El Al has regular international flights. Israel has a small, but growing, merchant navy.

Jerusalem's Old City is sacred to Jews, Christians and Moslems alike. In the foreground is the 'Dome of the Rock' built on a site where Mohammed is supposed to have ascended to heaven.

JORDAN

Area: 364:577 sq mi (944,254 km^2)
Population: 2,418,000
Population Growth Rate: 3·7%
Capital: Amman, pop 583,000
Language: Arabic
Religion: Moslem (Sunni) 90%, Christian 10%
Currency: Jordanian dinar = 1,000 fils

Jordan (officially the Hashemite Kingdom of Jordan) is a young Arab nation in an ancient, largely desert, land. Jordan's involvement in the Arab-Israeli conflict has cost the country dearly in lost territory, political unrest, a form of brief but savage civil war, and partial alienation from its Arab colleagues in the Middle East due to King Hussein's attempt to keep the peace with Israel.

The Land. Most of Jordan is desert. Only in the west are there more fertile regions, especially west of the River Jordan, where the Israeli-occupied limestone hill country of Samaria and Judea rises to more than 3,000 ft (914 m) near Hebron and Nablus. In the extreme north-west, the deeply incised valley of the Yarmuk, a major Jordan tributary, forms the boundary with Syria and the Israeli-occupied Golan Heights. The Jordan flows through the Ghor Depression which, at the surface of the Dead Sea, is 1,294 ft (394 m) below sea level. This rift valley continues to the Gulf of 'Aqaba, where Jordan has a few miles of coastline.

Beyond the fertile, narrow strip east of the river, the Jordanian highlands rise sharply, reaching their highest point in the extreme south near 'Aqaba, where Jebel Ram rises to 5,744 ft (1,754 m). East of the Hejaz Railway is the desert, which forms the greater part of the country.

The Wadi-el-Mojib is the Biblical 'Arnon' which formed the border between Moab and Israel.

Climate, Vegetation and Wildlife. The climate of the highlands is Mediterranean, with warm summers and cool winters, and some 35 in (889 mm) of rain annually. The desert is much hotter in summer and cooler in winter. Summer temperatures in the desert and the Jordan Valley regularly reach 95°F (35°C) and occasionally more than 115°F (45°C).

Eastern Palestine was known for its natural beauty and fertility in Biblical times. Some thin forests of evergreen oak and Aleppo pine still exist between Jerash and 'Ajlun. The hills of the West Bank territories have been cultivated and settled for so long that little of their natural vegetation remains, and most of the thorn scrub along the Jordan has been cleared. Farther east wormwood and other drought-resistant plants occur. Wildlife includes jackals, ibexes and wolves, found mainly in the southern highlands.

The People. In addition to the resident population of some 2,300,000, Jordan has about 550,000 Palestinian Arab refugees living in United Nations camps in East Bank Jordan.

Most Jordanians are Moslems of the Sunni sect. About 10% of the population is Christian and there are also a few Druzes. Only some 50,000 people still follow the nomadic Bedouin way of life.

Population is densest on the West Bank where most people live in stone-built villages clustered around a central mosque and square. Around the house are irrigated orchards and vegetable gardens. Conditions are much the same in the better watered but less populous highlands of the East Bank. Settlements in the semi-arid southern highlands are few. Amman, the capital, has grown very rapidly, and so have Zarqa and Irbid in the north.

Education has been impeded by lack of schools and teachers, and although it is theoretically free and compulsory, about 60% of the people are illiterate.

Government. Jordan is a constitutional monarchy with a Senate appointed by the king, and a 60-member House of Representatives elected by secret ballot and universal male suffrage.

Farming. Agriculture provides more than 20% of the gross national product and employs about 35% of the working population. Most farms are small, and many of the most fertile cannot be properly tended because they are near the tense border dividing Jew and Arab. Plans to extend irrigation on the eastern side of the Ghor Depression have been delayed for this reason.

Barley and wheat, the chief grain crops, are grown on unirrigated land and the yield thus fluctuates according to the amount of winter rain. Other important crops are pulses, tomatoes, fruit and vegetables. Large areas are devoted to vines and olives; and bananas, citrus and other fruits are grown. High-quality fresh fruit, vegetables and poultry are often air-lifted to places in the Persian Gulf. Sheep, goats and camels are reared by the Bedouin tribes of the drier areas.

Minerals and Industry. The only minerals at present worked commercially are phosphates (at Ruseifa and El Hasa), marble (Amman) and salt. Oil has not yet been found, and the refinery at Zarqa depends on supplies from the Trans-Arabian Pipeline (TAP-line) which crosses Jordan.

Despite lack of minerals, a wide range of industries has been established on the East Bank in recent years with government help, especially at Zarqa and Amman. Products range from cement and steel bars to textiles, electrical goods and beverages.

Tourism. Jordan suffered a serious loss of tourist income when Israel occupied the historic West Bank, but still has rose-red Petra with its Roman remains, and Jerash, a Graeco-Roman city, to attract the tourist.

Transport. Good roads link all the main towns, and there are desert highways to ‘Aqaba and Baghdad (Iraq). The Hejaz Railway at present ends at Ma‘an, but its reconstruction to Medina (Saudi Arabia) is in progress. There are internal air services to Ma‘an (for Petra) and ‘Aqaba. International services are provided by ALIA (Royal Jordanian Airlines).

International Trade. Phosphates account for nearly a third of exports by value; tomatoes, fruit, olive oil, cigarettes and asphalt are also exported. Imports include machinery, vehicles, cereals, livestock, foodstuffs and manufactured consumer goods. Trading partners include the United States, Britain, West Germany, Japan, Turkey and Jordan's Arab neighbours.

SAUDI ARABIA

Area: 850,000 sq mi (2,201,500 km²)
Population: 7,965,000
Population Growth Rate: 2·7%
Capital: Riyadh (Royal), Jeddah (admin), pop each 300,000
Language: Arabic
Religion: Moslem (Sunni)
Currency: Saudi riyal = 20 qursh

Saudi Arabia covers most of the Arabian Peninsula. This oil-rich desert kingdom nearly a quarter the size of the USA has special significance for Moslems everywhere, for it contains Mecca, the birthplace of Mohammed and Islam's most holy city, and Medina, another major place of Moslem pilgrimage where the Prophet was buried.

The nation was founded by Abdul-Aziz ibn Saud, a remarkable Bedouin leader of the strict Wahabi sect who established his authority over the warring tribes and opposing factions in the Arabian Peninsula (1902–33), welding Nejd, the Hejaz, Asir and Al-Hasa into a single kingdom. King Faisal ibn Abdul-Aziz, who took the throne in 1964 became one of the more conservative leaders in the Arab world.

The Land. Estimates of the area of Saudi Arabia vary greatly because its boundaries are largely undefined and disputed. In the north it shares a neutral zone with Iraq, established to facilitate the seasonal movements of nomads and their livestock.

Structurally Saudi Arabia is simply a great platform tilted downwards towards the Persian Gulf in the east. Most of Saudi Arabia is desert, but distinct regions are recognizable. The Red Sea coast in the west is bordered by the Tihama, a flat sandy plain from which mountains rise abruptly to elevations varying from 4,000 ft (1,220 m) north of Mecca to 8,000 ft (2,440 m) in the Asir highlands in the far south. Although isolated by poor communications, this south-western region is relatively fertile; only here is rain-fed agriculture possible.

Nejd, the central plateau region, has limestone escarpments, the largest of which is Tuwaiq, 500 mi (800 km) long and rising to 3,500 ft (1,050 m). Farther east, a narrow sand-belt called the Dahana runs for about 800 mi (1,300 km) between the North Arabian Desert or Nafud and the Rub 'al Khali Desert in the south. East of the Dahana is the barren Summan plateau, a hard rock plain about 100 mi (160 km) wide which rises to some 1,300 ft (400 m), but is replaced to the east by a low plain flanking the Persian Gulf. South-eastern Saudi Arabia consists of the great Rub 'al Khali or ‘Empty Quarter’. One of the hottest and driest places on earth, this desert is uninhabited except for a few hardy Bedouin living on its fringes.

Although streams occur in Asir, and dry *wadis* can become raging torrents during rare storms, there are no permanent rivers flowing to the sea. But water in abundance is known to exist below ground, and where it reaches the surface, oases have been developed. Spring and well water is relatively plentiful along the Persian Gulf coast, which is sandy and shallow. Large coastal areas are covered by salt flats (*sabhka*) and sometimes the salt concentration is rich enough for local mining.

Climate. Saudi Arabia has a hot desert climate notorious for the intense heat of the summer months when shade temperatures can exceed 120°F (49°C). On the coasts, particularly the Red Sea coast, temperatures are not so extreme, but high humidity makes the climate if anything more unpleasant. In Jidda, summer temperatures frequently soar to about 100°F (38°C), but humidity often exceeds 90%. The dryness of the interior, however, makes the higher temperatures more endurable. In winter, temperatures in north and central Saudi Arabia drop occasionally to freezing point.

Rainfall is low and very variable. Most of the country has less than 4 in (102 mm) annually and parts of the Rub ‘al Khali can be rainless for a decade or more. By contrast, parts of Asir affected by the monsoons of the Indian Ocean receive some 20 in (508 mm) annually. Winds also vary. In the east the prevailing wind is the *shamal*, which comes from the north or north-west, and which is notorious for its ability to whip up sandstorms.

Vegetation and Wildlife. With such a harsh environment, it is not surprising that plant and animal life is limited to only a few species. Large areas of desert are covered intermittently with saltbush. Sedges and grasses grow among the coastal dunes. There are no true forests, although parts of the Asir highlands are covered with juniper and olive.

Hunting has taken heavy toll of the wildlife. The oryx is now virtually extinct and the last ostrich was killed this century. Gazelles were once seen in large herds, but numbers are now greatly depleted. The ibex survives in Hejaz and troops of baboons frequent the mountains of Asir. While wolves and hyenas are fairly widespread, leopards are found only south of Jidda. Several species of large mammals face extinction in the near future at the hands of motorized hunting parties.

The People. No accurate census has ever been taken and estimates of the total population range from less than 4,000,000 to over 8,000,000, the largest figures coming from Saudi Arabia itself. The population has three typically Middle Eastern elements: town-dwellers, farmers living in villages, and nomads wandering in tribal groups. The number of nomads is probably about one million, although the Saudis themselves classify 65% of the population as nomadic.

Important towns lie in the west and north-west and include Mecca and Medina, the port of Yanbu, Jidda (the country's chief airport and commercial centre) and the town of Taif. Nejd, the central plateau region, contains Riyadh, the capital. The Eastern Province is the home of the vast oil industry, whose rapid development has given importance to places like Ras Tanura, Dhahran, Al Khobar and Dammam. This region also has the important oasis complexes of Qatif and Hofuf. Settled farming communities live in the mountainous south-west and nomadic Bedouin wander across the desert and semi-desert tracts.

Arabian travellers and scholars depict the Bedouin as a romantic figure, brave, proud and hospitable, yet leading a life of incredible frugality. Traditionally the nomad economy was based on the camel as a source of transport, merchandise, food, fuel, clothing and shelter; and on raiding and extortion. But nomadism is declining as the Bedouin are forced to settle or to become oil and construction workers by the disappearance of their traditional sources of income.

Beliefs and Culture. Islam is the state religion, most Saudis belonging to the Sunni sect. The *haj*, Islam's annual pilgrimage, brings more than one million Moslems from all over the world to Mecca every year.

Education is free, but not compulsory, and more than 80% of the population is illiterate. Adult literacy schools and other specialist schools have been established. There are universities at Riyadh and Jidda, and an Islamic university at Medina.

Mineral Riches. Saudi Arabia is an oil-rich state with nearly one-fifth of the world's known reserves. Production is confined to the Eastern Province, although oil has been discovered recently in the Rub'al Khali. The largest fields are Ghawar, Abqaiq (near Dammam) and the offshore Safaniya field, the largest of its kind in the world. Production is conducted mainly by Aramco (Arabian American Oil Co), with Getty Oil and the Japanese Arabian Oil Co operating in the joint Saudi-Kuwaiti zone in the north. Petromin, the Saudi Arabian government minerals agency, is playing an increasingly important role in the petroleum economy by various joint ventures. It handles internal marketing, has a refinery and steel mill at Jidda, is building another refinery at Riyadh and also has fertilizer interests. In 1969, 80% of Saudi

Pilgrims camping on the Plain of Arafat before entering Mecca, the birthplace of Mohammed and the spiritual centre of Islam. Mecca is visited by more than 300,000 pilgrims annually.

Arabia's oil was exported crude from the Ras Tanura tanker terminal, 13% went into the Ras Tanura refinery, and 5% was piped to the refinery on Bahrain Island. The long Trans-Arabian Pipeline (the 'TAPline') connects the oil fields with the Lebanese port of Saida (Sidon) on the Mediterranean.

Oil revenues, some £440 million in 1970, provide 80% of the national income and make Saudi Arabia one of the richest countries in the world. About 15,000 workers are employed in the oil industry, 80% of them Saudis, the rest being other Arabs (3%), Americans (10%), Indians (3·5%) and Pakistanis (2%). The Aramco settlements in Eastern Province are laid out like American suburbs. Aramco provides housing, health and education services, agricultural aid and a variety of other services.

The government is promoting diversification to lessen the overwhelming economic dependence on oil. A mineral exploration programme has revealed potentially workable deposits of iron, silver, gold, copper, sulphur, phosphates and uranium.

Agriculture. About 1,000,000 ac (404,686 ha) are under cultivation, of which 80% are irrigated and 20% (in Asir) rain-fed. Dates are the main crop, but their importance has declined since urbanization and increased wealth brought about dietary changes. Wheat, barley, millet and sorghum are the chief grains, and alfalfa is an increasingly important oasis crop. Cotton, rice, melons and fruit are also grown, and coffee in Asir. Animal husbandry, the main source of agricultural income, is largely practised by the Bedouin. The camel is still the most vital feature, although its transport role has been taken over by the motor truck. Sheep and goats pastured in the semi-desert areas are treated, like the camel, as all-purpose animals and provide meat, milk, wool, hair and skins. The donkey is the most important beast of burden in the oases. Falcons and the Saluki, a pure-bred hound highly esteemed in Arabia, are kept for hunting.

Oil revenues and the desire for greater self-sufficiency in food have led the government to make great efforts to develop agriculture. Water supply is crucial, and large sums of money have been spent on hydrological exploration. New areas are being brought under irrigation, dams built and experimental farms established. As recently as 1960 there was very little horticulture; today Saudi Arabia is self-sufficient in many vegetables and fruits. Progress is also being made in stock-breeding, dairying, pasturage and the conservation of grazing lands.

Fisheries. Cod, mackerel and other fish abound in the coastal waters, and the hitherto neglected fishing industry has considerable potential. Apart from one company freezing shrimp on the Persian Gulf, the industry is entirely in the hands of fishermen using traditional methods.

Transport. To create favourable conditions for industrial growth, large sums are being spent on road and port construction. In 1960, paved roads were limited to the few built by Aramco in Eastern Province. Today the network exceeds 4,000 mi (6,440 km) and is growing at a rate of 600 mi (9,660 km) per year. Major achievements include the trans-peninsular highway linking Dammam with Jidda via Riyadh and the spectacular Mecca-Taif, mountain road. A 1,000 mi (1,609 km) Red Sea coast road is under construction.

Against a background of rapidly increasing trade, special priority has been given to port expansion, notably at Jidda and Dammam. Yenbo, too, is being enlarged in the hope that it will be revitalized by the eventual re-opening of the Suez Canal.

The only railway operating in Saudi Arabia links Riyadh with Dammam. It has been losing both freight and passenger traffic as highway construction progresses. The old pilgrim railway between Damascus (Syria) and Medina which was partially destroyed by Lawrence of Arabia in 1918, is now being rebuilt by a British firm.

There are international airports at Jidda, Dhahran and Riyadh, and airfields for internal services at most other large towns. The state-owned Saudi Arabian Air Lines, managed by Trans-World Airlines and manned by American and Saudi pilots, operates internal and international services including special flights for Mecca's pilgrims.

International Trade. Oil and oil products account for 99% of Saudi Arabia's exports. The chief imports include foodstuffs, machinery, vehicles and electrical equipment. Exports exceed imports by a ratio of 8:2·5.

The United States is Saudi Arabia's chief trading partner. Other major trading partners include Great Britain, Italy, Lebanon, Japan and West Germany.

YEMEN ARAB REPUBLIC

Area: 75,000 sq mi (194,250 km²)
Population: 5,900,000
Population Growth Rate: 2·7%
Capital: San'a, pop 120,000
Language: Arabic
Religion: Moslem (Sunni 50%, Shi'a 50%)
Currency: Riyal = 40 bagsha

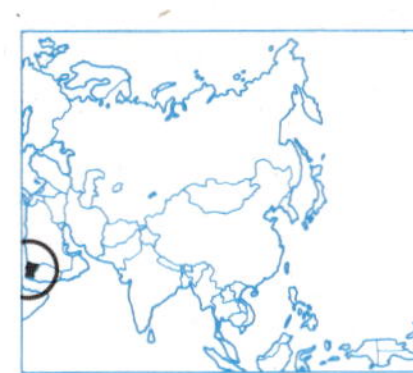

The Yemen Arab Republic (North Yemen), a small mountainous state in south-western Arabia, is the peninsula's most fertile and densely peopled country. Its fertility, and position on routes between India, Africa and the Eastern Mediterranean for many centuries gave it a special importance as an international trading centre.

History. Between about 750–100 BC what is now the Yemen Arab Republic was the Sabaean state of Saba, the Sheba of the Bible, which thrived on its trade with Syria and far-away places in India and Africa.

Later, Himyarites, Ethiopeans, Persians and Omayyads dominated but failed to crush the Yemenis, who from 897 AD were ruled by their own despotic Zeidi imams, although the country was nominally part of the Ottoman Empire from 1517 until 1918. In 1962 troops overthrew Imam Mohammed al-Badr and proclaimed a republic. Civil war followed, in which Saudi Arabia backed the deposed imam, and Egypt the new regime. By the early 1970s both Egypt and Saudi Arabia had withdrawn, and most of the country was in republican hands. Royalists were still holding out against republican forces, which meanwhile waged a sporadic war with Southern Yemen until Arab League mediation led to a truce.

The Yemen Arab Republic professes strict international non-alignment and its readiness to accept aid from any quarter provided no strings are attached. Britain, West Germany, the United States and the Soviet Union have given help. It is the Chinese, however, whose aid has made the deepest impression.

The Land. Coral reefs and monotonously low shelving beaches mark the Red Sea coastline. Inland lies the Tihamah, a coastal plain 20–45 mi (30–70 km) wide, covered with loose sand, shingle and alluvial deposits, and rising eastwards into an irregular foothill belt, through which *wadis* descend to the plain. Towering above it to the east, the scarped edge of mountains rises in gigantic steps to jagged crests often well above 10,000 ft (3,000 m), beyond which the rugged highland mass falls gradually to its less abrupt eastern scarp. Summits are highest in the central part of the crest, where Jebel Hadhur Nebi Shu'aib reaches 12,336 ft (3,760 m) west of San'a.

In the interior a maze of peaks and ridges encloses broad plateaux and rolling upland plains. Eastward the valleys draining to the desert open to wider, shallower basins covered with fertile loess. Around Arhab and Dhamar are the desolate plateaux of the *harra* ('burnt ground') with their barren unweathered lava and old volcanic cones (where hot springs and sulphurous fumes belie complete extinction). Away to the east, beyond the great basin crossed by Wadi Abrad and its tributaries, the ground falls away to the north-east where the dunes of the Rub'al Khali creep in across the frontier over the granite and sandstone surface. Nowhere do the *wadis* carry water continuously from source to mouth, though some have perennial reaches in the western highlands. Wells sunk in the gravels of the *wadi* beds and springs associated with fracturing and volcanic formations are important local sources of water.

Climate and Vegetation. Climatically, Yemen is the most favoured region of Arabia. Without the highlands to moderate tropical temperatures and draw moisture from the summer monsoon, the desert sands would

A Bedouin and tent in Southern Yemen. The nomadic Quabail are the indigenous people of the southern tip of the Arabian peninsula.

stretch unbroken to the Red Sea. The coastal lowlands are hot and arid, with high humidity and frequent dust storms. The highlands are much cooler and receive 16–32 in (406–812 mm) or more of rain, mainly between March and September. Snow is not unknown in the mountains. Eastwards, increasing heat and aridity mark the desert fringe.

Vegetation is denser and more varied than anywhere else in Arabia and ranges from the dwarf shrubs and palms of the desert to shrubby grasslands, woodland and the mountain juniper forests.

The People. Ethnically most Yemenis are tribesmen descended from the original South Arabian stock with a mixture of African elements on the coast and near the southern border. Mainly concentrated in the highlands, the population is almost equally divided between orthodox Sunni Shafa'i and the Zeidi sect of Shi'a Moslems. This division into two fundamentally opposed religious communities profoundly affects the social, economic and political structure and poses an urgent problem. The Shafa'i, probably a majority, control the country's trade.

Nearly all the major towns and cities are in the southern highlands: San'a, the capital, Ta'izz, a former capital, and Dhamar, Yarim, Ibb and Manakha, the centre for coffee production. Sa'ada is the only sizable town in the less productive north.

The Economy. Crops and livestock are still the basis of the economy. Farming is concentrated mainly in the highland plains and valleys which are watered by summer rain and intricately terraced. Elsewhere aridity limits cultivation to small irrigated oases in the *wadis* of the Tihamah and the eastern foothills.

Crops include cereals, vegetables and many kinds of fruit. Dates, cotton and tobacco are grown mainly in the Tihamah.

Mineral resources, which have still to be fully developed, include iron ore (near Sa'ada and elsewhere); copper, near Ta'izz and southwest of San'a; lignite, in the mountains east of Dhamar; and sulphur and uranium.

PEOPLE'S DEMOCRATIC REPUBLIC OF YEMEN

Area: 178,987 sq mi (463,576 km²)
Population: 1,425,000
Population Growth Rate: 2·9%
Capital: Aden, pop 225,000
Madinat as shaab, pop 10,000
Language: Arabic
Religion: Moslem (Sunni), Hindu, Christian (10%)
Currency: South Arabian dinar = 1,000 fils

The People's Democratic Republic of Yemen (Southern Yemen) emerged as an independent Arab state with the British withdrawal from southern Arabia in 1967. Government is now in the hands of the National Liberation Front. It consists of the port of Aden and a mountainous hinterland stretching north-eastwards, together with a number of islands: Kameran Island in the Red Sea, crescent-shaped Perim commanding the Bab al-Mandeb, and Socotra in the Indian Ocean, long known for its 'Dragon's Blood' (a red resin), myrrh and frankincense but now little cultivated. In ancient times the export of incense to Egypt, Greece and Rome helped to sustain a highly developed culture in this part of the Arabian Peninsula. Today it is strategically among the most significant areas of the Middle East.

The borders of Southern Yemen are still disputed, and although Arab League intervention succeeded in halting the frontier clashes with the Yemen Arab Republic (North Yemen) in 1972, when there was even talk of union, traditional rivalry between the two Yemens seems likely to continue. Also significant is Sino-Russian rivalry within the Yemens, expressed in economic and military aid.

Aden was an important port of call for ships bound through the Suez Canal. It is a part of the People's Republic of Yemen.

The Land. The contrast between the coast and the interior is very marked. The coastal plain, up to 40 mi (64 km) wide but sometimes – as at Mukalla – almost non-existent, is hot and practically rainless desert. Aden has an average temperature of 84°F (29°C) and an annual rainfall of 2 in (51 mm). Inland the terrain rises successively to ridges and plateaux, in places reaching 8,000 ft (2,440 m) before it slopes down to the great Rub'al Khali desert of southern Saudi Arabia. Rainfall increases with elevation to a meagre 10 in (254 mm), and is channelled into deep-cut *wadis* which have long been dammed to provide water for settled agriculture. Coconut palms thrive in coastal oases. Other vegetation includes myrrh bushes, frankincense trees, acacias and tamarisks.

The People of Southern Yemen are mostly Sunni Moslems. Their social structure is tribal, and inter-tribal strife (often over land) colours much of their history. True nomadism, however, is confined to the inhospitable northern border areas; elsewhere the people live for at least part of the year in permanent dwellings such as the extraordinary mud-built 'skyscrapers' of the Wadi Hadhramaut, the most important of the fertile *wadis*. Aden, the capital, stands on a volcanic crater harbour. Once a flourishing trading and refuelling port on the Suez Canal route to East Africa and the Far East, it has suffered greatly from the departure of the British and the closure of the Suez Canal.

Economy. Agriculture is the most important activity, and long-staple cotton, grown on the Abyan Delta east of Aden, the most important cash crop. Subsistence crops grown at higher elevations include millet, sesame and sorghum.

Like Aden itself, the British-built oil refinery at Little Aden has been hard hit by the closure of the Suez Canal. Lesser industries also initiated by the British include tanning, textiles, and cigarette-manufacture, but a major factor in the economy is the aid received from the Soviet Union and other communist countries. There are few surfaced roads and no railways.

Trade. Aden still serves to a limited extent as a transit port for neighbouring territories. Exports include cotton, fish, hides, skins and petroleum products; imports, rice and other foodstuffs. Kuwait, Britain, Iran and Japan are among the chief trade partners.

OMAN

Area: 82,000 sq mi (212,380 km²)
Population: 670,000
Population Growth Rate: 3·0%
Capital: Muscat, pop 6,200
Language: Arabic, English (commercial)
Religion: Moselm (Ibadhi 75%, Sunni 25%)
Currency: Rial Omani = 1,000 baiza

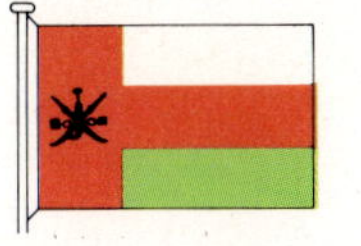

The Sultanate of Oman (formerly Muscat and Oman) occupies south-eastern Arabia as the easternmost outpost of the Arab world, and is the oldest independent sovereign Arab state in existence. Oman has remained outside the mainstream of Arab civilization; but for more than 150 years it has had close relations with Britain. The deposing in 1970 of the reactionary Sultan Said bin Taimur by his son, Qaboos, raised hopes of more rapid and liberal development in this remote corner of the world. But Dhofar in the south became the scene of a protracted war between government troops and Chinese-trained guerrilla forces who sought sweeping political and social changes.

Land and Climate. Oman is a land of desert, stony *wadis*, oases, scrub vegetation and barren, rocky mountains. It includes Masirah Island and the Kuria Muria Islands off the Dhofar coast. In the north, the Al Hajar Mountains rise to over 10,000 ft (3,000 m) in Jebel Akdar ('Green Mountain').

The climate is harsh. Masqat swelters in stifling heat and humidity. Inland, on the margins of the Rub'al Khali ('Empty Quarter'), the great Arabian desert, temperatures can reach 120°F (49°C). South-east facing coasts are cooler. Rainfall exceeds 5 in (125 mm) only in the mountains, but when it does occur, it is sudden and heavy.

The People and the Economy. The major areas of settlement are the northern coastlands and mountains, and the Salala district in the south. The intervening area is inhabited by nomadic tribes who raise the best riding camels in Arabia. In coastal areas up to half the population is non-Arabic (Indians, Baluchis, Persians and Negroes). The tribes of Dhofar, the land of frankincense and myrrh of antiquity, are among the most primitive in the Middle East.

Where water reaches the surface, as it does in the north, small but highly fertile areas result. The Batina coastal plain is famous for its dates, and fruit and grain are grown on the terraced slopes around Jebel Akdar. But the economic future lies with oil, discovered at Fahud and Natih in the interior and exported since 1967 from the Mina al Fahal terminal near Matrah. Oil revenues, some £30 million a year, have already brought great changes. A new £10 million port at Matrah is under construction, and plans have been prepared for improving roads, surveying water resources and expanding agriculture.

UNITED ARAB EMIRATES

Area: 32,000 sq mi (82,880 km²)
Population: 197,000
Population Growth Rate:
Capital: Abu Dhabi, pop 60,000
Language: Arabic. English (commercial)
Religion: Moslem (Sunni)
Currency: Bahrein Dinar = 1000 fils (Abu Dhabi); Qatar(Dubai riyal = 100 dirhams (elsewhere)

The United Arab Emirates (formerly the Trucial States) are a federation of seven sheikdoms strung out along the Arabian coast of the Persian Gulf between the Qatar peninsula and the Gulf of Oman, an area formerly known as Trucial Oman and in earlier times as the Pirate Coast. These sheikdoms came under British influence in the 1820s, and in a series of agreements they undertook to abandon piracy, suppress the slave trade and hand Britain responsibility for their defence and foreign relations.

When British forces were withdrawn from the Persian Gulf in 1971, six of the Trucial States – Abu Dhabi, Dubai, Sharjah, Ajman, Umm al-Qaiwain and Fujairah – came together as the United Arab Emirates under the presidency of Sheik Zayed bin Sultan al Nahayan, the ruler of Abu Dhabi. The seventh, Ras al-Khaimah, joined in 1972.

The seven states are similar in that each consists of a coastal town and a desert hinterland. The largest is Abu Dhabi, which includes parts of the Buraimi oasis and the Al Jiwa oases. Next in size is Dubai, through which most imports reach the emirates.

An Oil-Rich State. The importance and future prosperity of the emirates rests entirely on a single product – oil. Abu Dhabi is currently the largest producer, with an output worth £140 million in 1972 and production still rising. Dubai has been producing oil from offshore fields since 1969, but commercially worthwhile oil has still to be found in the other emirates.

Oil revenues have brought rapid expansion and development. Schools and hospitals have been built, ports enlarged and modernized, and new industries such as flour-milling and cement manufacture established. Agriculture has been encouraged by the opening of a trials station and agricultural school in Ras al-Khaimah, and by experimental farming in the fertile Buriami oasis (Al Ain), one of the few areas in the emirates where there is adequate water and good soil.

Dubai and Abu Dhabi are the chief seaports, and along with Sharjah have international airports.

QATAR

Area: 4000 sq mi (10,360 km²)
Population: 160,000
Population Growth Rate: 5·3%
Capital: Doha, pop 100,000
Language: Arabic. English (commercial)
Religion: Moslem (Sunni)
Currency: Qatar/Dubai riyal = 100 dirhams

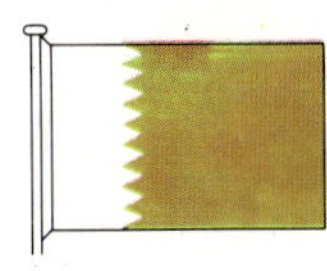

The sheikdom of Qatar, long a British

protectorate, declared its independence in 1971, when Britain withdrew from the Persian Gulf, but preserved the connection by a treaty of friendship. In 1972 the progressive Sheik Khalifa bin Hamad al-Thani became ruler after deposing his cousin in a bloodless coup.

The Land. Qatar is a low and mainly flat peninsula about 120 mi long projecting from the Arabian coast into the Persian Gulf. The highest point, 250 ft (76 m), occurs in the limestone outcrops near the west coast. The coast generally has low capes, narrow inlets, salt flats called *sabkha* and many small islands and coral reefs. Its pearl beds were a mainstay of the economy before the discovery of oil. Sand dunes predominate in the south, but the north has patches of vegetation suitable for nomadic grazing.

Climate. The climate is extremely hot and dry, summer temperatures sometimes reaching 120°F (49°C). Winter, with its occasional rain showers, is more pleasant. There is no surface water; shallow salty wells and three desalination plants provide the only supplies.

The People and the Economy. Most of the population lives in and around Doha (100,000), the capital. Immigrants from Pakistan, Iran, Oman and elsewhere now outnumber the native Qataris, who mostly lack the skills required by the oil industry, and who also are averse to manual labour.

Oil revenues amounting to some £50 million annually have transformed Qatar from an impoverished desert into a prosperous state. Oil was struck in 1940, and the Dukhan field went into production in 1949. A pipeline runs from Dukhan to Umm Said where tankers are loaded. Today Qatar's exports consist entirely of crude petroleum and refined products, except for shrimps and vegetables. The oil industry provides 90% of the national income, but employs less than 5% of the working population.

Natural gas from the oil fields makes possible the cheap generation of electricity and the production of fresh water by desalination. It is also the basis for the fertilizer and cement industries. A large flour-mill for processing imported wheat has been built at Umm Said, the oil terminal port. Lack of water and the poor soils were enormous obstacles to agriculture, but today Qatar is self-sufficient in vegetables and grows many other crops including tomatoes, eggplants and melons for export. The farms are Qatari-owned, but are operated by Palestinians with chiefly Baluchi labour; pesticides, seeds, windbreak trees and ploughing are provided free by the Ministry of Agriculture.

Qatar now has a road link with the rest of the Arabian Peninsula, and is thus no longer so geographically isolated. There is an international airport at Doha.

Oil rigs off the island of Halul in the Persian Gulf. Qatar is an important oil producer.

BAHREIN

Area: 255 sq mi (660 km²)
Population: 216,078
Population Growth Rate: 3·0%
Capital: Manama, pop 88,785
Language: Arabic. English widely spoken
Religion: Moslem (Sunni)
Currency: Bahrein dinar = 1,000 fils

Bahrain is a small desert Arab sheikdom in the Persian Gulf which was a British protected state from 1861 until 1971, when it proclaimed its independence and concluded a new treaty of friendship with Great Britain. Its name means 'Two Seas'.

The Land. The sheikdom occupies a group of low-lying islands west of the Qatar peninsula and 15 mi (24 km) from the coast of Saudi Arabia. Bahrain, the largest, is fringed with coral reefs and rises to 445 ft (136 m) in Jebel Dukhan. It contains the capital, Manama (88,785), the new port of Mina Sulman, the Dukhan oil field and one of the largest Middle East oil refineries. Linked with Bahrain by a causeway is Muharraq, with the international airport and the second largest town, also called Muharraq (49,387). The other main islands are Sitra, also linked with Bahrain; Nabi Saleh with its spring-watered date gardens; Jidda and Umm Nasan.

Climate. Annual rainfall is very low at 1·4–4·4 in (36–114 mm). There is also a high evaporation rate caused by summer (May-October) temperatures of over 100°F (38°C). Nonetheless Bahrain has abundant fresh water underground. Gazelles and jerboas are among the few wild animals that survive in Bahrain's arid areas.

The People. The population is preponderantly Arab, but includes Omanis (about 11,000), Iranian merchants, Indian clerks, Iraqis, Syrians, Lebanese and Egyptians, as well as some 3,000 Europeans and Americans. Important centres in addition to Manama and Muharraq include Awali, the oil workers' town, and the new town of Madinat Isa. The government provides free education and medical services.

Oil, Industry and Agriculture. The discovery of the Dukhan oil field in 1932 made Bahrain the Persian Gulf's first oil state. Since 1945, production has averaged 2–4 million tons annually, and while proven reserves may be exhausted by 1990, offshore developments are a possibility.

Only refined products such as petrol and turbine fuel are exported – from the 3 mi

(5 km) long terminal off Sitra. Only a third of the oil refined comes from Bahrain, the greater part being piped 34 mi (55 km) from the Damman field in Saudi Arabia.

Profiting from the rapidly-increasing prosperity of the Gulf, Bahrain continues to have a thriving entrepôt trade, especially with Saudi Arabia.

Oil revenues have been used to broaden the economy. Mina Sulman has been developed as a free port and industrial area. In 1971 a $968 million aluminium smelter was inaugurated, based on alumina from Western Australia and local natural gas. Ras Abu Jarjour, south of Manama, has a $4·8 million earth satellite station, the first in the Middle East. The well-watered northern fringe of Bahrain provides dates, fruits and vegetables. Fishing has prospered since a plant was opened to freeze prawns for export, but the once-important pearl fisheries have become insignificant.

KUWAIT

Area: 5,790 sq mi (14,996 km²)
Population: 831,000
Population Growth Rate: 9·0%
Capital: Al-kuwait, pop 295,000
Language: Arabic. English widely spoken
Religion: Moslem (Sunni) 95%; Christian 4%
Currency: Kuwait dinar = 1,000 fils

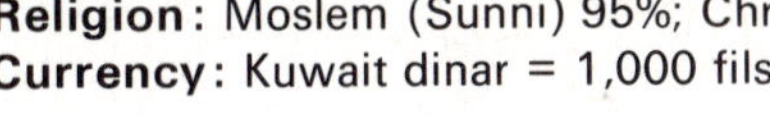

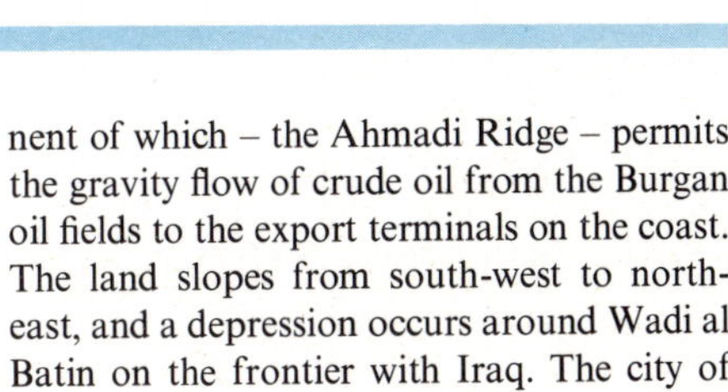

Kuwait, a small oil-rich nation at the northern end of the Persian Gulf, is the most prosperous and most progressive state in the Arab world. Indeed, in terms of per capita income, it is one of the richest countries in the world, and has more than once granted large loans to poorer members of the Arab League. In 1961 Kuwait became an independent state, but Britain promised military support if needed.

Kuwait's *amirs*, members of the Sabah dynasty which has ruled Kuwait since 1756, have used the enormous oil revenues for the benefit of the people. The modern state that has been built up has many remarkable features including free education, free medical services, subsidized housing, free telephone services, and no income tax. The symbols of an affluent society such as cars, television sets, radios, refrigerators and air-conditioning units abound. Here, at least, it seems that poverty has been virtually wiped out, except in a few remaining shanty villages.

However, although the foreigners in Kuwait, Palestinian Arabs, Saudi Arabians, Egyptians, Iraqis, Indians, Pakistanis and others, forming nearly half of the population and some 80% of the labour force, are among the wealth-makers, they have little standing in Kuwaiti society, and may not own property or participate in politics.

The Landscape. Kuwait consists of a mainland area on the north-west coast of the Persian Gulf and offshore islands including Būbiyān. South of the mainland area is a Neutral Zone shared with Saudi Arabia. The mainland is an undulating sandy plain broken occasionally by low ridges, the most prominent of which – the Ahmadi Ridge – permits the gravity flow of crude oil from the Burgan oil fields to the export terminals on the coast. The land slopes from south-west to north-east, and a depression occurs around Wadi al Batin on the frontier with Iraq. The city of Kuwait was originally founded on the blunt sandstone promontory on the southern shore of Kuwait Bay, a large and sheltered anchorage. The name Kuwait means 'little fort' and may have originated from the Portuguese castle that once stood here.

Climate and Vegetation. Kuwait is a land of extremely hot, dry summers and cool winters. Summer maximum shade temperatures often rise to 112°F (45°C), while winter temperatures drop below 45°F (8°C). Rainfall, about 4 in (100 mm) annually, is unreliable. Sandstorms are especially frequent from May to July.

The natural vegetation is thin and consists mostly of hardy bushes of the goosefoot family. In the extensive Tuffat al Adhami salt marshes in the Neutral Zone, only salt-resistant scrub survives. The date palms, acacias, tamarisks and other ornamental trees planted along the roads and in public parks are kept alive by regular irrigation.

The People. Almost half the population is concentrated in the city of Kuwait (295,000), which has had a carefully planned expansion since 1951. Al Ahmadi was developed as an oil town, and its ordered growth is reflected in a geometric street plan and well-landscaped gardens.

Kuwait is a Moslem state and most Kuwaitis belong to the Sunni sect. There are both state and private schools, and a university (founded 1966), but the illiteracy rate still exceeds 40%.

Oil Production. In 1971, with a total output of more than 140 million tons, Kuwait was the third largest oil producer in the Middle East. Burgan and Magwa Al Ahmadi are the oldest and most important oil fields, and are linked by pipeline with the export terminal at Mina al Ahmadi via Al Ahmadi, which also receives oil from the Raudhatain, Sabriya, Bahrah and other northern fields. There are other oil fields on land and offshore, in the Neutral Zone, such as Wafra, South Fuwaris and South Umm Gudair. Most of Kuwait's production is exported as crude oil, but some is processed before export at the Shuaybah refinery.

Industry. The Shuaybah refinery and petrochemical plant is but one of the growing number of manufacturing and service industries. An electricity generating station and sea water distillation plants have been built in the Shuaybah industrial zone, and use gas supplies from the Burgan oil field. State and private companies, backed by considerable government finance, now produce bricks, asbestos pipes and milled flour.

Farming and Fisheries. Kuwait has very little fresh water, and even the brackish supplies from Raudhatain and Sulaibiyah are insufficient to permit irrigated arable farming, although a government experimental farm has been established. Water and nearly all foods have to be imported. A long-standing project for bringing fresh water in from Iraq has still to be implemented. Meanwhile hydroponics has been investigated.

The government has also researched the suitability of various breeds of cattle for Kuwaiti conditions, and has encouraged the introduction of improved breeds of sheep and poultry. Privately maintained herds and goats have survived the oil boom though stock-rearing generally is in gradual decline.

Fishing is an important industry which has been modernized since 1959. Fish is an important part of the local diet. Every year large quantities of prawns are shipped from the deep freeze plant of the Gulf Fisheries Co to Europe and the United States. Since 1963 the state-owned Kuwait National Fisheries Co has been operating in the Persian Gulf, using a refrigerated supply and depot ship as its mobile headquarters.

Transport. Kuwait has a simple but excellent network of roads. Public transport does exist, but the private car is the most common form of transport. According to UN sources, 103,936 private cars – one for every seven persons – were registered in 1969.

The state-owned Kuwait Airways operates regional and international services.

Women dressed in their traditional black, veiled 'abiahs' in Al Jahra in central eastern Kuwait. Women enjoy a relative, though strictly limited, freedom in Kuwait.

International Trade. Because it has few resources except oil, Kuwait must import most of its food and consumer goods. The domestic market is limited by the small population, but the absence of strict government control on trade has encouraged the traditionally strong trading community to expand entrepôt trade. Kuwait acts as a merchant broker for much of it and trade is mainly with the United States, Japan and West European countries.

IRAQ

Area: 169,240 sq mi (438,332 km²)
Population: 9,750,000
Population Growth Rate: 3·2%
Capital: Baghdad, pop 2,969,000
Language: Arabic
Religion: Moslem (Shi'a 63%; Sunni 34%) Christian 3%
Currency: Iraqui dinar = 1,000 fils

Iraq is an Arab republic astride the great rivers Tigris and Euphrates. Today, Iraqi oil helps to provide the power upon which much of the industrialized western world depends: this oil pays for major improvements to Iraqi agriculture, but politically Iraq has shown no such advance: coups, counter-coups and rebellions have weakened Iraqi government since the 1950s. Internally, there was intermittent fighting between the government and its large Kurdish minority throughout the 1960s, and externally Iraq has had bitter disputes with Kuwait and Iran and is implacably opposed to monarchial Arab states.

The Land. Iraq has three contrasting regions: the deserts of the south and west, the Tigris-Euphrates alluvial plain, and the mountains of the north-east. The deserts, extending westwards into Jordan and Syria, and southwards into Saudi Arabia, are in fact largely arid limestone steppe bordered near the lower Euphrates by sand dunes. The Tigris-Euphrates plain stretches south-eastwards from Baghdad and Al-Falluja to end in lakes and marshes such as the Hor al Hammar as it approaches Iraq's short coastline on the Persian Gulf. The build-up of rich silts on the low-lying plain permits a high-yielding agriculture. North of Baghdad and Al-Falluja is the Iraqi *Jazira* ('Island'), a low plateau sloping south-eastwards between the two rivers and largely composed of desert except for a few better-watered valleys.

The Assyrian plains and foothills, rarely more than 3,000 ft (914 m) above sea level, separate the Mesopotamian region from the Kurdish mountains in the extreme north-east. Access to these rugged heights, which reach 12,229 ft (3,727 m) in the Algurd Dagh, is very difficult. From these jagged ranges flow four major tributaries of the Tigris: the Great and Little Zab, the Adhaim and the Diyala. The Euphrates has no tributaries in Iraq.

Climate, Vegetation and Wildlife. Summers in the plains are intensely hot with day shade temperatures often rising above 115°F (46°C) during July and August. Winters in the highlands are extremely severe, while even the plains experience frosts. In the plains about 7 in (178 mm) of rain falls annually, mainly between November and April. Winter precipitation of more than 40 in (1,016 mm) is common in the mountainous north-east, much of it falling as snow.

Vegetation ranges from the drought-resistant plants of the desert to the oak forests covering the slopes of the Kurdish mountains and succeeded at higher altitudes by stunted shrubs and thorn. In spring even the most poorly watered areas come to life with flowers and grasses following the winter rains, and north of the desert, steppe scrub lasts through the summer. Lower Iraq with its many streams, marshes and canals has perennial grasses and bulrushes and occasional poplars and willows.

Wildlife includes jackals, hyenas, desert rats and hares. Many snakes, lizards and waterfowl live along the great rivers.

The People. About 80% of the population is Arab and most of the rest Kurdish, although there are small numbers of Armenians, Syrian Christians, Jews and Yazidis. The large Iranian community concerned with trade and pilgrim traffic to the Shi'a shrines at Karbala and An Najaf has been nearly halved by recent expulsions, more than 89,000 having been driven out since 1969.

Most Iraqis, including the Kurds, are Sunni Moslems, but the Basra area has always been predominantly Shi'a. Education is free but not compulsory, and the illiteracy rate is about 80%.

The more densely settled parts of the country include the Assyrian plains in Mosul and Arbil provinces and the cultivated areas of Diyala, Baghdad and Basra provinces. More than half the population is concentrated in the cities and towns, including the capital, Baghdad (2,969,000), a rapidly growing metropolis; Basra, the chief port; Mosul, the commercial centre of the north; and Kirkuk, a Kurdish city and oil field centre. Some Kurds, like the Arab tribes of the desert, are pastoral nomads.

Oil and the Economy. Iraq is a major oil producer and its petroleum industry provides

The oilfields at Kirkuk, the starting point of the pipeline that runs across Iraq, Jordan and Syria to the Mediterranean coast at Lebanon.

about 25% of the gross national product. Output comes mainly from three areas: Kirkuk (Kirkuk, Bai Hassan and Jambur fields), the Ayn Zalah and Butmah fields north-west of Mosul; and the Al Zubair and Rumaila fields south of Basra. Minor producers are the old-established Naft Khaneh field and the Al Qaiyara field south of Mosul. Crude oil from the north is piped to the Mediterranean ports of Baniyas (Syria) and Tripoli (Lebanon), while southern oil is exported through Fao and the offshore terminal at Khawr al Amayah. Oil for use in Iraq is refined at Daura, near Baghdad, and other centres, and a petro-chemical industry is being developed. The Iraq Petroleum Company, the major oil producer, was nationalized in 1972.

The importance of oil has overshadowed Iraq's other mineral resources. These include the vast rock sulphur deposits at Al Mishraq, which are only just being exploited.

Industry. Baghdad is the chief manufacturing centre, although Basra and Mosul also have industrial importance. Products range from building materials to furniture, soap, dairy goods and beverages. The important textile (cottons, woollens and silks) and jute industries are concentrated in Baghdad, Mosul, Kut and Hilla. New industries are being established with Soviet technical expertise and machinery.

Farming, Forestry and Fisheries. About 43% of the land area is potentially cultivable, but only some 14% is actually under cultivation, about half of it being irrigated. Winter cereals, tobacco and fruit are the chief crops of the rain-fed north-east, while wheat, barley, millet, maize, vegetables and sesame are produced on the irrigated farms of the plains. The extreme south-east is important for its winter barley, summer rice, and also for dates, the chief export crop. Cattle, buffaloes, sheep, goats and horses are reared.

Since 1958 large areas of land have been distributed among the peasant farmers. New irrigation projects have been planned to supplement existing dams on the major rivers such as the Hindiya barrage on the Euphrates and the Kut barrage on the Tigris.

Neglect, indiscriminate cutting and persistent shifting cultivation have told heavily on the northern forests, which are mainly used as a source of charcoal and building materials. The extensive river and lake systems provide carp and other fish.

Transport. All major towns are linked by the state railway system, which also connects with the Syrian system at Tel Kotchek, thus providing a through route from Basra on the Persian Gulf to Turkey and Europe. An all-weather highway network also serves the major cities and provides links with Iran, Kuwait, Syria and Jordan. There are international airports at Baghdad and Basra. The chief seaport is Basra, but an increasing volume of traffic is now handled by the new port of Umm Qasr.

International Trade. Since 1958 international trade has been rigorously controlled by the government. Apart from oil, the only significant exports are dates and cement. Imports, including machinery, vehicles and foodstuffs, are provided mainly by the Soviet Union, Great Britain and West Germany. Among Iraq's other trading partners are Kuwait, Lebanon, Saudi Arabia, Italy, Japan and China.

IRAN

Area: 627,000 sq mi (1,623,930 km²)
Population: 30,159,000
Population Growth Rate: 2·9%
Capital: Tehran, pop 2,719,730
Language: Farsi (Persian. Kurdish, Turkic and Arabic spoken by minorities
Religion: Moslem, Shi'a (98%), Sunni (5%). Jewish and Christian minorities
Currency: Iranian rial = 100 dinars

Iran in South-West Asia is a kingdom three times the size of France. Economically it ranks as an emergent country, but its great central plateau and surrounding mountains

A Moslem shrine at Mahan on the edge of Iran's desert. Persia was converted to Islam after 671 A.D., when it was conquered by the Arabs.

have a historical tradition that goes back to the Persian Empire founded in the 6th century BC. Often invaded and occupied, Iran has nevertheless managed to retain its Persian character and traditions to the present day.

With the discovery of vast oil fields in south-western Iran in the early 1900s, the country was launched on an ambitious programme of modernization and reform. Since World War II, Shah Mohammed Reza Pahlavi, has followed a policy of political cooperation with the West, at the same time maintaining good economic relations with the USSR and other Eastern bloc countries. He achieved great political success in initiating revolutionary programmes for rapid economic and social development based on the profits from the massive oil industry, and has institutes land reforms and new building and road-construction projects.

The Land. Iran can be roughly divided into four regions: the mountains, the desert plateau, the coastal lowlands and the Khūzēstan Plain. The core of Iran lies on a high central plateau between the Elburz Mountains in the north, the Zagros Mountains of the south and west, and the lowlands adjacent to the Afghan and Pakistan frontier in the east. Much of this plateau is dry and barren, dominated by the Great Salt and Great Sand deserts, though there are several large and fertile basins. The Elburz mountain system runs at an average height of 10,200 ft (3,100 m) in an arc from west to east across northern Iran, attaining its maximum height in Mount Damāvand, 18,955 ft (5,780 m). The Zagros range, which borders the Iranian plateau to the south, also forms a mountain barrier between Iran and Iraq. It is deeply eroded by streams which drain into the Persian Gulf. Lowland areas are restricted to the agriculturally rich strip of coast along the Caspian Sea and the Khūzēstan Plain at the head of the Persian Gulf, an extension of the Mesopotamian region.

The country's land frontiers extend over 2,775 mi (4,400 km), most of which are established through international treaties. Border disputes exist with Iraq and Afghanistan. Iran has important sea boundaries including that along the Caspian Sea, which is shared with the USSR, and the Persian Gulf. Here, by virtue of its possession of offshore islands such as Khārk, Iran lays claim to more than half of the Gulf continental shelf.

Climate. The Caspian lowlands and the slopes of the Elburz Mountains receive a yearly average of 40–50 in (1,000–1,300 mm) of rainfall. The climate of the central plateau is one of extremes: very hot and dry in summer, and cold, with snow on higher ground, in winter. In the north-west and west, rainfall is fairly heavy and winters can be long and harsh with much snow. Like other areas of the country, this region experiences marked seasonal extremes of temperature.

In contrast, the southern lowlands have a hot and debilitating climate where rainfall is scanty.

Overall, more than a third of the land surface of Iran receives an average yearly rainfall of over 10 in (254 mm), while the heavy winter snowfalls in the mountains provide enough melt water to support spring cultivation and to refill underground reservoirs which supply water for irrigation throughout the year.

Vegetation and Wildlife. Man and his animal herds have done much to destroy and modify the pattern of natural vegetation. Nonetheless, more than 5,000 types of seed-bearing plants have been identified in Iran. Temperate and Mediterranean-type forests survive in the Elburz and Zagros mountains. Steppe vegetation grows on the plateau outside the deserts, thorns and salt-resistant plants persist locally in the deserts; date palms thrive in southern oases and tamarisks line dry southern watercourses.

Among a wide variety of animals the most famous wild species was the lion, adopted as the national emblem though now extinct. A few tigers may survive in the northern forests, and there are leopards, wolves, foxes and bears. Wild goats, sheep and a variety of birds of prey occur in the country's extensive mountain ranges, while gazelles roam isolated areas bordering the deserts.

The People. According to the census of 1966, the total population of Iran was 25,781,090 persons. Population distribution is extremely uneven. The massive mountain chains of the Elburz and the Zagros in the north and west, and the dry, barren deserts in the centre are very thinly populated. It is the fertile crescent of the west and north-west, the northern lowlands and the richer basins of the plateau that contain the bulk of the population. The population of Tehran, the capital, was estimated to have risen to 2·7 millions by 1970. An important chain of ancient cities lies along the line of the Zagros foothills from Hamadan through Eşfahān (575,000) and Shīrāz (300,000). The traditional Iranian cities are characterized by a fortified citadel (now often in ruins), a bazaar and caravanserai, and the quarters occupied by the permanent inhabitants of the town, often divided along ethnic or religious lines. The modern sections of the major cities are modelled on the Western pattern with wide streets, piped water, electricity supplies and mains sewage.

In marked contrast to the modern centres are the rural towns and villages of mud-brick houses and walls, surrounded by irrigated patches of cultivated land.

About 60% of the population is rural, including a declining minority of pastoral nomads such as the Bakhtiari, Kurds, Lurs, Qashqai and Shahsavan. About 49% of the total labour force is engaged in agriculture and 17·5% in industry. Conditions of health and education, formerly extremely poor in rural areas, have been improved considerably by crash programmes since 1962. The most widely spoken language is Persian (Farsi), written in Arabic script, but more than half of the population speaks different languages or dialects, including Turkish, Arabic, Baluchi, Kurdish and Gilaki. The state religion in the Shi'a branch of Islam to which more than 90% of the population adhere, though strong minority sects of Sunni Moslems, Christians, Jews and Baha'is are free to worship as they wish.

Agriculture. Agriculture forms the basis of the economy and directly and indirectly employs more than 60% of the Iranian labour force. The contribution by the agricultural sector to national income is estimated at 24%, larger by far than any other individual sector. Of the total surface area of the country, only 14% is used for agricultural purposes and of this only half is actively worked for field and orchard crops in any single year. Each year some 40% of cultivated land is irrigated, mainly by water from the ancient Persian system of underground water channels (*qanats*), or from the rivers. Large-scale modern irrigation schemes involving the building of dams are still being developed by the government, notably the Mohammed Reza Shah dam on the River Dez, the Empress Farah dam on the Safid Rud and the Amir Kabir dam on the River Karaj. Great care has to be taken to ensure that over-irrigation does not produce the twin evils afflicting agriculture in Iran: salination and soil erosion.

Farming methods are still primitive in many areas and oxen, donkeys and water buffalo are used as draught animals in conjunction with simple tools and implements. During the 1960s the government made great efforts to modernize agriculture through mechanization, and more than 8,500 tractors were sold to farmers by state agencies. A widespread impact has been made on agricultural productivity through the introduction of an extension and development brigade into most villages. Radical reforms in the land tenure system since 1963 have given all working farmers ownership rights to the plots they formerly worked as sharecroppers and tenants. A co-operative organization serves many villages, while large-scale farms are run by joint stock companies with peasant shareholders, and by agro-industrial units supported by foreign capital and largely export orientated.

Principal crops include wheat, barley and rice, though the varied climate permits the local cultivation of most vegetables and fruit. With the exception of wheat, quantities of which have occasionally to be imported, the country is self-sufficient in foodstuffs and even produces surpluses of cotton, rice, oilseed, nuts, dried fruit and gum tragacanth for export.

The raising of livestock is an important part of the agricultural economy, and it is estimated that the country supports some 27 million sheep, 12 million goats and 5·2 million cattle. Herding is carried on both by farmers and by nomads who spend the spring and summer grazing their flocks in the mountain valleys and in winter move to warmer areas along the Persian Gulf.

Forests and Fisheries. Despite heavy and indiscriminate cutting, large forest areas in Iran remain as natural woodland, of which the largest and commercially most important is the Caspian forest. All forests are owned by the state and are conserved as part of the national heritage.

The major source of fish in Iran is the Caspian Sea. Internationally famous, but now threatened with extinction through overfishing and pollution, is the Caspian sturgeon, from which comes the fine quality caviar sold on both domestic and export markets. The Caspian salmon is also taken from the sea off the northern coast and is regarded as a delicacy in Tehran. In the Persian Gulf, seer, mullet, sea bass, varieties of sardine and shrimps are caught on a small scale by local fishermen.

Pipelines are a relatively expensive way to transport oil and require a steady flow to be profitable. The huge oilfields in south-west Iran are linked, by pipeline, to the Persian Gulf.

Minerals. Iran is mineral rich, and in addition to being the major crude oil producer in the Middle East the country has large reserves of ferrous and non-ferrous metals and precious and semi-precious stones. Oil has been exploited since its discovery in southern Iran in 1908 and until 1951 was largely in British hands. Since that time the oil industry has been nationalized, though certain oil operations remain under control of a group of foreign oil companies known as the Consortium, of which the major shareholders are the British Petroleum Company and the Shell Oil Company. No less than 60% of the foreign exchange earnings of the country come from oil, a sum which totalled $938 million in 1969. In recent years there has been a growth of related industries, including petro-chemical and natural gas industries and metal pipe fabricating plants.

Minerals other than oil have a traditional importance but have taken on added significance following the discovery of copper ore deposits in the region of Sirjan in the Zagros. Rapid development of the iron ore and coal resources around Kermān has been stimulated by the demand for raw materials by a new steel mill set up at Eşfahān.

Industry. Since the 1930s, but especially since 1965, the government of Iran has pursued a policy of rapid industrialization with a view to reducing reliance on imports and to diversifying the economy which is over-dependent on oil and agriculture. Textiles, the main industry, is centred in Isfahan, Tabriz and Tehran. A multitude of plants processing such agricultural products as sugar, milk, cotton and rice have sprung up in recent years. The largest new industry produces transportation equipment, and the country now makes its own cars, trucks, buses, vans and tractors, though many components are still imported. Government or state-aided projects have been a decisive factor in the industrial programme.

Most industry is concentrated in the Tehran region where the nation's most prosperous market is to be found, but the oil fields of the south and the abundant labour supply in the regional cities has attracted industry to new areas.

Despite rapid modernization, cottage and bazaar craft industries still flourish. All Iranian cities boast a bazaar where potters, metal workers, jewellers and shoemakers ply their trades. Cottage industry produces high quality Persian carpets, a craft in which tribesfolk also participate.

Transport. A good network of roads has been built up in the present century, and an extensive rail system laid mainly during the 1930s links the capital with both the southern and northern ports over a distance of 2,175 mi (3,480 km). Chief ports are at Khorramshahr and Ābādān on the Persian Gulf and at Bandar-e Pahlevī on the Caspian Sea. Tehran and Ābādān airports serve international airlines including Iranian National Airways.

International Trade. The prime feature of Iran's trade is a consistent and large-scale deficit on accounts excluding the oil sector. The level of imports of capital and consumer goods has expanded rapidly in keeping with, and at times ahead of, the increase in income from oil. Deliveries of crude oil are directed mainly to Europe and Japan, while oil product exports from Ābādān refinery go to Africa and Asia, excluding Japan. Traditional exports are carpets, dried fruits, caviar and cotton. Goods other than carpets and cotton go mainly to Eastern Europe. Suppliers of capital, consumer goods and services are West Germany, USA and UK.

AFGHANISTAN

Area: 250,900 sq mi (649,831 km²)
Population: 17,087,000
Population Growth Rate: 2·4%
Capital: Kabul, pop 480,383
Language: Pushtu and Dari Persian
Religion: Moslem (Sunni) 90%
Currency: Afghani = 100 puls

Afghanistan is a landlocked, mountainous and conservative Moslem country of farmers and nomadic herdsmen. In the south and south-west, Pakistan and Iran isolate Afghanistan from the Arabian Sea. In the north, Afghanistan is bordered by the Soviet Union; in the east, by China. Border disputes have marred relations with Pakistan and Iran.

Afghanistan has a long history as a crossroads of political and religious forces and has known countless invasions. During the 19th century it served as a British dominated buffer state between Russia and British India and regained full independence only in 1921. Today, progressive reforms and foreign aid are slowly modernizing the country. In 1973 the constitutional monarchy came to an end when the king was deposed and a republic proclaimed.

Land, Climate and Vegetation. Afghanistan is dominated by the Hindu Kush, the great mountain range that slopes westwards across the country, diminishing in altitude from 22,000 ft (6,705 m) in the Wakhan corridor. The main ridge continues westwards in the Koh-i-Baba, extending to the Iran frontier along the Hari Rud River. The main geographical regions are the four great river valleys: the Hari Rud in the north-west; the Helmand draining the hill-and-desert south-west; the Kabul in the east; and the Amu Darya in the north.

The climate tends to great extremes, with many regional and seasonal variations. Mountains above 13,000 ft (4,000 m) tend to be permanently snow-covered. Areas above 8,200 ft (2,500 m) have more than six months of winter, while places above 4,000 ft (1,000 m), such as Kabul, have extreme variations in daily and seasonal temperatures. In the low river areas, summer temperatures reach 113°F (45°C), although winters can be cold, especially in the northern valleys. Summer heat is aggravated by strong hot winds (June-September). Annual rainfall averages 2–12 in (50–300 mm).

Afghanistan is an extremely beautiful country. The mountain areas are richly clothed in conifers, interspersed with numerous fruit and nut trees which give protection to many kinds of perennials, soft fruits and shrubs. By contrast the dry west has sparse shrubs and trees are rare.

Wildlife includes wolves, foxes, leopards, bears, ibexes, hyenas, wild dogs, wild cats and possibly tigers. Small mammals abound, and birds include pelicans, cranes, magpies, swallows, sparrows and many varieties of game birds.

The People. The Afghans are of diverse tribal origins but can be split into four main groups. The Pathans, related to the peoples of Persia and India, are the largest, comprising about half of the total population. Tajik tribes account for about 25%, while the Uzbeks and Hazares are also prominent.

There are two official languages, Pushtu and Dari (an Iranian dialect), but neither is nationally spoken. Most Afghans are Sunni Moslems; but some groups, notably the Hazares, are Shi'a Moslems.

The mountains and deserts are thinly populated, while the major valleys are densely settled. Kabul, once a small Pathan settlement, has expanded into a modern capital with 480,000 people. Other important centres are the oasis-town of Kandahār in the south and Herat in the west. Mazār-i-Sharīf is the rapidly growing administrative, defence and trading centre of the north.

About 2·5 million Afghans still live as nomads along the Pakistan border, but the government is encouraging them to settle by means of development schemes.

Despite compulsory education from 7 to 13 years of age, 90% of the population is illiterate, due in part to the shortage of schools and teachers.

Agriculture and Fisheries. Only a third of the country enjoys adequate rainfall. Irrigation is traditional and has been painstakingly and steadily extended by government schemes such as the Helmand Valley project. Just over half of the potentially cultivable land is used, about 13 million ac (5·3 million ha) being under irrigation. Wheat is a major crop, but maize, barley and rice are also important. Although Afghanistan is mainly self-sufficient in food stuffs, grain must be imported in drought years and sugar production is usually inadequate for domestic needs. Cotton thrives in the north around Mazār-i-Sharīf.

Livestock and related products are a mainstay of the economy and contribute about 80% of the exports. An estimated 6·5 million karakul sheep provide the valuable Persian lamb-skins. Some 15 million other sheep, 3 million goats and nearly 4 million cattle are also raised.

The commercial forests are mainly in the intermediate mountain areas of the east, where deciduous trees and the cedar predominate. Destructive exploitation has taken heavy toll of forest resources, but foreign aid has brought rationalized cutting and replanting to some areas.

Fish supplies come entirely from the rivers and lakes. The mountain streams teem with trout, and the lakes of Band-i-Amir, north-west of Kabul, are rich in fish.

Minerals and Industry. No complete survey of minerals has yet been made. Known deposits include silver, gold, sulphur and petroleum, but the terrain makes exploitation difficult and only coal, salt and chromite are currently worked. Large deposits of iron ore have been found at Hajigal, and lead, manganese, barite, beryl and gypsum await development. Afghanistan is the world's largest producer of lapis lazuli (at Badakhshan). Natural gas is extracted at Shorbaghan for domestic use and for export to the Soviet Union.

A view of Bamian in the north-western ranges of the Hindu Kush. Hundreds of man-made caves have been cut into the cliffs, together with two great Buddha figures, which probably date back to the 6th century A.D.

Traditional craft industries include metal-working, tiles (particularly at Herat), glass, leatherwork and jewellery. The famous Afghan rugs and carpets, woven in the rural areas, are a leading non-agricultural export. Modern industries include a sugar refinery (at Bāghlān), cement and textile mills (Pul-i-Khumri), cotton mills (Gulbahar), rayon and woollen mills, and a fruit-processing plant.

Transport. The rugged terrain and the long distances between the chief towns make transport a major problem. There are no railways, but an excellent system of all-weather trunk roads linking Kabul with Iran and Pakistan has been built with Russian and American aid. All major centres are linked by air services.

International Trade. Fruits, nuts and vegetables are exported to Pakistan and India, natural gas to the Soviet Union. Karakul sheep-skins, wool, cotton, carpets and rugs find ready overseas markets, and tourism is a growing source of foreign currency. Imports are mainly manufactured goods.

PAKISTAN

Area: 310,403 sq mi (803,944 km²)
Population: 59,000,000
Population Growth Rate: 2·1%
Capital: Islamabad, pop 50,000
Language: Urdu Bengali. English widely spoken. Other minority languages
Religion: Moslem (88%), Hindu, Christian and Buddhist
Currency: Rupee = 100 paisa

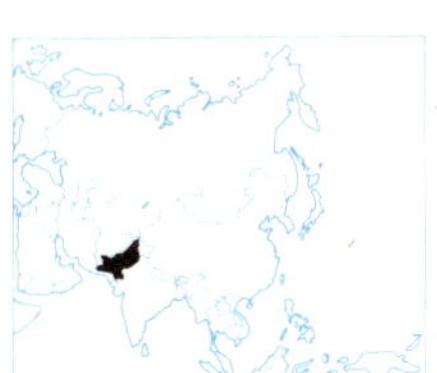

Pakistan, in the north-west corner of the Indian sub-continent, is a poor and largely undeveloped republic. It has few natural resources, little capital, and a considerable shortage of trained personnel. Its progress has been obstructed by wars, the loss of its eastern province (Bangladesh) – with more than half the country's population and valuable supplies of jute – and by the resistance of the mainly rural population to change. Such economic progress as has been made has been counteracted by the demands of the rapidly growing population, and population control is urgently needed.

The country was created amid immense loss of life and mass migration, in 1947, when British rule in India ended, to provide a homeland for the Moslem minority in India free from the threat of Hindu domination. But the physical division of Pakistan into two sections 1,000 mi (1,600 km) apart, and continuing antagonism with India caused endless unrest and in 1971 a war broke out between East and West Pakistan. India helped the former attain its desire for independence under the name of Bangladesh. The new socialist government of West Pakistan was faced with the task of restoring stability to a country damaged by war and with its economic base more or less halved.

The Land. Pakistan covers an area of 310,403 sq mi (803,944 km²). It includes the provinces of Sind and Punjab bordering India; Baluchistan, in the south-west, bordering Iran; and the famous North-West Frontier province, once the most troublesome outpost of the British Empire, along the border with Afghanistan. The north-eastern frontier lies along the cease-fire line in the disputed province of Jammu and Kashmir.

Pakistan is a land of great scenic contrasts. From the high peaks of the Himalayas in the north and north-west, the country falls gradually away to the south-east until, on the shores of the Arabian sea, it becomes low-lying plain, desert and swamp.

Huge mountain masses dominate the north. Some of the highest mountains in the world are found here, including Nanga Parbat at 26,660 ft (8,125 m). The River Indus emerges from the mountains, cutting deep gorges on its turbulent course to the plains. Chitral, west of the upper Indus, is an area of ranges and high plateaux. The Hindu Kush in the north-west forms a mountain wall separating Pakistan from Afghanistan. Further south, the famous Khyber Pass cuts through the mountains, joining Peshawar to Kabul. The western mountains branch out as they con-

The Khyber is a strategically important pass linking Afghanistan with Pakistan and the lands further to the east. In the past, it has been crossed by the armies of the Persians, Greeks, Tartars, Mongols and Afghans. In the 19th century it was vital to the English interests in India.

tinue southwards into several parallel ranges. The most important is the Sulaiman Range in northern Baluchistan, which reaches 11,000 ft (3,300 m) at Takht-i-Sulaiman.

Baluchistan, the western province, is a region of dry, high plateaux and mountain ranges, separated from the Sind region on the south-east by the limestone rampart of the Kirthar Range. A second plateau area lies around Rawalpindi in the north of the country separated from the vast Indus plain by the Salt Range. In the south, the province of Sind includes part of the Thar Desert and borders on the swampy Rann of Kutch.

But most of Pakistan is occupied by the huge alluvial plain of the Indus river basin. Sloping southward to the Arabian Sea, the plain is watered not only by the Indus but by the rivers of the Punjab ('land of five rivers'), the Jhelum, Chenab, Ravi, Beas and Sutlej. These rivers provide water for the extensive irrigation systems on which the high agricultural productivity of the area depends.

Climate, Vegetation and Wildlife. The climate of Pakistan is dominated by the monsoons, although considerable areas lie some distance from the sea and experience a more continental climate. In winter, from October to March, the cold, dry north-east monsoon blows off the land to the sea. Temperatures are low to very low in the north of the country and frost and heavy snow are frequent at higher altitudes. The north Indus plain is pleasant in winter. Lahore's temperature ranges from 40°F (4·5°C) to 68°F (27°C).

In summer, as the temperature of the land begins to rise, the south-west monsoon begins to blow inshore from the sea. For the first few months, the winds are dry and hot. Jacobabad, on the Indus plain, experiences temperatures of up to 113·9°F (45·5°C) in June – the highest in the subcontinent. On the coast the temperatures are lower, but, because of the humidity, the heat is more oppressive. Most rain falls towards the end of the summer months. The mean annual rainfall is 19·2 in (488 mm) at Lahore, 7·7 in (196 mm) at Karachi and 3·6 in (91 mm) at Jacobabad.

Much of Pakistan, including the Indus plain, has little natural vegetation but in the northern region, in the foothills and on the slopes of the Himalayas, great forests of deciduous trees and evergreens are found. The deodar is important for construction and oak, chestnut and walnut are all used for making furniture.

The northern mountain regions are also rich in wildlife. Leopards, wolves, Himalayan bears and the Siberian ibex are found and the waters of the Indus provide a home for what is probably the richest variety of water birds in the sub-continent. The sea off Karachi has one of the best fishing grounds in the world.

The People. In 1971 the estimated population of Pakistan was 55 million. The greater proportion lives on the Indus plain, particularly around Karachi. The northern mountains, the dry plateau of Baluchistan and the deserts are sparsely populated.

Invading Dravidians, Aryans, Greeks, Persians, Afghans, Arabs, Turks and others have all left traces of their blood in the present inhabitants of the country. The main group are the relatively light skinned Punjabis, a Mediterranean type. Other groups include the tall, fairer and often blue eyed Pathans, possibly of Semitic stock, and the Baluchi of sturdy Iranian stock.

Urdu is the official language, but English is widely spoken. Regional languages include Punjabi, Sindhi, Pushtu, Baluchi and Brahui. Only about 16% of the population is literate, but primary education became free in 1972 and there were plans to extend the school building programme. The vast majority of the population (about 88%) is Moslem but there are also important communities of Christians and Parsees.

Pakistan is still basically a rural country. Nearly 75% of the people get their living directly from the soil, working on subsistence farms and living in small villages where conditions are primitive, poverty ever-present and progress slow. Houses in the drier areas are mud-walled and have flat mud roofs. In the Baluchistan plateau, about half the population are nomads.

The largest cities are Karachi, Lahore, Lyallpur, Hyderabad and Multan. Islamabad, which replaced Karachi as the capital in 1959, has about 50,000 inhabitants.

Agriculture and Fisheries. Farming is easily the most important activity in Pakistan. Holdings are small and farming methods primitive. Fertilizers are in short supply and mechanization rare.

Over 50% of the cultivated land is under irrigation and the area is constantly extending. Important irrigation projects include the Guddu barrage on the Indus, the Rawal dam on the River Kurang, the Tanda dam on Kohat Toi and the Mangla dam on the River Jhelum, one of the largest earth fill dams in the world, which irrigates about 4 million acres and provides large quantities of hydroelectric power.

The staple food crop is wheat. Cotton has been cultivated in the Indus valley from prehistoric times and it is still a major crop, providing the country with much needed foreign exchange. Sugar-cane, rice, tobacco and oil seeds are among the other crops. The Punjab plain and irrigated parts of Sind grow superior quality rice and exports are increasing. Baluchistan produces fruit and dates. Livestock includes working animals such as oxen, camels, donkeys and mules as well as cattle and sheep. However, due to inadequate breeding facilities, lack of disease control measures and shortage of pasture lands, there are still far fewer animals than are required and the government has taken steps to improve the situation.

Fish are important, both as the major source of protein for nearly 40% of the population and as a major export commodity. The lakes of Sind abound in fish and many fish farms have been established. The waters of the Arabian Sea off the Karachi and Sind coast are particularly rich and tuna fishing is being encouraged.

Minerals. Pakistan's limited mineral sources have still to be fully developed. The Salt Range, Chitral, north-east Baluchistan and adjoining parts of Waziristan are the areas with the richest deposits. Coal is mined at Makerwal in the Trans-Indus region and several other places, but it is of low quality and seams are thin. Very small quantities of fairly good quality iron ore are mined in the northern mountain region. There are also reserves of gypsum, borite, chromite, rock salt, sulphur antimony and limestone.

Of more importance are oil and natural gas. There are indications of the existence of substantial quantities of oil but production from the six fields on the Potwar plateau meets only 20% of the country's requirements and there is a relentless search for further supplies. During oil exploration, vast deposits of natural gas were discovered. The gas field at Sui discovered in 1952 is considered one of the largest in the world. Supplies are piped to the industrial centres at Karachi and Multan.

Industry and Trade. Pakistan's major industry is cotton textiles. Nearly 50% of the cotton mills are located in Karachi, but Lahore, Lyallpur, Multan and Rahimyar Khan are all important centres. The export of finished cotton goods has become an important item of international trade. The country is also a major producer of good quality raw hides and skins; tanning materials are in plentiful supply and the dry climate of the Indus valley is ideally suited to tanning. The export of finished leather goods is increasing rapidly, and the Bata shoe factory at Lahore is one of the largest in the world. Oil refining and the manufacture of fertilizer are other large scale industries.

Pakistan has great hydro-electric power potential. The Mangla dam is already in action and the completion of the great Tarbela dam on the Indus River will substantially increase supplies of electricity.

Cottage industries employ about 10% of the population and provide fine hand-loomed textiles, carpets, lacquered goods, ceramics and metal ware. Methods are traditional and sometimes primitive, but the cottage industries are valuable because they require little or no investment and the brightly coloured rugs and fabrics are a popular export.

Transport. There are about 23,000 mi (37,000 km) of roads in Pakistan, about half of them surfaced, but the most important element in the transport system is the rail network. Since independence, a large part of the system has been modernized, and routes cover more than 5,000 mi (8,000 km). In many areas, however, bullock-, camel- and horse-drawn carts are still the main method of transportation.

Karachi is the country's main seaport and Pakistan International Airlines (PIA) serves major overseas routes from the international airports at Karachi and Lahore.

JAMMU AND KASHMIR

Area: 86,000 sq mi (222,740 km²)
Population: 4,615,176 (Indian side of ceasefire line only)
Capital: Srinagar, pop 285,257 Jammu, 102,738
Language: Urdu. Kashmiri and others
Religion: As for India and Pakistan
Currency: As for India and Pakistan

Since 1949 Pakistan has controlled Azad ('Free') Kashmir, Baltistan and Gilgit in the north-west of this former princely state, a mountain region abutting northern Pakistan. Since partition Pakistan has disputed this division with India and border troops have clashed in 1965 and 1969. Jammu and Kashmir (popularly called Kashmir) is therefore the focus of bitter hostility between India and Pakistan. This magnificently mountainous land lies between those two countries in the far north-west of the Indian sub-continent, and the dispute for it goes back to the 1947 partition of India into Moslem Pakistan and Hindu India: Kashmir remained apart but its Moslem majority encouraged Pakistan to invade. Kashmir's Hindu Maharaja appealed to India for help and declared Kashmir a part of the Indian Union. Fighting has continued sporadically ever since.

The Land. Kashmir is a beautiful country somewhat smaller than West Germany, composed of high mountain ranges and deep, fertile wooded valleys. To the north is the massive Karakoram Range and south-east lies the Zaskar Range of the Himalayas, containing some of the world's highest peaks including Mount Godwin Austen at 28,250 ft (8,543 m). The Indus, rising in the remote plateaux of Tibet, carves out a narrow gorge through the mountains. To its south, the Pangi and Pir Panjal ranges confine the Vale of Kashmir through which the River Jhelum flows.

Climate shows considerable variation. Srinagar, at 5,200 ft (1,585 m) has January temperatures of 30·7°F (−1°C) and a July average temperature of 73°F (23°C). The average rainfall is 25·7 in (653 mm) spread fairly evenly throughout the year.

Forest, mostly coniferous, covers 10·5% of the total area. Planes, poplars and rhododendrons are typical plants. Wild mammals include bears, deer and ibexes.

The People and their Work. The population of Indian Kashmir was about 4·6 million in 1971, with a further million or so in Azad Kashmir. The vast majority – some 80% – depends on agriculture, and the productive soils of the Vale of Kashmir support the densest population. Over two-thirds of the people are Moslems and most of the remainder are Hindus. Literacy is fairly low (18·3%) but education is free and well over half the children go to school. Indian controlled Kashmir has a Legislative Council and Legislative Assembly and sends six representatives to India's *Lok Sabha* (House of the People). Pakistan-controlled Azad Kashmir has a president and a nominated Ministerial Council.

Rice, maize and wheat are the leading crops, but apples, plums, peaches and apricots are important. Pastoralism is also practised and wool is used for a variety of local industries. Mulberry silk production, and gums and resins add appreciably to the country's earnings.

There are few mineral resources of value in Kashmir and industry is relatively underdeveloped. Apart from artistic craft industries such as papier mâché and wood carving, the chief source of earnings is tourism. Transport is fairly primitive. There are no railways and the most important means of communication, both internally and with the outside world, is road transport. Air transport connects Kashmir with other parts of India and there are airports at Srinagar (285,257) the summer capital, and at Jammu (102,738) the winter capital.

INDIA

Area: 1,127,345 sq mi (2,919,824 km²)
Population: 550,374,000
Population: Growth Rate: 2·5%
Capital: New Delhi, pop. 3,629,842
Language: Hindi. English. Fourteen other national languages
Religion: Hindu (83%), Moslem (10%), Christian, Sikh, Buddhist
Currency: Rupee = 100 paise

India is the world's seventh-largest nation and second-largest in population (with more people than South America and Africa combined). Its name comes from *Sindhu*, an old name for the River Indus. This river flows through Pakistan not India, but the Republic of India occupies most of the great Indian sub-continent of southern Asia.

India extends 2,000 mi (3,200 km) from the Himalaya Mountains in the north to Cape Comorin in the south, and over 1,500 mi (2,400 km) from the Gulf of Kutch in the west to Assam in the east. The country has 4,250 mi (6,840 km) of coast, and it is bounded on the west, south and east respectively by the Arabian Sea, Indian Ocean and Bay of Bengal. India has political borders with Pakistan in the west; Bangladesh and Burma in the east; and China, Nepal, Sikkim and Bhutan in the north. Half of India lies south of the Tropic of Capricorn and almost all of India has a form of tropical or sub-tropical climate.

The Emergence of India. Indians include all the main racial groups, divided by hundreds of dialects, two great religions (Hinduism and Islam) and the caste system which split Hindus into different socio-economic groups.

British rule ended in 1947, but British ideals of law and democratic government persist in the new Hindu-dominated nation of India that now comprises most of the subcontinent outside the two large Moslem areas of Pakistan and Bangladesh.

The Republic of India set out to raise its desperately low living standards by a series of economic five-year plans which began in 1951. Religious, political and linguistic divisions and old taboos have caused internal troubles, but the main national problem remains poverty worsened by a population explosion.

In international affairs India has clashed militarily with its neighbours Pakistan and China. Its leaders also gained a reputation as mediators between hostile major powers and as spokesman for the poorer nations of the so-called 'third world'.

Surface Features

India's 1,127,345 sq mi (2,919, 824 km²) represents an area about five times the size of France. This great slab of land has three main regions.

The Himalayas. Northern India is dominated by the Himalayas, although only their western and eastern ends are actually inside India's borders. These ranges effectively shut off the Indian sub-continent from Central Asia. With 92 peaks over 24,000 ft (7,315 m) and stretched out over 1,500 mi (2,415 km), the Himalayas are the tallest group of mountains in the world. Nanda Devi at 25,645 ft (7,817 m) is the highest peak in India itself.

The Himalayas are a complex region in which geographers recognize five parallel zones. The southernmost zone comprises the sandstone Siwalik Hills which run along the foot of the Himalayas proper. Behind the Siwaliks lie the Lesser Himalayas, with heights ranging from 6,000 ft to 10,000 ft (about 1,800 m to 3,000 m). Next comes a zone of deeply dissected spurs. This stands at about 15,000 ft (4,600 m), and is succeeded by the Greater Himalaya which fan out from their westernmost peak of Nanga Parbat 26,660 ft (8,131 m), in Pakistan-controlled Kashmir. Only a little of the fifth zone is in Indian-controlled Kashmir. This zone is the Indus–Tsangpo (Brahmaputra) Furrow, which at 12,000–14,000 ft (about 3,900–4,600 m) contains the headwaters of two of the sub-continent's largest rivers.

The Northern Plain. Immediately south of the Himalayas lies a great east-west lowland belt about 200 mi (320 km) wide. This is occupied by parts of three great river systems: the Indus, Ganges and Brahmaputra. The plain originated as a massive downfold in the earth's crust formed in the same buckling movements that thrust the Himalayas upwards. In places the depression is at least 6,500 ft (1,980 m) deep. But in the last few million years mud and stones dumped by the rivers flowing from the Himalayas have gradually filled this immense trench.

Today, the Northern Plain is a great, almost level, tract of low-lying land, whose deep layer of alluvium makes it among the world's most fertile farming regions. The great rivers that laid down this soil shift many hundred thousand tons of mud a day and provide water for irrigation and a useful annual silt deposit. The mud banks built up by these rivers are among the few natural features that break up the flatness of the land.

The Peninsula. The great central feature of the Indian peninsula is the Deccan, a triangular plateau with its apex in the south. North of the Deccan, the low Vindhya Mountains separate the plateau from the Northern Plain. To the west, the Deccan rises to the 5,000 ft (1,500 m) Western Ghats, a range that parallels the western coast from which it is only separated by a narrow coastal plain. The Deccan tilts gently down to the east where it ends in a low, broken range called the Eastern Ghats. In the south, the Eastern and Western Ghats converge to form the Nilgiri Hills which reach 8,648 ft (2,636 m). East of the Eastern Ghats, the land slopes down to a broad coastal plain.

The peninsula's biggest rivers (Cauvery, Godavari, Krishna, Mahanadi and Penner) flow east or south-east across this eastern plain and on through deltas into the Bay of Bengal. Unlike the Himalayan rivers, these receive no snow melt and have a greater seasonal variation in flow. Rivers over a mile wide during the rains may shrink to trickles a mere few feet across for nine months in the year.

Climate. Traditionally the Indian climate has been typified as 'monsoonal'. Derived from the Arabic word meaning 'season', monsoon in its Indian context refers to the rainy period which lasts over most of the land mass from June to September. Most of the sub-continent has four seasons, though these differ from those of countries in temperate latitudes. They can be called winter, spring, the rains and the retreat of the monsoon.

In winter (from mid-December to February) temperatures fall away the farther north you travel, and in the extreme north the Himalayas act as a cold air source. Over most of India, this period is relatively cool and dry; only in the extreme south do mean daily temperatures remain above 70°F (21°C), and while day temperatures in the north may reach 80°F (27°C) nights are often cold and frost can occur. Rainfall is slight.

In spring, heat 'lows', developing over the southern peninsula and moving northwards, reach the Punjab lowlands by June. Relatively cool and dry north-easterly winds are now replaced by warm moist southerly winds. When these low-level winds coincide with high level divergence in the westerly jet stream of air flowing above them south of the Himalayas, devastating storms can result. Most of the peninsula however remains dry, and through March and April temperatures rise rapidly particularly in the north. The middle Ganges plains reach temperatures of 100–110°F (38–43°C), while in the south temperatures remain somewhat lower, rarely exceeding 100°F (38°C).

The rainy season occurs from about the middle of June and the middle of September when winds moving in from the Arabian Sea and Bay of Bengal bring rain to most of India.

Temperatures over most of the peninsula drop somewhat after the onset of the rains. During the rainy season most of the land mass receives over 20 in (508 mm), but rainfall varies from under 10 in (254 mm) in the north-west's Thar Desert to over 400 in (over 10,000 mm) in Assam in the north-east, and from under 10 in (254 mm) in the southern central Deccan to over 200 in (over 5,000 mm) along the Western Ghats.

The large scale wind reversal marking the retreat of the monsoon (mid-September to December) is relatively slow. Low level cyclones continue to pass northwards over the peninsula. But rain-bearing winds are gradually forced farther south and alter direction to bring most of south-eastern India's rainfall (during the south-west monsoon season the rain-shadow effect of the Western Ghats keep the extreme south-east dry). Eventually, stable north-easterly airstreams push south and reassert themselves over the whole of the sub-continent.

Natural Vegetation and Wildlife. Under natural conditions most of India would be covered by trees, but thousands of years of human interference have removed or degraded vast areas of forest. Only 20% of the total land area (277,000 sq mi; 717,000 km²) is now classified as forest, but much of this is little more than scrub.

There are several kinds of natural forest. Moist tropical rain forest predominates between 1,500 ft and 4,500 ft (about 460–1,370 m) on the Western Ghats south of Bombay, and in Assam up to 3,000 ft (915 m). Where rainfall drops below 120 in (about 3,050 mm) this gives way to dry tropical cover, with sal forest in the north-east, and teak forest, particularly in Mysore in the south-west. Dry tropical forest covers more land than any other type. On the sea edge of most deltas, and particularly in Bengal, there are extensive mangrove swamps. In addition to these groups there are the sub-tropical and temperate hill forests of the south (the Nilgiri and Palani hills) and north (the Himalayas). Between 3,500 ft and 5,000 ft (about 1,070–1,525 m) in southern India, stunted rain forest gradually gives way to wet temperate forest. In the Himalayas there is a distinction between the wetter east and the drier west. While wet hill forest, with evergreen oaks and chestnuts, predominates in the east, further west sub-tropical pines, and at higher levels deodars and mixed conifers, are increasingly common.

Not all India is naturally forested. Only tough, drought-resistant bushes and smaller plants persist in hot desert areas and low-growing alpine plants replace trees high in the cold Himalayas.

As farms have replaced forests, some of India's big mammals have grown scarcer, among them several handsome kinds of deer

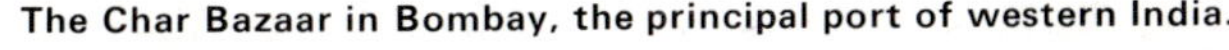

The Char Bazaar in Bombay, the principal port of western India.

and the gaur (the world's largest wild ox).

The Asiatic lion and Indian rhinoceros are now on the brink of extinction and fewer than 2,000 tigers remain in India. Indian elephants, leopards, bears of several species, many kinds of monkey and other smaller mammals appear to have more chance of survival. The more than 2,000 bird species recorded for the whole sub-continent is a variety perhaps unrivalled anywhere. Reptiles (with around 550 species) are also very varied and include three kinds of crocodile and the deadly cobra. Insects range from more than 1,400 decorative kinds of butterfly to the detested locusts, termites and disease-bearing mosquitoes.

The People

All of the world's major racial groups are represented in India, but much intermingling has occurred, making it difficult to place many individuals in any special category. Perhaps the people that have lived in India the longest are the Veddoids, broad-nosed, slightly built individuals including small forest groups such as the Kadars of the south. Some Veddoids are dark skinned and their ancestors may have interbred with Negroid peoples, although the only true Negroids still on Indian soil are the Negrito pygmies of the Andaman Islands. Some anthropologists consider the Veddoid peoples part of the Australoid group. Waves of Caucasoid peoples also entered India and evolved there in ancient times, among them the dark-skinned Tamils of the south, some perhaps displaced from the north by the Aryan Caucasoids who had seized the Northern Plains by 1200 BC and whose relatively fair-skinned descendants feature prominently in the north. There are many Mongoloid peoples in the mountainous north-east, in the Himalayas and Assam.

Culture and Religion. India is the home of nearly 200 languages, and this profusion naturally hinders communication and business. Ten of these languages are spoken by 90% of the population. In the north, Indo-Aryan languages predominate such as Bengali, Gujarati, Marathi, Urdu and Hindi (spoken by about 40% of Indians, and chosen in 1965 as the official language, although the former official language, English, went on being used). In the south, four main languages make up the Dravidian groups: Tamil, Teluga, Kanarese and Malayalam, spoken by a combined population of over 100 million people. The earliest written literature, the *Rig Veda*, dates from about 1000 BC, although the poems had been handed down possibly for 600 years before that.

India is constitutionally a secular state, but 85% of the population is Hindu. In 1961, however, there were nearly 47 million Moslems, as well as 11 million Christians, 3 million Buddhists and 112,000 Parsees.

Hinduism has largely shaped life in India, especially through the caste system which divides Hindus into some 3,000 different groups. Membership of a caste is determined largely by occupation and place of birth, and is inherited. Members of one caste seldom eat with or marry members of another. This system has helped to perpetuate craft skills and patterns of mutual aid, but it has prevented most low-caste individuals from ever bettering themselves and it condemned some 60 million so-called untouchables to performing none but the most menial tasks. New laws against caste discrimination and social changes linked with educational improvement and growing cities are only slowly altering this ancient social system.

The Hindu belief that cows are sacred and must not be killed or eaten has also influenced Indian life by aggravating widespread undernourishment. Poor sanitation is another disease-promoting factor that has helped to keep life expectancy low. By the 1960s, though, India's low average life expectancy of 42 years was 31% above the 1950 figure.

Poverty and Birth Control. The declining death rate is mainly due to a partial conquest of disease. By the mid-1950s, India had made huge strides against malaria and in other fields of public health. Mainly because better medicine kept more people alive, the population soared from about 439 millions in 1961 to about 547 millions in 1971 (a 25% rise in only 10 years). Unless the Indian government's birth control campaigns succeed, economists fear that the surging population will engulf any increase in farm and factory output, leaving India's traditionally poor peoples poorer than ever.

Sheer poverty partly explains the simple way in which most Indians live. Many men wear just a turban and a loincloth or a *dhoti* (a white cloth wrapped around the loins and legs) and many women wear a *sari* (a cloth wrapped around the body as a dress). Wealthy Indians often prefer such traditional dress; and rich women wear fine silk saris embroidered with gold thread. Most Indians eat a meagre starchy diet based on some kind of bread, often cooked in 'pancakes'.

Housing conditions are generally bad. Villagers occupy crowded mud huts with little furniture, no piped water and no electric light. They draw water from the local well. Cities contain frightful slums and shanty towns, while homeless thousands sleep on pavements.

Widespread lack of schools and teachers meant that by the mid-1960s 75% of Indians were still illiterate, but education standards were improving, and India had well over 60 universities.

Settlement Patterns. Over 80% of India's population lived in villages in 1961, and 350,000 of the total number of 564,258 villages held fewer than 500 people, and usually centred on the well or market. In the Himalayan regions village houses are often dispersed, and in some lowland areas they are strung out along river or canal banks.

Urban centres vary enormously from small traditional market towns to the sprawling cities that developed under British rule in the 18th and 19th centuries and expanded as manufacturing centres after India gained independence. These cities (for instance Calcutta, Bombay and Madras) reveal signs of their old role as colonial administrative centres in spacious barracks and administrative buildings that contrast with the narrow, tortuous streets of mixed commercial and residential buildings which have grown up around them.

Working elephants can understand up to thirty spoken commands from their particular 'Mahout'

Agriculture

Over 75% of India's population gains a living directly from the land, and farming yields about half the nation's income. Only four nations have more farm land than India and roughly half the country is under agriculture, but farming methods are largely primitive and crop yields per unit area are generally low. Moreover periodic floods and droughts have made India vulnerable to massive crop failures and resulting famines.

Types of Agriculture. There are three main kinds of farming in India, which may be labelled *shifting*, *sedentary peasant*, and *capitalist*.

Shifting agriculture is the most primitive type of farming and least important in the numbers of people it supports. It often involves clearing a patch of forest, cultivating food plants in the cleared ground, then moving on after a few years as soil exhaustion reduces crop yields. Shifting agriculture yields small harvests and has brought mounting soil erosion problems to Assam in the north-east, Orissa in the east and to parts of the south.

Sedentary peasant agriculture occupies the largest number of people and accounts for most land. It is also the most varied kind of farming with four sub-types: *wet* and *dry* cultivation, producing different crops and employing quite different techniques, and *kharif* (autumn harvest) and *rabi* (spring harvest) crops, occasionally grown on the same land. The kharif crops, sown immediately after the onset of the rains, are rice, jowar, bajra, sesame, cotton and jute. They are planted largely in flood-plain areas. Rabi crops on the other hand are sown after the rains have ended. Wheat, barley, gram, linseed, rape and mustard are the chief crops, grown mainly on uplands and *doabs* (the higher tracts between rivers). Of vital significance to all crops are the timing and amount of monsoon rains: thus rainfall that is too early or too heavy may wash away both seeds and freshly ploughed soil.

Capitalist farming is largely organized on the basis of plantations worked by hired labour. They tend to devote large areas to one or two profitable cash crops such as tea, rubber and coffee. Largely of foreign origin, plantations have expanded output rapidly since independence.

Types of Crops. Food crops occupy about three-quarters of the entire cultivated area of India, and a quarter of the entire cultivated area is under rice. Both in acreage and output rice is by far the most important single food crop. Over 4,000 varieties are grown. Most require a growing season mean temperature of 75°F (24°C) and 60–80 in (about 1,520–2,030 mm) of rain where irrigation is not available.

Rice is grown largely in the deltaic areas along the east coast – in Bengal, Andhra Pradesh and Tamil Nadu – and in Bihar, Madhya Pradesh and Uttar Pradesh in the centre and north. Yields per unit area are very low by world standards, but output of rice rose from approximately 39 million tons (39·6 million metric tons) between 1952–1956 to 62·7 million tons (63·7 million metric tons) in 1970.

Wheat, the second major food crop, occupies just over one-third of the area under rice and yielded just over one-third of the output in 1967. Wheat grows in drier areas than rice. The main wheat-growing states are Madhya Pradesh, Uttar Pradesh and the Punjab with two-thirds of India's total acreage. Jowar, a kharif crop, occupies a larger area than wheat but has a considerably lower output; it is grown principally in central and southern India.

Cash crops comprise some grown on big capitalist owned plantations and other crops mainly produced traditionally by peasant farmers. The main plantation crop is tea, which was an important export commodity long before independence. Production in 1967 had risen to 364,000 tons (375,000 metric tons), more than that of any other country. Coffee and rubber outputs were also growing, and India was the world's largest pepper producer.

Production of cash crops in the traditional peasant sector has also made rapid strides since independence, with increasing domestic demand. By 1967 India was the world's largest producer of raw cane sugar, annually producing over 90 million tons (over 91 million metric tons). Cotton (nearly 5 million tons; over 5 million metric tons) and peanuts (4·5 million tons; 4·6 million metric tons) were also mounting in importance. India is the world's biggest peanut grower.

Uttar Pradesh produced nearly half of India's sugar output; cotton and peanuts are more evenly distributed between Gujarat and Maharashtra in the west, Punjab in the north and Tamil Nadu in the south, where soils and climatic conditions are particularly favourable and an industrial market has been relatively long established.

Hindu veneration of the cow and a taboo on eating beef means that India's cattle are under-utilized, but they form an important element of the economy. In 1967 there were over 175 million cattle and 51 million buffaloes. Of these only 77 million were used for work, and the milk yield per cow is perhaps the world's lowest. There is no doubt that many of the cattle are an economic burden. Besides cattle and buffaloes there are 40 million sheep and 60 million goats but animals tend to suffer from poor breeding stock and sheer lack of fodder.

Darjeeling in West Bengal commands a fine view of Mt Everest and is a centre for the tourist trade and for the important tea trade.

Labour and Mechanization. The mechanization which has revolutionized Western farm-

The western Himalayas rise abruptly out of the Indian plain and form its border with China.

ing has had little direct impact on agricultural practices in India except in irrigation. Thus in 1961 of 40 million ploughs in use, 38 million were wooden; of 623,000 sugar-cane crushers, 590,000 were worked by oxen. The number of tractors in use was only 31,000 in 1961.

Hopes for raising crop yields are largely pinned to making more use of irrigation, fertilizer and selected seeds.

Traditional systems of irrigation have been in use for over 1,000 years. These are based on tanks, wells and canals. Salt accumulating in the soil and other watertable problems have not prevented a dramatic increase of 25% in the irrigated area since independence.

Indians have traditionally made relatively little use of natural fertilizers, but this situation is being changed by the development of modern fertilizer factories. In 1961 these produced half the total 259,000 tons (263,000 metric tons) of ammonium sulphate used in agriculture. Despite the great potential benefits of increased use of fertilizer, many technical and financial problems still hinder its effective widespread application.

New Seeds and Land Reform. Seed improvement is another innovation with enormous agricultural potential. Improved wheat strains have had a particularly beneficial effect, but there is no universally successful new strain of the chief food crop, rice, to replace the many local strains already grown. IR 8 ('miracle rice') and other improved strains are being introduced, but yields have remained obstinately low.

Land tenure systems also affect agriculture. Under old practices many peasants were the lifelong debtors of landlords whose rents rose steeply with rising land values. Moreover today farms still average a mere 5 acres (2 hectares), and most land is still farmed by peasants on a share-cropping basis. Land reform is crucial to agricultural progress, but recent legislation limiting the size of farm that one man may own remains ineffectual. The abolition of old landlord systems is little better than a legal fiction.

Forestry. We have already seen that population pressure on the land has cut India's once widespread forest cover to a mere one-fifth of the total area, and that much of this has been reduced to scrub. However, recent attempts at conservation have slowed and in some cases reversed deforestation. In 1964–65 the total forested area was 293,662 sq mi (752,982 km²), a rise of 2·5% since 1951–52.

About 7,750,000 cu yd (5,926,000 m³) of timber were produced in 1964–65, more than twice as much as in 1951–52. Minor produce such as bamboos, gums and resins, drugs and spices are also important. Although the Indian Government recognizes the urgent need to replenish forest resources, continuing pressure on the land makes any big increase in forested area unlikely.

Fisheries. Despite the extensive fishing grounds offered by India's continental shelf, fishing contributes relatively little to the Indian economy. But since independence there has been considerable growth in both freshwater fishing (notably in Bengal), and sea fishing. Between 1951 and 1966 the catch had grown by 82% to 1,367,400 (about 1,389,300 metric tons) of which more than a third were fresh water fish and the rest were mainly herrings, sardines and anchovies. Most of the total catch was marketed as fresh fish, and the rest was sold dried.

In theory there are great possibilities for expansion, despite local Hindu religious taboos on fish eating, but the development of a deep-sea fishing fleet is only just beginning.

Minerals. India has some of the largest deposits of iron ore of all qualities in the world. Total reserves are probably over 22,000 million tons of which more than one-third are highest grade ore. The eastern part of the peninsula is most richly endowed, its rocks containing several thousand million tons of haematite and lower grade ores. There are also large deposits in the south and elsewhere. Total iron ore production in 1969 was 21 million tons (21·3 million metric tons), a more than five-fold rise in 18 years. More than half this quantity was being exported in the late 1960s. But although India has perhaps a quarter of the world's iron ore reserves, only 2% of world output came from India in the late 1960s.

India is rather less well endowed with non-ferrous minerals, but manganese (important in steel making) is produced at a rate of over 1 million tons a year making India equal in output to South Africa and second only to the USSR. Magnesite, used for lining blast furnaces, also occurs in large reserves in the far south. There are big deposits of bauxite, the aluminium ore. Most comes from Bihar, Gujarat and Madhya Pradesh, three states straddling the Tropic of Cancer.

Rare and valuable minerals are also found in some quantity. As much as 90% of world mica extraction (for use with electrical devices) is carried out in the eastern states of Bihar, and in Rajasthan and the Punjab in the north-west. Gold production has slipped, from 14,117 lb (6,417 kg) in 1951 to 6,742 lb (3,058 kg) in 1969, all from the 2-mi (3·2-km) deep Kolar Gold Fields in Mysore. India is a leading salt producer; over 80% of its supply comes from evaporating sea water. Other minerals worked include apatite, asbestos, barytes, chromite, copper, corundum, diamonds, dolomite, fire-clay, ilmenite, lead, limestone, sillimanite, silver, steatite and zinc.

India has useful quantities of minerals used in power production. Two kinds – petroleum and the radio-active elements – are only recently exploited. While the geological signs indicate big oil deposits in the Northern Plains, reserves discovered so far total a relatively modest 45 million tons. By world standards output remains low but was rising by the 1970s.

Workable deposits of uranium have been found in Rajasthan, but sands on the south-west coast contain more valuable amounts of monazite (another source of nuclear energy). India has also some thorium.

India's most valuable exploited source of

energy is coal. Reserves of coal in India amount to over 70,000 million tons of medium and low grade coal, but a mere fraction of this quantity is coking coal. Almost all the coal is in the north-east, in West Bengal and Bihar. The Damodar Valley near Calcutta is the main source, where seams 80 ft (24 m) thick are relatively easily worked.

Coal output doubled between 1950 and 1970 to reach 70 million tons (71 million metric tons). Most of this came from West Bengal and Bihar. Lignite output has also increased rapidly and in 1967 reached nearly 3 million tons (about 3 million metric tons).

Energy Consumption. In 1967 electricity production was still less than that of Norway, a country with only 1 inhabitant for every 137 in India. Four main sources of electricity are now in use: coal, oil, flowing water and atomic energy. Thermal power stations account for about 60% of installed capacity, but these are largely placed to serve the cities. The development of a national grid is still far from completion.

There is hydro-electric power potential in three main zones: the Himalayan belt, the Western Ghats, and the southern hills. Schemes also operate in the south-east where rivers flow off the Deccan plateau. Seasonal variation in flow is one of the major problems facing hydro-electric developments. Even sub-Himalayan rivers vary enormously, for example during drought the River Tons may carry 8,000 times less water than it does in full spate. Moreover in the Himalayan area earth movements cause construction hazards. Despite these difficulties, installed hydro-electric power capacity increased nearly sevenfold between 1951 and 1966. Production remains about one-tenth that of the USA.

Industry

In the 17th century India was perhaps as industrially advanced as any nation. By the 19th century, British government restrictions and competition from British factories built in the Industrial Revolution were stunting India's own industrial growth. However, since India gained independence, industrialization has been rapid. Thus in the 10 years from 1956 to 1966 the industrial production index nearly doubled, largely through the help of foreign investment. But India remains industrially far behind the West, and only about 3% of its working population works in factories of any size.

India has two great groups of industries: (1) traditional small-scale consumer industries, with many products made at home; (2) new heavy industries, and consumer industries based upon them.

Since independence, the Indian government has stressed the need to expand old-style village industry and to expand the output of traditional products. Thus investment in small-scale industry was substantial in the first two five-year plans and helped to lift the total in employment from 11 million in 1951 to 16 million in 1961. The most important single product was *khadi*, a form of cotton cloth produced by 'home weaving' techniques. Khadi production rose from 843 million yd (778 million m) in 1951 to 1,865 million yd (1,721 million m) by 1959. Other products encouraged in the small-scale sector are sewing machines, fans, and bicycles, as well as numerous craft goods. However, small-scale industry rarely gives full-time employment; it is carried out largely in family units, and it is best developed in areas far from the largest cities and factories.

While cotton cloth manufacture is the most important of the traditional small-scale manufacturing industries, it is one that has most profited from expansion and modernization. Cotton manufacturing now employs over 25% of all factory workers and India usually ranks third or fourth in world production tables for cotton goods. The distribution of cotton mills reflects changes since mill production started at the end of the last century. The first big cotton milling centre was the port of Bombay, near cotton growing areas; convenient for importing the necessary stores, machinery and coal; and an outlet for exporting yarn to China. By 1950 Bombay still produced half the total output of finished goods, but since independence new centres have developed, notably Coimbatore in the southern state of Tamil Nadu, where spinning has expanded rapidly to meet local handloom demands. Nonetheless, by the early 1970s the leading cotton centres were still in the western cotton growing states of Maharashtra and Gujarat.

The second major traditional textile industry, jute, suffered a severe setback at Partition. For while all the processing factories were in West Bengal, largely in Calcutta, almost the entire production area was in East Pakistan (now Bangladesh). The woollen industry also suffered from Partition.

Besides some textiles, processed foods are among the traditional products that have forged ahead most strongly since independence. Output has steadily risen, both of food crops processed for home market (wheat, rice, sugar and salt) and of such processed export crops as tea.

The Modern Industrial Base. The modern iron and steel industry was only introduced to India in 1911 with the opening of the Jamshed-

The Secretariat buildings in New Delhi, designed by Sir Herbert Baker. The buildings form part of the ambitious government complex laid out by Sir Edwin L. Lutyens when the capital was transferred from Calcutta to New Delhi in 1912.

Gypsies are believed to have originated in India. Their language, Romany, is related to Sanskrit and Prakrit. These are Marwari Gypsies in Rajasthan in north-western India.

pur works. Iron and steel expansion was one of the major targets of the Second Five-Year Plan, and four new steel plants in the north-east (at Bhilai, Rourkela, Durgapur and Bokaro) helped to more than quadruple the output of steel ingots between 1951 and 1966. By the 1970s more steel plants were being built elsewhere. Yet in 1967 India was only the world's thirteenth largest steel producer, with one-eighteenth the output of the USA.

Expansion has been rapid in other modern industrial sectors. Heavy, light, and electrical engineering industries have all grown fast. In the 10 years to 1966, diesel-engine output, for example, rose from 7,246 to 87,876. Supplies of automobiles were still tiny, with 1 car per 1,000 people (against more than 402 per 1,000 in the USA). Production of engineering goods ranges from machine tools to jet aircraft. Demand from industry and agriculture has seen growth spurts in chemicals and cement. For instance, plentiful limestone helped to more than treble cement production between 1951 and 1966.

Most new industries have emerged near towns that grew up long ago as shipping centres for raw materials or by manufacturing products from agricultural raw materials, particularly jute and cotton. Thus Calcutta, Bombay and Ahmadabad are among the leading industrial centres, but new centres like Jamshedpur, based on coal and iron, have sprung up too. The main industrial areas are now in the north-east, on the west coast and in the south.

Despite massive capital investment, the increase in factory employment has been relatively slow, rising from 2,914,000 in 1951 to 4,682,000 in 1966, a mere 30% of the estimated number of unemployed. To create more jobs and decentralize industry, India has introduced planned industrial estates. By 1962, 71 estates had been established with a total of 138 factories. The best equipped and most successful were those near established industrial centres. Some industries including cotton continue to be privately financed, but heavy industry is under government control.

Transport and Communication. For centuries three means of transport prevailed in the Indian sub-continent – walking, bullock carts, and river craft. British rule saw the first major changes to this pattern, with the growth of a widespread rail network and the outlining of a trunk route road system. At independence India had over 30,000 mi (48,000 km) of rail, but still less than 100,000 mi (160,000 km) of surfaced road. The present day transport system still represents a mixture of the traditional and modern.

By the late 1960s India's more than 36,000 mi (nearly 60,000 km) of railways formed one of the largest rail systems in the world: but India still suffers major difficulties with all three types of road transport – inter-city highways, rural roads and urban transport. By the early 1970s nearly one-third of India's 500,000 villages were still more than 5 mi (8 km) from an all-weather road.

Coastal trade has expanded steadily since independence, and while the colonial ports of Bombay and Calcutta are still dominant, Madras and Vishakhapatnam are also important and Cochin and Kandla of increasing significance. Internal and external air services have developed greatly.

International Trade. India inherited a large reserve of foreign exchange as a result of British war spending. The position was altered dramatically by the heavy demand for capital goods sparked off by the second and third five-year plans, and by food shortages resulting from rapid population growth and poor harvests in the middle 1960s. In a decade (1956–66) the cost of these imports more than doubled. Exports on the other hand did not rise nearly so fast, and India devalued the rupee by more than one-third in an effort to resolve this crisis in its balance of payments. Difficulties continued, as India shipped in great quantities of raw materials yet made no major headway with its exports. Only iron ore exports showed a really big advance, more than doubling in value from 1960 to 1966.

Since independence the international structure of trade has also altered. North America is now by far the largest source of imports. Britain and other EEC countries are also large suppliers. Eastern Europe and Japan have greatly increased trade with India, and they now import Indian goods worth as much as those taken by North America and Europe together.

Possessions and Dependencies. India owns the Laccadive, Minicoy and Amindivi Islands in the Arabian Sea, and the Nicobar and Andaman Islands in the Bay of Bengal. India controls most of Jammu and Kashmir; the Kingdom of Sikkim is a protectorate.

MALDIVE ISLANDS

Area: 115 sq mi (298 km²)
Population: 114,469
Population Growth Rate: 1·8%
Capital: Male, pop 13,610
Language: Maldivian (a Sinhalese dialect)
Religion: Moslem (Sunni)
Currency: Maldivian rupee

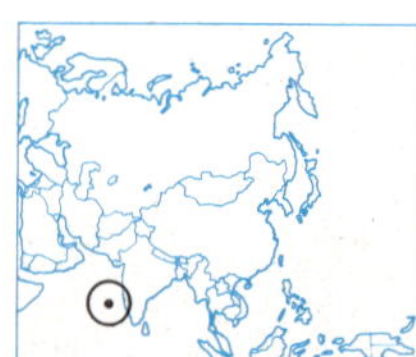

The 2,000 or so islands, of coral formation, which make up this group stretch in a chain of 12 atolls across nearly 500 mi (800 km) of the Indian Ocean south-west of India and Ceylon. The climate is hot and humid and the two monsoons that blow backwards and forwards over the islands bring an average of 60 in (1,500 mm) of rainfall every year. Grasses and tropical plants including coconut palms cover the land.

Only 220 of the islands are inhabited and the population of about 103,000 crowds densely on 115 sq mi (298 km²) of land surface. Half the population is engaged in fishing, and dried bonito, commonly known as Maldive fish, is the main export of the islands. Coconuts, copra, coir cordage and some fruits are also exported, but large quantities of rice have to be brought from India and Ceylon to supplement the local food supply. The Maldivians are great traders and navigators. They think nothing of making sea voyages of hundreds of miles – to Aden, Ceylon, Bombay or Calcutta – in their wooden *baggalas*. The capital is Male, with a population of over 12,000.

The Maldives were administered as part of Ceylon until 1948 and a republic was declared in 1968. In the early 1970s Britain still had facilities on the island of Gan in the Addu Atoll which had become a Royal Air Force staging post during World War II.

SRI LANKA

Area: 25,325 sq mi (65,592 km²)
Population: 12,747,755
Population Growth Rate: 2·3%
Capital: Colombo, pop 563,705
Language: Sinhalese. Tamil and English widely spoken
Religion: Hinayana Buddhist (64%), Hindu (20%), Christian and Moslem (15%)
Currency: Sri Lank rupee = 100 cents

Sri Lanka (Ceylon) is an island which hangs like a pear from the south-eastern tip of India. Sri Lanka, the republic's official name since 1972, means 'beautiful' or 'blessed country', a title well merited by its lush tropical forests, palm-edged beaches and prosperous plantations. Spices and other crops grown on the island were internationally famous 2,000 years ago, and today Ceylon grows more tea than any country other than India.

In 1948 Ceylon became the first British crown colony to achieve independence within the Commonwealth of Nations. Independent Ceylon underwent several changes of government, in which economic difficulties, and communal and religious differences, played an important role. In 1960, Mrs Sirimavo Bandaranaike succeeded her assassinated husband to become the world's first woman prime minister. The government survived a revolt by revolutionary students in 1971.

The Land. With an area of 25,325 sq mi (65,592 km²), Ceylon is a little more than half the size of England and slightly larger than the American state of West Virginia. Its position in the Indian Ocean has made the island important to shipping and international airlines. Proximity to southern India has created problems of illicit immigration and smuggling.

Physically, Ceylon can be divided into two distinctive parts: the south-central hill country or 'up country' averaging 3,000–7,000 ft (900–2,100 m) in elevation and surrounded by stepped upland; and the coastal plain, narrower in the west and south but broadening out in the north, its monotony relieved by isolated hills (monadnocks) of granite and other resistant rock. Between the gneiss or granite headlands along the low-lying coasts are sandy beaches, backed by lagoons, estuarine deltas and marshes.

The hill country is roughly anchor-like in shape. The shank running north-westwards contains Pidurutalagala (8,281 ft; 2,524 m), Ceylon's highest mountain; Kirigalpotta (7,857 ft; 2,395 m) and other peaks; and the high plains of Nuwara Eliya and Horton Plains. Plateaux flank the central ridge, the Hallon Plateau on the west and the Uva Basin (Welimada Plateau) on the east.

The western arm of the anchor ends in Adam's Peak (7,360 ft; 2,243 m) known throughout Asia as a place of pilgrimage. On its summit is the long 'footprint' which, according to Buddhist legend, was made by the Buddha himself, but which in Moslem tradition was made by Adam as he stood in penance after his expulsion from Eden. The eastern arm extends through Haputala and continues north-eastwards to form Namunkala in the Lunugala ridge. The two arms form the 'Southern Mountain Wall', so-called because of its sheer drop of more than 4,000 ft (1,200 m) on the southern side. To the south-west are the detached Rakwana hill country and the Bulutota Hills averaging 3,000 ft (900 m).

Ceylon's rivers radiate from the central massif. They are mostly short and narrow, and are subject to seasonal drought and flooding. The longest, Mahaweli Ganga (206 mi; 332 km) is tapped for irrigation along its tortuous course from the eastern mountains to the sea near Trincomalee. All the others, except Aruvi-aru, have lengths of less than 100 mi (160 km).

Climate, Vegation and Wildlife. Ceylon is near the equator and consequently experiences high temperatures, but temperatures in the lowlands are moderated by sea breezes and in the highlands by altitude. Colombo averages 77°F (25°C) in January and 82°F (28°C) in May; Nuwara Eliya, 6,199 ft (1,889 m) above sea level, 57°F (14°C) in January and 61°F (16°C) in May.

Rainfall is the most important factor. Ceylon has wet and dry zones. The Wet Zone, the south-western part of the island, benefits from the south-west (May-September) and north-east (December-February) monsoons, and also receives convectional and depressional rain during March and April. It is in this zone that most of the commercial crops are grown. The Dry Zone, the northern and eastern lowlands and the eastern hill country, suffers drought during the south-west monsoon. Average rainfall varies from less than

Panning for gemstones. Sapphires and rubies are found in the interior of Sri Lanka.

50 in (1,270 mm) in the Dry Zone to more than 200 in (5,080 mm) on the south-western hill-slopes in the Wet Zone.

Vegetation follows the climatic pattern. Thorn forest is typical of areas with less than 25 in (635 mm) of rain. Much of the Dry Zone has scrub jungle and forests containing ebony, satinwood and other valuable timber. The eastern Dry Zone has a few grassland areas (*talawa*). Equatorial forests are typical of the Wet Zone, while the hill country has temperate forests and localized areas of coarse grass (*patanas*). Palms, pandanus (screw pines) and mangrove swamps are found along the coasts. Ceylon has many beautiful flowering trees and shrubs including flame and tulip trees and the rhododendrons of the highlands, where orchids also thrive.

The island is rich in wildlife. The elephants, usually tuskless, have mostly been domesticated for haulage work in the forests. There are bears, leopards, jackals, water buffalo, monkeys, mongooses, fruit-eating and other bats, and innumerable birds. Poisonous snakes include the cobra, *tic polonga* (Russell's viper) and krait. Crocodiles live in the lakes and rivers, and there are several species of lizard.

The People. Ceylon has a multi-racial population of 12,747,755, a high birth rate and a declining death rate. The majority group, representing more than 70% of the population, are the Sinhalese. They are Buddhists, but unlike most Buddhist communities they have a caste system. Numerically, these are followed by the Ceylon Tamils and the Indian Tamils who have a Hindu culture and a caste system, and form about 22% of the population. The Ceylon and Indian Moors (Moslems), like the Malays, are Tamil-speaking, but learn parts of the Koran in Arabic. The Burghers are of mixed Sinhalese and European (Dutch or Portuguese) descent and like the Eurasians are mostly Christians. Christianity in Ceylon is mainly Roman Catholic, and is the faith of about 8% of the population.

Sri Lanka produces one-third of the world's tea. When the name of the country was changed it was agreed that the well known name 'Ceylon tea' would remain the same.

Sinhala (Sinhalese) is the official language, but Tamil was given limited recognition in 1966 and is used for some official purposes. English is the major second language. Some 80% of the population live in the 18,000 or more villages of the island. Nearly a fourth of the urban population lives in and around Colombo, the capital and chief port which, with the south-west coastal districts (Kalutara, Galle and Matara) and the Kandy district, is the most densely-populated part of Ceylon.

Mud and thatch are traditional building materials in the villages. In areas where tea and rubber are grown, the plantation workers live in 'lines' – contiguous one-room family quarters away from the village. The men wear sarongs and long shirt-like garments; the women, a sarong-and-bodice outfit called a *camboy*, or a sari. Rice is the basic food.

Agriculture. Ceylon has a predominately agricultural economy. More than half the total work force is employed in agriculture, which provides about a third of the national income. Some 4·5 million ac (1·8 million ha) are under cultivation, of which about 62% are in the Wet Zone. The plantation crops – tea (originally introduced by the British), rubber and coconuts – together earn nearly 85% of Ceylon's foreign exchange, tea alone providing some 55%. Ceylon is second only to India in world production of tea, and in 1970 produced 468 million lb (212 million kg) from 597,499 ac (241,799 ha). Because of its vital importance, tea receives special attention and many innovations have taken place.

Since Mrs Bandaranaike returned to power in 1970 the plantation industries, along with other industries, have been nationalized.

Paddy (rice), the main food crop, is grown on more than 1,870,000 ac (756,760 ha), about half of this area being irrigated. The ancient irrigation systems are now being restored, and the completion of the Mahaweli Ganga power and irrigation project will bring large new areas under irrigation. In 1970 Ceylon produced over 76 million bu (28 million hl) of rice, but large quantities have still to be imported.

Subsidiary crops include cocoa, cinnamon, citronella, spices and areca-nut. Gingelly (a type of sesame), chillies and grains are grown in the unirrigable highlands by the '*chena* system' (shifting cultivation). Ceylon is self-sufficient in fruits and vegetables.

Nearly 40% of the land area is covered by forests, a valuable source of timber and fuel; and schemes for conservation, reforestation and research are in hand.

Minerals. Commercially, the most important

mineral is graphite, mined chiefly for export. Ilmenite, which occurs along with rutile and monazite in the beach sands, is extracted at Pulmoddai on the north-east coast. The alluvial valleys around Ratnapura (the name means 'Gem City') contain a large variety of precious and semi-precious stones (sapphire, ruby, aquamarine, moonstone, topaz, chrysoberyl, zircon and others). The island also has deposits of china clay, iron ore and glass sand. The limestone of the north is the basis of the cement industry at Kankesanthurai.

Industry. Over the past few years industrial output has been increasing. Large-scale enterprises such as the Sapugaskanda oil refinery (opened in 1969) are controlled by State corporations. There are more than 20 such corporations responsible for a wide range of products including chemicals, fertilizers, textiles, paper, leather, ceramics, wood products, petroleum, oils and fats, steel, tyres, engineering products, sugar, salt and milk. Much of the output is absorbed by the domestic market.

Colombo is the main centre of medium and small-scale factories serving local needs. Improved transportation and electrification have encouraged industrial development in rural areas and the growth of cottage industries. The government has also sponsored the Ekala industrial estate and contemplates further estates at Boosa (Galle), Kundasala (Kandy) and Arialay (Jaffna).

Since there are no known deposits of coal or oil Ceylon relies on its water resources to provide electric power. Some hydro-electric plants are in operation, notably the Laxapana, and in mid-1972 work was proceeding on the Mahaweli, Maskeli Oya and Samanalawewa projects, which will ensure abundant supplies of power.

Transport and Trade. Ceylon has about 11,700 mi (18,829 km) of motorable roads and 932 mi (1,450 km) of state-owned railways. Colombo, on major sea routes, is one of the great ports of the East. Other ports include Trincomalee, from which most of the tea is shipped, Galle and Jaffna. Katunayaka is the island's international airport.

Exports include tea, cocoa, cinnamon and coconut products. Imports include rice, textiles, oil and gas. The United Kingdom, China and the United States are among Ceylon's leading trading partners.

NEPAL

Area: 54,362 sq mi (140,798 km²)
Population: 11,292,841
Population Growth Rate: 1·8%
Capital: Katmandu, pop 332,982
Language: Nepali. Tibeto-Burman and Indo-Aryan dialects
Religion: Hindu (75%) Buddhist (25%)
Currency: Nepali rupee = 100 pice

Nepal, a remote and mountainous kingdom, stretches across a 500 mi (800 km) arc in the central Himalayas. The calm and peaceful beauty of the country has attracted tourists and, recently, youthful escapists from the industrialized West. In the early 19th century, Nepal was larger than it is today, but it was gradually reduced by treaties with British India to its present limits between the Kanchenjunga range in the east and the River Kali in the west. In 1955 Nepal became a member of the United Nations.

The Land. Most of Nepal is mountainous. Along its borders with Tibet and Sikkim are some of the world's loftiest peaks, including Mount Everest (29,028 ft; 8,848 m), the highest mountain in the world. Makalu, Annapurna and Dhaulagiri are all over 26,000 ft (8,000 m). In the high Himalayas, the temperatures are sub-arctic and peaks are permanently covered with snow.

Below the high Himalayas and further to the south are the ranges of the middle Himalayas. Here, the rushing mountain rivers have cup deep gorges on their way to the plains and the River Ganges. In the river valleys, the climate is temperate. The original woodland vegetation of pine, oak, rhododendron and magnolia has given way to intensely cultivated or grazing land. The Katmandu valley, 30 mi (48 km) across, is one of the few pieces of level ground outside the plains or *Terai* in the south.

From the middle Himalayas, the ground continues to fall away, through the foothills, with their hot wet summers and warm winters, to the Terai itself. These flat, swampy plains, extend 20 mi (32 km) into the interior from the Indian border. Deciduous jungles flourish, full of wild game, and malaria is a serious hazard.

The People. In 1970, the population of Nepal was estimated at about 11,200,000 including about 7,000 refugees from Tibet. Nepal does not support everyone born in the country – many emigrate permanently or temporarily to India. There are many ethnic groups. Politically dominant are the Hindu, Nepali-speaking Parbatiyas of the middle Himalayas. This zone is settled also by a number of peoples speaking Tibeto-Burman languages, including the Gurkhas, long recruited into the British and Indian armies. The Katmandu valley is the ancient home of the Newars whose urban civilization is a relic of medieval Indian culture. Near the northern border are peoples of Tibetan stock, including the Sherpas, while in the Terai live groups similar to those in adjoining India. The population is presumed to be increasing despite the serious lack of medical resources.

Over 95% of the population is rural. People live in villages of thatched, mud-brick houses set amid the cultivated fields. More than half of them are Hindu, and the remainder are mostly Buddhists. Guatama Buddha, the founder of Buddhism, was born in Nepal. Only 12 out of every 100 Nepalese can read and write but the number of schools is increasing. Tribhuvan University at Katmandu was opened in 1960. The only large cities are Katmandu, the capital, Lalitpur (Patan) and Bhaktapur.

The Economy. Agriculture is easily the most important item in the Nepalese economy. About 10% of the country is cultivated and a third of this area, in the middle Himalayas, supports over two-thirds of the population. Irrigation is important, and the flat terraces rise high up the mountain slopes above the fertile valleys. Overcultivation and overgrazing have caused serious soil depletion in the middle Himalayas, but the flat and fertile Terai still has untouched reserves of land.

Subsistence cultivation of rice, maize, wheat and a few vegetables is the rule, although sugar-cane and jute are grown as cash crops in the Terai, and fruit and tea plantations have been started on a small scale in the middle Himalayas. In favourable years there is a small surplus of rice and potatoes for export. Many farms have cattle, buffaloes and goats which provide milk; otherwise, the diet is mainly vegetarian.

The forests, which cover perhaps two thirds of the country, are exploited to a limited extent. Wood from the sal forests of the foothills is exported to India for the construction industry and some Terai softwoods are cut for matches and plywood. The Terai forests hold reserves of many valuable hardwoods.

So far, Nepal has little industry, except for jute mills and the few industries, such as sugar, food and timber, which process local products. New industries (leather factories and chemical plants) are being established, and hydro-electric power is being developed with outside aid. Tourism is rapidly increasing.

The temple of Hamuman, the monkey god, at the Old Palace in Katmandu, capital of Nepal. Katmandu was an independent Newar Kingdom until conquered by the Gurkhas in 1768.

Deposits of many minerals notably coal, iron, dolomite and cobalt, have been found but have yet to be exploited.

A system of small airstrips links many far-flung parts of the country, but over large areas, transport is still by buffaloes and elephants or on foot. Several Indian railway lines touch the southern border, only one of them extending into Nepal at Birganj. In the late 1960s, roads were completed from Katmandu to Lhasa and Pokhara and a road was under construction along the whole length of the Terai. A network of local roads exists in the Terai, but Nepal has relied on foreign aid to build all major roads.

SIKKIM

Area: 2,823 sq mi (7,312 km²)
Population: 205,000
Population Growth Rate: 2·0%
Capital: Gangtok, pop 12,000
Language: Sikkimese, Nepalese, Lepcha
Religion: Hindu (75%) Buddhist (25%)
Currency: Rupee = 100 naya paise

Sikkim, a small kingdom in the eastern Himalayas, has been a protectorate of India since 1950. Like its eastern neighbour, Bhutan, Sikkim is of some strategic importance to India since it controls access to the Chinese province of Tibet. India provides Sikkim with aid for roads and other forms of development, and dictates foreign policy.

The Land. Sikkim covers 2,823 sqmi (7,312 km²), occupying an area of fast-flowing streams and rivers in the upper basin of the Tista River. The landscape ranges from perpetually snow-capped mountains like Kanchenjunga (28,146 ft; 8,585 m) on the Nepal border to lush tropical forests in southern river valleys less than 1,000 ft (300 m) above sea level.

The land slopes steeply from north to south and flat ground is rare. Rainfall averages range from over 130 in (3,300 mm) yearly in the south to less than 20 in (510 mm) in some northern valleys.

The People and the Economy. The population, an estimated 194,000 in 1971, is 75% Nepalese, the product of continued immigration over the last two or three hundred years. The rest are Lepchas, the original inhabitants, and Bhutias. The only city is the capital, Gangtok (12,000). English, Nepali, Hindi, Lepcha and Bhutanese are spoken. Buddhism is the state religion, but the Nepalese are Hindus.

Sikkim is ruled by a chogyal (maharaja) assisted by an Indian adviser and a state council. India controls defence, foreign relations and communications.

Sikkim is an agricultural country. The main crops are rice, maize, buckwheat and barley, but these are all consumed internally. The most important cash crops are cardamom (a spice), potatoes, fruit and apples. Sheep, goats, yaks and cattle are plentiful. Lead-zinc-copper ore is mined at Rhotang. Weaving and other handicrafts are widespread. Hydroelectric power is being developed; but the forests, covering about a third of the country, have still to be exploited.

Sikkim has poor roads and no railways or airports. India has helped improve road links, and is constructing strategic roads in the north.

BHUTAN

Area: 18,000 sq mi (46,620 km²)
Population: 1,100,000
Population Growth Rate: 2·2%
Capital: Thimbu, pop 8,000
Language: Dzongka. Nepali and tribal dialects
Religion: Buddhist (80%) Hindu (20%)
Currency: Indian rupee = 2 tikchung = 100 paise

Bhutan, a kingdom perched high in the remote mountains of the eastern Himalayas, has been less affected by 20th century progress than almost any other country in the world. The Bhutanese still follow their centuries-old way of life, raising crops and cattle in the fertile mountain valleys and centring their lives on the many Buddhist monasteries. However, despite its small size (18,150 sq mi; 47,000 km²), Bhutan occupies a strategic position between China on the north and India on the south. India controls Bhutan's foreign affairs and is providing financial aid for modernization. In 1971 Bhutan became a member of the United Nations.

The Land. Bhutan is extremely mountainous. Its northern border with Tibet roughly follows the edge of the Tibetan plateau and contains Kulhakangri (24,740 ft; 7,541 m) and other lofty peaks. The area is largely covered by ice and snow, but tundra vegetation supports deer and yaks.

From these great heights, the country falls away rapidly over a distance of about 90 mi (145 km) to the plain (the *Duars*) in the south, along the border with India. There are seven main river valleys in central Bhutan where the climate is mild, and even subtropical. The Duars are covered with tropical jungle and savanna grassland, and elephants, tigers, rhinoceros, and leopards are found.

Life and Work. Bhutan has never taken a

A fortified monastery or 'dzong', residence of the monk-governor of the surrounding area.

census, but in 1971, the population was estimated at a little over a million. About 75% are Bhutanese, descendants of early settlers from Tibet. The remainder are mostly Nepalese. Dzongkha, related to Tibetan, is the official language.

Bhutan is ruled by a monarch, King Jigme Dorji Wangchuk ('Fearless Thunderbolt Master of the Cosmic Powers'), and a National Assembly, which has the unique authority to depose the monarch by vote. But the real ruler of the country is still religion. Most of the people are Mahayana Buddhists and the fortified monasteries or *Dzonghs* in the valleys, the homes of the many lamas, are centres of administration and cultural life.

The people nearly all live in small villages in the valleys and on the mountain slopes and almost all are dependent on agriculture. Rice, wheat, barley, mustard, oranges, tea and jute are cultivated and in the mountains, the yaks and sturdy tangun ponies are tended. The mountain farmers are known for their fine embroideries, woodcarvings and metalwork. Coal, graphite and gypsum are found but are largely undeveloped. There is also considerable hydro-electric potential.

Until recently, Bhutan's roads were no more than mountain trails but paved roads are being constructed with Indian aid. India is the only trading partner and finances Bhutan's large trade deficit. Thimbu, the capital (about 8,000) is the only large town.

BANGLADESH

Area: 55,126 sq mi (142,776 km²)
Population: 72,000,000
Population Growth Rate:
Capital: Dacca, pop 915,000
Language: Bengali. Bihari, Hindi also spoken
Religion: Moslem (80%) Hindu (15%)
Currency: Taka = 100 paisa

Bangladesh ('the Bengal Nation') is a Moslem republic created from the Eastern Province of Pakistan by a bloody civil war and armed intervention by India. When the British rule of India ended after World War II, an independent Moslem nation was created at the same time as an independent, predominantly Hindu, India. Pakistan was the result, a divided country with two parts a thousand miles apart. Though the smaller part, East Pakistan had over half of the population, control lay predominantly in West Pakistan, and resentment gradually built up in the east.

When the central government thwarted the democratic election of the autonomy-seeking Awami league party in the East, they were forced to move in the national army to crush the ensuing revolt – about a million died and many million Bengalis fled into India. India then launched a lightning military campaign against West Pakistan and secured East Pakistan's independence (as Bangladesh).

The new nation has a few valuable assets to pit against its immense problems. Its Bengali majority gives it a greater cultural unity than India; dominating the fertile Ganges delta, its rich farmlands yield massive crops including, rice, sugar-cane, tea and jute.

Against this there are the burdens of dissident minorities and an enormous population (which increased from 42 million in 1941 to some 72 million in 1972). Moreover, the economy depends lopsidedly upon agriculture and suffers from periodic cyclones that sweep in from the Bay of Bengal, drowning low-lying areas and wrecking crops. There is no immediate hope of rapprochement with West Pakistan, and the country leans heavily on international good will for aid.

The Land. Most of Bangladesh is a vast, low and almost flat alluvial plain. This plain is in places less than 30 ft (9 m) above sea level. Though the country has scores of rivers, almost all are tributaries or distributaries of the three main rivers – the Ganges, the Jamuna (the main stream of the lower Brahmaputra) and the Meghna. These three rivers have formed here the largest delta in the world. Except for the south-western higher part and the old alluvial tracts, the entire plain is inundated every year during rains by the flooding of these rivers.

Due to the flatness of the plain, the small rivers change their course frequently during floods and sometimes even the big ones do so. The Ganges and the Brahmaputra have both changed their lower course in the last two centuries. The main tributaries of the Jamuna are the Tista and the Atrai, which drain the north-western part of the plain. A striking feature of the plain is the occurrence of a large number of marshes and swamps.

There are three old alluvial tracts: Barind, covering an area of 3,600 sq mi (9,300 km²) on the north-western part of the plain; the Madhupur tract, over 2,450 sq mi (6,350 km²) in the centre; and the Lalmai Tilas east of the Meghna, 20–100 ft (6–30 m) in elevation.

Low hills of 100–800 ft (30–240 m) are found in the north-eastern part of the plain, in Sylhet district.

Chittagong Hill Tracts are the only extensive hill area in Bangladesh. The 10 ranges run from north to south, parallel to each other. The average height is 2,000 ft (600 m). Between the sea and the hills, there is a narrow coastal plain.

Climate. The climate is basically monsoonal. As no part of Bangladesh is far from the sea, the range of temperature is small, though high humidity makes the heat oppressive.

Winters (November-February) are mild, dry and very pleasant. Maximum temperature at Chittagong, in January, is 78·6°F (25·9°C) and at Dacca 77·9°F (25·5°C). Summer (March-May) temperatures are moderately high. The hottest month is April. The temperatures fall during the rains and rise in September and October. The summer maximum is 90°F (32·2°C) to 95°F (35°C), generally attained in the northern districts. The minimum over the plain is hardly below 70°F (21·2°C).

Rainfall is heavy. The Bay of Bengal branch of the monsoon wind breaks over the plain in June and causes heavy showers till October. The annual rainfall over the plain is 50–100 in (1,040–12,700 mm) increasing towards the north-east. The hilly regions have 100–140 in (1,270 –3,560 mm). Humidity is never less than 75% rising to 95% during rain. The country is extremely vulnerable to cyclones sweeping in from the Bay of Bengal. In November 1970, a cyclone killed well over a quarter of a million people in the coastal areas.

Vegetation. The plain was probably once covered with tropical rain forest, but much has been cleared for cultivation. Only two natural forests exist. The Madhupur tract with an area of 1,600 sq mi (4,140 km^2) has sal as its main tree which is used for constructional work. In the south is the coastal Sundarbans tract containing probably the world's best example of tidal forest. The area covers 2,300 sq mi (5,960 km^2) and its marshy islands are clad in mangrove forest. The important trees are sundari nipa, and gewa. Sundari is used for boat construction and gewa for manufacturing match sticks and boxes. The Chittagong Hill Tracts and part of Chittagong district are also covered with dense tropical rain forest extending over an area of about 1,280 sq mi (3,310 km^2). The area has evergreen and deciduous forest. Trees grow as high as 200 ft (60 m). Teak and garjan produced here are mainly used for making furniture. Bamboo grows plentifully and is used for manufacturing paper. Banana, mango, jackfruit and coconut are the common fruit trees.

Wildlife. The dense tropical forest of Bangladesh is rich in wildlife. The Sundarbans is a renowned hunting ground where tigers and deer are found. Chittagong Hill Tracts are known for elephants, deer and bears. Crocodiles abound in shallow waters.

The People. Bangladesh's population of 72,000,000 is crammed into 55,126 sq mi (142,776 km^2) making it one of the most densely populated countries in the world. The average density is 1,369 people per sq mi (529 per km^2), over double that of Great Britain. The people are predominantly Bengalis, and 80% are Moslems, leaving a small Hindu minority. There are also 650,000 Biharis who are immigrants from Bihar. These are a source of friction, as when Bangladesh broke away from West Pakistan they showed sympathy for West Pakistan, whose language (Urdu) they speak.

An aerial view of an ancient temple standing next to a jute factory. Bangladesh produces 80% of the world's jute, a coarse fibre used for making burlap and other sacking.

Ninety per cent of the population live in the rural areas. Average farm plots are only 2·5 ac (1 ha). Houses on the plain are usually built on 20 ft (6 m) plinths as protection against the floods. Walls are of split bamboos, with reed matting or mud, and the conical roofs are thatched.

Towns are usually found at river junctions or on river banks. Dacca (pop. 915,000) is the capital and Chittagong (pop. 500,000) the chief port. There is no great movement into the towns. Only 20% of the people are literate, and although education receives high priority, the cost of building new schools and providing teachers is so great that any dramatic improvement is unlikely in the near future. The most serious problem, however, is that of population control; the birth rate has been running as high as 5% and every effort is being made to reduce it.

Agriculture. Bangladesh leans heavily on agriculture to help feed its enormous population. Three out of every four families work on the land. Of the total cultivated area of 22·5 million ac (9·1 million ha), nearly half is triple cropped. The three overlapping growing seasons are the *kharif* (April-August), the *haimantic* (June-December), and the *rabi* (October-March).

Rice is the most important food crop. It is the staple food and is grown on more than 60% of the total cultivated area. The area under rice was 24·1 million ac (9·71 million ha) in 1968–69, and production was over 11 million tons.

Jute, the golden fibre grown in the country from time immemorial and used for twine and making sacks, gained commercial importance in the early 19th century. Bangladesh now produces about 55% of the world's supply. The most important cash crop and the single largest source of foreign exchange, it is grown in almost all the districts of the province except Chittagong Hill Tracts and the Sundarbans. Rich soil, abundant and well distributed rain during summer and plenty of clean water rivers, provide ideal conditions. Jute was grown on about 2·2 million ac (0·9 million ha) in 1968–69. However, the yield per unit area has declined by about one-third since 1960. Good seed and fertilizer are being supplied by the government for increasing the yield. Bangladesh must successfully diversify its economy to reduce its dependence on jute, for which the world market is shrinking. Competition also comes from eastern India, another big jute grower.

Tea is the second most important cash crop of the province and is grown on the slopes of the Sylhet Hill and Chittagong Hill Tracts. The area of tea plantations is increasing. In 1968–69, the production was nearly 63 million lb (29 million kg). As domestic consumption has increased, tea is now imported from Ceylon. Sugar-cane, though an important cash crop, covers only 1·1% of the total cropped area, and is mainly produced in the northern and western part of the province. Wheat, tobacco and cotton are also grown in small quantities.

Minerals. Oil, peat, coal, natural gas and limestone offer the greatest potential. No coal mining has yet been undertaken, but several large gas fields are being exploited in Sylhet. Dacca receives gas by pipeline. Oil has been found in the Bay of Bengal. There is an oil refinery at Chittagong.

Industry. When the war of independence ended, what little industrial plant there was had been either disorganized or destroyed or deserted by the West Pakistanis who had financed and run them. Raw materials were short and the labour force volatile. Some re-establishment of small industries was reported by early 1973. Banks and some industries have been nationalized.

Jute manufacturing is the most important industry. In 1947 there was not a single jute mill. More than 20 have now been established and others are in construction. The main centres are Dacca, Narayanganj, Chittagong and Khulna. Almost half of the world's requirements of gunny bags and jute cloth is supplied by Bangladesh. Some 520,000 tons of jute goods were produced in 1968–69.

The Karnafuli paper mill at Chandraghana, 26 mi (42 km) from Chittagong, one of the largest in South Asia, was completed in 1953. The Khulna newsprint mill was established in 1959. Their total production in 1967–68, was 77,000 tons, soon to be exceeded by the North Bengal paper mills (Pabna) with 189,000 tons manufacturing capacity.

Cotton spinning and weaving are important cottage industries. The Karnafuli or Kaptai project is the only hydro-electric project in Bangladesh. Its installed capacity in 1966–67 was 80,000 kw.

International Trade. The magnitude of Bangladesh's difficulties is nowhere clearer than in the pattern of its foreign trade. In 1973 one-third of its estimated foreign exchange earnings was earmarked to pay for importing food grains, and this allows for huge gifts of food from friendly nations. Large supplies of sugar, edible oil, yarn, woollens and raw materials for textiles have to be imported. Plant machinery and equipment, agricultural equipment, fertilizers and fuel are among the other necessary and expensive imports.

Jute accounts for 80% of export earnings. Leather and fish exports are promising, but overall there is a prodigious gap between exports and imports.

Transport and Communications. The maze of rivers and creeks which breaks up much of the land, coupled with the frequent flooding of low-lying areas, hinders the growth of an effective overland transport system; however, the Ganges, Brahmaputra, Meghna and other inland waterways are important means of communication. Bangladesh's meagre transport facilities were naturally hit by the war in which Chittagong harbour and some 275 bridges were damaged, badly discolating the road network.

The railway system has 1,710 mi (2,736 km) of track. New lines are being opened to connect the industrial centres with Dacca and Chittagong. Bangladesh has its own small airline, Bangladesh Biman.

A market in Dacca, capital of Bangladesh. Umbrellas were introduced by English colonialists.

BURMA

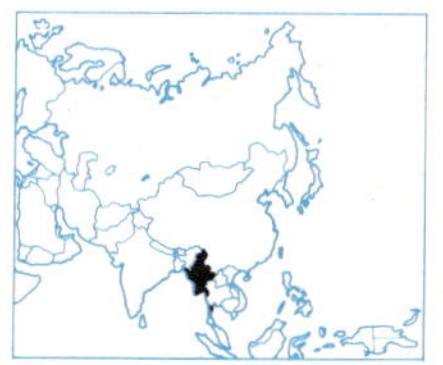

Area: 261,228 sq mi (676,580 km²)
Population: 28,201,000
Population Growth Rate: 2·2%
Capital: Rangoon, pop 1,851,159
Language: Burmese
Religion: Buddhist (85%) Hindu, Moslem, Christian, Animist (15%)
Currency: Kyat = 100 pyas

The Union of Burma came into existence as an independent democratic republic at 4.20 am on 4 January 1948 – a time ordained by astrologers. Despite modernization since independence, superstition and traditional religion are still powerful forces in Burmese life. Burma is well endowed with minerals and other resources; but political misdirection has hampered economic growth and Burma stands among the poorer nations.

Since the institution in 1962 of a military government, opposed to communism but dedicated to nationalizing the entire economy, Burma has concentrated on the improvement of agriculture (with mixed success, since rice production has fluctuated considerably) and the displacement of Indian and Chinese minorities from their dominant position in the country's commerce. The economic disruption caused by these measures, continuing internal unrest, and the threatening presence of communist China on the north-east frontier have helped to prevent Burma from achieving security and prosperity.

Rangoon, the capital of Burma, stands on the Rangoon river and is dominated by the 386 ft high Shwe Dagon pagoda.

The Land. Burma is dominated by a series of folded mountain ranges in the west and a block-faulted massif in the east. The western hilly zone, containing the Letha Range, the Chin Hills and the great Arakan Yoma reaches 10,030 ft (3,055 m) at Mount Victoria. Further north, in Kachin State, the country becomes even more mountainous, rising to 20,560 ft (5,885 m) at Hkakabo Ravi the highest peak in Burma. Between the mountains and the massif lie the central lowlands, extending from the deltas and valleys of the Irrawaddy and the Sittang to the interior plain of Mandalay. The extensive undulating Shan Plateau massif rises abruptly to the east of the lowlands, falling gradually as it runs southward into the long, narrow Tenasserim coastal strip.

More than half of Burma's area lies in the drainage basin of the Irrawaddy and its tributaries. Rising in the highest mountains in the north, the Irrawaddy flows southwards for over 1,000 mi (1,600 km) before emptying into the Bay of Bengal. The Salween enters Burma on the north-eastern border and traverses the Shan Plateau in a narrow incised valley. The Sittang and Mekong are other major rivers.

Climate, Vegetation and Wildlife. Like that of other countries in South-East Asia, the climate of Burma is monsoonal. The rainy season between May and October, when the area is under the influence of the warm, moist South-West Monsoon, is followed by the cool, dry season of the North-East Monsoon lasting until the middle of February. The hot, dry season, preceding the burst of the summer monsoon, is a period of transition.

Most places in Burma receive over 80% of their total annual rainfall in the wet season. In the interior, including the Shan Plateau, rainfall averages between 50 in (1,250 mm) and 75 in (1,875 mm) a year but the Mandalay Plain or Dry Zone, lying in the rain-shadow of the Arakan Yoma, receives less than 30 in (750 mm) while Moulmein and Tavoy on the Tenasserim coast average about 200 in (5,000 mm) a year.

The annual average temperature of Rangoon is 80°F (26·2°C), but the northern areas are considerably cooler. Maximum temperature is reached in April, while December and January are usually the coolest months.

Tropical rain forests and deciduous monsoon forests cover more than 60% of the country. In the Dry Zone, these give way, to scrub and semi-desert, and along the Irrawaddy delta extensive tracts of swamps and mangroves are found. Wildlife is abundant and includes elephants, tigers, leopards, monkeys, jackals, deer and bears. Snakes are common.

The People. The population of Burma was estimated to be nearly 29 million at the beginning of the 1970s and was increasing at a rate of 2·3% per annum. It is composed of several ethnic groups, with Burmans, Shan Karens, Kachins and Kayahs being the most numerous. Over 70% of the population is literate, but only two-thirds are Burmese speaking and more than 100 languages are spoken by the various hill tribes and minority groups.

More than 85% of Burmese people are Theravada (or Southern) Buddhists: the yellow-robed Buddhist monks or *pongyi* are the leaders of the community. Monasteries still play an important part in primary education, and education is free in all schools. Small Hindu, Moslem and Christian minorities also exist and animism is widely practised among the hill tribes.

For most Burmese, life centres upon the village. The houses of bamboo and thatch, often resting on stilts, cluster around the pagoda and the village may be fenced as protection against tigers and thieves. National dress for both men and women is a long wrap-around cylindrical skirt called a *lungi* made of finely coloured silk. Drama, music and classical dance are highly developed arts.

Rangoon, the largest city, is the commercial and administrative centre. Mandalay is the former royal capital and the cultural centre of the country, and Moulmein, on the Gulf of Martaban, is a busy seaport. Other urban areas are few and far between.

Farms, Forests and Fisheries. Farming is the most important economic activity of the Burmese, occupying no less than 65% of the country's population. By far the most important crop is rice, most of it grown in the hot wet areas of the Irrawaddy Delta. Burma is one of the few Asian countries to produce a surplus of rice and the crop is the most important export commodity. In drier areas, wheat, maize, millet, sesame, groundnuts, cotton, and pulses are grown. Cash crops include tea, tobacco, sugar-cane and vegetables. In the tropical Tenasserim Peninsula, plantations for the production of export crops such as rubber are important.

Agriculture in Burma is still backward. Slash and burn (shifting) cultivation is practised extensively by the hill tribes, and elsewhere, despite the nationalization and redistribution of land, there are many problems due to soil depletion and shortage of fertilizer.

Much of Burma's extensive forest cannot be commercially exploited because of transport difficulties, but the fine quality teak and pyinkado (ironwood) from the monsoon forests are widely extracted for construction and export. The logs are hauled to the river banks by trained elephants and floated downstream to the sawmills by bamboo rafts.

Fish is an important item in the Burmese diet. Much comes from fresh water but marine fishing is being encouraged.

Minerals. Burma is richly endowed with a wide range of minerals and the mining industry is an important sector of the economy. However, much of the country has not been systematically explored and known resources are as yet not being fully exploited. Petroleum is by far the most important mineral in Burma although production has not yet reached its pre-war peak. New strikes have increased the total known reserves substantially. Large natural gas reserves are located in the Chauk area. Tin, tungsten, antimony, lead, zinc and quicksilver are all important and extensive coal reserves have been discovered. Rubies, sapphires and other precious stones are found, as well as rich deposits of jade.

Industry. Although a million workers are employed in the manufacturing, power and construction sectors of the economy, the vast majority of industrial workers are engaged in the processing of agricultural produce and

A lion guarding the inner temple of the Buddhist Shwe Dagon pagoda in Rangoon.

minerals. Industries have traditionally been based on rice-mills, saw-mills and machinery repair shops. Of the various new industries, the fertilizer plants at Sale and Kynchaung are of the greatest significance to the country's predominantly agricultural economy. Practically all the industrial enterprises are now run by the state.

Transport and Trade. The Irrawaddy and the creeks on the delta are the traditional arteries of trade. The Irrawaddy itself is navigable up to Myitkyina, and the Chindwin tributary is navigable for another 400 mi (640 km). The Sittang is connected to the Irrawaddy system by 60 mi (96 km) of navigation canals. Surfaced roads are less important than the network of railways which spreads out from the main trunk route. The port of Rangoon which handles 85% of Burma's foreign trade is faced with serious problems of rapid silting, and Moulmein is potentially the major deepwater port. Internal air traffic links several of the provincial towns with the capital and regular flights are operated from Mingaladon International Airport to Bangkok and Calcutta.

Burma's main trading partners are in Western Europe and among the Asian nations. Leading exports include rice, timber, rubber and cotton. Major imports include machinery, paper, textiles and tobacco.

CHINA

Area: 3,690,546 sq mi (9,558,514 km²)
Population: 830,000,000 (own est.) 910,000,000
Population Growth Rate: 1·8%
Capital: Peking, pop 7,400,000
Language: Mandarin. Cantonese and Hakka dialects
Religion: Confuscianism, Buddhism and Taoism. Moslem 5%
Currency: Yuan = 100 cents

The People's Republic of China, the world's most populous and third largest nation, occupies more than one-fifth of Asia and dominates the continent's eastern mainland.

During its 3,500 years of recorded history, geographically isolated China pioneered such inventions as gunpowder, paper and printing. But its traditional way of life, based on peasant labour, loyalty to family and emperor, and a reverence for literary scholarship, discouraged great technological or social advance.

Since 1900 the backward giant has been politically and economically jolted into motion. The pace of change accelerated after the communists drove the nationalists on to the offshore island of Taiwan and seized power in 1949. In the next year Communist China invaded and absorbed independent Tibet and began the huge task of co-ordinating the exploitation of China's natural resources for the declared profit of the people as a whole. Ideological differences provoked a deep rift with Russia which still exists. Material progress has been hindered by problems including over-population, lack of capital, bad weather, harsh terrain, political unrest and government miscalculation. But much has been accomplished and China now ranks among the world's great powers, albeit as the poorest. In the 1970s China joined the United Nations and made friendly overtures to Western powers.

The Face of the Land

China is roughly the same size as the United States, occupying over 3·5 million sq mi (9·6 million km²). From the East China Sea west to the Pamirs, China's provinces and autonomous regions sprawl across 3,000 mi (5,000 km). From the Sino-Soviet Amur River frontier (54°N) south to Hainan Island (18°N) is about 2,500 mi (4,000 km). In the east, China's 14,000 mi (22,500 km) coastline faces the Yellow East China and South China seas.

China's land boundaries, totalling over 12,000 mi (20,000 km), touch 12 other countries: Afghanistan, Bhutan, Burma, India, Laos, Mongolia, Nepal, North Korea, North Vietnam, Pakistan, Sikkim and the Soviet Union. Sections of the Sino-Indian and Sino-Soviet borders are in dispute.

Surface Features. China is on the whole high in the west and comparatively low in the east. Some geographers divide it into three great zones: east, south-west, and (partly between the two) a west to south-east arc of plateaux and basins. Apart from a maze of mountains in the south and some big ranges in the north, much of the eastern zone is relatively low lying, especially the populous North China Plain, the North-East Plain, and the densely settled middle and lower Yangtze valley, the valley of the Hsi in the south and the south's narrow coastal fringe. The south-western zone is the great, cold waste of the 10,000 ft (3,000 m) high Tibet-Tsinghai Plateau, flanked on the south by the mighty Himalayas, on the west by the Karakorams, on the north by the Kunlun Shan, and on the east by ranges radiating from the plateau.

Immediately north and east of the plateau is the third zone, an arc of plateaux and basins averaging about 3,300 ft (1,000 m) to 6,600 ft (2,000 m) high. The arc sweeps round from the west through the Tarim and Dzungarian basins, and the Turfan Depression, China's lowest point at 505 ft (154 m) below sea level. It nexts takes in the Tsaidam Basin and the Inner Mongolian and loess plateaux, then turns abruptly south to include the Szechwan Basin and the Yunnan-Kweichow Plateau.

Of China's total surface area about one-third is classified as highlands, 26% as plateaux, 20% as basins, 10% as hill regions and 12% as plains.

China's major river systems are those of the Hwang Ho (Yellow River) and Yangtze Kiang which flow from west to east. Rivers often flow differently in the north-east, the south, and the great inland drainage basins in Sinkiang and Tsinghai, where rivers fed by melting snow and glaciers peter out through desert seepage and evaporation. From the 3,010 mi (4,845 km) Hwang Ho north to the Amur River, the northern rivers have relatively small flow, a short summer flood season and a high silt content, and they freeze in winter from one to six months. Southern rivers including the 3,600 mi (5,800 km) long Yangtze Kiang (the third longest river in the world) and the Hsi, tend to contain much more water even in winter, have longer flood seasons and less sediment and are always ice free.

The coast is of two types. North of Shanghai except for the Shantung and Liaotung peninsulas the coast is flat, low lying, relatively smooth and lapped by shallow waters. South of Shanghai, the coast is more moun-

A junk on the Hwang Ho, or Yellow River. The Hwang Ho, the 'cradle of Chinese civilization', has huge floodplains of yellow loess silt.

tainous and indented, with good natural harbours, and deep island-strewn waters.

Climate. Southern China experiences a distinctly monsoonal climate, while in the north and specially the north-west the pattern is more continental. In winter a great mass of cold dry high-pressure air over Central Asia sets up prevailing winds that blow out over the sea. In summer the pattern is reversed, with low pressure over the land, high pressure over the sea, and warm damp winds blowing on-shore.

The summer monsoon brings heavy rain and typhoons to south-eastern China, but rainfall drops off towards the north-west. Thus while parts of the south-east receive 118 in (3,000 mm) a year, north-western desert areas may get under 2 in (50 mm). The Ch'in Ling Shan watershed between the Yellow and Yangtze rivers acts as a major barrier to rain moving northwards across eastern China. Over 80% of China's annual rainfall occurs from May to October, and in much of the drier north it is concentrated into July and August.

Temperature patterns range from tropical maritime in the extreme south to cold continental in the far north-east. Generally July is the hottest month, averaging above 68°F (20°C), with Canton in the south only some 9°F (5°C) warmer than the north-eastern city of Harbin. In winter, however, this range increases to over 59°F (33°C) as the north-east, north-west, and Inner Mongolia experience harsh winters, while the extreme south remains mild.

China's huge size, and regional contrasts in altitude and distance from the sea produce some very different seasonal patterns. Thus the south-east has a long hot summer and short mild winter while the north-east and the Tibet-Tsinghai Plateau have a long winter and fleeting summer.

Natural Vegetation. There are two great zones of natural vegetation: woodland in the east and grassland and desert in the west. Both are about the same size.

Over the centuries most of the natural forests have been destroyed, causing disastrous soil erosion, particularly in northern China. Today only about 10% of the land is forested. Some 60% of China's timber reserves lie in the north-east in a horseshoe shaped belt running along the eastern edge of the Mongolian Plateau, the Greater and Lesser Khingan ranges and the Ch'ang-pai Shan massif. Other virgin forests are scattered about in the south and on the eastern flank of the Tibet-Tsinghai Plateau, but most of these reserves are more or less inaccessible.

There are six main kinds of forest. South of the Yangtze (from south to north) are tropical rain forest; evergreen broadleaved forest with evergreen oak and laurels; and mixed deciduous forest with ginkgoes. North of the Yangtze comes first deciduous broadleaved forest, mainly oaks; then mixed northern hardwood forest of maple, basswood and birch; and finally coniferous forest, largely of larch, spruce and fir. Each kind of forest may however contain plants from another kind.

The north-west consists mainly of steppe, semi-desert and desert. Here, fitfully flowing rivers peter out in inland drainage basins with a 'dead heart'. In the south-west are the cold deserts of Tibet, which support little more than scattered, low-growing cushion plants.

Wildlife. All told China possesses some 12% of the world's bird species, 10% of the mammals, and 9% of the fish. Remarkable species include the giant panda, found only on the Tibetan border, and such prehistoric survivals as the giant salamander and giant paddlefish. Nonetheless, intensive exploitation of land in eastern China has driven many species out of large areas or exterminated them.

The Great Wall, built in the 3rd century B.C. to repel the Huns and other nomads stretches 1,500 miles west from Peking. It is one of the greatest feats of construction ever undertaken.

The People

China has far more people than any other nation. Some estimates put its population in the early 1970s as high as 830 millions, roughly four times that of the USA and 20–25% of the world total. This means that China's population may well have doubled in a generation. By the early 1970s the annual growth rate was perhaps 20 millions a year, making a 1964 United Nations prediction of 1,000 million Chinese by 1980 seem very likely to come true. The Chinese government argues that its country's many millions represent a source of national strength. But the obvious results of overpopulation are relatively low living standards for many million Chinese.

One way of relieving population pressure might seem to be to shift surplus millions to thinly peopled areas, but these are generally too inhospitable for this to work. The result is that as much as 90% of the population lives on only 15% of the land. Thus China's average population density of about 200 per sq mi (77 per km^2) is misleading: for densities are far higher than that in the thickly settled central North China Plain, Yangtze Delta, middle Yangtze, Szechwan Basin, Canton Delta and the coastal plain between the Yangtze and the Hsi. There are over 2,500 people per sq mi (789 per km^2) along the lower Yangtze, for example.

These high densities are largely rural, and the huge majority of country dwellers live in small tightly knit farming towns. In the south these consist largely of traditionally built homes with bamboo and willow walls daubed with mud. In the north there are thatched, mud-walled huts and caves burrowed in the soft loess soil. In the semi-deserts of the north-west a few tent-dwelling shepherds pursue their old nomadic way of life, although new state farms are changing this.

Only about one in every five Chinese lives in a city, but around a quarter of the world's big cities are in China, concentrated in the north-east and Szechwan Basin. Shanghai, with over 10 million inhabitants, may well be the largest city on earth. China had around 20 other cities of about a million or more by the early 1970s of which the largest were Peking and Tientsin. Within the cities, the more fortunate people live in new blocks of flats, but millions are crammed into overcrowded shanty towns and riverside houseboats.

The Chinese differ from place to place not just in concentration but racially and ethnically. Almost all Chinese are basically Mongoloid. Some 94% are Han Chinese – eastern Chinese descendants of the founders of ancient China. The remaining 50 millions or so belong to other groups including Tibetans, Uighurs, Mongols, Manchus and Miaos. Most of their ancestors were absorbed as the Han Chinese expanded from their ancient heartland by the Yellow River. The 10 million or so Chuang who border Burma, Laos and North Vietnam are the biggest single national minority.

Most Chinese learn a widely spoken northern form of Mandarin, one of seven groups of Chinese dialects. More than half of all Chinese can now read this national language, thanks largely to an education programme which has reduced the traditional 40,000 Chinese characters to a more manageable 1,000.

Cultural differences involve religion as well as language. The main purely Chinese systems of belief are Confucianism (an ethic stressing individual morality and loyalty to family and state) and Taoism (proclaiming humility and involving magic). In the west, Islam and forms of Buddhism are traditionally strong. Non-Chinese religions have met varying amounts of official discouragements from the atheist communists.

In spite of their regional and cultural differences many Chinese share the same overall pattern of life. This is almost everywhere frugal by Western standards. Earnings are generally low. There are anyway few consumer goods to spend them on, and little variety in what there is. For instance, most Chinese now wear stereotyped workers' outfits of trousers, tunics and peaked caps.

Most families live in one- or two-roomed homes, often without piped water or electricity, or sharing sinks and lavatories with neighbours. Food tends to be monotonously based on cheap carbohydrates (rice in the south, wheat in the north) enlivened with various vegetables. Fish, pork and chicken are important protein foods but milk is lacking and few people can afford meat at every main meal.

The Hall of Annual Prayers was built during the Ming dynasty (1368–1644) and rebuilt after being destroyed by lightning in 1896. It is the centre of the Temple of Heaven complex in Peking.

Life traditionally centres on the enlarged family. Dozens of close relatives may share one household, but under government influence, loyalties have been shifted somewhat to the commune. This emphasis on service to the people is reinforced by a revised education system that makes young people spend less time at school or university and more at practical (particularly manual) collective work. However, life in China is far from dour. Leisure time activities include kite flying and wrestling. Traditional festivals feature fireworks and processions of mock monsters. The huge official displays also attract an immense following.

In many ways life for the masses is indeed better than ever. For instance, life expectancy is up; women at last enjoy parity with men; most Chinese are now literate. But to approach the living standards of the West, China's fast upsurge of numbers must be curbed. The government has backed birth control campaigns but in the early 1970s there was no nationwide drive to check population growth.

The Economy

Soils. The Chinese depend for food on the continuing fertility of their soils. Careful cultivation of these soils over the centuries has given a misleading impression of their great natural fertility. In fact, practices like stripping the natural vegetation have led to serious wind and water erosion in some places, while overcropping has caused soil exhaustion elsewhere. One recent mainland source states that 'poor' soils account for about 40% of the total cultivated area. Bad farming practices such as extensive deep ploughing may have been partly responsible for this percentage.

Agriculture. China ranks among the world's largest food producers. It accomplishes this by harnessing around one-third of the land as pasture and one-seventh (or less) as arable land. From these areas, the industrious Chinese peasants raise the world's biggest national rice crop and largest number of pigs; the second-largest harvest of vegetables; and the third-largest output of cotton, maize, tea, wheat and sheep.

Different climatic regions specialize in different products. Thus farmers raise millions of sheep and goats on the dry grasslands of the north and north-west. The fringes of north-western deserts and their oases also support grains and fruit. Much of the Tibetan plateau is too cold and dry to support crops, but the Chinese recently claimed to be growing barley at over 13,000 ft (4,600 m).

The east is devoted largely to arable farming. Here there are two major crops: wheat to the north of the Ch'in Ling Shan watershed and rice to the south. Most of the wheat grows between these mountains and the Great Wall of China. Other north-eastern crops include cotton, maize, millet and sugar beet.

South of the Ch'in Ling Shan farmers working irrigated paddy fields harvest at least two crops of rice a year. They also grow other food plants including sweet potatoes, maize and fruits. In the subtropical zone just south of the mountains cash crops include tea and tung oil. In the tropical far south

Textile workers. The Chinese government has expanded its work-force by breaking down traditional patriarchal family relations.

citrus fruits, sisal and rubber grow and farmers reap three rice harvests a year. In the south-eastern Canton Delta area a special pattern of farming features freshwater fish bred in artificial pools whose embankments support mulberry trees (needed for silkworm culture), sugar-cane or fruit trees.

Because cultivable land normally yields more food from plants than from animals the thrifty eastern farmers tend to limit their livestock to oxen and water buffaloes (used for ploughing and pulling carts) and chickens and pigs (which need little space and feed largely by scavenging).

South-west of the Ch'in Ling Shan, the rich Szechwan Basin province is a small but significant farming region, hemmed in by mountains. It produces more rice and mulberries than any other province.

Chinese farmers work largely on traditional lines. They laboriously terrace hillsides by hand, transplant whole fields of rice in the same way and irrigate crops with the help of foot-operated water wheels. They also rely heavily on traditional fertilizers: animal and human manure.

The face of Chinese farms and farming has changed remarkably since the communists came to power in 1949. Before this, millions of peasant farmers, like countless peasants before them for thousands of years, had eked out a bare existence, seldom fully fed, often poorly clothed and living in shacks. Many peasants were tenant farmers or landless labourers, heavily taxed and in debt for life to unscrupulous landlords. Lacking the cash to develop their land, they were often without the help of draught animals and were forced to overwork their soil, reeling under the unpredictable blows of nature, notably drought, flood and typhoon.

Understandably, where the Russian Revolution began in the cities, China's communist revolution came from the land. Mao Tse-tung's road to power lay in capitalizing on peasant discontent and placing farming resources in the hands of the people.

Agriculture Under Communism. The first steps toward the communist goals of a fully socialist society included a sweeping series of land reforms and drastic social reorganization. Official Chinese figures, possibly heavily weighted, claim that before 1949 less than 10% of the population owned more than 70% of the land. Under land reform the land was redistributed, largely to 300 million landless peasants, and by 1954 most peasant households were organized into *mutual aid teams* of about ten households which pooled labour, tools and animals permanently or during sowing and at harvest time.

Meanwhile larger units called *agricultural producers' co-operatives* had emerged and by the end of 1956 embraced 96% of the rural population, or 120 million households. In 1956 many co-operatives were enlarged and absorbed land, draught animals and large tools, although tiny 'private plots' were allowed for growing a few vegetables or keeping livestock. Full collectivization of labour was completed by 1957.

Full collectivization of labour was completed by 1957.

The Commune and its Influence. The following year the goal of an economic 'great leap forward' and the concept of the vast *people's commune* were proclaimed. The commune became a centre not only for agricultural, but for industrial, commercial para-military and social welfare activities. During the 1960s the original type of commune proved unworkable and was much modified, organizing its activities in smaller units. Lack of statistics prevents any useful evaluation of the success of the system. It continues to evolve and it remains to be seen what organizational developments there will be in the future.

By the early 1960s China's leaders realized that really extensive mechanization would take at least 20 years, and by 1965 China possessed only about 70,000 tractors. The livestock situation on the other hand appeared bright by the late 1960s when China had an estimated 180 million pigs; 100 million sheep and goats; and 65 million cattle and buffaloes. Livestock are a valuable foreign exchange earner. China's all-important grain output has also experienced vital improvements under communist rule. Since 1961 they have reported a row of rich harvests, culminating in 1970 in record total and per unit grain output: China at last seems to be beating the old bogey of famine, though there are no reliable statistics of agricultural production to prove it.

Forestry. Forests cover 10% of China, and in the mid-1960s China was the world's fourth largest producer of forest products. But the country's timber reserve is small when measured against the size of the population, especially since much of the timber grows in remote border areas such as Inner Mongolia. In fact China suffers from a very serious shortage of wood, and has to import it and make extensive use of bamboo as a timber substitute.

Most of the forests lie in remote border areas; Heilungkiang, Kirin and Inner Mongolia in the north-east together account for half the national production and this shows no signs of a significant rise.

In 1965 there were more than 3,500 state forest farms and 30,000 commune forest farms, but in many of these forestry was little more than a sideline. Mass afforestation schemes have become a regular feature of rural activities in the slack winter season but low sapling survival rates have hampered progress.

Fisheries. China possibly ranks as the world's second or third largest fish producer with fish production in the mid 1960s between 5 million and 7 million tons a year.

The seas off the China coast are less than 650 ft (200 m) deep and provide good fishing grounds. Seas off the coast of China contain over 1,500 varieties of fish. The northern waters are rich in cod; further south small and large yellow croakers, and hairtails are most important, although both hairtails and cuttlefish are found in northern and southern seas.

About 3,860 sq mi (10,000 km^2) of shallow seas are ideal for the production of mussels, oysters, seaslugs and kelp. The principal fishing centres are at Shanghai, Lüshun-Talien, Tsingtao, Canton and Chefoo; all have factories processing marine products.

Fish is one of China's most valuable sources of animal protein: there has been parallel scientific development of state deep-sea fisheries and the more traditional, commune-based, exploitation of shallow coastal and interior waters. The income from a carefully tended pond may be five times as great as that from a field of the same size.

Mineral Resources and Power Supplies. Western peoples have long believed China to possess almost limitless mineral riches. The country's mineral resources are undoubtedly great, but how great remains uncertain for prospectors have yet to examine great tracts of the land. Moreover even rich known resources remain largely untapped because the Chinese lack either the transport and mining equipment to extract the minerals or the factories to exploit them.

Certainly China is very rich in iron. By the 1960s China ranked as the world's fourth largest producer of iron ore. In 1969 and 1970 rapid growth of small iron mines was mainly responsible for the great increase in iron ore production, which was said to have jumped by 48% in one year. High-grade ore comes from central and northern China, but most ore is low grade and from the north-east.

China leads the world in its output of antimony and tungsten, and ranks second only to Malaysia in tin. Molybdenum, mercury and bismuth are also present in large amounts. There are only moderate supplies of copper, lead and zinc, and China is poor in chromium and nickel.

Similarly supplies of non-metals vary. Fluorite, graphite, magnesite and talc are important in world markets. There are abundant salt reserves. Cement is produced in comparatively large quantities; asbestos and barite deposits support considerable industries; the supply of sulphur is moderate; but phosphates are limited.

Many economists believe that China will remain self-sufficient in a wide range of mineral raw materials for the foreseeable future. But the exploitation of these resources seems unlikely to go ahead fast enough to place China among the world's industrial giants, at least until late in the century.

China's hopes for industrial advancement must lie largely in stepping up its supplies of industrial power. This in turn means raising the output of coal and oil. There was a steady rise in the extraction of both these fossil fuels between 1949 and 1960. China claimed to be the world's third-largest producer of coal by the mid 1960s, and the opening of many small coal mines in 1969 and 1970 almost certainly yielded considerably more than the 250 million tons (255 million metric tons) quoted by some economists.

The major coal mining centres are Fu-shun, Fou-hsin, Kailan (Hopeh), Huai-nan (Ànhwei), Chi-hsi, Ho-kang and Ta-t'ung (Shansi). This suggests a heavy concentration of coal reserves north of the Yangtze. The Chinese now claim that large coal deposits have been found south of the Yangtze.

Economists believe that China possesses rich oil reserves. In the interior, large quantities of oil are now flowing from wells from the Dzungarian Basin in the north-west to the Szechwan Basin in the south. Since the mid-1960s China has claimed self-sufficiency in oil-production, yet its per capita output is extraordinarily low. Prospecting continues in the interior and presumably off shore as well.

Although China possesses an enormous hydro-electric power potential, much of this lies in inaccessible areas and is untapped. Electricity generation depends mainly upon

The 20th anniversary of the founding of the People's Republic of China in October of 1969 was marked with parades and demonstrations. The statue is of Mao-Tse-Tung. In the right rear, is the Gate of the Heavenly Peace in Peking from which Chinese leaders often review parades.

The Potala in Lhasa was the palace-monastery of the Dalai Lamas.

coal-fired thermal electricity plants and there are only a handful of large hydro-electric plants, the ratio being 4:1. However, output is rising. The new giant San-men Gorge multi-purpose project should make a large impact when it is fully operational. The widespread building of local, and often very primitive hydro-electric power plants in 1969 and 1970, must also measurably increase electricity supplies. While much of the nation's electricity is harnessed to urban industry, supplies are increasingly reaching out into the countryside.

Industrialization: the Key to Material Progress. After the communists came to power they backed programmes of rapid industrialization in order to build a balanced, self-sufficient modern economy. Economic and political crises have hampered the actual rate of China's industrial growth, but this has at times been impressive.

Before 1949 though traditional art and craft industries produced an unsurpassed list of masterpieces in porcelain, jade, lacquer, silk and bronze, China's industrial growth had been stunted. What modern industry there was, was the result of the penetration of China by Western nations, and tended to be restricted to the production of consumer goods. China badly lacked modern heavy industrial plant.

After they gained power the Chinese communists decided on a major relocation of industry. There was a strong emphasis on the development of the northern half of the country, especially the north-east. Away from this area, Pao-t'ou, Tientsin, and Peking grew rapidly. Iron and steel works were developed at Shihching-shan (Peking) and T'ai-yüan. In the north-west, Sian and Lan-chou emerged as industrial nuclei and production at Kansu's Yü-men and Sinkiang's Karamai oil fields expanded. At the same time, the growth of light and handicraft industrial centres further south proved to be much more gradual. By the early 1970s most expansion had been in existing industrial areas.

Probably one half of the country's modern industrial capacity remains in the three eastern provinces of Kiangsu, Hopeh and Liaoning, and two-thirds of the total capacity when southern central Szechwan and Hupeh are added. In particular the Chinese were far more dependent on the established industrial metropolis of Shanghai than their government cared to admit.

The uncomfortable predominance of Shanghai is entirely against current Chinese thinking. The aim is not to build big cities but to disperse newly built factories in the rural areas. This enables members of the rural people's communes to take part in industrial construction and the families of workers and staff members to participate in agricultural production. The movement to establish small rural areas so that workers can experience both industrial and agricultural life.

Recent Industrial Progress. After an initial great surge forward in growth rates under the communist regime, there was a difficult period after 1960 when the Russians withdrew technical aid: there were crop failures in the early 1960s and economic dislocation appears to have been caused by the 'Cultural Revolution' in 1967 and 1968. Money and scientific talent are constantly drained from the economy by massive investment in military projects. Nevertheless, in terms of technological progress China has made impressive gains. Chinese scientists have developed and tested atomic and hydrogen explosive devices; a transistorized computer appeared in 1967; and China's first satellite was launched in 1970. For the masses, government priorities on advanced research and necessities like lorries and tractors mean that such consumer goods as cars and washing machines are almost all unobtainable however, and the industrial way forward will not be easy.

Rice planting. Four-fifths of the Chinese work the land, and arable land is used to its limit.

The inscription on this bridge reads 'Mao-Tse-Tung's thought is the guiding principle for all the work of the party and the country'.

Transport

Although the railway system links the major centres, motor vehicles are uncommon and many people move by boats, bicycles, ox- or horse-drawn carts, or simply walk, bearing their loads on their backs or slung from poles.

Two factors have conditioned the growth of China's transport network under communist rule. The first factor was economic. An improved communications grid promised to help the opening up of pioneer settlement areas, the exploitation of natural resources, and industrialization. The main effort in railway building was in the south. In fact the less developed regions received 80% of the new railways up to 1964.

The second factor conditioning transport development was strategic, involving both national defence and internal stability. For instance the road and railway system has been strengthened opposite Taiwan, and highway construction in Tibet has been undertaken ostensibly for reasons of 'national defence'. Early on, priority also went to building rail links with the USSR but these were not extended after relations deteriorated between both countries.

Railways. By the end of 1965 about 22,000 mi (35,400 km) were open to traffic. In 1949 all western China had been without railways while there were only limited, temporarily out of action routes in three southern provinces. Since then many major connections have been effected, for instance the Paoki-Chengtu and Chengtu-Chungking lines and lines in the far south joining up with Vietnam. Railway construction has included throwing three major bridges across the Yangtze, Wu-han, Chungking and Nanking.

China's engineers face some formidable construction and maintenance difficulties, especially in the harsher climatic areas: sand-dune encroachment is a particular hazard in the north-west. Economic difficulties hit the building of trunk lines in the 1960s. Also, China cannot make enough locomotives and rolling stock and must import the balance (from Eastern Europe and Britain) or in some mining and forest areas make do with simple home-made engines run on wooden tracks.

Roads. A recent non-Chinese source states that China has 190,000 mi (300,000 km) of 'predominantly all-weather' roads usable by motor vehicles, or only one-fifteenth the American total. There are also 310,000 mi (500,000 km) of unsurfaced secondary roads, and countless mountain trails between remote villages. Local roads, rather than railways and highways, provide the main lines of communication for most people. Vehicles are rare but the narrow local roads suffer heavily from traffic congestion caused by a tangle of travellers carrying traditional carrying poles, pushing wheelbarrows and handcarts, and riding bicycles or driving pack animals.

Inland Waterways. These provide a third important transport system, used by countless small barges, and larger freighters and passenger steamers. These routes include both rivers and canals, notably the 1,100 mi (1,700 km) Grand Canal between Peking and Hangchow began about 600 AD. In 1960 there were reportedly 104,000 mi (168,000 km) of navigable waterways, a quarter of them open to river-going steamers. The Yangtze and its tributaries account for about 40% of the total length of inland shipping routes. Since 1949 the Chinese say they have built or enlarged nearly 100 ports and harbours along the banks of the Yangtze, and claim that loading and unloading is now mechanized at big ports. In terms of freight carried by modern means, China's inland waterways lead the roads and rank second only to the railways. But China's seagoing merchant fleet is small.

Air Services. By 1966 China had an internal air service, and in the early 1970s the Chinese were contemplating an international air line. Meanwhile foreign airlines served Peking, Shanghai and Canton.

International Trade. The total value of China's trade reached a record $US 4,300 million in 1966. This was a mere one-thirteenth of the American figure for that year. Moreover the volume of trade fell in the two troubled years that followed, but since then trade has been expanding. Also, in almost every year since 1955 the Chinese have had a favourable balance of trade – a situation which some wealthy Western nations have been unable to achieve.

During the 1950s China traded mainly with socialist countries, mostly with the Soviet Union. Then, as relations with the USSR worsened, United Nations statistics show that between 1960 and 1968 Soviet exports to China fell from 735 million roubles to 53 million roubles, while imports from China declined from 763 million roubles to 53 million roubles.

Meanwhile, the Chinese began to do more business with non-communist Western Europe, Canada and Australia. By the late 1960s Japan, Hong Kong and West Germany had become established as China's biggest trading partners. By the early 1970s growing tolerance between China and the USA suggested that there could be further changes.

Barges on the Soochow River in Shanghai, China's largest port. Shanghai is located on the delta of the Yangtze river.

MONGOLIA

Area: 604,090 sq mi (1,564,593 km²)
Population: 1,230,200
Population Growth Rate: 3·2%
Capital: Ulan Bator, pop 282,000
Language: Khalkha Mongolian. Turkic (7%)
Religion: Tibetan Buddhist Lamaism (officially suppressed)
Currency: Togrog = 100 mongo

The Mongolian People's Republic (Mongolia) is the largest landlocked state in the world. Greater in area than Alaska and nearly three times the size of France, it is an enormous sparsely populated buffer state between the two giants of the communist world, the Soviet Union and China. Both of these nations have guaranteed the independence of the republic. Mongolia was under Chinese rule from the 17th century until Russian support led in 1924 to the independence of Outer Mongolia, which became the Mongolian People's Republic, and the Mongols have looked to Moscow ever since. Visible expressions of their alliance with the Soviet Union are the continuing flow of Russian financial and technical aid, and the presence of large Soviet forces in the country.

The Land. Everywhere the land stands high, the average altitude above sea level exceeding 5,000 ft (1,500 m). The north and west are dominated by mountains: the Altai range has snow-capped peaks well above 15,000 ft (4,600 m), and the lofty Hangayn and Hentiyn ranges enclose the fertile Selenga-Tuul basin, where the old Khans had their capital, Karakorum.

The Selenga, flowing northwards to Lake Baikal, is one of many rivers descending from the mountains, and is joined by the Orhon. The Kerulen flows to Lake Hu-lun (Dalai) in China, and the Onon eventually joins the Amur. Many large and deep lakes, some salt and others of fresh water, lie between the mountains; Lake Hövsgöl, for example, covers more than 1,000 sq mi (2,590 km²) and is about 780 ft (238 m) in depth. East of the mountains the Mongolian plateau falls away to the south and east embracing part of the sandswept Gobi Desert.

The climate is harsh and rather dry. Winters are bitterly cold, the short summers hot. Rainfall, occurring mainly in summer, averages only 10–20 in (254–508 mm) a year and decreases to below 4 in (102 mm) in the extreme south-east. Rivers and lakes freeze over in winter, but there is little snow. Strong winds and dust storms are frequent. Much of the country is steppe, and nearly half is semi-desert or desert. Less than one-tenth of Mongolia is forested, the most common tree being the Siberian larch.

Wildlife includes spotted deer, elk, bears, boars, lynxes, wolves, sables, beavers, reindeer, wild sheep and asses.

The People. The Mongolian People's Republic is one of the most thinly-populated countries in the world. Most of the people are Mongols of the Khalkha group, but there is a Turkic-speaking Kazakh minority in the west. Most Mongols have now abandoned their Lamaist form of Buddhism, and only a few still live in the traditional, round and movable felt-covered tent called a *ger* or *yurt*. Nearly a quarter of the population lives in Ulan Bator (282,000). This city is not only the capital, but the chief industrial, transport and cultural centre, where Soviet-inspired buildings and Western ways of life are found. Choybalsan and Darhan, both situated on railways, are growing industrial towns.

Education has made great strides and most Mongols are literate. Their language, quite unlike Russian or Chinese, is now written in a modified form of Cyrillic script. Ulan Bator has a university, an Academy of Sciences and many technical institutes.

The Economy. The Mongols have always been great herdsmen, and stock-breeding is still the mainstay of the economy. Using all-year pastures in the wetter highlands, and seasonal steppe pastures, the state and collective farms raise millions of sheep, goats, cattle, horses and Bactrian camels.

The Soviet system of state and collective farming, despite such problems as aridity, spring floods and soil erosion, has also had success in growing grain, mostly in the wetter north. Some scattered areas are irrigated for fruit cultivation.

Mongolians are turning increasingly to mining and manufacturing to meet domestic needs and provide exports. Indeed industry now provides nearly two-thirds of national production. There are large coal-mines near Ulan Bator and Darhan; soft coal from opencast workings at Sharyn Gol is railed to Darhan to fuel a new thermal electricity plant.

Oil is produced at Dzüünbayan in the Gobi Desert. Gold is mined, and wolfram and fluorspar are exported to the Soviet Union. Large deposits of copper, tin, phosphates and fluorite have recently been found, as well as silver and uranium.

Transport. Side by side with the camel caravan and the ox-cart, still widely used even in the capital, are the most modern forms of transport. Soviet-built aircraft, operated by Mongolians, link Ulan Bator with Moscow and Irkutsk. Some river bridges, a few metalled and many improved roads now permit the use of lorry and *gazik* (the Russian equivalent of the jeep). There are also a few broad-gauge railways, constructed originally for Soviet strategic purposes, and some narrow-gauge lines serving the mines.

International Trade. About 75% of the republic's foreign trade is with the Soviet Union, and a further 15% with the republic's East European partners in COMECON. Fluorspar, wool, furs, hides, livestock and meat are the chief exports. Imports include aircraft, vehicles, machinery, petroleum, cloth and foodstuffs.

DEMOCRATIC PEOPLE'S REPUBLIC OF KOREA

Area: 46,780 sq mi (121,160 km²)
Population: 14,281,000
Population Growth Rate: 2·8%
Capital: P'yongyang, pop 1,500,000
Language: Korean
Religion: Officially discouraged. Buddhism, Confucianism, Shamanism and Ch'ondoko
Currency: Won = 100 jun

The People's Democratic Republic of Korea (North Korea), is a communist state occupying the northern part of the Korean peninsula. Bordered on the north by Manchuria (part of Communist China), it stretches roughly to the 38th parallel, where it is divided from South Korea by the 487 sq mi (1,261 km²) demilitarized zone.

The division, which has existed since the end of World War II, created a serious imbalance in the peninsula's economy: the North possessed most of the country's mineral resources and industries, while the South was primarily agricultural. In North Korea, this has led to strenuous efforts to increase agricultural production coupled with extensive industrial redevelopment. On the political front, hostility between North and South Korea showed the first signs of cautious relaxation in July 1972, when the two countries agreed to seek peaceful reunification.

Geographical Features. North Korea is predominantly mountainous. Along the border with Manchuria, running from north-east to south-west, a series of high, rugged mountain ranges rises to over 6,000 ft (1,829 m), culminating in the country's highest peak, P'ai-t'ou-shan (9,003 ft; 2,744 m). A further mountain chain extends along the east coast and into South Korea. With the exception of the Tumen, which runs 323 mi (517 km) eastwards from P'ai-t'ou-shan, most of North Korea's rivers are in the west. The Yalu rises in China and is navigable for most of its 491 m (786 km) course west of P'ai-t'ou-shan. The Ch'ongch'ŏn and the Taedong water the country's sole plain, in the west. The east coast is mainly steep and unindented, but has several good harbours, such as Ch'ŏngjin and Wŏnsan.

Climate, Vegetation and Wildlife. Korea has a monsoon climate which is influenced mainly by continental elements. In winter, cold, dry air masses move across the country from Siberia. Winter may last as long as six months. January temperatures sink as low as 15°F (−9°C), and even lower in the mountains. In summer, temperatures average 68°F (20°C). Annual precipitation, most of which falls in summer, ranges around 25 in (635 mm).

Much of the country's once extensive forest has been decimated, but there are still large areas of larch and spruce in the north. Here, too, are most of the remaining large wild mammals – tigers, lynxes, wolves and bears.

The People. Traditional modes of life in North Korea have steadily given way beneath the authoritarian pressures of the communist regime, which have nevertheless meant the virtual elimination of poverty and illiteracy. Agriculture has been collectivized, while the focus of urban life is the factory.

The largest urban centre is the capital, P'yongyang (1,500,000). Education in North Korea is free and compulsory from the age of 8 to 17. There are some 3 million students at 8,690 primary and secondary schools; and about 214,000 students in higher education.

The Economy. In 1971 North Korea had an estimated gross national product of $2,800 million. Of this figure, about 70% comes from industry and mining, which employ over half of the country's labour force. In recent years, the emphasis has been on the development of heavy industry. Most manufacturing industries are concentrated in the south-west, particularly around Pyŏngyang, and produce chemical fertilizers, cement, steel, pig iron and textiles. Mineral resources are extensive. North Korea is a leading producer of graphite, tungsten and magnesite, and has good deposits of coal, iron ore, lead, zinc and other minerals. Hydro-electric power is well developed.

Agriculture has undergone extensive changes under the communist regime. By 1959, the country's one million farm households had been reorganized into 3,800 peasant

A village in North Korea, a mountainous still largely agricultural land.

co-operatives. But 5% of the cultivable land is reserved for private plot farming. Improved irrigation and increased mechanization have raised agricultural output. Between 1956 and 1971 grain production increased from 2·8 million tons (2·9 million metric tons) to 5·9 million tons (6 million metric tons). By far the most important crop is rice, followed by millet, barley and wheat.

North Korea's fisheries are being developed and modernized.

Transport and Trade. By the late 1960s North Korea claimed 6,200 mi (10,000 km) of railway, but roads were poor and unspecified. Waterborne traffic was increasing substantially. Most of North Korea's trade is with other communist countries, particularly China and the USSR. The main exports are metal ores and metal products, minerals and chemicals. Imports include machinery and fuel. The main non-communist trading partner is Japan.

REPUBLIC OF KOREA

Area: 38,027 sq mi (98,490 km²)
Population: 33,524,000
Population Growth Rate: 1·2%
Capital: Seoul, pop 5,536,000
Language: Korean
Religion: Buddhism, Confuscianism, Shamanism, Chondokyo
Currency: Won = 10 hwan

The Republic of Korea (South Korea) occupies that half of the Korean peninsula south of the 38th parallel. It is bounded by the Yellow Sea to the west and the Sea of Japan to the east. In the south-east, it is separated from Kyūshū, the southernmost island of Japan, by the 120 mi (192 km) wide Korea Strait. South Korea was declared a republic on 15 August 1948.

South Korea's economic difficulties stem from the fact that it has only limited mineral wealth and industrial capacity compared to the North, while its population density is very high (819 persons per sq mi; 316 per km²). Since the Korean War (1950–53), which severely damaged the economy, South Korea has concentrated on industrial growth.

Geographical Features. Rough, mountainous terrain prevails throughout South Korea, though only a few peaks rise above 5,000 ft (1,500 m), the highest being Chiri-san (6,283 ft; 1,915 m). From the northern border, one mountain chain extends roughly south-west while another runs south along the east coast, which is steep and rocky. The west and south coasts are deeply indented, with many natural harbours, including Inch'ŏn in the west and Pusan in the east. There are over 3,000 offshore islands, the largest being Cheju, about 50 mi (80 km) off the south coast.

Lowland occurs mainly in the form of small coastal plains, the most extensive and fertile being in the south-west. South Korea's largest rivers are the 326 mi (522 km) Naktong, which waters the south-eastern region, and the 320 mi (512 km) Han, which empties into the Yellow Sea near Inch'ŏn.

Climate and Vegetation. South Korea has relatively dry, mild winters, and monsoonal summers. January temperatures average about 26°F (−3°C). In July the average is about 75°F (24°C). Warm, moist, maritime air masses from the south-east move across Korea in the summer, bringing rather rainy conditions. Annual precipitation is around 50 in (1,270 mm). The south is vulnerable to typhoons, particularly during September.

The country's once extensive forest cover has been ruined by excessive felling, burning and disease, but there remain scattered stands of pine and bamboo.

The People. South Korea's population has grown rapidly since 1945, swollen by successive waves of refugees from the North. The estimated population in 1945 was 16,500,000; by 1972 the figure had more than doubled to 33,524,000. Some 32% of the people live in urban areas, the largest city being the capital, Seoul (5,536,330).

Korea has often served as a cultural bridge between China and Japan. Nevertheless, the Koreans evolved a distinctive culture. In rural areas, Koreans live and dress much as they did centuries ago, but in the cities Western influences are increasingly evident.

Most South Koreans are Buddists or Confucians. However, the principal organized religion is Christianity: there are some 2 million Protestants and about 500,000 Roman Catholics. Chondogyo – an eclectric religion founded in the 19th century – claims some 1·5 million adherents.

Primary education – from the age of 7 to 13 – is compulsory. South Korea has over 8,000 schools of all grades, with more than 7,300,000 students. There are 27 universities and 783 colleges and technical schools. More than 172,400 students are enrolled in higher education.

The Economy. In 1971 South Korea's gross national product was estimated at $8,100 million. Of this figure, some 28% came from agriculture and some 20% from manufacturing industries. In recent years, the state has encouraged land reclamation schemes, irrigation projects, and the use of improved varieties of grain.

Farmers have been given grants to purchase agricultural machinery. As a result, agriculture has made considerable progress. Output of rice (which accounts for about 60% of the grain grown) rose from an annual average of 3·5 million tons (3·6 million

Sokcho is a port just above the 38th parallel on South Korea's mountainous east coast.

metric tons) in the years 1952–56 to over 5·4 million tons (5·5 million metric tons) in 1970. Over the same period, barley output doubled and wheat output trebled. Other important crops are sweet potatoes, soybeans and tobacco. Fisheries are a vital sector of the South Korean economy: the catch in 1969 was 849,174 tons (862,800 metric tons).

Industrialization has made rapid progress since 1965. Light industries lead the manufacturing sector. Textiles (including silk) are the most important product, followed by plywood, plastic goods, paint, fertilizers, cement and electrical equipment. Japanese assistance, in the form of loans and co-operation in the development of heavy industrial projects, has been increasingly important.

South Korea's mineral resources are much smaller than those of North Korea, with the exception of tungsten – South Korea provides 6% of the world supply of this metal. Other mining products are coal, iron ore, fluorspar, limestone, graphite, gold, copper ore, lead, kaolin and talc.

Transport and Trade. There are some 1,800 mi (2,900 km) of railway and 22,000 mi (35,500 km) of largely unsurfaced road. Korean and other airlines operate internationally and there are 15,000 vessels (largely fishing boats).

South Korea's major trading partners are the United States, Japan and Hong Kong. The main exports are textiles and clothing, plywood, electrical equipment, fish and fish products. Imports include machinery, transportation equipment and food grains.

JAPAN

Area: 142,885 sq mi (370,072 km²)
Population: 105,220,000
Population Growth Rate: 1·1%
Capital: Tokyo, pop 11,408,000
Language: Japanese
Religion: Buddhism and Shinto
Currency: Yen = 100 sen = 1,000 rin

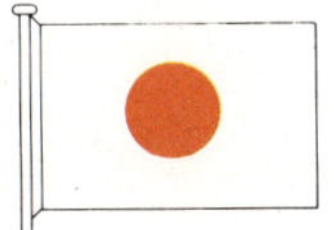

Japan consists of four principal islands, Hokkaidō, Honshū, Kyūshū and Shikoku, about 500 less important islands and about 3,000 minor ones. The insularity of this mountainous nation off the East Asian mainland did much to develop its national consciousness, but the country's sheer lack of land is a problem for its expanding population. In the 1930s Japan's military leaders sought to solve the problem by aggressive expansion culminating in World War II. Despite the ruinous devastation of the ensuing defeat, and a lack of natural resources, the people of Japan have since made such good use of their skills that their country has become one of the leading industrial nations of the world and indeed Asia's sole economically developed major nation. Japan's industrial growth now enables the islands to support far more people and far higher living standards than ever before.

Economic and cultural changes along Western lines has happened so fast that Japan now exhibits a confusing mixture of ancient and modern. In the country, patchworks of paddy fields surround traditional thatched villages emblazoned with brilliant signboards advertising the latest gadgets. In the cities ceremonial geisha teahouses adjoin modern coffee bars.

In strong contrast to the aggressive policies of imperialist pre-war Japan, post-war Japan has relied largely on American military support, and the resulting saving on defence expenditure has helped make the economic growth rate the highest in the world.

The Land. The islands of Japan form one of the archipelagoes which extend around the rim of the Pacific Ocean, and which are characterized by frequent volcanic activity and crustal movement. Each year, over a thousand earth tremors are recorded in Japan; but really severe earthquakes, which cause damage to buildings, are less frequent and occur on average once every five years.

While the Japanese islands are mountainous throughout, several arcs of folding converge within Japan to produce particularly conspicuous knots of highland. The Japan Alps, which rise to over 10,000 ft (3,050 m), provide the highest terrain in the country, though the highest single peak, Mt Fuji (12,388 ft; 3,770 m), is an extinct volcano unrelated to the fold mountains of the Alps.

Japan contains 192 active volcanoes, and volcanic deposits and ash cover nearly 25% of the surface area of the country. Major zones of active volcanoes and hot springs occur in Hokkaidō, in northern and central Honshū, and in southern Kyūshū.

Japan has a number of swift-flowing rivers and streams which are good sources of hydro-electric power, though of little use for navigation. Seasonal variations in the amount of precipitation, however, produce marked fluctuations in the rate of flow of the mountain torrents, and thus hamper the efficient operation of hydro-electric plants throughout the year. There are also many inland lakes, the largest being Lake Biwa, near Kyōto. The Inland Sea is also virtually a lake, since it is almost completely enclosed by the three islands of Honshū, Shikoku and Kyūshū. The surrounding scenery is strikingly beautiful, and the whole country's landscapes of mountains, rivers, lakes, waterfalls and sea views have won acclaim for their beauty from many travellers.

The acute shortage of lowland is one of the salient characteristics of the geography of Japan. While the Japanese population currently exceeds 105 million, cultivable land amounts to only 16% of the total area. The small and constricted coastal plains of Japan, which contain most of the cultivated land and all of the major concentrations of population and industry, thus assume vital importance.

None of the Japanese plains is extensive, but three bays of the Pacific coast of Honshū are flanked by comparatively wide alluvial plains. The largest of these, the Kanto, occupies an area of 5,000 sq mi (12,950 km²) and includes the urban conurbations of Tokyo and Yokohama.

The coastal fringe of the typical Japanese plain is usually low-lying, and in the larger plains, particularly on the Pacific coast, extensive tracts lie below sea level, and are protected from inundation by sea dikes. Along the Japan Sea coast, by contrast, high sand dunes are often prominent features of the coastline.

Climate. The climate of Japan, characterized by a marked seasonal alternation in the direction of the prevailing winds, may be described as monsoonal. Winter weather in the Japan Sea coastlands is characterized by a heavy cloud cover and by frequent blizzards. On the Pacific side of the country, however, humidity is lower, and cold, clear, invigorating weather predominates. Within a narrow zone lying along the Pacific coast, winter temperatures are ameliorated by the influence of the warm Japan Current.

Winter temperatures vary according to the contrast between the two sides of the country, and also according to latitude. Thus, north of latitude 38 N, January temperatures fall below 32°F (0°C), and reach 20°F (−6·7°C) in Hokkaidō. By contrast, average January temperatures in Kyūshū, the southernmost island, can reach 44°F (6·7°C).

Towards the end of March, the first season of maximum rainfall occurs. The Bai-u rains, which begin in the middle of June and last for up to three weeks, produce advantageous conditions for the transplanting of rice seedlings into the flooded paddy fields.

By early July, a complete reversal of the winter pattern is well-established, and the south-easterly monsoon begins. During the summer months, heat and humidity are intense throughout most of the country, and living conditions in the low-lying plains bordering the Pacific coast are particularly uncomfortable. As in winter, temperatures vary according to latitude. Whereas average August temperatures in western Honshū and in Kyūshū rise to 80°F (26·7°C), northern Japan experiences cooler summers. Thus, the average August temperature at Sendai in northern Honshū is 75°F (23·9°C), and at Sapporo in Hokkaidō it is 70°F (21·1°C).

In late August and early September, the arrival of successive typhoons from the equatorial regions of the Pacific brings the second annual rainfall maximum. Coastlands exposed to the south and south-east are particularly vulnerable. Typhoon damage, however, is not confined to these areas, and throughout the country the maturing rice crop can be devastated by wind and rain.

Owing largely to insularity and rugged relief, precipitation is generally high throughout Japan. Average annual precipitation varies from 33 in (840 mm) in eastern Hokkaidō to 62 in (1,550 mm) in Tokyo and over 120 in (3,050 mm) in the mountains of central Honshū and in the mountainous parts of the Pacific coastlands.

Vegetation and Wildlife. The natural vegetation of the Japanese islands consists almost entirely of forest. Owing to the prevalence of rugged relief, which has greatly restricted the spread of cultivation, forests cover extensive areas and occupy about 70% of the surface area of Japan.

The boundaries of the forest zones of Japan are determined largely by temperatures, and vary according to altitude and latitude. Thus the sub-tropical forest zone, which is today characterized by such species as oak, camphor, wax and bamboo, predominates in south-western Japan. North of the sub-tropical zone stretches a region of broad-leafed deciduous and mixed forest, which is characterized by a variety of planted conifers as well as maple, birch, beech, poplar and oak. This zone extends over northern Honshū, south-west Hokkaidō, and the mountains of central Japan. Where average annual temperatures fall below 43°F (6·1°C), boreal forest predominates. Characterized by coniferous trees such as fir and spruce, together with birch, alder and aspen, it covers most of Hokkaidō east of the Ishikari plain.

Animal life in Japan also shows a great diversity. Japanese animals are often smaller than, but closely resemble, those of the Asian mainland. There are 140 or so mammal species. Larger mammals include deer, monkeys, bears, wild boars and wolves. Among the many smaller mammals are badgers, ermines, foxes, hares, mink, otters and squirrels. About 450 species of birds have been observed over Japanese territory. The teeming waters near the islands offer salmon, sea-bream, sardine and tuna. Trout are abundant in the rivers and lakes.

For the occasion of the Olympic games in 1967, Japan opened the first urban monorail service between its centre and the airport.

The People. The present Japanese population of over 105 million means that in terms of arable land per person Japan is the most densely populated nation of the world. Over two-thirds of the people now live in cities of 10,000 and over. The most thickly populated areas are south and south-east Honshū and north Kyūshū – areas of strong industrial concentration.

The Japanese are a mixture of racial elements from North-East and South-East Asia, but many centuries of isolation and inbreeding have produced a homogeneous national stock. The aboriginal inhabitants are some 15,000 Ainus, a primitive people of Caucasian (white) origin, and there are about 500,000 foreigners (mostly Koreans) living in the country. The Japanese are basically a Mongoloid people, but with more facial and bodily hair than other Mongoloids, possibly due to slight mixtures of Ainu blood. The Japanese language is the universal tongue of the country – except among the Ainus (whose language is not related to any other) and the Koreans.

Religion and Education. Buddhism, coming to Japan in the 6th century, has remained the religion of the common man. In all, there are perhaps 70,000,000 Buddhists in modern Japan, most of whom also observe Shinto, which is based on the worship of nature and one's ancestors and attempts to maintain the traditions of national life. Christians in Japan comprise less than 1% of the population.

Japanese schools are overcrowded, but the literary rate of 98% for the adult population is the highest in Asia. Education is free and obligatory for the first nine grades, after which about 60% of the students go on to senior high schools, most of which are co-educational. There are many adult education and vocational institutions. The higher education system was remodelled on American lines after World War II and includes 439 colleges and universities (of which seven are major national universities) and 473 junior colleges. Much university work is in English, now the leading foreign language in Japan.

Culture. Although the Japanese borrowed much from the Chinese (and the Koreans) over the centuries, Japan now differs widely from China in literature and art. There is a unique Japanese drama form, and typically Japanese graphic arts, more exotic than the Chinese.

Since World War II and the American occupation, Japanese culture has undergone far-reaching changes. Western influences deeply affect Japanese youth, and because many of the older generation retain their prewar traditions almost intact, there is a widening age gap in ways of living, working and dressing. Many in the cities have moved to modern apartment houses, but the majority still prefer the traditional small, low house with reed mats on the floor instead of chairs.

Agriculture. Although less than one-sixth of Japan's land area is arable, agriculture was the major source of income until after World War II. A high level of technology combined with intensive cultivation has resulted in regularly increasing harvests. The average farm has only about 2·5 ac (1 ha), but is heavily fertilized and often irrigated.

The leading crop and staple food is rice, of which annual yields reached about 14 million tons by the late 1960s. Rice is grown throughout the country, including locations where relief and climatic conditions are far from ideal. Only in the northern and eastern extremities of Hokkaidō, and at altitudes above 3,000 ft (900 m) is rice not cultivated. Technological changes and government price supports have brought about a steady increase in rice output, despite stagnant consumption. By the end of 1970 Japan's rice surplus amounted to 8 million tons.

Before World War II, silk cocoons constituted the second most important category of agricultural production. With the collapse of the American silk market following the stock market crash of 1929, and the subsequent gradual substitution of synthetic fibres for silk, output declined. In postwar years, in conjunction with growing affluence and gradual westernization of the Japanese diet, livestock products have replaced silk as the second most important agricultural sector, contributing about 18·6% of the value of agricultural production. Vegetables, including wheat, barley, potatoes and soybeans, now occupy third place, with about 12·6% of the value of agricultural production. But in spite of intensive farming, Japan must still import the bulk of its food.

Forestry. The forests of Japan, which cover about 60% of the total land area, are more extensive than those of Sweden or Finland. Forestry and timber-related industries (paper, furniture, etc.) employ about 3·7 million people. In contrast to the Nordic countries, a very high proportion of the Japanese forests lie in inaccessible mountain areas, and only 27% of the forest cover is commercially exploited. In many parts of Japan, the value of this exploited area has been greatly reduced by poor management and excessive felling.

Since the beginning of rapid economic growth in 1955, the demand for timber from the Japanese construction industry has risen swiftly. In 1969, domestic forests supplied 55% of the demand for timber, and the remainder was imported.

Fisheries. From earliest times, fish has constituted the main source of protein in the Japanese diet. With an annual catch of about 7 million tons, Japan ranks immediately after Peru as the most important fishing nation of the world. Well-equipped vessels range as far afield as the Antarctic, Africa and the west coast of North and South America. However, both cold water and warm water species of fish are present in abundance in close proximity to Japan. Bonito are caught in the warm water of the Japan Current, and saury are caught in the cold Oya Shio current. Cold water species such as cod, salmon and sea trout are fished off Hokkaidō. Japan's annual whale catch has been declining in recent years. In fact the fishing industry as a whole, despite the switch in emphasis away from inshore fishing to deep-sea fishing and government support of fish farming schemes, has experienced a gradual decline. Today, despite the size of the annual catch, fishing represents only 2% of the value of Japan's national income, employs only 2% of the civilian labour force, and provides a mere 3% of Japan's exports.

Minerals and Power. Japan is poorly endowed with mineral resources and its manufacturing industries rely heavily on imported fuels and raw materials. The most valuable mineral is a low-grade coal, but good quality coking coal is in particularly short supply and the steel industry relies heavily on imports from the USA and Australia. The two main coal-fields in Hokkaidō and northern Kyūshū are remote from the main industrial regions. Small amounts of iron ore, crude oil, gold, copper and sulphur are produced, but in general Japan is almost entirely dependent on imports.

Since 1955, oil has rapidly replaced coal as the major source of energy. In the electricity generating industry, oil-fired thermal electric power stations, mostly located on the coast, have now become more important than hydro-electric plants.

Japan's first nuclear power station at Tokaimura began full power generation in 1967, and it is estimated that by 1985, nuclear power will provide 10% of total energy supply.

Industry. Japan, like Germany, has experienced an 'economic miracle' since 1945 that has turned the nation into one of the world's leading industrial powers. The economy experiences real growth of roughly 10% a year – the highest in the world. Japanese electronic equipment, cameras, textiles, steel and automobiles have flooded many markets and radically changed the trade pattern of many nations.

The industrial boom has altered the economic structure of the country. Since 1955 the contribution of the secondary sector has risen from 30·2% to 37%. About 26% of the labour force work in manufacturing industries; 22% are engaged in trade and finance. The last decade and a half has seen a growing shortage of workers in industry, especially of skilled workers.

Government support of industry is still a salient feature of the Japanese economy. In the postwar years, the state has provided subsidies, low-interest loans, and highly effective economic plans, while successfully implementing monetary policies to control the rate of economic expansion.

Since 1955 the manufacture of machinery, metals and chemicals has steadily increased; by contrast, the share of the total value of factory shipments contributed by the textile industry has dropped. Since 1965, the machinery industry has shown the fastest rate of growth and now accounts for over 30% of the total value of factory shipments. Growth has been particularly dramatic in general machinery, including turbines, engines, bearings, and machine tools; and in electrical machinery, including radio and television sets. Output of passenger cars rose from 696,000 in 1965 to 2 million in 1970, by which year Japan ranked as the world's second largest producer. In shipbuilding, especially in construction of super tankers and bulk

Above: **Geishas play an important role in Japanese social and business life. The girls are trained in such arts as singing, dancing and conversation to help Japanese men relax from the cares of the day.** ***Right:*** **The Kanan Temple at Isakusa. The overhanging, curved roof is typical of Japanese architecture.**

Inexpensive, mass produced Japanese items, such as these Honda cars, have radically altered world trade patterns since World War II.

carriers of over 300,000 tons, Japan leads the world.

Expansion in the steel industry, upon which the growth of all manufacturing industries ultimately depends, has been equally spectacular. Japan's output of crude steel rose from about 40 million tons in 1965 to over 90 million tons in 1970, a rate of growth which, if maintained, would enable Japanese steel production to outstrip that of the USA by the end of the decade.

In chemicals, dramatic gains have been made in petroleum refining and in the production of petro-chemicals. Refined petroleum output rose from 2,470 million gal (11 million kilolitres) in 1965 to 4,619 million gal (21 million kilolitres) in 1970, while during the same period, polyethylene production increased from 400,000 tons to over 1 million tons.

Most of Japan's industrial output is concentrated in a belt stretching from the Kanto plain in the east to northern Kyūshū in the west. Within this belt are three conspicuous bay-head concentrations of industry: the Keihin region, which includes the cities of Tokyo, Yokohama and Kawasaki and is by far the most important; the Hanshin region, contained in the prefectures of Osaka, Hyōgo and Kyōto; and the Chukyo region, centred upon the city of Nagoya and the lowlands surrounding Ise Bay. Owing largely to the twin factors of land shortage and reliance on imported raw materials, postwar industrial location in Japan has been characterized by the establishment of many industrial complexes built on land reclaimed from the sea.

Transport. The development of the Japanese railway network began in the 1870s and 1880s. Since the beginning of this century, the strategic trunk lines, including the Tokaido line linking Tokyo and Osaka, have been operated by the state, while the development of feeder lines has been left mainly to private enterprise. Until very recently, the railways carried a high proportion of passenger traffic, as well as the bulk of freight traffic. Now, however, more Japanese are able to afford cars and the increase in road traffic has taken business away from the railroads.

Despite recent advances, roads in Japan are poorly developed and are not comparable with roads in Europe and America. Although the Tokyo-Kōbe motorway, which was fully opened in 1969, represents the beginning of an ambitious programme to construct a nation-wide network of similar highways, only 30% of Japan's roads were paved in 1968, in comparison with 70% in Italy and 100% in Great Britain.

Coastal shipping is an important means of transport in Japan, especially between ports on the Pacific coast and those on the Inland Sea. Domestic air services are well developed, with daily passenger flights between the cities of Sapporo, Tokyo, Nagoya, Osaka and Fukuoka.

International Trade. Since Japan lacks most natural resources, trade is a vital component of the national economy. Both imports and exports run at about 13,000 million dollars a year. Since World War II by far the most striking change in the pattern of Japan's foreign trade has been the dramatic decline in the importance of mainland China as a trading partner. China today is well-placed to supply Japan with industrial fuels and raw materials, and to provide Japan with a potentially enormous market. However, political factors, particularly the alliance between Japan and the USA as demonstrated by the Security Pact of 1960, have been responsible for the relative lack of trade between Japan and China. As a result, Japan now depends on the USA for 30% of all exports and imports. About half of the 200 largest US corporations are involved in the Japanese market – selling, distributing or manufacturing. American policy toward trade is therefore of crucial importance to Japan.

Exports consist primarily of ships, machinery and other iron and steel products, textiles, chemicals and electronic equipment. The USA is by far the greatest market for Japanese exports. Imports include foodstuffs, machinery, metal ores, petroleum and raw cotton and wool for the textile industry. Although about one-third of the value of all imports comes from the USA, it is significant that Japan has now become Australia's biggest customer. In this connection, Japanese participation in the development and exploitation of mineral deposits in Western Australia has been of paramount importance.

As the efficient Japanese economy has grown and penetrated world markets, trade relations with many countries have become strained – especially those with the USA. There have been bitter complaints about the

Mt Fujiyama is an inactive volcano and, at 12,385 ft, Japan's highest peak. It last erupted in 1707 and has long been a source of inspiration to Japanese artists and poets.

'flood' of inexpensive Japanese textiles, shoes and electronic equipment. The Japanese have also been widely criticized for their tariff protection on imported goods, though in fact formal trade barriers have been largely lifted.

Minor Islands. The main outlying Japanese islands are two chains extending southwards in the Pacific Ocean – the Rynkyn Islands (including Okinawa) in the west, and an eastern group including the Bonin and Volcano Islands. Marcus Island and the Daitō Islands are among more isolated possessions south of the major islands of Japan.

TAIWAN

Area: 13,885 sq mi (35, 962 km²)
Population: 15,249,000
Population Growth Rate: 2·4%
Capital: Taipei, pop 1,897,803
Language: Mandarin Chinese. Amoy and Hakka dialects
Religion: Buddhism and Taoism. Christian and Moslem minorities
Currency: New Taiwan dollar = 100 cents

Taiwan (Formosa or Nationalist China), the seat of the National Republic of China, is an island separated from the mainland of China by the Taiwan Straits, which are only 90 mi (140 km) wide at the narrowest point. Taiwan province includes several groups of islands: the Taiwan, Penghu (Pescadores), Quemoy and Ma-tsu groups.

Taiwan became the refuge of China's nationalist government under Chiang Kai-shek after the Chinese communists drove it off the mainland in 1949. Since then this crowded island has made very substantial economic progress. But the nationalists have lost ground internationally as more countries have politically recognized Communist China. This trend culminated in Taiwan's expulsion from the United Nations in 1971.

The Land. Structurally, the island is composed of a tilted fault block, running roughly north-east to south-west, with a steep face looking eastwards, and sloping more gently towards the west. Taiwan is dominated by the Central Range, a ridge of high mountains occupying almost half the surface area of the island. More than 60 peaks are about or over 10,000 ft (3,000 m), the highest being Hsinkao at 13,113 ft (3,994 m). The Tatun Shan area has prominent volcanic peaks, hot springs and fumaroles.

Rather over half of Taiwan is classified as mountainous, while 11% is made up of foothills and terraced tablelands, between about 1,600 ft (500 m) and 300 ft (100 m) high. However, by far the most important of the land form categories are the alluvial plains, which cover only 24·3% of the surface area. The plains are most extensive in the west.

Rising in the Central Range the major rivers flow mainly east to west and are often used to feed a vitally important irrigation network.

Climate, Natural Vegetation and Wildlife. Generally Taiwan's monsoonal climate is tropical in the south and sub-tropical in the north, although strongly modified by oceanic and relief factors. Apart from the mountainous areas, summer temperatures average 80°F and winter temperatures about 60°F. The summer monsoon (early May to late September) brings heavy rains to the south, whereas during the winter monsoon (late October to late March), with the reversal of wind direction from south-west to north-east, the rainy season comes to the windward side of north-east Taiwan.

Typhoons are frequently serious, especially from July until September. At the same time the tropical climate allows the extension of multiple cropping systems of two or even more harvests a year.

Forests cover over 55% of the surface of Taiwan, and include conifers and bamboo. Acacia is ubiquitous in the lower hills.

The People. Taiwan's population doubled in the 20 years before 1968, due to the influx of population from the mainland and high rates of natural increase.

The people of Taiwan are predominantly Chinese but there are distinctions between those recently arrived from the mainland after the communist takeover and the native Taiwanese. The former tend to group in the cities and the latter, who have adopted some Japanese styles of living, favour the country. With a population of 15,249,000, Taiwan is one of the most densely populated countries in the world, has the highest population density of any sovereign country in Asia apart from Sigapore, and one of the highest in the world.

Around half the population is under 20 years of age, and so there is a high growth potential. Although the population is predominantly rural, the urban population is increasing with the expansion of industry. Of the population aged 15 and over, 50% are engaged in agriculture, forestry and fisheries; manufacturing industries account for 12%

The climate of Taiwan is temperate enough to permit a twice-annual rice harvest. This rice is necessary to feed the island's enormous population of over 1,000 people per square mile.

The manufacture of electronic components, in this transistor factory in Taiwan, requires a delicate and precise craftsmanship in which both the Taiwanese and Japanese excel.

of the population. Illiteracy rates have dropped from 42% in 1953, to 16% in 1968.

At the present rate of growth the population would double before the year 2000 AD. Even with an effective birth control programme, however, the population will be over 20 millions by 1989.

Agriculture. The very considerable economic advances achieved by Taiwan during the last few years are largely based on successes in the agricultural sector. A quarter of the surface area is arable, and the climate encourages the production of crops such as rice, sweet potatoes, sugar-cane, vegetables, peanuts, soybeans, bananas, pineapples, and citrus fruits. The natural fertility of Taiwan's soils is comparatively low, however, so they require heavy application of fertilizers.

Methods of improvement include land consolidation designed to bring together tiny, fragmented, random-shaped plots, so that rectangular fields between 0·5 ac (0·2 ha) and 0·7 ac (0·3 ha) are served by irrigation and drainage ditches, and farm roads. The government aims to consolidate roughly half of the total cultivated area of 2,260,800 ac (914,900 ha) by the late 1970s.

Agricultural advances have resulted largely from substantial increases in yields per unit area. In 1968 2·48 million tons (2·52 million metric tons) of rice was harvested, over 60% more than in 1952, yet on more or less the same acreage. In the same period unit yields of crops such as pineapples and bananas more than doubled.

Forestry. Taiwan's forest reserve covers over half the island. It has increased from 266 million cu yd (203 million m³) in 1952 to 315 million cu yd (241 million m³). Bamboo reserves in 1969 stood at 460 million poles. Taiwan is the world's second largest exporter of composition board and plywood (after Japan), and in 1969 timber exports were valued at over US$110 million. Windbreaks are being planted in coastal areas to protect crops from typhoon damage.

Fisheries. In recent years Taiwan's deep-sea fishing fleets have been modernized and have grown considerably. In 1968 there were over 25,000 vessels, of which 10,000 were powered; the remainder were classified as sampans (4,600), and rafts (10,600).

Minerals. Taiwan is comparatively poor in economically useful minerals. Coal is the most important mineral resource, but the industry suffers from fragmentation and shortage of workers, and the remaining reserves are deep, with thin, irregular seams.

Among minerals with sufficient reserves to meet domestic requirements are: gold, copper, sulphur, salt, silver, limestone, serpentine, building stone and glass sand. The rich reserves of natural gas are important both as fuel and for the petro-chemical industry.

Industry. During World War II Taiwan's industry was badly damaged, and output declined by over two-thirds compared with pre-war peaks. Nevertheless Taiwan provides a striking example of an Asian country which has undergone a remarkable expansion in its industrial sector, largely financed by American government aid and (since this virtually ended in 1965) by big business investment.

Unlike that in many developing countries, Taiwan's industrial growth rests firmly on a sound agricultural base.

The trend in industry has been towards more durable products and industrial intermediates geared increasingly to the export trade. Examples include the setting up and expansion of cement, chemical products such as detergents and pharmaceuticals, glass, household appliances, machinery, metal products, plywood, rayon, and rubber tyres and tubes. Textiles have become Taiwan's leading export.

The electronics industry is a recent arrival. Some 67 new approved plants built with American, Dutch and Japanese investment are producing components such as capacitors, memory systems, semi-conductors, tuners and transformers; and also records, radios, television sets and tape-recorders.

As in the case of Hong Kong the tourist industry is a growing source of foreign exchange, which is capable of considerable expansion.

Transport. The government-run railways operate over 575 mi (1,000 km) of track. The west line links the northern and southern ports of Keelung and Kao-hsiung, and then continues via Pingtung as far south as Fangliao. Most major cities and the key economic areas are served by this line.

The provincial government aims to complete the around-the-island railway before 1980. Taiwan had approximately 10,600 mi (17,000 km) of 'highways' in 1970.

China Airlines (CAL) operate 6 scheduled domestic routes, and 11 international flights connecting with other Asian cities.

International Trade. Both the rich agricultural base and the buoyant industrial development of Taiwan are reflected in the country's trade statistics. The value of foreign trade increased, at an accelerating rate, about ten-fold between 1950 and 1969.

HONG KONG

Area: 398 sq mi (1,031 km²)
Population: 4,077,400
Population Growth Rate: 2·2%
Capital: Victoria, pop 671,000
Language: English. Cantonese and Mandarin widely spoken
Religion: Confucianism-Buddhist-Taoist
Currency: Hong Kong dollar = 100 cents

Hong Kong has been a British Crown colony since 1843 and now enjoys a large measure of autonomy. This tiny holding at the edge of mainland China developed an impressive transit trade and a major port to handle it. But this trade has been overtaken in economic importance by the export of manufactured goods as a result of the colony's remarkable

A junk sailing in Kowloon bay against a background of modern skyscrapers. Kowloon is a mainland peninsula and a part of Hong Kong. The colony is a major port and manufacturing centre.

industrial expansion since World War II. In 1967, during China's 'Cultural Revolution', there were serious riots and terrorist bombings, the work of local communists. However, China itself evidently regards Hong Kong as too valuable an outlet to be forcibly taken over.

Island and Mainland. Lying on and off the south-east coast of China about 90 mi (145 km) south-east of Canton, Hong Kong consists of Hong Kong Island, Kowloon Peninsula, and the New Territories, a part of the mainland leased to Britain in 1898 for a period of 99 years, together with Deep Bay and Mirs Bay, and over 230 small islands. Hong Kong Island is about 11 mi (18 km) long and some 2–5 mi (3–8 km) wide. Its hills rise sharply from the sea, some exceeding 1,000 ft (300 m) in height; Victoria Peak, the highest, reaches 1,809 ft (551 km). Kowloon and the New Territories are also hilly, Tai Mo Shan rising to 3,144 ft (968 m).

Flat land is generally scarce and consequently reclamation of land from the sea has been gradually achieved throughout Hong Kong's history. Reclaimed land is used for housing and factories, and for such projects as the extension of Kai Tak international airport. The coastlines are rugged and deeply indented. The magnificent harbour lies between Kowloon Peninsula on the mainland and the city of Victoria on the northern shore of Hong Kong Island.

Hong Kong has a tropical monsoon climate, with temperatures averaging 82°F (27·8°C) in July and 59°F (15°C) in February. Rain, some 85 in (2,159 mm) annually, falls mainly between June and September, but water has to be piped from Kwangtung in China to supplement local supplies. Occasional devastating typhoons occur in summer.

The People. The population is almost entirely Chinese and mainly Buddhist, although Confucianism and Taoism are practised. There are more than 400,000 Christians and small Moslem, Hindu and Jewish communities.

Overcrowding has been worsened by large numbers of refugees from China, and despite vigorous rehousing schemes, several hundred thousand people still live in shanty-towns.

Industry and Tourism. The post-war industrial expansion was made possible by Hong Kong's rich commercial heritage, and more particularly by the arrival of great numbers of immigrants who were ready to learn and to work hard. Some 750,000 arrived from China in 1949, and in 1950 the border had to be closed.

Hong Kong's success is based on light industry, especially cotton textiles for which experts from Shanghai were available. Next comes plastics, followed by electronic, photographic and optical equipment, transistor radios and light industrial machinery. Heavy industry includes small-scale steel production and shipbuilding and repair. Small quantities of iron ore, wolframite, graphite, feldspar and quartz are mined.

Hong Kong also has a highly lucrative tourist industry. In 1970 more than 927,000 tourists visited the colony.

Agriculture and Fisheries. Only about 13% of the land area is farmed. More profitable activities such as growing fruit and vegetables for the dense urban population are increasingly replacing rice cultivation. Pigs and poultry are the chief livestock.

The marine fisheries, based on Aberdeen and Shau Kei Wan on Hong Kong Island, and on Tai Po and Sai Kung in the New Territories, are an important source of food. The sea harvest is supplemented by fish from countless ponds along the coast of Deep Bay, also noted for its oysters.

Transport. Hong Kong has been relatively slow in adapting to modern shipping methods, although a container terminal is to be built at Kwai Chung. A £24 million cross-harbour road tunnel linking Hong Kong Island with Kowloon was opened in 1972, and a £230 million transport network with underground railways and hovercraft is planned.

International Trade. Textiles, clothing and other manufactured goods are the chief exports, more than half of which go to the United States and Britain. Re-exports include diamonds, pharmaceutical products, and textile fabrics. The major imports are foodstuffs and raw materials.

MACAO

Area: 6·17 sq mi (16 km²)
Population: 321,000
Population Growth Rate: 1·7%
Capital: Macao, pop 321,000
Language: Portuguese. Cantonese widely spoken
Religion: Confucianist-Buddhist-Taoist
Currency: Pataca = 100 avos

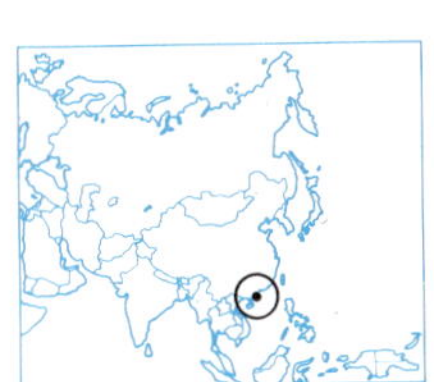

Macao (Macau), is a tiny overseas province of Portugal at the mouth of the Canton River in southern China, about 40 mi (64 km) west of Hong Kong. It has been Portuguese since

1557, and Portugal's continued possession reflects its importance to China as an entrepôt. In 1967, following communist-inspired riots that were an overspill of China's 'Cultural Revolution', China forced Portugal to admit responsibility for the riots, pay an indemnity, expel Chinese refugees and eliminate all Nationalist Chinese (Taiwan) organizations.

Macao consists of a mainland peninsula on which stands the city of Macao, and the adjoining islands of Taipa and Colôane. In this small area of some 6 sq mi (16 km²) live 321,000 people, nearly all of them Chinese.

Long known for its free gold market, its smuggling and its gambling, Macao is host to more than 2 million tourists annually, mainly from Hong Kong. Major attractions include the Hotel Lisboa casino and the Cheoc Wan resort centre on Colôane Island, opened in 1970. Industry, once limited to matches and fireworks, now includes small factories making plastics, textiles, cameras and binoculars. Some rice and vegetables are grown.

PHILIPPINES

Area: 115,600 sq mi (299,404 km²)
Population: 39,769,000
Population Growth Rate: 3·5%
Capital: Quezon City, pop 585,000
Language: Filipino. English widely spoken
Religion: Roman Catholic (80%), Protestant (10%), Moslem (5%)
Currency: Philippine peso = 100 centavos

The Republic of the Philippines consists of over 7,000 islands in the Western Pacific Ocean, lying north of Indonesia and south of Taiwan. The total area of the islands is slightly smaller than that of the British Isles, but the irregular coastline of 14,400 mi (18,543 km) is nearly as long as that of the USA. Eleven of the largest islands contain most of the land area and population. The Philippines is the only predominantly Christian country in Asia.

Standing at the eastern fringe of South-East Asia, it straddles major trade routes from the USA, East, South and South-East Asia, and thus lies at a crossroads of East and West. It has a remarkably complex culture containing elements of both, and since attaining independence from the USA in 1946, it has played an active role as a mediator in South-East Asia affairs. Despite a communist-led rebellion by the 'Huks', which lasted into the 1950s, the economic condition of the republic gradually improved, helped by American aid.

The Land. The total area of the Philippines is 115,601 sq mi (299,404 km²). The archipelago is 1,150 mi (1,851 km) long, and it is roughly triangular in shape with its apex in the north and its base to the east of Borneo. Only about 730 of the over 7,000 islands are inhabited. The country is generally divided into three main groups centring upon the 11 largest islands. In the north is Luzon, the largest island, with Mindoro and other islands. In the centre is the Visayan group, including Bohol; Cebu, the most heavily populated of the islands; Leyte, of World War II fame; Negros; Panay; Samar; Masbate; and long, thin Palawan stretching way to the west. In the south is the Mindanao group, including the islands of the Sulu Archipelago which reach towards Borneo. The republic is bounded on the west by the China basin, some 14,000 ft deep (4,200 m) and on the east by the Philippine deep, at 36,000 ft (10,973 m) one of the deepest troughs in the world.

The islands of the Philippines are of volcanic origin and are for the most part mountainous, with many active volcanoes; coral reef formations fringe most of the islands. The mountain ranges generally run along the main axes of the islands, parallel to each other and close to the coasts. Flatlands are rare, the largest lowlands being on Luzon and Mindanao, with other large plains on Panay and western Negros. The tallest mountain peak is Mt Apo, a quiescent volcano 9,690 ft high (2,953 m) near Davao on Mindanao. Active Mt Mayon in Luzon has the most perfect volcanic cone in the world. The river systems of the Philippines trend generally northward. The rivers are short and swift, seasonal in their flow, and seldom navigable.

Climate. The climate is warm, with high humidity. Temperatures range very little from north to south, averaging about 80°F (26·7°C) at sea level. Rainfall, however, varies a great deal from place to place and according to season, ranging from less than 40 in (102 cm) to over 200 in (508 cm). The eastern side of the archipelago is very wet, the wettest period coinciding with the north-east monsoons. In addition the islands are lashed by damaging typhoons in the summer and autumn. The western parts, however, have a dry season during the winter months because they are sheltered from the prevailing north-east monsoons.

Vegetation and Wildlife. About 42% of the Philippines land area is covered by forests, most of which are government-managed. Not many are open to lumbering. The tropical rain forests contain over 3,000 species of trees, including *Shorea* species whose wood is known as Philippine mahogany. There are also heavy groves of bamboo and rattan and thousands of flowering plants. About 18% of the land area is covered in grassland, difficult to cultivate, and a smaller percentage in mangrove swamps and marshes. Fruits and flowers, including many rare orchids grow abundantly, and bananas are becoming commercially important.

Mammals of the Philippines include monkeys, bats, wild boar, Palawan's dwarf mouse deer and the tamaraw, or pygmy buffalo of Mindoro. There are over 750 species of birds, and reptiles include pythons, cobras and many lizards. Crocodiles were once abundant, but have been almost exterminated by skin hunters. The encompassing seas probably have a richer variety of fishes and molluscs than any comparable area elsewhere.

The People. The population of the Philippines, estimated at 39 million (1972), is very unevenly distributed. Cebu and parts of Luzon have over 1,000 persons per sq mi (386 per km²), but Palawan, Mindoro and Mindanao have less than one-tenth of this density. Panay and Negros are also heavily populated. The larger cities are growing fast, although 70% of the population is still rural. Basically, the Filipinos are of Malaysian stock, with an admixture of Mongoloid, and a more recent blending of Caucasian. There are sizable minorities of Americans and Spanish, and some 500,000 Chinese. There have been strong cultural influences from the Americans and Spanish, especially in the larger cities and around Manila and south through the Visayan Islands. Mindanao has remained stubbornly Islamic and apart. Although there are many local languages of the Malayo-Polynesian family, Pilipino (based on Tagalog) is the official language. English and Spanish are also widely spoken. Most Filipinos are farmers and fishers, who subsist on rice and fish and live in large families in *barrios* (small villages) of raised huts made of rattan or bamboo. Women enjoy equal rights.

Nearly 80% of Filipinos profess Roman Catholicism. The Moslems of the south form the next largest religious minority, with about 5% of the population. Literacy is widespread in the Philippines, with over 5 million children attending elementary schools, and a smaller number going on to high schools. There are also many colleges and universities.

Agriculture. With 57·3% of the working

Rice terraces, built into the steep mountainsides of north-eastern Luzon, rise as high as 4,000 feet out of the valleys. The dry rice of the upland areas is distinct from the lowland variety.

population employed in it, agriculture is the basic industry of the Philippines. Nearly three-quarters of the area planted is given to food crops (principally in Luzon and the western Visayas), the chief of which is rice, with maize a staple for nearly 25% of the population. Sweet potatoes, manioc and other vegetables, and many fruits, including the banana, are also grown. Cattle and dairy products are scarce, but hogs, poultry and fish make up for this deficiency.

The chief commercial (export) crops are coconut, sugar, abaca and tobacco. Coconut is an important subsistence and commercial crop, and 5 to 6 million people depend upon it directly. The Philippines has nearly 30% of the world's acreage, and coconut palms grow widely throughout the islands. About 95% of the coconuts is converted into copra, or dried meat (from which oil is extracted), and the Philippines has long been the world's chief exporter of copra. Sugar-cane is grown on plantations on Negros and Luzon. Abaca (exported as the fibre known as Manila hemp) is a Philippines monopoly. Cigar tobacco is grown chiefly on Luzon. There is also some coffee, pineapples, and a developing rubber industry.

Agricultural patterns range from much shifting cultivation, and terraced-irrigation farming, to large mechanized farms and plantations. Most farms are small, averaging 7·7 ac (3 ha). Continuous cropping, soil depletion, erosion and lack of fertilizing presents a perennial problem. Agricultural reform and diversification of exports are urgent necessities.

Forests and Fisheries. The Philippines takes sixth place among the world's lumber producers, and lumbering accounts for a seventh of the total value of exports. Every year the forests yield over 430 million board ft (over 130 million m) of lumber, as well as logs, most of which now comes from the Dipterocarpaceae tree family.

Fish is second only to rice in the Philippine diet, yet much fish has to be imported, and commercial fishing is underdeveloped. Moreover the shelves around the islands where fish breed are too narrow for much commercial fishing. Nevertheless, anchovies, mackerel, sardines, tuna and similar fish are caught, and trawling accounts for 35% of the catch. Inland fishing in swamps, marshes and fishponds is also important.

Minerals and Industry. The islands are rich in minerals, which one day may become the basis of a new industrialization programme. At present the Philippines ranks eighth in world gold production, and is a leading copper producer. Chromite, iron, manganese, lead and zinc are also mined, as well as coal, asphalt, sand, gravel and salt.

Most Philippine industry is concerned with processing farm crops and minerals. Heavy industry is lacking, although in recent years factories have been established in cement, textiles, chemicals, tyres and aluminium. New hydro-electric projects should aid further development. Market-oriented local manufacturing (textiles, cosmetics, shoes, beverages, etc.) is growing.

Transport and Trade. The Philippines, thanks partly to the American occupation, has one of the best transportation systems in Asia. There are about 35,000 mi (56,300 km) of surfaced roads, at least half of these first-class. Some 800 mi (1,300 km) of railways operate chiefly on Luzon, with smaller lines on Cebu and Panay. The efficient Philippine Air Lines (PAL) flies international routes, as well as from island to island within the country. Water transport is all-important in an island country like the Philippines. About 150 vessels ply regular routes among the islands, and about 20 sail on international routes. In addition there is a large, heterogeneous mass of shipping for the inter-island trade and communications.

The Philippines exports abaca fibres, coconut products, sugar and wood; and imports metals, food and machinery. Most trade is with the USA.

DEMOCRATIC REPUBLIC OF VIETNAM

Area: 61,294 sq mi (158,751 km²)
Population: 22,038,000
Population Growth Rate: 2·6%
Capital: Hanoi, pop 920,000
Language: Vietnamese
Religion: Confucianist-Buddhist-Ancestor Worship. Catholic (4%)
Currency: Dông = 10 hào = 100 xu

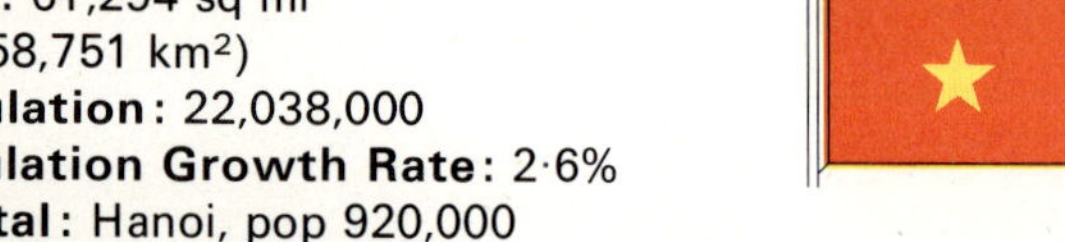

The Democratic Republic of Vietnam (North Vietnam) is a small communist state between communist China and non-communist South Vietnam. Its struggle to be independent and then to absorb South Vietnam tied down French then American and South Vietnamese forces in decades of largely fruitless endeavour. By the early 1970s, North Vietnam's expansionist aims seemed on the point of at least partial success after a cease-fire agreement left North Vietnamese troops deep inside South Vietnam.

Vietnam was one of the states which emerged after World War II with the break-up of French Indochina. Vietnam became independent within the French Union in 1949, but the French were defeated by the revolutionary nationalist (and communist) Vietminh. Victory at Dien Bien Phu and the Geneva Conference (1954) divided Vietnam at the 17th parallel. The north became the Democratic Republic of Vietnam and the south the Republic of Vietnam.

North Vietnam, backed by China and working through the Vietcong guerrilla movement in the south, now strove to master the whole of Vietnam while America began sending massive aid to South Vietnam to prevent this happening. There was soon complete American involvement in full-scale war not only against the Vietcong in South Vietnam, but spilling over into Cambodia and Laos. By 1969, the Americans had had enough of the war they could not win and began withdrawing their forces from South Vietnam. This encouraged the North to renew full scale invasions against the South. The resulting stalemate ended in the cease-fire of 28 January 1973. By this time, American bombing had badly damaged North Vietnam's industry, while South Vietnam had been devastated by explosives and defoliants, and sizable parts were held by the enemy.

The Land. North Vietnam is a mountainous country. The East Tonkin Highlands, drained by the rivers Chay, Gam and Cau, are mostly over 3,250 ft (9,906 m) above sea level, while the West Tonkin Mountains, drained by the Black River (Song-Bo) system, are the highest and most rugged part of the whole Indochinese peninsula. Here Fan-si-pan, Vietnam's highest mountain, rises to 10,312 ft (3,143 m). A contrast is provided by the delta of the Red River (Song Koi). This densely populated and fertile delta covers some 5,600 sq mi (14,502 km²) and is subject to frequent flooding; around Hanoi the protective dykes are some 40 ft (12 m) high and about 150 ft (46 m) thick.

Running almost the entire length of the Indochinese peninsula is a mountain backbone, the Annamite Chain, containing several peaks above 7,650 ft (2,332 m) and throwing off spurs to the narrow coastal plain. This chain, together with the Red River delta in the north and the Mekong delta in South Vietnam, provides the popular description of Vietnam as 'a pole balanced by two rice baskets'.

Climate and Vegetation. North Vietnam has a tropical monsoon climate. Temperatures are uniformly high during summer, averaging 84°F (29°C) on the plains. Winter temperatures are around 60°F (16°C). Most areas receive more than 60 in (1,524 mm) of rain annually, mostly in summer, but parts of the Annamite Chain and the West Tonkin Mountains sometimes receive more than 160 in (4,064 mm).

The dominant type of vegetation is tropical evergreen forest, which, however, may give way to deciduous forest or wooded savanna in areas with a more pronounced dry season. Mangroves fringe the coasts near Mong Cai.

The People. About 85% of the population speak Vietnamese and live in the lowlands, the most densely populated province being Thai-binh on the Red River delta. There are also some 175,000 ethnic Chinese and various tribal minorities such as the Thai, Muong, Nung, Miao and Man. Buddhism is the prevailing religion, but the tribal groups are mostly animist. There are about 500,000 Roman Catholics. Basic education consists of a 10-year course, and it was claimed in 1970 that illiteracy had been eliminated. There are two universities at Hanoi, and North Vietnamese students also study in China and the Soviet Union.

Agriculture. North Vietnam is a primarily agricultural country. More than 80% of the working population is engaged in farming. About 97% of the farms belong to the 28,000 agricultural co-operatives and there are also some 59 state farms. Rice is the major crop, but the annual production (about 4·5 million tons in a normal year) is barely sufficient to feed the people. Other crops include groundnuts, cotton, sugar-cane and jute. Animal husbandry (water buffaloes, cattle, pigs, poultry) is becoming increasingly important.

Foresty and Fisheries. North Vietnam has valuable reserves of timber, mainly teak, ebony and rosewood.

The Gulf of Tonkin abounds in tuna, bonito and mackerel. Along Bay is noted for its prawns and crayfish. The numerous ponds, lakes, rivers and canals, and even the wet paddy fields, provide freshwater fish.

Minerals. At Quang Yen, north of Hanoi, North Vietnam has the largest anthracite deposits in South-East Asia. With smaller deposits at Phan Me and Tuyan Quang, the total reserves are about 20 million tons – enough to meet all domestic needs and to provide substantial exports as well. Iron ore is mined at Thai Nguyen, also north of Hanoi, and elsewhere. Other important minerals include phosphates (Cao Cai), zinc (Tuyan Quang and Thai Nguyen), tin (Tinh Tuc), antimony (Mong Cai) and graphite (Lao Cai).

Industry. Modern industries have been developed, with Russian and Chinese help, but more than 50% of the industrial output still comes from the handicraft sector. The chief industrial centre is Hanoi, which has a wide range of light industries as well as engineering and machine plants. Haiphong comes second with its shipbuilding, cement, glass, porcelain, and cotton textile industries. Thai Nguyen has an iron and steel complex, while Nam Dinh is known for its textile mills. Viet Tri has paper mills and chemical plants, and Vinh has engineering works.

Transport. The railway network, severely damaged during the fighting, focuses on Hanoi, whence lines run to China (via Lang Son and Lao Cai), to Dong Hoi in the south, and to Haiphong, the outport of Hanoi. The road network follows the railways fairly closely, and because of its greater flexibility carries much of the freight.

Inland waterways are of special significance. Hanoi, the administrative, commercial and cultural hub of the country, stands at the confluence of the major rivers. The Red River is navigable from Hanoi to Yen Bai; the Black River, from Hanoi to Cho Bo; the Song Lo, to Tuyan Quang. Haiphong is the only port for ocean-going ships. The Chinese provide regular air services between Gia Lam (Hanoi's airport) and Peking.

International trade is conducted mostly with other communist countries; major non-communist trading partners include Japan, Hong Kong, Malaysia and countries of the European Economic Community. Exports include coal, ores, cement, handicrafts, sugar and timber. Imports include machinery, transport equipment, steel, plastics, cotton textiles, rubber, petroleum and petro-chemicals.

REPUBLIC OF VIETNAM

Area: 67,107 sq mi (173,807 km²)
Population: 18,809,000
Population Growth Rate: 2·6%
Capital: Saigon, pop 1,763,692
Language: Vietnamese
Religion: Buddhist-Taoist (81%) Hoa Hao, Cao Dai (15%)
Currency: Dông = 100 centimes

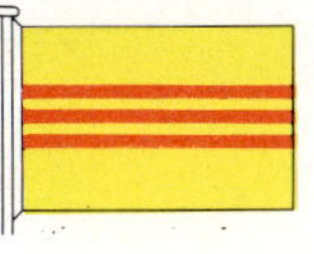

The Republic of Vietnam (South Vietnam), a small agricultural state in South-East Asia, emerged in 1955 as the non-communist half of a divided Vietnam. For most of the next 18 years, South Vietnam was ravaged by war as the Vietcong (communist guerrillas) and Vietminh (North Vietnamese troops) fought to impose communism upon the country which was defended by South Vietnamese troops bolstered by massive American military aid. Victory eluded both sides and in 1973 the USA pulled out its disheartened troops on the promise of peace. But the peace terms left pockets of communist troops controlling parts of South Vietnam, whose political and economic future seemed as doubtful as ever. (See also the Democratic Republic of Vietnam).

The Land. South Vietnam consists of the southern two-thirds of Annam, formerly a French protectorate, and the former French colony of Cochin-China. Entering from the north is the Annamite Chain, a mountain backbone containing Ngoc Linh (8,523 ft; 2,598 m), Chu Yang Sin (7,900 ft; 2,405 m) and other peaks, and with spurs thrusting seawards across the narrow coastal plain. In the south is the great alluvial delta of the Mekong River, very similar to the Red River delta in North Vietnam, but much larger and less densely settled. The Mekong delta, a flat

Many inhabitants of the Mekong delta live in river-dwellings supported on piles.

The Market place in Hue. Hue is the ancient walled capital of Annam, a kingdom which once extended over North and South Vietnam.

and fertile plain, covers about 14,600 sq mi (37,814 km²).

Climate. Like North Vietnam, South Vietnam has a tropical monsoonal climate. Temperatures vary little, from around 86°F (30°C) in summer to around 80°F (27°C) in winter. Annual rainfall exceeds 60 in (1,524 mm) in most areas and occurs mostly in summer. But coastal areas exposed to the north-east monsoon, which blows across the South China Sea in winter, may receive up to 10 in (254 mm) in January.

Vegetation. Tropical evergreen forest predominates, especially on higher ground, but drier areas usually have deciduous forest or even woodland savanna. The Mekong delta has extensive freshwater swamps, and its coast is fringed by mangroves.

As a result of the war, about one-twelfth of South Vietnam has been defoliated. Areas so affected may take seven years or more to recover. Fortunately the paddy fields of the Mekong delta, the country's rice basket, have not been seriously affected.

The People. The population is predominantly Vietnamese, but includes more than 1,000,000 Chinese, numbers of Khmers (Cambodians), about 800,000 refugees from North Vietnam, and a profusion of tribal groups such as the Bahnar and Jarai. More than 75% of the population normally live in villages, although South Vietnam has a number of sizable towns.

Many South Vietnamese practise Taoism and its variants. Buddhism, too, is widespread and there are some one million Roman Catholics. Caodaism, a blend of Buddhism, Christianity and Confucianism, claims many followers, as does the Buddhist-influenced Hoa Hoa sect. About 40% of the population is illiterate. The traditional university centres are Saigon, Hue and Da Lat.

Agriculture. South Vietnam, like the North, is primarily an agricultural country. Its two million farms are mostly in the Mekong delta and the narrow coastal plain, and in small alluvial areas at the foot of the Annamite Chain. Unlike the North, where two (and in some areas, three) crops of rice are grown in a year, South Vietnam depends mostly on a single annual crop. War ravages interrupted South Vietnam's traditional role as a major rice exporter, but resumed production of an export surplus seemed likely after the peace settlement of 1973. Other important crops include rubber, sugar-cane, tobacco and tea. Animal husbandry (water buffaloes, cattle, pigs and poultry) is underdeveloped.

Forests and Fisheries. About two-thirds of Vietnam is forested. War and the difficulty of terrain have prevented the full exploitation of the valuable tropical hardwoods.

The coastal harbours are advantageously near some of the richest fishing grounds in the South China Sea, which are harvested by some 250,000 South Vietnamese operating about 60,000 small craft. There are also many large ponds providing freshwater fish.

Minerals. In contrast to North Vietnam, the South has only limited minerals – a little gold at Bong Mieu and small coal deposits at Nong Son north of Da Nang. Proper investigation, however, may reveal a wide range of minerals.

Industrial Development, due to lack of mineral and power resources, is not so advanced as in North Vietnam. The war has also compelled suspension of such projects as the An-Hoa complex near Da Nang. Food-processing and textiles are important, and glass, paper, cement, jute bags, bicycles and cigarettes are manufactured. Saigon is the chief industrial centre.

Transport. South Vietnam has a small railway network consisting of a coastal route from Dong Ha southwards to Saigon and My Tho, and its branch line from Thap Cham to Da Lat. The road network, focused on Saigon, is of much greater economic significance. The Mekong and its branches are navigable. Long Xuyen, Can Tho, My Tho and Saigon are all river ports.

International Trade. France, the traditional trading partner, has largely been supplanted by the United States, with Japan, Taiwan and Great Britain following. Exports still consist of agricultural raw materials such as rubber, tea, groundnuts and copra, but in much reduced quantities. Imports include rice, cotton, sugar, milk products, frozen foods, machinery, vehicles and consumer goods.

LAOS

Area: 91,428 sq mi (236,798 km²)
Population: 3,106,000
Population Growth Rate: 2·4%
Capital: Vientiane, pop 162,000
Language: Lao. French widely used
Religion: Theravada Buddhist (80%)
Currency: Kip = 10 bi = 100 at

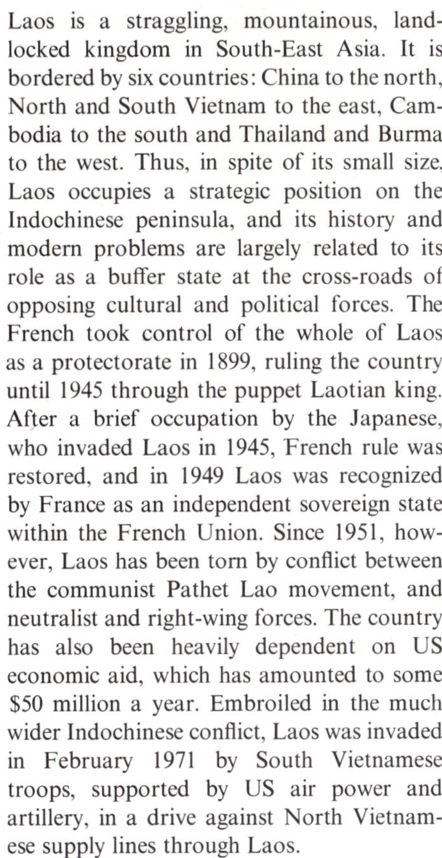

Laos is a straggling, mountainous, landlocked kingdom in South-East Asia. It is bordered by six countries: China to the north, North and South Vietnam to the east, Cambodia to the south and Thailand and Burma to the west. Thus, in spite of its small size, Laos occupies a strategic position on the Indochinese peninsula, and its history and modern problems are largely related to its role as a buffer state at the cross-roads of opposing cultural and political forces. The French took control of the whole of Laos as a protectorate in 1899, ruling the country until 1945 through the puppet Laotian king. After a brief occupation by the Japanese, who invaded Laos in 1945, French rule was restored, and in 1949 Laos was recognized by France as an independent sovereign state within the French Union. Since 1951, however, Laos has been torn by conflict between the communist Pathet Lao movement, and neutralist and right-wing forces. The country has also been heavily dependent on US economic aid, which has amounted to some $50 million a year. Embroiled in the much wider Indochinese conflict, Laos was invaded in February 1971 by South Vietnamese troops, supported by US air power and artillery, in a drive against North Vietnamese supply lines through Laos.

Land and Climate. Almost the entire length of Laos is bound by the Mekong River on the west and the summits of the Annamese chain in the east, but four major physiographic regions can be recognized. (1) Folded ranges in the north run north-east to south-west with deeply dissected plateaux. The relief here is rugged, with mountains rising steeply above the narrow valleys. Phu Bia, at an elevation of 9,240 ft (2,817 m), is the highest peak in Laos. Xieng Khouang Plateau, on which the Plain of Jars is located, is the only relatively flat area. (2) The greatest mountain range in Laos is the Annamese Chain in the east, along the Vietnam border. (3) Limestone plateaux in the south, including the Plateau des Bolovens, slope westward toward the Mekong River. (4) The Mekong lowlands are cut deeply by the fast-flowing tributaries of that great river, and small pockets of lowlying land are found only where the tributaries enter the main stream.

The river system of Laos consists of a series of tributaries of the Mekong, including the Nam Ou, Nam Ngum, Nam Ca Dinh and Nam Bang Hieng. Only Houa Phan province and part of Xieng Khouang lie outside the Mekong basin, where the Nam Het Song Ma and the Nam Het Song Chu flow into the Gulf of Tonkin via North Vietnam.

The climate of Laos is tropical monsoonal. Annual average temperatures range from 78°F (24°C) in the north to 82°F (26°C) in the south. The amount of annual precipitation varies considerably according to elevation and exposure to the south-west monsoon, but most areas receive between 60 in (1,500 mm) and 100 in (2,500 mm) per year. There are three seasons in Laos: the hot, rainy season lasts from June to October, when the area is under the influence of the south-west monsoon; the cool, dry season between November and February occurs when the monsoon comes from the landward direction; and a transitional hot, dry season lasts from March through May.

Vegetation and Wildlife. More than two-thirds of Laos is covered with dense sub-tropical forests. High-altitude montane woodland predominates in the north and is gradually replaced by deciduous broad-leafed forests on the lower slopes in the south. In the drier areas, forests give way to savanna, as on the Xieng Khouang and Bolovens plateaux. There are an estimated 15,000 ac (6,000 ha) of valuable teak forests in Sayaboury province on the Thai border and some 8 million ac (3·2 million ha) of commercially exploitable open forests in the south. Extensive areas of forest are being cleared for agricultural purposes.

Animal life in Laos includes elephants, tigers, leopards and gaurs (wild oxen). Elephants are used as draught animals, especially in lumbering.

The People. The people of Laos belong to a number of different ethnic groups, the largest being the Lao, who total over half the population. They are a Thai people who migrated from China many centuries ago. Chiefly poor farmers, they live in isolated valley communities. Most practise Hinayana (Theravada) Buddhism, the official religion, but with animist variations. The rest of the population consists of mountain tribes, which include the Mans, Thais, Miaos and Hos, most being of Chinese origin, and the Khas, who are Indonesian aborigines. There are minorities of Chinese, Indians and Pakistanis who live mainly in the cities.

Over 90% of the population of Laos live in rural village communities with up to 120 families. Many such settlements are situated in the middle of dense bamboo groves and are surrounded by extensive paddy fields. Laos has few towns and these are generally small and resemble overgrown rural markets. Only five centres had a population exceeding

The Meo are a Laotian mountain tribe of Chinese origin. Many of the Meos make their living from the cultivation of opium.

An aerial view of a remote village in the hilly, forested terrain of Laos' northern provinces. It is here that the Pathet Lao guerrillas struggle to overthrow the pro-western government.

10,000 in 1966. These include Vientiane, the administrative and commercial centre; Pakse, the marketing centre of the south; Savannakhet, a trans-shipment point along the Mekong and commercial centre of south-central Laos; Luang Prabang, the royal capital; and Thakhek, another market centre with a cement works. The other provincial towns had populations ranging from 2,000 to 4,000.

Agriculture, Forestry and Fisheries. The total agricultural area in Laos is estimated at 2·5 million ac (1 million ha), most of which is devoted to rice cultivation. The annual production of approximately 570,000 tons of rice is normally barely sufficient for local consumption, and in recent years, substantial quantities have had to be imported for urban and military use. Other agricultural produce of some significance includes small quantities of maize from Xieng Khouang and Houa Phan provinces, tobacco from the Vientiane Plain, coffee from the Plateau des Bolovens and cotton from various southern localities. In addition to rice, bananas, coconuts, papaya and other tropical fruits are grown in the lowland villages for domestic consumption. With the introduction of modern irrigation facilities around Vientiane, the importation of fresh vegetables from across the Mekong is no longer necessary. Livestock is an important source of supplementary farm income and there are about 400,000 head of cattle, 500,000 water buffalo and 1 million pigs.

Of the 58,600 sq mi (150,000 km^2) of forests in Laos, some 27,350 sq mi (70,000 km^2) are considered to be commercially workable, but less than 60% of this amount is actually being exploited. Inaccessibility caused by the rugged terrain and the inadequacy of transport facilities is responsible for the low level of exploitation. Of the total amount of timber felled annually, 70% is for making charcoal, the predominant form of local fuel. Apart from timber, a few tons of gums, resins and balsam are extracted from the forests.

Commercial fishing in Laos has yet to be developed, but a large quantity of fish is harvested annually along the Mekong for local consumption. Most of the catch consists of carp, catfish, mullet and perch. The major centres of fishing are Vientiane, Borikhane, Sithandone, Sendone, Savannakhet, Thakhek and Houei Sai.

Minerals, Industry and Transport. Laos is believed to be rich in mineral resources but exploration is far from being completed. There are known reserves of about 70,000 tons of tin in the Phong Tiou area, and considerable quantities of high quality iron at Xieng Khouang, copper in Champassac province, coal near Luang Prabang, Vientiane and in Saravane province in the south, and lead in Vientiane province. The limestone deposits of Thakhek support a small cement-making industry.

The major part of Laotian industry is of the cottage type and consists mainly of pottery work, the weaving of fine silks and brocades, silversmithing and leather curing. Development of modern industry has been hampered by a general fuel shortage, a limited domestic market and inadequate transport. Under such conditions, the only organized industries in Laos consist of rice-mills and saw-mills of small capacity.

In recent years, in addition to the small cement factory at Thakhek, a number of small factories manufacturing light consumer goods such as small plastic articles, cigarettes and soft drinks have appeared in the capital city, Vientiane.

Laos has poor supplies of thermal electricity and few towns have adequate electricity supplies. But hydro-electric developments linked with the Mekong River promise improvements, and the new Nam Ngum dam should raise the generating capacity tenfold.

Adequate means of transport is perhaps the most serious problem facing economic development in Laos. The Mekong River has traditionally been the principal highway in the country. However, the Laotian section of the river is interrupted by no less than four formidable rapids. At Vientiane, the river is so shallow as to permit only boats of less than 50 tons in the dry season and 100 tons in the wet season.

There is no railway in Laos and road traffic remains under-developed. The 860 mi (1,370 km) long north-south trunk road runs from Luang Prabang through Vientiane to Khong on the Cambodian border. The total road mileage in 1966 was 3,500 (5,600 km), but most roads are passable only in the dry season.

International Trade. In recent years, Laotian foreign trade has produced large deficits in the balance of trade, with the value of imports 20 or 30 times that of exports. In addition to a wide range of consumer goods, including transportation equipment and petroleum products, rice and foodstuffs constitute about 25% of all imports.

Major export commodities include tin ore, timber, green coffee, benzoin, sticklac, leather and hides.

The principal trading partners of Laos include Thailand, the USA, Japan, Singapore, Malaysia and France.

KHMER

Area: 69,897 sq mi
(181,033 km²)
Population: 6,994,000
Population Growth Rate: 2·3%
Capital: Phnom Penh, pop 600,000
Language: Khmer. French.
Religion: Hinayana Buddhist (90%)
Currency: Riel = 100 sen

The Khmer Republic (formerly the Kingdom of Cambodia) occupies the south-western portion of the peninsula of Indochina. Its well-watered lowlands yield rice and various tropical crops, but living standards are low and civil war has damaged the beginnings of industrial growth. After gaining independence from France in 1953 Cambodia tried to keep out the Vietnam War that embroiled the rest of the peninsula. But in 1970 Lieutenant General Lon Nol deposed the pro-communist head of state Prince Norodom Sihanouk, renamed the Kingdom of Cambodia the Khmer Republic, and aligned it with the United States. US and South Vietnamese troops entered the country to fight the Vietcong and the North Vietnamese, who had long established hide-outs and supply routes in Cambodia. Following the withdrawal of these troops, the small and ill-equipped Cambodian army was unable to stop large sections of the country being overrun by communist forces.

Geographical Features. Cambodia is rather like a saucer, with the great plain of Tonle Sap in the centre, and a mountainous rim comprising the Phanom Dong Rak range to the north, the Cardamom Mountains to the south-west and the plateaux of Mondalkiri and Ratanakiri to the east. These ranges are generally between 2,500 ft (760 m) and 3,000 ft (914 m) in elevation.

Cambodia is dominated by the Mekong River (flowing north to south through the east-central part of the country) and Tonle Sap Lake, and their seasonal changes are of immense importance. The lake is drained by the Tonle Sap River, a tributary of the Mekong. Between May and October the Mekong receives a tremendous volume of flood water, which raises its level and reverses the flow of the Tonle Sap, inundating an enormous area around the lake. The surface area of Tonle Sap Lake thus expands from 1,200 sq mi (3,100 km²) to 4,000 sq mi (10,360 km²) at the height of the floods. This phenomenon is of great significance for rice cultivation and inland fisheries.

Cambodia has a 350 mi (560 km) rocky shore-line along the Gulf of Siam and a few small off-shore islands. The coastal plain is for the most part very narrow and is interrupted by the Elephant Mountains.

Climate, Vegetation and Wildlife. Cambodia has a tropical monsoon climate with a distinctive dry season from November to May. Temperatures range from 68°F (20°C) to 97°F (36°C). Annual rainfall varies from less than 60 in (1,524 mm) on the Great Plain to as much as 250 in (6,350 mm) in the Cardamom Mountains.

About half of Cambodia is forested. On the mountain ranges and plateaux, where rainfall is very high, tropical evergreen forests are extensive, but in the drier lowland areas they rapidly give way to deciduous forests and wooded savanna areas. There are some mangroves on the coast, and a small grass savanna area in north-western Battambang.

There is abundant wildlife, including elephants, water buffaloes, tigers, leopards, bears and crocodiles. Cambodia is rich in freshwater fish.

The People. About 85% of Cambodians are Khmers – a homogeneous group with a rich history of cultural and racial admixture. The principal racial minorities are the Chinese and Vietnamese. The Chinese traditionally play an important part in the country's economic life. There are also small numbers of Chams, Europeans, Indians and Thais, and several primitive hill tribes. About 89% of the population live in rural areas, most of them in villages in the Mekong and Tonle Sap basins. The only sizeable city is the capital, Phnom Penh (600,000).

The illiteracy rate is 42%, but education is being expanded. There are some 5,880 schools with over 1·1 million pupils, and nine universities with 14,560 students. Many primary schools are still run by Buddhist monks, reflecting the state religion, Theravada Buddhism, to which most Cambodians adhere.

Farming, Forestry and Fisheries. Cambodia is primarily an agricultural country. Some 850,000 families are engaged in farming. The total cropped area amounts to 4 million ac (1·6 million ha) and a further 15 million ac (6 million ha) are considered potentially cultivable. The two main crops are rice and rubber.

Over 80% of the farm area is devoted to rice cultivation on a self-subsistence basis. Rice holdings are very small and yields are low. The national average yield of just under 1 ton per acre is considered low even by South-East Asian standards. However, in particularly fertile areas, as in Battambang and Kampong Cham, output per acre can rise above the 3-ton mark. Annual total rice production is over 2 million tons, but less than 18% of this is available for export. Rice production can be greatly enhanced by double cropping, made possible by better water control and irrigation facilities.

Rubber was first introduced in 1921. From a modest start of 3,500 ac (1,400 ha) in the Chup Plantation, the planting area steadily

The central market in Phnom Penh, capital of the Khmer Republic. The name derives from the 'phnom' (hill) crowned by a Buddhist temple which rises north of the Royal Palace.

A typical fishing craft on the Mekong. The Khmers are mainly fishermen and rice farmers whose lives follow the cycles of the Mekong river.

increased to about 50,000 ac (20,000 ha). Practically all the plantations are on the 'red' soils of Kampong Cham province. Annual production is about 50,000 tons and further increases will come from newly established areas at Pailin in Battambang province and around Kampong Som.

Pepper is an important export crop, with a major centre of production in Kampot province. However, production of pepper has drastically declined, owing to plant disease and a drop in foreign demand. Sugar palms are particularly numerous in the provinces of Takeo, Kandal and Kampong Chhnang, but production is insufficient for local consumption and fairly substantial quantities of foreign cane sugar have to be imported. Other cash crops include maize, cassava, tobacco, kapok, and cotton. Since most of the population are Buddhists (and therefore vegetarians), livestock are raised mainly as work animals, though some are raised for export.

The lumber industry is still underdeveloped. Annual production is about 300,000 cu yd (230,000 m³), of which 26% are exported. Elephants provide the haulage labour to the river bank, from whence the logs float down the Mekong to the saw-mills at Phnom Penh.

Fish constitute an important part of the Cambodian diet and 90% of the annual catch is consumed locally. The country is famous for its production of fish sauce (*tuk trei*) and fish paste (*prahoc*). Three-quarters of the catch comes from the Tonle Sap area, where the main fishing season is between June and November. Marine fishing is of only limited importance. A small surplus is exported.

Industry. More than half the 3,788 industrial establishments are associated with food processing and another 20% with saw-milling. Industrial businesses are mainly small family concerns. Most of the medium scale factories are in the state sector and these include an oil refinery (put out of operation by military action in 1971), sugar refinery, textile mills, cement works, tyre factories, glassworks and tractor and lorry assembling plants. The major industrial centres are Phnom Penh, Kampong Som, and Kampong Cham.

Mineral resources are largely unexploited. There are deposits of iron ore, phosphates, manganese, gold, gemstones and coal.

Transport. Cambodia has 2,620 mi (4,200 km) of national and provincial roads, in addition to 7,500 mi (12,000 km) of motorable tracks. The national network focuses on Phnom Penh, the capital, which has an efficient modern road link with Kampong Som, the major port. The country is also served by 400 mi (655 km) of metre-gauge railway, with one branch linking Phnom Penh with Poipet on the Thailand border and the other with Kampong Som. A considerable proportion of the internal traffic moves along the navigable sections of the Mekong, and Phnom Penh, Kampong Cham and Kampong (Chhnang are the most important river ports. Kampong Som can be used by ocean-going ships.

International Trade. Rice, rubber, maize, timber, livestock and pepper, in that order of importance, are the main commodities exported. Imports include machinery, minerals, textiles and other manufactured goods. Cambodia's main trading partners are Hong Kong, France, Communist China, Singapore and South Vietnam.

Exports have been badly hit by the Vietnam War: rice exports, in particular, have been curtailed and rubber production has been stopped by communist occupation of the plantations.

THAILAND

Area: 198,455 sq mi (513,998 km²)
Population: 34,200,000
Population Growth Rate: 3·1%
Capital: Bangkok, pop 2,840,000
Language: Thai. Chinese, Malay and Vietnamese used
Religion: Buddhist (90%) Moslem, Confucianist, Christian (10%)
Currency: Baht = 100 satangs

The Kingdom of Thailand (formerly Siam), flanked on the east by Laos and Cambodia, on the west by Burma, and on the south by Malaysia, covers a total land area of 198,455 sq mi (513,998 km²), only slightly less than that of France.

This ancient monarchy is the only country in South-East Asia never to have been ruled by a European colonial power. Appropriately, its official name *Muang-Thai* means 'Land of the Free'. Since World War II there have been several military coups and changes in government. Thailand is a member of the South-East Asia Treaty Organization (SEATO) and supported the United States in Vietnam. Although there has been some sporadic communist terrorist activity in the northern border areas since 1965, Thailand has remained one of the most prosperous and stable nations in South-East Asia.

Land and Climate. Thailand can be roughly divided into five physiographical regions: (1) The Northern Mountains, a rugged highland region, include all peaks of more than 6,000 ft (1,825 m) above sea level. Doi Inthanon (8,512 ft; 2,595 m) near Burma is the highest. (2) The Central Plain, a triangular depression, extends 300 mi (485 km) from north to south and can be subdivided into an upper and lower zone. The latter zone, as the delta of the Chao Phraya, is the most densely settled and agriculturally developed area of the country. (3) The Korat Plateau, an upland area east of the Chao Phraya, has a gently rolling surface and lies at a general elevation of 500 ft (150 m) above sea level. A range of low hills, the Phu Phan, separates the smaller tributaries of the Mekong River from the eastern Chi-Mun system. (4) The ranges and valleys of the South-Eastern Hills represent the northward extension of the ranges of western Cambodia and run in a general south-east-north-west direction. (5) Peninsular Thailand forms a region of low hills which extend from the Tenasserim Ranges southwards into Malaysia.

Most of Thailand's rivers belong either to the Chao Phraya system, as in north and central Thailand, or to the Chi-Mun system in the north-east. The Nan, which flows 620 mi (1,000 km) before it joins the Chao Phraya, is the longest river.

Most of Thailand has a tropical monsoonal climate, marked by a very distinct alternation of wet and dry seasons. The wet season is the time of the south-west monsoon, which lasts from May to September. Most areas receive more than 70% of their annual rainfall during this season, but the total amount varies with the exposure and distance inland. Inland areas generally have an average rainfall of 50 in (1,250 mm) a year.

The average daily temperature remains above 77°F (25°C) throughout the year. January temperatures average 60°F (15·6°C) in the north and 80°F (26·5°C) in the south, while in the hot, dry season temperatures normally exceed 80°F (26·5°C).

Vegetation and Wildlife. The natural vegetation of Thailand is characteristically that of a monsoonal area. Much of the original cover, especially in the valleys and plains, has been cleared for agricultural purposes. Nevertheless, many mountainous areas of the north are still heavily wooded and contain valuable timber of tropical hardwoods like teak and yang. The western slopes of the Peninsular ranges are also densely forested. Mangroves fringe much of the Gulf of Siam.

Thailand's forests are rich in wildlife such as tigers, leopards and monkeys, and many species of birds. The water buffalo is a widely used draught animal, and the elephant has also been domesticated for work in the forests.

The People. According to the 1970 census, the total population of Thailand was 34·2 million. About 80% of the total population are Thais, there are, however, large Chinese communities in the urban areas and a significant number of Vietnamese refugees have settled in the north-east. Laotian and Cambodian minorities are also numerous in that region. In the southernmost provinces, a large proportion of the population is of Malay origin.

More than 80% of all Thais live in the 44,600 villages situated in the midst of the rice cultivating areas. Many of the 7 million people living in the urban areas are concentrated heavily in the capital, Bangkok. Other large urban centres include Chiang Mai, Nakhon Ratchasima and Thon Buri.

Agriculture, Forestry and Fisheries. Approximately 35% of the gross national product of Thailand is derived from agriculture.

Many of the people of Bangkok live in small houseboats and conduct much of their trade without going ashore.

Rice is by far the most important crop and the annual harvest of over 12 million tons comes from some 20 million ac (8 million ha). Thailand is the world's third largest producer of natural rubber, which comes from the southern region. The annual production of maize, another important export crop, is today about 1·5 million tons. Thailand also exports about 250,000 tons of upland jute, grown in the north-east, and an increasing amount of tapioca from the south-east. A wide range of tropical and subtropical crops are also produced for local consumption.

There are appoximately 106,000 sq mi (275,000 km^2) of forests in Thailand, and annual timber production amounts to about 3·4 million cu yd (2·6 million m^3). Important varieties of wood include yang, teak, and the more valuable teng-rang, takian and daeng.

Fish and related products are important sources of protein in the Thai diet, and over a million tons of marine and freshwater fish are harvested each year. Marine fishing in the Gulf of Siam is concentrated in Samut Prakan, Samut Sakhon, Chumphon, Chanthaburi and Songkhla provinces.

Tin, in the form of cassiterite, is Thailand's most important mineral resource. As one of the world's major suppliers, Thailand annually produces about 30,000 tons of tin concentrates, most of which come from the island of Phuket, Phang-nga and Ranong provinces. Apart from tin, however, the mineral resources of Thailand are relatively poor. Other

Stilt-mounted dwellings are common in Thailand where canals and rivers are, in many places, the only means of transport. Waterways carry more than half of Thailand's freight.

minerals produced include modest amounts of wolfram, lead, manganese, iron ore, fluorite, gypsum and lignite.

Industry. Most of Thailand's industries are related to the processing of raw materials; rice-mills and saw-mills are particularly numerous. However, with the expansion of the domestic market, the increase of foreign capital investment and the general development of the economy, the country has in recent years acquired an impressive range of modern industries. These include textile mills, chemical plants, cement factories, motor assembly plants, petroleum refineries, metal works and steel rolling plants. Electricity generating capacity is increasing rapidly and is expected to exceed 1 million kw by 1972.

Transport. Each year about 48 million passengers and 5 million tons of freight are carried on the railway network of 2,340 mi (3,765 km). The road network has been expanded to 7,820 mi (12,586 km). The construction of rural feeder roads is making a significant impact on the economic development of the previously isolated interior areas.

The Port of Bangkok at Klung Toey handles over 90% of Thailand's foreign trade. Other coastal ports include Phuket, Songkhla and Pattani, while Nong-Khai and Mukdahan are important river ports on the Mekong.

International Trade. Thailand's chief exports include rice, maize, tin, timber and rubber. In terms of total volume, Japan and the United States are the most important trading partners. The rapidly developing tourist industry is contributing substantially in closing the trade gap. Major imports include machinery, motor vehicles, iron and steel products, building materials and petroleum. These are supplied mainly by Japan, the United States, West Germany and the United Kingdom.

MALAYSIA

Area: 127,581 sq mi (330,435 km²)
Population: 12,324,000
Population Growth Rate: 2·9%
Capital: Kuala Lumpur, pop 451,810
Language: Malay. English and Chinese also spoken
Religion: Moslem (99%)
Currency: Malaysian dollar = 100 cents

The Federation of Malaysia occupies the southern part of the Malay Peninsula, called West Malaysia, and much of northern Borneo – the states of Sarawak and Sabah, collectively called East Malaysia. The federation was formed on 16 September 1963, by the union of the old British dominated Federation of Malaya with Singapore and the British protectorates of Sarawak and North Borneo (now Sabah). In 1965 Singapore seceded by mutual agreement and became an independent republic.

The country has rich natural resources (notably rubber and tin), a soundly based economy and a satisfactory balance of trade. It is one of the most prosperous and progressive countries in South-East Asia.

Nevertheless, Malaysia faces some perennial political and social problems. Foremost among these is the fear of the ruling Moslem Malays that Malaysia's enterprising Chinese, who control commerce and are prominent in some of the professions, will eventually dominate the political life of the country. In addition, the long and costly struggle against Chinese communist terrorists in the late 1940s and 1950s has not been forgotten, and there are fears that the surviving terrorists in jungle hide-outs along the Thailand border might again prove troublesome.

Internationally, Malaysia has restored friendly relations with Indonesia, whose former President Sukarno waged guerrilla warfare against the federation (1963–65). Relations with the Philippines are marred by Filipino claims to Sabah.

Geographical Features. Nearly three-quarters of West Malaysia has an elevation over 600 ft (183 m). The terrain is dominated by a central mountain core, surrounded by narrow coastal plains which broaden into a rolling landscape in the south. The Central Mountains are composed of a series of roughly parallel ranges. Peaks exceeding 6,500 ft (1,980 m) include Gunong Tahan (7,182 ft; 2,189 m), the highest peak on the peninsula. The major lowland areas are the Kedah Plain in the north, the Selangor Plain in the centre, the Johore Lowlands in the south, the Pahang Plain in the east and the Kelantan Plain in the north-east. The three longest rivers of West Malaysia are the Pahang, the Perak and the Kelantan.

East Malaysia is predominantly mountainous. Over 90% of Sabah and 80% of Sarawak lie above 1,600 ft (500 m). Sabah has Malaysia's highest peak, Mt Kinabalu (13,455 ft; 4,101 m). The highland areas are the northern end of the great ranges that form the backbone of central Borneo. The coastal plains are very narrow, particularly in Sabah. Rajang, Baram and Kinabatangan rivers form the main constituents of the drainage pattern of Sabah and Sarawak.

Climate. Malaysia has an equatorial maritime climate, characterized by constant high temperatures and heavy rainfall, though considerably influenced by the reversal of wind directions associated with the monsoon regime. Average temperatures in West Malaysia are fairly uniform, with a mean of 81°F (27°C). In East Malaysia temperatures are higher, reaching 88°F (31°C), but vary rather more: lower temperature limits are 66°F (19°C) along the coasts of Sabah and 72°F (22°C) in Sarawak.

Variations in rainfall are, however, much greater. Although most places receive between 100 and 160 in (2,540–4,000 mm), annual totals depend very much on altitude and exposure to the monsoons. Annual averages can rage between 60 and 175 in (1,520–4,500 mm) in West Malaysia and Sabah, while in Sarawak averages are generally higher, ranging from 120 to 160 in (3,000–4,000 mm). Relative humidity remains high throughout the year and Mt Kinabalu and the Cameron Highlands (West Malaysia) are often shrouded in cloud and mist.

Vegetation and Wildlife. The natural vegetation of Malaysia is equatorial evergreen forest. Heavy rainfall and high temperatures have led to the development of a luxuriant forest cover. Three-quarters of West Malaysia and over 90% of East Malaysia are still forest covered. Swamp vegetation dominates in the low-lying localities and mangroves are extensive on the more sheltered coasts. However, the natural vegetation of the river deltas and the valleys has long been cleared for rice cultivation, especially in the more densely settled areas. The foothills on the Peninsula are now extensively planted with rubber trees, while alluvial tin mining in West Malaysia and shifting cultivation in East Malaysia have brought about significant changes in the vegetation cover. Coarse *ladang* grass and hardy shrubs have replaced the rain forests in many such areas. Malaysia is rich in jungle wildlife, including tigers, leopards, elephants, wild cattle (gaur), crocodiles, wild pigs, apes and monkeys. There are several species of deer, two species of rhinoceros (near extinction by 1970), and a large variety of snakes, birds, insects and fish.

The People. The most outstanding characteristic of the Malaysian population is its multiracial nature. Malays (about 45% of the population) and Chinese (about 35%) form the major ethnic groups. About 9% are Indians and Pakistanis.

At the end of the 19th century, when the mines and rubber plantations were undergoing rapid development, Malaya experienced mass immigration of Chinese and Indian labourers. The tide began to ebb towards the end of the 1930s, and the admission of female immigrants introduced significant changes. Today, most of the minority peoples were actually born in Malaysia. The two states of East Malaysia, however, were much less affected by these trends. Indigenous tribesmen form a substantial proportion of the population of the Borneo territories. The most important groups include the Kadazans, Bajaus, Bruneis and Muruts in Sabah, and the Sea Dayaks, Land Dayaks and Melanaus in Sarawak.

Population density in West Malaysia is considerably higher than in East Malaysia. Several areas on the west coast of the peninsula are feeling the strains of population pressure in relation to available agricultural land, while East Malaysia has an obvious demand for manpower to meet the needs of an accelerated economic development programme. The paradox cannot be resolved for some time since free flow of population between the two wings is still fraught with political difficulties.

The population of Malaysia is predominantly rural (61%), but the urban population is proportionally more numerous than in most South-East Asian countries. As would be expected, most important towns owe their origins and significance to the earlier expansion of tin mining and rubber plantation. The largest centre in West Malaysia is the federal capital, Kuala Lumpur (451,810). It is conveniently located in an area important both for its tin and rubber. With the added advantages of being the major road and rail hub, and its special role as the administrative, cultural and commercial centre, Kuala Lumpur developed the country's first industrial estates.

The traditional settlement in peninsular Malaysia is the *kompong*, which is usually located at a river bank or on the coast. Settlement density is highest on the small deltas.

In East Malaysia urban centres are usually small in size and few in number, as Sabah and Sarawak have not yet reached the same degree of economic development as the West. Towns with populations over 10,000 are limited to Sandakan, Kota Kinabalu (formerly Jesselton), Kuching, Miri and Sibu. Rural settlements are typified by the long-house communities of the non-Moslem indigenous tribes.

Religion and Education. Islam is the state religion and is practised by most Malays. Most of the Chinese are Buddhists, Confucians or Taoists. The majority of the Indian populations are Hindus, though 5% of their number are Christians. The tribal peoples of Sabah and Sarawak follow animist beliefs.

Education is complicated by the fact that separate school systems have to be maintained for the different ethnic groups. Illiteracy runs at an overall rate of 57%. West Malaysia has 5,325 schools with some 1,934,860 pupils, three universities with a combined enrollment of over 8,500 students, and some 6,000 students in other forms of higher education. Sabah has 760 schools with 137,430 pupils, and Sarawak has 1,334 schools with some 179,470 pupils.

Stalls outside the market in Kuala Lumpur. A modern city which dates its development from 1896 when it became capital of Malaya, Kuala Lumpur is more than 60% Chinese.

Agriculture. Agriculture is the most important sector of the Malaysian economy. The agricultural area of Malaysia is about 7 million ac (2·8 million ha) with some 85% in West Malaysia and only 320,000 and 500,000 acres in Sabah and Sarawak respectively. The most important form of land use is rubber planting. Nearly two-thirds of the agricultural land on the Peninsula and a slightly larger proportion in the Borneo territories are devoted to this crop. Coconuts are another major product. Palm oil is a relatively new cash crop and occupies third place in agricultural exports. This crop has experienced a sustained boom in recent years and is thought to be of great potential in future development. The traditional method of producing agricultural raw materials in Malaysia has been the organization of large-scale plantations, but in recent years an increasing percentage of the producing area has been devoted to small holdings. Nevertheless, the slightly smaller proportion of agricultural land in the plantations is producing more than half the total output of the principal crops.

Production for the world market has several problems. Although demand for agricultural raw materials is increasing, exporting countries have to face ever keener competition from producers elsewhere. An even greater threat comes from synthetic products. It has therefore become essential for the exporters to lower the cost of production through labour-saving techniques and through increasing yields, and Malaysia has made remarkable progress in these directions. On the other hand, production of rice for local consumption has remained comparatively low, with about 1 million ac (400,000 ha) in West Malaysia and 260,000 ac (105,000 ha) in East Malaysia. In Sabah and Sarawak more than 40% of the rice needed still has to be imported from other countries. Malaysia has a number of other significant export

Mt Kinabalu, the highest peak in Borneo.

crops, including sago, pepper and tea.

Forestry and Fisheries. Timber production is an important industry in Malaysia. There are some 9,000 sq mi (14,500 km²) of commercially exploited forests in West Malaysia, contributing some 2% of the total exports. In Sabah, timber from 7,300 sq mi (12,000 km²) accounts for over half of the area's exports. A third of the export earnings of Sarawak are derived from the forestry sector. Total Malaysian timber output is over 6 million tons annually.

Although fish is a major item in the Malaysian diet, the fishery industry is relatively underdeveloped, employing only a small proportion of the work-force, though it is being steadily developed and modernized. The annual catch is just under 3 million tons.

Minerals. Malaysia is the world's leading tin producing country. Production is primarily from the alluvial deposits of cassiterite, containing 75% tin. The major concentrations are on the foothills of the west coast of the Peninsula, extending from Perak in the north to Negri Sembilan, with the Ipoh and Kuala Lumpur areas being the most significant. Annual production of tin concentrates amounts to about 70,000 tons.

Iron ore of some importance and about 5 million tons are mined annually. The major iron mines are located at Ipoh, Semiling, Bukit Besi and Tanah Abang. The Telok Ramunia bauxite mine produces about a million tons each year. There are also limited deposits of oil (in Sarawak) and gold.

Industry. The economy of Malaysia and the former British territories in Borneo has traditionally been based on the production of agricultural raw materials for export, and until recently the manufacturing sector had been limited to the processing of agricultural produce and minerals. Tin smelting, with plants at Penang, Butterworth and Kelang; and oil refining, with refineries at Port Dickson in West Malaysia and Lutong in Sarawak, are the most important large-scale establishments. Rubber processing; copra milling; saw milling; and cement, brick and soap manufacturing; machinery maintenance, locomotive engineering and foundry work are locally important. A variety of factories process foodstuffs and produce consumer goods for the domestic market. Investment has promoted new industries such as plants for motor vehicle assembly, steel milling, aluminium rolling and the manufacture of fertilizers.

To date, the manufacturing sector contributes some 11% of the gross national product of Malaysia (estimated at $4,360 million in 1971), and employs 8·5% of the economically active population. A total of 2,500 factories are already in operation and production in new plants (many of them in new industrial estates and in economically depressed areas) was expanding in the 1970s.

International Trade. The chief export commodities of West Malaysia include rubber, tin, iron ore, and palm oil. Much of these items is handled by Singapore but direct exports to the United States, Japan, United Kingdom and USSR are steadily increasing and are now over 40% of the total. The major imports are foodstuffs, livestock, machinery and transport equipment, manufactured goods, chemicals, fuels and crude materials, and these are mainly supplied directly by the United Kingdom, Japan, Australia, China and Thailand. Exports from Sabah are primarily composed of rubber and timber, which traditionally go to Singapore and Japan. Rice, machinery and manufactured goods form the bulk of the imports and are supplied mainly through Singapore. Sarawak exports rubber, timber, sago and pepper. Crude oil imported from Brunei is either refined locally or shipped to West Malaysia.

The railroad station in Kuala Lumpur reflects the strong Moslem influence in Malaysia.

Transport. As the transport network of Malaysia was primarily designed to serve the 'Tin and Rubber Belt', it tends to concentrate on the west coast of the peninsula, focusing on the country's few ports and inland production centres. There are about 14,600 mi (23,320 km) of roads, of which 8,500 mi (13,600 km) are surfaced. The main road network in West Malaysia runs along the western coast of the Peninsula. The pattern of railways closely follows that of the roads. There are just over 1,000 mi (1,600 km) of railway in Malaysia, but only 96 mi (154 km) are in East Malaysia (Sabah).

Port Swettenham, the largest port, now handles over 2 million tons of cargo annually. Penang (George Town), with its special free-port status, is Malaysia's second-largest port, and plans are at hand for the development of deep-water port facilities at Butterworth on the mainland. The major ports of East Malaysia include Sandakan in Sabah and Kuching in Sarawak.

SINGAPORE

Area: 225·6 sq mi (584 km²)
Population: 2,147,000
Population Growth Rate: 2·1%
Capital: City of Singapore, pop 1,239,500
Language: Malay, Mandarin, Tamil, English
Religion: Confucianist-Buddhist-Taoist (75%) Moslem (20%)
Currency: Singapore dollar = 100 cents

Singapore is a tiny island republic at the southern tip of the Malay Peninsula. The country is about as small as England's Isle of Wight yet it supports one of South-East Asia's greatest manufacturing and trading cities whose prosperity sprang from its position at a cross-roads of major Asian sea routes. In World War II Singapore was a vitally important British seaport and naval base. Singapore joined the Federation of Malaysia in 1963, but in the face of Malay distrust of its large Chinese population withdrew in 1965 to stand on its own.

Although Singapore is now the most prosperous state in South-East Asia, it has its problems. Some stem from the multiracial character of its increasing population; others from economic losses following the British military run-down.

Physical Geography. The republic consists of one large island, diamond-shaped Singapore Island, and a number of very small and sparsely populated outer islands. Singapore Island is low and undulating with a few hills in the north-west and rather extensive swamps in the south-east. Its highest point is Bukit Timah (581 ft; 177 m). Short streams descend from the hills to the coast, which is mostly sandy or marshy but which in the south, where Singapore city stands, has one of the finest natural harbours in the East.

Climate and Vegetation. The climate is usually hot and humid with annual temperatures averaging 78°F (25°C). The annual rainfall, which is heaviest in December, is about 100 in (2,500 mm). August is the driest month.

Development and urban growth have robbed the island of much of its original mangrove and tropical rain forest, but some of the finest specimens of these types of vegetation are preserved in the Catchment Area Reserve and the Pandan Nature Reserve.

The People. Singapore has a multi-racial society partly based on peoples attracted from nearby areas in the recent past. In 1970 it included nearly 1,600,000 Chinese, about 311,000 Malays and more than 145,000 Indians and Pakistanis. With a density of over 9,000 people per sq mi (3,500 per km^2), there are inevitably problems of employment and housing. However, industrialization has created many new jobs, and a vigorous family planning campaign has more than halved the annual growth rate of the population. There has also been much government-sponsored slum clearance and rehousing.

Industrial Development. Industrialization began with the rise of Singapore as an international entrepôt, and the first industries were based on imported raw materials such as rubber, copra and timber. Since 1965 the government has been transforming the economy from one based on commerce and foreign military expenditure to a bustling industrial economy. Faced with a land shortage, the government reclaimed the swamps of Jurong and there built a thriving industrial complex with its own deep-water berths. Further industrial estates of this kind are planned.

The Singapore river is continually congested with barges and other port traffic. Singapore is a free port and much of its traffic is transit cargo.

To older industries such as tin smelting (Palau Barani) and oil refining (Palau Bukom, Pasir Panjang and Jurong) have been added a wide range of new industries such as textiles, shipbuilding and repair, chemicals, tyres, plastics and cables. Finance for these developments has come from Hong Kong, Japan, Britain, the Netherlands and the United States.

Transport. Singapore has a complex road network and a railway. A mile-long causeway carrying a road, railway and water pipeline links Singapore Island to the Malay Peninsula. International communications are served by one of the world's largest and busiest ports, and an international airport. This is an important scheduled stop for most leading airlines and the arrival point for over half of the important tourist trade.

International Trade. Re-export of processed rubber and petroleum occupies an important position in Singapore's trade, and Malaysia and Indonesia are thus the most significant trading partners. Exports of textiles and manufactured goods to South-East Asian countries are increasing. Rice and other foodstuffs, machinery and capital equipment are among the chief imports and come mainly from Japan and Britain.

BRUNEI

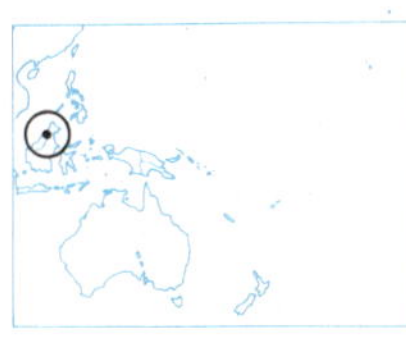

Area: 2,226 sq mi (5,765 km^2)
Population: 136,000
Population Growth Rate: 3·7%
Capital: Bandar Seri Begawan, pop 37,000
Language: Malay, Chinese, English
Religion: Moslem (60%) Confucianist-Buddhist-Taoist (35%)
Currency: Brunei dollar = 100 cents

The Sultanate of Brunei in northern Borneo was once a large and powerful state with a thriving trade with southern China. In 1888 it became a British protected state, a status confirmed by an agreement in 1971 by which Brunei retains full internal self-government while leaving foreign relations in British hands.

Land and People. Brunei looks out on the South China Sea, but is otherwise completely surrounded by Sarawak which, in fact,

The Seria oilfield in Brunei was discovered in 1929. Borneo's oilfields are some of the world's finest. Here, natural gas, found with the oil, is being burnt off.

divides the sultanate into two separate parts. Unlike Sarawak, Brunei does not belong to the Federation of Malaysia. Most of the sultanate is low-lying coastal plain. Proximity to the equator ensures a hot and humid climate throughout the year with temperatures around 80°F (26·7°C) and more than 100 in (2,540 mm) of rain yearly. Where the ground has not been cleared for cultivation it is densely covered with luxuriant equatorial forest containing much valuable timber.

Brunei Malays, mostly Moslem, form more than half the population; Chinese, living mainly in Bandar Seri Begawan, the capital and chief port, and in the oil town of Seria, more than a quarter. Dayak tribes inhabit the interior.

The Economy. Oil and natural gas are Brunei's most valuable resources. The oil industry, mainstay of the economy, employs about 10% of the working population. The Seria oil-field, discovered in 1929, is now past its peak and rather more than half the current production (6,900,000 metric tons in 1971) comes from the recently developed Ampa field offshore. Some of the output is refined in Brunei, but the bulk is piped to Lutong and Miri in Sarawak.

Other export products are rubber, pepper and buffalo hides, but quantities are small. Rice is the chief food crop. Other crops include coconuts and sago. Sarawak, Britain, Singapore and Japan are Brunei's chief trading partners.

INDONESIA

Area: 735,267 sq mi (1,904, 342 km²)
Population: 125,812,000
Population Growth Rate: 2·8%
Capital: Djakarta, pop 5,849,000
Language: Bahasa Indonesian. Over 200 other languages and dialects
Religion: Moslem (90%) Christian and Buddhist
Currency: Rupiah = 100 sen

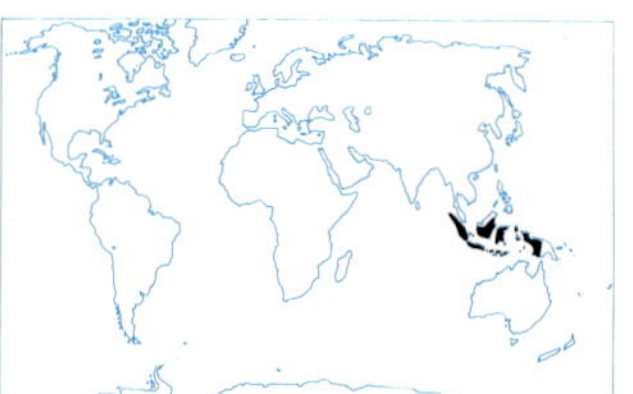

Indonesia (formerly the Netherlands or Dutch East Indies) is the largest country in South-East Asia, and in population the fifth largest in the world. Lying along the equator, it consists of more than 3,000 islands stretching 3,400 mi (5,500 km) from the Malay Peninsula to Australia. The total land surface of Indonesia is rather less than that of Western Europe.

Although the republic occupies major parts of New Guinea and Borneo, the second and third largest islands in the world, 66% of Indonesia's population and all its largest cities are crowded into the small, fertile island of Java, which represents only 7% of the country's land area. This imbalance, which increased under Dutch colonial rule, has been a major factor in Indonesia's history, and presents enormous problems for the present regime which, after the stagnation of the war years and the mismanagement by the postwar Sukarno regime, is attempting to give an essentially underdeveloped country a modern industrial economy. Not the least of its problems is maintaining a national identity in the face of divisive regional and ethnic difference throughout this sprawling island nation. Indonesia was declared independent on 17 August 1945, and a constitution was drawn up which called for a strong central government, a 'unitary republic' with power vested in a 'People's Congress'. Independence was finally recognized by the Dutch in 1949. Sukarno, Indonesia's first president, attempted to institute 'guided democracy', a left-ward-oriented, authoritarian regime and increasing reliance on Communist China not only enfeebled the economy but led to mounting discontent. In 1965, after an abortive coup, there was a massacre of Communists (some 300,000 to 500,000 died). Sukarno gradually thereafter lost power to General Suharto, who reversed all Sukarno's policies, and by 1968 had become Indonesia's second president.

The Land. Standing between the Pacific and Indian oceans, the islands of Indonesia occupy a great maritime crossroads, exposed to a wide range of external influences. It is said that in Indonesia 'the land divides but the sea unites', for the islands are mountainous, including more than 100 active volcanoes and much impassable swampland, but they are closely linked by the sheltered waters of the narrow Java and Banda seas, which act as maritime highways. This geographical fact largely explains the national unity of Indonesia.

Indonesia, the largest island group in the world, has a total land area of 735,267 sq mi (1,904,342 km²). It is generally divided into the Greater Sunda Islands, containing most of the population and wealth, and including Sumatra, Java, Borneo and Sulawesi (Celebes); the Lesser Sundas (Nusa Tenggara), stretching from Bali to Timor; and Maluku (the former Moluccas or Spice Islands). In addition, there is West Irian, the western half of New Guinea. The Indonesian two-thirds of Borneo is called Kalimantan. Both of these areas, but especially West Irian, are covered by tropical rain forests, undeveloped and thinly populated.

In general the Indonesian islands have mountainous interiors with much volcanic activity, testifying to the youthful geological structure of the archipelago. There is a basic distinction, however, between those islands such as Sumatra, Java, Borneo and New Guinea which represent extensions of the Asian or Australian mainlands, separated only by shallow seas, and those like Sulawesi and several Nusa Tenggara and Maluku islands which rise abruptly from the very deep seas lying between the two continental shelves. Thus mountainous Sulawesi is strikingly deficient in lowland, whereas those island areas (Sumatra, Kalimantan) facing the shallow inland seas are apt to be extremely low-lying and swampy.

Swamps and mountains make the four largest islands largely unsuitable for human occupation. Java on the other hand has little swampland and its interior mountains, though high, are discontinuous and do not impede communications. Moreover it has a high percentage of land suitable for cultivation, and in the east (and in Bali) which has the best soils, the long dry season provides ideal conditions for the ripening of grain.

Climate. Most of Indonesia has a maritime equatorial climate. Because of the equalizing effect of the all-encompassing seas, the mean monthly temperatures throughout the year vary by only two or three degrees from an annual average of around 78–80°F (25·5–26·6°C), except for the higher uplands where temperatures drop about 3°F (1·7°C) for every 1,000 ft (295 m) above sea level. Rainfall, on the other hand, shows considerable variation, both in amount and in seasonal incidence (the average for Indonesia as a whole per year is about 75 in (190·5 cm)), owing to the fact that the islands lie across the path of the two great monsoonal air streams which alternately flow in opposite directions between East Asia and Australia.

Thus in December-March the winds blow from the north-east over the northern part of Sumatra, Kalimantan and Sulawesi, but swing round in crossing the equator to become west/north-west over Java, Nusa Tenggara and beyond. In June-October the direction is reversed, with an extremely dry air stream flowing from the Australian deserts over the southern half of Indonesia, before changing direction to south-west across Sumatra, by which stage it has become heavily charged with moisture. During this season, therefore, there is a marked dry season from Nusa Tenggara through eastern Java. In most parts of the country except the south-east rainfall is fairly well distributed throughout the year, with the exception of areas, such as the south-west-facing slopes of Sumatra, which are particularly exposed to moisture-laden monsoons.

Vegetation and Wildlife. Since Indonesia is a region of overlap between species derived from tropical Asia and Australasia, and lies in the humid tropics, it has an exceptionally luxuriant diversity of plant life. Mangroves and other swamp plants cover the lower lying coastal fringes, giving ground inland to tropical rain forest over the greater part of Sumatra, Kalimantan, Sulawesi and West Irian. Most of western Java has long been cleared for cultivation, and in the east the forest cover thins out to savanna in the dry Nusa Tenggara, with teak, casuarina and eucalyptus in between. Teak and other hardwoods are used for building, along with the ubiquitous bamboo. There are many palms and fruit trees, including the banana, mango, durian and several kinds of citrus.

Wildlife is equally varied, and the profusion of bird life is proverbial. Indian elephants are found in Sumatra and Kalimantan. Tigers, leopards and apes including the great orang-utan, as well as crocodiles, snakes of all sizes, and lizards (including the huge Komodo Dragon) are found in various parts of the country. But some animals, including the Sumatran and Javan rhinoceroses, are becoming very rare.

The People. Of Indonesia's estimated population of 125,812,000, about two-thirds live on Java, where the population density is over 1,507 persons per sq mi (605 per km^2), whereas the average density for the country as a whole is about 158 per sq mi (61 per km^2). Moreover, 7 out of Indonesia's 10 largest

The temples at Borobudur were built around 800 A.D. when the Buddhist Saliendra dynasty dominated the island of Java. They contain 3 miles of stone carvings of Buddhist texts.

A bull burning in effigy. Balinese religion is a mixture of Hinduism and animism.

cities are in Java, including the capital and largest city, Djakarta. Since 70% of the population of overcrowded Java still lives by agriculture, the increasing pressure on cultivable land presents a growing problem, aggravated by Indonesia's high rate of population growth, which is nearly 3 million persons per annum. A 'trans-migration' policy, the large-scale removal of people from Java to the outer islands, is being attempted by the present government, but so far with little success – particularly since it is difficult to check the influx into the large cities from the countryside as well as from the other islands.

Most Indonesians are of Malay stock, although there are Arab and Papuan (West Irian) minorities and some 3 million people of Chinese descent, concentrated mainly in the larger towns. The Chinese generally make up the commercial middle class. Over 250 languages of the Malayo-Polynesian group are spoken in Indonesia, including Javanese, which is widespread on Java. However, the official language, *bahasa Indonesia*, which is descended from the trading language of colonial Indonesia, is now increasingly known in all the islands. English has replaced Dutch as the principal Western language, and is often taught in schools and universities.

The majority of Indonesians are farmers and live predominantly in villages, where their lives are governed by a body of local custom known as *adat*. Villages in densely-populated Java are large, compact and closely spaced, with well-built individual family houses. In the outer islands village and house types reflect the great diversity of regional cultures. Among many people, including the Malays of central Sumatra and the primitive Dayaks of interior Kalimantan, villages consist of communal long houses. Traditionally, men and women wear either a *kain* or a *sarong*, a colourful strip of cloth wrapped around the body. The main food of most Indonesians is rice, cooked and served in many ways and with different foods.

There is an enormous diversity of cultures in Indonesia, from the Stone Age Papuans of West Irian and the Dayaks of Borneo to the Hindu-Buddhist cultivators of inland Java and Bali, and the Moslem coastal people whose main business is trade. In the larger cities the wealthier classes ape Western ways. Despite this cultural diversity, however, about 90% of the people are Moslems, and only 10% profess Christianity, Hinduism, Buddhism or other faiths. Basic and traditional Indonesian beliefs underlie all religion, especially on Java, where there is a pervading belief in an unseen world of spirits, good and bad, and an emphasis on ancestor worship. On Bali, a religion called Bali-Hinduism emphasizes worship of the spirits of mountains, trees and other natural features. The Moslem majority on all the islands is noted for its devoutness.

An aspect of the deep-rooted Hindu-Buddhist cultures of Java and Bali is the famous village and court dances, as well as the 'shadow play' puppet dramas. *Gamelan* orchestras, mostly percussion, accompany both puppet shows and dancers. The most famous Indonesian handicraft is the making of exotically-coloured and designed *batik* cloth.

One of the great successes of the new nationalist government has been in education. Elementary schools and a high school system are provided by the government and children are required by law to go to school between 8 and 14. Private schools are also aided, and there is a strong programme of adult education. As a result, perhaps three-quarters of the people can now read and write, whereas at the end of World War II illiteracy stood at around 90%. There are many colleges, and some 14 new universities were established between 1945 and 1960 alone. The programme as a whole is rapidly creating a sophisticated educated class, free of tradition and ready to help turn Indonesia into a modern state.

Agriculture. Agriculture, employing some 70% of the population of Indonesia (though on only 7·4% of its land surface) is by far the most important single occupation in the country. It also provides about 70% by value of the country's exports. The principal subsistence crop is rice; however maize and cassava are also important, while soybeans, peanuts, green beans and sweet potatoes are also grown. Subsistence agriculture has markedly increased its share of the Gross National Product in recent years, a reflection of the huge increase in population and of the effort needed simply to keep the people fed. Pressure of people upon the land means that the average size of family holdings, particularly in Java, has declined and is now below 3 ac (1·2 ha) and the number of wholly landless peasants has continued to rise. Despite the Suharto government's stress upon agricultural improvement, and the encouraging introduction of new strains of high-yield rice, the situation is still dangerous.

Indonesian agriculture can be divided roughly into two main categories, subsistence and commercial. The former accounts for much the greater part of the total labour input. Subsistence agriculture is of two sorts, closely correlated with areas of high and low population density. *Ladang*, or slash-and-burn cultivation, is the older type and is still the most widespread in the less-populous islands. Freshly cleared forest land is planted for a year or two, without manure or fertilizer, and is then left fallow for up to 18 years in order to restore the fertility of the soil. The main crops grown by this wasteful method are dry rice, cassava (tapioca) and peanuts, together with green vegetables and fruit trees.

The second type of subsistence agriculture is the more advanced permanent cultivation, or *sawah*, consisting mostly of the cultivation of rice in wet fields. *Sawah* covers much less land than *ladang*, but produces far more food, often involving double-cropping. Most *sawah* cultivation (74%) is in Java, with negligible amounts in other islands – 9% in Kalimantan and 7·5% in Sumatra. Two-thirds of the *sawah* lands are watered through irrigation.

Indonesia as a whole is not a land of exceptionally fertile soils. There is much land capable of supporting plantations of commercially important trees, but the really rich soils are generally confined to central and eastern Java, Bali and parts of northern Sumatra. Because of population growth and pressures on the land, Indonesia still does not grow sufficient rice for its population and must import much rice, which at times costs over one-fifth of its foreign-exchange earnings.

Commercial agriculture in Indonesia has a long history, dating back to the cultivation of spices in Maluku which first attracted the European traders. Under the Dutch, commercial cultivation was intruded into the already intensive cultivation system of Java, with sugar (an annual) being grown in rotation with rice on lowlands, and perennial coffee on uplands. At the same time the Dutch began to set up plantations in Sumatra, beginning with tobacco, then establishing estates of coffee, rubber, tea, palm oil and sisal. Soon the outer islands began to overtake

Indonesian dancers are trained from childhood to interpret the traditional epics of love and adventure with graceful, rhythmic movements of the fingers, hands, limbs and body.

Founded as a Dutch East India trading post in 1914, Djakarta, on the north-west coast of Java, is the modern capital of Indonesia.

Java in commercial production, while small-holder cultivation of cash crops – especially rubber and copra – rapidly expanded in all the major islands outside Java.

Since independence there has been an acceleration in the shift of emphasis in commercial cultivation both from Java to the outer islands and from the estates to small-holders. This has resulted partly from the prolonged neglect of the commercial estates during the war and the Sukarno period. Since the Suharto regime came to power, most of the foreign-owned estates (Dutch, British and American) which had been siezed by the government have now been restored to their former owners on conditions which should encourage their return to effective production. Rubber and copra production already show substantial increases over prewar levels. In order to remain competitive with the highly efficient Malaysian production and synthetic rubber, Indonesia must replant its rubber with high-yielding varieties. But the main emphasis in commercial agriculture will remain in the small-holder sector. With efficient organization, this should begin to prove increasingly competitive.

Livestock play a small part in Indonesia's agriculture, being raised mainly as part of the subsistence *sawah* system in Java, with both oxen and buffaloes used with the plough. The fodder problem is becoming acute. Pigs, sheep, goats and especially poultry are widely raised. The possible development of commercial stock raising in the drier eastern islands of Nusa Tenggara is being considered.

Forestry. Approximately two-thirds of Indonesia's total land surface, or about 470,000 sq mi (1,218,000 km^2) is under forest, but 97·5% of this is outside Java. Nevertheless, Indonesian timber production is concentrated in Java, largely because about half of it is destined for firewood and charcoal. All forest lands are under government control. Even if 40% of the total, mainly in upland areas, is protected as a conservation measure and another 18% is to be reserved for future cultivation, there still remain vast areas available for exploitation. However, only some 20% of the total acreage is considered close enough to sea or river transport to make exploitation commercially worthwhile. At present Java is the main producer of teak, and Sumatra of other timber and by-products, but the potential of Kalimantan far exceeds that of Sumatra. Forestry could be a very important industry for Indonesia, and the economic Five Year Plan operating in the early 1970s was geared to a production of 500,000 cu yd (400,000 m^3) of teak and a major expansion in that of other timbers, amounting to a total of 2 million cu yd (1·5 million m^3).

Fisheries. In 1968 Indonesia's total fish catch was over 1 million tons, of which three-fifths was salt-water catch and the rest from rivers, inland fish ponds, and the flooded rice paddies of central and eastern Java. At present commercial fishing of more than local importance is not extensive, but the potential in both lakes and seas is great, and the government is encouraging joint fishing projects with foreign – particularly Japanese – capital. The prime need is motorized vessels to facilitate deep-sea fishing. The Five Year Plan in hand in the early 1970s anticipated raising Indonesia's total fish catch to nearly 2 million tons.

Minerals. Tin, extracted with the use of Chinese labour, was the basis of Indonesia's first important mining industry, but early in the present century tin was overtaken by petroleum. By 1938 the total production of crude oil amounted to 7·3 million tons (7·4 million metric tons), and accounted for 24% of the country's exports. Since World War II oil production has increased more than four-fold to 29·5 million tons (30 million metric tons) in 1968, while in the same year tin had fallen to 16,600 tons (16,900 metric tons). Nevertheless, Indonesia stands with Malaysia and Bolivia among the world's most important suppliers of tin, with the United States as its most important consumer.

During the Sukarno period the country became steadily more dependent on oil. By this time Indonesia had attained the rank of the world's eleventh oil producer, though it produced only 3% of the global totals, and its known reserves amounted to only 2·3% of the total. Since only about 5% of Indonesia has been geologically mapped in detail, the oil reserves may indeed prove to be greater than this; however the important fact is that Indonesia is the largest oil producer between the Middle East and the American Pacific coast, and its main producing areas are readily accessible by sea. Most of Indonesia's oil today comes from eastern Sumatra, with Kalimantan as the second centre, and some small production in Java and West Irian. There are good prospects for further developments on the continental shelves off Sumatra and between Java and Kalimantan. Foreign prospectors are at present encouraged to undertake projects in which they

Lake Toba is the largest of a chain of mountain lakes in the Sumatran highlands.

receive payment in the form of a percentage of the oil they produce. Similar policies are also favoured in respect of other minerals, of which Indonesia is known to possess a great variety including nickel, bauxite, copper and manganese. Coal and iodine are among other Indonesian mining products. Salt, a government monopoly, is mined for internal use.

Manufacturing Industry. During their last years in Indonesia the Dutch were very slow in recognizing the need for industrialization, and little was done. Under the nationalistic Sukarno regime after the war the expulsion of foreign technicians and managers hindered what little development was taking place in industry. The contribution of manufacturing to the national product actually declined from 13·2% in 1958 to 12% in 1967. Factories, which are concentrated in Java, are therefore small, and the product, with the exception of the oil refineries in Sumatra and Java and a large cement factory at Padang, is mostly for local consumption. In Java the main emphasis is on consumer goods, notably textiles, clothing, food, drink, tobacco and household goods including furniture. Approximately one-third of the Indonesian labour force is involved in the manufacture of food and tobacco products, and one fourth in the manufacture of clothing and related industries. The Suharto government's Five Year Plan begun in 1969 envisaged trebling the old annual textile output of about 330 million yd (300 million m), and building up heavier industry such as iron and steel and electrical equipment, as well as pulp, paper, rayon and the petro-chemical industry.

The role of the existing small-scale domestic industries in absorbing labour and maintaining living standards in overcrowded Java should not be disparaged. But expansion in this regard is hampered by the lack of electric power, despite the increase of the installed capacity of the State Electricity Company from 157,500 kw in 1945 to 541,605 in 1969. The addition of a further 125,000 kw from the Djatiluhur electrical project in the latter year has greatly improved the position in the two key centres of Djakarta and Bandung on Java.

Batik making in Java. Using wax, Indonesians are able to print cloth with beautiful designs.

International Trade. Indonesia's trade, both export and import, has changed little in its overall pattern since colonial times. The export trade still consists almost exclusively of unprocessed or partially processed primary produce of many kinds. Raw rubber has headed the list, followed by oil and oil products, then such items as copra, palm oil, spices, tobacco and sugar. However, the share of petroleum and rubber, items which headed the 1938 list and together accounted for 47% of the total, has since risen to over two-thirds, while the proportions formed by tea and tobacco have dropped sharply and sugar has disappeared from the list. Tin and tin ores have always held a small place in Indonesia's total exports (never higher than 7%). As early as 1938 oil accounted for 24% of exports and tin only 5%. Export of oil to Japan today is of particular importance because of Japan's fast expanding oil consumption and the low sulphur content of Indonesian oil. (Japan's growing atmospheric pollution is partly due to burning fuels that give off toxic sulphur compounds.)

Consumer goods make up a large percentage of imports, with capital goods second. Thus rice, textile fabrics, wheat flour, dried or salted fish and similar products head the list. Trading partners have changed markedly, with the role of the Netherlands declining. Japan now holds first place, with the United States second.

Transport and Communications. Roads, inter-island, coastal and river shipping services and air transport are of particular importance in an island state like Indonesia. The Dutch, particularly on Java, instituted excellent systems, which seriously deteriorated during the war and under the Sukarno regime. They are only now beginning to be renovated and enlarged, although population growth and increasing use make the task difficult. About 85% of the nation's railway service is still on Java, and these lines are slowly being improved. So too is the new national airline, Garuda Indonesian Airways, which is also being extended. Similarly the privately-owned Dutch network of coastal and inter-island shipping services was allowed to deteriorate but has now formed the nucleus for a new merchant marine whose tonnage today is more than double that of the old Dutch company.

In telecommunications Java again is far better served than any of the other islands, although there is no normal telephonic communications between the islands. However microwave and similar systems are now being developed, and radio broadcasting provides an important link throughout the nation. Television is also rapidly spreading both on Java and the other islands.

PORTUGUESE TIMOR

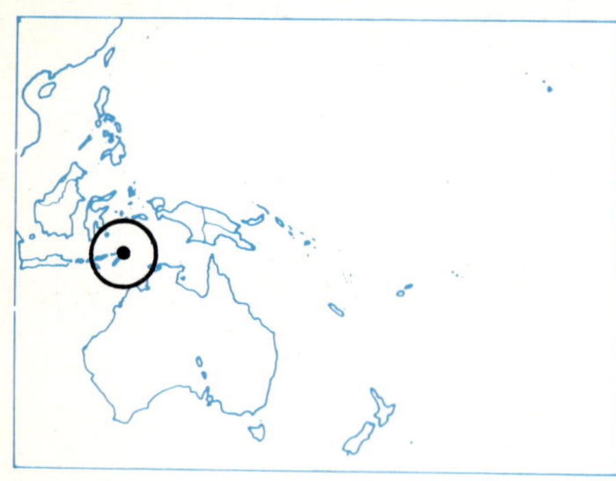

Portuguese Timor, in the Malay Archipelago, mainly consists of the eastern part of Timor Island (the western, formerly Dutch, portion is part of Indonesia). Portugal's holding occupies 7,240 sq mi (18,752 km²), ranks as an overseas territory of Portugal, and includes the enclave of Ocussi Ambeno on the northern coast of Indonesian Timor and the islands of Pulo Cambing and Pulo Jako. The Portuguese first arrived in Timor about 1512.

Land and People. Timor is part of the southern or outermost arc of the Malay Archipelago which extends from the Mentawai Islands along the submarine ridge south of Java to Sumba and other islands. Timor is rugged and mountainous, with small plateaux and fertile valleys, although its soils cannot compare in fertility with those of Java and Bali. It is also less favoured climatically, for it has a prolonged dry season during the south-east monsoon (April-October) which blows directly from the Australian deserts. During the rest of the year the pattern is set by the north-west monsoon with heavy and often torrential rains. The vegetation reflects the island's position between South-East Asia and Australia, and ranges from coastal mangrove swamp to savanna, open grassland and the interior monsoon forest.

Portuguese Timor has a population of 610,541, including about 1,700 Europeans. The capital and port is Dili (10,753). The Timorese proper, the Atoni, are mainly Melanesians and live in the interior, whence they were driven by the descendants of later Malay/Indonesian arrivals, who occupy the better coastal areas. The people of the interior are still largely animist, while the coastal zone contains both Moslems and Christians.

The Economy. The indigenous people practise shifting cultivation, growing maize, dry rice and sweet potatoes. Some wet rice is grown on irrigated terraces, and cattle, pigs, goats, sheep and poultry are reared. Export crops include coffee, copra, rubber and wax, but the quantities involved are small. Sandalwood is also exported. The chief road follows the north coast, with extensions through the interior to the south and a link with Indonesian Timor.

Timor is the easternmost island of the Sunda group of the Malay archipelago. It suffers from a difficult climate with a long dry season and considerable over-farming.

CHRISTMAS ISLAND

Christmas Island lies in the Indian Ocean about 250 mi (402 km) south-west of Java. The island covers 52 sq mi (135 km²), most of which is plateau – the top of a submerged mountain. The island was annexed by Britain in 1888 and transferred to Australia in 1958. The population of 3,361 consists mainly of Chinese and Malays brought in to work the large phosphate deposits under the control of a joint Australian-New Zealand commission. Phosphates are shipped to Australia and New Zealand for the most part, but also to Singapore, Malaysia and other countries.

This island should not be confused with Christmas Island in the Line Islands in the Pacific which was the scene of British and American nuclear tests (1957–62).

COCOS (KEELING) ISLANDS

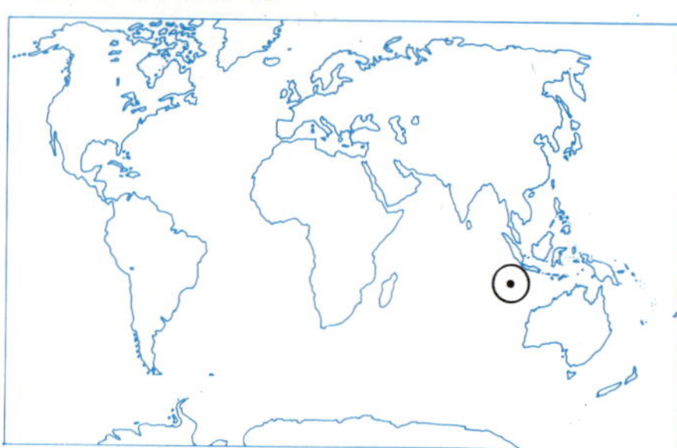

The Cocos (Keeling) Islands are 27 low-lying, palm-covered coral islands in the Indian Ocean. They were settled in the 1820s by Alexander Hare and John Clunies-Ross, annexed by Britain in 1857 and transferred to Australia in 1955. Land rights were granted in perpetuity to the Clunies-Ross family in 1886. The largest island is West Island, which has the airfield and most of the 128-strong European community. Home Island, the headquarters of the Clunies-Ross Estate is where most of the 483 Cocos Islanders live.

Until 1967 the islands were a refuelling point on the Australia-South Africa air route. Today the main activity is copra production.

AFRICA

Africa, the second largest continent, is famous for its vast tropical forests and grasslands; big game reserves; and diverse Negro peoples. Yet although Africa has nearly one-quarter of the world's land and one-third of its independent nations, it possesses only some 10% of the world's population, producing a mere 2% of the world's total services and goods. No other continent contains so many desperately poor nations as Africa. Moreover, their prospects of growing richer dwindle as rapid population increase threatens to absorb the benefits of economic growth. The main exceptions are nations (like Libya and South Africa) in the extreme north and south where industrial growth is bringing at least relative prosperity.

Africa in History

The Dark Continent. Africa is one of the oldest known homes of man, and contained in Egypt one of the great ancient civilizations, but its size and daunting physical obstacles made it the last of the earth's great land masses to be explored and developed by civilized peoples. It was regarded for centuries as 'the Dark Continent', mysterious and isolated: the northern fringe, tied to Asia by the Suez isthmus, and narrowly separated from Europe by the Strait of Gibraltar, was for long the only area known to Europeans.

The wide, inhospitable wastes of the Sahara discouraged contact between Mediterranean Europe and tropical Africa. Entry elsewhere was hampered by surfbound coasts, impassable rivers, dense mangrove swamps and forests. European travellers found climatic conditions harsh and easily fell victim to fatal diseases. The survivors had to contend with the hostility of the peoples of the interior – especially in areas where organized slavery and the slave trade flourished such as East Africa, where the Arabs were active, and West Africa, where the Europeans themselves had promoted the shameful traffic.

The Evolution of Modern Africa. From the later 18th century interior Africa was gradually opened up by geographically curious Europeans. Western governments became interested in their discoveries in the late 19th century, when nearly every part of Africa was claimed by Belgium, Britain, France, Germany, Italy, Portugal or Spain. Almost simultaneously, economic development was stimulated by the discovery of minerals, notably diamonds and gold.

During the present century, economic and social development has continued, but political progress has taken a new course. With the rise of African nationalism most territories have become politically independent. Now there are more than 40 independent countries, some as huge as Zaire, others as small as Gambia. Only in the south do white minorities still govern.

Nearly all the newly independent countries are republics, but progress has been chequered. Coups have created authoritarian military governments. There have been costly and disastrous civil wars and tribal strife, notably in the Congo (now Zaire) in 1960, Nigeria in 1967 and Burundi in 1972. Many nations depend heavily on foreign aid.

Africa, however, has a sense of unity forged from misfortunes and exploitation. Pan-Africanism, a belief in the unity of all African peoples, is supported throughout black Africa. The Organization of African Unity (OAU), founded in 1963, works to this end; its members include every African country except those ruled by white minorities.

Over: **The remote Hoggar mountains in the southern Sahara.** *Below:* **A typical oasis village at the foot of the Atlas mountains.**

The Face of Africa

Location and Size. Africa covers nearly 12 million sq mi (30 million km²), about one-fifth of the earth's total land surface. It is three times the size of Europe and half as large again as North America. Together with Eurasia and Australia, Africa makes up the great land mass known as the Old World. This immense continent stretches about 5,000 mi (8,000 km) from Cape Bon (Tunisia) to Cape Agulhas (South Africa), and 4,600 mi (7,500 km) from Cape Verde (West Africa) to Ras Hafun (Somalia). It is the only continent to be crossed by the equator and both tropics (Cancer and Capricorn). More than 75% of its area lies within the tropics, although two-thirds of Africa lie north of the equator.

Bordered by the Atlantic and Indian oceans, and by the Mediterranean and Red seas, Africa has few large gulfs and inlets apart from those of Guinea in the west, Aden in the east and Tripoli in the north. Its comparatively few islands include Madagascar, the world's fifth largest island.

The Land. The African plateau, in which some very ancient and hard crystalline rocks are exposed, is dominated by granites and highly metamorphosed schists and gneisses. Millions of years of erosion are largely responsible for the uniform surface features occurring over extensive areas. Nearly everywhere the plateaux rise sharply from the usually narrow coastal plains. In general the plateaux of southern Africa, sometimes called High Africa, are more extensive than those in the north and west (Low Africa). In the south, there is a particularly steep edge to the plateaux in the Drakensberg Mountains of South Africa, which overlook the narrow coastal plain along the Indian Ocean. North of the Congo Basin, the plateau is lower and much more broken. In West Africa, the higher parts are widely separated from one another, with the Fouta Djallon Plateau in the extreme west, the Jos Plateau in central Nigeria, and the Adamawa and Cameroon mountains in the east. The highest parts of the Sahara are the Aïr Plateau and the mountains of Ahaggar and Tibesti.

In the extreme north and south of the continent are high folded ranges. In the north, the Atlas Mountains, rising in places to 14,000 ft (4,000m) are associated with the great Eurasian mountain system that includes the Alps and Himalayas. In the south, east of Cape Town, are the lower yet very prominent Cape Ranges rising above 3,000 ft (915 m).

Africa's most spectacular physical feature is the East African Rift Valley, which forms a

AFRICA
vegetation
high mountain flora
dry mixed woodland and forest
irrigated areas
tropical rain forest
savannah and bush woodlands
steppe grassland
desert and semi-desert
cultivated areas
oases
swamp
scale 1 : 40,000,000
0
500
1000 st. miles
MEDITERRANEAN SEA
BLACK SEA
CASPIAN SEA
Aral Sea
PERSIAN GULF
RED SEA
GULF OF ADEN
DAHLAK ARCH.
Bab el Mandeb
Socotra
Cape Guardafui
AZORES
Madeira
CANARY ISLANDS
Bay of Biscay
Corsica
Sardinia
BALEARIC ISLANDS
Tyrrhenian Sea
Adriatic Sea
Sicily
Malta
Crete
Cyprus
Str. of Gibraltar
Tangier
Casablanca
Algiers
Tunis
Cape Bon
TELL ATLAS
GREAT ATLAS
SAHARAN ATLAS
ANTI ATLAS
Chotts
Marrakech
Gulf of Gabes
Tripoli
Gulf of Sidra
Gulf of Salum
Alexandria
Suez Canal
Cairo
Qattara Depression
CYRENAICA
Siwa Oases
Gr. Eastern Erg
Gr. Western Erg
Erg Iguidi
Erg Chech
Tidikelt
FEZZAN
AHAGGAR
TASSILI-N-AJJER
Kufrah Oases
Kharga Oases
Libyan Desert
Nile
L. Nasser
Nubian Desert
Tropic of Cancer
Cape Blanc
ADRAR
S A H A R A
AIR OR AZBINE
TIBESTI
TÉNÉRÉ
BORKOU
Senegal
EL DJOUF
Tombouctou
Cape Verde
Dakar
Gambia
Niger
Lake Chad
KORDOFAN
Blue Nile
White Nile
FOUTA DJALLON
S U D A N
Chari
Benue
SUDD
AMHAR
Addis Ababa
PLATEAU
Freetown
UPPER GUINEA
ASHANTI
Accra
Lagos
Ivory Coast
Gold Coast
Slave Coast
Bight of Benin
Mt. Cameroon 13353
Fernando Poo
Bight of Biafra
Principe
São Tomé
GULF OF GUINEA
Cape Lopez
Annobon
Equator
Lake Rudolf
Lake Albert
Lake Kyoga
Mogadiscio
Ubangi
Stanley Falls
Mt. Margherita 16821
Lake Edward
Lake Victoria
Mt. Kenya 17040
CONGO BASIN
Lake Kivu
Zaire (Congo)
LOWER GUINEA
Brazzaville
Kinshasa
Kasai
Mt. Kilimanjaro 19590
Mombasa
Pemba
Zanzibar
Dar es Salaam
TANGANYIKA PLATEAU
Lake Tanganyika
Ascension
A T L A N T I C
O C E A N
KATANGA
Lake Mweru
Lake Malawi
COMORES ISLANDS
Benguela
LUANDA PLATEAU
Lubumbashi
Mozambique
MOZAMBIQUE CHANNEL
St. Helena
Katima Falls
Zambezi
Salisbury
Victoria Falls
Beira
MADAGASCAR
Mauritius
Réunion
Namib Desert
Okovanggo Basin
Limpopo
Tropic of Capricorn
Walvis Bay
Kalahari Desert
Pretoria
Johannesburg
Witwatersrand
Delagoa Bay
Lourenço Marques
Orange River
Bloemfontein
Durban
St. Helena Bay
Table Bay
DRAKENSBERG
Cape Town
Cape of Good Hope
Cape Agulhas
Port Elizabeth
I N D I A N
O C E A N
0°
10°
20°
30°
40°
50°
60°

tremendous gash of land from the Red Sea to the Shire River in southern Africa. A series of mostly extinct volcanoes lies along the edges of the valley and many lakes have formed in its lower parts. Lake Tanganyika, 4,700 ft (1,570 m) deep, is the world's second deepest lake.

The Rift Valley and its branches can be traced out by its lakes. Across central Tanzania, the rift is less conspicuous until Lake Malawi (formerly Nyasa) is reached. Beyond, the rift continues down the Shire almost to Beira on the Indian Ocean. From the northern end of Lake Malawi, a great arm extends north-westwards and westwards. Lake Victoria, Africa's largest lake, occupies a fairly shallow depression on the surface of the plateau.

The volcanic peaks near the Rift Valley are among Africa's highest mountains. They include Kilimanjaro (19,340 ft; 5,900 m), the grandest of all. In the Virunga or Mfumbiro Mountains on the Uganda-Zaire border, there is still occasional volcanic activity. By contrast, the mountains of the great Ruwenzori Range, also between Uganda and the Congo basin, and extending from Lake Edward to Lake Albert, owe their height to the hardness of their constituent rocks.

Coastal lowlands less than 1,000 ft (300 m) above sea level are generally few, and are restricted in extent. Most are associated with the lower sections of river basins.

Rivers. Africa has many rivers, some very long with extensive basins. Only the Congo among Africa's largest rivers reaches the sea through an estuary rather than a delta. The longest river, the Nile, kept the secret of its headwaters from explorers until the 1860s. More than 4,000 mi (6,400 km) long, it has important tributaries like the Blue Nile and the Atbara. The Congo, though shorter (2,918 mi; 4,600 km), drains a much larger area: its basin is second in size only to that of the Amazon.

The Niger (2,600 mi; 4,100 km) in West Africa rises on the Sierra Leone-Guinea border and flows in a great arc eastwards, and then southwards, before reaching the Gulf of Guinea through its vast delta in southern Nigeria. Its seasonal flow varies greatly. Upstream it is navigable for some 400 mi (650 km), but its left bank tributary, the Benue, can be navigated for another 600 mi (960 km) in the rainy season.

Rapids and falls impede navigation on many rivers. The most famous and magnificent, the Victoria Falls on the Zambezi, are one of the natural wonders of the world, 355 ft (108 m) high – more than twice as high as North America's Niagara Falls. The Orange River (1,300 mi; 2,092 km), the longest river in South Africa, resembles the Nile in flowing through desert and semi-desert regions in its middle and lower courses. Although its chief tributary, the Vaal, rises in the well-watered areas of the Drakensberg, so much water evaporates or is used for irrigation that at times there is almost no flow in its lower reaches.

Man-Made Lakes. Several of Africa's great rivers are dammed for hydro-electric power and irrigation, and some of the world's largest man-made lakes have resulted. The lake reservoir formed by the Akosombo Dam on the Volta River in Ghana covers 3,275 sq mi (8,482 km²). Lake Nasser, on the Nile in Egypt, 300 mi (483 km) long by 1980, was designed to irrigate some 6 million ac (over 2·4 million ha). Lake Kariba, on the Zambezi, 175 mi (373 km) long and 12 mi (19 km) wide, is the third largest man-made lake in the world.

Climate. Because of the predominantly plateau nature of the continent, climatic conditions are uniform over wide areas. Most of Africa is hot for most of the year, with clear wet and dry seasons (except in the equatorial lowlands), and only small seasonal changes in temperature, not always as high as latitude suggests owing to the effects of altitude. Near the equator, snow lies permanently on the upper slopes of some higher mountains such as Kilimanjaro and Kenya.

Climate is dominated by the sun's movements. It is overhead at the equator in March and September, at the tropic of Cancer in June, and at the tropic of Capricorn in December. Thus northern and southern Africa have appreciable seasonal changes in temperature. For example, in July much of northern Africa averages about 80°F (27°C), and parts of the Sahara average up to 100°F (38°C). Meanwhile, temperatures average 64°F (18°C) or less in extreme southern Africa. In January the situation is reversed. Temperatures in equatorial areas are less extreme due to the cooling influence of rain, and considerable cloud cover.

More marked than regional temperature contrasts are differences between the rainy months and those with little or none. Near the equator there is generally no dry month, and total rainfall ranges from 60 in (1,550 mm) to as much as 400 in (10,000 mm) annually. North and south of the equator the rain generally comes in the hottest months. The dry season becomes longer and fiercer near the virtually rainless Sahara and Kalahari deserts. In the extreme north and south, the climate is Mediterranean, with hot, dry summers and warm rainy winters.

Vegetation. Broadly speaking Africa has four main vegetation zones: a central equatorial zone of rain forest sandwiched to the north and south by successive zones of grassland, desert, and drought-resistant Mediterranean-type trees and scrub (found in the extreme north and south of the continent). On the whole, the vegetation zones north of the equator are much more extensive than their southern counterparts.

Vegetation varies with rainfall. Dense tropical rain forest covers 10% of the continent in a belt across western equatorial Africa. Growing to heights of over 150 ft (46 m), the trees form a canopy almost excluding sunlight from the ground below. These rain forests

The Kariba dam on the Zambezi River has created an artificial lake of more than 2,000 square miles. It was completed in 1959 and provides electricity for Zambia and Rhodesia.

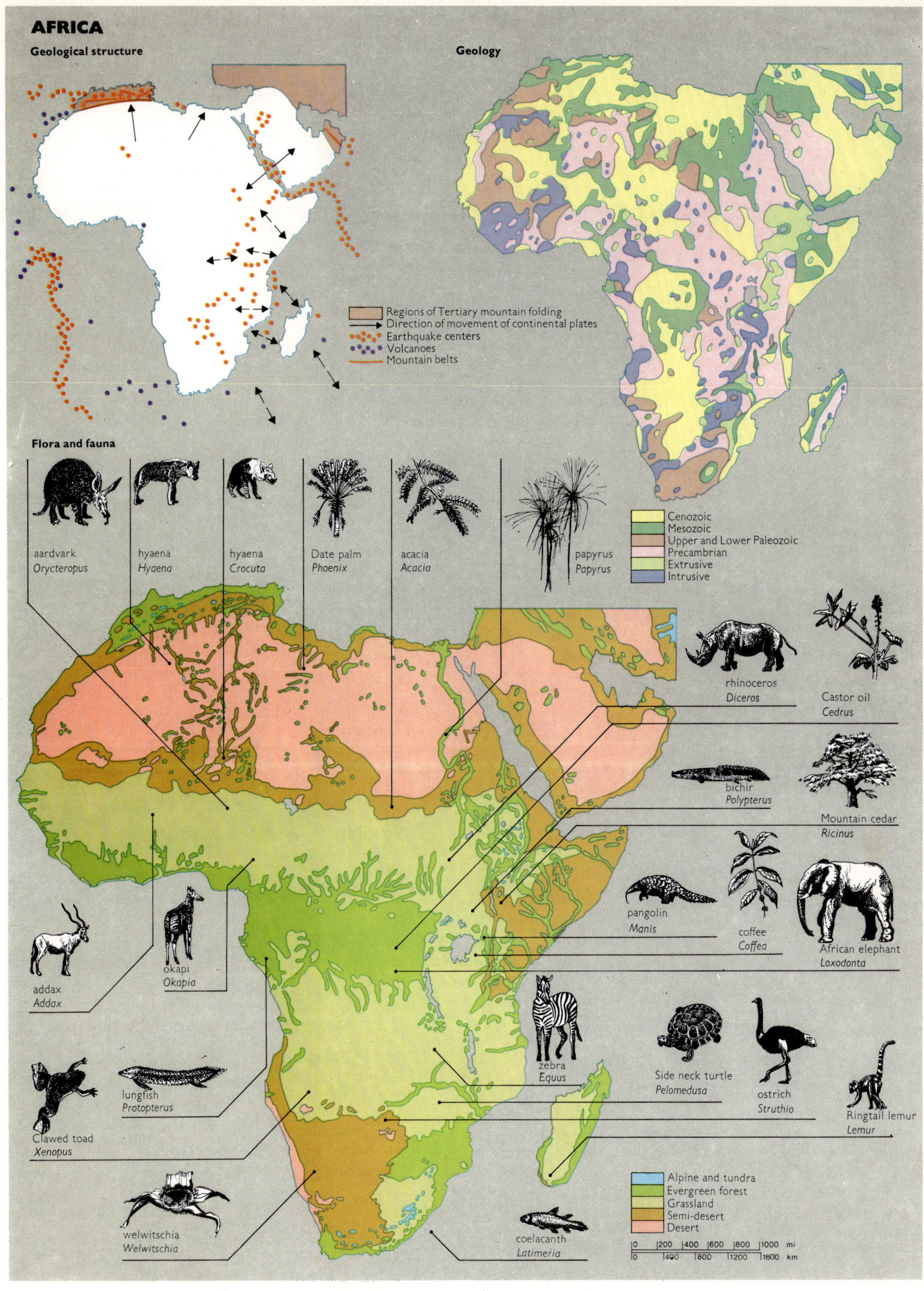
AFRICA
Geological structure
Regions of Tertiary mountain folding
Direction of movement of continental plates
Earthquake centers
Volcanoes
Mountain belts
Geology
Cenozoic
Mesozoic
Upper and Lower Paleozoic
Precambrian
Extrusive
Intrusive
Flora and fauna
aardvark
Orycteropus
hyaena
Hyaena
hyaena
Crocuta
Date palm
Phoenix
acacia
Acacia
papyrus
Papyrus
rhinoceros
Diceros
Castor oil
Cedrus
bichir
Polypterus
Mountain cedar
Ricinus
pangolin
Manis
coffee
Coffea
African elephant
Loxodonta
addax
Addax
okapi
Okapia
zebra
Equus
Side neck turtle
Pelomedusa
ostrich
Struthio
Ringtail lemur
Lemur
lungfish
Protopterus
Clawed toad
Xenopus
welwitschia
Welwitschia
coelacanth
Latimeria
Alpine and tundra
Evergreen forest
Grassland
Semi-desert
Desert
0 200 400 600 800 1000 mi
0 400 800 1200 1600 km

The Fulani are cattleherders and nomads whose territory stretches from the Upper Nile to Senegal. They are an Islamic people who have been at times a great power in Northern Africa.

provide valuable hardwoods such as mahogany and African walnut.

North, south and east of this belt are vast tracts of drier tropical grassland (savanna). As these become drier, trees are more scattered, giving way to dry brown grassland which puts forth new, tall, green growth during the rains. Acacias and the drought-resistant baobab tree are common.

Savanna is replaced by thorn-bush and semi-desert which finally degenerates into true desert in such arid areas as the Kalahari Desert and the Sahara Desert – the largest desert on earth.

Areas with a Mediterranean climate have typically Mediterranean vegetation, which features tough-leaved trees and shrubs adapted to conserving moisture during the long dry summers.

Natural vegetation varies with local climate. For example, in mountainous East Africa and Ethiopia, savanna gives way at about 6,000 ft (1,830 m) to warm temperate forest. Higher still are bamboo forests, then cool grasslands, and lastly afro-alpine vegetation.

Wildlife. North of the Sahara, Africa has a Mediterranean-type fauna. Here, mammals include porcupines and Barbary apes. Many animals have found it impossible to cross the formidable desert barrier of the Sahara, but some species are actually adapted to life in the desert, like the fernec, a desert fox.

The open savanna south of the Sahara offers excellent feeding grounds for enormous herds of antelopes, giraffes, zebras, and other herbivores; and for the lions, leopards and hunting dogs which prey upon them. Here too, live rhinoceroses and African elephants – the world's largest land mammals. But big game hunting and poaching have slashed herd sizes and the pressure of human population has helped to confine big game largely to 60 or more national parks and game reserves in Kenya, Uganda, Tanzania and South Africa.

The equatorial forests possess hippopotamuses and an arboreal fauna, including gorillas, chimpanzees and monkeys.

Africa is most famous for its big mammals, but it is also rich in birds (including the ostrich, flamingo, and stork) and in reptiles, including the Nile crocodile.

The People

Negroes and Caucasians. With an estimated population of 336 million, Africa has nearly 10% of the world's population. The continent is the original home of the world's Negro peoples and has an enormous variety of languages and cultures. Thus although 95% of the population is classed as African there is no such thing as a typical African.

Africa's peoples mostly fall into two broad racial divisions: the Hamites, Semites and other mainly Caucasian groups north of the Sahara, and those of Negro origin to the south. Most of the northerners have narrow noses and lips and wavy hair. The majority are Moslem, but have had close contact with Mediterranean Europeans.

Negro peoples account for about 70% of the population. They tend to have dark skin, frizzy hair, thick lips and broad noses, but there are marked regional differences. East African Negroes have relatively thin lips and noses through centuries of interbreeding with Caucasian peoples from North-East Africa. Sudanese Negroes from West Africa have more pronounced Negroid features. The many different Negro ethnic groups include the Masai, the Kikuyu and the Zulu of eastern and south-eastern Africa who are generally described as Bantu-speaking.

Three peoples of central-southern and southern Africa, are now in decline. The Pygmies of the Congo forest now number only some 100,000; the 50,000 or so Bushmen of the Kalahari Desert, primitive hunters, and the 40,000-odd Hottentots of South West Africa now face extinction. Bushmen and Hottentots have yellowish skin, high cheekbones and kinky hair. Malagasy's people are largely of Indonesian descent.

The descendants of European colonists form important minority groups and number in all over 5 million, most prominent in northern and southern Africa and in higher and healthier parts of tropical Africa (Kenya, Zambia, Rhodesia). Asians, especially Indians, are found in East Africa and in Natal in South Africa.

The position of many non-African groups has been affected by the emergence of politically independent states. In the north and east, the number of Europeans has declined considerably, while the economic standing of Asians in East Africa, and Lebanese and Syrian traders in West Africa, has worsened. At the same time, Europeans in the Republic of South Africa and more recently in Rhodesia have strengthened their position by implementing *apartheid*, the doctrine of 'separate development' of the white and coloured populations.

Settlement. Though the overall population density is very low, averaging only 30 per sq mi (11 per km^2), it is unevenly distributed. Huge areas like the Sahara are virtually uninhabited. Yet there are extraordinary concentrations of people in limited areas, as in the Nile Valley in Egypt, the home of 34 million people. The population explosion in such fertile but often economically undeveloped areas is putting an intolerable pressure upon the land.

Villages and towns are the main areas of population increase. There have long been important cities like Cairo and Alexandria in Egypt, and Timbuktu and Kano in West Africa, as well as Arab trading centres like Zanzibar off the East African coast. But such historic centres are relatively few; the impetus to urban growth came with the Europeans and, to a lesser extent, the Asians. For example, Nairobi, now the capital of Kenya with a population of over 500,000, was established less than a century ago on an uninhabited site as a railway headquarters for East Africa. Economic development, especially of minerals has resulted in large towns in such areas as Katanga (Zaire) and the adjacent copper belt of Zambia.

The Economy

Africa's economy suffers from two superficially contradictory problems – too many people and too few. Many poorly developed African countries face the prospect of doubling their populations before the end of the century without being able to double their food supply. Nothing can avert disastrous famines except a rapid acceleration in food

AFRICA
political
CITY population more than 1,000,000
CITY population more than 500,000
CITY population more than 100,000
City population less than 100,000
INDEPENDENT AFTER 1945:
former British territory
former French territory
former Italian territory
former Belgian territory
former Spanish territory
railways
roads
tracks
airport
scale 1 : 40,000,000
0
500
1000 st. miles
MEDITERRANEAN SEA
BLACK SEA
CASPIAN SEA
Aral Sea
RED SEA
PERSIAN GULF
GULF OF ADEN
GULF OF GUINEA
ATLANTIC OCEAN
INDIAN OCEAN
MOZAMBIQUE CHANNEL
Tropic of Cancer
Tropic of Capricorn
Equator
MOROCCO
ALGERIA
TUNISIA
LIBYA
EGYPT
SPANISH SAHARA
MAURITANIA
MALI
NIGER
CHAD
SUDAN
SAHARA
SENEGAL
GAMBIA
PORT. GUINEA
GUINEA
SIERRA LEONE
LIBERIA
IVORY COAST
UPPER VOLTA
GHANA
TOGO
DAHOMEY
NIGERIA
CAMEROON
CENTRAL AFRICAN REPUBLIC
ETHIOPIA
FR. TER. OF AFARS AND ISSAS
SOMALIA
KENYA
UGANDA
EQ. GUINEA
GABON
CONGO
ZAIRE
RWANDA
BURUNDI
TANZANIA
ANGOLA (PORT.)
ZAMBIA
MALAWI
MOZAMBIQUE (PORT.)
RHODESIA
NAMIBIA
SOUTH WEST AFRICA
BOTSWANA
SOUTH AFRICA
SWAZILAND
LESOTHO
MALAGASY REP.
MADRID
MARSEILLE
ROME
ATHENS
BAKU
TEHRAN
JERUSALEM
MECCA
Aden
AZORES (P.)
Madeira (P.)
CANARY ISLANDS (Sp.)
LAS PALMAS
Cape Verde
Ascension (Br.)
St. Helena (Br.)
Fernando Poo (E.G.)
Santa Isabel
Príncipe (P.)
São Tomé (P.)
Annobon (E.G.)
Cabinda (P.)
ALDABRA ISL. (Br.)
FARQUHIR ISL.(Br.)
COMORES ISLANDS (FR.)
Mauritius
Réunion (Fr.)
Socotra
Cape Guardafui
Cape of Good Hope
Corsica
Sardinia
Sicily
Malta
Crete
CYPRUS
BALEARIC ISLANDS
Gibraltar (Br.)
Ceuta (Sp.)
TANGIER
RABAT
CASABLANCA
FEZ
MEKNES
MARRAKECH
ORAN
ALGIERS
ANNABA
CONSTANTINE
Bizerte
TUNIS
Sfax
TRIPOLI
Zuwārah
BENGASI
Salûm
ALEXANDRIA
CAIRO
PORT SAID
SUEZ
ASYUT
ASWAN
Nile
Sidi Ifni
Béchar
Touggourt
Tindouf
Adrar
Aiun
Villa Cisneros
F Dérick
Marzūq
Nouadhibou
Nouakchott
Senegal
DAKAR
Bathurst
Bissau
CONAKRY
FREETOWN
Monrovia
Tombouctou
Niger
BAMAKO
Ouagadougou
Niamey
Agadès
KANO
ABIDJAN
KUMASI
ACCRA
Lomé
Porto-Novo
IBADAN
ABEOKUTA
LAGOS
PORT HARCOURT
Lake Chad
Fort Lamy
Abéché
Wadi Halfa
PORT SOUDAN
OMDURMAN
KHARTOUM
Kassala
Massaua
ASMARA
El Fasher
El Obeid
White Nile
Blue Nile
ADDIS ABABA
Djibouti
Berbera
Wau
Juba
MOGADISCIO
DOUALA
YAOUNDÉ
BANGUI
Bata
Libreville
Zaire (Congo)
Mbandaka
KISANGANI
Entebbe
KAMPALA
Kigali
Bujumbura
Lake Victoria
NAIROBI
MOMBASA
Pemba
Zanzibar
DAR ES SALAAM
BRAZZAVILLE
KINSHASA
Pointe Noire
Matadi
KANANGA
Kalemie
Ujiji
Tabora
Lake Tanganyika
LUANDA
Benguela
Nova Lisboa
Moçâmedes
LIKASI
LUBUMBASHI
NDOLA
Kabwe
LUSAKA
Zambezi
Lake Malawi
Lilongwe
Zomba
BLANTYRE
Mozambique
Maramba
SALISBURY
BULAWAYO
Beira
Limpopo
Walvis Bay (S.A.)
Windhoek
Lüderitz
Orange River
Gaberone
PRETORIA
LOURENÇO MARQUES
Mbabane
Mafeking
JOHANNESBURG
Kimberley
BLOEMFONTEIN
Maseru
DURBAN
CAPE TOWN
EAST LONDEN
PORT ELIZABETH
Diego Suarez
Majunga
Tamatave
TANANARIVE
Tuléar

production, and this calls for radical changes in Africa's socio-economic structure.

A major factor limiting economic growth, however, is under-population. Many countries could develop economically if they had more people to provide mass markets, labour and services. Indeed, economically successful Rhodesia and South Africa are large-scale importers of labour. The Witwatersrand district of South Africa, for example, has drawn heavily on other parts of southern Africa. Adverse physical conditions and the lack of resources explain the general sparsity of population. Vast areas seem unlikely ever to support large numbers. Where populations increase more rapidly than prospects for economic development, serious problems such as soil exhaustion arise. Agricultural education has an important role to play in such areas.

Agriculture. Farming has long been the main occupation and is likely to remain so. It was traditionally of two kinds: crop-growing on well watered soils, and nomadic stock-raising on the dry grasslands. There are now increasing attempts to combine the two. A more significant division is that between subsistence and commercial farming.

Subsistence cultivation using simple and even primitive methods persist widely. Shifting cultivation (bush fallowing) is practised over vast areas of tropical Africa. At the beginning of the rainy season, a small clearing is made in the forest by cutting down and then burning the undergrowth and all but the largest trees. The ash is spread over the soil and mixed crops such as yams, sweet potatoes and bananas are grown. Since the conditions are favourably hot and humid, the crops are ready for harvest within weeks. Similarly in grassland areas, plots are cleared and sown with root-crops such as yams or cereals like maize or, in drier areas, millet. With declining fertility, the small farms are abandoned after one or two seasons, and the farmers move on to repeat the whole process.

Some African peoples, such as the Fulani of interior West Africa and the Masai of East Africa, rely for food entirely on stock-rearing. Cattle, sheep and goats are all raised, and Africa has nearly one-third of the world's goats. But domestic livestock are locally subject to such debilitating diseases as rinderpest, tuberculosis and trypanosomiasis, spread by the tsetse fly.

Despite the continued dominance of subsistence agriculture, commercial farming is of increasing importance in tropical Africa. Among the chief crops, usually grown for export, are coffee, cocoa, cotton, peanuts and sisal. Forest trees whose fruits are collected for domestic use and export include the oil-palms of West Africa and the Congo Basin. Some pastoral tribes have become commercially-minded and now sell their surplus animals to centrally located meat-canning plants.

Outside the tropical lowlands, farming has followed rather different patterns and Europeans have played important roles in its development, especially in northern Africa, the Kenya highlands and parts of Tanzania, Zambia, South Africa and Rhodesia. Livestock can often be raised successfully and cereals like wheat and maize, and crops such as tobacco and tea, have done well. South Africa, like the Mediterranean countries (Morocco, Algeria and Tunisia), has developed wine-producing and fruit-growing.

Forestry. About 20% of Africa is covered by forest; this represents 25% of the world's total forest area, but the continent as yet produces only 1–2% of the world's timber. However, countries like Ivory Coast and Gabon may be among the world's major sources of timber by the end of the century.

Extensive areas of virgin forest remain. Their preservation, especially in tropical areas, is encouraged to protect water supplies and prevent soil erosion, and also because of the influence of forest upon climate. The potential value of the timber and such forest products as gums and resins is also recognized.

Since the 1940s, there has been greatly increased output of tropical hardwoods such as mahoganies used as cabinet woods, and of veneer woods like okume. The leading producers include Ivory Coast, Nigeria, Ghana and Gabon.

The savanna woodlands are of little commercial value. Only relatively small areas of forest have survived in non-tropical Africa because of ruthless clearance and burning. Here, as elsewhere, new policies of conservation, afforestation and scientific management are having important economic consequences.

Fisheries. Although Africa has a long coastline and many lakes and rivers, it has only a limited area of continental shelf, and so fishing is relatively unimportant. Africa produces only 6% of the world's total catch and methods are mostly simple and inefficient with commercial fishing only a recent development. The chief sea-fishing grounds are the Agulhas Bank off the coast of South Africa, the banks off South West Africa and Angola, and those off Morocco and Mauritania in the north-west. The Republic of South Africa is tenth among the world's fish-producing countries. Catches consist mainly of anchovy, pilchard, sardine, tuna and mackerel.

Africa's extensive inland waters, with at least 2,000 distinct species of fish, are increasingly exploited, and fishponds have been established in many areas, notably in Zaire.

Minerals. Mining now ranks only after agriculture in value of production, with at least a million workers, including more than 100,000 non-Africans; it is the most highly organized sector of the economy.

The industry is concentrated mainly in the Republic of South Africa, Rhodesia, Zambia, the Shaba (Katanga) district of Zaire, Nigeria, Ghana, Liberia and other countries of West Africa and along the Mediterranean. Minerals account for more than a third of the

Bantu women in Transvaal, South Africa. The Bantu migrated to South Africa from the north at about the same time that Europeans began settling the coast.

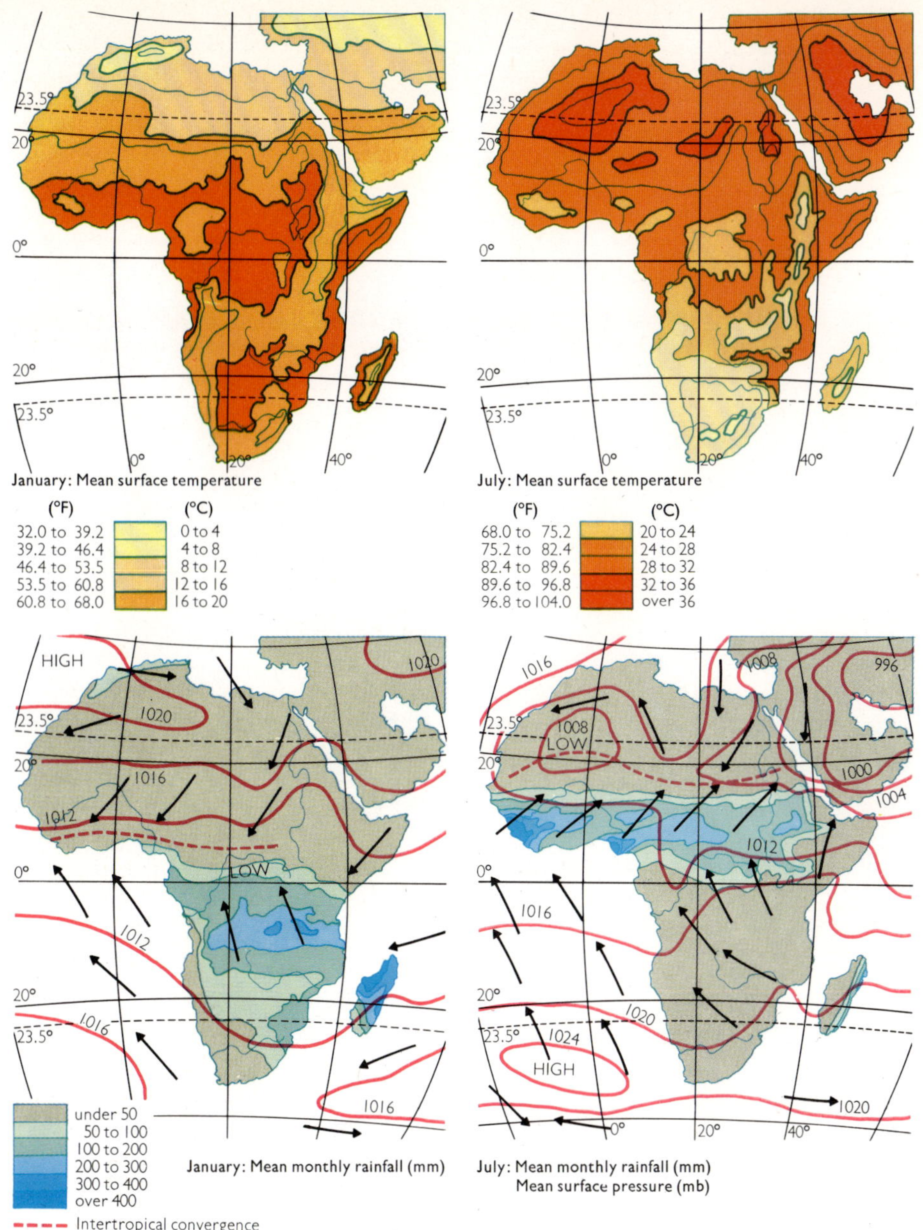

continent's total value of exports, but Africa's share of total world mineral exports is under 10%. South Africa dominates production, accounting for over 40% of the total.

Until recently, gold, diamonds and copper represented a very high proportion of Africa's output and it yielded nearly all the world's diamonds and two-thirds of its gold. Other important minerals included asbestos, chromite, cobalt, iron ore and phosphate. Since the mid-1950s, oil reserves in such countries as Algeria, Libya and Nigeria have made Africa an important producer of petroleum.

Manufacturing. Many traditional crafts thrive throughout Africa, but commercial manufacturing is generally a recent development usually associated with the processing of minerals or agricultural products. Development was greatly stimulated by the reduction in supplies from other parts of the world during World War II, especially in South Africa. Since then, political independence has promoted the establishment of manufacturing enterprises in many countries.

Transport. Outside northern and southern Africa, there is no real railway network but rather a pattern of unconnected lines joining ports to inland areas of minerals or cash crops. Africa has 42,100 mi (68,000 km) of railway track – about 5% of the world total – with much congestion in places since the volume of freight traffic has trebled since 1940 while the railway mileage has barely altered.

Since the early 1900s, many roads have been built. But of a possible 1 million mi (1·6 million km) of motorable roads less than 10% are all-weather roads. In general, the roads supplement the railways. Air services have been developed considerably since the early 1950s. Africa now has many international airports and a growing internal network including flying doctor services.

International Trade. Africa is less involved in international trade than most other continents and accounts for well under 10% of world trade, though it is heavily dependent on imports.

For the most part the old associations with European countries have been maintained and even, in some cases, strengthened. The pattern is largely an exchange of African primary products for the manufactured goods and expertise of Europe, the Soviet Union and the United States. In recent years, there has been a remarkable increase in trade with Japan. Chinese involvement in Africa is apt to be exaggerated.

Africa Hall in Addis Ababa, the headquarters of UN economic commission for Africa. Addis Ababa, the capital of Ethiopia, has emerged as a major centre of the movement towards African unity.

AZORES

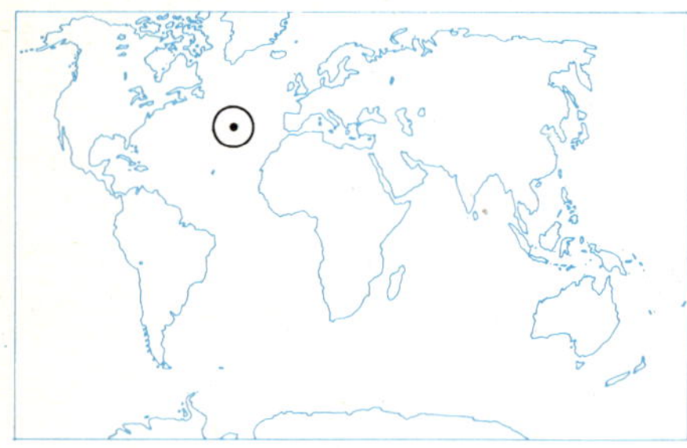

Azores. This mid-Atlantic Portuguese archipelago comprises nine islands divided into three separate groups. The population (291,000) lives mainly by farming and fishing. Ponta Delgada is the capital and chief port.

MADEIRA

Madeira is a Portuguese island group south-east of the Azores known for its wine and embroidery. Funchal, on Madeira, is the capital and port. The population is about 253,200.

MOROCCO

Area: 166,000 sq mi (429,940 km²)
Population: 15,379,259
Population Growth Rate: 3·0%
Capital: Rabat, pop 460,000
Language: Arabic. Berber, Spanish and French also spoken.
Religion: Moslem (Sunni)
Currency: Dirham = 100 Moroccan francs

Morocco in north-west Africa stands at the westernmost limit of the Arab world, remote from the mainstream of Arab nationalism. Although a member of the Arab League, this mountainous kingdom has never been violently militant in its cause, but has favoured close economic links with the United States and Western Europe, only a few miles distant across the Strait of Gibraltar. The French and Spanish are particularly influential in Morocco. It has the largest French community in North Africa, which helps to provide much-needed managerial expertise and teaching skills but is resented by those wishing to completely Arabize the country. Spain has small outposts on Morocco's Mediterranean coast comprising Spanish North Africa.

Among Morocco's most urgent problems are the rapid growth of population, the modernization of the economy, providing work for the increasing number of well-educated Moroccans, and the establishment of stable democratic government.

European Penetration. Once part of the great Almoravid Empire, and later a pirate centre, Morocco had become weak and lawless by the 19th century. In 1906, the great powers of Europe guaranteed the country's independence, but France was given special rights which enabled her to establish a protectorate, part of which she ceded to Spain. Tangier was later given international status.

In 1955, the French were compelled to restore the Sultan, whom they had exiled in 1953, and to grant independence in 1956. At the same time, Tangier and the Spanish zone (except for Ceuta, Melilla and a few small islands) became part of Morocco. Spain handed over the small Ifni enclave in 1969.

The new Morocco emerged as a supposedly constitutional monarchy: but King Hassan II attempted to extend his powers in the late 1960s and this emperilled his position. Although Hassan commanded wide support, especially in the rural areas, military coups and his assassination were attempted in 1971 and 1972, plots in which the revolutionary government of Libya may have been involved.

Mountains and Plains. Morocco is a mountainous country. In the north, an arc of high ridges and broad valleys extends from Ceuta to Melilla. These are the Er Rif Mountains, where Abd-el-Krim long defied the French and Spanish in the 1920s, and where Mt Tidiquin rises to 8,054 ft (2,455 m). South of these mountains, the Atlas Mountains sweep across central Morocco, embracing in the east large plateaux and high plains that extend into Algeria, and descending abruptly in the south-west of Morocco's Atlantic coast. South-west of the Middle Atlas is the great central range, the High Atlas containing Jebel Toubkal (13,664 ft; 4,165 m), Morocco's highest mountain.

In the south, the Anti-Atlas, more rugged tableland than mountain chain, fall away to the Wad Dra. Morocco's Saharan border with Algeria was not determined until 1971.

Many streams and rivers descend from the Atlas. Some, like the Dra and Sous, flow to the Atlantic; others, like the Moulouya, Morocco's longest river, to the Mediterranean. Many of the small streams disappear in the desert.

Climate and Vegetation. The climate is Mediterranean in the north, with hot summers and mild winters, but Saharan in the south-east with long intensely hot summers and hot dry winds. Between both areas is the relatively cool zone of the Atlas ranges. Along the narrow Atlantic coastal strip, the hot Saharan weather is moderated by ocean breezes; and mild temperatures, high humidities and falls of dew permit the cultivation of wheat in places that get very little rain throughout the year.

Rainfall varies from place to place. The plains of the Rharb, in the north-west, have an annual rainfall of about 20 in (500 mm), while the high ridges of the Er Rif sometimes receive 71 in (1,800 mm) yearly. Towards the south and east rainfall is unreliable. Parts of the Moroccan Sahara average under 4 in (100 mm) of rain a year, but land that is drought stricken for many months can be flooded by a sudden downpour.

Forests of evergreen oak, Atlantic cedar, juniper and thuja clothe many mountain slopes, although centuries of felling and goat-grazing have thinned or otherwise impoverished many stands of trees. The plains are covered mostly with rough grass and bushes, with thorny scrub in the east and south. The plain of Sous is known for its thorny argan trees. The desert has date palm oases such as Tafilet, the original home of the present royal dynasty (the Filalian dynasty of the Sherifs).

The People. This north-western corner of Africa has been the halting place for several peoples and cultures, and Moroccan culture is a blending of influences from South-West Asia, Africa and Europe. The ancient indigenous Berber culture still survives in modified forms, mainly in isolated groups in the remoter mountain and desert areas. Most of the Berbers, however, have merged with the Arabs who invaded the country between the 8th and 14th centuries. Andalusian Moslems and Jews who took refuge in Morocco after the Christian reconquest of Spain in the 1400s have also left their mark on city life. About

100,000 Frenchmen live in the country, a legacy of early 20th-century French colonialism.

Today most Moroccans are Arabic-speaking, although some 35% speak a Berber tongue and secondary-school lessons are in French. Islam is the established state religion. Education is compulsory from 7 to 13. Some 1·17 million children attend primary schools; 300,000 attend secondary schools; and there are about 13,000 students at the three universities.

Government takes the form of a constitutional monarchy where the king has the power to appoint ministers (including the prime minister), Veto any legislation and dissolve Parliament. But royal power has been slowly eroded during the last few years.

Towns and Countryside. Morocco features many picturesque old Moslem cities, whose walls embrace three invariable elements: a large 'Friday' mosque, a *Kasbah* (the fortress-palace of the sultan) and a *souk* or market-place. The French and Spanish tended to build their suburbs and new towns apart, as at Casablanca, leaving the old as it was. These new suburbs and cities retain much of their French or Spanish character. The cities have attracted an increasing flow of people from the countryside, but lack of proper housing has led shanty-towns to spring up in the shadow of the skyscrapers, and the drift to the cities presents a big social problem, for perhaps 20% or more of the urban population is unemployed.

In the countryside, the typical farmer's house is the *nouaïel*, a round structure of wood and straw-thatch, or the adobe *diour*. Wandering herdsmen use tents made from goats' hair, wool, or wild palmetto fibres.

Moroccan Farming. Agriculture is by far the most important activity, supporting about 70% of the population. Its patterns vary from the traditional husbandry of the Berbers, whose crops include fruits, nuts, vegetables and the hard wheat used in *Kous-Kous*, the most common North African dish, to the large farms, originally laid out by the French, which produce soft wheat, citrus fruits, vines, and the early vegetables that fetch a high price in the French market. Some Berbers are semi-nomadic, following their flocks of sheep and goats from the high pastures in summer down to the lowlands in winter. The true nomad, wandering perhaps 60 mi (96 km) a year in search of pasture for his camels, survives only in the more arid areas.

While the land was less extensively colonized than in neighbouring Algeria, foreigners still produce most of Morocco's export crops. But land held by foreigners is now being taken over for co-operative management and mechanization. Other agricultural projects include damming perennial rivers of the Middle and High Atlas for irrigation (and power).

Forests and Fisheries. Cedar and other timber, cork and tannin are the main forest products. One of the chief woodworking factories is on the Rharb plain behind Rabat.

The cool Atlantic waters abound in fish, especially sardines, which are harvested by fleets operating from Agadir, Safi, Casablanca and other ports where there are fish canneries. The fishing industry has great potential.

Minerals and Industry. Morocco is one of the world's leading producers of phosphates, worked at Khouriba and Youssoufia, and shipped out through Safi and Casablanca. Small quantities of coal, lead, zinc, manganese and cobalt are also mined, and Morocco now has a share in Algeria's Gara-Djebilet iron ore deposits since a 1971 agreement for their joint exploitation and export by way of Morocco.

Morocco's small oil fields, opened many years ago, have a declining output, but natural gas has been found in the Sebou valley and offshore oil prospecting goes on. Three-quarters of the power used by Morocco comes from hydro-electric installations.

Industry mainly concerns the processing and packing of agricultural products for export. To this is being added the production of consumer goods for the local market and such industries as car and truck assembly, and cement, chemicals and textiles, the manufacture of which offer new job opportunities for the unemployed.

Transport. Morocco has roughly 15,500 mi (25,000 km) of surfaced roads. The chief standard-gauge railway links Casablanca with Rabat, Tangier, Fez and Oujda, and then continues through Algeria to Tunis; another links Casablanca with Marrakech, whence two important branches serve the phosphate and other mining areas. The chief ports are Casablanca, known in Arabic as Dar el Beida, and Safi, Mohammedia and the free port of Tangier. There are four international airports.

International Trade. Morocco, like Algeria and Tunisia, still has strong trade links with France. To France in particular goes the produce of the commercial farms which forms nearly half of Morocco's exports. Morocco's trade horizon is widening, however, especially in Western Europe. The kingdom also trades with the United States, and some American capital has been attracted.

Two items of growing importance are the export of labour to Europe, especially France, which helps the Moroccan balance of payments, and the expanding tourist industry.

A city gate in the ancient city of Fez, showing its characteristic Moorish architecture. Fez, situated in a valley of the Atlas mountains in Morocco, is a centre of Islamic study.

SPANISH NORTH AFRICA

Spanish North Africa comprises small Spanish toeholds on the Moroccan coast. The main ones are the port of Ceuta, under Spanish rule since 1580 and formerly capital of Spanish Morocco; and the port of Melilla, Spanish since 1496. Both cities are regarded as parts of metropolitan Spain. Also under Spanish sovereignty are Alhucemas, an island off the Er Rif coast; the Chafarinas Islands near Melilla; and the small island of Peñón de Vélez de la Gomera south-east of Ceuta.

These are all that remain of larger holdings. From 1912 until 1956, Spain held part of northern Morocco as a protectorate, despite fierce opposition from the Rif leader Abd-el-Krim, who was finally overcome by combined French and Spanish forces in 1926. From 1923 Spain also participated in the administration of the then international zone of Tangier, and had held the small enclave of Ifni, on Morocco's Atlantic coast, since 1860.

When Morocco became independent in 1956, Spain relinquished its protectorate. The international status of Tangier was abolished that same year, while Ifni was returned to Morocco in 1969.

ALGERIA

Area: 952,198 sq mi (2,466,193 km²)
Population: 14,769,000
Population Growth Rate: 3·2%
Capital: Algiers, pop 1,200,000
Language: Arabic, Berber, French
Religion: Moslem
Currency: Algerian dinar = 100 centimes

Algeria is the third largest country in Africa. From its Mediterranean coastline between Morocco and Tunisia, it extends southwards for some 1,190 mi (1,915 km) embracing a large section of the Sahara Desert. Its name derives from the capital Algiers, a Europeanization of the Arabic Al Jazair ('the islands'). Berbers lived in Algeria long before the Phoenicians, Romans, Vandals and Byzantines ruled Algeria. Today a Berber language is still spoken by 20–30% of the people.

French rule was terminated by the bloody and bitterly-fought Algerian Revolution (1954–61), and the independence of Algeria was formerly recognized by France in 1962.

Contrasting Landscapes. Northern Algeria and the Sahara are quite distinct from each other. The north is, in fact, the central section of the Atlas Mountains, and has four regions: the eastern and western zones of coastal ranges which Algerians call the *Tell* (the hill); a succession of interior plains; and further ranges to the south.

The road leading westwards from Algiers into the Tell in Oran and neighbouring *wilayas* (the former French departments) passes through the semi-arid Chéliff valley and then crosses the fertile Sahel d'Oran, a plain of vineyards, grain fields and olive groves, to reach Oran itself. South of the Chéliff valley rise the dry ranges of the Tell, which presently give way to a few high basins and interior tablelands.

By contrast, roads winding into the Tell east of Algiers pass through a far more verdant countryside where the plains and basins are less extensive, and where wheat has been grown on the slopes since Biblical times. This area is dominated by Constantine, a large city (243,558) spectacularly perched on an almost isolated chalk rock, 330 ft (91 m) high. The eastern Tell also contains some high and remote mountain regions such as the Great and Little Kabylia, where ancient villages nestle below the woods and winter snows of the higher slopes.

Behind the hilly coastlands lies a broken belt of high and wide interior basins whose streams mostly drain to inland salt lakes called *Shotts*, or *Chotts*. Beyond these grassland plains is the last mountain barrier to the Sahara. In the west, this range forms the folded hills of the Saharan Atlas; only the highest peaks have perennial streams, but large springs on the dry lower slopes water populous valleys. In the east, the massive block of the Aurès Mountains, often snow-covered in winter, glowers over the Sahara.

South of the Saharan Atlas, the wind blows warm and dry from the desert and active sand-dunes appear, cutting across the road in places. Several streams and springs from the Atlas water the northern edge of the desert, but few of them reach out far; only the Wadi Saoura sends occasional floods up to a few hundred miles from the mountains. A number of populous oases depend on these northern streams, which are used to irrigate large palm groves.

To the south, the desert presents two faces: the monotonously flat, blackened and pebble-covered plateaux contrast vividly with the red sand seas – some as large as France – which contain great mountains of sand, perhaps 1,000 ft (300 m) high.

In the extreme south, the central Saharan plains have been disturbed by a very large uplift, the Ahaggar (Hoggar) Mountains, which culminate in the volcanic massif of Mt Tahat (9,573 ft; 2,918 m), the highest peak in Algeria.

Climate and Vegetation. The most striking characteristic of the Algerian climate is aridity. In the central Sahara, there are climatic stations which do not see rain for several years on end, but farther north is a climatic zone affected, in winter, by the frontal storms of temperate latitudes. But the rains are never reliable, so that there may be several successive years of drought followed by a few with winter rain. Only in the mountains north of the Sahara and along the Mediterranean coast is rainfall certain. Annual precipitation averages of more than 55 in

The Kasbah in Algiers is surrounded by Turkish fortifications and still retains its ancient Arab flavour.

(1,400 mm) are recorded in some mountain areas.

Interior Algeria has extremes of temperature. Frost is not at all unusual during the Saharan winter, but temperatures frequently soar to 120°F (49°C) in summer. The north also has a hot, dry summer, although coastal and mountain temperatures are lower than elsewhere. Winters are mild along the coast and cool to cold in the Atlas Mountains, where snow falls.

Vegetation in uncultivated northern areas is typically a sparse scrub that has suffered generations of grazing by sheep and goats and cutting for firewood. On the higher mountains, the scrub becomes open woodland or, occasionally, denser forest of evergreen or deciduous oak. Vegetation thins southwards, first to the grassy steppe of the high plateaux and then to sparse, heath-like scrub on the northern Saharan plains. About 300 mi (480 km) south of the Saharan Atlas even this disappears.

The People. Most Algerians are descended from an ancient indigenous Berber stock which was in time leavened by other peoples including Negroes in the south. Today, a Berber language is still spoken by 20–30% of the people. From the 7th century on, a lasting cultural imprint was provided by the Moslem Arabs, who established themselves as administrators and merchants in the towns. Between the 11th and 13th centuries, pastoral Arabs also occupied most of the drier northern areas.

In isolated areas there are still small Berber communities living in the age-old manner. A Berber group with a distinctive desert culture are the Tuareg people of the Ahaggar mountain region, whose camels and nomadism were adopted from the Arabs, and who still use an ancient alphabet reputedly of Phoenician origin.

The French have also left an indelible cultural, and perhaps linguistic, mark upon Algeria. They not only subdued the country, but politically integrated it with France. Settlement was encouraged, and in time more than a million Europeans were living in Algeria, many of them Algerian-born. The picture was changed by the nationalist revolution which began in 1954; it caused perhaps a million deaths, the resettlement of about two million peasants, and the emigration to Tunisia and Morocco of about 250,000 refugees. Since independence (1962) all but about 100,000 Europeans have left.

Settlement Patterns. About 60% of the people live in the countryside, mainly in the fertile mountainous north. Some rural Algerians (largely Arab) are nomads living in small groups and sleeping in tents. Others (mainly Berber) live in villages or hamlets, some fortified. The French introduced a new design element into this pattern. Their farms were usually dispersed groups of substantial buildings retaining the French architectural style.

Algeria's large towns are limited mainly to the fertile Mediterranean coast area (site of the largest cities, Algiers, Oran and Constantine). Smaller trading centres like Ghardaïa lie as far south as the northern edge of the Sahara Desert.

Many of Algeria's ancient towns suffered under French replanning; in Algiers only the quarter known as the Kasbah retains some of its former charm. In appearance, the cities of today are mainly a French creation and until 1962 they were also largely French in population. Oran (325,000), the second largest city in Algeria, had a European (mainly Spanish) majority until the late 1950s and the largest city, Algiers (1,200,000), had a bare Moslem majority.

Although sand dunes, continually changing their shape with wind direction, characterize the desert, they only occupy about a tenth of the Sahara's 3·5 million square miles.

Young men from the countryside increasingly drift to shanty towns on the edges of the cities in search of work, often to be disappointed. Enormous numbers (perhaps one in three of the working population) are unemployed. A social relief system based on the extended family maintains these people but the usual resort of the unemployed is to seek work in Western Europe. About 500,000 Algerians now find employment in French, Belgian or German factories. Their earnings are an important part of the Algerian foreign exchange balance.

Agriculture. This provides about a third of the gross national product. Wheat, barley, grapes (for wine-making), olives, citrus fruits and dates are all important crops and stock-raising (sheep, goats, cattle, donkeys and camels) makes a significant contribution to the economy. The agricultural landscape of Algeria is largely a product of its history. The ancient Berbers of North Africa evolved a subsistence farming pattern whose primitive methods persist in the remoter highlands and oases. In the mountains, peasant farmers cultivate small fields of barley and hard wheat and a number of tree crops; in the oases their crops are chiefly dates, cereals and vegetables.

During French rule, Europeans appropriated large parts of the countryside, particularly fertile regions near the coast where they commercially exploited parcels of land that ranged in size from small vineyards to vast estates. The economy developed almost entirely in European hands, since most of the exports – at that time predominantly agricultural – were produced on their farms.

The swift changes in politics, population and the economy since the 1950s have made their mark even in the remotest parts of the country. The rapid growth in population has been accompanied by near stagnation in agricultural production. Over-use of the land has led to serious soil erosion in many places; and ancient techniques, lack of capital, and outmoded land tenure systems has made it impossible for farmers to meet the new challenges. Also, resettlement during the revolution disrupted agriculture in many areas.

The government is now trying to introduce collective farming, withdraw marginal land from cereal cultivation, and plant more tree crops in the hope of preventing soil erosion. The former lands of the Europeans and some other parcels of better land – amounting to a

The oil field at Hassi Messaoud in the Sahara was one of Algeria's first and was opened in 1957. It is connected by oil pipeline to Bougie and by gas pipeline to Oran.

quarter of the country's agricultural land or 7·5 million ac (over 3 million ha) – have been nationalized and placed under worker self-management schemes. These 'auto-gestion' schemes have surprised many by their high levels of production, particularly of citrus fruits and cereals. But the initial levels of production of processed farm products have not been maintained.

Minerals. Although less richly endowed in phosphates than neighbouring Tunisia and Morocco, and not so far ahead in the development of the iron resources of the western Sahara as Mauritania, Algeria is one of the richer oil and gas countries of the world.

The first oil discoveries were made in 1956 around Hassi-Messaoud in the eastern Sahara, gas being found around Hassi-R'Mel farther west. The fields were rapidly developed and pipelines laid to the Algerian and Tunisian coasts. Some remote fields remain to be tapped.

Small towns have mushroomed around the principal oil fields and the oil ports of Arzew (east of Oran) and Skikda have also been growing rapidly. Highly-skilled labour is needed, and until recently this has had to be mainly foreign. Gas and petroleum deposits mean supplies of cheap fuel for Algerians, and a valuable income in taxes, royalties and profits.

Industry. To diversify the economy and combat unemployment, the government has focused its attention increasingly on industry. Plants for producing fertilizers, plastics and chemicals are now operating at Arzew and Skikda. Jobs are also being created by developing heavy industry such as the large iron and steel complex at Annaba, opened in 1970 and based on local low-grade iron ore.

There is also room to expand the labour force in the established industrial sectors that were disrupted by the struggle for independence and its immediate aftermath. By the 1970s, Algeria's main industry was still raising farm produce for the local market. Other industries, such as car and truck assembly, encouraged by the government, were the subsidiaries of foreign concerns, especially Russian and American.

Algeria aimed to be fully developed and economically independent by 1980, a target date called *Horizon Quatre-Vingt* (Horizon Eighty). The programme involved three stages: the pre-plan, which ran to the end of 1969, intended to decolonize and to reduce Algeria's dependence on France; the 4-year plan, 1970–73, concentrating large investments in primary industry; and a 7-year plan, beginning in 1974, intended to complete the full economic development and independence of Algeria. Among the ambitious second stage projects were dams to control the scarce water supply and provide irrigation and hydro-electric power. Large areas were also re-forested. In order to achieve these goals, the government pegged salaries and wages at 1962 levels.

Transport. Algeria was bequeathed a relatively good system of metalled roads by the French, especially in the north. The trans-Saharan metalled road reaches only to El Goléa, one third of the way to the southern border. In the eastern Sahara a metalled road runs as far as the southern limit of the Grand Erg Oriental. (Erg is the term used for an extensive sand desert in the Sahara.)

The railway network has some 2,548 mi (4,100 km) of track and is being slowly modernized. The chief seaports are Algiers, Oran, Bejaïa, Annaba, Skikda and Arzew. The national airline, Air Algeria, provides domestic and international services from the main airport at Algiers.

Trade. Algeria is striving to break its international trade out of the French mould. The pre-1962 agricultural economy was developed solely with the French market in mind. In 1969, France provided more than 44% of Algeria's imports and took more than 54% of its exports.

The French government has kept its agreement to buy Algerian wines in spite of strong opposition from French wine-growers. France, too, has been the main source of capital and technical know-how for the development of Algerian oil. Saharan gas, on the other hand, was first bought by the British gas industry. The slow weaning from France during Algeria's first decade of independence has made other West European customers, and to some extent the Soviet Union, increasingly important. The United States also is showing a keen interest in liquefied natural gas. Algeria's primary exports are now oil, gas and wine.

TUNISIA

Area: 63,362 sq mi (164,108 km²)
Population: 5,137,000
Population Growth Rate: 2·2%
Capital: Tunis, pop 642,000
Language: Arabic. French widely used.
Religion: Moslem (95%)
Currency: Tunisian dinar = 1,000 millimes

Tunisia is a young republic in a land with a long history. The site of Carthage, the powerful seat of empire which Phoenicians founded about 814 BC, can still be traced near the modern city of Tunis. The Romans made the area one of their granaries.

In 1956 the French granted Independence and in 1957, with its hereditary monarchy abolished, Tunisia became a republic; Habib Bourguiba became its president. Modern Tunisia is largely his creation.

In contrast to some Arab states, Tunisia has emerged as a practical force, concerned with self-improvement rather than grand international Arab causes. The nation has indeed had a moderating influence upon its more jingoistic fellow members of the Arab League.

Under Bourguiba's firm but enlightened rule Tunisia has made considerable social and economic progress. Some projects, such as the co-operativization of agriculture, have proved of doubtful benefit, but others have been notably successful. Great advances have been made in education and public health. Women's status, traditionally low in Arab countries, has also been improved in such matters as marriage and divorce.

The Land. Tunisia lies at the eastern end of the Atlas Mountains, which enter from Algeria to form a plateau called the Dorsale ('backbone'). This plateau extends to the Gulf of Tunis at an average height of 1,600 ft (488 m); its highest points are Jebel Chambi (5,000 ft; 1,524 m) and Jebel Zaghouan (4,260 ft; 1,298 m) at the western and eastern ends respectively. Two important rivers, the Medjerda and the Milyan, flow from the Dorsale through extensive alluvial lowlands where their waters are used for irrigation.

North of the Dorsale, the Tell Atlas enter from Algeria, swinging north-east to form the Northern Tell. Formed of limestone, sandstone and clay, these hills do not rise above 3,000 ft (914 m), but nonetheless slope steeply to the sea, limiting the northern coastal lowlands.

South of the Dorsale, the land takes on an increasingly Saharan character, with featureless lowlands and coastal and inland *sebkha* (salt flats) such as the Chott Djerid south of Gafsa. The extreme south is a region of poor steppe pasture and sandy desert. The south has no perennial streams. Those that flow occasionally sometimes cause much damage as they turbulently course through the dry valleys; but they also provide sufficient moisture for the shifting cultivator to raise occasional crops of cereals.

Climate. Tunisia has a Mediterranean climate with hot, dry summers and mild, wet winters. The coastal areas are subject to maritime influences and have smaller annual and daily variations in temperature than the interior where extremes are the rule. In winter, temperatures may fall below freezing point in the highlands. In summer, they can soar to 120°F (49°C) in the desert areas of the south.

Rainfall is almost everywhere low and unreliable, except in the north-eastern hills which receive more than 30 in (762 mm) annually. In the Medjerda and Milyan valleys, rainfall is considerably lower and may drop below the minimum required for reasonable wheat yields. Rainfall declines sharply towards the south, where 4 in (100 mm) is the annual average. There is a marked summer drought everywhere; and relatively high temperatures, even in winter, make the low and unreliable rainfall less effective. Over all but the north-eastern hills evaporation rates far exceed the total precipitation.

Vegetation and Wildlife. An estimated 6% of the country is woodland, much of which has degenerated to scrub through cutting and overgrazing. Cork oaks predominate in the Northern Tell, and Aleppo pines in the drier Dorsale. There are steppe grasslands where rainfall is less than 16 in (400 mm) yearly, and esparto grass covers large areas of the west and south. In the desert, Saharan species such as acacia and saltbush, together with the date palm in the oases, are the only vegetation. Wild animals include wild boars and semi-desert species such as antelopes, sandgrouse and locusts.

The Medina of Tunis, the central and ancient part of the city. Tunis stands next to the site of the Phoenician city of Carthage which was razed in 698 AD.

The People are predominantly Arabic-speaking. Only about 2% of the population speaks the Berber language, but most Tunisians are of ancient Berber stock although ethnologists have traced local groups descended from the Vandals and from Negro slaves. The European population, which numbered 255,300 in 1956, had declined to 40,000 by 1963 and has continued to decrease.

Tunisia is more densely populated than other North African countries. The most populous areas are the lowlands around Tunis and those extending eastwards through the Medjerda valley. Very few people live in the south. More than 40% of the population lives in the towns, the most important of which are on the coast. Tunis, the capital (642,000), is a fast-growing modern city with an ancient centre (the *medina*) and a colourful market (*souk*). It is the seat of government and a major manufacturing centre and port. Sousse is also a port and manufacturing centre and, like Sfax, engages in fishing. Bizerte is a naval base with manufacturing industries and an oil refinery. Kairouan is the major Moslem shrine in North Africa with numerous mosques and an Islamic university.

Islam is the state religion, but there are small Roman Catholic and other Christian communities.

The drive to reduce illiteracy has had considerable success, and the rate is now below 35%. There is a university at Tunis.

Agriculture. Agriculture produces just over 18% of the gross domestic product and employs more than 60% of the working population. A little more than half of Tunisia is suitable for agriculture, but most of this land is grazed and only 27% of the country is under cultivation. Dry farming is most common, and yields fluctuate because of the uncertain rainfall.

The main arable area consists of the Medjerda valley, the plain of Tunis and the Cape Bon peninsula, from which comes nearly all Tunisia's wheat, barley, wine and citrus fruits. Olives are grown on the south-east Dorsale plateau and on the east coast lowlands from north of Sousse to beyond Sfax in the south. The southern oases provide high quality dates for export, those of the Tozeur Oasis being especially esteemed.

Traditional 'gardens' are a feature of coastal east Tunisia. Field crops such as winter grain and vegetables, as well as tomatoes, melons and other summer crops, are grown between the olive and fruit trees, which are in turn shaded by date palms.

Livestock are raised mainly on marginal land that is unsuitable for crops. Sheep and goats are the most important stock, but cattle are increasing in number. On the steppes of the south, semi-nomadic people rear sheep, goats and camels, and grow crops whenever rainfall is sufficient.

French and Italian farmers changed the pattern of cultivation in Tunisia by developing large areas of the best land, especially in the Medjerda valley, and by efficient dry farming and extensive mechanization achieved high yields. The French also initiated the Medjerda Valley Scheme. Major dams now control the river and provide water storage on the Mallas and at Ben Metir and El Aroussia. The increase in farm incomes in the irrigated areas has been phenomenal.

Currently agriculture is developing mainly through careful seed selection and the intro-

In north-western Tunisia, between two spurs of the Atlas mountains, flows the Mejerda (Majardah) river which provides water for Tunisia's major wheat-producing area.

duction of new varieties, the object being to ensure Tunisia's self-sufficiency in cereals. New crops such as sugar-beet and cotton have been introduced.

Forestry and Fisheries. Cork is the main forest product and some 8,000 tons are produced annually from more than 8 million trees. Pine, eucalyptus and other varieties provide raw materials for extractive industries – for tannin, gums and eucalyptus oil, for example. Soil erosion is being checked by reforestation.

Some 60% of the population lives within 30 mi (48 km) of the coast, and there is thus a ready market for the large fishing industry centred on Sousse, Sfax and Tunis, all of which have canneries.

Minerals. Tunisia is one of the world's major producers of phosphates. These are extracted mainly in the Gafsa region. Expansion should be encouraged by the new processing plant at Gabès and port facilities at Sfax. High quality iron ore is mined at Djerissa and Douaria, and exploitation of deposits at Jebel al'Urq south of Gafsa began in 1968.

Sufficient oil for local needs and modest export is produced at El Borma and refined at Bizerte. Natural gas is available and fields in the Cape Bon area supply the city of Tunis. Small quantities of lead and zinc are mined, and there are minor deposits of potash and mercury. The development of mineral resources has stimulated port and railway construction.

Industry based on the processing of local raw materials such as wool, leather, esparto and food crops continues to expand. Olive oil is prepared for export at all the major eastern centres, and factories for processing farm produce are an integral part of the Medjerda valley development. The range of industry includes sugar refining (Béja), chemical processing (Gabès, Sfax, El Metlaoui), cellulose and paper production (Kasserine), oil refining (Bizerte), steel (Menzel Bourguiba), and the assembly of vehicles and electronic equipment (Sousse and Tunis).

Tourism is the fastest growing industry and a major earner of foreign currency. The climate is comfortable throughout the year at such coastal resorts as Monastir and Hammamet, and charter flights to local airports make these centres readily accessible to European tourists. A major tourist complex has been built on Djerba, the legendary island of the lotus-eaters and once a corsair haven.

Transport. Most of Tunisia's roads and railways were built by the French. The state-owned railways include the electrified line from Tunis to La Marsa, the capital's most favoured coastal suburb. The chief seaports are Tunis-Halq el Wadi, which handles about 60% of Tunisia's trade, Bizerte, Sousse and Sfax. Tunis has an international airport and there are airfields at Djerba, Sfax, Gabès, Monastir and Tozeur.

International Trade. Machinery and equipment are major imports, and in some years grain has had to be imported following bad harvests after low winter rains. The main exports are phosphates, olive oil, wine, citrus fruits and iron ore.

LIBYA

Area: 679,358 sq mi (1,759,537 km²)
Population: 2,010,000
Population Growth Rate: 3·1%
Capital: Tripoli, pop 247,000
Language: Arabic. English and Italian in commerce.
Religion: Moslem (Sunni) 97%. Christian
Currency: Libyan dinar = 1,000 dirhams

Libya is an oil-rich Arab republic in North Africa, lying between Tunisia and Egypt. Since 1959 oil has transformed this thinly peopled desert nation from a poor and, agriculturally, largely sterile land into one of the wealthiest in Africa with the continent's highest per-head Gross National Product. By the late 1960s Libya still had perhaps the lowest population density and one of the lowest populations in Africa, but food and medicine partly financed by oil were helping to support Africa's fastest population growth rate.

In 1969 army leaders deposed the king and proclaimed a socialist republic; they then closed American and British bases, nationalized foreign businesses, raised the price of Libyan oil, and worked for a union with Egypt.

The Land. Libya is a country of huge arid and semi-arid basins, except for narrow Mediterranean coastal areas in the north-east and north-west. Over 90% of the land is desert and semi-desert. There are rock and gravel deserts, and sand 'seas' cover 30% of the Eastern Muhafadat (Eastern Provinces, formerly Cyrenaica) and 20% of the Southern Muhafadat (Southern Provinces, formerly Fezzan).

The highest mountains are in the extreme south-east and reach 6,200 ft (1,892 m). There are coastal hills in the north-east and north-west, and although these rise to only 2,000 ft (650 m) they are economically important as they attract the rainfall essential to agriculture. Between the Jebel Nefusa and the coast is the triangular lowland area of the Jefara Plain, which broadens towards Tunisia. This is the site of Libya's most intensive and productive agriculture. There are few permanent streams in Libya and all are short.

Climate. Only at the coast is rainfall sufficient for settled agriculture. Tripoli receives 14·3 in (362 mm) and Benghazi 10·5 in (266 mm) per year. Coastal Libya has a Mediterranean climate with rainfall in winter and dry summers. Winter rain is unreliable, sometimes 50% above or below the mean figure, and such fluctuations can have disastrous effects on dry farming. Rainfall in south and central Libya and on the coast of the Gulf of Sidra is negligible.

The very high temperatures cause the inadequate rainfall to evaporate quickly and even near the coast only one-fifth of it remains available for transpiration, run-off and recharge of underground water.

Temperatures reach their highest in July and are lowest in January. In any one year, however, the absolute maximum may be recorded at any time between May and September when the southerly hot dry wind from the desert, called the *Ghibli*, causes temperatures to rise within an hour by 27°F (16°C).

Vegetation and Wildlife. Little natural vegetation remains in the coastal areas, where as much land as possible is devoted to agriculture. Heavily overgrazed steppe occurs in uncultivated areas, giving way to acacia, saltbush and other drought-resistant plants as rainfall decreases away from the coast. On the uplands up to 60 mi (100 km) from the coast there are poor Mediterranean woodlands, with juniper predominating. These and the steppe areas are used for grazing by nomads and semi-nomads.

There is little animal life, gazelles, rabbits, vipers, scorpions, wild sheep and birds are found in the desert and massif areas.

The People. Most Libyans are either Berbers or Arabs. Nearly all are Moslems. It was the Arabs who overran the country in two waves in the 7th and 11th centuries. There are also small numbers of Italians (35,000) and Jews (4,000).

Three-quarters of the population is concentrated in the north-west corner of the country on the Tripolitanian coastal plain.

The waterfront of Tripoli, the main city and, with Benghazi, the capital of Libya.

One-quarter of the population is in the north-east. In the Fezzan (the south-west) there are 55,000 nomads who largely live in scattered oases.

The oil revenue has helped a rapid change in styles of life and population distribution. By 1970 most Libyans were reportedly working outside agriculture and there has been an increased flow from the land to the towns. European clothes are replacing traditional Arab garments. As much as 16% of the total budget is being applied to education and literacy is fast increasing from 40%.

Over 25% of the population lives in the two main cities of Tripoli and Benghazi on the coast, where nearly all the important cities are found.

Agriculture. Less than 2% of the total area can be used for agriculture. Nevertheless, agriculture was the base of the economy before oil exports began in 1961. At that time nearly 80% of the population lived off the land. Apart from a few modern farms owned mainly by Italians, crops were grown largely on a subsistence basis, although cereals and livestock occasionally yielded marketable surpluses.

Agricultural production is largely determined by the unpredictable rainfall, and yields vary. The most important crops are barley, olives, peanuts and tobacco, grown on the Jebel Nefusa and near Tripoli. Citrus fruits, almonds, potatoes, melons, peppers and salad vegetables are grown throughout the coastal area of the Jefara Plain.

The government was allocating between 10% and 30% of the development budget to agricultural development in the late 1960s and early 1970s, and agricultural production is rising slowly.

Yields from irrigated land are normally five times those from other farms. Thus the most productive farming is carried out on the irrigated Jefara Plain and especially within 30 mi (50 km) of Tripoli. Both public and private investment has been high in this region and there is much new irrigation equipment. Unfortunately the increase in the number of wells and pumping equipment has led to a serious drop in the water table, which is falling at more than 3 ft (1 m) per year at points 12 mi (20 km) from the coast.

There is some agriculture in oasis areas in the south. Water is available from wells in wadi bottoms and irrigated cereals and vegetables are grown throughout the year, but these regions are remote and development potential very limited. 'Fossil' underground water, suitable for agriculture, has been found in the Kufra areas of south-east Libya, and experiments are being carried out to utilize the water to raise fodder crops for stock fattening.

Meanwhile imports of agricultural products are rising rapidly.

Petroleum. The exploration which led to the dramatic discovery of Libya's underground wealth began in 1955. Oil was discovered four years later and exports began in 1961. In 1962, the Libyan government received US $39,200,000; in 1965 output was nearly 60 million tons and by 1969 around 150 million tons, having increased at an average rate of 29% in the five years up to 1969. Production, already increasing rapidly, was given a boost in June 1967 with the closure of the Suez Canal. By 1972 Libya was receiving some US $2,000,000,000 per annum in revenue. Favourably placed for the important West European market, Libya is the largest oil producer in Africa and second only to Iran in the Middle East.

Most producing fields are in the Sidra basin. Pipelines up to 320 mi (500 km) long link fields with oil terminals such as Sidra, Ras Lanuf, Marsa Brega and Marsa Harega (Tobruk).

Other mineral resources are little known and discovered resources unimportant. The government is taking a strong interest in surveys of the country to discover economic deposits, aiming to provide an alternative source of revenue when petroleum resources run out.

Industry. Manufacturing industry is confined largely to the processing of agricultural products. The most important industries are olive oil pressing, the manufacture of building materials and woodworking. Both these last industries service the building boom which is affecting both city and country. The main cities have many new public buildings, and private housing developments are evident in all towns and especially on the farms.

Industries located in or near Tripoli include the manufacture of soap and detergents, soft drinks, tomato products, fish products, matches and macaroni, and the processing of dates, salt and gypsum. There are printing works and the factory of the state tobacco monopoly is in Tripoli. Libya produces about half its leaf requirements which are made up with imported tobacco. Cement is processed at Homs, 65 mi (105 km) to the east of Tripoli, and at Benghazi which has a number of workshops and flour milling.

Transport and Communications. There have been no railways in Libya since 1964 but the road system is being rapidly extended and improved. The coastal road (1,140 mi; 1,820 km) is of international standard, and roads extend over 500 mi (800 km) into the southern deserts to Sebha and beyond, and to Kufra (under construction). Tripoli is the main port and alone handles over 50% of Libyan import trade. The other main port is Benghazi, and there is some trade at Derna, Tobruk and Misurata.

EGYPT

Area: 386,198 sq mi (1,000,253 km²)
Population: 34,383,000
Population Growth Rate: 2·5%
Capital: Cairo, pop 5,384,000
Language: Arabic
Religion: Moslem (Sunni) 92%. Coptic 7%
Currency: Egyptian pound = 100 piastres

The Arab Republic of Egypt (formerly the United Arab Republic) occupies the north-eastern corner of Africa, where its neighbours are Libya to the west, Sudan to the south and Israel to the north-east. Two seas – the Mediterranean to the north, and the Red Sea and its associated gulfs of 'Aqaba and Suez to the east – form the other boundaries.

Cairo is the largest and oldest city in Africa. This view of the old town shows the Sultan Hasan mosque (left rear) which was built in 1361 and, to its right, the Al Rifa'i mosque (1912).

In 1953 revolutionary Egyptian army leaders deposed Egypt's last royal ruler and made Egypt a socialist-inclined republic.

Later a costly Arab confrontation with Israel left Egypt dependent on expensive foreign military aid, for which it looked both to the West and the USSR.

Today, Egypt ranks as the world's leading Arab state. It is only Africa's eleventh-largest nation, but supports Africa's second-highest population, herded mainly into the Nile Valley and Nile Delta.

Geographical Features. The Nile Valley is one of Egypt's five distinct physical regions; the rest being the Nile Delta, the Western and Eastern deserts, and the Sinai Peninsula. The Nile Valley separates the Western and Eastern deserts, and comprises a narrow, sinuous strip extending some 800 mi (1,280 km) from the Sudanese frontier to the Mediterranean. From the southern frontier, for some 200 mi (320 km) downstream, the narrow valley winds between high walls of Nubian sandstone. Much of this section is now flooded by the waters of Lake Nasser, penned back by the massive Aswān High Dam. Below Aswān the valley broadens and is densely settled. Further downstream the valley averages 12 mi (19 km) in width as the River Nile flows between white limestone bluffs. Just south of Cairo the limestone walls diminish and part to form the apex of the Delta.

The Nile Delta, which holds the bulk of Egypt's population, is a triangular alluvial plain stretching some 100 mi (160 km) from Cairo to the Mediterranean and 155 mi (250 km) from Alexandria to Port Said. It consists of Nile silt which has filled in an old gulf of the Mediterranean. There are now only two distributaries of the Nile, reaching the Mediterranean at Rashīd (Rosetta) and Dumyat (Damietta); but in classical times there were seven. Just behind the coast lies a zone of marsh and brackish lagoons – Lakes Maryūt, Idku, Burullus and Manzala – parts of which are being reclaimed for agriculture. The coastline itself is sculpted into a series of curves by the west-to-east longshore drift of sediment from the Nile's two distributaries.

The Western Desert covers three-quarters of Egypt (260,000 sq mi; 673,000 km²). This is the eastern part of the great Libyan Desert. It is an arid plateau area mostly averaging 700–800 ft (210–250 m) in height, with 7,000 ft (2,130 m) high outcrops of rock in the extreme south-west. The surface presents wide vistas of stone-covered wastes and occasional sandy plains. In the centre and west occur longitudinal belts of sand dunes, some nearly 200 ft (60 m) in height. A number of depressions in the plateau surface permits shallow wells to reach underground water and to support the oasis settlements of El Khārga, Dakhla, Farafra, Baharīya and Siwa; otherwise the desert is empty of people.

The Eastern Desert stretches eastwards

from the Nile Valley to the Red Sea. To the east, paralleling the coast, are the Red Sea Mountains – a prolongation of the Ethiopian Highlands. These are flanked to the west and north-west by deeply dissected limestone and sandstone plateaux. The rugged Red Sea Mountains have many peaks over 5,000 ft (1,520 m). The highest, at 7,150 ft (2,175 m), is Jebel Shāyib. Water is found only in sheltered hollows, beneath dry stream beds and in occasional springs.

The Sinai Peninsula is an irregular triangular plateau separated from Egypt proper by the isthmus of Suez. The highest parts are in the south where the ancient crystalline rocks of Jebel Katherina attain 8,651 ft (2,585 m). Here the Sinai mountains are a mass of sharp peaks, ridges and gorges. The northern two-thirds of the peninsula is a heavily dissected limestone plateau. In the far north the land becomes lower and more open and drains by shallow wadis to the flat, dune-covered Mediterranean shore. Most of the very sparse population is found in the north.

Climate, Vegetation and Wildlife. Egypt lies within the zone of great tropical deserts. Except for the Mediterranean fringe it is a dry country with only two clearly recognizable seasons: a hot summer from May to October and a cooler winter from November to April. In the desert, summer daytime temperatures exceed 100°F (37°C), but heat escapes into the cloudless sky at night, when temperatures fall by 25° or 30°F (15°–18°C). Winter temperatures are substantially lower, the January mean being 55°–60°F (13°–16°C). Years may pass without any rainfall and then a sudden storm might give one or two inches. In spring, occasional depressions from the Sahara cross Egypt towards the Delta, bringing the *khamsin*, a dry scorching wind notorious for dust and sand storms.

Unlike the interior, the Mediterranean coast receives fairly regular, if small, winter rainfall (4–8 in; 100–200 mm). The coast also has milder winters and rather lower summer temperatures than the interior through the tempering effect of the sea.

Sparse scrub finds a roothold on wadi floors within the desert, especially east of the Nile, but most of the desert's sandy, gravelly or rocky surface is bare of vegetation. Desert plants include coarse grass, tamarisks and dwarf mimosas. Date palms flourish in oases.

Large desert mammals include gazelles, hyenas and jackals. Kites, hawks and vultures are common, and the Nile and the Delta lakes attract waterbirds.

The People. Egypt's population has more than doubled in little over a generation. Roughly two-thirds of it is packed into the fertile Delta, whose towns attract migrants from the poor, overcrowded farm lands of Upper Egypt.

The famous pyramids at Giza near Cairo are the most celebrated in the world. They were built between 2600 and 2500 BC to honour the pharaohs of the Fourth Dynasty.

Some 40% of Egyptians now live in towns or cities. Chief among these are Cairo and Alexandria, Africa's two largest cities, whose modern districts contrast with old narrow streets and bazaars. Cairo (pop 5,384,000 in 1972) is the capital and the major centre of banking and industry. Alexandria is Egypt's chief port and second only to Cairo in industrial growth.

The *fellahin* (peasants), who make up the bulk of the population, live in small, mud-brick homes huddled in tightly-knit villages scattered every 2 mi (3 km) or so over the Delta and up the Nile. A few nomadic Bedouin, living in black goatskin tents, tend herds and flocks in the Eastern Desert.

Most of the fellahin men wear the traditional *galabiyah*, a long cotton 'shirt', while women favour long brightly coloured cotton dresses and often veil themselves before strangers. Both sexes often go barefoot.

Egyptians comprise three main groups: the black Nubians of the south; the white but dark-complexioned descendants of the so-called Hamites, founders of ancient Egypt; and the Arabs. Ninety-eight per cent of Egyptians speak Arabic (the official language) but local minorities speak Beja, Berber and Nubian. In the late 1960s over 70% of Egyptians remained illiterate, and illiteracy will remain a problem as long as the government fails to fully enforce its policy of compulsory education for children aged 6 to 12. Most Egyptians are Sunni Moslems, but a few belong to Egypt's ancient Coptic Church.

In the time that Egypt's population has doubled, its agricultural land has increased by only 26%, putting Egypt's farm-based economy under great strain and causing unemployment. Many Egyptians thus live in poverty, and the average per capita earnings are around one-tenth of those for some industrialized nations.

Irrigation: the Key to Agriculture. Almost all agriculture in Egypt depends upon irrigation from the Nile. The growing demand for food crops has combined with improved technology to produce an increasingly complex water control system. This now features dams and barrages (weirs) that raise the water level to feed a broadening network of distributary canals. A major factor in this system is the Aswān High Dam (completed in 1969), four miles south of an older dam. The High Dam has a vast capacity (170 milliard cu yd; 130 milliard m^3) and permits control of the Egyptian Nile, so that there is now no wasteful seasonal flooding. Instead, the water collecting in Lake Nasser, behind the dam, means that large tracts of once arid land at the edge of the desert are being brought into profitable cultivation. Other, and more limited, sources of agricultural water in Egypt are its natural underground reservoirs, especially important in the oases of the Western Desert.

Egyptian agriculture is now highly productive thanks to a sunny climate, well regulated use of chemical fertilizers and a new approach to labour-intensive methods of production. Before the land reforms begun in the early 1950s, most Egyptian farm land was owned by a few thousand rich men and farmed in tiny plots by a share-cropping servile peasantry. Now no family may possess more than 100 *feddans* (1 feddan = 1·038 ac; 0·42 ha). Larger areas have been expropriated and redistributed to the landless peasantry in small plots averaging 2 to 3 feddans. Each of these holdings forms part of a large field, perhaps of 50–70 feddans, which grows one crop. Such a unit is large enough to be economically viable, but the peasants retain the satisfaction of working their own plots.

Many of Egypt's fields yield two or even three harvests a year. The chief seasonal crops are (winter) wheat, beans, onions and clover, and (summer and autumn) cotton, millet and maize. Rice (largely for export) is grown towards the north of the Delta and sugar-cane is an important crop of upper Egypt. But high quality, long-staple cotton is Egypt's chief export crop. Yields, at about 500 lb per ac (567 kg per ha) are among the world's highest. Livestock rearing is a minor occupation, for little land can be spared for fodder crops. However, cattle are widely kept for draught purposes.

Mining and Industry. In recent years, there have been many mineral surveys in Egypt, and crude oil, iron ore, phosphates, salt and manganese are now extensively worked. Oil comes from the Qattara and El 'Alamein area of the Western Desert. Phosphate of lime comes from quarries on the Red Sea coast near El Quseir and, before the Israeli seizure, manganese was mined at Umm Bugma near Sinai's west coast. Rich hematite reserves lie just east of Aswān. Gypsum, talc, lead, wolfram and asbestos are also mined in small quantities. Building materials, in the form of limestone, alabaster, marble and basalt are widely available.

Egypt is now the second most heavily industrialized country in Africa. Most industrial development has occurred since 1939 and substantial expansion took place during the 1960s. Some two-thirds of all industrial establishments are concentrated in and around Cairo and Alexandria, and most of the rest lie in the Delta and the Suez Canal Zone. Textile manufacturing is the chief industry, and includes the spinning, weaving, dyeing and printing of cotton, wool, silk and rayon. Egypt now makes and exports a wide range of fabrics, knit wear and furnishings. The main textile centres are the Delta towns of Mahalla el Kubra and Kafr el Dauwar.

Other major industries stem from agricultural or mineral raw materials and include brewing, sugar refining, vegetable and fruit processing and the manufacture of leather goods, alcohol, soap, cement and glass. A small iron and steel plant operates at Helwān just south of Cairo, producing about 220,000 tons (223,520 metric tons) of steel a year. There are cement works at Cairo and Alexandria, and artificial fertilizer plants at Aswān and Suez, where petro-chemical industries are based on the oil refineries. Other chemical products include sulphuric acid, caustic soda, ammonia, glycerin and sodium carbonate. Egypt possesses no coal and its oil resources are only moderate. Thus the vast hydro-electric power capacity of the Aswān High Dam (2,100 MW) will prove a boon to the nation's expanding industrial sector.

Transport and Communications. Habitable Egypt has an efficient system of communications. A railway network serves the Delta and the Nile Valley as far south as Aswān. There are 2,800 mi (4,510 km) of railway, all state owned and operated. Considerable road improvements have been carried out since 1952. Egypt has about 13,830 mi (22,140 km) of surfaced roads.

The sandy isthmus joining the Delta to the Sinai Peninsula is breached by the Suez Canal, which links the Mediterranean with the Red Sea. After the Arab-Israeli war of 1967 the canal remained closed to shipping. In 1966, 21,250 vessels passed through, carrying cargoes of 274 million tons (278 million metric tons).

International Trade. Since the 1950s much trade has been diverted to the communist countries, which now take about 49% of Egypt's exports (mainly raw materials). Western Europe and North America take about 21%; the rest of the world, mainly Arab countries, takes about 30% (foodstuffs and manufactures). The Eastern Bloc supplies about 40% of imports (iron and steel products and timber); Western Europe and North America about 42% (machinery, vehicles and fertilizers). Egypt's principal exports include raw cotton, cotton piece goods, rice, onions, phosphates, cement and mineral oils. Imports include wheat and flour, chemicals and fertilizers, industrial machinery, electrical apparatus and vehicles.

SUDAN

Area: 967,500 sq mi (2,505,825 km^2)
Population: 16,489,000
Population Growth Rate: 2·8%
Capital: Khartoum, pop 475,000
Language: Arabic. English widely understood.
Religion: Moslem (70%) Animist and Christian
Currency: Sudanese pound = 100 piastres

The Sudan, with an area of some 967,500 sq mi (2,506,000 km^2), is the largest country in Africa. Its name comes from the Arabic *Bilād-es-Sudan*, meaning 'Land of the Blacks'. Most of the republic is a vast plateau crossed and watered by the Nile and its tributaries. The nation has a 400 mi (640 km) coastline on the Red Sea. Extremes of climate produce desert in the north and areas of rain forest in the south.

Since the Sudan gained independence from England in 1956, military governments have alternated with periods of civilian rule, and there has been no agreement upon a permanent constitution. Sporadic civil war with tribal southerners led in 1972 to an agreement giving the south regional autonomy.

Physical Features. The Sudan, lying south of Egypt and west of Ethiopia, consists largely of high plains (1,000–2,000 ft; 300–600 m) crossed by the White Nile flowing from the south and the Blue Nile from Ethiopia in the south-east. They meet at Khartoum to form

Picking cotton in Gezira. The quality of cotton is determined by the length and fineness of the lint. The Egyptian cotton grown on the irrigated plains of Sudan is among the world's finest.

the Nile. High mountains (up to 10,000 ft; 3,000 m) border the country in the west and south and along the Red Sea coast. Temperatures are generally high, ranging from around 100°F (38°C) during the summer, to 60°F (16°C) in the north and 80°F (27°C) in the south during the winter.

Big local rainfall variations help to produce regional differences in vegetation.

Desert (a part of the Sahara) covers a third of the country in the north, shading off in the central Sudan into increasingly wet savanna which gives way to remnants of equatorial rain forest in the south, which has as much as 50–60 in (1,270–1,524 mm) of annual rainfall. The seasonal floods of the White Nile and its tributaries as they cross the southern plains create a papyrus swamp called the *Sudd*, one of the largest swamps in the world.

Wildlife is richest in the well-watered south – a haunt of lions, elephants, hippopotamuses, rhinoceroses and other big game.

The People. The Sudan's 16 million people are unevenly distributed. About one half of the land is almost uninhabited, and over 50% of the people live on about 15% of the area. Two-thirds of the people live in the north and the central savanna area, largely in the three cities at the confluence of the White and Blue Niles: Khartoum, the capital; Omdurman, the second largest city; and Khartoum North, an industrial area. North of the cities most people live in the Nile Valley although nomadic herdsmen range westward across the fringes of the desert. These people are Arabic-speaking Moslems of mixed Negro, Nubian and Arab ancestry. A third of the population is Negro and lives in the south. Most southerners follow tribal religions (there are also Christian and Moslem minorities), and speak languages including Dinka, Nuer and Zande.

Most Sudanese are farmers growing sorghum and millet; the nomads of the drier parts herd camels in the north and cattle in the south.

A large proportion of Sudanese still cannot read or write, although illiteracy is being reduced by elementary schools and adult education. There are two universities, at Khartoum and Omdurman, and a Khartoum branch of Cairo University.

The Economy. Crops and livestock are the basis of the economy. The nomads in the north follow the rain with their camels, goats and sheep, and in the south drive their cattle away from the Nile floods. Only about 5% of the land is cultivated but the Sudan is largely self-sufficient in food, the farmers raising sorghum, millet, cassava, maize and pulses for local use.

Modern irrigation schemes are essential to Sudanese agriculture. The parent project was the Gezira scheme south of Khartoum which uses water from the Sennar dam on the Blue Nile. Here long-staple Egyptian cotton is grown, which forms the Sudan's principal export, along with some short-staple cotton, groundnuts, sesame, hides and skins. Over half a million ac (200,000 ha) are irrigated in the Gezira scheme, and other irrigation projects are under construction.

Outside agriculture the economy is generally weak, although the Sudan is the leading world supplier of gum arabic, a forest product used in making perfumes and sweets. Fishing along the Nile and the Red Sea coast is of only local importance. Mining and industrial development is strictly limited, and Port Sudan, on the Red Sea, is the only major port. The country exports cotton, gum arabic, peanuts and sesame (largely to Western Europe) and imports metals, machinery, vehicles, petroleum and textiles (mainly from Great Britain, India and Japan).

Transport and Communications. Transport systems are generally inadequate, especially in the south. However, there are some 3,000 mi (4,800 km) of railways connecting the major centres, and river boats operate on about 1,500 mi (2,400 km) of the Nile and its tributaries. There is a large network of roads, but most are dirt tracks. Sudan Airways operates internal airlines as well as international routes.

ETHIOPIA

Area: 471,778 sq mi (1,221,905 km²)
Population: 25,248,000
Population Growth Rate: 1·9%
Capital: Addis Ababa, pop 796,000
Language: Amharic. English widely used.
Religion: Ethiopian Orthodox (50%) Moslem (33%) Animist.
Currency: Ethiopian dollar = 100 cents

Ethiopia (Abyssinia) is the oldest independent African nation, with more than 2,000 years of distinctive cultural traditions. It covers a mountainous area of 471,778 sq mi (1,221,905 km²) between the Red Sea in the north and Kenya in the south. Ethiopia's per capita Gross National Product is one of the lowest in Africa and no other African nation has fewer doctors or cars per 1,000 inhabitants.

Under the continuing leadership of Haile Selassie, emperor since 1930, Ethiopia helped to found the Organization of African Unity (OAU), with Addis Ababa as its headquarters. In 1952 Eritrea, a former Italian colony on the Red Sea, joined Ethiopia, providing important coastal outlets.

Plateaux and Valleys. Most of Ethiopia is covered by two great plateaux, separated by part of the Great Rift Valley. The Ethiopian or Central Plateau, the larger of the two, lies west of the Great Rift Valley. Its 7,000–14,000 ft (2,100–4,300 m) high surface is broken by deep river valleys and dotted with mesa-like *ambas*. The most fertile part of the country, the Ethiopian Plateau, is also the most densely populated.

East of the Great Rift Valley is the Somali Plateau, which reaches over 14,000 ft (4,267 m) in the Urgoma Mountains. The rift valley separating the two plateaux is a long, narrow cleft dotted with lakes and broadening in the north to form the Danakil Depression.

Lake Tana, the largest lake in the country, lies in north-western Ethiopia outside the Rift Valley and is the source of the Blue Nile. Ethiopia has many other rivers, most of which rise in the Ethiopian Plateau, but only the Baro River is navigable.

Climatic Zones and Wildlife. Ethiopia has three climatic zones. The so-called *Quolla* (regions below 6,000 ft; 1,828 m) are mostly hot, parched grasslands, supporting some cattle and sheep.

Above the Quolla is the *Woina Dega* (6,000–8,000 ft; 1,800–2,400 m), a cooler, well-populated, malaria-free zone. It contains Ethiopia's most typical highland crops of teff, barley and wheat grown over large areas where the original forest vegetation has been removed.

The yet higher *Dega* zone is one of truly temperate climate with average temperatures below 60°F (16°C). Wheat and barley are cultivated but there is also much natural pasture for cattle, sheep and horses.

At over 10,500 ft (3,195 m) is the cool to cold *Wirch* zone of afro-alpine vegetation.

Rainfall ranges from less than 4 in (100 mm) along the Red Sea coast to over 40 in (1,000 mm) in the highlands and over 80 in (2,000 mm) in the south-west.

Ethiopia's wealth of wildlife includes elephants, lions, antelopes, giraffes, zebras, rhinoceroses, hippopotamuses and baboons. Birds include ostriches, parrots and eagles.

The People. The Semitic Amhara and the Hamitic, or Cushitic, ethnic groups are Ethiopia's largest – they are descendants of peoples who originally came from the Arabian Peninsula. The Amharas comprise about one-third of the population and make up most of the ruling class. They live chiefly in the Central Plateau region. The Hamitic peoples (who include the Galla and Somali), form the largest ethnic group in Ethiopia. They once practised tribal religions, but many have adopted Christianity or Islam. Negro tribes, found mainly in western and southern Ethiopia, comprise a significant minority (5% to 10% of the population). To the north of Lake Tana live communities of Falashas (Ethiopian Jews). Greek, Italian, Armenian, Arab and Indian minorities all play an important part in the country's commercial life.

About 90% of the people live rurally and work in agriculture. The main centres of population are Addis Ababa, the capital and chief commercial centre; and Asmara, the second-largest city and capital of Eritrea province. Other cities include Gonder, Mesewa, Aseb and Harer.

Agriculture. Farming and livestock-raising form the basis of Ethiopia's economy and occupy about 90% of the total population. The most important cash crop for export is coffee. Crops grown for local consumption include wheat, oats, barley, millet, fruits and vegetables. Enough sugar to supply domestic markets is grown in the Awash Valley, and cotton plantations are being developed on irrigated lands in the lower Awash Valley.

Much of Ethiopia is ideally suited to livestock-raising, and livestock products are more important than crops to the domestic economy. Among the livestock raised are cattle, goats, sheep, horses, donkeys, and mules. Milk and meat is sold locally, but hides and skins are one of Ethiopia's chief exports.

Forestry and Mining. Much of the highland forests have been cleared for agriculture, for firewood, or by overgrazing. The principal forest resources remain in the inaccessible rain forests of the south-west. The shortage of firewood in the highlands has been met by the introduction of *Eucalyptus globulus* from Australia, plantations of which are seen around towns and villages.

There is a variety of mineral deposits but relatively little mining. Some gold, platinum and manganese is produced, and salt and potash are exploited in the Danakil Depression. Prospecting for oil along the Red Sea coasts has been encouraged by finds of natural gas and by the development of oil fields on the Egyptian coast.

Industry and Power. The government is encouraging industrial development and foreign investment, but manufacturing is mainly limited to light industries. Factories process food and manufacture textiles, leather goods, furniture, tobacco and cigarettes, chemicals, and building materials including cement. Ethiopia still lacks adequate power supplies for industrial development, though the establishment of new hydro-electric plants on the Awash River has boosted industry in the Addis Ababa-Nazareth region.

Transport. Surfaced roads in Ethiopia total 4,700 mi (7,600 km). An uneven terrain with deep gorges and escarpments makes road construction difficult, and the major roads include stretches of spectacular engineering, some of which were constructed during Italian colonial occupation (1936–41).

There are about 680 mi (1,094 km) of railways in Ethiopia, serving areas of commercial agriculture and industry. The difficulties of terrain and the inadequacies of ground transport services have favoured internal air transport, and Ethiopia Air Lines also provide an extensive international service. The principal ports are Mesewa and Aseb. Ethiopia also uses Djibouti in the neighbouring French Territory of the Afars and the Issas.

International Trade. The volume of international trade is small. Coffee is by far the most important export and the amounts exported largely influence the balance of trade. Other important agricultural exports include oilseeds, pulses, hides, skins and cereals. Chief imports include transportation equipment, textiles, machinery, yarns and fabrics, petroleum products and foodstuffs. Ethiopia's major trading partners are the USA, Italy, West Germany, Japan and Saudi Arabia.

At Tis Asat, the Blue Nile cascades 138 ft where its course was blocked long ago by a lava flow.

FRENCH TERRITORY OF AFARS AND ISSAS

The French Territory of the Afars and the Issas (formerly French Somaliland) was a French colony from 1896 to 1967. In that year it chose continued association with France in preference to total independence and took a new name reflecting its two major ethnic groups. Opponents of the change were mostly Somalis hoping for eventual union with Somalia.

This is a land of stony desert, with extensive lava sheets and volcanic hills which sometimes exceed 5,000 ft (1,500 m) in height. Faulting parallel to that of the Red Sea has created the troughs filled by the Hanlé Plains and by the saline lakes Alol and Assal. Lake Assal is 492 ft (150 m) below sea level. The coastline is deeply indented by the Gulf of Tadjoura, where the hot and humid climate is at its most oppressive, and by the bay of Ghoubbat-el-Kharab. Vegetation is mostly thornbush and desert scrub, except for thin patches of forest on some hills, the natural groves of doum palm along dry river courses, and stretches of mangrove along the coast.

The People. The indigenous peoples, the Afars and the Issas, are Moslem and have been rivals ever since they began competing for grazing long ago. Most Afars live in the north (and in neighbouring Ethiopia, including Eritrea); the Issas, a Somali clan, in the south where they also extend into adjoining parts of Ethiopia and Somalia. Many Somalis of the Issa and other clans live in the capital Djibouti, but are not regarded as permanent residents. In addition to more than 58,000 Issa Somalis and 48,000 Afars, the Territory has more than 8,000 Arabs living mainly in Djibouti and the coastal settlements, and over 10,000 Europeans, including many French and Greeks.

The Economy. Agriculture is possible only in very restricted localities with irrigation such as Ambouli and Dikhil, where dates and vegetables are grown. The rural population depends mainly on its goats, sheep, camels and a few cattle which provide milk, meat, blood and hides. Seasonal grazing means that the numbers fluctuate with nomadic movements, especially by Somalis, into and out of neighbouring Ethiopia and Somalia. Salt is produced from the sea by evaporation and from huge deposits at Lake Assal, but the output is very small.

International trade is based almost entirely on the capital and port of Djibouti and its transit traffic. As a port of call, Djibouti depended heavily upon traffic through the Suez Canal until its closure in 1967. Linked by the Franco-Ethiopian Railway with Addis Ababa, the port is the most convenient outlet for most of Ethiopia, including the Harer region, and plans have long existed for an extension to Sidamo province.

Somalia is typically a sparsely vegetated land of dry mountains, plateaux and savannah. Its people are almost entirely nomadic.

SOMALIA

Area: 246,201 sq mi (637,661 km²)
Population: 2,864,000
Population Growth Rate: 2·2%
Capital: Mogadishu, pop 200,000
Language: Somali. English, Italian, Arabic spoken.
Religion: Moslem (Sunni)
Currency: Somali shilling = 100 centesimi

Today this arid, north-eastern corner of Africa, the Horn of Africa, is the Somali Democratic Republic, an independent state born in 1960 when the former British Somaliland Protectorate and the UN Trust Territory (and former Italian colony) of Somalia were merged.

The Somalis are a pastoral people organized in clans. Their traditional grazing areas include parts of neighbouring Kenya, Ethiopia and the French Territory of the Afars and the Issas, and there have consequently been border disputes and clashes, and territorial claims. Uncontrolled nepotism and corruption led to a military coup in 1969.

Land and People. Somalia is mainly an arid land of plains and low plateaux with a highland block in the north running parallel to the Gulf of Aden and rising to some 7,000 ft (2,134 m). The prevailing dryness gives added significance to the country's only two permanent rivers flowing from the Ethiopian highlands: the Juba, which enters the Indian Ocean near Kismayu and which is navigable in its lower reaches: and the Webbe Shibeli, which is deflected to flow parallel to the coast and only rarely reaches the sea.

The climate is hot, with temperatures averaging more than 80°F (27°C) and annual rainfall of about 3 in (76 mm) in the north and about 20 in (508 mm) in the south. Scrub and dry thorn vegetation predominates, and wildlife, including big game, abounds.

The Somali people are divided into clans such as the Dir, Isaq, Hawiye, Darod, Digil and Rahanwin. These clans are regionally distributed, affect land tenure, are influential politically and inspire strong loyalties. In the more fertile southern region between the two rivers are people of Bantu or other African origin.

Most of the people are nomadic or semi-nomadic Moslems, and nearly all are illiterate, although there is a rich heritage of oral legends and narrative poems. Three of the four leading towns are ports; Hargeisa is pleasantly situated in the highlands. Despite the long history of trade with the Arabian Peninsula, the numerous coastal settlements have little more than unimpressive ruins.

The Economy. The nomadic pastoral system with its millions of camels, sheep, goats and cattle is the basis of the economy. Cultivation is widespread only in the Benadir area between the two rivers, where maize, sorghum, sesame and groundnuts are the chief crops. Bananas, citrus fruits and sugar cane are grown under irrigation along the rivers.

Minerals, including iron ore, gypsum and uranium, have still to be developed. Industry is limited to a few small enterprises such as the meat packing plant at Kismayu and the fish cannery at Ras Koreh, both built by the Russians.

Transport and Trade. Somalia has no railways and only a limited road network. New roads are being built, mainly to serve the chief ports. New ports have been built at Berbera and Kismayu.

International Trade. The chief exports are livestock, exported to the Arabian Peninsula, and bananas, exported to Italy and other European countries. Imports include foodstuffs, vehicles and consumer goods.

CHAD

Area: 495,752 sq mi (1,283,998 km²)
Population: 3,800,000
Population Growth Rate: 2·3%
Capital: Fort-Lamy, pop 179,000
Language: French. Arabic spoken.
Religion: Moslem (50%) Animist (40%) Christian (5%)
Currency: CFA franc = 100 centimes

The Republic of Chad in northern-central Africa is a landlocked country, part grassland, part desert. It is nearly as large as West Germany, France and Spain combined, but has fewer people than the conurbation of Paris.

The small population of this former part of French Equatorial Africa is bitterly split between the Moslem pastoralists of the north and the Negro farmers of the south. In the past, the Moslem northerners were slave traders, and the southerners their victims.

After Chad became fully independent in 1960, the south dominated the north, with a one-party régime. By 1969 the north was in revolt and 3,500 French troops were called in to help subdue the rebels.

The Land. Chad takes its name from the large freshwater lake which straddles the south-western border. Shallow and marshy, Lake Chad has been contracting gradually over the years. Its main, but intermittent, feeders are the Chari and Logone rivers. Flowing across the south-west of the country, and uniting at Fort-Lamy before entering the lake, they are the only rivers in Chad of importance for irrigation and seasonal navigation. From the lake the rest of Chad spreads out as a large inland basin whose rim is formed by the Tibesti Mountains in the north, where Emi Koussi rises to 11,204 ft (3,415 m), and by the Ennedi Plateau and the Quaddai Massif in the north-east and east.

Lake Chad is a shallow freshwater lake which varies in size between 4,000 and 10,000 square miles depending on the rate of evaporation.

Climate and Vegetation. Temperatures vary from season to season but are generally high, averaging over 80°F (27°C) in the centre and south. Rainfall varies from 47 in (1,194 mm) in the extreme south to 1 in (25 mm) at Largeau (Faya) in the north. The southern third of the country has grasslands with scattered trees, and is the main farming area. The open grasslands of central Chad favour pastoralism and their abundant wildlife and game reserves could be the basis of a tourist industry.

The People. Only 10% of the population is urban, and the two largest towns, Fort-Lamy, the capital, and Sarh (formerly Fort-Archambault), are in the south, where most of the population is concentrated, the largest southern ethnic group being the 1,000,000 negroid Sara people. By contrast, the three northern provinces – Borku, Ennedi and Tibesti – have only 100,000 people. The northern peoples are of Arab and Hamitic origin and lead a nomadic or semi-nomadic life.

Islam, introduced by the Arabs, is the dominant religion in north and central Chad. The south is mainly animist, but has a few Christians. Schools are few and mainly in the south; little more than 5% of the population is literate.

Agriculture. Chad lives mainly by farming and stock-breeding. Except for oases, the northern half of the country is too arid for cultivation, and the people are mostly nomads moving from place to place with their goats, sheep, camels and cattle. Cattle and livestock products are Chad's second export by value.

The southern third of the country provides Chad's two leading cash crops, cotton and groundnuts, and various food crops such as millet, guinea corn and rice. The cotton crop is bought and processed by COTONFRAN (*Société Cotonnière Franco-Tchadien*), whose small ginneries are located to reduce the distance over which the growers have to carry their head-loaded harvest. Sugar cultivation is of increasing importance along the Chari and Logone rivers. These rivers, and also Lake Chad, are a valuable source of fish, which is dried or smoked for local sale or for export to Nigeria and Cameroon.

Minerals and Industry. Natron, won from shallow workings in the Kanem region east of Lake Chad, is the only mineral of export significance. Traces of tungsten have been found in the Tibesti area, and there may be oil in the Lake Chad depression. Industry is mostly limited to processing farm and pastoral products.

Transport. Remote from any port, Chad has to pay heavily for imports and its exports are less competitive because of the high transport costs to markets. The country has no railway, and roads are the links with rail and river trans-shipment points in neighbouring countries, but are of little use during the rains. A rail link with the Nigerian or trans-Cameroon railway is under consideration. Meanwhile Chad relies heavily on air-lifting imports and exports. Fort-Lamy is one of the most important freight-handling airports in Africa.

International Trade. Cotton normally accounts for about 75% of the value of exports, meat products for a further 15%. Consumer goods form more than 50% of the imports. France is the chief overseas trading partner, while the Congo Republic and the Central African Republic are Chad's chief customers in Africa.

NIGER

Area: 489,190 sq mi (1,267,002 km²)
Population: 4,243,000
Population Growth Rate: 2·7%
Capital: Niamey, pop 102,000
Language: French. Fulani and Hausa (commercial)
Religion: Moslem
Currency: CFA franc = 100 centimes

Niger covers a similar area to that of its eastern neighbour Chad. Much of this vast landlocked and remote country is desert, and it has only a small but ethnically diverse population. The republic (a former territory of French West Africa) is young, becoming fully independent only in 1960. Its political inexperience, ethnic divisions and lack of natural resources make it difficult to see how Niger can develop a prosperous economy, or indeed create a strong sense of national unity.

President Hamani Diori, the dominant political force from the 1950s, gained French support and made Niger a one-party state.

The Land. The northern two-thirds of Niger are part of the Sahara Desert, but by no means flat, for the Aïr massif rises to well over 5,900 ft (1,798 m). The lowest parts of the country lie in the south-west along the valley of the River Niger, and in the south-east around Lake Chad. The seasonal flooding of the Niger sustains the valley pastures and crops like rice.

Climate. This is a hot and arid land. Temperatures at Niamey, the capital, average 91°–100°F (33°–38°C). Rainfall here, and along parts of the Nigerian border, averages 22 in (559 mm) annually. Agadès in central Niger has an annual average of only some 7 in (179 mm), and even less falls further n rth.

Vegetation and Wildlife. There are tropical grasslands in the south, but these gradually give way to semi-desert and desert in the north and north-east. Wildlife includes lions and elephants.

The People. The main ethnic groups are the Hausa, numbering over two million, the Djerma (948,000), the Fulani or Peuls (426,000), the Beriberi-Manga (365,000) and the nomadic Tuareg (120,000). Most people are of mainly Negro origin. About 85% of the population are Moslems and some 14% animists. Only about 11% can read and write, and the help of France and UNESCO has been sought to increase the number of schools and teachers.

Most of the people live in the south and south-west. Niamey, the capital (102,000), stands on the banks of the Niger; it was once the capital of the Songhai Empire. Eastwards lies Zinder, the chief commercial centre and the local capital of the Hausas.

The Economy. Two-thirds of the Gross National Product come from farming and stock-breeding. The south is the truly productive part of the country, providing sorghum, millet, maize, rice, groundnuts and cotton, with cattle, sheep and goats on the drier margins. Northwards, cultivation becomes increasingly rare, and nomadic pastoralists exist precariously.

Currently the most significant mineral is uranium, mined since 1971 by a French-controlled company (SOMAIR) at Arlit about 200 mi (322 km) north of Agadès. Small amounts of tin are mined in Aïr, but the low-grade iron ore at Say, and Tahoua's phosphates and gypsum have still to be exploited.

Transport is a major problem. Roads are few, railways non-existent. The Niger is at present navigable from Niamey to the Dahomey frontier, but only from October to May; measures to improve navigation and extend its season are being planned.

Logically the two main outlets from Niger are either through Nigeria, via Zinder, or south-westwards through Dahomey to the port of Cotonou. But even from Jerma Ganda, Cotonou is more than 500 mi (805 km) by road and rail.

International Trade. Cattle and groundnuts are the chief exports; vehicles, cotton, textiles and foodstuffs, the chief imports. Salt from the small oasis of Bilma, in the desert east of the Aïr massif, is still exported by camel caravan. France and Nigeria are among Niger's leading trade partners.

UPPER VOLTA

Area: 105,838 sq mi (274,120 km²)
Population: 5,485,971
Population Growth Rate: 2·1%
Capital: Ouagadougou, pop 110,000
Language: French. Mossi spoken
Religion: Animist (80%) Moslem (18%) Christian
Currency: CFA franc = 100 centimes

Upper Volta, so named because it straddles the upper reaches of the main headstreams of the Volta river system, became a fully independent republic in 1960 after decades as a French colony. It is a poor, undeveloped and landlocked country, larger than West Germany, and has relied heavily on aid from France, the World Bank and the European Economic Community. There was a military coup in 1966 but in 1970, a new constitution was approved which foreshadowed a full return to civilian government. Surprisingly, Upper Volta is more influential in current African politics than its size or economy would suggest. The republic has particularly close links with Ivory Coast and these seem to offer Upper Volta's only real chance of economic progress.

The Land. Most of Upper Volta is a plateau about 1,000 ft (300 m) above sea level, incised by the Black, Red and White Volta rivers which rise near the north-western border with Mali. In the south, the river valleys are infested by the *Simulium* fly, whose bite can bring on blindness, while the Gourma swamps are the home of the tsetse fly which indirectly causes sleeping sickness. Much of the western part of the country is sandy and infertile, but there are no true deserts.

Upper Volta has a hot, dry climate. Temperatures average 89–100°F (32–38°C), and rainfall 35 in (889 mm). Little rain is retained by the poor soil and vegetation is rarely more than poor savanna.

The People. The dominant ethnic group are the Mossi, whose city of Ouagadougou, once an independent kingdom, is now Upper Volta's capital. Bobo-Dioulasso, the chief market town, is in the south-west, the home of the Bobo people, the second largest ethnic group. Other groups include the Gourma in the east, and the nomadic Fulani in the north. Most of the people are animists, but there are about one million Moslems and more than 200,000 Roman Catholics. Not more than 10% of the population are literate.

Upper Volta is an arid, impoverished land. Its main source of wealth is its livestock.

Upper Volta is over-populated, and thousands of young workers migrate, temporarily or permanently, to Ivory Coast and Ghana.

The Economy. Agriculture is the most important activity. Cash crops include peanuts, cotton, sesame and karité nuts. The chief food crops are sorghum in the south and bulrush millet in the north. Cattle, sheep and goats are reared in the north and east of the country.

Iron ore is known to exist, and traces of copper and uranium have been found. Japanese interests are developing the extremely large manganese deposits at Tambao. There is little industry as yet, although Banfora has a sugar refinery and Koudougou a small textile mill.

Upper Volta exports mainly hides, groundnuts and cotton. Textiles, foodstuffs, vehicles and machinery are among the chief imports. Trade is mainly with France and countries of the Franc Zone.

Transport. There is an extensive road network, and Ouagadougou is the terminal of the Abidjan-Niger railway, which may be extended to Bamako in Mali. There are airports at Ouagadougou and Bobo-Dioulasso.

MALI

Area: 478,767 sq mi (1,240,006 km²)
Population: 5,257,000
Population Growth Rate: 2·1%
Capital: Bamako, pop 282,000
Language: French. Malinké, Bambara and Dyula
Religion: Moslem (60%) Animist (40%)
Currency: Franc Malien = 100 centimes

The Republic of Mali, which is well over twice the size of France, was formerly the colony of French Sudan. After a brief period of autonomy (1958), it joined with Senegal in forming the Federation of Mali. This collapsed in 1960, when Sudan became the independent republic of Mali and withdrew from the French Community. All governmental powers are now vested in the military National Liberation Committee.

The Land. Mali is a landlocked country of vast proportions. Much of it is monotonously flat, although the south-west is mountainous and the Adrar des Iforas dominates the extreme north-east. Much of the country is desert or semi-desert, but the River Niger cuts a great arc across the south, with a swampy middle section called the Inland Niger Delta. Watering the south-west is the upper Senegal River, and Mali is associated with Mauritania and Senegal in its development.

Climate, Vegetation and Wildlife. Mali is a hot and arid country, with temperatures ranging between 85°F (29°C) and 105°F (41°C) according to season. Annual rainfall is only 2 in (51 mm) in the extreme north, but increases to some 55 in (1,397 mm) at Sikasso in the southern mountains. A dry, dusty wind called the *harmattan* blows in the dry season (November-April). Tall grasses and trees grow in the south, and scrub, thorn and acacia north of this savanna. Antelopes, hippopotamuses and lions are among the country's larger mammals.

The People. Its long history as a West African cultural cross-roads has given Mali an ethnically complex population. Some of its peoples, such as the Bambara, Marka, Songhai and Dogon, are sedentary; others, like the Fulani and Tuareg, are nomadic. Most of the people are Moslems, but there are many animists and about 20,000 Roman Catholics. Only some 5% are literate.

The main towns are mostly along the Niger, from the 1,000-year-old trading town of Timbuktu upstream to Bamako, the capital. Seasonal migration takes many workers into Senegal, Gambia, Guinea, Ivory Coast and Ghana.

The Economy. Because of its aridity Mali is largely a livestock producing country, rearing cattle, sheep and goats, and indeed exporting cattle to Senegal, Ivory Coast, Ghana, Liberia and Nigeria. Crops irrigated by water tapped from the River Niger are important in the south. They include cotton, rice, millet, sorghum and maize.

Mineral resources, including iron ore, bauxite, gold and salt, have still to be developed. Some factory industry – food processing, textiles and tobacco – has been established.

Transport. Mali suffers from its isolation and poor internal transport facilities. Bamako has an 805 mi (1,296 km) rail link with Dakar (Senegal). In association with Guinea and China, Mali has planned a further rail link to Conakry, the Guinean capital. More all-season roads, and improved navigation on the Niger are urgent needs.

International Trade. Livestock, dried fish, peanuts and cotton are the chief exports; foodstuffs, textiles, iron and steel products, machinery and vehicles, the chief imports. Mali trades mainly with France and countries of the Franc Zone, and with East European countries and China.

Schoolchildren outside the mosque at Djenné on the river Bam in the fertile south-central lowlands. In the early middle ages, Mali was the centre of a powerful Negro-Islamic empire.

MAURITANIA

Area: 397,956 sq mi (1,030,706 km²)
Population: 1,200,000
Population Growth Rate: 2·2%
Capital: Nouakchott, pop 48,000
Language: Arabic. French. Wolof and Tukolov also spoken.
Religion: Moslem
Currency: CFA franc = 100 centimes

Mauritania is a large, sparsely-populated, Islamic republic linking Arab/Berber north-west Africa and Negro West Africa. When full independence from France came in 1960, the strong economic links were retained.

The Land. Most of Mauritania is a monotonous series of plateaux edged by westward-facing scarps. Only in small areas of western Adrar does the land rise above 1,600 ft (488 m). At least 40% of the land surface is covered with shifting sand.

Climate and Vegetation. Mauritania is a hot and arid country. The northern two-thirds of the country have an annual rainfall of less than 4 in (102 mm); and even in the south, rainfall – about 25 in (635 mm) annually – is only marginally adequate for cultivation. While the south has some grasslands and the fertile flood plain of the Senegal River, some areas are virtually without vegetation.

The People. About 75% of the population is of Moorish (Arab/Berber) stock. All are basically nomadic pastoralists. Negro peoples live in the south, mainly as sedentary cultivators. Not more than 10% of the population is urban-dwelling. Nearly all inhabitants are Moslems, and more than 90% are illiterate.

Farming, Fishing and Mining. Crop-raising and stock-breeding are the chief economic activities. Millet is the chief crop, other crops include dates, maize, rice and red beans. Cattle are raised in the south-west; camels, goats and sheep farther north.

There are rich offshore fishing grounds, and fish-meal factories and a freezing plant have been built at Nouadhibou, which is also the port for the Zouerate iron ore deposits. These large deposits are being exploited by MIFERMA, an international (but predominantly French) company. An Anglo-American group has the largest interests in SOMIMA, the company exploiting the large copper deposits near Akjoujt, while yttrium and other rare minerals are being produced by a Franco-Mauritanian company. Titanium and gypsum exist in exportable quantities.

Mauritania's population is largely moorish and its way of life still mostly nomadic.

Transport and Trade. Apart from the iron ore railway linking Nouadhibou with Akjoujt, internal transport is by a poor road system and by air. Nouadhibou is an international airport and has deep-water facilities.

International Trade. Iron ore is by far the most important export. Other exports include livestock, fish, gum arabic, dates and salt. Petroleum products, vehicles, machinery and electrical equipment are among the chief imports.

SPANISH SAHARA

Spanish Sahara is an overseas province of Spain that occupies some 102,680 sq mi (266,000 km²) of coastal north-west Africa. It is divided into two regions: Saguiet el Hamra in the north (capital, Aaiuń) and Río de Oro in the south (capital, Villa Cisneros). Mainly desert, the province has rich phosphate deposits. The Sahrawi people, the bulk of the total population of about 76,000, are mostly nomads. Spain has promised to hold a referendum to determine the future of the province.

CANARY ISLANDS

The Canary Islands lie some 60 mi (97 km) off the north-west coast of Africa, but politically constitute two Spanish provinces: Santa Cruz de Tenerife and Las Palmas. There are two groups of islands: Tenerife, Gran Canaria, Palma, Hierro and Gomera, all mountain peaks rising directly from the ocean depths; and Lanzarote, Fuerteventura and six barren islets. Pico de Teide on Tenerife rises to 12,162 ft (3,706 m). The Canaries have an equable climate, and grow bananas, citrus fruit and sugar-cane. Tourism is important. The population numbers 908,718, of whom some 287,000 live in Las Palmas, the capital.

CAPE VERDE ISLANDS

Cape Verde Islands off West Africa are a Portuguese overseas territory comprising 10 islands and 5 islets, divided into two groups called Barlavento (Windward) and Sotavento (Leeward). All but three of the islands are mountainous, and Fogo has an active volcano. Temperatures are always high and rainfall is unreliable. The whole area amounts to 1,557 sq mi (4,033 km²) and has a mainly mixed Afro-Portuguese population of over 272,000. Only emigration prevents overpopulation. The capital is Praia (6,000) on São Tiago, but Mindelo on São Vicente is the chief port and commercial centre. Products include bananas, coffee, peanuts and castor beans. Maize and beans are the chief food crops.

SENEGAL

Area: 75,750 sq mi (196,192 km²)
Population: 3,822,000
Population Growth Rate: 2·4%
Capital: Dakar, pop 581,000
Language: French. Wolof, Fulani and Mende spoken.
Religion: Moslem (80%) Christian (10%) Traditional beliefs (10%)
Currency: CFA franc = 100 centimes

Senegal, Africa's westernmost nation, is roughly the same size as South Dakota but has fewer people than Detroit. This low, dry, sparsely peopled country became a self-governing republic within the French Community in 1958, and gained full independence two years later after the failure of its federation with what is now Mali. This left Senegal with little more than its peanut crop and an elaborate and expensive administration. The country became heavily dependent upon French and other foreign aid and still labours under heavy economic burdens, despite economic diversification encouraged by its first president, Léopold Sédar Senghor whose authoritarian rule remained unbroken by the early 1970s.

A Senegalese festival dancer from the Wolof tribe in a gold brocade dress. The port of Dakar in Senegal was once the administrative centre of French West Africa.

A Land of Plains. Senegal consists of great plains less than 600 ft (180 m) above sea level in which are cut the broad valleys of the rivers Senegal, Saloum, Gambia and Casamance. In the central northern area, the low-lying, dry and sandy plains of the Ferlo 'desert' are remnants of a former southward extension of the Sahara. Between them and the coast plain are the even lower and rather wetter Western Plains. In the far north, the left bank flood-plain of the Senegal River is known as the Fouta where it forms the boundary with Mauritania. The only highlands are in the east, where the Bambouk Mountains enter from Mali, and in the south-east, where the northern edge of the Fouta Djallon protrudes from Guinea. The coast is characterized by sand bars, delta swamps and low cliffs.

The independent state of Gambia, stretching west-east for 200 mi (320 km) along both banks of the River Gambia, cuts western Senegal in two, although its divisive effect is lessened by the Trans-Gambian Highway from Kaolack to Ziguinchor.

Climate, Vegetation and Wildlife. Senegal has a Sudanic climate with temperatures ranging from 65°–88°F (18°–31°C) and rainfall rising to more than 60 in (1,524 mm), but in many areas averaging less than 30 in (762 mm). Temperatures and rainfall are high inland, the coast being surprisingly cool for its latitude.

The vegetation is generally low tree and bush or grass savanna. Lions, antelopes, warthogs and other large mammals persist, but mainly in reserves.

The People. The people include groups of both Negro and at least partly Caucasian stock, and most live along or near the coast. The two most numerous ethnic groups, the Wolofs and Fulani, are also found in Gambia. Other major groups are the Serer, Tukulor and Diola. About 80% of the population is Moslem, and some 10% Christian. Less than 10% can read and write.

Dakar, the capital, still has a large transit trade, and by African standards is a large, modern and highly sophisticated city. The contrast between life in Dakar and in the poor and limited hinterland is striking. Kaolack, on the estuary of the Saloum River, is the export centre for peanut products.

Agriculture. Peanuts are the major cash crop and account for more than 90% of exports by value. Falling world prices have hit the country hard. The chief producing areas are along the Dakar-Saint-Louis railway and along the lines to Kaolack and Kayes. A small experimental rice and cotton scheme is operating at Guede on the edge of the flood plain. In the (southern) Casamance area, where rainfall is somewhat higher, oil-palms, maize, yams, cassava, sisal and citrus fruits are grown inland from the coastal swamps. Bulrush millet and sorghum are the staple food crops, but rice and other foodstuffs have to be imported. Sheep, goats, cattle and other livestock are raised, especially in the drier regions.

Minerals and Industry. Phosphates are quarried near Thiès and exported, while the Saloum estuary area yields titanium. Limestone is quarried and used in the cement factory at Rufisque, east of Dakar. Industry is centred on Dakar, which has peanut-oil mills, sugar and oil refineries, fish canneries, breweries, and soap, textile and chemical factories.

Transport and International Trade. The railway network is vital to agricultural development, and an improved road network is being extended from Dakar. Rivers such as the Senegal are important routes.

Senegal's chief exports are peanuts and peanut products, followed by phosphates and salt. Imports include wheat, sugar, petroleum products, machinery and textiles. Most of the country's trade is with France.

GAMBIA

Area: 4,261 sq mi (11,035 km²)
Population: 378,730
Population Growth Rate: 1·9%
Capital: Banjul, pop 54,000
Language: English. Malinke and Wolof
Religion: Moslem (55%) Animist (35%) Christian (10%)
Currency: Dalasi = 100 butut

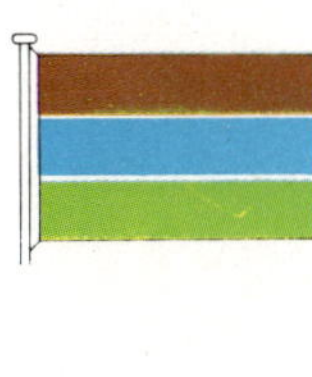

Gambia, a small West African republic within the Commonwealth, is a narrow ribbon of land running inland from the Atlantic Ocean astride the River Gambia. Only 15–30 mi (24–48 km) wide and some 290 mi (464 km) long, it is Africa's smallest independent state, flanked on three sides by the much larger French-speaking state of Senegal. It gained independence from Britain in 1963, and since 1970 has been a republic.

The Land. Gambia is low lying and vulnerable to river flooding. Mangrove swamps line the River Gambia's banks; behind these lie grasslands subject to seasonal flooding and called the *banto faros*; and behind the *banto faros* rise low sandy plateaux.

Gambia has a predominantly dry tropical climate, with rain in summer months.

The country's tropical wildlife includes crocodiles and hippopotamuses; but away from the rivers few animals larger than the numerous monkeys persist, for the land is largely cleared for agriculture.

The People mainly consist of different largely Negro tribal groups. About 40% are peanut-farming Mandingo. The cattle-raising Fulani (13%) are the next largest group, but the few hundred British, Syrians and Lebanese control trade. Only some 10% of Gambia's 379,000 inhabitants live in towns, notably in Banjul (54,000). In spite of free primary education, illiteracy remains high.

The Economy. Over 90% of the total value of exports comes from peanuts, grown on the sandy soils of the higher, flood-free areas. Peanuts already occupy about two-thirds of the cultivated uplands, but there is great scope for increasing yields. Basic food crops include rice, millet and maize. Attempts to expand food output by mechanizing rice farming and developing poultry farming and fisheries have so far more or less failed.

There is little major industry apart from peanut processing. Gambia lacks important mineral resources, but its palm-fringed beaches are proving a tourist attraction. Moreover food-processing industries are emerging around Bathurst, and there is small-scale cotton spinning and weaving. Gambia's limited trade is chiefly with Great Britain, Italy, Japan and Portugal.

The entirely navigable River Gambia could prove one of West Africa's most valuable waterways but so far carries little besides peanuts. Apart from the Trans-Gambia Road, completed in 1958, the land transport system is little developed, with few all-weather roads and no railway.

With a population of well under 500,000, heavy dependence upon one cash crop, and few natural resources, Gambia would seem to have poor potential as an independent state. Largely because its fine inland waterway could provide Senegal with a valuable outlet, Gambia's long-term progress may well depend upon working closely with Senegal.

Palm oil production in the Gambia. Palm oil is extracted by pounding the pith which surrounds the palm nut until it yields its oil. The Gambia's principal resource is peanuts.

PORTUGUESE GUINEA

Portuguese Guinea lies on the West African coast where it forms a wedge between Senegal to the north and Guinea to the east and south. Like other Portuguese overseas territories in Africa, this low-lying tropical land (largely of swamp and forest) is the scene of guerrilla warfare between a revolutionary independence movement and the Portuguese army. By the early 1970s, the African rebels were pinning down some 30,000 troops.

Land and People. Covering about 13,948 sq mi

(36,125 km²), tiny Portuguese Guinea is highest in the south-east where low sandstone outliers of the Fouta Djallon plateau rise to about 990 ft (300 m). Flowing westwards from this plateau are the main rivers – the Corubal (Rio Grande), Gêba and Cacheu. They often overflow during the rainy season, and the coastal plain is therefore swampy and forested, whereas the interior is savanna. Temperatures are always high, and rainfall is torrential on the very humid coast. Leopards, elephants and crocodiles are among the country's wild animals.

The population of 532,000 is composed of local tribes like the Balante, Fulani and Mandyako; *assimilados*, people of African or mixed blood who have taken Portuguese citizenship; and non-Africans, mainly Portuguese. Bissau (65,000) is the capital and chief port. Nearly half the population is Moslem. The Portuguese are encouraging the use of their language, but the *lingua franca* is a Cape Verde-Guinean *crioulo*.

The Economy. Most of the people are peasant farmers, but the war has played havoc with what was always a shaky economy. The main food crop is swamp rice, grown on cleared mangrove swamps and in flooded river valleys by the Balante. Other food crops include coconuts, cassava, sweet potatoes and maize. Peanuts are almost the only cash crop and are grown inland along with upland rice. The Fulani rear cattle on the plateau. Oil, bauxite and other minerals remain unexploited. There is little industry except food processing.

GUINEA

Area: 95,000 sq mi (246,050 km²)
Population: 4,010,000
Population Growth Rate: 2·2%
Capital: Conakry, pop 197,000
Language: French. Soussou and Manika
Religion: Moslem (65%) Animist (30%) Christian (1%)
Currency: Sily = 100 corilles

The Republic of Guinea proclaimed its independence in 1958. Under President Ahmed Sékou Touré, the republic rejected membership of the French Community and declared for 'positive neutralism'. Aid without strings was accepted from communist and non-communist countries. Attempts to overthrow the regime provoked savage reprisals. Yet progress has been made, and Guinea's rich soils and mineral reserves may eventually bring prosperity to the poor inhabitants of this mountainous West African nation.

The Land. Beyond the Atlantic estuaries, mangrove swamps and narrow coastal plain, the Fouta Djallon plateau rises steeply to more than 3,300 ft (1,006 m). Its deeply dissected slopes are densely forested, and there are many waterfalls, some capable of providing hydro-electric power. In the extreme south-east, the Guinea Highlands rise to 5,800 ft (1,768 m) in Mt Nimba on the Liberian border. In the east, grassy plains are drained by the River Niger and its tributaries.

Climate, Vegetation and Wildlife. The climate is hot and humid, and rainfall (April-November) is especially heavy along the coast. Conakry has average temperatures of more than 80°F (27°C) and an average annual rainfall of 169 in (4,293 mm).

Mangroves and oil-palms flourish on the coast. Creepers and tropical hardwoods thrive in the rain forest areas below the uplands, where savanna is predominant. Wildlife is richly varied and includes lions, leopards, hippopotamuses, chimpanzees and crocodiles.

The People. Of the 16 or so ethnic groups, the most important are the Fulani (Fulbe or Peuls), the Malinké or Mandingo, the Soussou and the Kissi. Most of the people are Moslems; 33% are animists and 1% Christian (mainly Roman Catholic). Education is free.

Most of the people live in villages. Conakry, the capital and chief port, stands on Tombo Island, which is linked to the Kaloum Peninsula by a causeway. It is isolated from the south-east and the interior, where Kankan is the main centre.

The Economy. Minerals are becoming increasingly important with the development, by an American company, of the vast bauxite deposits at Boké and Fria. The Fria deposits are converted into alumina with hydro-electricity from falls on the Konkouré River, and an aluminium plant is to be built. Sangaredi, the new centre of the Boké operation, has been linked by rail with the new deep-water port of Kamsar.

Iron ore is mined near Conakry, and rich deposits extending into Guinea from Liberia in the Mt Nimba area can be exploited when communications are improved.

Agriculture remains highly important. Rice, maize, oil-palms, bananas and pineapples are grown in the wetter, settled areas of the south-west, and the Guinea Highlands have coffee and cinchona plantations. On the Niger plains rice is grown where irrigation is possible, but elsewhere the chief crops are millet, cassava and peanuts.

Since independence, manufacturing industries have been established with foreign aid and are run as state enterprises. They include a British-built textile mill using imported cotton; a Chinese cigarette factory; a Russian sawmill, and a Yugoslav furniture factory.

Transport and Trade. Roads link Guinea with Senegal, Ivory Coast and Mali. New transport developments include the extension by Chinese engineers of the Conakry-Kankan railway to Bamako in Mali.

Alumina accounts for about 60% of Guinea's exports by value. Bananas and pineapples are the chief agricultural exports. Imports include rice, textiles and machinery.

SIERRA LEONE

Area: 27,925 sq mi (72,325 km²)
Population: 2,600,000
Population Growth Rate: 1·5%
Capital: Freetown, pop 178,600
Language: English. Krio, Mende, Temne
Religion: Animist (62%) Moslem (35%) Christian (5%)
Currency: Leone = 100 cents

Sierra Leone is a small and compact West African republic that gained independence from Britain in 1961, but remained within the Commonwealth. It is bounded to the north and east by Guinea and to the south by Liberia. The country has relatively few natural resources, and a problem of uneven development – the south is richer than the north.

The Land. Sierra Leone comprises two main regions. Broad coastal lowlands (broken only by rocky Freetown Peninsula) stretch inland for about 50 mi (80 km), and are succeeded by a mountainous plateau over 2,000 ft (600 m) above sea level. Rainfall ranges from over 80 in (2,000 mm) in the north-west to over 200 in (5,000 mm) in the south-east. The heavy rainfall once helped to support extensive forest cover, but most of this has been cleared for agriculture. Mangrove and other swamps fringe the coast. Outside the swampy areas savanna is widespread on poor soils that are liable to erosion. Big game animals still live in the remoter areas.

The People. The country's 2·6 million people comprise at least 13 different tribes, of which the Temne in the north-east and the Mende in the south are the two most numerous. But perhaps the most distinctive and influential inhabitants are the Creoles, numbering some 120,000 and concentrated in the Freetown area; they are descendants of slaves repatriated from America between 1787 and 1870 and form an elite whose presence arguably militates against true national unity.

Freetown (178,600) is the capital and has a fine harbour which was once used by the British as their main naval base for the suppression of the slave trade in West Africa.

Education is not compulsory and illiteracy is high.

Agriculture and Fisheries. Almost all the major West African crops are grown, ranging from the perennial tree cash crops of palm oil, cocoa, coffee and kola nuts to such food crops as rice, sorghum-millets, maize and cassava. More recently such crops as tobacco, rubber and bananas have been developed. Since the 1920s, the mangrove swamplands have been increasingly cut and drained to make way for the development of swamp rice and to the harvesting of such swamp products as piassava, or the leaf stalks of the Raphia palm.

New tools and techniques have helped to raise the largely canoe-based catches of fish including sardines and bonga. But by the late 1960s annual totals of under 30,000 tons remained below the nation's needs and additional supplies were being imported.

Mining and Industry. Most of Sierra Leone's export revenue comes from mining, especially for diamonds and iron. Alluvial diamonds are now mined chiefly in the south-east and east of the country. Iron ore is mined near Larampa at Lunsar and taken by mineral rail to the ore port of Peppel in the Rokel estuary. Other iron deposits are found further inland in the Sula Mountains. Sierra Leone also has alluvial gold, platinum, bauxite, molybdenum and the world's largest known deposits of rutile (titanium oxide).

Other industrialization has developed little in the country, but recent progress includes completion of a new oil refinery and power station at Koidu and Kenema.

Freetown, the capital of Sierra Leone, was founded in 1787 by an Englishman, Granville Sharp, to settle slaves freed from ships captured by the British navy.

Transport and Trade. Over 40% of public investment has been earmarked for transport and communications in recent development planning. Ambitious road-building programmes are opening up remoter parts of the interior and forging a link with Liberia. River and coastal traffic has increased. Sierra Leone Airways runs domestic flights between several airfields. The country's major international outlets are Lungi Airport near Freetown and the port of Freetown.

The United Kingdom is still Sierra Leone's chief trading partner, accounting for about 77% of Sierra Leone's exports and over 30% of its imports by value. The country's main imports include food, fuel oils, textiles, vehicles, building materials and machinery. The chief exports include diamonds, iron ore, bauxite, palm kernels, coffee, cocoa and piassava.

LIBERIA

Area: 43,000 sq mi (111,370 km²)
Population: 1,571,000
Population Growth Rate: 1·7%
Capital: Monrovia, pop 180,000
Language: English
Religion: Animist (90%) Christian (7%) Moslem (3%)
Currency: Liberian dollar = 100 cents

Liberia, the world's second oldest Negro republic and the oldest independent state in West Africa, grew out of settlements made by freed slaves sent there from America by colonization societies, beginning in the 1820s and 1830s.

Broken terrain, tropical diseases and a sparse largely primitive population have hampered development of the interior. But iron ore and rubber production are now financing the start of social improvements that could gather momentum late in the century.

The Land. Lying on the West African coast between Ivory Coast and Sierra Leone, Liberia is essentially a hilly, wet, and largely forested country. Behind a coastal plain 10–50 mi (16–80 km) wide rise the low hills and plateaux which cover much of the land and lead up into the smaller areas of high mountains, notably the Guinea Highlands, many of which are edged by dramatic scarps. Rainfall on the coast varies from 100–200 in (2,540–5,080 mm), but inland areas have lower totals. Average temperatures range from 76–80°F (24–27°C). Liberia's large wild

Brightly decorated fishing boats in the Liberian interior. Over 90% of Liberia's population is indigenous. The 5% who are descended from American slaves live mostly in coastal cities.

animals include leopards and pygmy hippopotamuses.

The People. Descendants of the nation's American Negro founders comprise only about 5% of the total population but they occupy most of the positions of social and political leadership, and are concentrated in the cities.

The majority of the population comprises some 20 indigenous Negro tribes, divided by language into four major groups: Mendetan (notably the Vai people); Mande-fu; West Atlantic; and, easily the largest, the Kru. Most of these groups, especially in the forested mountainous interior, have been little touched by modern life. Literacy is very low, and only 1 in 10 children of school age attend school. The three major centres in Liberia are the capital, Monrovia (with a population around 180,000), Harper and Buchanan.

Agriculture. Liberia features elements of both the rice and cassava (root crop) zones of West Africa. It is far from being self-sufficient in food, and must import a good deal of rice. The country suffers from a land tenure system which inhibits the development of co-operative marketing. In this difficult environment, too, there are few livestock.

Yet there has been considerable development in cash crop farming, often on a plantation basis, notably for coffee, cocoa, kola and, especially, rubber. The Firestone rubber plantations, begun in 1926, brought much American investment to the country, and in 1945 accounted for 97% of the value of Liberian exports. Now, some 60% of the rubber is grown on private plantations owned by Liberians. Even so, the Firestone rubber plantations having all the services of a large town have had a great impact on the economic and social development of Liberia.

Minerals and Industry. Rich deposits of iron ore have been exploited since 1951, notably in the Bomi Hills some 45 mi (72 km) north-west of Monrovia, and at Mt Nimba near the Guinea border. Liberia exported 17·2 million tons (17·5 million metric tons) of ore in 1967. Gold and diamonds are also mined. What little manufacturing industry there is includes an iron-ore washing and treatment plant, textile factories and a tyre factory. Liberia has the potential for a thriving timber industry which may be developed in the future.

Transport. Internally Liberia has until recently possessed very poor and limiting transport facilities indeed. The three railway lines – to the Sierra Leone border past Bomi, to the Bong Mountains and Mt Nimba – were built to carry minerals. Shipping registered in Liberia under 'flags of convenience' account for some 10% of Liberia's national budget. Monrovia is a good deep-water port.

IVORY COAST

Area: 124,502 sq mi (322,460 km²)
Population: 4,420,000
Population Growth Rate: 2·3%
Capital: Abidjan, pop 500,000
Language: French. Dioula also spoken
Religion: Animist (60%) Moslem (23%) Roman Catholic (14%)
Currency: CFA franc = 100 centimes

Ivory Coast lies on the Guinea coast of West Africa, with the Atlantic Ocean to the south, Ghana to the east, Liberia and Guinea to the west and Mali and Upper Volta to the north. The roughly square republic covers almost 125,000 sq mi (323,000 km²) and is thus rather larger than Italy.

Ivory Coast has developed rapidly since gaining full independence from the French in 1960 and is now the wealthiest of the West African nations. Under the leadership of Houphouët Boigny: there has been successful diversification of the economy, especially agriculture, and foreign investment has been enthusiastic. Ivory Coast's rich farm and forest output and its role as an outlet for other West African nations indicate increasing prosperity – provided the government remains stable and controlled by economic realists. Ivory Coast is a one-party state.

The Land. The southern third of the country consists of low, fertile plains covered in thick rain forest. Much of the coastline is fringed by sand-bars and extensive lagoons. The northern two-thirds of the country is savanna, averaging about 1,000 ft (300 m) above sea level, and rising in the Western Highlands to summits above 3,000 ft (900 m). In these mountains of the west, rainfall totals reach 80 in (2,000 mm), but the national average is about 60 in (1,250 mm). Temperatures hover around 80°F (27°C) in the equatorial south.

The People and the Government. With a population of about four million, Ivory Coast has far fewer people than the smaller country of Ghana lying to the east, and the need to enlarge the national labour force has led to immigration from neighbouring countries, especially Upper Volta. Ivory Coast's inhabitants are nearly all Negroes. Perhaps a quarter of the population is 'foreign' – comprising African immigrants and white residents. All told, over 60 ethnic groups are represented, notably the Baoule, Kina-Kouas, Krounen, Mande and Voltaic peoples.

The population is remarkably evenly distributed over the country, though the highest densities are in the south-east and centre.

Harvesting cocoa in a Yaou village in the Ivory Coast. Chocolate is made when dried and ground cocoa beans are mixed with cocoa butter and sugar.

The capital and chief port of Abidjan has a population of half a million, and other major towns are Bouaké, Man and Grand-Bassam.

Just over half of all school-age children receive education. There is a university at Abidjan.

Agriculture. By African standards Ivory Coast has a prosperous and robust economy, largely due to agricultural wealth. The chief food crop in the south-west is rice, whereas in the east yams and cassava are more important; and in the north maize and sorghum dominate. Cassava, plantains and fonio are widely dispersed. The wealth of Ivory Coast, however, has been largely built up on two tree crops: cocoa and coffee. Coffee is grown throughout the southern third of the country. Cocoa is grown chiefly in the south-east where it is characteristically grown in close association with coffee – cocoa on the hill tops or valley bottoms and coffee on slopes. Attempts to diversify have been successful; palm oil, rubber, sugar, cotton and pineapples now form increasingly important elements of the export structure.

Minerals and Forestry. Diamonds and manganese are both mined, the former in the alluvial workings of the Bou tributary of the Bandama River, and manganese near Grand Lahou. There are iron ore deposits in the north-west. Ivory Coast is the major African timber producer and probably has greater stands of useful timber than any other West African territory.

Industry and Power. The first paper factory in tropical Africa is at Bimbresso, just west of Abidjan, but the majority of industrial establishments are in and around the capital and chief port of Abidjan. Here the many industries include the manufacture of margarine, soap, motor-vehicle assembly, fruit canning and brewing. A chemical industry, too, has recently been established. Power for industrial development comes from a number of hydro-electric centres, notably at Ayame on the Bia River and at Kossou on the Bandama River.

Transport. Ivory Coast has a relatively highly developed network of communications. The great Vridi Canal, completed in 1950, has made possible the growth of Abidjan as a major port, for this canal connects the city through to the sea via the lagoon. Abidjan's role as the nation's chief transport centre is further assured by its location at the southern end of the railway to Upper Volta and of road links with five neighbouring countries. Abidjan also has an international airport.

GHANA

Area: 92,100 sq mi (238,539 km²)
Population: 8,545,561
Population Growth Rate: 3·0%
Capital: Accra, pop 848,825
Language: English. Twi, Fanti, Ga, Hausa and others.
Religion: Animist (46%) Christian (42%) Moslem (12%)
Currency: New cedi = 100 pesawas

Ghana is a West African republic on the Gulf of Guinea, sharing boundaries with the Ivory Coast, Upper Volta and Togo. With an area of 92,100 sq mi (240,000 km²) it is about the same size as the United Kingdom but has less than one-sixth as many people. In 1957 Ghana became the first European colony in Negro Africa to secure independence.

After independence from Britain a one-party socialist state gradually evolved with an ambitious programme of state welfare and economic development. Vigorous suppression of the opposition, controversial foreign policy and economic problems led to military coups in 1966 and 1972.

Land and Climate. Ghana is in general a low-lying country. Beyond a narrow coastal plain, the Kwahu Plateau extends inland, giving way to rolling savanna in northern Ghana. In the east a range of hills rises to little more than 2,000 ft (600 m) above sea level.

The Volta River system covers two-thirds of the northern part of the country, with its tributaries, the Black and White Voltas and the Oti. Since the completion in 1965 of Akosombo Dam, about 70 mi (113 km) from the sea, the Volta has been flooded to form the world's largest man-made lake, the 3,275 sq mi (8,482 km^2) Lake Volta created for hydro-electric power, irrigation and fisheries.

The climate of the country is dominated by the meeting of two airstreams. The *harmattan* wind from over the Sahara is dry, hot and dustladen; the monsoonal wind from the Atlantic is cool and moist. Dry or rainy seasons occur according to which wind is dominant. Annual mean temperatures vary from 79°F (26°C) on the coast to 84°F (29°C) inland, with the highest temperatures between February and April and the lowest in January. Total rainfall varies from more than 75 in (1,900 mm) in the extreme south-west, through 58 in (1,470 mm) at Kumasi to 43 in (1,010 mm) at Navrongo. The south-eastern coast is semi-arid.

Vegetation and Wildlife. Much of the Kwahu Plateau is covered with tropical rain forest, with trees up to 200 ft (60 m) high. Species such as African mahogany, wawa and sapele provide valuable commercial timbers. South of this region is lowland scrub and grassland, bordered in many places along the coast by mangroves. Northern Ghana is a region of dry savanna grassland.

Animal life in Ghana, less abundant today than it once was, includes elephants, leopards, antelopes, buffalo, monkeys and many kinds of reptiles and birds.

The People. A UN estimate for 1971 put Ghana's total population at 8,860,000. About 66% of this number dwell in rural villages, though urbanization is developing rapidly. Accra is the fastest-growing urban area and the country's main administrative, commercial and industrial centre.

Of the many Negro groups in Ghana, the most numerous is the Akan family, which includes the Fanti, living mainly along the coast, and the Ashanti, predominant in the forest regions of central Ghana. Other large groups include the Ga, Ewe and Moshi-Dagomba tribes. The official language of Ghana is English, spoken by most people, though tribal languages are also used.

Ghana has a high level of education, with 10 years of free and compulsory schooling. There are two universities.

Agriculture. Farming occupies over 60% of the population. Cocoa, introduced successfully into the country in 1879, has for over 50 years been the leading export. Other leading food crops are maize, yams, cassava, millet, guinea corn and plantain. Rice cultivation is found in parts of the south-west and tobacco is gaining increasing importance as a cash crop.

Much of Ghana is tsetse infested, and although sheep and goats are raised throughout the country, cattle are restricted to the savanna areas of the south-east and north.

Forestry and Fisheries. Forest products account for about 10% of the value of Ghana's exports. Logs still make up nearly two-thirds of timber exports but efforts are being made to increase the export of sawn timber, plywood and veneers from factories in Takoradi and Wiawso.

Ghana has a well developed fishing industry. Fishing from traditional native canoes, now assisted by outboard motors, still accounts for 43% of the catch, almost equal in tonnage to that from deep-sea vessels. The total catch of the fishing industry in 1970, including large amounts of herring and tuna, was about 65,000 tons.

The Volta Dam, started by President Nkrumah, allows Ghana to operate her own heavy industry. Lake Volta is the world's largest man-made lake.

Industry. After agriculture, mining is the most important source of Ghana's income. Gold, which originally attracted Europeans to the coast in the late 15th century, still accounts for 8% of export value. Diamonds (mostly industrial), bauxite and manganese are also exported. Other minerals include low-grade deposits of iron ore, and deposits of nickel, graphite, ilmenite and chromite.

Since independence in 1957, Ghana has pursued an active policy of developing the economy by diversifying its agriculture and by industrialization. In January 1966 the Volta River Project was opened, and as a part of the project, aluminium smelting at Tema has become Ghana's main industry, though the raw alumina is imported. Most industry, however, is still centred upon the processing of agricultural products such as cocoa, coffee, palm oil, coconuts and rubber, though there is some oil refining and steel smelting. The largest concentration of industrial development has been at Accra-Tema, Sekondi-Takoradi and Kumasi.

Transport and Trade. Ghana has about 800 mi (1,300 km) of railways, most of which runs between Accra, Kumasi and the coastal ports of Tema and Takoradi, with branch routes. There are over 20,000 mi (32,000 km) of roads, though many become impassable during the rainy season. Lake Volta provides a new waterway. There is an international airport at Accra.

Cocoa still provides over 60% of Ghana's export earnings, representing a dangerous overdependence on a crop whose value varies greatly from year to year with violent fluctuations in price on world markets. Other exports include timber, gold, diamonds, aluminium, bauxite and manganese. Machinery, transport equipment and foodstuffs still make up the bulk of Ghana's imports, along with raw materials such as alumina and crude oil.

Despite uncontrolled hunting, some elephants and lions occur in the north.

TOGO

Area: 21,621 sq mi (55,998 km²)
Population: 2,022,000
Population Growth Rate: 2·5%
Capital: Lomé, pop 141,000
Language: French
Religion: Animist (66%)
Currency: CFA franc = 100 centimes

Togo is the second smallest state in West Africa, forming a narrow strip between Ghana and Dahomey with Upper Volta to the north and the Gulf of Guinea to the south. This is a poor republic, heavily dependent upon subsistence agriculture, with largely undeveloped mineral assets, little industry and an unstable government.

French Togoland became a fully independent republic in 1960. In 1967 a military coup suspended the constitution and suppressed all opposition.

The Land. The coast is fringed with sandbars and lagoons. Togo's narrow coastal plain rises inland to a broad low plateau whose southern slopes are drained by the Mono River. On the northern edge of the plateau are the Togo-Atakora Mountains, rising to over 3,500 ft (1,000 m). North of the mountains is a thinly-populated low plateau drained by the Oti River, where floods often alternate with droughts.

Temperatures average 81°F (27°C) and the humidity is generally high, especially in the south. The coastal lands, however, are relatively dry: Lomé, the capital, averages less than 39 in (990 mm) of rain yearly. Inland, rainfall increases to about 70 in (1,778 mm) annually, diminishing again in the north.

Savanna woodland predominates, with tropical forest along rivers and in the mountains. Mangroves fringe the coastal lagoons. But crops have replaced much of the natural vegetation.

The People. Togo has an almost completely Negro population of 2,022,000, made up of many different tribal groups. In the relatively densely-populated south are the Ewe, Ouatchi and Mina. The Kabre and Losso predominate in the north-east, the only other densely populated region. More than 90% of the population live in rural villages.

The largest town is Lomé, the capital and chief port; other urban centres include Sokodé in the north and Palimé in the south-east.

Some 160,000 children attend primary schools but only 13,000 receive secondary education. This is in French, the official language.

The 1963 constitution established government by an elected president and National Assembly with ministers chosen by and answerable to the president.

The Economy. Nearly half of the Gross National Product of $205 million is provided by agriculture. Many of the small farms produce food crops such as cassava, corn, yams, millet and sorghum. The chief export crops – cacao, coffee and palm nuts – are grown on commercial plantations in the south. Cotton and peanuts have been introduced in the north.

Later deposits of phosphates are worked north-east of Lomé. Iron ore, bauxite and chromium deposits await exploitation. Industry is largely limited to processing agricultural products and manufacturing a few consumer goods.

Togo has 3,021 mi (4,860 km) of variable roads. Railways (305 mi; 490 km) link Lomé with Anécho, Palimé and Blitta. Lomé is the main seaport and airport. Togo trades mainly with France and other countries in the EEC and with Japan and the United States.

DAHOMEY

Area: 43,480 sq mi (112,613 km²)
Population: 2,760,000
Population Growth Rate: 2·5%
Capital: Porto-Novo, pop 74,000
Language: French. Fon, Mina, Yoruba, Dendi
Religion: Animist (68%) Moslem (15%) Roman Catholic (14%) Protestant (3%)
Currency: CFA franc = 100 centimes

Dahomey is a West African nation the size of Tennessee with fewer people than metropolitan Detroit. It is a poor country with a near-stagnant economy and high unemployment. Economic disaster is averted only by foreign aid, notably from France.

Dahomey, once a slave exporting country, formed part of French West Africa until becoming an independent republic in 1960. Since then the country has been plagued by government crises and military coups, largely provoked by the inability of successive regimes to cope with the country's severe economic problems.

Behind Dahomey's coastal strip lie a number of lagoons settled by fishing people.

Land and Climate. Dahomey is a long and narrow country aligned north-south. The short southern coastline on the Gulf of Guinea consists of sand-bars and lagoons which obstruct the outflow of the Ouémé and other rivers to the sea. The interior is low plateau country with small highland areas, such as the Atacora Mountains in the north-west.

The climate of the south is hot and humid with an average annual rainfall of about 50 in (1,270 mm). Inland, rainfall decreases and the dry season increases in length and intensity. Vegetation changes from the forests of the inland south to the grassland savanna of the north where lions, elephants and antelopes live.

The People. Most of the predominantly negroid people live in the south, where the two major groups are the Fon, with Abomey as their ancient centre, and the Yoruba, who are much more numerous in neighbouring Nigeria. Among the many interesting northern groups are the Somba, who live in large fortified compounds.

Most Dahomeans are animists, but there are sizable Moslem and Christian (mainly Roman Catholic) communities.

Less than 30% of the people are literate, but the need to find jobs for skilled people encourages gross overstaffing of the civil service.

Economic Production. Dahomey is primarily an agricultural country, self-sufficient in food (cassava, maize, peas, sweet potatoes, chillies, beans and peanuts) and with large oil-palm plantations in the south.

Iron ore and chrome are known to exist, but there is yet no significant mining. Great hopes are pinned on the recently discovered offshore oil. There is little industry; most of it, such as food processing, is found in Cotonou and Porto-Novo.

Transport and Trade. The best roads are in the south, and the Benin-Niger railway runs up country to Parakou. Cotonou has an international airport and a fine new artificial harbour.

Oil-palm products provide nearly 70% of Dahomey's exports by value. Livestock, dried and smoked fish, peanuts, coffee and cotton are also exported. Imports include textiles, machinery and vehicles. France and other countries of the Franc Zone are the leading trade partners.

NIGERIA

Area: 356,669 sq mi (923,773 km²)
Population: 66,174,000
Population Growth Rate: 2·5%
Capital: Lagos, pop 875,417
Language: English. Hausa, Yoruba, Ibo, Edo.
Religion: Animist (43%) Moslem (38%) Christian (19%)
Currency: Naira = 100 kobo

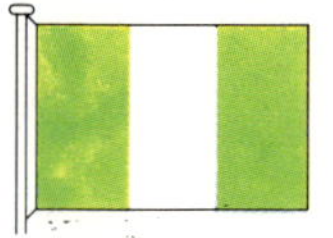

Nigeria is the fourteenth-largest country in Africa but the most populous. With its enormous diversity of ethnic groups, the federal republic has been dogged by problems since independence, chief among them being the creation of national unity and the stimulation of economic development. Into the 1970s Nigeria remained mainly an agricultural nation, but mineral and manpower resources were being increasingly harnessed to satisfy international needs, and the large home market. Given internal peace and skilled leadership, Nigeria has better prospects of prosperity than most other African nations.

Emergent Nigeria. In 1914 the British government formed the colony and protectorate of Nigeria from the former protectorates of Lagos and Southern and Northern Nigeria. It became a federation – under a governor-general – in 1954, a sovereign member of the Commonwealth of Nations six years later, and a republic in 1963.

On independence, in 1960, Nigeria was a federation of three regions – Western, Northern and Eastern. A fourth region, the Mid-West, was created in 1963. The Northern Region, however, covered two-thirds of the country's area and had just over half the total population. The imbalance led to frequent crises: in 1966 the constitution was suspended and a military government took control, and a year later the country was re-divided into 12 states, to provide a more balanced federal structure. This did not prevent the secession of the eastern states as 'Biafra' under Lieutenant Colonel Ojukwu. The bloody civil war which followed continued until Biafra's defeat and reabsorption in January 1970.

Since the end of the civil war, Nigeria has begun to tackle the problems of reconciliation and reconstruction in the east. Economic recovery has been boosted by the rapidly increasing exploitation of abundant oil deposits.

The Land. The Niger River flows south-eastwards through western Nigeria, bearing due south below Lokoja, where it is joined from the east by its chief tributary, the Benue. The Niger empties into the Gulf of Guinea through an extensive swampy delta. Inland from this low coastal region is a hilly belt pierced by river valleys. North of the broad valleys of the Niger and the Benue lies the Jos Plateau, which rises to over 5,000 ft (1,520 m) and then falls away farther north to sandy plains averaging 2,000 ft (610 m) in height. To the south-west lie the Yoruba Highlands and in the east, along the border with Cameroon, is another highland area (also over 5,000 ft). The Jos Plateau and the Yoruba Highlands form important watersheds, with many streams flowing north and south.

Climate. Nigeria's climate is influenced by its location just north of the equator. In the southern parts of the country annual temperatures average between 70°F (21°C) and 90°F (32°C). In the north temperatures are more extreme, ranging from 50°F (10°C) in winter to as high as 110°F (43°C) in summer.

Most of Nigeria is favoured with abundant annual rainfall. Precipitation varies from over 100 in (2,540 mm) in the coastal areas to about 20 in (508 mm) in the extreme north. In the south there are rains for between 8 and 11 months of the year, with two dry seasons occurring some time between November and March and during July-August. In the north, the rainy period lasts between 4 and 7 months, and there is drought from October to June.

Vegetation and Wildlife. There are several distinct vegetation zones. Mangrove swamps line much of the coast. Behind these, rain forests of sapele, obeche and other valuable timber trees extend up to 150 mi (240 km) into the interior. These are succeeded by savanna with tall grasses and scattered broadleaved trees. In the dry far north the grass is shorter and interspersed with thorn trees and bushes. Much of the vegetation has been cleared for farming.

Nigeria has a wide range of tropical wildlife, though some species – including giraffes and elephants – are increasingly threatened by the spread of agriculture. Several species of apes and monkeys inhabit the forest regions, while crocodiles, turtles and hippopotamuses occur in the rivers. Leopards are found in most parts of the country, and lions, cheetahs, caracals, and various antelopes range over the savannas.

The People. Nigerians belong to some 250 different but predominantly Negro ethnic groups, each with its own language or dialect (though English is the widely spoken official language). The four principal ethnic groups are the Yoruba (about 13 million) in the south-west, the Ibo (7·8 million) in the east, and the Hausa (6·8 million) and Fulani (5 million) in the north. Other important groups are the Edo in the west; the Kanuri, Tiv and Nupe in the north; and the Ibibio and Ijaw in the east.

In terms of population density, Nigeria can be divided into three broad zones. The three northern states (North-Western, North-Central and Kano) have a fairly high density, with Kano having 347 persons per sq mi (134 per km²). The three middle states (North-Eastern, Benue Plateau and Kwara) have a relatively low density, while the six southern states (Lagos, Western, Mid-Western, East-Central, South-Eastern and Rivers) have a high population density, reaching 1,045 per sq mi (403 per km²) in Lagos State.

Over-population and the inadequacy of local resources in certain regions have led to increasing migration within the country, particularly to the urban centres. Nigeria is highly urbanized, some of its towns and cities dating from medieval times. The capital, Lagos, is Nigeria's largest city, with a population of 875,417.

Education and Religion. Although education has made considerable progress since the 1940s, the country still suffers from a shortage of schools (particularly secondary schools) and trained teachers. Almost two-fifths of Nigeria's children do not attend school. The overall illiteracy rate is still very high, at around 75%. Only a few states have free primary school education, and many primary and secondary schools are still run by Christian missionary societies, except in the Moslem areas of the north. Nigeria has about 16,330 primary and secondary schools, with some 3,252,350 pupils. There are some 40,200 students in higher education, the five universities having a combined enrolment of over 7,000.

Northern Nigeria is predominantly Moslem and about half the total population is Moslem. Christianity is strongest in the south and west, and its adherents constitute roughly a third of the population. However, a very large proportion – including many who have adopted Christianity – still follow animist beliefs and traditional religions.

The walled inner city of Kano, in northern Nigeria, is made mostly of mud.

Agriculture. About 56% of Nigeria's gross

Lagos, Nigeria's chief port and capital, occupies three islands in the Gulf of Guinea.

Groundnuts are stacked in great pyramids while awaiting shipment from Kano.

national product – estimated at $9,900 million in 1971 – comes from agriculture, which employs about 70% of the country's labour force. Nigeria grows almost all its own food. The southern forest areas grow mainly root crops of yams, cassava and coco-yam, with maize and rice in some areas. In the grassland areas and in the north cereals predominate, especially millet and sorghum. Animal husbandry is confined to the north, where it is a major occupation.

The production of major cash crops shows strong regional specialization: oil-palm products come principally from eastern Nigeria; rubber is produced in the mid-west, cocoa in the west, soybeans in the Benue Valley, cotton and peanuts in the north, and tobacco in the north and west. Southern Nigeria is an important source of hardwood timber.

Minerals. Mining plays an increasingly important role in the development of Nigeria's economy and contributes 5% of the Gross National Product. There are large deposits of petroleum in the south-east, particularly in the Niger Delta. Crude oil represents the country's major export commodity, and by late 1972 output – which had tripled in six years – reached 1·8 million barrels per day.

Next in importance are columbite (Nigeria is the leading producer of this ore, providing some 95% of the world supply) and tin, both mined on the Jos Plateau. The Enugu area has large deposits of coal which are exploited for domestic use. Other minerals include limestone, gold, iron, zinc and wolfram.

Industry. Though Nigeria has mostly secondary industries, this sector is expanding rapidly and is having a growing impact. In the 1960s, major efforts were devoted to the importation of machinery, skilled manpower and semi-processed raw material necessary for the domestic manufacture of previously imported consumer goods. Because this kind of industrial development was in itself dependent on imports, it brought special significance to the port cities, particularly Lagos. Consequently, Greater Lagos has more than 30% of the country's industrial establishments. Some 16 urban centres account for most of Nigeria's industrial output. This concentration of industrial activity has had considerable impact on the pattern of population movements within the country.

Manufacturing now contributes 6% of the Gross National Product. The chief products are processed foods, refined oil, textiles, cement, peanut oil, soap, cigarettes, ceramic goods, plywood, beer and soft drinks.

Industrialization gained an added impetus from the completion in 1969 of the Kainji Dam on the Niger River, which provides cheap hydro-electric power. The dam was also intended to increase irrigation, improve navigation on the Niger and create opportunities for new fisheries.

International Trade. With the exception of crude oil – which represented 58% of Nigeria's exports in 1970 – the main exports are agricultural products. Nigeria is the world's leading producer of peanuts and the second-largest producer of cocoa. It is also a leading exporter of oil palm products, rubber and cotton. The United Kingdom takes about 28% of the country's exports, followed by the Netherlands, the United States, France and West Germany. Nigeria imports mainly machinery, motor vehicles, chemicals, foods and various manufactured goods. Some 31% of the imports come from the United Kingdom, other sources being the United States, West Germany, Japan and Italy.

Transport. Nigeria's transport system forms a network running roughly south to north, linking the interior with the ports in the south. There are 2,356 mi (3,770 km) of railway track, and some 55,000 mi (88,000 km) of roads, of which 9,500 mi (15,200 km) are surfaced. The waterways provided by the Niger and Benue rivers give some limited, though highly seasonal, facilities. There is a growing network of internal air routes linking the state capitals, and Lagos and Kano have international airports. The two main ports of Lagos and Port Harcourt have proved inadequate, and there has been vigorous development of new ports, notably at Warri in the Niger Delta and at Calabar in the extreme south-east.

Marua village, Cameroon. As with many west African countries, its north is arid.

CAMEROON

Area: 183,569 sq mi (475,444 km²)
Population: 5,836,000
Population Growth Rate: 2·2%
Capital: Yaoundé, pop 178,000
Language: French (East), English (West) African languages
Religion: Animist (52%) Christian (33%) Moslem (15%)
Currency: CFA franc = 100 centimes

Cameroon, which lies between West and Central Africa, is one of the continent's most varied countries, with contrasting climatic regions, two official languages (French and English), and hundreds of different tribes and local tongues. It was formed in 1961 when the small British Southern Cameroons joined the much larger Republic of Cameroon (formerly French Cameroons) to make the present federal republic. In that year the northern part of the British Cameroons voted to join Nigeria. The two states of the Federal Republic are now called East Cameroon (with 80% of the people and 90% of the land area) and West Cameroon.

Land and Climate. Cameroon is divided into five different regions. Along the coast runs a hot and humid plain covered with marshes and mangrove forests. Inland is the southern region, a plateau smothered in dense rain forest, which shades into the Adamawa

Plateau, running east-west with an average elevation of 3,400 ft (1,036 m), a transitional area with rain forest in the south changing to savanna in the north. Beyond this plateau, the northern savanna, covered in grass and shrub, runs down to the Lake Chad basin. Finally, the western mountains, an irregular chain of plateaux, hills and peaks, runs along the western border almost to Lake Chad, with Mt Cameroon (13,350 ft; 4,070 m), West Africa's highest point, rising at its lower end above the coastal plain. In the coastal area, especially around Mt Cameroon, rainfall rises to over 160 in (4,000 mm) a year, and here are some of the wettest places on earth. Rainfall diminishes to the north, but the countrywide average is 40–70 in (1,000–1,800 mm) a year.

The People. Of Cameroon's some 6 million basically Negro population 80% live in East Cameroon. The population is widely scattered. The only major concentrations are around the capital, Yaoundé (178,000) and the chief port, Douala (250,000). Over 100 different tribes live in the country, speaking as many different tongues. In addition, the influence of English in West Cameroon contrasts with the influence of French in the rest of the country. Most of the population are farmers. The northern people raise cattle while those in the south and west grow crops. Small groups of pygmies live in the southern forests. The people grow millet, manioc, cassava, tuberous vegetables and palm oil. Most worship tribal spirits, but there are strong Roman Catholic minorities in the south, and Moslems in the north. State primary schools, supplemented by Christian missions, are well attended in the south, less so in the north. There is a university at Yaoundé, with over 1,000 students.

The Economy. The area around the capital at Yaoundé and the principal port, Douala, is the economic heart of Cameroon. The two towns are linked by rail and the region is well developed agriculturally and has the country's only industrial activity. Most agriculture is of the subsistence type. The region's chief export crop (and that of the entire country) is cocoa, followed by coffee, cotton and rice. The region also produces large tonnages of hardwood timber. A major hydro-electric generating plant on the Sanaga River supplies power for a number of small factories but especially for the large Edéa smelter producing aluminium, Cameroon's most important industrial product. Outside this region, the north produces peanuts and cotton and, with the Adamawa Plateau, is an important cattle raising area. Plantations are well developed in West Cameroon, which produces bananas and rubber for export, as well as palm nut oil.

Cameroon's principal trading partners are the countries of the Common Market (especially France), and the USA and Japan. Guinea provides most of the bauxite smelted at the Edéa aluminium plant.

Transport. With Cameroon's diversity of language, historical origins and economy, good transport links are all-important in developing political unity, stimulating export production and encouraging economic interdependence between the regions. Important in this respect has been the recent extension of the railway (only about 294 mi; 483 km) both north and west, and improvement and extension of the road network of about 8,500 mi (13,680 km). Traditionally, many northern products are traded through Nigeria. If these can be diverted south through Douala by means of improved communications, that port may become a major outlet for the northern section of Central Africa. Both Douala and Yaoundé are already international airports of some importance.

CENTRAL AFRICAN REPUBLIC

Area: 240,535 sq mi (622,986 km²)
Population: 1,637,000
Population Growth Rate: 2·2%
Capital: Bangui, pop 302,000
Language: French, Somgho
Religion: Animist (60%) Christian (35%) Moslem (5%)
Currency: CFA franc = 100 centimes

The Central African Republic is a landlocked country in the heart of Africa, forming a link between the Sudan and the Congo basin. It is one of Africa's most thinly populated states and suffers from its isolation and lack of natural resources. The country was part of French Equatorial Africa until achieving independence in 1960. General Jean-Bédel Bokassa seized power in 1966 and in 1972 had the sole political party appoint him president for life.

Geographical Features. The country consists mainly of undulating plateau surfaces at about 2,500 ft (760 m) which form the watershed for drainage southwards to the Ubangi-Congo system and northwards to the Chari River and Lake Chad. In the north-east the Bongo Massif reaches a height of 4,593 ft (1,400 m). Most of the country is grassland, grazed by antelopes. The grassland becomes drier and treeless in the north. A dense tropical rain forest inhabited by gorillas and chimpanzees covers the southern part of the country, which becomes virtually inaccessible during the rainy season (June-October), when the numerous rivers become flooded.

Annual rainfall averages from 64 in (1,625 mm) in the south to 44 in (1,117 mm) in the north. Temperatures average about 80°F (27°C).

The People. There is no dominant tribal group in the Central African Republic, which has about 80 different ethnic but basically Negro groups. The principal groups are the Ubangi in the south, the Mandia and Baya in the west, the Sara in the north-west and the Banda in the centre. The most common spoken language is Sangho, though there are many other languages and dialects.

Overall population density is 6 persons per sq mi (2·7 per km). The only sizable city is the capital, Bangui (302,000). While 25% of the population is classed as urban, many town dwellers actually live by farming.

Only about 40% of the children attend school, and the rate of illiteracy is 82%. Most of the people are animists, though one-quarter are Christians and there are Moslems in the north.

Economic Production. Agriculture is the predominant element in the republic's economy, even though only about 2% of the land is actually under cultivation. The great majority of the population are engaged in subsistence farming, growing millet, sorghum and maize. In 1971 output of the main cash crops was: cotton, 53,000 tons (54,000 metric tons); coffee, 12,164 tons (12,359 metric tons) and peanuts, 7,280 tons (7,400 metric tons). Other products include rubber, tobacco and palm oil. The tsetse fly prevents any significant expansion of livestock husbandry. In the South-west there are forest resources.

The country's mineral resources are few, though diamonds have become the chief export (493,605 carats were produced in 1970), representing over 40% of the total exports. Some gold is also mined, and there are plans to exploit known deposits of uranium.

Industry is restricted to the processing of local raw materials and some light manufacturing. The one large industrial complex is the textile plant at Bouali (40 mi – 60 km – north-west of Bangui), which uses hydro-electric power. There are a number of plants for such activities as the ginning of cotton and the extraction of oils from cotton seed, peanuts, sesame and palm nuts.

International Trade. The republic's balance of trade has shown a deficit for some years, and the country relies on a considerable amount of aid from France. France is the main trading partner and accounts for over 50% of both imports and exports. The Central African Republic also trades with other Common Market countries, and with Israel and the United States. The chief exports are diamonds, cotton and coffee. The republic imports machinery, electrical equipment, textiles, vehicles, chemicals and metals.

Transport. The focal point of the country's transport network is the capital, Bangui, which is a river port on the Ubangi. The Ubangi River is the chief link with other countries and much of the republic's trade is handled by river traffic between Bangui and Brazzaville in the Congo Republic.

Of the 13,300 mi (21,280 km) of roads and tracks, only 3,750 mi (6,000 km) are all-weather roads. There are no railways, but there are domestic and international air services to and from Bangui.

CONGO

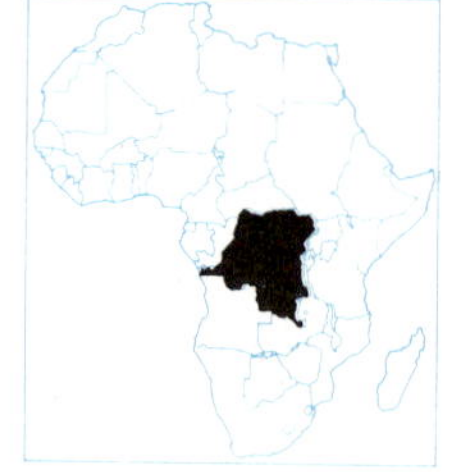

Area: 128,127 sq mi (331,849 km²)
Population: 958,000
Population Growth Rate: 2·1%
Capital: Brazzaville, pop 200,000
Language: French, Lingala and Kongo
Religion: Animist (50%) Roman Catholic (49%) Moslem (1%)
Currency: CFA franc = 100 centimes

The People's Republic of the Congo, also known as Congo (Brazzaville), in equatorial west-central Africa is a hot, humid and largely forested country about the size of Finland but with a population a little larger than that of the English city of Birmingham. Poor in natural resources, the Congo has had to rely on foreign aid since this former part of French Equatorial Africa became independent in 1960. Its history has been turbulent. A change of government in 1963 led to the adoption of generally left-wing policies and the Congo has vigorously courted the communist countries, particularly China. The Congo's relations with neighbouring Zaire have been poor.

Geographical Features. The eastern boundary of the country is formed for the most part by the Congo and Ubangi rivers while in the west there is a short Atlantic coastal strip. The equator crosses the northern part of the country. The coastline is relatively smooth and the immediate coastal zone is sandy in the north but more swampy below the mouth of the Kouilou River. There are numerous lagoons backing the coast. From the narrow coastal plain the forested Mayoumbé Mountains rise abruptly to heights of 2,600 feet (790 m). To the east, the Niari Valley is a broad depression of woods and savanna formed in softer sandstones and with superior soils. A further upland area to the north of the Niari Valley provides the main watershed between drainage to the coast and to the Congo Basin. The hilly Brazzaville area gives way northwards to the Batéké Plateau, fairly level surfaces of savanna deeply dissected by forested tributary valleys of the Congo. The Sangha region of the north is true Congo Basin country with dense equatorial forests.

The climatic pattern is essentially equatorial, with average temperatures ranging between 70° and 80°F (21°–27°C). There is rainfall throughout most of the year, averaging 80–100 in (2,000–2,500 mm).

The ridges and rain forests of the Mayombé Escarpment divide coastal from inland Congo.

The People. There are four main ethnic groups in the Republic of Congo: the Kongo, the Batéké, the M'Bochi and the Sangha. There are some 12,000 Europeans, mostly of French descent. The population is concentrated in the south. There has been rapid urbanization since the early 1960s and some 39% of the population are found in the urban centres, the largest of which are the capital, Brazzaville (200,000), and Pointe Noire (100,000).

About 48% of the population are Christians. There are about 4,500 Moslems, and the rest are animists. The literacy rate is only about 22%, but is steadily rising. There are about 950 primary and secondary schools, with a total of over 237,370 pupils.

Agriculture and Forestry. The Congo Republic is essentially an agricultural country. Oil-palm plantations are found mainly in the Likouala and Sangha river areas in the east. Coffee and cocoa production is scattered in the forest areas and tobacco is grown on scattered holdings in the Batéké region and on larger units in the Niari Valley. The Niari Valley is particularly favoured by soil, climate and water supply, and has extensive areas under sugar-cane, bananas, peanuts, tobacco, maize, manioc, cotton, vegetables and rice. Cattle ranching has been established on the drier margins of the valley.

Timber is the country's principal export in terms of value, and forest reserves occupy an estimated 42 million ac (17 million ha), or roughly half the country. About a quarter of this forest area is found in the Mayoumbé and Chaillu areas and the rest is in the north. Annual timber exports are usually in excess of 400,000 tons.

Minerals and Industry. Annual petroleum production had declined to less than 50,000 tons by the early 1970s. There are small quantities of tin, lead, zinc, copper and gold. A more significant development for the future is the increasing production of high-grade potash.

Industry is concerned mainly with the processing of timber and food products. The south has saw-mills, plywood factories, sugar refineries, breweries and flour-mills. There is hope that a hydro-electric scheme on the Kouilou River will provide power for further industrial development.

Transport and Trade. The Congo-Ocean railway between Pointe Noire and the river port of Brazzaville, with the navigable sections of the Congo, Sangha and Ubangi rivers, constitutes the main routeway for interior Chad and the Central African Republic as well as the Congo, and handles much of the neighbouring countries' trade. The road network is poorly developed and there are no surfaced roads outside the main towns. Pointe Noiré is the main seaport, and Brazzaville has an international airport.

The Congo exports mainly timber, industrial diamonds and palm products; and imports machinery, textiles and chemicals. Chief trading partners are France, West Germany, the Netherlands, Britain and the United States.

A street market in Libreville, capital of the newly independent Gabonese Republic.

GABON

Area: 103,088 sq mi (266,998 km²)
Population: 500,000
Population Growth Rate: 1·3%
Capital: Libreville, pop 105,080
Language: French. Fang (North) and Bantu
Religion: Roman Catholic (50%) Animist (49·5%) Moslem
Currency: CFA franc = 100 centimes

Gabon is a thickly forested West African republic on the equator, a little larger than the United Kingdom. Since gaining independence from France in 1960, Gabon's small and poor population has increasingly profited from some of Africa's richest forest and mineral resources. Now, international loans offer hope of rapid economic advancement.

Land and Climate. Gabon's surface is largely sculpted by the westward flowing Ogooué River and its tributaries, which have carved basins and valleys out of a broad inland plateau whose highest point is Mt Iboundji (5,167 ft; 1,849 m). The coastal lowland is narrow in the south and north but broadens in the lower Ogooué region. Gabon's equatorial climate features uniformly high temperatures, high relative humidity and heavy rainfall. Rain forest vegetation covers 75% of the country and only in the drier south-east and south is there savanna.

The People. With only 500,000 people (5 per sq mi; 2 per km²), Gabon is one of Africa's most sparsely populated nations. Gabonese who live in rural villages cultivate forest plots of cassava, yams and bananas. The shy, primitive forest pygmies live by hunting and food gathering. Many Gabonese have moved to the towns, and the capital, Libreville (pop 105,080), and Port Gentil (pop. 30,000) together contain about 25% of the population.

The northern Fang tribe is the politically dominant group and Fang is the leading northern language. Bantu dialects are predominant in the south, and French is taught in the schools. Today, nearly 90% of all children of primary school age attend schools, and secondary school attendance is rapidly increasing. Christianity is supplanting animism as the chief form of religious belief.

Agriculture and Forestry. The country's large forest area is potentially productive, but small-scale subsistence farming is the rule and production of cash crops is poorly developed. Coffee and cocoa are grown in the fertile north-west, and oil-palms are cultivated around Kango, Lambaréné and Moabi. Rice has been successfully introduced in the southern Tchibanga region.

Until the late 1950s timber provided as much as 90% of the value of exports. By 1968 its share had dropped to 34% but overall production had increased. The most important wood is okoumé (*Aucumea klaineana*), used in making veneers and plywood. The main forestry concessions are in north-central Gabon.

Minerals. Since the 1950s, mineral exploitation has transformed the economy of Gabon: by 1968 minerals provided 61% of the value of Gabon's exports. In 1955, the discovery of an important reserve of heavy oil in the central coastal zone foreshadowed later finds to the south and a rapid growth in annual production: from 1·5 million tons in 1966 to over 5 million tons in 1969.

Moanda, in the south-east, has one of the world's largest manganese deposits. Mining commenced in 1962 and COMILOG (Compagnie Minière de l'Ogooué) now exports around 1·5 million tons a year. At Mounana, also in the south-east, uranium has been mined since 1961 and 1,500–2,000 tons of uranium concentrate are now exported.

There are huge proven reserves of iron ore and active planning is in hand to exploit them.

Industry. Although Gabon is essentially a primary producer, it is now developing industries based on local raw materials. Most of the timber cut is exported; about 15% is processed in Gabon. An oil refinery opened at Port Gentil in 1967. Coffee and palm oil are processed in the producing areas, while manufactures, located mainly at Libreville and Port Gentil, include cement, paint, furniture, soap, textiles, beer and mineral waters.

In 1969 work started on the Kinguelé hydro-electric power scheme, but by the early 1970s electricity was still supplied by the natural gas-fired plant at Port Gentil and diesel plants elsewhere.

Transport and Trade. Transport as yet is poorly developed, since most of Gabon's 1,200 mi (1,900 km) of roads are impassable to heavy transport. The new Owendo-Belinga railway, Gabon's first, and the first deep-water port at Owendo (under construction) will vastly improve transport and shipping facilities.

Gabon's trade produces a considerable balance of payments surplus. In 1968 unprocessed minerals and petroleum accounted for 55% of the total value of exports, followed by timber (27%), processed timber (8%) and processed minerals (6%). In terms of value Gabon's main customers are France (34%), the USA (12%), and Curaçao (12%).

EQUATORIAL GUINEA

Area: 9,828 sq mi (25,454 km²)
Population: 290,000
Population Growth Rate: 1·7%
Capital: Santa Isabel, pop 60,000
Language: Spanish. Fang, Bubi and Ibo
Religion: Roman Catholic (66%) Protestant, Moslem, Animist.
Currency: Guinea peseta = centimos

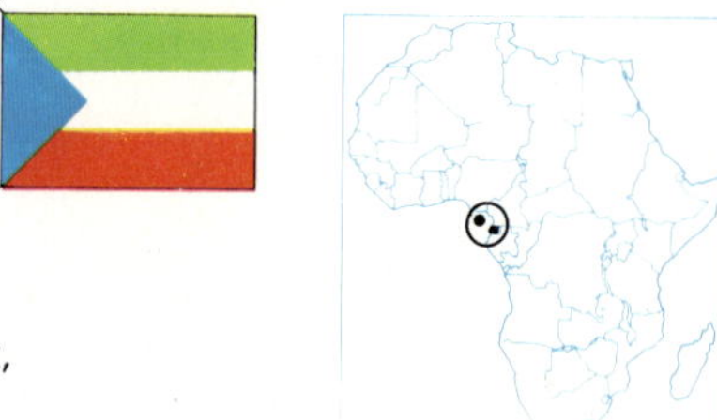

Equatorial Guinea, formerly known as Spanish Guinea, was one of the last of the colonial territories of Africa to become independent (in 1968). The republic consists of two provinces: Río Muni and Fernando Po. Over 10,000 sq mi (26,000 km²) of the new state are in Río Muni, comprising the west coast of equatorial Africa between Cameroon and Gabon and the tiny offshore islands of Corisco, Great Elobey and Little Elobey. Fernando Po, 779 sq mi (2,030 km²), lies in the Gulf of Guinea 149 mi (240 km) off the coast of Cameroon and the province includes Annobón, a tiny island of only 7 sq mi (18 km²), 440 mi (644 km) farther south-west.

Fernando Po, in spite of its small size, has most of the population and resources of the new republic. Fernando Po island is volcanic, rising steeply from a narrow coastal plain to twin peaks: Santa Isabel (9,350 ft; 2,850 m) and San Carlos (6,580 ft; 2,010 m). The lowland climate is equatorial, and natural vegetation varies with altitude, from tropical forest, to woodland savanna, montane woodland, grassland and moor. Soils are fertile with plantations occupying most of the cultivated area. Cocoa and *liberica* coffee are the main cash crops. Above 3,000 ft (900 m), cattle are grazed and temperate vegetables are grown.

The indigenous population are the Bantu Bubi people, but commerce is largely run by the coast-dwelling Porto (of mixed Negro, Spanish and Portuguese ancestry), and most of the 25,000 migrant workers on the plantations are Nigerian Ibos. Few Spanish settlers remain. The port of Santa Isabel is the federal capital, handling cocoa and coffee exports.

Annobón is only 20% cultivable. Palm oil, coconuts, cocoa and coffee are exported, but Annobón is over-populated, and many men leave to work on the plantations of Fernando Po or in Río Muni.

Río Muni features a plateau which drops steeply to a coastal plain which occupies about a third of the province. The main river is the Río Benito. Temperatures hover around 78°F (26°C), and rainfall is heavy. Most of the land is forested; timber concessions employ immigrant labour, and okoumé and mahogany are exported from Bata, the provincial capital. Some coffee and cocoa are grown by Fang tribesmen and there is considerable offshore fishing. Río Muni was neglected by the Spanish in favour of the more fertile island of Fernando Po, and remains backward today.

SÃO TOMÉ e PRINCIPE

São Tomé e Principe in the Gulf of Guinea, with the adjacent islets of Pedras Tinhosas and Rôlas, have been under Portuguese rule since 1522, and now constitute Portugal's smallest overseas province. Their total area is 372 sq mi (964 km²), the largest island being São Tomé (330 sq mi; 854 km²), on which stands São Tomé (7,000), the capital and chief port.

Both São Tomé and Principe are of volcanic origin, and are rugged, mountainous and jungle-covered. Pico de São Tomé rises to 6,640 ft (2,024 m). Lying just north of the equator, São Tomé and Principe have a hot, steamy climate with temperatures around 80°F (27°C) and heavy rainfall.

Most of the population of 76,218 live on São Tomé, mainly in compounds on the plantations. Apart from some 1,500 Portuguese and the native-born islanders of mixed but mainly African stock, the population consists of contract workers from the Cape Verde Islands.

Large, Portuguese-owned cacao and coffee plantations have been carved out of the jungle on the eastern slopes and coastland of São Tomé. Other commercial products include coconuts, coffee, cinchona and palm oil.

SAINT HELENA

St Helena is a British crown colony in the South Atlantic, best known as the island of Napoleon I's exile and death (1821). Of volcanic origin, it covers 47 sq mi (122 km²). Its uplands, rising to 2,700 ft (823 m), have forest and grassland and moderate temperatures; the lowland areas, reaching out to formidable coastal cliffs, have higher temperatures and less rain, and cactus scrubland.

The population of over 5,000 is of mixed origin, with elements derived from the original British East India Company settlers, slaves from Madagascar and the East Indies, Chinese and Africans. The capital and port is Jamestown (about 1,600). The people speak English and are mostly Anglicans or Baptists. They grow potatoes, maize, oats and barley, raise livestock and fish. The closure of the government flax mills in 1966 was a severe blow. Flax is still grown, but now largely as an anti-erosion crop. The island retains some importance as a port of call and communications centre.

Dependencies. St Helena administers other British islands in the South Atlantic: the Tristan da Cunha group and Ascension Island.

Tristan da Cunha comprises several islands: volcanic Tristan da Cunha, the home of some 270 people who live by farming, fishing and working in a South African crawfish-freezing plant; the three small uninhabited Nightingale Islands, the haunt of seals and seabirds; Inaccessible Island; and Gough Island, with its meteorological station and guano deposits.

Ascension Island, with some 1,230 inhabitants, is lush and volcanic, and has been noted for its turtles. The main settlement is Georgetown.

ZAIRE

Area: 895,348 sq mi (2,318,951 km²)
Population: 22,477,000
Population Growth Rate: 4·2%
Capital: Kinshasa, pop 1,623,760
Language: French. Swahili, Kiluba, and many others
Religion: Roman Catholic (50%) Protestant and Animist.
Currency: Zaire = 100 makuta = 10,000 sengi

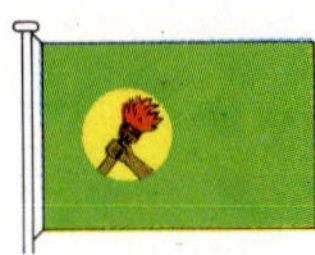

The Republic of Zaire (formerly the Belgian Congo, and then the Democratic Republic of the Congo) is the largest country in Africa after the Sudan, about a quarter the size of the United States. The nation is almost completely landlocked and it stands astride the equator. Zaire has immense mineral wealth, yet most of its people live lives of rural poverty.

A Turbulent Birth. Unrest and nationalist demands forced Belgium to grant the Congo its independence in·1960. The early years of independence were bedevilled by bitter rivalries between the different political leaders and by the continued involvement of Belgian troops and Belgian business interests. A series of army mutinies and other disturbances led to the virtual disintegration of the government and the secession of mineral-rich Katanga. United Nations troops were eventually called in to restore order, and the secession of Katanga ended in 1963.

General Joseph Mobutu came to power in 1965 and President Mobutu proceeded to establish strong central government and attract foreign investment. In 1971, Mobutu implemented an Africanization policy which involved the adoption of African place names and the expulsion of foreigners.

Geographical Features. Zaire covers a large part of the Congo (or Zaire) River Basin, which is a shallow crustal depression rising gradually from about 1,000 ft (300 m) at its centre to rather more than 3,000 ft (914 m) on the surrounding plateau rim.

The highland areas in the west and southeast rise above 6,000 ft (1,830 m), while the Ruwenzori Range on the Uganda border rises well over 10,000 ft (3,000 m). In the extreme west, there is a short 25-mi (40-km) coastline on the Atlantic Ocean. The edge of the Great Rift Valley in the east contains lakes Albert, Edward, Kivu and Tanganyika.

The most striking single feature is the great Congo River system. Known as the Lualaba along its upper course, the Congo flows some 2,900 mi (4,660 km). It rises in the south-east and flows northward, turning westward in a large arc, then south-westward to empty into the Atlantic. The Congo is fed by a great network of tributaries, including the Lomami, Ubangi, Kasai and Kwango rivers.

The Congo Basin has a characteristically equatorial climate, with a fairly constant mean temperature of 77°F (25°C) and two rainy seasons. Annual rainfall averages 60 in (1,524 mm). There tends to be a wider range of temperatures towards the south.

Vegetation and Wildlife. Most of northern and central Zaire is covered with tropical rain forests, while in the south there are extensive areas of grassland. Bamboo and scrub forests are found in the mountain areas, and the rivers near the Atlantic are surrounded by swamps and marshes.

There is an abundance of wildlife typical of Central Africa – including monkeys, gorillas, elephants, giraffes, lions, leopards, zebras, antelopes and hippopotamuses. There is a great variety of reptiles, birds, fish and insects.

The People. The great majority of Zaire's 200 ethnic groups belong to the Bantu peoples, with Sudanese and Nilotic peoples in the far north and east. There are many groups of sharply contrasting characteristics, like the Zande (Azande) in the north-east, the Luba in the south-east, or the Kongo in the lower Congo. The original inhabitants of the Congo Basin, the pygmies, have been reduced to about 100,000. The most important of the many African languages spoken are Lingala, Kingwana, Kongo, Luba and Swahili.

The population is unevenly distributed, the greatest concentration being around the edge of the Congo Basin. While about 70% of the people still live in rural settlements, urbanization is proceeding fairly rapidly. The largest city is the capital, Kinshasa (1,623,760).

Education and Religion. Zaire has about 11,000 primary and secondary schools, with some 2,185,000 pupils. The National University of Zaire has about 4,400 students at faculties in Kinshasa, Lubumbashi and Kisangani. There are also 16 establishments of higher education. The overall rate of illiteracy, at 60%, is still high.

Soldiers strolling in the market in Léopoldville (now Kinshasa), capital of Zaire. Zaire has been under military rule since General Mobutu seized power in 1965.

Many of the people still practise traditional religions involving ancestor worship and animist beliefs, but about half the population belongs to the Christian faith (mainly Roman Catholicism). There is also a Moslem minority.

The Economy. The main food crops of the near-subsistence agricultural economy vary from one part of the country to another, but cassava is most widespread. Maize occupies a smaller area, and in the drier savanna areas the staple food is often millet. Rice and bananas are cultivated mainly in the eastern provinces. The chief cash crops are cotton, coffee, rubber, palm oil, sugar and tea. Cotton is the most extensive cash crop. Zaire's forests have valuable woods such as teak, ebony, cedar and mahogany, but much of this great potential is unexploited.

Though the majority of the labour force is engaged in agriculture, Zaire's rich mineral resources are the mainstay of its economy. The world's leading producer of industrial diamonds, cobalt and pitchblende, Zaire also mines large amounts of copper, gold, tin, tungsten, manganese and zinc. Most of the deposits are in Shaba (formerly Katanga) province. Industrial diamonds come from the neighbouring province of Kasai-Oriental.

Most of the industries are related to exploitation of the mineral resources. Owing to the great distance from the coast and the consequent high cost of rail haulage Shaba ores are smelted at the mines. Manufacturing industries produce processed foods, cement, chemicals, textiles, and consumer goods.

Zaire, with its great rivers, has immense potential for the production of hydro-electric power. The possibilities of development on the lower Congo have as yet hardly been touched, though the Inga project is potentially one of the greatest hydro-electric power schemes in the world.

International Trade. The country's chief exports are copper (which accounts for more than half the total value of exports), industrial diamonds and coffee. Palm oil, cotton, rubber, tea and cocoa are also exported. Imports consist mainly of machinery, vehicles, foodstuffs, textiles and petroleum products. Belgium, with its long history of involvement in the area, is by far the leading trading partner, accounting for over 20% of Zaire's imports and exports. Other trading partners include West Germany, Italy, France, the United States and Britain.

Transport. Zaire's vast size and the fact that major areas of economic development occur around its edge make transport and communications a major problem. There is only one important ocean port, Matadi, and the most important railway line is that linking Matadi with Kinshasa. Most of the rest of the 3,622 mi (5,795 km) of railways serve the mining regions of Shaba.

The major transport network is formed by the Congo River and its tributaries and the lakes: there are about 8,590 mi (13,744 km) of navigable waterways. The chief river port, Kinshasa, is linked by water with Kisangani and Port Francqui.

Road transport is little developed other than on a local scale. Of the 87,500 mi (140,000 km) of roads, only some 1,250 mi (2,000 km) are surfaced.

RWANDA

Area: 10,166 sq mi (26,330 km²)
Population: 3,724,000
Population Growth Rate: 2·8%
Capital: Kigali, pop 20,000
Language: French and Kinyarwanda, Kiswahili
Religion: Roman Catholic (60%) Animist and Moslem
Currency: Rwanda franc = 100 centimes

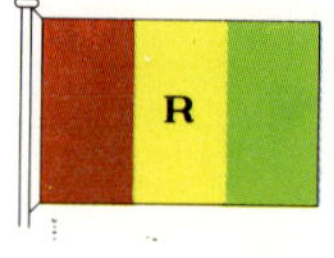

Rwanda is a poor, small, disease-riddled Central African republic. Once part of the UN trust territory of Ruanda-Urundi, it became independent in 1962.

Rwanda includes part of the East African Rift Valley, here largely occupied by Lake Kivu, from which the Ruzizi River flows southwards into Burundi. The high volcanic mountains forming the eastern edge of the valley include the Virunga Mountains, where Karisimbi, the country's highest peak, reaches 14,786 ft (4,507 m). Rivers flow eastward from the mountains across high plateaux, gradually descending towards Lake Victoria. Temperatures in the Rift Valley are usually over 68°F (20°C), and rainfall is everywhere over 40 in (1,016 mm).

The People. Rwanda is one of Africa's most densely populated nations. The Hutu (Bahutu), a farming people, form about 90% of the population. Other groups include the Tutsi (Watutsi) pastoralists, and the Twa, hunters and pygmies. Most live in small villages. Only some 10% are literate.

The Economy. The per-head value of economic output is Africa's lowest. Coffee, tea and pyrethrum are grown as cash crops. Small amounts of tin are mined east of Lake Kivu. Coffee and tin account for 80% of exports by value, but trade is limited since Rwanda's remoteness makes transport difficult and costly. Roads are very poor and there is no railway system.

BURUNDI

Area: 10,747 sq mi (27,835 km²)
Population Growth Rate: 2·3%
Capital: Bujumbura, pop 100,000
Language: French and Kirundi. Swahili (commerce)
Religion: Roman Catholic (50%) Animist (49%) Moslem (1%)
Currency: Burundi franc = 100 centimes

Burundi in Central Africa shares its northern neighbour Rwanda's standing as one of Africa's smallest and poorest states. Both were once part of German East Africa, later became a UN trust territory (Ruanda-Urundi) administered by Belgium, and achieved independence in 1962.

Burundi's monarchy survived independence until 1966 when a one-party republic was proclaimed. Since then, Burundi has seen political assassination, attempted coups and internecine tribal warfare. In 1972 there was widespread slaughter of Tutsis by Hutus, and of Hutus by Tutsis, and thousands fled to Zaire and Tanzania.

The Land. Stretching along Burundi's western border is part of the East African Rift Valley, here represented by the Ruzizi River flowing southwards from Lake Kivu, and by Lake Tanganyika. On its eastern edge, the Rift Valley rises steeply to a narrow range of mountains reaching an average height of 6,000 ft (1,830 m) and rich in fertile volcanic soil. Successive plateaux, separated by east-facing escarpments, fall away eastwards to Tanzania. Many streams flow down through deep valleys in the higher areas to join the Kagera River on its way to Lake Victoria.

Climate. The Rift Valley is warm, with average temperatures of 73°F (23°C) and a low rainfall for its latitude, of about 30 in (762 mm). Temperatures are lower in the mountains, where rainfall increases to almost 60 in (1,524 mm). This heavy mountain rain often comes in violent storms and causes much leaching of the fertile soils and serious erosion.

The People. No other African country is so densely populated. The average density of 350 per sq mi (135 per km^2) is approached only by Rwanda, which has 335 per sq mi (129 per km^2). The rapidly increasing population is expected to pass the 7 million mark by the end of the century.

There are three main ethnic groups: the Hutu, a farming people who form about 84% of the population; the pastoral Tutsi, a tall Hamitic tribe, forming about 15%; and the Twa, an aboriginal pygmy people who live by hunting. More than half the population is Christian (mainly Roman Catholic), the rest being mostly animists. Only 10% are literate.

The only large town is the capital, Bujumbura (formerly Usumbura), on the north-eastern shore of Lake Tanganyika. Its many fine buildings were erected during the period of Belgian administration.

Economic Production. Subsistence agriculture is the chief activity. Because the Rift Valley is so hot and arid it is unfavourable for settlement and agriculture, and the people have naturally been attracted to the fertile mountain slopes on its eastern edge, which are densely populated and intensively cultivated. Food crops like rice, yams, cassava, beans and sweet potatoes grow on the lower slopes, but higher up give way to millet and, above 6,000 ft (1,829 m), to wheat and barley. Cash crops range from *robusta* coffee, cotton and oil-palm to *arabica* coffee, tobacco, tea and pyrethum at higher levels. Soil erosion remains a problem in spite of terracing and eucalyptus planting. Attempts have been made to promote mechanized and co-operative cultivation of rice and cotton in the Ruzizi valley: but the overflow population from the scattered mountain farmsteads prefers temporary work in the copper mines of Zaire or in Tanzania to moving down permanently into the traditionally unhealthy Rift Valley. Cattle are reared by the Tutsi on the plateau grasslands.

Lake Tanganyika is rich in fish, and large quantities are caught and dried, finding a ready market in a country which is short of protein foods.

Significant mineral deposits have still to be found. Small quantities of gold, tin and kaolin are produced, and oil may yet be found in the Ruzizi valley. Bujumbura is the embryonic industrial centre producing blankets and domestic utensils, and processing local coffee and cotton. Power comes from a hydro-electric plant on the Ruzizi River.

Trade and Transport. Some cotton is exported, but the main export is coffee. About 90% of the crop goes to the United States. The distance of Burundi from the sea, and its poor internal communications, limit foreign trade. The republic has no railways and only poor roads. From Bujumbura on Lake Tanganyika international traffic goes by water, either south to Kigoma (Tanzania) and thence by rail to Dar es Salaam, or west to Kalemi (formerly Albertville) in Zaire and thence by rail and the River Congo to Matadi.

Burundi's main export is coffee. High quality Arabica is grown in valleys such as this one.

UGANDA

Area: 91,134 sq mi (236,037 km^2)
Population: 10,127,000
Population Growth Rate: 4·4%
Capital: Kampala, pop 331,000
Language: English. Luganda
Religion: Christian (50%) Moslem (6%) Traditional remainder
Currency: Uganda shilling = 100 cents

Uganda lies mostly north of the equator, to the north and west of Lake Victoria. Its great range of scenery, wildlife parks and fertile farmlands make this republic perhaps the most beautiful country of east-central Africa.

Much of the country's history has been turbulent. When Uganda became independent from Britain in 1962, it was still plagued by rivalries between the four traditional kingdoms, among which Buganda was dominant. The federal system was abolished in 1966 and the country became a republic the following year. In 1971 Major-General Idi Amin established a military regime.

His impulsive one-man rule and lightning drive to Africanize the economy created serious problems for his country. In particular the expulsion of thousands of non-citizen Asians in 1972 left a serious vacuum since Asians had played a prominent part in the country's commerce.

Geographical Features. Some 15% of Uganda's area is water: parts of lakes Edward, Albert and Victoria, all of Lake Kyoga, and the Victoria and Albert branches of the River Nile.

More than 80% of the land area is a plateau between 3,000 and 5,000 ft (900–1,500 m) above sea level, with highlands to the east and west. In the east, on the border with Kenya, Mt Elgon rears its volcanic mass to 14,178 ft (4,321 m). In the south-west, the Ruwenzori Range rises to 16,763 ft (5,110 m) with Mt Margherita.

The plateau slopes gently to Lake Kyoga in the centre. The western highlands fall away sharply to lowlands along lakes Edward and Albert and along the swamp-fringed Albert Nile River.

Uganda's climate is ameliorated by its altitude and by its extensive bodies of inland water. Midday temperatures average 80°F (27°C) throughout the year. Rainfall varies from 30 to 60 in (762–1,524 mm) on the plateau, and is heaviest around Lake Victoria

Uganda is famed for the range of its scenery, its wildlife parks and its fertile farmlands.

and in the mountains. Only the extreme north-east and some southern areas have less than 30 in (762 mm) rainfall yearly.

Vegetation and Wildlife. Northern and central Uganda are dominated by wooded savanna. The vegetation changes to dry steppe in the east and open grassland in the south and south-west. There are areas of mixed forest and elephant grass near lakes Victoria and Albert. Forests and moorlands mark the mountainous regions.

Uganda has the typical wildlife of central and eastern Africa – including lions, leopards, elephants, rhinoceroses, hippopotamuses, antelopes and giraffes. Uganda has three national parks which help to conserve its wildlife.

The People. Uganda has some 20 different African tribes. The Baganda in the southern and central regions are the largest and most prosperous tribe, numbering over one million. These and other Bantu peoples are contrasted with the Nilotic northern groups, such as the nomadic Karamojong. About two-thirds of the African population speak Ki-Swahili and other Bantu languages; the rest speak various languages belonging to the Sudanic, or Nilotic, group.

Population density is high by African standards – averaging about 127 inhabitants per sq mi (49 per km²). The south is the most densely peopled area, and the largest city is the capital, Kampala (331,000). The once substantial Asian population has been considerably depleted since 1972. There are about 10,000 Europeans. Less than 8% of the population live in urban centres.

Uganda has some 2,800 primary and secondary schools with about 745,640 pupils. Makerere University College, in Kampala, accounts for a considerable percentage of the more than 8,600 students in higher education. The rate of illiteracy, however, is still high, approximating 80%. About half of Uganda's population are Christians, some 6% are Moslems and the remainder are animists.

The Economy. Uganda is primarily an agricultural country. Almost 53% of the gross national product (an estimated $1,060,000,000 in 1970) is derived from agriculture. In regions of high rainfall, planted crops – notably plantains, sweet potatoes and cassava – provide the food supply; while in the rest of Uganda sown crops – finger millet, maize and pulses – predominate. The main cash crops are coffee, cotton and tea, followed by tobacco, peanuts, maize and sugar-cane. Livestock are raised in the north, and the forests provide hardwoods. The many lakes and rivers constitute one of the world's largest freshwater fisheries. Fish-farming of carp and other species is becoming increasingly important.

Uganda's chief mineral resource is copper, mined at Kilembe in the Ruwenzori Range and smelted at Jinja. Tin ore and small amounts of beryl and wolfram are also mined.

Most of the country's industries are connected with the processing of agricultural goods: cotton ginning, sugar milling, cigarette manufacture, vegetable-oil extraction and so on. Most of these are located in the producing areas near Lake Victoria, particularly in the Kampala-Jinja area. However, other industries are being developed with the continued exploitation of hydro-electric power from the Owen Falls dam at Jinja. Jinja has textile, plywood and brewing industries and a steel rolling mill. Cement manufacturing is well established at Tororo.

International Trade. Uganda depends heavily on export earnings from coffee and cotton, which together account for over 75% of its exports. Copper and tea are also important. Imports consist mainly of machinery, transportation equipment, manufactured goods, chemicals and fuels. The chief trading partners are Britain, Kenya, the United States, West Germany, Japan and Italy.

Transport. Uganda has 766 mi (1,226 km) of railways, which are part of the East African railway system. Together with waterway services operating on the lakes and on the Albert Nile River, they carry much of the country's trade. There are some 15,200 mi (24,300 km) of roads, including 3,870 mi (6,200 km) of all-weather roads. Entebbe has an international airport.

KENYA

Area: 224,960 sq mi (582,646 km²)
Population: 11,694,000
Population Growth Rate: 3·0%
Capital: Nairobi, pop 535,200
Language: English and Swahili. Kikuyu and Luo widely spoken
Religion: Animist (55%) Christian (33%) Moslem and Hindu
Currency: Kenya shilling = 100 cents

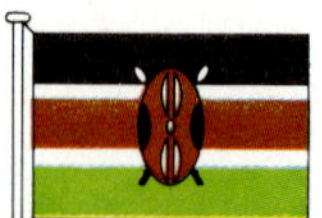

Kenya is an East African republic with a richly varied landscape and people. It is one of the most prosperous and politically stable states in Africa, although progress has been affected by the country's birth rate, the highest in East Africa.

Kenya's lack of mineral resources has been offset by the skilful development and diversification of agriculture and by the encouragement of tourism. Every year, thousands of visitors are drawn by Kenya's big game reserves and beautiful scenery. Annual earnings from tourism now exceed $50 million.

Until its independence in 1963, Kenya was a British colony. In 1964 it became a republic with Jomo Kenyatta as its president. An early problem was maintaining national unity with some 40 different tribes. Kenyatta's unifying influence proved strong, but there seemed no guarantee of unity under Kenyatta's successors.

Geographical Features. Straddling the equator, Kenya has a landscape of great variety. Bordering the Indian Ocean in the south-east is a narrow, fertile coastal strip with scattered rain forest, mangrove swamps and coconut palms. Inland, from the coastal strip to the northern border, stretch vast, dry plains which cover more than half the country. The plains are about 150 mi (240 km) wide in the south, where they are crossed by Kenya's two main rivers, the Tana and the Athi. The north-western and southern sections are at an altitude of about 4,000 ft (1,200 m), but most of the plains are below 1,200 ft (360 m).

The highlands, which have most of the country's population, are mainly in the south-west. They are cut from north to south by the Great Rift Valley, in the northern part of which is Lake Rudolf. The highlands are mostly 5,000–8,000 ft (1,520–2,440 m) above sea level. Two lofty cones have been raised by volcanic action: Mt Kenya (17,058 ft; 5,200 m), Kenya's highest mountain, and Mt Elgon (14,178 ft; 4,321 m) on the Uganda border. Most of the country's crops are grown on the rich volcanic soils of this area.

The western plateau (sometimes called the Nyanza Plateau) stretches from the highlands and the Great Rift Valley to Lake Victoria, which Kenya shares with Uganda and Tanzania. At an average altitude of 4,000 ft (1,220 m), this well-watered plateau has productive farmlands, forests and grasslands.

Climate. Temperatures are largely determined by altitude, with few seasonal variations. The low coastal strip has an average temperature of 80°F (27°C). The highlands are generally much cooler, with averages ranging around 60°F (16°C). Temperatures climb to between 70° and 80°F (21°–27°C) towards Lake Victoria.

Rainy seasons vary considerably from region to region. Annual averages range from

5 in (127 mm) in the drier parts of the north, to 40 in (1,016 mm) on the coast and in the highlands, and over 50 in (1,270 mm) in the west.

Wildlife. Kenya is world famous for its big game, which is particularly abundant in sparsely peopled areas in the south and east. Reserves hold large numbers of lions, elephants, hippopotamuses, giraffes, antelopes, zebras and other large mammals.

The People. More than 95% of the population is African, consisting of about 40 different tribal groups. The Kikuyu, numbering over 1,600,000, are the largest group, followed by the Luo, Luhya and Kamba – each numbering about a million. Other large groups include the Kisii, Meru, Turkana and Masai. The Asian community, which numbered about 192,000 in mid-1968, has been considerably reduced by the 'Kenyanization' of commerce and industry, which drove thousands of Asians out of Kenya from 1968 onwards. There are some 40,000 Europeans, mainly British, and about the same number of Arabs.

Overall population density is about 44 inhabitants per sq mi (17 per km²), but because of the aridity of large parts of the country some 75% of the total population is concentrated in the south-west, on only 10% of the land. Urban centres account for only 8% of the population. The capital, Nairobi (535,200), is the largest city.

Kenya's social structure is very varied. The people range from Europeanized professional people, businessmen, government officials and so on, to merchants, plantation owners, farmers and simple nomadic herdsmen. Some tribes, such as the Kikuyu and Luo, mainly raise crops. Others, like the Masai in the south, breed cattle. Some Kenyans have westernized urban homes or modern farmhouses, but many more live in traditional tribal homes with clay walls and thatched roofs.

Education and Religion. The rate of illiteracy, at approximately 75%, is still high. There are almost 7,000 primary and secondary schools with a combined enrolment of about 1·5 million pupils. Some 4,000 students study at the Kenya Polytechnic in Nairobi and the Mombasa Technical Institute, and over 1,220 at the University of Nairobi.

Over half the population are animists and about a third are Christians – the Roman Catholic Church being the most prominent group, with over 1·5 million adherents. The rest are Moslems or Hindus.

Production. Agriculture is the dominant factor in Kenya's economy, employing most of the labour force, in spite of the fact that over half the land is fit only for grazing. Local rainfall deficiencies largely restrict cultivation to the highlands, the region of Lake Victoria and the coast. Within these regions are areas of intensive production supporting dense rural populations. Much of the farming is traditional, small-scale and family-based. The cultivation of cash crops is increasing, especially with the introduction of co-operatives and the growth of urban markets.

The Parliament building in Nairobi. Nairobi has developed extensively since World War II.

The leading export crop is coffee, which provides a quarter of Kenya's export earnings. Next in importance is tea. Sisal is also exported, but has declined in importance. Kenya is a major world producer of pyrethrum (used in making insecticides).

Maize is the chief food crop and is grown, along with wheat, mainly in the highlands. Other important crops include cassava, cotton, sugar-cane, coconuts and cashew nuts. A significant livestock industry is based on high-grade highland pastures and on a number of ranches.

In the highlands, there are commercially important forests providing softwoods, hardwoods and bamboos. Fisheries are of only limited importance, though Lake Victoria supplies considerable quantities of fish for local consumption.

Most of Kenya's industries are devoted to processing agricultural raw materials or to the production of light consumer goods. Nevertheless, industries include oil refining, motor vehicle assembly and the production of cement, metal goods and textiles. Industrial employment is concentrated in Nairobi and Mombasa (the chief port). There are few known mineral resources, but soda ash is mined, together with small quantities of gold and copper.

International Trade. Exports are largely agricultural, with coffee and tea accounting for some 40% of the total. Other exports include pyrethrum, sisal, wattle bark extract, tinned meat, hides and skins. The principal industrial exports are petroleum (from imported crude oil), cement and soda ash. Imports include machinery, transportation equipment, manufactured goods, metals and chemicals. Since 1967, Kenya has been a member of the East African Community. The chief overseas trading partners are the United Kingdom, West Germany, the United States, the Netherlands and Japan.

Transport. The chief axis of transport is formed by railways and roads from Mombasa through Nairobi and across the highlands to Uganda. The 1,300 mi (2,080 km) of railways are part of the East African system linking Kenya, Tanzania and Uganda. There are some 26,250 mi (42,000 km) of roads, of which about 1,875 mi (3,000 km) are surfaced. The major port is Mombasa.

TANZANIA

Area: 362,821 sq mi (939,706 km²)
Population: 13,630,000
Population Growth Rate: 2·6%
Capital: Dar es Salaam, pop 343,911
Language: Swahili and English. Bantu dialects
Religion: Animist (44%) Moslem (31%) Christian (25%)
Currency: Tanzanian shilling = 100 cents

Tanzania is the largest East African nation, renowned for its wildlife and magnificent lake and mountain scenery. Tanzania was formed in 1964 by the union of Tanganyika,

Lake Manyara in Tanzania, showing (rear) the ridges that border the Great Rift Valley.

until 1961 British trusteeship territory, and the island of Zanzibar, until 1963 a British protectorate. It has pursued a strict policy of non-alignment in an attempt to create a socialist economy in which private, public and co-operative enterprises have their place.

The Face of the Land. Mainland Tanzania (formerly Tanganyika) covers 362,821 sq mi (939,706 km²), including 20,650 sq mi (53,484 km²) of inland water. It forms part of the Great Rift Valley of East Africa. The coastal belt, fringed by sand beaches, coral reefs and mangrove swamps, leads inland to plains and plateaux 2,000–4,000 ft (600–1,200 m) above sea level, with grasslands and open woodlands, and isolated hills and hill ranges. Rising above the plateaux are the Usambara, Livingstone and Pare mountains, which have Kilimanjaro, Africa's highest peak at 19,340 ft (5,895 m), clothed in forests and afro-alpine plants.

Much of the south is drained by the Rufiji, Tanzania's largest river. Within Tanzania's boundaries are parts of two important African lakes – Lake Victoria, which extends into northern Tanzania, and Lake Tanganyika, which forms part of the western border.

The fertile, coral-formed islands of Zanzibar and Pemba, covering 1,021 sq mi (2,644 km²), are separated from the mainland by the 22 mi (35 km) wide Zanzibar Channel. Their climate, like that of the mainland coastal belt, is equatorial.

On the mainland, Dar es Salaam, with a mean annual temperature of 79°F (26°C), is typical of the tropical humid coast, while Tabora, at 72°F (22°C), is representative of the interior plateau. Rainfall is variable, and generally lower than would be expected for these latitudes. About 79% of the country receives below 30 in (750 mm) a year. In general there is one rainy season between December and May.

A traditional safari area, Tanzania supports an abundance of wildlife, including elephants, lions, giraffes, leopards, zebras, antelopes and buffalo. Many species are protected in game reserves on the Serengeti Plain and the Ngorongoro Crater.

The People. Tanzania has a population of about 13·6 million, of which about 380,000 live on the islands of Zanzibar and Pemba. The capital and largest city is Dar es Salaam (344,000). Other large centres include Tanga, Arusha and Mwanza. On Zanzibar the largest centre is the city of Zanzibar (68,000). Tanzania is predominantly rural, however, with 94% of the population living in areas of low population density between the large centres.

The people of Tanzania are mainly African, of more than 100 Bantu tribes each with its distinctive dialect and traditional customs. There are also Arab, Indian and European minorities. In line with Tanzania's particular form of African socialism, there are kibbutz-like rural co-operatives engaged in communal farming.

Agriculture and Forestry. Agriculture provides more than 60% of the Gross National Product and more than 80% of Tanzania's exports. A wide range of crops is grown; maize or millet is the most common food crop grown over most of the country. The main cash crops are coffee, cotton, sisal, cloves (from Zanzibar and Pemba) and cashew nuts. Other agricultural exports include tea, tobacco and oilseeds. Rice, sugar-cane, wheat, fruits and peppers are also grown, and livestock is important, though it could be made more so. Timber trees include camphor and types of African mahogany.

Mining. Tanzania's mineral resources thus far have been only partially exploited. Diamonds, mined at Mwadui in Shinyanga Region, account for nearly 90% of mineral sales. The production of gold, once an important mineral export, has declined in recent years, but there is increasing production of mica, salt, tin and lead. There are large reserves of iron ore and coal in the south-west which as yet are unexploited.

Industry and Transport. Industry in Tanzania is still at an early stage of development, and is centred on the processing of agricultural products and light industries serving local markets. Textile factories are found in Dar es Salaam, Arusha and Mwanza; other industries produce soap, cement, farm tools, shoes, soft drinks and beer, cashew nuts, bicycles, fertilizer and petroleum products. Tourism is an important and growing industry, concentrated on Dar es Salaam and Arusha in the north.

The republic has 2,940 mi (4,730 km) of surfaced roads and about 1,800 mi (2,900 km) of railways. The new Tanzam Railway, built with substantial aid from Communist China, is designed to link Dar es Salaam with the copper belt in northern Zambia. Dar es Salaam, Tanga, Mwanza (on Lake Victoria) and Mtwara are the chief ports.

International Trade. Tanzania is a typical developing country in that its exports are chiefly primary products. These include cotton, coffee, sisal, cloves, tea, pyrethrum, cashew nuts, diamonds, gold and mica. Major imports include machinery, transportation equipment, manufactured goods and textiles. Trade, traditionally with Europe, Uganda and Kenya, is now being diversified. Two-way trade links with the Middle East, East Asia (including China) and Eastern Europe are being developed.

SEYCHELLES

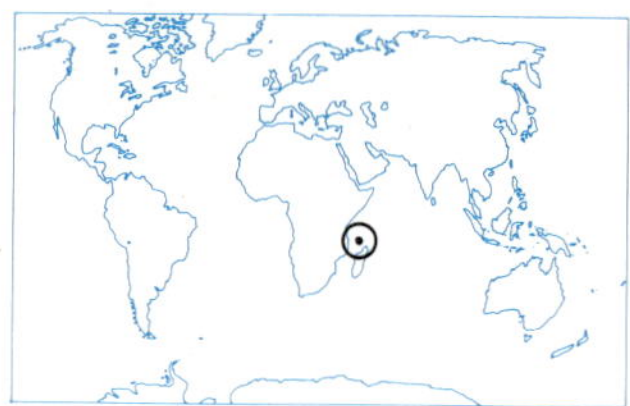

The Seychelles are a British island colony in the Indian Ocean about 600 mi (970 km) north-east of Madagascar. For many years this beautiful archipelago remained unspoilt and little known. Today it is being 'discovered', and Mahé, the largest island, now has an international airport and modern hotels to receive the growing number of tourists.

Annexed by the French in 1756, the Seychells were captured by the British in 1794 and have been a British colony ever since. Both political parties in the colony cherish the British connection.

Land and People. The Seychelles comprise about 85 scattered islands with a total land area of about 107 sq mi (277 km²). They form two distinct physical groups: the Granitic, including Mahé, Praslin and adjacent islands, with mountainous interiors rising to almost 3,000 ft (over 900 m); and the outlying dependencies of the Coralline group, including the Amirantes and such islands as Assumption, Astove, Cosmoledo, Providence, Coetivy and Platte Island, principally atolls and reefs just above high water. During the main wet season (November-March), temperatures and humidity are high, and rainfall is heavy. The main islands have lush tropical vegetation, the atolls mainly coconut palms. Wildlife includes many seabirds and turtles.

The population of over 52,000 is mainly of African or Afro-French descent, and predominantly Roman Catholic. English, French and a Creole *patois* are spoken. More than half the population lives on the island of Mahé, where Victoria, the capital and chief port, now has more than 13,600 inhabitants.

The Economy. Until 1955 the Seychelles required little outside aid. Imports of foodstuffs, petroleum and manufactured goods were more than balanced by exports of copra, cinnamon oil and bark, dried fish, guano, tortoiseshell and timber. The growth of population and the need for social services, however, have made financial aid necessary since that date. Tea planting has been successfully introduced, and farming co-operatives have been formed to reduce dependence on imported food. Improvements in water and power supplies, the road network, and the harbour and airport serving Victoria have encouraged private investment and tourism.

BRITISH INDIAN OCEAN TERRITORY

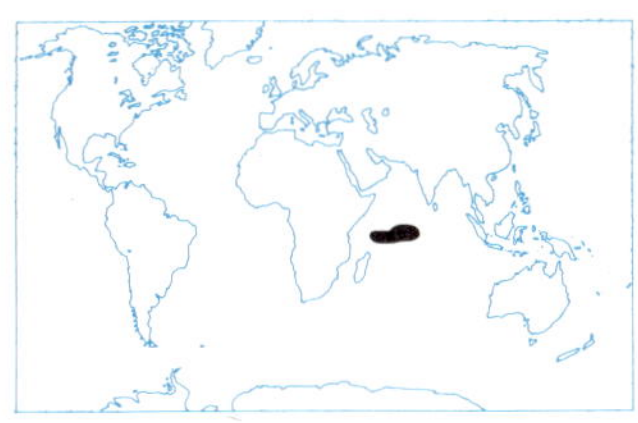

British Indian Ocean Territory comprises some small, widely scattered coral islands and atolls politically linked as a new colony in 1965 to facilitate the establishment of British and United States defence bases in the Indian Ocean. The territory consists of the reef-girt islands of the Chagos Archipelago, some 1,200 mi (1,931 km) north-east of Mauritius, and the Aldabra and Farquhar groups and Desroches Island lying north and north-east of Madagascar. Their total land area is only about 176 sq mi (456 km^2) and most of the racially mixed population of about 1,000 live in the Chagos Archipelago. Copra is the chief product of the archipelago, but the largest island, Diego Garcia (11 sq mi; 28 km^2), could become strategically important.

By far the most interesting of the islands is Aldabra, an elongated atoll with a remarkable wildlife. It is the home of the giant tortoise, found elsewhere only in the Galápagos Islands, and such rare birds as the flightless rail and the pink-footed booby. It is a breeding ground of the green turtle. In 1971 Aldabra was leased to the Royal Society until 1985, with the option of a further 20 years, for scientific research and conservation studies of its unique fauna and flora.

COMORO ARCHIPELAGO

The Comoro Islands, a French overseas territory, lie at the northern end of the Mozambique Channel between Africa and Madagascar. The archipelago consists of four small islands (Grande Comore, Anjouan, Mayotte and Mohéli) and a number of islets and coral reefs, their total area being 878 sq mi (2,274 km^2). The main islands are volcanic, and Grande Comore, the largest, is dominated by Mt Karthala (7,746 ft; 2,361 m), an active volcano with densely forested slopes which is visible 100 mi (161 km) out to sea. The climate is hot and rainy from November until April, and cooler and dry throughout the rest of the year.

The People. About 275,000 people, including 1,500 Europeans, live in the islands. Arabs, Africans, Malagasys, Persians, Indians, Indonesians and Europeans have contributed to their tangled racial background. African and Arab influences are strongest. Most of the people are Moslems. Grande Comore is the most populous island and has the capital and principal town, Moroni (15,000).

The Economy. Agriculture, inhibited by an archaic system of land tenure, employs most of the working population and provides nearly all the exports. Sugar-cane was once the chief cash crop, but the chief products today are vanilla, copra, sisal, cloves and essential oils such as citronella and ylang. The islands cannot pay their way, however, and the large annual deficit has to be covered by subsidies from France, the Comoros' chief trading partner.

MOZAMBIQUE

Area: 303,070 sq mi (784,951 km^2)
Population: 8,233,034
Population Growth Rate: 1·2%
Capital: Lourenço Marques, pop 444,000
Language: Portuguese. Bantu dialects.
Religion: Moslem, Roman Catholic, traditional religions (majority)
Currency: Mozambique escudo = 100 centavos

Mozambique (Portuguese East Africa), is an overseas province of Portugal stretching about 1,500 mi (2,400 km) along the shores of the Indian Ocean. Its total area is 303,070 sq mi (784,951 km^2). Its position gives it an importance in the economic and political life of southern Africa unjustified by its natural resources. For not only do its ports and railways serve its landlocked neighbours (Malawi, Zambia, Rhodesia and Swaziland) but its chief port, Lourenço Marques, is the nearest port to the great South African mining and industrial area of the Witwatersrand.

As one of the few remaining dependent territories of Africa, Mozambique's relations with independent Tanzania and Zambia are strained. The new oil pipeline from Dar es Salaam to the Zambian copper belt and the new Tanzam Railway will divert Zambian trade northward instead of through Mozambique. In addition, since Rhodesia's unilateral declaration of independence in 1965, UN economic sanctions have aimed at preventing Rhodesian trade from using the ports of Beira and Lourenço Marques.

Geographical Features. Mozambique's coastal plain is much wider than that of most parts of Africa. It occupies about two-fifths of the country and is widest in the south. Large areas are covered with infertile sandy soils. Where rivers like the Zambezi, Pungwe, Buzi, Save

The Zambezi is Africa's fourth largest river. It was explored from its source by David Livingstone.

and Limpopo cross the plain to the sea there are more fertile alluvial soils, but seasonal floods and the tsetse fly limit their usefulness.

The plateaux of the Low Veld descend to the plain in a series of steps. The granite country of the north is higher than the sandstones of the south, but nowhere does it rise to more than 8,000 ft (2,400 m).

The coastal plain is hot and humid, with average annual temperatures ranging from 72°F (22°C) to 85°F (29°C). The plateaux are not high enough to cause any appreciable cooling and in the interior – at Tete in the Zambezi Valley for instance – temperatures can rise as high as 113°F (45°C). Rainfall averages between 30 in (762 mm) in the south and 56 in (1,422 mm) in the north.

The People. The majority of Mozambique's population (8,233,034 in 1970) belong to various Bantu ethnic groups. The standard of living among the African population is low and most people subsist on agriculture or on the various labour schemes which compel Africans to work on European-owned agricultural or industrial concerns. Some 500,000 Africans a year are recruited as contract labour for farms and mines in South Africa and Rhodesia. Political rights are limited: in theory, there is no colour bar, but qualifications are so stringent that only about 5,000 Africans have become Portuguese citizens. The languages are Portuguese and Bantu.

There are over 120,000 Europeans in Mozambique. There is also a substantial Asian minority. The capital, Lourenço Marques (pop 444,000), is an important port with factories producing consumer goods of all kinds and an oil refinery.

Literacy among Africans is very low, with a rate of about 1%. There is no compulsory schooling for African children and most African education is provided by missions.

Agriculture. The people of the plateau country are mainly subsistence farmers growing cassava, yams and maize. Some farmers in the north also grow cotton. Cattle are reared only on the Angonia Highlands in the south. Climatic conditions have deterred Portuguese settlers from coming to Mozambique, but a few Europeans own tea plantations.

The poor soils over most of the coastal plain are a handicap to agriculture and the main products are cashew nuts (Mozambique is the world's chief source of this product), coconuts (mostly north of the Zambezi), and sisal. Maize, cotton, sugar and pineapples are grown on a small scale by African farmers on pockets of better soil in the south. There are also some European plantations, mainly in the river valleys, where water is available for irrigation. There are sugar plantations near the lower Zambezi, Buzi and Incomati; and rice, sugar, cotton, maize, millet and wheat are grown in the southern area irrigated by the Limpopo.

Minerals. Thick seams of coal occur in the Karroo Beds at Moatize, near Tete, and are mined for domestic use. Iron-ore deposits have been discovered at Narampa north of Nacala, and oil has been discovered at Inhambane, south of Beira.

Industry. Apart from the processing of agricultural products there are few industries. Sugar and tea factories are situated on the estates and there are processing plants for cashew nuts along the coast. Cotton is ginned in the growing areas and there is a textile factory at Vila Pery on the railway from Beira to Rhodesia. At Cabora Bassa on the Zambezi, a huge dam is being constructed which on completion will be Africa's largest single source of hydro-electric power.

Trade and Communications. Over four-fifths of the country's exports by value come from sugar, cotton, copra, sisal, cashew nuts and tea, but the total income is small and Mozambique depends largely upon revenue from transit trade and on money sent home by the large numbers of migrant workers employed in Rhodesia and South Africa.

The pattern of roads and railways reflects their main function as feeders to the ports of Lourenço Marques, Beira, Mozambique and Nacala. There are about 2,500 mi (4,000 km) of railway line, but no north-south rail routes. Railways leading to Lourenço Marques carry all Swaziland's iron ore and, by agreement with the Republic of South Africa, 47·5% of the trade of the Transvaal. Mozambique has about 25,000 mi (40,000 km) of roads.

MALAWI

Area: 36,350 sq mi (94,146 km²)
Population: 4,549,000
Population Growth Rate: 3·0%
Capital: Zomba, pop 20,000
Language: English. Nyanja, Tumbuka and others
Religion: Christian (50%) Moslem (33%) Animist remainder
Currency: Kwacha = 100 tambala

Malawi (formerly Nyasaland) lies in the south-east of Central Africa, west and south of Lake Malawi (formerly Lake Nyasa). This is Africa's third largest lake, most of which belongs to Malawi. Lake Malawi is among Africa's most beautiful areas. But Malawi is small and its per capita Gross National Product is one of the continent's lowest.

From 1953 to 1963 Malawi was part of the Federation of Rhodesia and Nyasaland, and in 1964 it became the independent state of Malawi, remaining within the British Commonwealth. It was proclaimed a republic on 6 July 1966, and unlike many African nations, has maintained close diplomatic and economic relations with South Africa.

Land and Climate. Malawi is situated near the southern extremity of the East African Rift Valley system. Although completely landlocked, it is at its nearest point only 120 mi (193 km) from the Indian Ocean. Lakes Malawi (Nyasa), Chilwa and Chiuta fill most of the valley, covering nearly one-fifth of Malawi's entire territory.

Malawi is a country of high grassland and savanna. The lowest point, only 120 ft (36 m) above sea level, is near Nsanje in the south, and the highest, Mlanje Peak in the south-east, reaches 9,848 ft (3,002 m). Between these extremes are areas of high plateau at 5,000–8,000 ft (1,500–2,400 m); plains at about 3,500 ft (1,060 m); and rift valley floor at 120–2,000 ft (36–609 m). These areas are often divided by steeply-sloping and dissected scarps.

Malawi's climate can be broadly classified as tropical continental, though differences in altitude and aspect produce wide variations of temperatures and rainfall within short distances. Temperatures are generally high, reaching a maximum in October or November, immediately before the rains. The rainy season is over by April and temperatures fall during the dry season, making July and August the coolest months. Altitude affects temperatures, and whereas Nsanje in the Shire River plain has a July mean of 69°F (21°C) and an October mean of 84°F (29°C); Dedza at over 5,000 ft (1,500 m) in the highlands has a July mean of 57°F (14°C) and an October mean of 69°F (21°C).

Vegetation and Wildlife. Variations of relief, climate and soils account for a great diversity of vegetation. Dry thorn savanna with succulents, mopane and baobab characterize the lower Shire valley and other rain shadow areas. These contrast with the rain forests found on the southern slopes of Mlanje and on the shores of Lake Malawi around Nkhata Bay. On the higher plateaux and scarps open woodland predominates, occasionally broken by *dambo* (seasonal marsh) grasslands. Areas above about 6,000 ft (1,830 m) are covered with montane grassland.

Five game reserves in Malawi protect a wide range of species once common throughout the country. Mammals include the African elephant, lion, leopard, cheetah, jackal, honey badger, great anteater, ground pangolin, hippopotamus, zebra and a variety of monkeys and antelopes. Over 600 species of birds have been recorded.

The People. Except for some 11,300 Asians and about 7,400 Europeans, the population of over 4,500,000 is made up entirely of Africans belonging to many different Bantu tribes. Each has its own language and cultural traditions, although many people have embraced Christianity. Education is also imposing a measure of uniformity. There are around 1,800 primary schools, over 40 secondary schools, a university and adult education facilities.

The largest centre is the city of Blantyre, the hub of commerce and industry and seat of Malawi University. But the main settlement units are the villages, in which more than 90% of Malawi's people live. The residents of a typical village are traditionally linked by ties of kinship. Physically, a village usually consists of a loose cluster of huts whose thatch sweeps nearly to the ground, protecting the mud walls from driving rain.

Agriculture and Fishing. In the absence of minerals, farming is the most important activity in Malawi. Tobacco, grown mainly on the plains of the central and southern regions, is the most important cash crop. Tea is also grown in the highlands, and cotton in the lowlands. Other cash crops include sugar, cotton, peanuts and rice. Maize, cassava, millet, fruit and vegetables are grown widely for subsistence.

Malawi has about 9,000 sq mi (23,300 km²) of forest, of which about 3,000 sq mi (7,800 km²) are incorporated into state forest reserves. Important woods produced include mahogany, African ebony, Mlanje cedar and eucalyptus.

Industry and Transport. Manufacturing industries in Malawi are concerned mainly with processing agricultural products either for export or to satisfy the growing local market. They include tea and tobacco processing, meat and fish canning, and the manufacture of cigarettes, cotton textiles and garments, blankets, knitwear, matches, bricks, pesticides, timber and wood products and radios. The demand for electricity created by these light industries has trebled since 1964. As a result, hydro-electric stations have been built at Nkula and Tedzani Falls on the middle Shire River.

A British knitwear factory in Malawi. Malawi has the lowest per capita income in Africa.

Malawi is linked by rail to two ports in Mozambique, Beira and Nacala. The overall road network covers more than 6,000 mi (9,600 km), and good quality bitumen roads link Mlanje, Thyolo, Blantyre, Zomba, Lilongwe and Mchinji. At present there is an international airport at Blantyre, but Lilongwe airport will be enlarged to take over this function when the capital moves.

International Trade. Malawi trades chiefly with Great Britain, South Africa, Rhodesia and the United States. Exports include tobacco, tea, corn, peanuts and cotton. Major imports include transportation equipment, textiles, petroleum products, pharmaceutical products and food products.

ZAMBIA

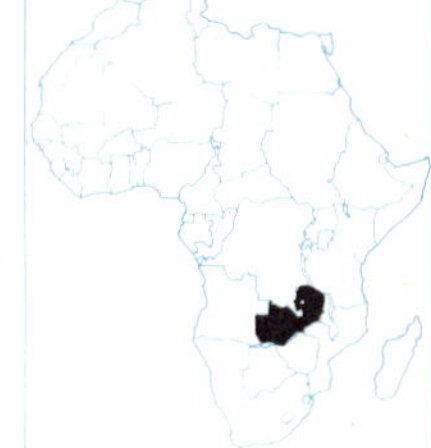

Area: 290,586 sq mi (752,618 km²)
Population: 4,515,000
Population Growth Rate: 3·0%
Capital: Lusaka, pop 347,900
Language: English. Nyanja, Bemba, Afrikaans
Religion: Christian (16%) Moslem, Hindu and traditional remainder
Currency: Zambian Kwacha = 100 ngwee

Zambia (formerly Northern Rhodesia) is a landlocked upland republic in south-central Africa. It is the largest producer of copper for the world market and consequently one of the wealthiest of the new African nations. However, Zambia has suffered from its heavy dependence upon other countries for outlets and in particular upon white-ruled Rhodesia.

Loading copper, Zambia's major export. Zambia has enormous mineral wealth.

Political friction with Rhodesia has threatened to hamstring Zambia's copper-dependent economy, but new outlets and economic diversification promise to solve the country's worst problems.

The area was first explored by David Livingstone in the 19th century. Between 1953 and 1963 the region was part of the British Federation of Nyasaland and Rhodesia. A year later it became the independent Republic of Zambia, named after the Zambezi River which runs through the country and along its southern border. Later in the 1960s plans were launched to give Zambia control of the companies exploiting its minerals.

Land and Climate. Zambia lies on a plateau about 3,000–5,000 ft (over 900–1,500 m) above sea level, dissected by broad river valleys and rising to the Muchinga Mountains (7,000 ft; 2,100 m) in the north-east. The Kafue and Luangwa rivers flow south to join the Zambezi which is punctuated by the dramatic Victoria Falls and one of Africa's largest man-made lakes, Lake Kariba, formed by the Kariba Dam.

Although Zambia lies entirely in the tropics, the climate is mild because of the altitude. A cool dry season from April to August may bring frosts at night and day temperatures of 70°F to 87°F (21°C to 31°C). Day temperatures in the hot dry season, from late August into November, may rise to 95°F (35°C) or higher. During the warm wet season, from November to April, temperatures fall and rainfall is heavy, especially in the north.

Vegetation and Wildlife. Zambia's vegetation is savanna with light tree growth and bushes and interrupted occasionally by unbroken grassland, swamp or trees almost dense enough to rank as forest. The once abundant wildlife has been decimated by hunting and is now largely concentrated in game parks, such as the large and well-stocked Kafue National Park. There is an exceptionally rich variety of birds.

The People. Zambia's population of about 4,515,000 is 98% African Negro, comprising Bantu-speaking peoples. Over 70 tribes are represented, and many dialects. The four main languages are Tonga, Bemba, Nyanja and Lozi. English is widely spoken. The chief city and capital, Lusaka, had a population of 347,900 in 1972. Other large cities are Kitwe in the copper belt, Ndola and Livingstone. In many parts of Zambia the people still live a tribal life under their chiefs, growing food crops such as maize, millet and cassava around their circular, primitive huts. There is an increasing drift to the towns, especially those in the mining areas, and the old tribal ways are rapidly disappearing.

Although some tribal beliefs persist, most Zambians are Christian. There are Hindu and Moslem minorities. At least half the population is illiterate, but primary education is now available for all children aged 7, secondary schools are numerous, and the University of Zambia was opened in 1966.

Agriculture. Nearly three-quarters of Zambians subsist on crops like millet, peanuts and maize raised by the traditional 'slash and burn' method. The diet is supplemented by fishing, hunting and food-gathering, as well as cattle-herding. Commercial crops, grown in the better soils along the line of the railway, consist of maize (corn), tobacco, peanuts, cotton, fruit and vegetables. Here, too, are found nearly all the beef and dairy herds of modern cross-breeds.

Mining, Power and Industry. Copper mining dominates Zambia's economy, accounting for about 90% of exports. The copper is extracted from the copper belt area along the northern border, which is partly supplied with electricity from one of Africa's largest hydro-electric installations, at the Kariba Dam. The state holds a 51% controlling interest in mining and monopolizes mineral rights. Lead, zinc and coal are also mined.

There are developments in manufacturing or processing food and beverages, metal products, clothing, textiles and vehicles; but manufacturing industry as a whole is still weak and unbalanced.

Transport and Trade. Zambia's original single railway line, built 1905–09 and linked to all neighbouring systems by 1931, covers 650 mi (1,050 km). This line traditionally carried the bulk of exports and imports through neighbouring Rhodesia. The imposition of sanctions on Rhodesia after 1965 and subsequent disputes with Rhodesia harmed Zambia's trade, and in 1970 spurred the start of a new Tanzania-Zambia railway from Dar es Salaam on the coast.

There were about 21,200 mi (34,000 km) of roads in Zambia in the mid 1960s, but only about 840 mi (1,350 km) of these were tar-surfaced highways. There are international airports at Lusaka, Ndola and Livingstone. A microwave link provides telephone and television services between Lusaka and the copper belt.

Zambia exports mainly copper, maize and tobacco; and imports wheat, fertilizers, textiles, vehicles and fuel oils. Sterling area and Common Market countries are among Zambia's main trading partners. Trade with Rhodesia and South Africa has declined.

A Tonga village in a remote part of Zambia. The Tonga are one of five main Bantu tribes making up Zambia's population. Zambia was formerly known as Northern Rhodesia.

ANGOLA

Area: 481,351 sq mi (1,246,699 km²)
Population: 5,673,046
Population Growth Rate: 1·3%
Capital: Luanda, pop 347,000
Language: Portuguese, Bantu and other languages
Religion: Roman Catholic (Majority). Protestant and traditional remainder
Currency: Angolan escudo = 100 centavos

Angola (Portuguese West Africa), an overseas province of Portugal, is one of the last remaining colonial territories of Africa. This slab of land facing the Atlantic Ocean is more than twice the size of France, but has fewer inhabitants than metropolitan Chicago. Zaire separates Angola proper from Cabinda, a tiny coastal exclave to the north.

Since 1960, guerilla warfare waged by African nationalists has ravaged the country. Nonetheless after Portugal lifted its restrictive trade policies in 1961 the economy began a remarkable expansion.

The Land. A narrow plain rises gently from the Atlantic coast in the north and more steeply in the central and southern sections to reach the Benguela-Bié Plateau (5–6,000 ft; 1,500–1,800 m). In the north-east, rivers drain into the Congo Basin; elsewhere they flow into the Atlantic. Where the drop from the plateau is gentle as in the north, the rivers are navigable. Farther south, however, the rivers flow over a steeper plateau edge and the falls have been harnessed for hydro-electricity.

Temperatures along the coastal plain are moderated by the cold offshore Benguela Current. On the plateau, rainfall increases from only 15 in (380 mm) in the south to 50 in (1,250 mm) in the Cabinda district. Temperatures vary with altitude, with a larger annual range in the south than in the north.

There is a gradual transition northward from desert on the border with South-West Africa (Namibia) through dry savanna with euphorbia herbs, acacia and baobab (monkey-bread trees) to woodland savanna. The isolated Cabinda area north of the River Congo is tropical rain forest.

Lions, elephants, antelopes and ostriches figure among Angola's larger wild animals.

The People. The population (estimated in 1970 at only 5·7 million), mostly comprises Bantu Negroes. There are also about 300,000 Europeans and 30,000 *assimilados* or Africans who have taken Portuguese citizenship. This privilege is open to all who choose to live under Portuguese law, are literate and reach certain income levels. Only a few Africans have applied for citizenship, because of the obligations of higher taxes and military service. Nine out of ten of the people are illiterate and most cherish animist beliefs.

The largest cities are the ports of Luanda and Lobito. Luanda, the capital, has a population of 347,000, nearly a quarter of them Europeans. Its deep-water harbour is protected by an island linked by road with the mainland, and it is Angola's major port by value, handling much of the nation's exports. The city's industrial activities include oil refining, brewing and textile and cement manufacture.

Lobito, a port with a population of 35,000 (7,500 Europeans), is the terminus of the Benguela Railway. It is an outlet for copper from Zambia and Zaire, for iron ore from Cuima, and for Angolan maize and other agricultural products. Moçâmedes (Port Salazar) was originally a fish-processing port, but it has grown in importance with the development of the iron ore deposits at Cassinga.

Agriculture and Industry. Most people remain subsistence farmers who grow millet in the south, and yams, maize and cassava farther north. But Angola has ranked as Africa's leading coffee producer since 1962. *Robusta* coffee is the main cash crop, grown primarily north of the Cuanza River. Cotton and oil palm are also grown commercially, and in the Cabinda district the forests supply valuable hardwoods. Portuguese settlements, especially on the high Benguela-Bié Plateau, grow maize, citrus fruit, pineapples and bananas. On the coastal plain some sisal is produced and used for the manufacture of sacks. Where irrigation water from the rivers is available, sugar plantations have been established. Nonetheless, in some districts agricultural progress is slow mainly because of limited capital and poor communications.

Victoria Falls, on the Zambezi between Rhodesia and Zambia, plunge 355 feet into a narrow gorge at right angles to the river's course. It was discovered by David Livingstone in 1855.

Excellent fishing grounds exist in the waters off Angola. Drying and canning factories, mostly run by Portuguese, have been built along the coast.

Industrial development was accelerating by the 1970s, and factory output was climbing by an annual 15%, partly assisted by electricity from various hydro-electric projects. Industry is concentrated in Luanda, Lobito, Benguela and Moçâmedes and includes the refining of oil and sugar, textile manufacture and brewing.

Minerals. Restrictions on foreign investment long hindered mineral development, and until fairly recently diamonds were the only important product. These are obtained by open-cast methods in the north-east, in an extension of the Congo diamond fields of the Kasai Valley. About one-tenth of the world's gem stones come from Angola.

Valuable deposits of iron ore are worked at Cuima and Cassinga. The ore from Cuima goes by a new railway to the Benguela Railway at Vila Robert Williams and so to Lobito. Krupps, the German steel firm, has a contract for the ore from Cassinga which is linked by rail to the line to Moçâmedes. Other deposits are mined near the Luanda Railway.

Petroleum is extracted at Benefica and refined in Luanda. Valuable new fields have been discovered offshore from Cabinda. Some manganese and copper are produced.

Transport and Trade. Poor communications have hindered development. Railways run inland from the four main ports, but the lines are not linked with each other. The Benguela Railway, completed in 1928, is the only line to reach a border. It connects with Zaire to the coast at Lobito, and the value of the freight made it worthwhile overcoming difficulties in supplying water and fuel and the steep gradient up the plateau edge. The transit traffic is still an important source of income but it may decline as independent Zaire diverts copper to an all-Congo route. Roads have been improved especially in the north, but many areas remain isolated.

Angola's trade expanded sharply in the 1960s. Half of the export income came from coffee, followed by products including diamonds, iron ore, fish, sisal and timber. Imports included machinery, vehicles and textiles. Portugal is Angola's main trading partner.

RHODESIA

Area: 150,820 sq mi (390,624 km²)
Population: 5,550,000
Population Growth Rate: 3·2%
Capital: Salisbury, pop 463,000
Language: English. Sindebele and Chishona
Religion: Christian (15%) Animist and traditional remainder
Currency: Rhodesian dollar = 100 cents

Rhodesia (formerly Southern Rhodesia) is a landlocked country in south-central Africa; one of the two nations in Africa governed by a white minority. To perpetuate white supremacy, the British colony's white-dominated government unilaterally declared itself independent in 1965, successfully defied British and world opinion and economic sanctions, and proclaimed a republic in 1970. In the early 1970s, this beautiful highland country remained one of Africa's most prosperous states, but with an African majority far poorer than the 4% white minority.

In 1964, Great Britain denied Rhodesia independence after Rhodesia's white leaders had refused to guarantee rule by the (black African) majority. After the failure of negotiations, Rhodesian Prime Minister Ian D. Smith unilaterally declared Rhodesia independent (11 November 1965).

Britain declared these moves illegal and prohibited trade with Rhodesia. The United Nations also imposed economic sanctions (1966), but on 2 March 1970 Rhodesia proclaimed itself a republic. In the Parliament, which consists of a 23-member Senate and a 66-member Assembly, Europeans now elect 13 members of the Senate and 50 members of the Assembly.

Land and Climate. Rhodesia is a landlocked country of south-central Africa and has an area greater than that of Japan. It lies astride the high plateaux between the Zambezi and Limpopo rivers and may be divided into four geographical regions. The High Veld at above 4,000 ft (1,220 m), extends across the country from south-west to north-east and is most extensive in the north-east. The Middle Veld between 3,000 and 4,000 ft (915 and 1,220 m) flanks the High Veld and in turn gives way to the Low Veld land below 3,000 ft (915 m) which comprises the Zambezi basin in the north and the more extensive Limpopo and Sabi-Lundi basins in the south and south-east. These three regions consist predominantly of gently undulating plateaux. The fourth physical region, the Eastern Highlands, is a narrow, mountainous belt along the Mozambique border. It marks the uplifted, folded edge of the great tableland of south-central Africa.

The Rhodesian landscape generally is rugged, but beautiful. Natural vegetation ranges from open savanna to the lofty evergreen forests of the more mountainous, high-rainfall eastern areas. The valleys of the Zambezi, Limpopo and Sabi rivers have fringing forests, mainly of hardwoods.

Although Rhodesia lies entirely within the tropics, the climate is moderated by the comparatively high altitude. Mean monthly temperatures range from 72°F (22°C) in October and 55°F (13°C) in July on the High Veld to 86°F (30°C) and 68°F (20°C) in the low-lying Zambezi valley.

Rainfall occurs mainly from November to March and, except in the eastern highlands, is variable in amount and in distribution. Mean annual precipitation ranges from 55 in (1,400 mm) in the Eastern Highlands, to 30 in (800 mm) on the north-eastern High Veld and to less than 15 in (400 mm) in the Limpopo valley.

There are some 12 national parks and game reserves, including Wankie National Park (5,249 sq mi; 13,595 km²), with its elephants, buffalo, zebras, giraffes, lions and other large mammals.

The People. More than 95% of the population of Rhodesia is African. Whites, mainly of Rhodesian, South African or British stock, number only 234,000. There are also 25,000 Asians and Coloureds (persons of mixed descent). The Africans mostly belong to the Matabele (Ndebele) and Mashona (Shona) tribes.

The largest cities are Salisbury, the capital (463,000), Bulawayo, Umtali and Gwelo. About 42% of the non-African population lives in Salisbury, and a further 23% in Bulawayo.

More than 60% of the Africans live on over-populated tribal lands in the rural areas.

Africans who work in the cities live in their own townships. Others work on the large farms operated by some 6,000 white farmers or farming companies. From 1965 to 1969, African employees numbered 655,000 and their average annual earnings in 1969 amounted to R$296, as opposed to R$2,980 for non-Africans.

Some Africans are Christians but most are animists. Among whites, Anglicans form the largest religious group.

There are African and non-African education systems; most African children get five years of elementary education. Africans account for over one-third of the students at the country's university college.

Agriculture. Agricultural activities have become increasingly diversified since economic sanctions were imposed by the United Nations in 1966. Dependence on the production of Virginia tobacco has been reduced, and today the chief crops are maize, tobacco and cotton, together with sown grasses and fodder crops. Wheat, sugar-cane, citrus and deciduous fruits, tea and coffee also contribute significantly to local needs and to export. Beef is produced also for domestic and export markets and Rhodesia is self-sufficient in all dairy products.

Mining. Rhodesia is rich in mineral resources, and mineral wealth provides a substantial part of the Gross National Product. The most important minerals are gold, asbestos, copper, chrome, nickel and tin; these are chiefly exported. Iron ore, limestone, iron pyrites, phosphates and several other minerals provide raw materials for local industries. Coal from Wankie and, since 1960, hydro-electricity from the Kariba Dam are the main sources of power.

Industry and Transport. Oil refining and vehicle assembly were halted by UN economic sanctions, but there have been significant increases in the manufacture of consumer goods such as clothing, furniture and foodstuffs. The production of textiles and paper has also expanded, and a large fertilizer plant has been constructed at Que Que.

Landlocked Rhodesia's chief outlets to the sea are the ports of Beira and Lourenço Marques in Mozambique. Its railways (2,019 mi; 3,250 km) also provide access to ports in the Republic of South Africa and link with the Zambian system. There are 5,259 mi (8,460 km) of surfaced roads.

International Trade. Great Britain, South Africa and the United States were traditional trading partners before the sanctions imposed in 1966. Since then Rhodesia has been able to obtain necessary imports through such co-operative neighbours as Mozambique and South Africa. Major exports include asbestos, tobacco, copper, chrome, meat and sugar. Imports are chiefly machinery and transportation equipment, petroleum products, textiles, iron and steel products and foodstuffs.

Salisbury, the capital of Rhodesia, was founded in 1890 by Cecil Rhodes' Pioneer Column. It is situated in a rich agricultural plain, 4,830 feet above sea level.

BOTSWANA

Area: 222,000 sq mi (574,980 km²)
Population: 625,898
Population Growth Rate: 2·7%
Capital: Gaborone, pop 18,436
Language: English. Tswana and Khoisan spoken
Religion: Christian (15%). Remainder tribal
Currency: Rand = 100 cents

Botswana is a poor landlocked republic larger than France located in southern Africa. It is bordered on the north and west by South-West Africa; on the north-east by Rhodesia; and on the south and south-east by South Africa. Formerly the British Protectorate of Bechuanaland, it achieved independence in 1966.

Plateau and Desert. Botswana lies an average 3,300 ft (1,000 m) above sea level. Much of the country is a dry and bush-covered plateau called the Kalahari Desert. The only perennial watercourses are in the wetter north, where the Okavango River forms the Okavango Swamp and the intermittent Lake Ngami. Eastwards, the Makarikari Pan is alternately wet and dry according to the character of the season.

Botswana is hottest in January, temperatures averaging 85°F (29°C), and coolest in July, when temperatures average 55°F (13°C). Rain falls mostly in the six summer months, but very erratically in the south. Most of the country is covered by mixed tree-and-grass savanna. The abundant wildlife includes lions, elephants, antelopes and giraffes. Some 30,000 sq mi (77,700 km²) have been set aside as game sanctuaries.

The People. The population of 626,000 is almost entirely African. Europeans and people of mixed descent number less than 8,000. About 80% of the people live in the more favoured eastern third, which contains the capital, Gaborone, and other main towns.

The Batswana (people of Botswana) are grouped in eight major tribes and about 20 dependent tribes. The largest and most important tribe is the Bamangwato. Many Batswana live in scattered settlements of round, thatched huts, tending their cattle; others live in concentrated tribal villages such as Serowe, the largest.

Botswana is also the principal home of the Bushmen, a primitive hunting people of the Kalahari famous for their rock paintings. Their numbers (at present about 20,000) are declining.

The Economy. Traditionally Botswana has lived by cattle-raising. By the 1970s some 35,000 male Batswana were also finding work in South Africa or Rhodesia. But the country may be transformed into one of Africa's leading mineral producers by the development of large deposits found since independence.

Most of Botswana is too dry for successful commercial crop raising which explains why the main farming operation is stock-breeding. Livestock products account for more than

The Bushmen are one of the oldest African races, and only a few thousand true Bushmen survive, mostly in the Kalahari. It is thought that they evolved locally from now extinct stone age men. The Bushman is about 4½ ft tall.

90% of Botswana's exports.

Minerals include gold, silver, manganese and asbestos. Large deposits of nickel and copper at Pelikwe and in the Tati area promise substantial revenue. The Orapa diamond pipe, opened in 1972, is the second largest in the world. Large deposits of medium-grade coal have also been found.

Transport and Trade. A single 394 mi (634 km) railway, part of the South Africa-Rhodesia line, passes through eastern Botswana. Roads are still poor and much of the country is not easily accessible. Botswana exports livestock products and imports cereals, sugar, petroleum products, textiles and iron and steel products. Although Botswana is joined with South Africa in a customs union, its chief trading partners include Great Britain, Zambia and other African countries.

SOUTH AFRICA

Area: 471,445 sq mi (1,221,043 km²)
Population: 21,447,230
Population Growth Rate: 2·4%
Capital: Cape Town, pop 1,096,257
Pretoria, pop 516,703
Language: English and Afrikaans, tribal
Religion: Christian (mainly Protestant) majority, Bantu and tribal
Currency: Rand = 100 cents

South Africa is over twice the size of France and covers nearly half of southern Africa, whose other states it largely dominates. South Africa totally surrounds Lesotho, and has extensive borders with Botswana and Swaziland. Much of the republic occupies great grassy plateaux with a cooler, less enervating climate than most of Africa – which encouraged European settlement and development.

Endowed with rich mineral resources, South Africa now boasts the strongest economy of the African continent. It is the world's leading producer of gold, one of the chief sources of gem diamonds and a leading producer of uranium. It is also one of the world's wealthy industrialized nations.

The land is ruled by a white minority which accounts for only one in five of the population. The white government's policy of *apartheid* is based on separate development of the white and non-white communities.

This policy has drawn criticism from other world powers, especially through the United Nations, which in 1962 called for a boycott of trade and an arms embargo on South Africa. *Apartheid* has also created economic problems. An acute manpower shortage in skilled jobs reserved for whites has prevented domestic industry from keeping up with the demand for goods. Imports have soared and inflation has continued unchecked.

The Land. Most of South Africa lies on a series of high plateaux, 3,000–4,000 ft (900–1,200 m) above sea level. Between the plateaux and the narrow coastal lowlands is a rim of mountains which reach their highest in the south-eastern Drakensberg Range where Mont aux Sources reaches 10,822 ft (3,293 m). From the rim, flanked by steep escarpments, the land falls away steeply to the sea.

A diamond mine in Kimberley in north-eastern Cape Province, the diamond centre of South Africa. South Africa produces about 66% of the world's gem diamonds.

The inland plateaux make up the vast central area of South Africa. To the west are the dry *karroos*, as these plateaux are called; to the east is the highest and most fertile plateau, the Northern Karroo or High Veld. North-east of the Northern Karroo the land descends to the Transv al Plateau Basin, a hot and humid region. In the north-west is the southernmost part of the Kalahari Desert, and along the west coast lies part of the arid Namib Desert, which is continued in South-West Africa. In the extreme south occur one of the very few folded mountain ranges of Africa – the Cape Mountains, which run north and east of Cape Town.

The South African coastline is generally smooth with very few deep bays and projecting capes, so that sites for good harbours are rare. There are few natural lakes in South Africa, and none of the rivers is navigable. The Orange River, which flows 1,300 mi (2,092 km) to the Atlantic, is the most important. Others are the Limpopo and the Vaal.

Climate. Because South Africa lies almost wholly outside the tropics and none of it is more than 500 mi (805 km) from the sea, it enjoys a relatively temperate climate. Moreover, the high altidude of the plateaux makes summers warm rather than hot like those below the escarpment. Mean annual temperatures range from 64°F (18°C) at Port Elizabeth on the coast to 61°F (16°C) at Johannesburg on the High Veld. In the comparatively dry atmosphere of the plateaux, cold nights follow warm days and, in winter, frosts can occur for up to 150 days. The coastal ocean currents have a considerable effect on climatic conditions. The cold Benguela Current has a cooling effect as it flows north along South Africa's west coast, but the warm Mozambique Current heats eastern coastal areas.

Over most of the country, rainfall occurs mainly in the summer six months of the year and steadily decreases from about 45–50 in

(1,143–1,270 mm) per year in the east to less than 5 in (127 mm) in the west. The extreme south-west, however, experiences maximum rainfall in winter, and has a Mediterranean climate. Drought is always a serious threat to farming in South Africa, where about half the country receives less than 15 in (381 mm) of rainfall per year and evaporation rates are high. Rainfall generally occurs on the plateaux in thunderstorms, often accompanied by hail, but the south and east coasts also receive rainfall derived from low pressure centres which move eastwards across the southern tip of the continent. In winter, strong currents of cold Antarctic air occasionally move in over the plateaux and bring temporary cold and wet conditions.

The Drakensburg mountains lie along the south-east of the country parallel with the coast.

Natural Vegetation and Wildlife. Most of the wetter eastern half of the South African plateau region is covered by rolling grassland, little varied by indigenous bush or forest. As rainfall decreases in the west, the grasses become shorter and ultimately give way to a semi-arid scrub vegetation in the dry *karroo* region. On the west coast lies the Namib Desert, a sandy expanse of dunes. After spring rains these dry areas are often carpeted with bright orange Namaqualand daisies.

Plant life similar to that of the Mediterranean is found in the south-west Cape area. Here, tough-leaved species are adapted to survive the warm dry summers. The warmer east coast is flanked by natural bush, and indigenous forest occurs in ravines leading down from the plateau edge. Valuable trees such as the Cape mahogany and Cape ebony which were formerly part of the sub-tropical forest in the south-east have been so heavily exploited that they have virtually disappeared. Further north, in the low eastern Transvaal country, is dry savanna bush.

Once rich in wildlife, South Africa has had to provide special reserves for many of the much-hunted larger mammals. The best-known reserve is Kruger National Park, in the eastern Transvaal, which provides a refuge for lions, elephants, antelopes, hippopotamuses, rhinoceroses, giraffes and leopards. Other game sanctuaries in Natal and the north-west Cape Province give protection to the rare white rhinoceros and the gemsbok.

The People. The whites in South Africa consist mainly of people of Dutch, British, French or German descent. The first Europeans to settle in South Africa were the Dutch, who became known as *Boers* (farmers). Their modern descendants are called Afrikaners. They speak Afrikaans, a language developed from Dutch. The whites have a style and standard of living similar to the British and Americans, helped by the abundance of cheap non-white labour.

The non-whites are classified as Bantus, Coloureds and Asians. In line with the official apartheid policy, the government's aims are separate development, separate residential areas and ultimate political independence for the whites, Asians, Bantus and Coloureds. In 1959 acts were passed which created a structure for the eventual creation of 9 Bantu states. The Transkei became the first of these semi-autonomous territories in 1963. The Bantus (Africans) are divided according to their 'nationalities' (tribes), the most important being the Xhosa (3,900,000), Zulu (3,800,000), Tswana (1,700,000), Bapedi (1,600,000) and Sotho (1,400,000). About 100,000 Bantus work in the mines.

The Coloureds are descended from the Hottentots and other African peoples who mixed with the early white settlers; they live mainly in Cape Province. The Asians are of Indian origin and live mainly in Natal, where Indian labour was imported in the 19th century to work the sugar plantations.

South Africa's largest cities are Johannesburg; Durban; Cape Town, the legislative capital; Pretoria, the administrative capital (516,703); and Port Elizabeth (386,577). There are 11 other cities with populations of over 100,000. The official languages, Afrikaans and English, are spoken by most whites and Coloureds, and by many Asians. The Africans speak Bantu languages.

Each group has its separate schools. Nearly all non-white schools are government-controlled. There are 11 universities for whites, and another 5 universities for Bantus, Coloureds and Asians. The overall illiteracy rate is about 11%, but is much higher among the Bantus (50% in 1969).

The Afrikaans-speaking people, white and non-white, mainly belong to the three Dutch Reformed sects, which together have over 2,600,000 members. The Bantu Churches claim more than 2,000,000 members; the Methodists over 1,700,000 and the Anglicans more than 1,400,000.

The Economy. South Africa has great and varied mineral resources and the republic has about 66% of the world's gem diamonds. South Africa is also a major wool producer, but manufacturing is the chief economic activity, providing nearly 25% of the Gross National Product. Mining contributes about 12%, and agriculture about 10%.

Most of the cultivated land is owned by white farmers and operated with non-white labour. The chief crop is maize, grown mainly in the southern Transvaal and eastern Orange Free State. Wheat and other grains, sugar-cane (Natal), tobacco and potatoes are important. Citrus fruits are grown mainly in the Transvaal and Natal. Cape Province is noted for its grapes, wine and orchard fruits. White and Bantu farmers raise cattle, and dairying flourishes, but much of the country provides relatively poor grazing due to low rainfall, the high evaporation rate, poor soil and soil erosion. Wool, produced mainly in the drier western part of the country, is the most important farm product and a leading export. The problem of drought will be partly solved by the long term Orange River Project, begun in 1966.

South African gold is mined mainly in the southern Transvaal (Witwatersrand) and in the Orange Free State. Uranium is also extracted from the gold-bearing rock. Diamonds are mined at Jagersfontein (Orange Free State), Kimberley (Cape Province) and near Pretoria. South Africa's large deposits of coal occur mainly in the Transvaal and Natal. Other minerals include iron ore, asbestos,

The famous Table Mountain looms behind the harbour at Cape Town, South Africa's principal port and one of its two capitals. Founded in 1652 by a Dutch naval surgeon, Jan van Rierbeeck, as a stopping-off point for Dutch East India Company ships, Cape Town is today a thriving commercial and governmental centre, with fine public buildings and some picturesque gabled houses in the 17th-century Dutch style. It has excellent beaches and a sunny pleasant climate which make it a popular holiday resort.

copper, nickel, chrome and phosphates.

Manufacturing ranges from the production of iron and steel (Pretoria and Vereeniging), and oil-from-coal (Sasolburg) to engineering, textiles, food-processing and a wide variety of consumer goods. There are four main manufacturing areas: the southern Transvaal (the most important), western Cape Province, Natal (Durban-Pinetown area) and the Port Elizabeth-Uitenhage area of Cape Province. Because South African industry is so highly concentrated, the government plans to attract industry to the less developed parts of the country, including the African reserves.

Transport and Communications. South Africa has built up a very efficient system of railways, roads and airways. The railways, which are government operated, carried nearly 500 million passengers and over 120 million tons of freight in 1968–69. South Africa has 50,000 mi (80,460 km) of surfaced roads penetrating all parts of the country. Its airways have expanded remarkably.

Since the closing of the Suez Canal in 1967, sea traffic via the Cape route has increased considerably.

International Trade. South Africa's trade with the rest of the world is substantial and is steadily increasing. Chief trading partners are Great Britain, the United States, Japan and West Germany. The other nations of southern Africa provide markets for many South African goods, and these countries are economically linked to South Africa by a customs union.

The single most important import is machinery and transport equipment. Gold is the chief export and accounts for about one-third of the total value of exports. Other exports include diamonds, wool and fruits.

Dependencies. South Africa controls Namibia (South-West Africa) in defiance of a UN ruling declaring its mandate ended. South Africa also includes remote Prince Edward and Marion islands in the Indian Ocean.

NAMIBIA

Area: 317,827 sq mi (823,172 km²)
Population: 746,328
Population Growth Rate: 3·7%
Capital: Windhoek, pop 60,000
Language: English and Afrikaans, German widely spoken
Religion: As South Africa
Currency: Rand = 100 cents

Namibia (South-West Africa) on the west coast of southern Africa has been administered by South Africa since 1920. But the United Nations regarded the territory as a UN trusteeship, and declared South Africa's mandate ended in 1966.

Geographical Features. Namibia comprises three regions: the Namib, an arid desert coastal belt; a central upland some 4,000 ft (1,200 m) above sea level; and the Kalahari, a desert plateau running eastwards into Botswana. Much of the country is desert or semi-desert, but desert changes to grassland and tree savanna as you travel north-eastwards. Antelopes, hyenas, jackals and leopards are widespread. Rainfall is very low in the coastal and southern part, but increases steadily north-eastwards. Temperatures range from 55°F (13°C) in winter to 75°F (24°C) in summer.

The People. The population includes several ethnic groups, chief among them being the Ovambos and Hereros (two Bantu groups), Bushmen, Hottentots and Basters (people of mixed Hottentot and European descent). There are also some 96,000 people of European descent, living mainly in the drier central and southern districts.

The Economy. Namibia's principal resources – providing most of its revenue – are its minerals. Diamonds are obtained along the coast and from the bed of the Orange River, and copper is mined in the interior. There are also deposits of lead, zinc and vanadium.

A main railway from South Africa runs northwards through the country, totalling about 1,645 mi (2,632 km). There are over 19,900 mi (31,840 km) of roads. Walvis Bay is the chief port.

LESOTHO

Area: 11,716 sq mi (30,344 km²)
Population: 1,043,000
Population Growth Rate: 4·0%
Capital: Maseru, pop 14,000
Language: English and Sesotho
Religion: Christian (70%)
Currency: Rand = 100 cents

Lesotho (formerly Basutoland) adopted its present name in 1966 when this British High Commission territory became a fully independent kingdom. It is a black enclave in white South Africa but linked with South Africa in a customs and monetary union, and as the country has little industry, thousands of Lesotho's men work in South Africa; indeed South African co-operation is essential to Lesotho's economic survival.

The Land. Except for the lower western quarter, about 5,500 ft (1,676 m) above sea level, Lesotho is a mountainous country rising to over 11,000 ft (3,353 m) in the Drakensberg (Dragon Mountains). Southern Africa's largest and longest river, the Orange, rises in Lesotho and flows westwards towards the Atlantic Ocean.

Partly because it lies high above sea level, Lesotho has a mild and moist climate. Snow occurs in the mountains in winter, but the lowlands have warm, wet summers.

The People. Two-thirds of the Basotho (the people of Lesotho) live on the western lowlands in scattered rural homesteads under a system of communal tenure in which the land belongs to the people as a whole. The largest town is Maseru, the capital (14,000). At nearby Roma is the main campus of the three-state University of Botswana, Lesotho and Swaziland. Education is Lesotho is largely conducted by the three leading Christian missions.

The Economy. Although good quality land is limited, most of the people are dependent upon agriculture. Wheat and maize are the chief crops, and large numbers of cattle, sheep and goats are reared both for subsistence and for the export of meat, hides, wool and mohair. Diamonds are mined for export, but as yet no other significant minerals have been found. The manufacturing industry is still in its infancy. The ambitious Malibamatso project would generate much hydroelectricity and divert water from the upper Malibamatso River for sale to the industrialized Witwatersrand area of South Africa. All imports come from or through South Africa. Exports, mainly of livestock and animal products, are mostly sold to South Africa or by that country to overseas buyers on behalf of Lesotho.

SWAZILAND

Area: 6,705 sq mi (17,366 km²)
Population: 421,000
Population Growth Rate: 2·7%
Capital: Mbabane, pop 131,803
Language: English and Siswati
Religion: Christian (55%) Traditional beliefs (43%)
Currency: Rand = 100 cents

Swaziland, formerly one of the three British High Commission territories in southern Africa, became fully independent in 1968. Wedged between Mozambique and the Republic of South Africa, the kingdom is an outpost of black independence in a region of white supremacy. Like two other such outposts, Botswana and Lesotho, it benefits from a customs and monetary union with South Africa.

The Land. Swaziland is a country half the size of the Netherlands. It consists of a series of parallel steps descending from the African plateau towards the Indian Ocean: from the fairly mountainous High Veld averaging 4,000 ft (1,220 m) above sea level to the Middle Veld (2,000 ft; 610 m) and the Low Veld (900 ft; 270 m). The low Lebombo Mountains form the country's eastern boundary. The chief rivers run at right-angles to these steps, breaking through the Lebombo in deep gorges.

Climate and Vegetation. The climate ranges from cool and wet on the High Veld to hot and dry on the Low Veld. Rainfall occurs predominantly during summer. Vegetation ranges from short grass in the higher regions, through taller grasses to broad-leaved savanna in the lower and drier parts of the country.

The People. Except for some 12,000 people of mainly European descent, the population is entirely Swazi and predominantly rural, living in widely-scattered homesteads of mud, grass and thatch. About 55% are Christians.

The king rules constitutionally through a prime minister and parliament. The *Libandla*, the council of chiefs, and the *Liqoqo* or royal council have great influence over the political and economic affairs of the country.

The Economy. Swazi subsistence farming combines cultivation (especially of maize) with cattle-breeding. Commercial agriculture, introduced by the Europeans, ranges from sheep and cattle rearing to intensive irrigated farming of sub-tropical crops such as sugar, citrus fruits and rice. Forestry has made great strides, and there are large softwood plantations on the High Veld.

The considerable mineral wealth has still to be fully developed, but large quantities of asbestos and iron are mined for export, and a little coal is obtained for local use. New developments include an iron ore processing plant and the exploitation of silicates.

Transport. The railway from Kadake which links with the Mozambique network was built in the 1960s for transporting iron ore to the port of Lourenço Marques (Mozambique) for export to Japan. A bituminized road traverses most of the country from west to east. The chief airport is at Matsapa.

International Trade. Swaziland's chief trading partners are Britain, which buys most of its sugar, asbestos and woodpulp; Japan, which purchases all its iron ore; and South Africa.

MALAGASY REPUBLIC

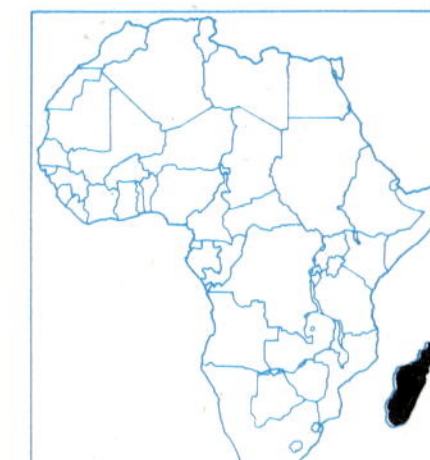

Area: 229,233 sq mi (593,713 km²)
Population: 7,655,134
Population Growth Rate: 2·3%
Capital: Tananarive, pop 322,000
Language: Malagasy and French. Hova spoken
Religion: Animist (50%) Christian (45%) Moslem (5%)
Currency: Franc Malagache = 100 centimes

The Malagasy Republic comprises Madagascar (the world's fourth largest island) and small offshore islands. It lies about 240 mi (390 km) off the east coast of Africa.

Racially and culturally Madagascar is more reminiscent of South-East Asia than nearby Africa. The inhabitants are mainly of mixed Asian and African backgrounds, being descended from Indonesian colonists from the Malay Archipelago who began arriving over 2,000 years ago; and Negroes from Africa, many of them brought in as slaves by Arab merchants. In 1958 the Malagasy Republic was proclaimed a state within the French Community. In 1960 it became entirely independent, but the ties with France remained close.

The Land. The topography of Madagascar is extremely varied. In the west a sandy, swamp coast is backed by broad desiccated plains crossed by fertile valleys containing the island's longest (often navigable) rivers. In the centre the plains rise to a plateau at 2,500–4,500 ft (760–1,370 m) from which mountain massifs climb to over 6,500 ft (1,980 m) in the north, centre and south. The eastern side of the highlands falls away in steep escarpments through which short, swift rivers descend to a narrow coastal plain, about 30 mi (48 km) wide. This terminates in a straight, largely lagoon-fringed eastern coast. The island's many lakes are largely of volcanic origin.

Climate. The coastal lowlands experience mean annual temperatures ranging from 80°F (27°C) in the north and west to 73°F (23°C) in the south and east. In the highlands, in the Tananarive area, the mean annual temperature is under 64°F (18°C).

Vegetation and Wildlife. The natural vegetation ranges from tropical rain forest on the eastern coast through scrub and grassland, giving way to savanna, on the plateau regions which have been extensively deforested and eroded. True forest now covers only about 7% of the island. Many of the plants are unique to the island. So, too, are many of the animals. These include most of the world's known lemurs (among them the rare aye-aye).

The People. The estimated population of the country in 1970 was over 7,650,000. Having doubled in 50 years, it is now increasing even more rapidly, putting heavy pressure on the island's limited area of fertile land and other resources. The capital and by far the largest city is Tananarive (about 320,000). Most of the people live in this area and on the east coast where Tamatave is the nation's main port. However, the rest of the island is sparsely populated. Though the inhabitants are of mixed descent, those of predominantly African origin live in the coastal regions, those of Indonesian origin in the central and southern highlands. Most are peasant farmers, though the educated Merinas tend to become merchants or officials.

The official languages are French and Malagasy. Malagasy, spoken everywhere, is closely related to Malay and Indonesian. With about half the children attending primary school, illiteracy is declining, though it still stands around 50%. A minority of the inhabitants of the island are Christian; the rest practise animistic tribal religions.

The Economy. Most people live by agriculture. Rice is the main food crop, along with millet, cassava, maize and potatoes. Oxen are raised, and cattle outnumber the people. Coffee, grown on the east coast, is the most important export. Madagascar is the world's chief producer of vanilla, and is also one of the largest exporters of cloves.

Industry is little developed and largely involves processing agricultural products. It includes sugar refining, the preparation of rice, and meat preserving and tanning. There is some mining for graphite, phosphates and mica. Tourism, especially from South Africa, is being developed and there are increasing signs that Madagascar's almost traditional isolation is being reduced. There is now an American satellite-tracking station on the island, and there are plans to build cement, rubber and plate glass factories on the north-west coast, and to exploit bauxite deposits.

There are some 20,000 mi (32,000 km) of passable roads. Some 535 mi (860 km) of railways include two that link the eastern coastlands with the plateau. The navigable western rivers are supplemented by the Canal des Pangalanes running behind the sand dunes of the east coast.

Madagascar mainly exports agricultural products, minerals and textiles; imports include mineral products, vehicles, machinery, and petroleum products and other manufactured or processed goods. France is by far the main trading partner.

REUNION

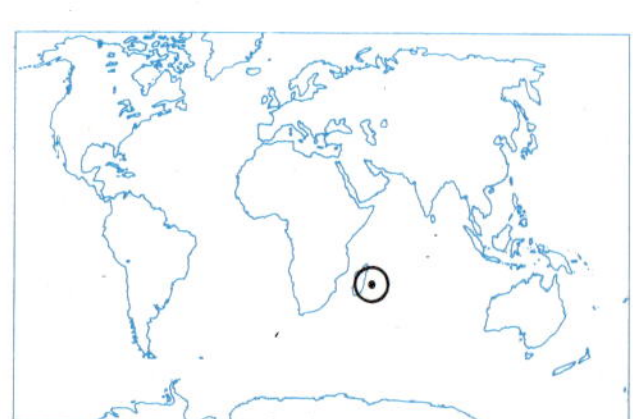

Réunion, an island about 500 mi (800 km) east of Madagascar, was settled by the French in 1642, and it has been a French overseas department since 1946. Covering 969 sq mi (2,519 km²), it is oval, of volcanic origin and mountainous (Piton des Neiges, 10,069 ft; 3,069 m). It has a tropical climate and abundant rainfall.

The population of 457,900 consists of the descendants of French colonists and the Malagasys, Indians, Annamites and others brought in to work the sugar plantations. St Denis (87,000), the capital and largest town, is linked by railway with other coastal towns. Sugar is the most important crop and chief export. Rum, vanilla and essential oils are also exported.

MAURITIUS

Area: 720 sq mi (1,865 km²)
Population: 830,606
Population Growth Rate: 2·2%
Capital: Port Louis, pop 141,000
Language: English, French, Creole
Religion: Hindu (50%) Christian (40%) Moslem (10%)
Currency: Mauritian rupee = 100 cents

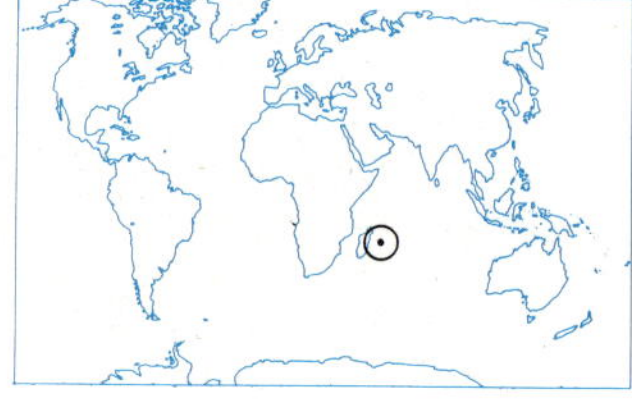

Mauritius is a small island republic within the Commonwealth of Nations, situated in the Indian Ocean about 500 mi (800 km) east of Madagascar. Since independence from Britain (1968) the two most pressing tasks have been installing a sense of national unity in the fast-increasing, multi-racial population, and reducing the island's dependence upon a single crop – sugar-cane.

Land and People. Mauritius is a hilly, volcanic island girdled by a coral reef. The central uplands rise to more than 2,000 ft (609 m) but the highest peaks are in the south-west, where the Black Mountains rise to 2,713 ft (827 m). The climate is dominated by the south-east trade winds, which bring nearly 200 in (5,080 mm) of rain to the higher areas, but only some 30 in (762 mm) to the sheltered north. Hurricanes are frequent during the main rainy season (December-June). Temperatures range from 68°F (20°C) to 80°F (27°C).

Until the abolition of slavery (1835), the island was peopled by descendants of European settlers and their African slaves. Then Indian labourers were brought in, and some Chinese also came. Today's population (one of the densest on earth) is nearly 70% Indo-Mauritian. Hinduism is the leading religion.

The Economy. Sugar plantations cover nearly half the island, and sugar and its by-products account for 95% of the exports. Secondary crops include tea and tobacco. Industry is mainly concerned with processing agricultural products. Tourism is on a modest scale.

Most of the island's export trade is with Britain; imports largely come from Britain, Burma, South Africa and Australia.

NORTH AMERICA

North America, the third largest continent in the world after Asia and Africa, is situated in the Western Hemisphere between the Atlantic and Pacific oceans. Its area is generally taken to include not only the United States, Canada and Mexico, but also Greenland to the north and Central America to the south, and the islands of the Caribbean Sea. Included within its vast area are arid deserts, towering mountain ranges, rich farming land and such striking natural features as the Great Lakes, Niagara Falls and Grand Canyon.

Although North America covers 16% of the land area of the world, it has only 7·5% of the world's total population. Its people are grouped into 15 independent nations, from the smallest, Barbados, to Canada, the second largest country in the world after the USSR. Canada and the USA enjoy the highest standard of living in the world.

North America in History

The Indians were the first to settle in North America, moving out of Asia some 25,000 years ago. The Eskimos came about 6,000 years ago, and settled in the northern areas.

The first Europeans to reach North America were the Norsemen. They settled in Greenland, and about 1000 AD Leif Ericsson reputedly landed in Nova Scotia. The first permanent contacts, however, were made in the 1490s by Christopher Columbus in the Bahamas, and the Cabots along the Newfoundland coast. During the following centuries, the leading sea powers of Europe and Russia began to found permanent colonies, mostly along the coasts and fought to extend their spheres of interest.

Britain lost its 13 seaboard colonies during the War of Independence (1773–1785) and they combined to form the Republic of the United States which rapidly emerged as the most powerful nation in North America. Beginning in 1810, the Central American colonies in turn shook off Spanish rule. Canada gained dominion status from Britain in 1867. By the 1970s, independence had replaced colonial rule everywhere except Greenland, which remained a Danish province.

By the 1970s a continent that had supported 4 million Indians was now easily maintaining about 80 times that number.

Face of the Continent

The Land. North America is shaped like an inverted triangle, roughly 4,000 mi (6,440 km) maximum north to south and 4,900 mi (7,889 km) east to west, with its apex to the south connecting with South America. A number of large gulfs break into the pattern, notably Hudson Bay to the north and the Gulf of Mexico to the south. The most prominent feature of the continent is perhaps the broad belt of the Cordillera, or Western Highlands, which is divided into the Rocky Mountains, the backbone of the continent, and the Pacific ranges to the west. Between the two is a broad interior plateau region of mountains, deep canyons and deserts. The Cordillera extend from Alaska into Mexico, where the Eastern and Western Sierra Madre flank the high Mexican plateau. The highest mountain in the Pacific ranges (indeed in North America) is Mt McKinley in Alaska at 20,300 ft (6,192 m).

Complementing the Cordillera in the east is the Appalachian system, or Eastern Highlands, paralleling the Atlantic coastline and running from Alabama into Canada. Separating the Appalachians from the Atlantic Ocean are the coastal plains, which run unbroken from Long Island to the Mexican coast and reach considerable widths, especially in Texas and Louisiana. The western coastal plain is virtually non-existent, the Pacific mountains rising steeply from the Pacific Ocean.

North and west of the Appalachians is the vast Canadian, or Laurentian, Shield stretching half way across Canada, an area of hard, very old rocks which has been heavily glaciated. Below the Shield and south of the Great Lakes is an area of rich soils, the Middle West Corn Belt, which received the materials scraped off the Shield by the advancing ice sheets. The Corn Belt forms part of the great Interior Plain, or Lowlands, of North America which stretch over 1,000 mi (1,600 km) east and west between the Appalachians and the Rocky Mountains, and a maximum 2,000 mi (3,200 km) north and south. The western, drier part of the Interior Plain is known as the Great Plains and in Canada as the Prairies. The core of the Interior Plain is the Mississippi Valley. Since about 70% of the continent was covered by the four major advances of the ice sheets, there has been extensive erosion and deposition. The Great Lakes were of glacial origin, and Great Salt Lake in Utah is the remnant of a vast glacial lake, Lake Bonneville.

Of the North American river systems draining the Interior Plain to the oceans, the Mississippi-Missouri system is by far the greatest, draining virtually the whole of central USA. Those rivers draining to the Pacific and those to the Atlantic are determined by the Great Divide at the crest of the Rockies. The Yukon, Columbia and Fraser rivers empty into the Pacific, the Colorado (which carved the Grand Canyon) into the Gulf of California. The Mackenzie, which flows from Canada's Great Slave Lake, empties into the Arctic Ocean. The St Lawrence River flows from the Great Lakes to the Atlantic, and with the Great Lakes makes up the St Lawrence Seaway, one of the great inland waterway systems of the world. The eastern flanks of the Appalachians are drained by a number of short rivers emptying into the Atlantic, such as the Hudson, Delaware, Potomac and James rivers.

Numerous islands, of which Greenland is the largest (it is also the largest island in the world), lie off North America's coasts. These include the West Indies, with Cuba and Haiti/Dominican Republic, Bermuda, Newfoundland, Long Island, and Vancouver and other islands off the western coast, as well as the Aleutians off Alaska.

Over: The skyline mid-town Manhattan, New York City. To the rear is the Chrysler building built in 1929. *Below:* A Navajo Indian and his horse overlooking Monument Valley Utah.

Climate. The climate of North America varies enormously, from the sub-tropical climate of southern Florida and southern Mexico to the dry cold climate of the Arctic regions, and from the abundant rainfall of some of the mountain areas to the virtually rainless deserts of Arizona and New Mexico. Thanks to the Pacific Ocean and the Alaskan Current, the west coasts of Canada and the

NORTH AMERICA
vegetation
snow-and icecaps
tundras and high mountain flora
mixed forest and northern coniferous forest
tropical rain forest
monsoon forest and thorn scrub
steppe- and mountain grassland
desert and semi-desert
cultivated areas
irrigated areas
swamp
scale 1 : 40,000,000
0
500
1000 st miles
ARCTIC OCEAN
North Pole
PACIFIC OCEAN
ATLANTIC OCEAN
BERING SEA
BEAUFORT SEA
HUDSON BAY
BAFFIN BAY
GREENLAND
DAVIS STRAIT
DENMARK STRAIT
FOXE BASIN
LABRADOR
GULF OF ALASKA
GULF OF MEXICO
GULF OF CAMPECHE
GULF OF CALIFORNIA
CARIBBEAN SEA
SARGASSO SEA
ASIA
EUROPE
SOUTH AMERICA
ICELAND
NEWFOUNDLAND
Wrangel Island
Jan Mayen
Tropic of Cancer
Equator
Aleutian Islands
Alaska Peninsula
Alaska Range
Mt McKinley 20320
Anchorage
Mt Logan 19850
Queen Charlotte Isl
Vancouver Island
Vancouver
Mt Rainier 14410
Mt Shasta 14162
Mt Whitney 14495
San Francisco
Los Angeles
Mt Elbert 14431
Blanca Peak 14317
Mt Mitchell 6684
Great Slave Lake
Great Bear Lake
Lake Athabasca
Reindeer Lake
Lake Winnipeg
Lake Manitoba
Winnipeg
Churchill
Lake Superior
Lake Michigan
Lake Huron
Lake Erie
Lake Ontario
Detroit
Chicago
Quebec
New York
Washington
Kansas City
St Louis
New Orleans
Mexico City
Popocatepetl 17887
Citlaltepetl 18696
Mexican Plateau
Central Plains
Coastal Plain
Bermuda Islands
Bahama Islands
Havanna
Cuba
Jamaica
Hispaniola
Puerto Rico
Greater Antilles
Lesser Antilles
Grand Cayman Islands
Revillagigedo Islands
Galápagos Islands
Lake Nicaragua
Pico Blanco 12861
Panama
Isthmus of Panama
Gulf of Panama
Caracas
Guayaquil
Trinidad
Barbados
Grenada
Nova Scotia
Long Island
C. Hatteras
C. Cod
C. Sable
C. Farewell
Hudson Strait
Yucatan Channel
Thule

Gary, Indiana was laid out and planned in 1905 by the United States Steel corporation and named for Albert H. Gary one of its founders.

USA have a warm rainy climate with dry summers, with a concentration of heavy rainfall in the north-west. The Gulf Stream and the Atlantic similarly modify the climate of the east coast. Since the principal mountain ranges are parallel to the Atlantic and Pacific oceans, they tend to reduce the effect of the oceans and the ocean currents on the interior, which therefore has a harsher climate than the coastal regions. Rainfall is unreliable in the interior lowlands, especially the western parts; while a semi-arid region, shading into near-rainless desert beyond the Rockies, covers most of the western half of the continent and runs from Canada to Mexico. Summer hurricanes are a regular hazard of the southern Gulf states and the Atlantic seaboard of the USA, while tornadoes, violent hailstorms, blizzards and electric storms menace the Interior Plains region from southern Canada to central Texas.

Vegetation and Wildlife. The widely differing belts of vegetation in North America are determined largely by the climate from the low ranging desert scrub of the south-western USA and Mexico to the Arctic tundra of Greenland, Canada and Alaska or the tropical rain forests of Central America. The largest tree in the world, the sequoia, grows in the rainy forests of the Pacific Northwest. Though much of the Interior Plain is now farmed, vast areas of grassland still support large herds of sheep and cattle. Coniferous forests grow over much of Canada, the south-eastern USA and the northern Pacific areas, as well as the central mountain areas of the USA. Mixed forests of hardwood and softwood trees, and some conifers characterize the Appalachians, while deciduous forests are found on the eastern coastal plain and the eastern areas of the Interior Plain.

Animal life is also varied, ranging from the moose and elk of the western forests to the polar bears, seals and walruses of the far north, and the alligators, monkeys and armadillos of Central America. The former number of buffaloes has been vastly reduced.

The People

North America as a whole with 320 million people is far less densely populated than either Europe or southern or eastern Asia. Yet the north-eastern USA and south-eastern Canada has one of the most densely populated areas in the world. Westwards to Chicago and southwards to the Gulf of Mexico the population is still moderately dense, while other concentrations occur around Mexico City, Los Angeles, San Francisco, and the Pacific Northwest. The rest of the continent is very lightly populated, the interior of Greenland and northern Canada virtually empty. Canada's population is confined to a narrow belt in the south along the border with the USA.

The rapid growth of population in North America has been in large part a result of immigration from various parts of Europe. The main influx of European settlers, particularly into the USA and Canada, has taken place since 1860, but since 1900 immigration has declined as a result of increasingly strict controls. The major groups of immigrants, which today still retain some traces of their national origin, were from England, Ireland, France, Spain, Germany, Scandinavia, the Baltic, Russia and Southern Europe. In Mexico and Central America many claim Spanish descent, although in Mexico today the bulk of the population is now mixed European and Indian (mestizo). In the United States about 12% of the population is non-European, and this is largely Negro, descended from the large numbers of imported slaves. Only about 600,000 of the original inhabitants of North America, the

NORTH AMERICA
political
CITY population more than 1,000,000
CITY population more than 500,000
CITY population more than 100,000
City population less than 100,000
INDEPENDENT AFTER 1945
former British territory
railways
roads
airport
scale 1: 40,000,000
0
500
1000 st. miles
North Pole
ARCTIC OCEAN
ASIA
Ust Kamchatsk
Anadyr
Arctic Circle
Wrangell Island
East Cape
C. Lisburne
BERING SEA
Dutch Harbor
ALEUTIAN ISLANDS
Nome
ALASKA
Yukon
Fairbanks
Anchorage
Pt. Barrow
BEAUFORT SEA
Aklavik
Great Bear Lake
Port Radium
Whitehorse
Skagway
Juneau
ALEXANDER ARCHIPELAGO
Great Slave Lake
Uranium City
Hines Creek
Prince Rupert
CANADA
EDMONTON
Saskatchewan
CALGARY
Vancouver Island
Victoria
VANCOUVER
REGINA
Medicine Hat
L. Manitoba
L. Winnipeg
WINNIPEG
Churchill
HUDSON BAY
SVERDRUP ISLANDS
PARRY ISLANDS
Ellesmere
Banks I.
Victoria I.
Prince of Wales Island
Somerset I.
Devon I.
Baffin Island
BAFFIN BAY
DAVIS STRAIT
GREENLAND (DENMARK)
Thule
Angmagssalik
Godthaab
Cape Farewell
Jan Mayen (Norway)
ICELAND
Reykjavik
EUROPE
Bergen
Nain
Schefferville
Labrador
Battle Harbour
Sept Îles
Anticosti
Gander
NEWFOUNDLAND
St. John's
C. Race
Miquelon (Fr.)
St. Pierre (Fr.)
Charlotte Town
HALIFAX
QUEBEC
St. Lawrence
Cochrane
Thunder Bay
L. Superior
MONTREAL
OTTAWA
Sherbrooke
Augusta
BOSTON
NEW HAVEN
Long Island
NEW YORK
PHILADELPHIA
BALTIMORE
WASHINGTON
RICHMOND
NORFOLK
C. Hatteras
TORONTO
HAMILTON
BUFFALO
LONDON
CLEVELAND
PITTSBURGH
L. Huron
L. Michigan
DETROIT
MILWAUKEE
CHICAGO
INDIANAPOLIS
COLUMBUS
DULUTH
ST. PAUL
MINNEAPOLIS
Bismarck
Pierre
Missouri
DES MOINES
LINCOLN
KANSAS CITY
ST. LOUIS
Ohio
NASHVILLE
RALEIGH
CHARLOTTE
COLUMBIA
ATLANTA
BIRMINGHAM
MONTGOMERY
SAVANNAH
JACKSONVILLE
MEMPHIS
LITTLE ROCK
TULSA
OKLAHOMA CITY
AMARILLO
Wichita Falls
FT. WORTH
DALLAS
JACKSON
Mississippi
MOBILE
BATON ROUGE
NEW ORLEANS
HOUSTON
SAN ANTONIO
CORPUS CHRISTI
Tallahassee
Orlando
C. Canaveral
TAMPA
MIAMI
C. Sable
Key West
UNITED STATES
SEATTLE
Columbia
PORTLAND
Helena
Boise
Great Salt Lake
C. Mendocino
SAN FRANCISCO
SACRAMENTO
Reno
Carson City
FRESNO
SALT LAKE CITY
Cheyenne
Colorado
DENVER
Arkansas
LAS VEGAS
LOS ANGELES
LONG BEACH
SAN DIEGO
Santa Fe
ALBUQUERQUE
TUCSON
EL PASO
MEXICALI
CIUDAD JUAREZ
GULF OF CALIFORNIA
Guadalupe (Mex.)
Tropic of Cancer
CHIHUAHUA
Rio Grande
TORREON
CULIACAN
La Paz
MEXICO
MONTERREY
NUEVO LAREDO
MATAMOROS
AGUASCALIENTES
TAMPICO
GUADALAJARA
MORELIA
MEXICO CITY
PUEBLA
VERACRUZ
GULF OF CAMPECHE
MERIDA
REVILLAGIGEDO ISLANDS (MEX.)
GULF OF MEXICO
BRITISH HONDURAS
GUATEMALA
GUATEMALA
HONDURAS
TEGUCIGALPA
SAN SALVADOR
EL SALVADOR
NICARAGUA
MANAGUA
COSTA RICA
SAN JOSE
Colón
PANAMA
PANAMA
CANAL ZONE (U.S.)
BERMUDA ISLANDS (BR.)
BAHAMA ISLANDS
Nassau
HAVANNA
CUBA
CAMAGUEY
SANTIAGO DE CUBA
GREATER ANTILLES
KINGSTON
JAMAICA
HAITI
PORT-AU-PRINCE
DOMINIC. REP.
SANTO DOMINGO
PUERTO RICO (U.S.)
SAN JUAN
St. Martin (Neth., Fr.)
LESSER ANTILLES
Guadeloupe (Fr.)
Dominica (Br.)
Martinique (Fr.)
St. Lucia (Br.)
St. Vincent (Br.)
Grenada
PORT OF SPAIN
ANTILLES (NETH.)
CARIBBEAN SEA
CARACAS
MARACAIBO
MEDELLIN
BOGOTA
CALI
SOUTH AMERICA
Equator
QUITO
Iquitos
Rio Branco
GALAPAGOS ISLANDS (ECUADOR)
PACIFIC OCEAN
ATLANTIC OCEAN

Bayfront Park backed by resort hotels along Biscayne Boulevard in Miami Florida. Miami is one of the nation's most popular tourist resorts.

Indians, remain in the USA, and Canada has some 200,000 Indians and Eskimos. Only in Mexico is there still a marked Indian group (about 30%).

English is generally spoken throughout the United States and in most of Canada. French is spoken in parts of Canada and in some of the Caribbean islands. Spanish is the language for most of Mexico and Central America, though many Indian (and Eskimo) dialects are still spoken in all parts of the continent.

Christianity is the dominant religion of the continent. The majority of people in Mexico and Central America are Roman Catholics, while about 60% of the population of the United States is Protestant, and most of the rest Catholic. Canada is about half Protestant, half Catholic. Significant minority religious groups include Eastern Orthodox Christians, and Jews.

The dominant feature of the population of North America today is that it has become predominantly urban (well over 70%), especially in the USA and Canada. More accurately, it is becoming suburban, for in recent years there has been a marked outward spread of the cities. Between 1960 and 1970, the population of the major cities of the USA increased by 3 million, but that of their suburbs by 15 million. There is also a tendency to move from the smaller cities to the larger ones. In major areas of North America the village pattern of Europe is virtually non-existent.

Another characteristic of the population is its increasing mobility, rooted in a pioneering past. There is also a continual westward movement of population, so that, for example, California is the most populous state of the USA.

The Economy

With the exception of the enormous land mass of Eurasia, North America produces more corn and wheat, copper and iron, coal and petroleum and cattle and timber than any other continent, and surpasses all of them in the production of natural gas and hydroelectric power. This wealth, however, is most unevenly spread. Farms in the USA and Canada supply most of the farm products; in Mexico and Central America farms are small and generally poorly mechanized. Moreover, Cuba and some other southern countries depend dangerously on the market and climatic vagaries affecting a single major commercial crop – sugar, coffee, or bananas.

Although Mexico has abundant lead, silver and zinc, the most important industrial

The city of Oaxaca in Southern Mexico is situated on a high plateau and surrounded by mountains.

minerals, such as coal, copper and oil, again are concentrated in the USA and Canada. The areas of greatest industrialization on the continent reflect this distribution. The industrial centres of the USA are found east of a line drawn from Minneapolis through St Louis, and in southern California, while another area is found in the southern part of the Mexican plateau.

Atitlan is a Guatemalan lake of spectacular beauty on the Pacific edge of the highlands.

Farms, Forests and Fisheries. Because of the wide climatic variations within North America, almost all temperate crops and many sub-tropical ones can be grown, although additional supplies of coffee, tea, rubber and similar crops have to be imported. The USA has the largest yield of crops and the greatest variety of products, especially fruits and vegetables: most crops are commercial, and for a national market. Canada comes next, and the two countries together are leading world suppliers of wheat, corn, cotton, and also barley, rice and meat products.

Because of mechanization, fewer people are being employed in farming while more and more is being produced from a relatively static or declining area of farm land. Limits of production have been pushed further into the drier zones, and, in Canada, zones with shorter growing seasons.

Mexico is primarily an agricultural nation, but production is hampered by relatively poor soils, and inefficient land use.

The development of the forest industry in North America is most marked in the southern parts of the Canadian forests, in the luxuriant forests of the Pacific Northwest, and in the coniferous forests of the south-eastern USA. There is a great deal of commercial fishing, especially on the more northerly coasts of the continent, and the Great Lakes area, but fishing takes a very minor place in the vast economy of the continent.

Mining. Nearly half the world's known coal reserves are in North America, and the largest part of these are in the USA, with some in Canada. Anthracite coal is practically an American monopoly. North America also has substantial reserves of petroleum, and there have been spectacular new discoveries since the early 1960s notably in Alberta and Alaska. Nevertheless, the USA now supplies only about 25% of the total world production compared with 50% at the beginning of the 1950s. This change is largely due to big discoveries elsewhere. North America, especially again the USA and Canada, is also richly endowed with iron and nickel (yielding most of the world's production), as well as sulphur, vanadium and molybdenum. Among the major metal ores it principally lacks is tin. One of the world's richest sources of uranium ore is found in Canada.

Economic, political and strategic factors all combine to complicate the mineral picture. For instance, the USA both imports and exports oil and iron ore. Depletion of existing reserves, as in the great Mesabi Range in Minnesota, is very dangerous, since mineral reserves cannot be replaced. So far exhaustion of one deposit has been compensated for by the discovery of new deposits elsewhere, but this cannot last indefinitely.

Industry. The North American continent as a whole is one of the major industrial regions of the world. The United States leads the world in industry, and because of its better developed transport system, larger home market and earlier start, it has developed a very much larger and more complex industrial economy than Canada. However Canada has made spectacular advances over the last few years.

The continent is the world's major producer in virtually all fields except shipbuilding and marine engineering. Today one can talk of a North American manufacturing belt stretching from Chicago and St Louis in the west through the American and Canadian lake shore areas and extending in the east through Toronto, Montreal and Quebec to Boston, New York and Baltimore. However the boundaries of this belt are becoming less rigid as increasing ease of transport, increased mobility of the population and the development of electrical supplies combine to create new industrial centres, especially in California and British Columbia, in the prairie region, and in Texas and the Gulf states of the USA. Industry is now beginning to be more market-oriented than dependent upon material resources and fuel.

Transport and Communications. North America boasts the most extensive and complex

An elk drinking from Bow River, Alberta.

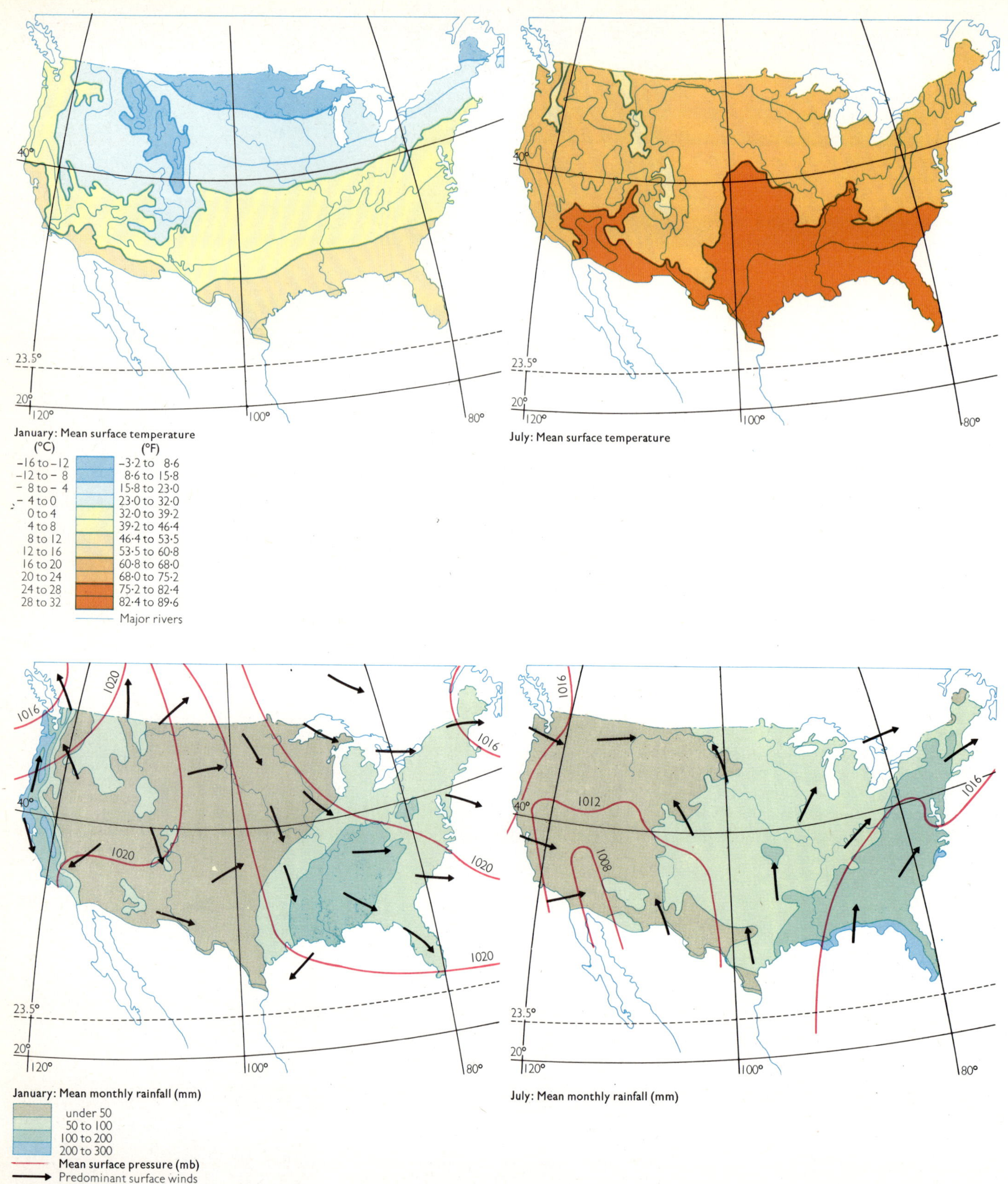

transport network in the world. Again, most of the development has been in the USA, and next in Canada: although the Pan American Highway leads from Mexico down into South America, Mexico and Central America are far less advanced.

Waterways, one of the earliest forms of transportation to be developed, are still extensively used for heavy freight, and thanks to the St Lawrence Seaway-Great Lakes waterway system, and the huge Mississippi-Missouri system, North America has the largest inland waterways system in the world. This also includes intracoastal waterways along the east coasts of the United States, and along the west coast from Seattle to Alaska. At the southern tip of the continent the Panama Canal, built between 1904 and 1914, links the Atlantic and Pacific oceans.

The railways of North America, especially those of the USA and Canada, reached their maximum extent in 1920 with over 290,000 (464,000 km) route miles. Since then the system has suffered severe competition from

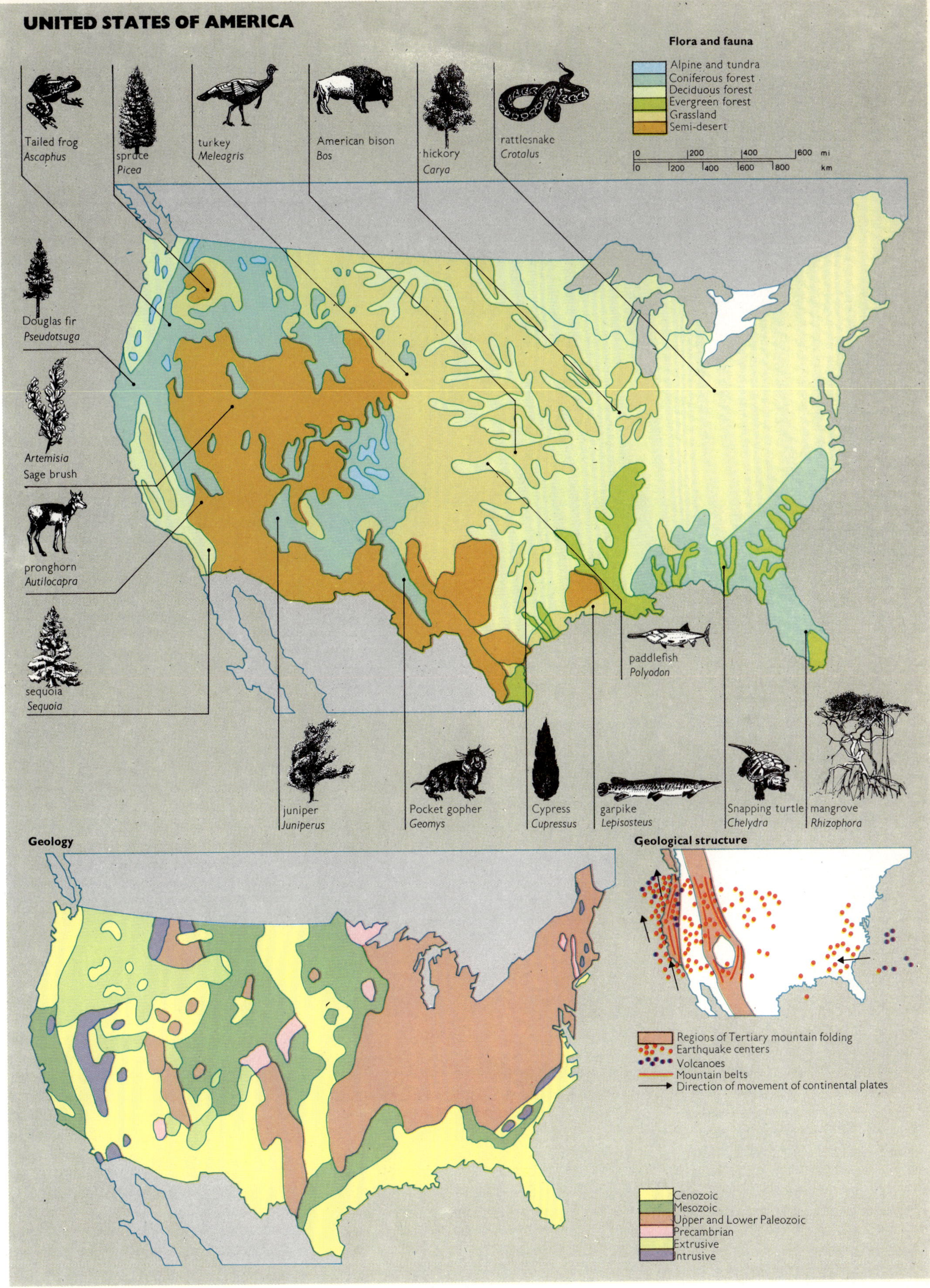
UNITED STATES OF AMERICA
Flora and fauna
Alpine and tundra
Coniferous forest
Deciduous forest
Evergreen forest
Grassland
Semi-desert
0 200 400 600 mi
0 200 400 600 800 km
Tailed frog
Ascaphus
spruce
Picea
turkey
Meleagris
American bison
Bos
hickory
Carya
rattlesnake
Crotalus
Douglas fir
Pseudotsuga
Artemisia
Sage brush
pronghorn
Autilocapra
sequoia
Sequoia
paddlefish
Polyodon
juniper
Juniperus
Pocket gopher
Geomys
Cypress
Cupressus
garpike
Lepisosteus
Snapping turtle
Chelydra
mangrove
Rhizophora
Geology
Geological structure
Regions of Tertiary mountain folding
Earthquake centers
Volcanoes
Mountain belts
Direction of movement of continental plates
Cenozoic
Mesozoic
Upper and Lower Paleozoic
Precambrian
Extrusive
Intrusive

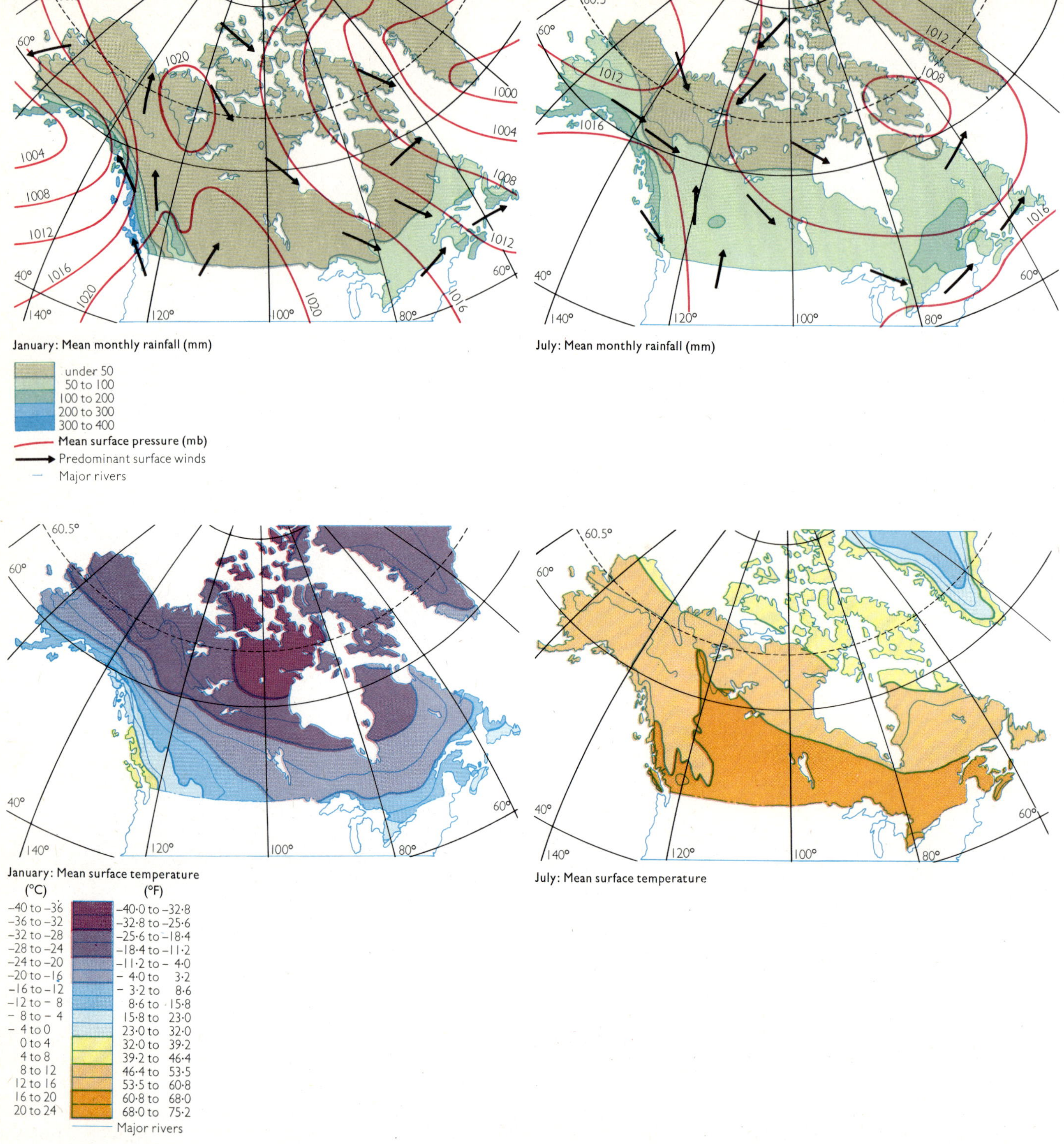

January: Mean monthly rainfall (mm)

July: Mean monthly rainfall (mm)

January: Mean surface temperature

July: Mean surface temperature

other forms of transport, notably the automobile and the aeroplane, and has lost both passenger and valuable freight traffic. By 1966, the US-Canadian network was down to less than 254,000 route miles (about 400,000 km). Since then rail modernization and increasing congestion on the roads have seen some return to the railways.

Most towns and cities in the USA and Canada are linked by good roads and highways. Road systems, following the geography of the continent, are especially well developed in the eastern and central parts. In the USA the network thins in the Great Plains area and through the Cordillera to the west. In west-central Canada the network is again rather thin. Today the USA has nearly completed its 42,000 mi (107,000 km) interstate highway system, and Canada has recently completed its 4,860 mi (7,700 km) Trans-Canada Highway, the first coast to coast metalled highway link.

As a means of moving liquids and gases, the pipeline has been well developed in North America, especially in the USA. Complex networks join the major oil and gas fields to the principal consuming areas.

Airways are particularly appropriate to the large distances of North America, and take the long distance traffic away from the railways. About 70 million passengers a year now go by air, and an increasing (though still small) amount of freight.

Communications are more highly developed in North America, especially in the USA and Canada, than in any other area of the world. Some 2,000 daily newspapers are published, and there are more telephones and television sets per person than in any other part of the world.

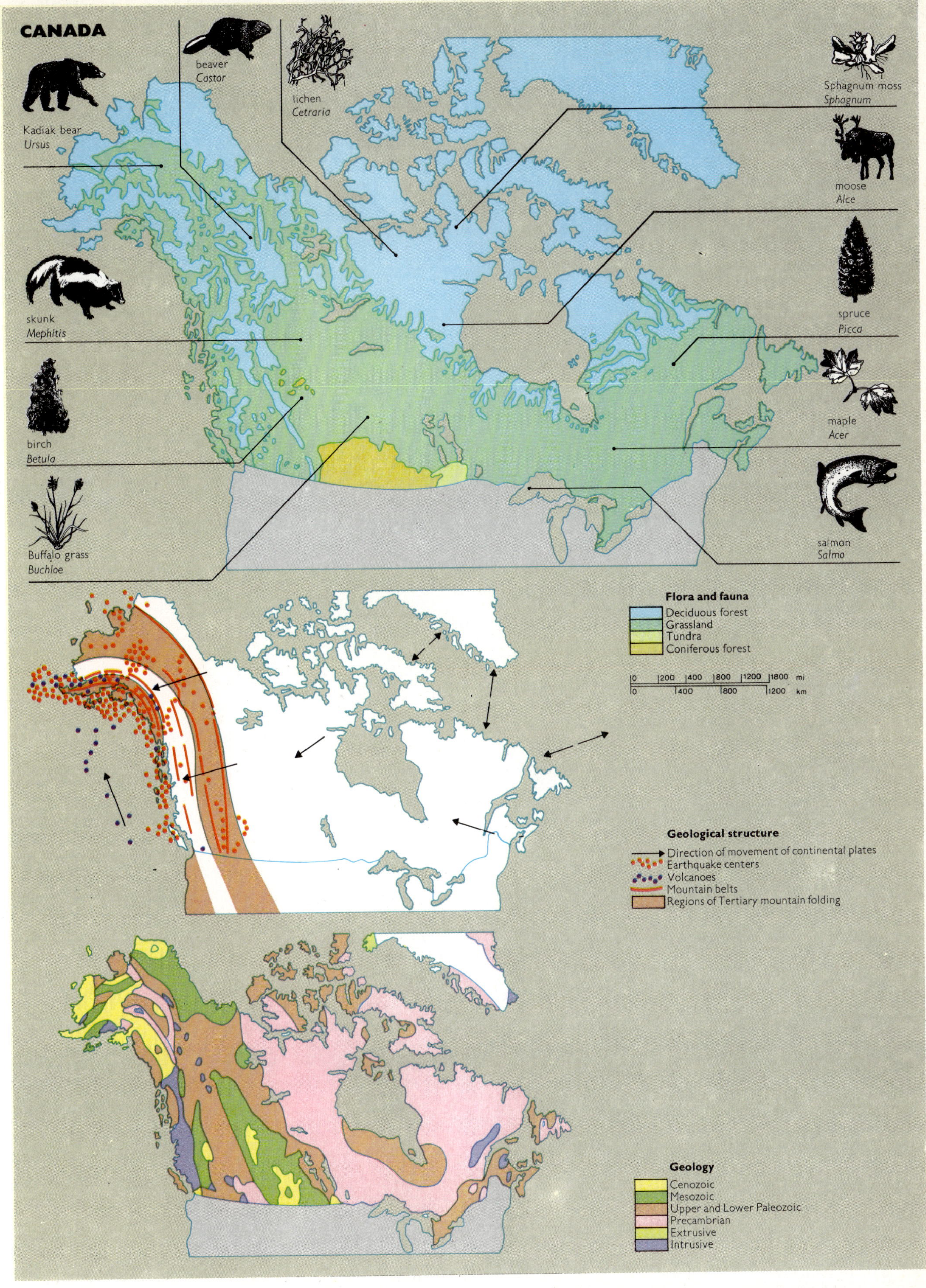
CANADA
Kadiak bear
Ursus
beaver
Castor
lichen
Cetraria
Sphagnum moss
Sphagnum
moose
Alce
spruce
Picca
maple
Acer
salmon
Salmo
skunk
Mephitis
birch
Betula
Buffalo grass
Buchloe
Flora and fauna
Deciduous forest
Grassland
Tundra
Coniferous forest
0 200 400 800 1200 1800 mi
0 400 800 1200 km
Geological structure
Direction of movement of continental plates
Earthquake centers
Volcanoes
Mountain belts
Regions of Tertiary mountain folding
Geology
Cenozoic
Mesozoic
Upper and Lower Paleozoic
Precambrian
Extrusive
Intrusive

CANADA

Area: 3,560,238 sq mi (9,221,016 km²)
Population: 21,788,000
Population Growth Rate: 1·7%
Capital: Ottawa, pop 602,510
Language: English (80%) French (31%)
Religion: Roman Catholic (45%) United Church of Canada (20%) Anglican (13%)
Currency: Canadian dollar = 100 cents

Canada occupies the entire northern half of the North American continent, with the exception of Alaska and the tiny French islands of Saint Pierre and Miquelon. From its southernmost point at Middle Island on Lake Erie (41° 41′ N), Canada stretches north to Cape Columbia on Ellesmere Island (83° 06′ N), some 1,150 mi (1,840 km) within the Arctic Circle. It is the second-largest country in the world – only the USSR is larger. Within its vast area, Canada has tremendous natural resources, including immense mineral deposits. It is one of the world's leading industrial and trading nations.

Canada is a sovereign state and federal parliamentary democracy, but preserves its ties with Great Britain as a constitutional monarchy and as a member of the Commonwealth of Nations. It is composed of 10 provinces and two territories: the Atlantic or Maritime Provinces (New Brunswick, Newfoundland, Nova Scotia and Prince Edward Island), Quebec, Ontario, the Prairie Provinces (Alberta, Manitoba and Saskatchewan), British Columbia, and the Yukon and Northwest Territories.

The majority of Canada's population lives within 200 mi (320 km) of the country's 3,000-mi (1,875-km) border with the United States. This has inevitably coloured the nation's way of life and its people's attitudes. To the casual observer, there seems little difference between culture, social customs and speech in Canada and the USA. There are, however, many significant differences. European origins still exert a strong influence on the Canadian way of life. Canada is also distinguished by lower population densities and relative slowness in developing its great natural resources. It is still searching its vast land areas for deposits of exploitable minerals.

Nevertheless, Canada's economy is closely linked with that of the United States. Many industries are owned by US concerns and US investment capital helps to feed the nation's economy. Some 65% of Canada's exports go to the USA, while 71% of its imports come from that country.

There have been increasing demands in the 1970s for a lessening of cultural and economic dependence on the USA, stemming partly from a reaction to developments in US affairs and partly from a desire for greater national identity. Prime Minister Pierre Elliott Trudeau – who came into office in 1968 and narrowly retained power in 1972 – initiated policies which went some way towards answering these demands.

A major problem of the 1960s and 1970s was the demand by French Canadians in Quebec for greater autonomy. Extremists demanded complete independence for French Canada, and backed their demands with terrorist tactics. Most French Canadians, however, did not support the extremists, although they did feel they were discriminated against in some fields. The federal government was forced to recognize the need for measures to promote true equality between French- and English-speaking Canadians.

Vancouver, the principal city of British Colombia and largest Canadian seaport on the Pacific.

The Land

Nearly half of Canada comes within the Canadian Shield, also known as the Laurentian Plateau. This continental core, covering some 1,771,000 sq mi (4,568,889 km²), is composed mainly of extremely old and hard rocks. It spreads in roughly the shape of a shield around Hudson Bay, extending from the coast of Labrador along the St Lawrence River and Lakes Huron and Superior, reaching into the United States before swinging north-westward through the larger western lakes to the Arctic Ocean near the mouth of the Mackenzie River.

The Shield has been compared to a saucer because its rim is generally higher than its centre, Hudson Bay. In the south-east, along the St Lawrence River and the Gulf of St Lawrence, the Shield rises abruptly from the river plain. Around the rim, streams have cut deeply into the rocks, creating a landscape of smoothly-rounded rocky hills. Some valleys have been over-deepened by the action of glaciers; many others, blocked by glacial debris, have been filled by small lakes. Within the rim, especially in areas draining into Hudson Bay, there is little surface relief and drainage is relatively poor; the result is shallow basins containing lakes, marsh or muskeg.

The Great Lakes – St Lawrence Lowlands. This area, south-east of the Shield, consists of peninsular southern Ontario and the gently rolling area south-west of the city of Quebec. Running south-eastwards from the Bruce Peninsula in Lake Huron is a layer of hard limestone, the Niagara escarpment, through which the Niagara River has cut the dramatic gorge of the Niagara Falls. The St Lawrence Lowlands, between the Shield and Appala-

Quebec's first nuclear generating station is the small 250,000 Kw facility at Gentilly. Quebec has an abundance of hydroelectric power.

chian Canada, were covered by the sea after glaciation and are now a featureless plain of marine sediments, except for the eight Monteregian Hills rising abruptly in a line running eastwards from Montreal. The lowlands are separated from the Ontario peninsula by a south-eastern extension of the Shield, which is crossed by the St Lawrence River near the outlet of Lake Ontario.

The Appalachian Region. Part of the Appalachian mountain system, this region comprises the eastern borderlands of Quebec, the island of Newfoundland and the provinces of New Brunswick, Nova Scotia and Prince Edward Island. It has a general pattern in Quebec of gently rolling uplands separated by deep valleys. The original Maritime Provinces are a basin lying between the Canadian Shield on the north and another area of hard rocks in southern Nova Scotia. The typical north-east/south-west alignment of the basin has been modified by glacial erosion and deposition.

The Interior Plains. West of the Canadian Shield, a vast wedge-like extension of the United States' Great Plains tapers northward from a base of some 800 mi (1,280 km) along the United States border to an Arctic apex about 200 mi (320 km) wide. The Interior Plains are clearly stepped at different levels. The first rise, to about 1,600 ft (490 m), comes with the Manitoba escarpment (Duck, Riding and Porcupine mountains). The second, to an average 3,000 ft (910 m), with the Missouri Coteau in Saskatchewan. From Alberta, the plains fall away to the Mackenzie River valley. Deposits from ancient glacial lakes have made the plains mostly featureless, but in Saskatchewan and Alberta rivers have cut deeply through them. Their southern grasslands are the famous Prairies.

The Cordilleran Region. The area west of the Interior Plains is composed of much younger rocks than the Canadian Shield and is often called the Rockies after its easternmost mountain range. Basically, it consists of a central plateau flanked by twin mountain ranges broken by deep troughs. The western longitudinal trough has been invaded by the sea, and the westernmost mountain range is represented by Vancouver Island and the Queen Charlotte Islands. The granitic Coast Mountains rise precipitously east of the trough to heights of more than 9,000 ft (2,740 m), preventing access to the interior except where the Fraser River cuts through, east of Vancouver. North-west of the river are rugged alpine peaks, including Mt Waddington (13,260 ft; 4,042 m). In the south-west Yukon are the St Elias Mountains with Mt Logan (19,850 ft; 6,050 m), the highest peak in Canada.

The central plateau is considerably dissected in the south, but northwards becomes gently rolling upland until, in northern British Columbia, the Skeena and Hazelton Mountains, rising to 8,000 ft (2,438 m), bar the way to the Yukon plateau beyond. The eastern rampart of the Cordillera is the Rocky

A farm in Saskatchewan, the 'breadbasket of Canada'. Saskatchewan has over a third of Canada's farmland and produces 60% of its wheat.

Mountains, containing at least 30 peaks more than 10,000 ft (3,050 m) high, the loftiest being Mt Robson (12,972 ft; 3,954 m). Near the British Columbia-Yukon border the Rockies end, the comparatively level Liard plain and plateau linking the Yukon with the Interior Plains and providing the route for the Alaska Highway. Farther north the Selwyn, Mackenzie and Richardson mountains continue the alignment of the Rockies almost to the Arctic Ocean.

The Innuitian and Arctic Regions. The eastern fringe of the Arctic Archipelago is characterized by the high mountains of Baffin and Ellesmere islands, which rise in places to 9,000 ft (2,740 m). There are many ice-caps and glaciers, and the coasts have numerous fjords. Many of the north-central islands rise sheer out of the sea to heights of 500–800 ft (152–244 m) and, although scarred by canyons and ravines, are mainly level plateaux. The action of sea and glaciers on the south-central islands has produced low-lying, lake-strewn landscapes. The shallow lakes rest on a permanently frozen layer (permafrost), which on some islands is more than 1,500 ft (457 m) thick.

River Systems. All inland waters in Canada connect eventually with the sea; there are no areas of inland drainage. There are three major river systems: the Great Lakes–St Lawrence system drains central North America and eastern Canada into the Atlantic off Newfoundland. The southern Interior Plains, from the Rockies to eastern Manitoba, are drained by the Saskatchewan–Red–Nelson river system into Hudson Bay. Canada's longest river, the Mackenzie (2,635 mi; 4,241 km) drains the northern Interior Plains and much of the western Arctic into the Arctic Ocean.

Climate. Most of Canada has extremes of climate. Winters are generally long and cold, but mildest on the west and south-east coasts. Summers in settled Canada are generally warm, with July temperatures averaging 60°F (16°C) or more. The main factors influencing climate are distance from the ocean and distance north. A comparison of St John's, Victoria and Winnipeg shows the effects of the sea. Victoria is mild in winter (39°F; 4°C) and cool in summer (60°F; 15°C) with relatively little change from one season to the other. St John's has comparable summer temperatures (60°F; 15°C), but in winter cold westerlies blow from the interior and lower the temperature (24°F; −4°C). Winnipeg lacks the moderating effects of the sea, and warm summers (68°F; 20°C) are quickly followed by very cold winters (0°F; −18°C). A similar though colder pattern occurs at Eureka in the high Arctic, where winter temperatures average −34°F (−37°C) and summer temperatures 42°F (6°C).

Precipitation, rain or snow, also depends on distance from the sea, although the Gulf of Mexico to the south plays an important role. The west coast is wet because moist air can readily move in from the Pacific. The western slopes of the Coast Mountains, the wettest part of Canada, have a rainfall of 100 in (2,540 mm) or more, most of which falls during the winter months.

But on the prairies the air is dry, having lost its moisture in the mountains, and central and north Canada are generally dry. Winnipeg averages 20.35 in (516 mm) annually, and Regina 15.73 in (394 mm). East of the Prairies, moist air from the Gulf of Mexico brings precipitation throughout the year, and in the Maritime Provinces storms moving up the Atlantic seaboard add more moisture.

Soils and Vegetation. The climatic pattern is closely followed by soils and vegetation. The treeline runs from the mouth of the Mackenzie River to the southern shores of Hudson Bay and around the Ungava Peninsula. North of this line poorly drained soil above a permafrost layer supports only tundra vegetation; to the south, forest is everywhere, except on the prairies. Canadian forests, growing on thin and infertile soils, are mainly coniferous.

The main productive forest area, called the Boreal Forest, stretches in a broad belt from the Atlantic coast to Alaska and consists mainly of spruce, balsam fir and pine, but also includes deciduous trees like white birch and poplar. Mixed broad-leaf and coniferous woodlands have developed on the better soils of southern Ontario, Quebec and the Maritime Provinces. In coastal British Columbia conifers predominate. In the dry southwestern corner of the prairies, the fairly fertile brown soil supports short grass. In the damper north-east are the most fertile black soils. The soil colour deepens to black, grasses are taller and there are scattered clumps of aspens.

Wildlife. Canada has a rich and varied wildlife. Polar bears and musk oxen live in the Arctic; the ocean is the home of seals, walruses and whales. Caribou graze the tundra and the fringing forest-lands. Farther south, deer, elk, moose, black and brown bears, wolves and foxes abound, along with beavers, muskrats, martens and other fur-bearing animals. The western mountains are the realm of the grizzly bear and the Rocky Mountain goat. Settled Canada has mainly smaller animals such as squirrels, chipmunks, skunks, weasels and otters, but deer are also common.

There are many species of birds – from golden eagles and other birds of prey to game birds such as grouse and partridge, and the ubiquitous sparrows and starlings. Most birds, especially ducks, geese and cranes, migrate south for the winter. Canada's coastal waters, lakes and rivers teem with fish.

The People

Canada's 1966 census recorded a population of 20,014,880. This had increased to an estimated 21,788,000 by 1972, but the birth rate has been decreasing since 1960. Immigration has always been important, though fluctuating considerably. In 1971, immigrants numbered 121,900 – a far cry from the record 400,870 in 1913. The largest groups came from the UK and the USA. These gains have been offset to some extent by emigration.

Ethnic Patterns. In Canada, the population's diverse ethnic origins are much more evident than they are in the United States. In the past, groups of immigrant settlers in less populated rural areas tended to preserve many of the customs of their mother countries. Today, the pattern is one of two linguistic cultures, English and French, and the differences between the two have led to considerable tension.

About 80% of the population speak English, but only some 44% are of British stock. People of other European origins – Germans, Ukrainians, Italians, Dutch, Poles, Scandinavians, Hungarians and others – account for about 23% of the population. There are also Asian and Negro groups. Most of these people have adopted the English language and Anglo-Saxon cultural patterns, and are greatly affected by the proximity of the United States.

French Canadians, however, representing some 30% of the population, form a much more distinct group. Living mainly in Quebec province, with significant French-speaking minorities, they preserve their own special culture and way of life.

In recent years, French Canadians have become more outward-looking, but have also begun to demand recognition of their special identity. They resent the emphasis placed on the English language and have protested

A rock arch and lake, Ontario. In the interior of Ontario are some 250,000 lakes most of them dotting the ancient surface of the Canadian shield.

against discrimination, particularly in employment. Not more than 12% of Canada's population speak both French and English. Today, Canada has about 230,000 Indians and 16,000 Eskimos.

Settlement Patterns. The pattern of Canada's population densities shows a clear demarcation between south and north. Most Canadians live in the south, and 74% in urban areas. Of Canada's 21 cities with populations over 100,000, only Edmonton and Sudbury are more than 300 mi (480 km) from the southern border. In the rest of the country population density drops to less than two persons per sq mi (0·8 per km^2); some parts are uninhabited. Even in the south, settlement is not continuous.

In the Maritime Provinces, as in the Gaspé Peninsula of eastern Quebec, settlement is largely confined to the coasts. The St Lawrence Lowlands and eastern townships are densely populated, but the Quebec section of the Canadian Shield is empty except for a few southern valleys. In Ontario, most people live south of the Shield, from Sault Ste Marie to the prairies and almost the only settlements are logging camps and small mining towns.

The Interior Plains are completely, though in places thinly, settled. To the north, within the woodlands and on the Shield, settlement is almost non-existent, though there is some movement northwards. The most densely populated area of British Columbia is around Vancouver and Victoria, and the southern inland valleys are well populated. The Yukon and the Northwest Territories have only 53,000 people collected together in a number of small towns.

Canadian Cities. Montreal and Toronto vie for leadership among Canada's cities. In the early 1970s Montreal was the larger, but Toronto, the leading industrial city, appeared likely to overtake it. Vancouver is growing rapidly as Canada's trade becomes more oriented towards the Far East. Edmonton and Calgary are both experiencing a population explosion due to oil.

Ottawa, the federal capital, (metropolitan area, 602,510; city, 302,341), on the right bank of the Ottawa River, has been conscientiously developed as a capital city, and has many impressive public buildings, Roman Catholic and Anglican cathedrals, and two universities (Ottawa and Carleton). Most of its people are employed by the government.

Educational and Religion. Teaching is mainly in English, except in Quebec. The educational system reflects the religious beliefs of the people. In predominantly French Quebec, for example, most of the schools are Roman Catholic. Most provinces have 'separate schools' for minority groups. The federal government maintains schools for Indians and Eskimos, and is responsible for education in the Territories. Illiteracy is negligible. The 60 or so universities range from small colleges to large bilingual institutions like the University of Ottawa.

All major, and many minor, Christian denominations are represented in Canada. More than 45% of the population are Roman Catholics. Some 20% belong to the United Church of Canada, a union of Methodists, Congregationalists and Presbyterians, and over 13% to the Anglican Church. Other groups include Lutherans, Greek Orthodox, Ukrainian Catholic, Mennonites, Salvation Army and Mormons. There are some 275,000 Jews.

The Economy

Canada has natural resources on a gargantuan scale – vast expanses of fertile soils, great forest belts spanning the country, immensely rich fishing grounds and mineral wealth astonishing in its extent and variety. All these are highly important to the national economy, and provide most of the raw materials for Canadian industries, which also have the benefit of readily available water and electric power.

The St. Lawrence at Montreal. The St. Lawrence Seaway connects the Great Lakes to the sea.

Industry. One of the most significant developments of the present century has been the emergence of Canada as a major manufacturing country. Leading products include motor vehicles (about 900,000 per year), pulp and paper, petroleum products, iron and steel, motor vehicle parts and accessories, dairy products, sawn timber, machinery and factory equipment, industrial chemicals and aircraft. Other important industries involve smelting and refining, meat-packing, printing and publishing, communications equipment and processed foods. In 1970 manufacturing accounted for about 25% of the Gross National Product.

Industry is concentrated mainly in the provinces of Quebec and Ontario, which have all but 4 of Canada's 12 leading manufacturing cities, and is centred mainly on Montreal and Toronto. Montreal has a wide variety of industries ranging from the preparation of foods and beverages to the building of ships and aircraft. Steel plants, chemical works, oil refineries and clothing factories are all part of the industrial scene, to which motor vehicle assembly has recently been added. Textile industries are characteristic of many small towns in the St Lawrence Lowlands, where their development has been encouraged by the availability of water power. Eastern Quebec has pulp and paper mills and aluminium plants.

Toronto has varied industries. Like Windsor, on the Detroit River opposite Detroit (Michigan), it is particularly associated with the motor vehicle industry, and the Canadian light aircraft industry is based just north of the city. Toronto also manufactures electrical and automotive equipment, chemicals, clothing, knitwear and many other products. Among other centres in Ontario are Cornwall, which manufactures chemicals, rayon and paper; Sarnia, at the southern end of Lake Huron, another chemicals centre and terminal of the oil pipeline from the west; London, known for its buses and railway locomotives; and Hamilton, a hub of Canadian iron and steel production. Another large iron and steel works is at Sault Ste Marie, and Thunder Bay makes buses and railway rolling-stock.

The only major industrial area in the Maritime Provinces is the iron and steel complex on Cape Breton Island, with the nearby town of Trenton using local steel to make railway rolling-stock. But Newfoundland has one of the world's largest pulp and paper mills, at Corner Brook.

Prairie industry is closely connected with the agricultural products of the region, but surprisingly does not include agricultural

A weather station at Eureka on Ellesmere Island, Northwest Territories. Cape Columbia on Ellesmere Island is the continent's northernmost point.

engineering, although half the farm machinery produced in Canada is bought by Saskatchewan. Farther west industry is associated with oil fields, refineries and petro-chemical plants, either serving these enterprises or using their products.

British Columbia has seen considerable industrial growth since World War II through its desire to meet local needs rather than import manufactured goods from the east. A great variety of goods is produced near Vancouver and Victoria. Forest products and tinned salmon are sent east, and the province has a major Canadian aluminium smelter at Kitimat, based on cheap local hydro-electric power and imported raw bauxite and alumina from the Caribbean.

Electric Power. Until 1950 Canada relied almost entirely on hydro-electricity. Since then, thermal and atomic generating stations have been built, and hydro-electricity's share of power production has fallen to 70%. In Ontario, all the immediately practicable hydro-electric sites have been developed, and current construction is concerned with thermal stations.

By contrast, hydro-electricity is a comparative innovation in Saskatchewan, where the Island Falls, Squaw Rapids and Gardiner Dam hydro-electric plants are now in service. Manitoba's 437,000-kilowatt Grand Rapids development on the Saskatchewan River was completed in 1968. Other large developments include the recently completed Manicouagan River scheme in Quebec and the large Churchill Falls project on the Hamilton River in Labrador, which began operating in 1972.

Agriculture. Although Canada is now a predominantly manufacturing country, agriculture continues to have an important role in the economy. In 1969 it accounted for 12% by value of commodities exported, and it provides work for more than 7% of the labour force. Canada's farmlands occupy 272,070 sq mi (704,661 km^2), less than 8% of the total land area. Poor soils and poor drainage make the Canadian Shield unsuitable for farming, and the length of the growing season has largely limited agriculture to areas south of latitude 55°N.

Certain areas tend to specialize in certain crops. The best example is the Prairie Provinces, world famous for grain crops such as wheat, barley, oats and rye, although cattle ranching is important in the drier areas. Crops of 1,000 million bushels (363,680,000 hl) are usual, but these far exceed Canadian requirements and enormous quantities are available for export. Overseas markets have been so depressed, however, that by 1970 there was two years' supply of grain in store and Canadian grain-growers were being paid to keep their land out of production. Since then there have been indications of a recovery in world demand. Many farmers are now diversifying with mixed stock and grain farms.

British Columbia has three specialized areas. The most important is the Okanagan Valley in the southern plateau region which has about 15% of British Columbia's agricultural land, devoted almost exclusively to fruit production. Apples are the chief crop, but pears, peaches and plums are also important. Rainfall in this area is less than 12 in (305 mm) annually, and irrigation is essential. Market uncertainties have led to some diversification (vegetables and dairying), but the essential character of the area remains unchanged. The northern plateaux have cattle ranches and sheep farms, and fodder crops are grown under irrigation to provide winter silage.

Good soils, the favourable climate and easily accessible markets give southern Ontario a prosperous and diversified agriculture.

Extreme western Ontario is Canada's 'Corn Belt', specializing in maize, while the sandy area along Lake Erie provides most of the Canadian tobacco crop. Fruit and vegetables are grown in a belt around Lake Ontario, the north producing apples and vegetables for canning; the Toronto-Hamilton area, fresh vegetables and fruit; and the Niagara peninsula, peaches and grapes.

Ontario has two dairying areas, one embracing the major urban hinterland around eastern Lake Erie and western Lake Ontario, the other being an extension of the Quebec dairying region. The western area provides mainly fresh milk; the eastern Canadian cheddar cheese. The rest of southern Ontario is a mixed farming region deriving 70% of its income from the sale of livestock or animal products. Farther north, farming is limited to a few favourable basins where milk is produced for nearby centres.

In Quebec, agriculture has only recently been supplanted by manufacturing as the main form of employment. Dairying is most important, especially around Montreal where about half of the milk is made into butter, a third being sold in liquid form and the rest converted into cheese or evaporated milk. On the St Lawrence Lowlands the breeding of meat animals is widespread, and vegetables and fruit are important in some areas. Near Three Rivers (Trois Rivières) tobacco has been introduced as a cash crop. Beyond the lowlands, poor land and remoteness from major markets make farming less economic and many farmers augment their income by forestry work in winter.

In the Maritime Provinces agriculture accounts for less than 20% of the net value of production. In all provinces hay and oats are the leading field crops, but in the St John River valley and on Prince Edward Island potatoes are the main cash crop. Dairying is important everywhere, but especially near the larger towns and cities. In Newfoundland agriculture is almost entirely a subsistence and part-time activity.

Forests and Fisheries. Forest industries are important throughout Canada except on Prince Edward Island, where few trees remain, and in the Territories, where few suitable trees exist. Sawn timber, plywood, and pulp for newsprint are the most important products. The industry employs less than 1% of the total labour force but makes a significant contribution to export earnings. Quebec has the largest area and volume of usable timber, but much is only suitable for pulping. Forestry is the most important industry in British Columbia, whose Douglas firs provide sawn timber and plywood.

In many areas forestry is a winter activity. The logs are transported to the nearest river and dumped on the ice until the spring break-up, when they are driven down river to the saw or pulp mills. Some logs are towed as great rafts or booms by powerful tugs, or even floated down artificial streams called flumes. Others are transported by road or rail.

Both the Atlantic and Pacific coasts have valuable fisheries. Salmon are the most important fish on the west coast, where they are caught as they leave the sea to ascend the rivers to their spawning grounds. In recent years the industry has suffered from over-fishing, and from interference with fish breeding-grounds by new dams and other works affecting the flow of the rivers.

There is some offshore fishing in British Columbia, mainly for halibut, but offshore fishing is more characteristic of the Atlantic coast, where the catch is mainly cod. The ocean currents there, and large areas of shallow seas such as the Grand Banks of Newfoundland, attract vast shoals of not only cod, but hake, pollock, herrings and sardines as well. The salted dried cod that was once the major export of the Atlantic fisheries is no longer in such high demand and most fish is now sold fresh or as frozen fillets, or is canned.

Mineral Riches. Mining and the processing of ores are widespread. Canada is world leader in the production of nickel, asbestos, zinc and platinum, and second only to the United States in uranium and molybdenum. No other country exports more iron ore. Minerals usually provide 30% of Canada's exports.

In the Maritime Provinces mineral resources are limited to the copper-lead-zinc-silver complex at Buchans in west-central Newfoundland, nickel at Bathurst in New Brunswick, and coal in Cape Breton. Coal production was declining until the increase in thermal electricity generation boosted demand in eastern Canada. In the Labrador-Ungava area are the immense iron ore deposits of Schefferville, Labrador City and Gagnon: ore and concentrated pellets are sent to the north-eastern United States.

South-eastern Quebec is the largest asbestos-producing area in the world. The main deposits are at Asbestos and Thetford Mines, but recent discoveries near Ungava

Toronto from the Dominion Tower. Toronto has 30% of Canada's manufacturing capacity.

Bay suggest that these centres may soon have a rival. Canada's so-called 'Gold Belt' runs from north-western Quebec across northern Ontario almost to Manitoba, with the chief mines concentrated in the east around the towns of Val d'Or, Rouyn-Noranda, Timmins and Kirkland Lake. The gold industry has been very prosperous, but rising costs have reduced profitability and made government subsidies necessary.

Canada's richest mining area is the Sudbury basin north of Lake Huron which produces nickel and copper. Poisonous fumes from the smelters and refineries have killed all natural vegetation in large areas around the towns, and soil erosion has created a grim landscape of bare rock. Farther west there is a base metals complex at Manitouwadge, and iron ore is mined at Michipicoten and at Steep Rock Lake, where a lake 15 mi (24 km) long was pumped out to reach the deposits.

Mining centres in the Prairie Provinces include Flin Flon on the Manitoba-Saskatchewan border (silver-copper-zinc), Lynn Lake and Thompson in Manitoba (nickel) and Uranium City in northern Saskatchewan (uranium). Uranium mining has run into difficulties due to over-production throughout the world, and the Canadian government is forced to stockpile the output. In the south, fossil fuels are the most important minerals. At one time this meant mainly the bituminous coal of south-western Alberta and the lignite of south-eastern Saskatchewan, but now all three Prairie Provinces have oil and natural gas.

The deposits in Manitoba and Saskatchewan are small compared to those in Alberta, where the famous Leduc, Redwater, and Turner Valley fields have made the province second only to Ontario in the value of minerals produced. Pipelines carry oil and natural gas to eastern Canada from Alberta and Saskatchewan, while others supply Vancouver and the western United States. Potash, used in chemical fertilizers, has become important in central Saskatchewan, but world over-production has depressed prices.

Half the mineral output of British Columbia comes from the Kootenay district in the south-east, where the Sullivan Mine is one of the world's largest lead and zinc ore producers. The ores are refined at Trail, the chief city in the area, which also has ancillary industries including a small iron and steel plant. High production costs have made most of British Columbia's gold mines idle. Copper and some iron ore are mined along the coast, and asbestos deposits are being exploited at Cassiar in the north.

Within the Territories, mineral exploration and expectations always exceed production. The Yukon has a silver-lead-zinc mine at Keno Hill north of Dawson City, and a natural gas field on the plateau; but further ventures are discouraged by distance from markets. In the Northwest Territories, gold

Above: A catch of herring at a New Brunswick dock. Fishing is a long established industry in the region. *Below:* Pulpwood jamming the Ottawa river. Logs are rafted from the interior.

is mined at Yellowknife, the territorial capital, and silver-lead-zinc at Pine Point on Great Slave Lake. Norman Wells has a small oil field, but despite extensive surveys no new fields have come to light.

Canada produces far more minerals than it can use and is thus at the mercy of fluctuations in world prices. Nickel and copper command good prices on the world market, but in the case of gold, uranium and potash, production has to be cut back or government-assisted. Recent Japanese requirements have revitalized the coal industry in western Canada and may well increase exports of iron ore. In the east, interest is centred on northern Quebec and Baffin Island, where large iron ore deposits await exploitation. In the Prairies, one of the world's largest petroleum reserves has been found in the Athabasca tar sands.

International Trade. Canada is one of the great trading nations of the world, ranking fifth among Western nations after the United States, West Germany, Great Britain and France. Primary products account for 47% by value of Canadian exports. Countries dependent on exporting raw or semi-processed materials are usually at the mercy of fluctuating world prices, but Canada exports such a broad range of commodities that a fall in the price of one is often offset by a rise in the price of another.

Some 65% of Canada's exports go to the United States, about 9% to Great Britain and 5% to Japan. About 71% of Canada's imports come from the USA, 5% from Great Britain and 3% from Japan. In the early 1960s Canada imported much more than it exported, but since 1966 imports and exports have been more or less equal. There is considerable concern over the dominant position of the USA in Canadian trade, and even more over the free-trade areas elsewhere in the world from which Canadian exports are excluded by tariff walls.

Transport. Canada has 45,240 mi (72,384 km) of railways and 518,190 mi (829,100 km) of roads, of which 396,100 mi (633,760 km) are paved.

The railways consists of two transcontinental systems, the privately owned Canadian Pacific Railway and the government-owned Canadian National, and 21 other lines privately or provincially owned. Since 1959 the closure of unprofitable lines has exceeded the construction of new routes. The railways' share of passenger and freight traffic has declined since 1920, and at present both the major railways are trying to abandon much of their remaining passenger services. Road transport and pipelines have reduced rail's share of freight, but the development of 'piggy-back' (truck trailer-carrying) services and containerization has enabled the railways to fight back.

The car is the preferred means of passenger transport and freight movement over distances of less than 300 mi (480 km) are the prerogative of trucks. More than half of Canada's roads are in the Prairie Provinces. The completion in 1965 of the Trans-Canada Highway (4,860 mi; 7,780 km) made it possible to drive from coast to coast on paved roads.

Canada has one of the world's greatest commercial waterways in the Great Lakes – St Lawrence Seaway system, opened in 1959. Since the Seaway was opened, the tonnage using the system has doubled, but the number of ships has remained fairly constant. Elsewhere water transport is used to move supplies to the Yukon from Vancouver by way of Skagway, and to float supplies to the Arctic down the Mackenzie River.

As a result of increasing trade with Japan and China, Vancouver had become Canada's leading seaport by 1967. The iron ore port of Sept-Iles in eastern Quebec ranks second.

Canada is served by two major airlines, the government-owned Air Canada and the privately owned Canadian Pacific Airlines. Long-distance passenger travel in the south is mostly by air; air freight has not yet become of national importance. In the north, however, air transport is usually the only way of transporting goods and people, and services are provided by many companies and private operators ('bush pilots'). A start has been made on building a road network in the north, but this is unlikely to diminish the importance of aircraft.

The Canadian side of Niagara Falls is receding nearly 5 ft a year as the soft rock under the hard dolomite river bed is cut away.

GREENLAND

Drying fish in south-western Greenland. Gradual warming of Greenland's seas has meant that cod, halibut, shrimp and wolf-fish have moved their range north to Greenland.

Greenland is the largest island in the world. Lying in the North Atlantic, and for the most part within the Arctic Circle, it covers 840,000 sq mi (2,175,590 km²), of which some 727,000 sq mi (1,934,720 km²) are ice covered. Its population is small, its resources few, but its geographical position makes Greenland strategically important. There are radar stations of North America's DEW-line (Distant Early Warning Line) on the island, and American air bases such as Thule. Greenland is an integral part of Denmark and thus comes under the protection of NATO; a Danish-American agreement for common defence of the island was signed in 1951.

Difficulties of Settlement. The far north of Greenland is only 25 mi (40 km) from Ellesmere Island in the Canadian Arctic Archipelago. Greenland's first settlers, the Eskimos, came from the North American mainland in the 8th or 9th century. Today, however, only two major Eskimo settlements remain. The first arrivals from Europe were the Norsemen in the 10th century. In 1953 the island was made a county of Denmark. It has two representatives in the Danish parliament and also has its own democratically elected council, the *landsraad*.

The Great Ice Sheet. Geologically Greenland belongs to the Laurentian Shield and is mainly a plateau of predominantly Pre-Cambrian rocks, especially granites, bordered on the west, north and east by mountain ranges which rise in places to more than 12,000 ft (3,658 m). The coast has many deep fjords, the largest being Scoresby Sound, about 185 mi (298 km) long.

The interior of the island is covered by the largest ice sheet in the Northern Hemisphere. In east-central Greenland it is over 11,000 ft (3,350 m) thick. Glaciers flow from the ice sheet to the sea, the largest being the Humboldt Glacier, 60 mi (97 km) wide where it terminates on the north coast.

Greenland has a polar climate. Temperatures are above freezing for only a few months of the year. Winter temperatures are relatively higher on the south coast, influenced by the North Atlantic Drift. The east coast, washed by the Labrador Current, is much colder than the west. Precipitation is heavy in the south-west, but lighter elsewhere.

The Greenlanders and Their Work. Greenland has a population of 46,531 of which only 7,417 were born outside the island. More than half of the population lives in the west coast towns and settlements, the largest of which is Godthaab (8,209), the capital. Most Greenlanders are of mixed Danish-Eskimo descent. Eskimo is commonly spoken, but the use of Danish is spreading.

Greenlanders once lived by sealing and trapping, but today fishing is far more important with Godthaab and Frederikshaab as the main centres. Fish and shellfish are frozen, salted or canned for export. Some farming is only possible along parts of the south-west coast, where sheep and cattle are raised and small summer crops of hay and vegetables grown.

Greenland is the sole source of natural cryolite, used in making aluminium and ceramics, which has been mined near Ivigtut since 1865. A concession for working lead and zinc near Umanak was granted to a Danish company in 1971, and there is growing interest in oil and other prospecting. Lying on a number of polar air routes, Greenland has become an important centre for weather and radio stations. The Søndre Strømfjord airport serves polar flights linking Denmark with the Pacific.

ST PIERRE AND MIQUELON

Saint Pierre and Miquelon (linked by a shingle isthmus to the former separate island of Langlade), are the main islands of an archipelago off the southern coast of Newfoundland. They cover about 93 sq mi (242 km²) and are part of the French Overseas Territories. Saint Pierre (4,362), on the island of Saint Pierre, is the capital. The islanders are chiefly of Breton, Norman and Basque descent. The economy of the islands is based primarily on fishing.

UNITED STATES OF AMERICA

Area: 3,536,855 sq mi (9,160,454 km²)
Population: 204,765,770
Population Growth Rate: 1·1%
Capital: Washington, pop 756,510
Language: English
Religion: Protestant (66%) Roman Catholic (25%) Jewish (3%)
Currency: US dollar = 100 cents

The United States is the world's fourth-largest nation, in area as well as in population. In nearly every branch of its economy, it surpasses all other countries. Although it accounts for only 6% of the world's population, it possesses 50% of the world's wealth.

Without its rich farmland, plentiful mineral resources and abundant water supplies the United States could never have become the great economic power it is today. But equally responsible for America's economic pre-eminence has been the ingenuity and resourcefulness of its citizens. The explorers and settlers from Europe who followed after Columbus' discovery of the New World in 1492 faced with a wild and hostile continent, transformed it in a mere four centuries into the richest and most powerful nation on earth. And from these diverse transplanted Europeans, at first beset by national rivalries in their explorations, gradually evolved the American – inventive, optimistic and sturdy. The settlers from England had proved the most successful in establishing their colonies and the American was therefore a creature of predominantly British stock. But he became fully independent of his national origins after the Revolutionary War with Britain (1775–1781). The mastering of the continent produced an unequalled social and geographical mobility, and a remarkably open political and cultural life which also contributed to the nation's economic growth.

The United States is a federal republic of 50 states and the District of Columbia (the site of Washington, the nation's capital). The 48 contiguous states together form a broad band stretching 2,800 mi (4,500 km) across the North American continent from the Atlantic to the Pacific. With the additional outlying states of Alaska at the northwest corner of the continent, and Hawaii, an archipelago in the Pacific Ocean (described in a separate article), the United States extends to the north above the Arctic Circle, and to the south into the tropics.

Swiftcurrent Lake lies among the towering peaks of the Rockies in Glacier National Park, Montana. Some 60 glaciers remain from the Ice sheet that once covered the area.

America's Recent Interest. After the outbreak of World War II in 1939, the United States was gradually drawn into the conflict, then decisively thrown into it when Japan attacked the Pearl Harbour naval base in December 1941. The next four years saw the United States fighting on two fronts – Europe and in the Pacific – and on land and sea and in the air. The war ended in August 1945, after the first use of the atom bomb on Japanese cities. In the post-war years the United States came to the aid of ravaged Europe with a series of imaginative aid programmes, both civilian and military, and warmly supported the founding of the United Nations in 1945. But the post-war years also saw the development of the so-called Cold War with Russia and the other countries in the communist bloc.

The 1960s, years of unparalleled prosperity, nevertheless saw drastic internal disturbances and overcommitment abroad. The Negro civil rights movement developed from peaceful demonstrations in the South to violent riots in the Northern ghettoes and on the nation's campuses. In foreign affairs, the chief source of discontent there was the growing involvement in the futile and costly Vietnam War. Richard M. Nixon, who became president in 1969, sensing the public mood, gradually withdrew American troops from Vietnam and finally brought the war to an uneasy end in 1973. His policy of detente of the Cold War situation including a historic visit to Communist China and another to Russia, was widely acclaimed. These foreign preoccupations, however, precluded more active policies on the all-important domestic front and accusations of presidential complicity in political corruption tarnished Nixon's image.

The Land

The vast area of the continental United States can be roughly divided into six geographical regions: the Atlantic and Gulf Coastal Plains; the Appalachian Highlands; the Central Plains; the Rocky Mountains; the Western Plateau and Basin Region and the Pacific Coastlands. We shall here consider Alaska's varied terrain as a seventh region.

The Atlantic and Gulf Coastal Plains extend from Long Island in New York State south to Florida and westward to the Mexican border. Apart from Cape Cod in Massachusetts and part of the Rhode Island-Connecticut coast, the coastal plain is virtually

The deserts of south-west Arizona where the annual rainfall is only 3 in are among the hottest and most arid regions of the US.

non-existent in New England. Rough, rocky coasts are characteristic of this region, and the stony soils of the hilly interior make farming on a large scale difficult.

The coastal plain itself reaches its greatest extent in South Carolina, Georgia and northern Florida, where in places it is up to several hundred miles wide. It was here that the great plantations of the South flourished. Small industrial centres like Richmond, Virginia, and Raleigh, North Carolina, grew up in the interior along the fall line, where the rivers from the Appalachian uplands tumbled down to the coastal plain, furnishing water power for manufacturing. Goods were then shipped down the navigable portion of the rivers to the sea.

Apart from important bays and inlets such as Long Island Sound, New York Bay, Chesapeake Bay and Delaware Bay, the Atlantic coast south of New England can be characterized as one of low marshes with sand dunes, and in the South, fringed with a chain of islands, lagoons and sandbars. Important rivers that drain the Atlantic coastal plain include the Hudson, Susquehanna, Potomac and Delaware.

The Gulf coastal plain runs from Florida to Mexico and includes much of the area covered by Florida, Alabama, Mississippi, Louisiana and eastern Texas. Perhaps the most important feature of the Gulf Coast is the delta of the great Mississippi River, which drains most of the Central Plains, and a large part of the Gulf Coastal plain. The rich alluvial soils of the delta region make it a prosperous agricultural lowland; rice and many sub-tropical fruits are grown. The Rio Grande, which rises nearly 2,000 mi (3,220 km) away in the Rocky Mountains, also drains into the Gulf of Mexico. Important port cities along the Gulf Coast include Jacksonville, Florida; Mobile, Alabama; New Orleans, Louisiana; and Galveston and Corpus Christi, Texas.

The Appalachian Highlands run in a south-westerly direction from Nova Scotia in Canada to Alabama. In the north they include such ranges as the White Mountains of New Hampshire, the Green Mountains of Vermont and the Adirondack Mountains of New York. Further south, they form a narrow chain of mountains flanked on the west by a ridge and valley belt, and west of t at, by the Allegheny Plateau. The mountain chain includes the Allegheny, Blue Ridge, Black and Great Smoky mountains. The Allegheny Plateau, cut deeply by rivers, is not flat, but rises sharply in steep hillsides in many places. Apart from a few fertile valleys, there is little good farmland in this region, for the soil is easily eroded from the rocky hillsides. Rich deposits of coal have made mining one of the traditional occupations of the region, though some subsistence farming also takes place. However, a large part of this region in eastern Kentucky, Tennessee and West Virginia is an economic and cultural backwater, and the local dialects, customs and way of life of the hill people have remained virtually the same since the early days of settlement.

The Central Plains stretch westward from the Allegheny Plateau in the east to the Rocky Mountains in the west. This vast region, created to a large extent by glacial action, is the drainage basin of the huge Mississippi-Missouri river system and its tributaries, of which the Ohio River is one of the most important. These rivers have provided a natural water route through the interior since the earliest days of exploration. East of the Mississippi the land seldom rises to over 1,000 ft (300 m), but to the west it rises gradually to as high as 4,000 ft (1,200 m) in Colorado and Wyoming. The western portion of the Central Plains is often called the Great Plains; states in this region, such as Iowa, Kansas and Nebraska, can boast some of the world's most productive farmland. Within the Central Plains are also a number of upland masses including the Black Hills of the Dakotas, the Colorado Plateau in Texas, the hills fringing the southern shores of Lake Superior, and the Ouachita Hills and Ozark Mountains of southern Missouri, Arkansas and eastern Oklahoma.

Encroaching on the plains in the north-east are the five Great Lakes, covering a total area of 95,170 sq mi (247,000 km^2). They have played a major part in the development of the Central Plains, and such cities as Chicago, Detroit, Duluth and Milwaukee owe their importance to their position on the Great Lakes, today connected to world shipping routes by the St Lawrence Seaway.

The Rocky Mountains form the backbone of

Multnomah Falls 30 mi east of Portland Oregon, tumble 620 ft down the walls of the Great Gorge of the Columbia River.

North America, running from Alaska into Canada and through the United States into northern Mexico. In the American portion, especially in the highest range of the Southern Rockies in Colorado, many peaks reach elevations of over 14,000 ft (4,270 m). Between the southern and northern sections of the Rocky Mountains lies the Wyoming Basin, a plateau about 7,000 ft (2,175 m) high with isolated low mountains. Rivers such as the Bighorn, North Platte and Laramie cut deep canyons into the basin floor. Much of this intermontane basin region is semi-arid, treeless grassland, suitable for grazing.

The Northern Rockies stretch across northern Utah and Wyoming, Montana, Idaho and north-eastern Washington. Although altitudes are generally lower than those of the Southern Rockies, there are fewer passes through these ranges. Glacial features are most pronounced in Glacier National Park in Montana. Yellowstone National Park, in Wyoming, Idaho and Montana, is a lava basin surrounded on three sides by mountains. The hot springs and active geysers that have made this region famous are among the most extensive in the world.

The Continental Divide, which passes through the Rockies, marks the line between the streams that flow eastward toward the Atlantic Ocean and those that flow westward to the Pacific. The longest rivers west of the Continental Divide are the Snake-Columbia system of the north-west and the Colorado River in the south. Despite the existence of other, smaller streams, the total amount of water that reaches the Pacific is small compared with the area to be drained.

Most of this mountainous region is sparsely populated; settlements in the area originally grew up around mineral deposits during the great mining boom of the mid-19th century. However, the Rockies afford some of the most spectacular scenery in the United States: high, snow-capped peaks are mirrored in ice-blue glacial lakes; forests support an abundance of wildlife; and above the tree line, green alpine meadows bloom in spring with carpets and bright flowering plants. This region attracts large numbers of tourists and vacationers each year.

The Western Plateau and Basin Region is an extensive complex of high tablelands, plateaux, intermontane basins, canyons, deserts and mountain ranges that lie to the west of the Rocky Mountain chain. Covering parts of Oregon, Idaho, Utah, Nevada, New Mexico, Arizona and south-eastern California, it includes such features as the Columbia Plateau of southern Idaho, Oregon and Washington; the Grand Canyon of Arizona, cut out of the Colorado Plateau by the Colorado River; the Great Basin, the largest area of internal drainage in North America; and numerous small, roughly parallel mountain ranges. The Great Basin is essentially a desert region, fed sporadically by streams which appear after violent rainstorms and flow out of the upland areas onto the flat, sandy sinks or *playas*, where they are absorbed into the ground. A number of shallow salt lakes also characterize this region, the largest of which is Great Salt Lake in northern Utah, about 80 mi (129 km) long and 50 mi (81 km) wide. Death Valley in California, which lies 282 ft (85 m) below sea level, is also a feature of the Great Basin.

To the west, the Plateau and Basin Region is separated from the Pacific Coastlands by two mountain ranges, the Cascade Range of northern California, Oregon and Washington, and the Sierra Nevada Range of eastern California. Taken together, these ranges run north-south from Canada to Mexico and are comparable to the Rockies in height and ruggedness. A popular recreation area, they are also rich in mineral deposits, including the gold that sparked the great California Gold Rush of 1849. The Cascade Range includes a number of extinct volcanoes, among them Mt Shasta and Lassen Peak in northern California and Mt Rainier in Washington.

The Pacific Coastlands lie to the west of the Cascade and Sierra Nevada ranges. A series of broad lowlands and rich valleys including the Puget Sound lowland, the Willamette River Valley and the long Central Valley of California, are sheltered from the Pacific by the Coast Ranges, which in many places drop sharply to the sea. Important cities of this lowland and valley region include Seattle and

The Matanuska River near Anchorage in south-central Alaska.

Skyscrapers in New York. Skyscrapers were pioneered in New York and Chicago.

Tacoma, Washington; Portland, Oregon; and Fresno and Sacramento in California. A natural break in the Coast Range in central California, providing easy access to the Central Valley, was in part responsible for the growth of the port of San Francisco. The Pacific coastal plain is virtually non-existent; it is broadest around Los Angeles, one of the West's great metropolitan centres.

Alaska, bounded by Canada, the Arctic and Pacific oceans and the Bering Sea, is largely dominated by northward extensions of the Rockies and related mountain systems. Part of the Pacific mountain system follows the great curve of Alaska's southern shoreline which ends westwards in the Aleutian Islands. Glaciers and jagged peaks including Mt McKinley (20,300 ft; 6,192 m), North America's highest point, tower above the coast. North of Mt McKinley's Alaskan Range lies an area of low hills aligned east-west and drained by the westward-flowing Yukon – Alaska's longest river. North of this region rise the steep flanks of the Brooks Range (the northern end of the Rockies). North of the Brooks Range the low Arctic coastal plain extends to the Arctic Ocean.

The Climate. The climate of the main continental area of the United States is heavily influenced by its geographic position, with extensive oceans to the east and west; a rather shallow, warm sea to the south with winter water temperatures that average as high as 60°F (15°C); and the extensive landmass of Canada to the north. On the west coast (which, with the south-east has some of the country's highest annual precipitation levels), westerly winds blow inland from the cool Pacific and are forced to rise over the Coast Ranges. This air is warmed and dried on the eastern slopes, so that annual rainfall tapers from over 70 in (1,780 mm) in some west coast areas to under 8 in (200 mm) locally inland. To the north and south of the country no such physical barriers intervene and there is no obstruction to general air movement. On the east coast, the prevailing winds are offshore and the warming effect of the Atlantic Ocean is of little significance in winter. As a result, many parts of the country experience a harsh continental climate with great extremes in temperature. Temperatures of below −40°F (−40°C) in winter and above 100°F (38°C) in summer are not infrequently recorded in the northern interior. Even New York City on the Atlantic Coast experiences relatively severe winters and hot summers.

Alaska differs from the rest of the continental United States mainly in its generally much lower temperatures.

Winter and Spring. In winter, maximum precipitation occurs on the west coast, when about 50% of the mean annual rainfall (70 in) is recorded. Further inland, the significance of these winter rains decreases and in the Great Plains winter precipitation is only about 15% of the annual amount, most of it falling as snow. The other source of moisture is the Gulf of Mexico, and depressions that originate here frequently increase in intensity as they move north-east into the colder air. Thus while the south may receive mild air and rain, the north will experience much colder air and snowfall can be heavy.

Temperatures in winter vary greatly over the country and depend on the frequency of cold and warm air streams. Since much of the west is sheltered on the east by the mountain systems, temperatures are determined by the maritime air from the Pacific: they are relatively high and the mean daily temperature rarely falls below freezing. In the lowland interior, cold Arctic air can sweep down as far as the Gulf Coast, bringing temperatures of −50°F (−46°C) to northern Minnesota and 0°F (−18°C) to north-west Florida. Conversely, warm air can penetrate far northwards, and winters are characteristically variable, with 'cold waves' and blizzards in the north and 'northers' in the south. In contrast to the west coast, the east coast experiences a continental climate with extremes of temperature because of the prevailing offshore winds. In Alaska, January temperatures are everywhere below freezing. Prospect

San Francisco is famous for its hills and for its cable cars.

A Jazz Festival in New Orleans, the city which, more than any other, is the home of jazz.

Nevada's principal source of income is from the tourism which results from the liberal divorce and gambling laws of its two largest cities, Reno and Las Vegas. Half of all Nevadans live in the Las Vegas area, above, whose streets are lined with nightclubs, casinos and bars.

The Pueblo Indians of Arizona and New Mexico live in permanent towns of adobe and mud.

Creek Camp in northern Alaska holds the record for the nation's lowest temperature: −78·8°F (−61·5°C), recorded in January 1971.

In spring, with the northward movement of the sun, the southern areas begin to warm up. Mean daily maximum temperatures reach 80°F (26°C) in the south-western deserts, south-west Texas and Florida and may reach 50°F (10°C) in the north. In some areas, precipitation patterns also change. Southern California receives much less rain, and in Oregon and Washington, although precipitation is still fairly high (over 15 in; 380 mm), spring is one of the drier seasons. East of the Rockies, precipitation is rather more evenly distributed. In Wyoming about 35% of the mean annual rainfall comes in spring, when there is a greater influx of moist Pacific or Gulf air. Elsewhere precipitation distribution is largely similar to the winter pattern. Tornadoes in the Mississippi Valley most often occur in spring, when cold Arctic air meets warm Gulf air to create an unstable situation.

Summer and Autumn. Precipitation in summer follows a different pattern. Over most of the west rainfall is at a minimum. A large area stretching across the Great Plains and the south-east has a summer maximum of precipitation. In the south-east this is the result of hurricanes moving eastwards through the Caribbean and then curving northwards over the United States, giving prolonged and heavy rain.

Temperatures in summer are generally high, with mean July daily maximums of above 80°F (26°C) for most of the country and up to 110°F (43°C) in the almost cloudless deserts of the south-west. But in Alaska coastal temperatures reach only around 55°F (13°C) and inland temperatures rise little above 40°F (4°C). Other relatively cool areas include the extreme north-east – Maine and Vermont – where depressions are more frequent, giving greater cloud cover; the western coastal zone, where the cool California Current and onshore sea-breezes keep maximum temperatures below 70°F (21°C). In San Francisco, the mean daily maximum is only 65°F (18°C), but in the Sacramento Valley, only a few miles inland, it reaches 89°F (32°C). Coupled with these low temperatures on the coast, relative humidities are high. In the east and south, relative humidities are also very high, rarely falling below 50% in July. Combined with the high temperatures, this makes working conditions difficult unless air conditioning is available.

Annual rainfall begins to reach a peak in autumn in the north-west. In the south and east, hurricanes and thunderstorms decrease in frequency but autumn still produces 20–25% of the annual average rainfall.

Natural Vegetation. The wide variety of climates and geographical regions of the

Above left: **the Okeefenokee Swamp in Georgia is a federal wildlife reserve.** ***Below left:*** **The Rio Grande is a shallow river practically dry in summer, which forms the border between Texas and Mexico.** ***Right:*** **Sequoia trees grow on the wet western slopes of the Sierra Nevada.**

continental United States have produced a dramatic range of natural vegetation, from Alaskan tundra to the hardwoods of the Great Lakes region and the cacti and mesquite of the south-western deserts.

From Maine to Minnesota in the north, and running south along the Appalachian ranges are extensive coniferous forests of pine, hemlock, balsam fir and spruce, together with smaller numbers of alder, birch, tamarack and poplar. At lower altitudes, mixed forests of coniferous and deciduous trees contain maple, oak, hickory, sycamore, beech, ash, locust and walnut. On the sandy coastal plains of Georgia and the Carolinas pines are common, and in the southern swamps, bald cypress, white cedar and tupelo may be found. Live oak, palmetto, tropical palms, satinwood and mangrove occur along the Gulf Coast and in southern Florida.

Grassland vegetation is found principally in the Great Plains region. Here a variety of hardy grasses, including buffalo, wheat, needle, grama and bunch grass, low plants and sagebrush make up the ground cover. With the rise in altitude in the Rocky Mountains, grassland gives way to forests of conifers – white pine, Douglas fir, red fir, hemlock, redwood, spruces and yellow pine – which cover the high plateaux and mountains below the tree line. The densest of these forests are in the humid regions west of the Cascade and Coast ranges in Washington, Oregon and northern California, where the magnificent redwood may grow to over 300 ft (90 m).

Above the tree line alpine meadows bloom in spring with small flowering plants; between the mountains sagebrush, juniper and piñon dot the arid, semi-desert basins and low ridges. The south-western desert is noted for numerous varieties of cactus, notably the picturesque giant saguaro, which may grow up to 60 ft (18 m), and many smaller varieties which produce brilliant blooms in spring.

The forests of Alaska are primarily hemlock, Sitka spruce and cedar, with smaller stands of balsam, poplar, alder, birch, aspen and cottonwood. Wild flowers include alpine types, and on the treeless Aleutian Islands, lush grasses and small shrubs provide the ground cover. To the north, in arctic Alaska, tundra vegetation predominates. Here, mosses and lichens are prominent, though small shrubs and various grasses can also withstand the severe climate.

Wildlife. The great variety of animals of the United States can be roughly grouped with natural vegetation (and hence climatic) zones. Most species of the continental United States are those of temperate or arctic climates, with the exception of such tropical animals as the

Oil drilling derricks and pumps in Texas. The first oil was struck near Beaumont in 1901. Texas now produces a third of the nation's oil.

alligator of southern Florida and the armadillo of the south-central states.

The streams and forests of the north-east and east support a variety of fishes, reptiles and amphibians, including catfish, salamanders, tree frogs, minnows and water and land snakes. Mammals and birds include the beaver, black bear, Virginia deer, raccoon, opossum, wild turkey, woodpeckers, tanagers, humming-birds and orioles.

Of all the animals of the grassland interior the buffalo, or American bison, is probably the most famous. Herds of these magnificent animals once roamed the Plains, but during the 19th century their numbers were dangerously reduced through indiscriminate slaughter. Since the early part of the 20th century, however, the buffalo has been protected by law, and very gradually the herds are being built up once again.

Smaller mammals of the plains include the coyote, ground squirrel, prairie dog, American badger and pocket gopher. Grouse species known as prairie chickens are characteristic of the region, though many other bird species come to the plains from other areas.

The desert of the south-west is far from lifeless, though many species are nocturnal, an adaptation enabling them to escape the daytime heat. Reptiles abound; probably the most characteristic of the region are the poisonous rattlesnake and Gila monster, only two of a wide range of species of snakes and lizards. Mammals include the kangaroo rat, peccary and ring-tailed cat.

The vast mountain ranges of the west afford a home for a number of rock-dwelling species, including the Rocky Mountain goat, bighorn sheep, pika and marmot. Other mammals include the wapiti deer, beaver, mule deer and mountain lion.

True seals inhabit both east and west coasts, but sea lions are found only in the Pacific. On the east coast, the manatee, or sea cow, can be found in certain large rivers in Florida.

The People

In 1970, the population of the United States was recorded by the official census at 204,765,770. In 1790, when the first official census was taken, the population was 3,929,214, with fewer than 5% living west of the Appalachians. In less than 200 years, the American people have expanded across an entire continent, have developed vast natural resources and have industrialized to become the richest nation on earth.

The large cities and industrial areas of the east still make up the most densely-populated part of the United States, though in the last 20 years the south and west experienced the highest growth rates in the entire country. New Jersey has the highest density of population at 953 persons per sq mi (367·8 per km^2); Wyoming the lowest at 3·4 persons per sq mi (1·3 per km^2).

In the mountainous areas of the west, the average density is rarely more than 8 per sq mi (3·1 per km^2) and in places it falls below 1.

The New Immigrants. Until 1840 it was primarily the English and Scots who came in the wake of the Industrial Revolution; then came the Irish, after the potato famine of the 1840s. Successive waves of Germans and Scandinavians, who helped develop the rich farmland of the Midwest, followed, but until 1880 most immigrants were from northern Europe. Then a new type of immigrant appeared: from the Mediterranean came Greeks, Southern Italians and Sicilians; from eastern Europe came Czechs, Poles, Slovaks, Russian Jews and Armenians.

The peak decade was from 1900–10, which brought over eight million immigrants to the large eastern ports of entry, in particular New York City. American business was booming and factories were crying out for labour. Many of the newcomers settled in urban neighbourhoods crowded together with others of the same national background. Out of these settlements came the 'Little Italys' and other ethnic communities which still exist to some extent today in most major cities.

The high rate of immigration began to alarm some of the older-established Anglo-Saxon Americans, and after World War I a series of progressively more stringent laws restricting immigration were passed and introducing a variable quota system to limit the number who might enter from the 'less desirable' nations. During the Depression of the 1930s emigrants actually exceeded immigrants.

Today the foreign-born comprise less than 5% of the total population. Their major countries of origin are Italy, Germany, Canada, Britain, Poland, Russia and Mexico.

Black America. Negroes first entered the United States as slaves, brought in during the 17th and 18th centuries to work the large plantations in the South. By the outbreak of the Civil War in 1860 there were nearly 4·5 million Negroes, of whom 4 million were slaves. Between 1790 and 1860 the proportion of Negroes in the total population dropped from just under 20% to 14%, and today the Negro population is 22·7 million (1970 census), 11·8% of the total.

The distribution of Negroes has changed markedly since 1900, however. In the early part of the century, lack of capital to buy land on any scale, lack of opportunities to rent land in the settled areas to the north, lack of certain agricultural skills and disenchantment with the rural South led many Negroes to look northwards for opportunities. Today the conurbations of Chicago and New York contain more Negroes than any southern state, with over one million each. The urban Negro tends to be confined by economic and social pressures to the older property within the city, from which the former residents have moved out to better conditions in the outer suburbs. Overcrowding, unemployment and the difficulty of escape from these urban 'ghettoes' has contributed to the increase in racial tension, crime and violence in recent years.

The Indian. When the white man first came to settle in America in the 17th century, there were tribes of Indians scattered over the whole country, descendants of the Asian wanderers who had crossed the Bering isthmus over 20,000 years earlier.

Gradually the tribes were deprived of their ancestral lands as the frontier of settlement moved westwards. By 1890, after the final succession of wars between the whites and Indians, the entire Indian population had shrunk almost to a sixth.

The US government, awakening to the fact that the Indians were in danger of extinction, finally intervened to improve conditions on the reservations, and gradually the Indian population ceased to decline. By 1970 their numbers had reached 791,830. Today Indians are scattered throughout all the states, but they are mainly concentrated on large reservations in the middle and far west: Arizona (5%), Oklahoma (12%), New Mexico (10%), California and North Carolina (about 7·5% each).

Housing, education and unemployment are perennial problems on the reservations. However, the increasing self-awareness of Indian leaders had made them more forceful in seeking their rights as a group.

Other National Groups. While nearly every nationality in the world is represented to a greater or lesser degree in the United States, there are several fairly large non-European groups that deserve mention. Many Chinese and Japanese, for instance, came originally to the west coast in the rush for gold in 1849. They stayed to work on the railways, to farm, and to carry on certain trades. Today, most of the Japanese live in California, while the Chinese are more or less evenly split between California and the remainder of the country, the largest community being Chinatown in New York City. Over 100,000 other Asians from Indonesia, Polynesia, Korea, Indo-china and other places, live on American soil.

Spanish-Americans. Another large concentration of non-European immigrants are Spanish-Americans from Central and South America. The greatest number in the south-west are Mexicans, many of whom are employed as seasonal labourers picking crops in Texas and California. How many Mexicans can be actually termed immigrants

The $18 million Atlanta Stadium, built in the mid-60s and the modern city behind are demonstrations of the city's new prosperity.

is difficult to assess, for many of them move back and forth between Mexico and the US – often illegally.

Puerto Ricans comprise a second sizable group of Spanish-Americans. Having had US citizenship conferred on them in 1917, they are free to emigrate to the US as they wish. During the 1950s and 60s, over 700,000 settled there, most of them in New York City.

Patterns of Living

The Cities. Since 1790, the United States has undergone a remarkable transformation from a predominantly rural country to a largely urban one. In 1790 only 5% of the population lived in urban centres; by 1970 that percentage had risen to 70%. This phenomenon was brought about in large part by the industrial and agricultural revolutions, which on the one hand established the factory system and provided the economic stimulus for the development of cities; and on the other mechanized farm methods and increased the productivity of the farmer, releasing farm workers for jobs in industry. In 1790 there were no centres of over 50,000 inhabitants; today there are over 150 cities of over 100,000, 26 of which have over 500,000 and 6 of which have over 1 million.

Despite the number of large centres, however, the United States is economically and politically a decentralized nation, and not all the big cities hold the same national importance. New York, for example, stands alone as the nation's cultural, financial and commercial centre. Chicago and Philadelphia are distribution and marketing centres for huge industrial areas; cities like Detroit (automobiles) and Pittsburgh (steel) have developed through the growth of a particular industry.

Spreading suburbs. In recent decades, the population of the central portion of many cities has dropped sharply as more and more people have migrated to the suburbs. Between 1950 and 1960 the number of people living in standard metropolitan areas increased by about 24%; at the same time the population in the central cities increased by only 10·7% while suburban populations increased by 48·6%.

Several factors were responsible for this growing exodus. For one thing, facilities for commuting greater distances have improved. At the same time overcrowding, the high cost of living comfortably, the increasing crime rate and the effects of pollution have made the city a difficult place to live in. Thus urban decay has become an increasing problem in a number of large cities, where the exodus of a predominantly middle-class population has left the cities to the very rich and the very poor. As the tax base is eroded there is a breakdown in services, welfare and education programmes, and some cities, New York in particular, have appealed to the federal government for aid.

Small Town and Rural Life. It may be said that the real America is found in the thousands of small towns with 10,000 to 50,000 inhabitants. Many towns grew up from settlements of only one or two families, and for a long time a strong community spirit, a legacy of pioneer days, dominated these scattered centres and gave them their distinctly American character. 'Main Street, USA' is often used to describe the uniformly laid out centre of many small towns, usually a single street lined with shops, the town hall, post office, church and school.

In rural areas people are rarely concentrated in villages, for the traditional unit of settlement was the large farm or plantation which was relatively self-sufficient. The only exception to this is in New England, where settlements were more directly patterned on those of Europe, and where fishing and mercantile enterprises necessitated close co-operation with others.

Religion. The early settlers in America brought with them their individual religious beliefs and formed churches dedicated to their own particular creeds. Subsequent immigration and the development of a number of completely new sects added to the great variety of practised religions. Furthermore, the strict separation of church and state as set out in the Constitution has fostered the equality of all religions before the law, and all forms of worship are recognized.

The largest individual religious body in the United States today is the Roman Catholic Church, with nearly 48 million members. The largest Protestant groups are the Baptists, Lutherans and Methodists. There are nearly 5 million Jews and over 2 million Mormons Many Americans would admit to no distinct religious tie, and a small number belong to various sects that have evolved from both Eastern and Western religions.

Education. As far back as the early 19th century, American educators preached the need for a widespread system of public education in order to develop the country's potential. Today over 90% of all children remain in school until the age of 13, and about 70% graduate from high school. Nearly 20% receive a degree from a four-year college or university course.

The state and federal governments are playing an increasing part in financing public schools systems. Much of this aid is provided in federal and state-sponsored programmes set up to deal with problems such as the teaching of under-privileged children in areas where local resources cannot cope with overcrowding or inadequate facilities. Additional federal funds have been provided in the field of higher education for new univer-

The Capitol building is by William Thornton, founder of the US Patent Office who won a contest for its design in 1792. It was rebuilt after British troops burnt it down in 1817.

The Moyie River Dam in The Rockies of northern Idaho provides both hydroelectric power and water storage for irrigation.

sity facilities and low-interest loans to students.

The Economy

In 1971 the US Gross National Product was the highest ever, totalling $1,050,356,000,000. The previous year, when the figure was $976,445,000,000, the gross national product of the Soviet Union was about $434,900,000,000, that of all the Common Market countries $491,300,000,000 and of Japan $198,800,000,000.

Agriculture. Until 1910 the United States was a predominantly agricultural nation. In 1920 there were some 32 million people living on farms, and 13·4 million farm workers were employed on over 6 million farms. Since then the numbers in these categories have diminished, and today less than 10 million people live on some 3 million farms. Large-scale mechanization and reorganization of farming have enabled a much smaller labour force to work almost as many acres as in 1920. In addition, this acreage is now divided into far fewer farms and produces more of most farm commodities. Average farm size grew from 147 ac (59 ha) in 1920 to 213 ac (86 ha) in 1950 and 389 ac (157 ha) in 1968.

The range of farm production includes practically all the temperate and sub-tropical field crops, vegetables, fruits and nuts. Types of produce can be roughly divided by region: the Dairy Belt runs across the north and north-east, with butter and cheese particularly important in the Great Lakes region; feed grains and livestock come from the Corn Belt of the mid-west; the spring-wheat belt dominates the north-western plains, with winter wheat growing to its south-east; and the range livestock belt embraces not only parts of the High Plains but also much of the intermontane basins and plateaux of the west. South of the Corn Belt lies a mixed farming region, and south of that the traditional Cotton Belt, which is now much altered from its appearance 50 years ago.

Most cotton now comes from Texas, Arizona, New Mexico and California rather than from the South. Instead, the South has concentrated on livestock, dairying and peanuts and other crops, as well as forestry, which has provided a partial remedy for badly eroded soils. On the north-eastern edge of the cotton zone lies the fragmented tobacco belt. Finally, there are widely scattered speciality crops: fruit and vegetables in the truck-farming areas of Florida, California and the Great Lakes' shorelands; rice in Louisiana and California; citrus fruits in south-west Texas, California and Florida; and sugar-cane on the Gulf coast.

The main problem facing the farmer, apart from the hazards of the climate, is that of surplus production. In itself, this can be a reflection of climatic factors, but it is also caused by the use of fertilizers, increasingly prolific crop strains and improved farming practices. In an attempt to stabilize farm prices, the federal government has instituted acreage restriction schemes and certain minimum price guarantees, and has established the Commodity Credit Corporation to take excess produce off the market. Attempts to reduce acreage devoted to crops in surplus have, nevertheless, been only partially successful, since productivity on the remaining land has tended to increase.

Above: Harvesting tobacco in North Carolina; about 40% of the nation's crops is produced in the state. The tobacco plant is native to North America but has since spread around the world. *Below:* A ranch in the California mountains. Two mountain ranges parallel the California coast, the Coast Ranges and the Sierra Nevada. Two-thirds of the state is mountainous.

The long-serious problem of soil erosion, especially severe in the 'Dust Bowl' region of the south-central states, has been tackled by state and federal agencies. The Soil Bank, conservation and land retirement programmes and other schemes have done much to help. Agricultural education is being conducted through schools, colleges, 4H clubs and agricultural field offices.

Minerals. The United States ranks first in the world in the volume and diversity of its mineral production. Nevertheless, since some mineral reserves are dwindling and others are uneconomic to exploit, the United States has to import appreciable amounts of nickel, tin, tungsten, copper, industrial diamonds, manganese, platinum, bauxite and other minerals. As in agriculture, the mining industry's labour force has shrunk over the years, while its output has increased. In 1920 the mining labour force was over 1·2 million, by 1968 the figure had dropped to 585,000.

The chief coal-field areas are the Appalachian field, with West Virginia, Kentucky and Pennsylvania the principal producers; the interior field of Indiana and Illinois; the smaller fields of Alabama; and fields in the western half of the country which have important local significance. The main production areas for petroleum are in the south-central states, led by Texas and Louisiana. Outside this region, only California was once of any importance in petroleum production, though production is increasing in the Great Plains states and Alaska's immense oil reserves are now being opened up. The pattern of natural gas production is similar to that of petroleum, although while the USA imports an increasing percentage of its petroleum requirements, it is largely self-sufficient in natural gas.

The main region of non-ferrous metal production is the West. Iron ore is mined in the Lake Superior fields of Minnesota, Michigan and Wisconsin, and in New York and Alabama.

Manufacturing Industry. Industry first developed in New England, where it was associated with an abundance of water power and with the commercial centres in the coastal ports. The growing use of coal eventually led to the development of heavy industry in Pittsburgh and the surrounding areas of Pennsylvania and Ohio. Changing patterns of resource supply and markets, as well as the development of much larger units of production over the past century, have led to changes in the location of industries. While Pittsburgh, for instance, is still a major centre of the steel industry, it is now equalled by the Chicago-Gary area. Steel production is also important in Cleveland, Detroit and Duluth; on the west coast, the eastern seaboard and the Gulf coast.

The advent of the internal combustion

engine had considerable impact on the American way of life and led to a vast industry which is the country's major user of steel, rubber and certain other products. Detroit was the cradle of this industry and remains the motor-car industry's chief centre. Surrounding towns in Michigan, Ohio and Indiana are important in component manufacture and assembly. Current annual production is around 8·5 million passenger vehicles and 2 million commercial vehicles.

Whereas the motor-car industry is concentrated in the southern Great Lakes states, the aircraft industry is located in the west and the Great Plains, in such centres as Wichita, Fort Worth, San Diego and Seattle. Engine manufacture, however, is centred in New England, New York, New Jersey and Michigan.

The textile industry was initially concentrated in New England. But in the course of time, manufacturing plants moved closer to their sources of raw material and took advantage of lower fuel, labour and plant costs in the South. Most of the textile industry has moved to the Piedmont region of the Carolinas and Georgia. Less than 4% of cotton textiles capacity remain in New England.

Electronics and the manufacture of mechanical electrical equipment have now become prominent in New England. Electronics research industries are particularly important around Boston, New Haven and Worcester, where they have been joined by branches of chemical industry. An important factor in this development has been the proximity of the research institutions of the region's major universities.

The consumer goods industries are widely scattered, since they are particularly market orientated. Food-processing is largely located near particular supply areas: for instance, flour milling at Buffalo and Chicago; meat packing at Omaha and Kansas City; and fruit-juice and canning industries in southern Florida and California.

Service Industry. The provision of services (wholesale, retail, professional and so on) is of increasing importance in the US economy. This group is by far the most important employer, and the ability to attract employees in this category is very important for an area's wellbeing. Good examples of cities whose populations work largely in service industries are Washington, DC, Denver (Colorado) and Lincoln (Nebraska).

International Trade. United States foreign commerce is of great significance to the world economy. With only 6% of the entire world population, America produces about 30% of the total output of goods and services. Each year US exports amount to about 15% and imports to about 12% of the world total. America's external trade can be looked at in two ways, however. From the American standpoint, it appears relatively unimportant. Imports of goods and services accounted for only 6·2% of the gross national product in 1971 and exports for the same proportion. This is a tiny percentage by comparison with the world's other major trading nations. However, the sheer scale of the American economy makes it the most important single country in international trade, and fluctuations in the US economy have world-wide repercussions.

The balance of payments crisis that disrupted the international monetary system in 1971

The patchwork patterns of the Kansas wheat belt stretch as far as the eye can see. Kansas produces 20% of the nation's wheat.

and resulted in the effective devaluation of the dollar stemmed from a complex interrelationship of causes. Investments abroad, aid to developing countries, military spending, foreign tariffs and escalating costs all contributed to a massive flight from the dollar in and after 1971, but no one factor could be taken in isolation.

The United States' chief trading partner is Canada, which takes more than 25% of all US exports. Other important trading partners include Japan, West Germany and Great Britain.

Transportation

When Henry Ford established the Ford Motor Company, the automobile revolution was just around the corner. By 1920 there were 8 million private-owned cars in the United States; by 1930, 23 million; and, by 1972, 96 million, or nearly one car for every two Americans. To cater for these and the equally spectacular increase in trucks and other commercial vehicles, considerable road-building and improvement schemes had to be undertaken. In 1969 there were 3·7 million mi (5·9 million km) of surfaced roads in the US. Since the Federal Highway Act (1921) the government has been playing a part in the construction and improvement of major routeways; more recent acts have widened the scope of this aid. Acts passed in 1944 and 1956 called for a highway network to connect, by the most direct routes possible, the major urban and industrial centres, and to connect suitable border points on important continental routes. This established the vast National System of Interstate and Defense Highways, of which 42,500 mi (68,300 km) had been completed by 1972.

Los Angeles is a remarkably decentralized city and sprawls over an area of 458 square miles. It has more cars per inhabitant than any other city in the United States.

Waterways. Rivers provided the first means of inland transportation and by the early 19th century Americans were designing canals to extend navigable river systems. Today major canals include the New York State Barge Canal, which replaced the Erie Canal in 1918, and the Soo Canal, a series of progressively larger locks and canals linking lakes Superior and Huron.

Although canals have lost their old pre-eminence, inland waterways remain important. The Mississippi-Missouri system has been regularized, straightened and improved to such an extent that a 9-ft (2·7-m) channel is now maintained to Omaha and Minneapolis, Knoxville and Pittsburgh, and is in the process of being extended. Improvements have also been made to other rivers to make them navigable for some distance so that ocean-going vessels can reach such important interior centres as Portland, Oregon, and Sacramento, California. The St Lawrence Seaway gives ocean-going vessels access to remote inland cities of the Great Lakes area, including Chicago, Duluth and Milwaukee, and has been a major factor in the development of large industrial cities along the lake shores including one Canadian and five American cities of over one million inhabitants. But the Great Lakes and the northern half of the Mississippi river system are largely closed by ice in winter except where ice-breakers keep ferry routes open.

Another major waterway neared completion in the 1960s when construction started on the Cross-Florida Canal, designed to link the Atlantic and Gulf intracoastal waterways (systems of rivers and sea inlets linked by canals). This canal completes 3,000 mi (4,800 km) of sheltered waterway along the country's southern and eastern seaboards.

Railroads. In the last 40 years of the 19th century the American railroads were an instrument of frontier expansion, pushing the line of settlement westward. In addition to providing the means of movement for goods and passengers, especially in areas west of the chief navigable river systems, they also provided actual incentives for people to settle in the new territories.

Initially the railways were financed by private, institutional, company and sometimes state funds. This worked while the lines were relatively short and linked established commercial centres. But once the lines ventured west of the Mississippi, financial aid was not so readily available. Help from the federal government was largely provided in the form of land, and land grants

A grain boat passing through the locks at Sault Sainte-Marie. The Soo canal links Lakes Superior and Huron and handles 90 million tons of cargo a year.

became a feature of railroad financing.

New York, Pennsylvania and Massachusetts each had 1,000 mi (1,600 km) of railway line by 1850. Illinois and Pennsylvania were the first states to reach 10,000 mi (16,000 km) in 1900; since then Texas has also passed this total.

Today much is heard of the declining importance of the railways. Since the peak period of the 1920s, railway route mileage has shown a 20% drop due to economic pressures. To try and save failing companies, the federal government in recent years allowed a number of mergers to take place.

The greatest decline has taken place in passenger services: the annual number of passengers carried dropped from 1,270 million in 1920 to 289 million in 1970. Competition from aircraft, especially for long distance travel; from buses, for cheap travel; and from the automobile, for convenience, have been major factors in this decline. Although it is little more than a century since a golden spike was driven to complete the first transcontinental railway line, today there are no transcontinental passenger trains and few through services from the Pacific coast to the areas east of the Rockies. In 1970 the US Congress authorized the establishment of the National Railroad Passenger Corporation (Amtrak) to operate inter-city passenger trains.

Airways. The vast distances within the United States have encouraged the development of a complex network of internal airways. Today there are nearly 4,000 public and 6,300 private airports. Over 250,000 route mi (400,000 km) are in operation, serviced by 40 domestic and 9 international operators employing nearly 250,000 people. Costs of air travel have declined while speeds have increased. However, the problems of airport congestion and of the time taken to get from airports to city centres has made some short-haul routes only marginally advantageous over rail in terms of time. The increasing size of runways required for larger planes and the increasing popularity of air travel have necessitated the search for newer and larger airport sites. By the early 1970s New York, with three international airports, was looking for a fourth site. Chicago's O'Hare International Airport, the busiest commercial airport in the world, remained unable to cope with air traffic despite two other airports, and Chicago was therefore also looking for an additional site. During the economic slump of the late 60s and early 70s, however, air traffic did not increase as fast as the airlines had predicted, and a serious problem of over capacity has developed on some routes.

Pipelines. Although experiments have been carried out successfully in transporting pulverized coal, wood chips and even steel capsules by pipeline, virtually the only products transported in this way up to now have been crude oil, refined petroleum products, and natural gas. Today there are over 160,000 mi (258,000 km) of crude oil pipeline in the United States, of which 116,000 mi (187,000 km) are trunk lines and the rest feeder lines.

The densest network of pipe is that linking the southern oil fields to local refineries and to the chief areas of consumption in California and the north-east. A natural gas pipeline system similar to the network of oil pipelines links up the main consumer and producer areas.

Dependencies

Caribbean dependencies are Puerto Rico; American Virgin Islands; Panama Canal Zone; Corn Islands; Guantánamo Bay military base in Cuba; Swan Islands (disputed with Honduras) and Quita Sueño Bank, Serrana Bank and Roncador Cay (jointly held with Colombia). Pacific Ocean holdings are Guam; American Samoa; Trust Territory of the Pacific Islands; Midway Islands; Wake, Howland, Baker, Jarvis, Johnston and Palmyra Islands; Canton and Enderbury Islands (jointly held with Britain) and Okinawa military base. (Separate articles describe the larger holdings.)

BERMUDA

Area: 20·6 sq mi (53·3 km²)
Population: 53,000
Population Growth Rate: 2·1%
Capital: Hamilton, pop 3,000
Language: English
Religion: Protestant
Currency: Bermuda dollar = 100 cents

Bermuda, a British possession, is a group of some 300 small coral islands in the western North Atlantic nearly 700 mi (1,127 km) east of Cape Hatteras, North Carolina. The islands are said to have been discovered by the Spaniard Juan de Bermúdez in 1503. By 1684 Bermuda had become a British crown colony. A new constitution, ratified in 1968, introduced a large measure of internal self-government, while leaving the governor responsible for external affairs, defence and the police.

The island's sunny climate, warm seas, and tax concessions attract tourism and foreign investment, giving Bermuda a prosperity unwarranted by its otherwise limited resources.

Island Landscape. Bermuda's low, coral islands cover an oval area about 22 mi (35 km) long from north-east to south-west, and some 5 mi (8 km) wide. The largest island is Great Bermuda, about 14 mi (23 km) long.

Throughout the year Bermuda enjoys a mild climate influenced by the warm Gulf Stream. Peak temperatures of about 86°F (30°C) are reached in July and August. Even in February, the coolest month, daytime temperatures rarely fall below 60°F (16°C). Annual rainfall averages 60 in (1,524 mm).

The People. With an area of only 20.6 sq mi (53 km²) and an estimated population of 53,000 (1970) crowded on only about 20 of the islands, Bermuda is one of the most densely populated areas in the world (2,558 per sq mi; 981 per km²).

About two-thirds of the population are Negroes, descendants of African slaves brought to Bermuda in the 17th and 18th centuries. The remainder are largely of British or Portuguese stock. Hamilton (3,000) is the capital and chief port.

The Economy. Tourism is Bermuda's major source of income, contributing about £27·5 million (about $66 million) annually to the economy.

Bermuda has little scope for agriculture. The chief crops are early vegetables and flowers (especially Easter lilies), bananas and citrus fruits. Small-scale industry involves ship repairing, small-boat building, pharmaceuticals, perfume and essential oils. But most products, including foodstuffs, have to be imported. The islands are well served by international air and shipping lines.

BAHAMAS

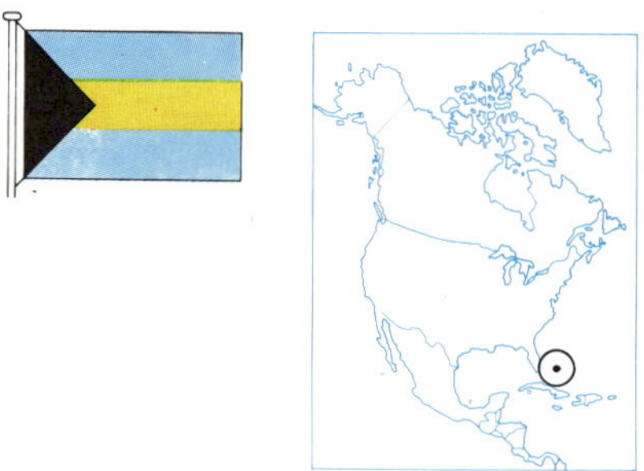

The Commonwealth of the Bahama Islands, formerly an internally self-governing British colony, gained independence on 10 July 1973. This many islanded archipelago lies off the south-east coast of Florida, stretching some 800 mi (1,300 km) south-eastwards. It is the north-eastern section of the West Indies.

There are more than 700 islands and 2,000 rocks and cays but only some 30 of the islands are inhabited. The largest is Andros; the most important and most populous, New Providence, on which stands Nassau, the capital. Of lesser importance are Abaco, Grand Bahama, Cat Island, Long Island, Eleuthera and Exuma. Watling's Island or San Salvador is traditionally the first land in the New World to have been reached by Christopher Columbus (1492). The total area of the Bahamas is 5,382 sq mi (13,938 km²).

Land and Climate. The Bahamas rest on a submarine shelf of coralline limestone and are mostly low-lying and flat. Their beautiful coastlines are justly famous.

The Tropic of Cancer bisects the archipelago, and the sub-tropical conditions have made the islands a favourite and expensive holiday resort. The warming influence of the Gulf Stream helps to keep the average winter temperature at 70°F (21°C) and summer temperatures at 80–90°F (27–32°C). Rainfall averages 40–50 in (1,020–1,270 mm) annually. The rainy months are May, June, September and October.

Blacks and Whites. The population is about 85% Negro, the remainder being of British, American or Canadian stock. Since the 1600s the islands have been subject to mainly British cultural influences. The Bahamas were long the haunt of pirates and slave-traders. Nassau

The Bahamas is made up of almost 700 islands in the Caribbean only about 30 of which are inhabited. The economy of the Bahamas is largely dependent on its sunny beaches.

prospered during the American Civil War as a base for blockade-runners and during the Prohibition era was the rendezvous of boot-leggers. English is the language of the islands, and education is compulsory between the ages of 5 and 14. Internal self-government was introduced in 1964. Island politics were dominated by the white minority until 1967, when an all-Negro government took office.

The Economy. The tourist industry, backed by American capital, is the mainstay of the economy. The constant sunshine, inviting beaches and warm seas attract large numbers of tourists to such centres as Nassau and the more recently developed Freeport on Grand Bahama Island. Tax concessions encouraged investment in Freeport and also made the Bahamas an attractive place for retirement. Real estate and finance became important factors in the economy.

As yet, there is little industry, apart from the oil refinery at Freeport and a rum distillery on New Providence. Fishing is one of the major activities of the Negro population. Vegetables and fruit are grown on the limited arable land, and some livestock is raised.

In 1968 a constitutional amendment transferred control of the police and internal security to a minister designated by the prime minister. There have also been attempts to remove white, and especially expatriate, predominance in the Bahamian economy, symbolized by Freeport.

TURKS AND CAICOS ISLANDS

Turks and Caicos Islands are a British colony in the south-eastern group of the Bahamas, about 575 mi (880 km) south-east of Florida. They cover an area of about 166 sq mi (432 km²). Only a handful of the small islands are populated, including Grand Turk, Grand Caicos and Salt Cay. The climate is hot, rainfall slight and the islands are vulnerable to hurricanes. The population of 6,000 are predominantly of African descent, and depend on fishing and salt production.

The islands are governed by a Crown-appointed administrator, helped by executive and legislative councils.

MEXICO

Area: 761,530 sq mi (1,967,183 km²)
Population: 50,830,000
Population Growth Rate: 3·5%
Capital: Mexico City, pop 8,206,000
Language: Spanish
Religion: Roman Catholic (96%)
Currency: Mexican peso = 100 centavos

Mexico, the third largest of the Latin American countries, is one of the few to have made impressive economic progress. Its industrial development since World War II has excited comparison with that of Japan, and the capital, Mexico City, now covers a larger area than Paris. But this great industrial and cultural centre has its share of the appalling slums to be found in and around most large Latin American cities. Moreover rural living standards are largely low and farming methods primitive. Mexico's picturesque peasantry, relics of great Pre-Columbian cultures, and towering volcanoes annually help to attract nearly two million tourists – mainly from Mexico's northern neighbour, the United States. Tourism is indeed Mexico's main source of dollar income. But only a continuing broad economic growth and a drop in the high birthrate will finance the social improvements still badly needed.

Economic progress accelerated under benevolent dictatorships that nonetheless failed to halt rising peasant discontent with the dominant upper classes. A revolution which radically changed Mexico's history came in 1910. Since then a relatively stable government has gradually come to grips with the problem of land distribution and industrialization.

The Land. The heart of Mexico, and three-quarters of its area, is the central plateau bounded by the mountain ranges and escarpments of the Sierra Madre Occidental and the Sierra Madre Oriental. This central plateau is interspersed with hills, mountain chains and basins, and its elevation ranges from 8,000 ft (2,440 m) in the south to 4,000 ft (1,220 m) in the north. The main plateau surface consists of low hills and broad flat basins which cover one-third of the area.

The southern part of the central plateau, the Mesa Central, is a highly dissected region and many of its features provide evidence of recent volcanic activity. Here the southern plateau edge consists of a chain of snow-capped volcanoes. These include the still active Popocatépetl 'the Smoking Mountain' (17,900 ft; 5,450 m). Crater lakes and lava flows add distinctive landscape features to this mountainous region.

South of the central plateau is a rugged mountainous region ranging between 7,000 and 8,000 ft (2,130–2,440 m), with some districts rising to over 10,000 ft (3,000 m). Steep sided valleys and mountains surround the basins of central Colina, central Oaxaca and the Balsas depression.

The most spectacular surface features in Mexico are the rugged steep eastern and western escarpments of the central plateau, descending from heights of 7,000–8,000 ft (2,130–2,440 m) to the coastal plains of the Gulf coast and the Pacific. The Gulf coast plain is up to 100 mi (160 km) wide, extending from the Rio Bravo (Rio Grande) to the Yucatán peninsula. Much of this lowland consists of swamps and lagoons, with the plain merging inland into low ridges of hills towards the escarpment. The Pacific coast lowlands are much less extensive and consist of low hills and plains. Southwards the lowlands form a narrow coastal strip.

The northern half of Mexico including part of the central plateau is arid. There are few rivers, but after heavy local rainfall the desert basins hold intermittent lakes. Further south, the central plateau is drained by three main river systems – the Santiago–Lerma; Pánuco–Moctezuma; and Balsas. These rivers have cut deep valleys through the plateau escarpments and built alluvial flood plains and deltas on the coastal plains.

Climate. The country has three main climatic zones: the *tierra caliente, tierra templada* and *tierra fría*.

The *tierra caliente* includes the coastal lowlands and foothills of southern Mexico. Day-time temperatures are 85–90°F (29–32°C). This is the region of tropical agriculture. The *tierra templada* (2,500–6,000 ft; 760–1,830 m) covers most of the central plateau. Day-time temperatures of 75–80°F (24–27°C) are less enervating than those of the lowlands and much of the population and farming is found in this climatic zone.

The *tierra fría*, at 6,000–12,000 ft (1,830–3,660 m) is mainly found in the basins and mountain slopes of the Mesa Central. Temperatures during the day are 60–70°F (16–21°C). Frosts are common during the cooler months (December to February). A fourth climatic zone, the *tierra helada* (frozen land), covers only the highest mountain peaks (over 12,000 ft; 3,660 m). Temperatures are below 50°F (10°C) all year round.

Precipitation varies greatly from one region to another. Parts of the arid north have less than 4 in (100 mm) of rain a year. This is partly because the plateau's flanking mountains shut out rain-bearing winds. Over most of the central plateau and the Pacific coast lowlands the rainfall is seasonal, and three-quarters of the rain falls between May and October.

The eastern and southern regions lie directly in the path of rain-bearing trade winds which sweep in from the Caribbean and the Gulf of Mexico and are the wettest part of the country with an average rainfall of 40–120 in (1,020–3,050 mm). In southern Mexico, during the summer months, tropical air masses bring thunderstorms in their wake and intense tropical disturbances affect the whole of the south. Spectacular hurricanes from the Gulf of Mexico rush over the eastern coastal region during the hurricane season (July–October).

Vegetation and Wildlife. Southern lowlands are covered by tropical rain forest and savanna, with temperate rain forest at higher altitudes. In south-eastern Mexico, large stretches of the rain forest have been cut down to provide land for pastures and plantations. North of the tropical rain forest, the southern part of the central plateau is covered by grassland and steppe, giving way northwards to scrub and barren desert.

Mexican animals include North and South American species. Thus wolves and bears roam remote plateau areas, while jaguars, monkeys and tapirs inhabit the flanking lowlands, and parrots and toucans are among the many colourful birds. Reptiles include caymans and iguanas and many species of poisonous snakes. Ticks and termites are perhaps the most noxious of Mexican insects.

The People. Most Mexicans are mestizos (of mixed white and American Indian ancestry), although there are 3 million Indians in the population of 48 million. Increasing at its current rate, the population should reach 85 million by the year 2000 (equal to nearly half that of all South America in 1970). Over 40% of the population is less than 14 years old. Half the population lives in towns, and half the population is concentrated on the central plateau. Around Mexico City, densities exceed 300 persons per sq mi (116 per km²). Sixteen per cent of the population lives in Mexico City, compared with 4% in 1900. This gives some idea of the scale of rural-urban migration.

Life in the country is generally different from life in the towns. In most villages hygiene and sanitation are primitive and illiteracy is common. Food is based on maize (the basis of the pancake-like tortilla), beans, potatoes, tomatoes, chilli-peppers and fruit. Goods are often bartered, not sold, and the market place is normally the centre of village social activity.

Women in the market in the city of Pueblo south-east of Mexico City. It has been estimated that Mexico's population is 10% white, 30% Indian and 60% Mestizo (of mixed background).

Life in the towns and cities is on the whole more sophisticated. Many towns have old Spanish-colonial buildings, with heavy wooden gates and courtyards, but suburbs are now frequently reminiscent of the United States. However, there are also disfiguring slums whose poverty matches that in rural areas. The largest cities are Guadalajara, Monterrey, Netzahnalcoyott and León. Mexico City is the centre of the country and has a population of 8,206,000.

Despite the prolonged campaign against illiteracy, in which 20 million free textbooks have been handed out, 30% of children do not attend school.

Most Mexicans are at least nominally Roman Catholics, but there are small Protestant and other minorities. No Church is allowed to own land and since 1917 the government has claimed land formerly held by the Roman Catholic Church.

Farms, Forests and Fisheries. Agricultural production is no longer keeping pace with food requirements, and the gap between supply and demand has been increasing. This is partly through sheer lack of fertile land. Two-thirds of the country is ruggedly mountainous, the northern half is arid, and only 10% of the land is under crops, concentrated mainly in the central plateau which has fertile volcanic soils and reliable summer rain.

Mexican agriculture long suffered from a socially inequitable system of land ownership. Before the revolution of 1910 most of the fertile land was divided into *haciendas* (large private estates) and 96% of the population were landless peasants. Redistribution of land on a fairer basis was started in 1922 and is still continuing. The expropriated land has been redistributed as *ejidos*, farmed by 20 heads of families either as a collective or more commonly in individual plots of 10 to 50 ac (4–20 ha). Now almost half of the former large estates have been redistributed in this way. The *hacienda* owner was allowed to retain a part of the estate as private land and this *pequeña propiedad* forms a second type of land ownership under the new system.

The bulk of the peasant population practise subsistence farming, particularly in central and southern Mexico. Both tools and crops have barely changed since the Spanish conquest. In the tropical forest regions, particularly on the Indian lands, slash and burn shifting cultivation is still common.

Commercial agriculture has nonetheless greatly expanded. The main crops are wheat, maize, beans, cotton, coffee and sugar-cane. Cotton is the chief export crop. Its production is now highly mechanized and Mexico is the seventh largest producer in the world. The three main regions of commercial farming are the sub-tropical and tropical south, the central plateau and the new irrigated regions in the north.

The Gulf Coast lowlands are the main region of tropical agriculture and plantation crops are grown on a commercial scale. Coffee, sugar-cane, bananas and many tropical fruits are the main crops. The lowlands on the Pacific coast form a similar but less important area of plantation agriculture. Expensive government attempts are being made to develop new plantations.

Northern Mexico is the most rapidly growing farming region. In all there are 10 main irrigation districts covering over 3·5 million ac (1·4 million ha). Cotton, wheat, alfalfa, tomatoes and melons are the main crops, and with this expansion Mexico is now the most important cotton-producing country in Latin America.

Cattle rearing occurs on a small scale in the mountainous region south of the Mesa Central, and on large commercial ranches on the semi-arid pastureland of the north. In the south the small farmers usually graze a few sheep and cattle, chiefly for wool and milk. On large ranches that have survived in spite of land reform in northern Mexico, there are now over 6 million cattle, 3 million goats and 2 million sheep.

Dairying is a major agricultural development in Mexico and is almost all concentrated around the large cities of the central plateau, particularly around Mexico City. One-fifth of Mexico is covered by forest. The main areas of commercial forestry are the Sierra Madre Occidental and western Chihuahua state, where the main product is pinewood, used in the building industry and for railway sleepers.

The fishing industry has greatly expanded since the 1940s. The main fishing grounds lie off the Pacific coast. Sardines and tuna are the chief catch, and shrimps found in the shallow inshore waters form the basis of a lucrative inshore fishing trade.

Minerals. Mexico is the largest producer of silver in the world and many mines have been worked since the Spanish conquest. As some of these silver veins were worked out, zinc, lead and copper ores were discovered and these three metals now account for over three-quarters of the mineral production. A recent development has been the discovery of sulphur in southern Veracruz.

Coal, oil and natural gas deposits have also been found and exploited over the past century. The Salinas Basin is the main coal-producing area, yielding high grade coking coal. Large oil and natural gas deposits have been found along the Gulf coast and these now supply power to over three-quarters of Mexican industry. In 1938 the oil fields were nationalized and since then oil has been an important factor in the country's rapid industrial growth.

The Temple of the Magician at Uxmal in Yucatan. Uxmal rose to prominence about 1000 AD.

Industry. During the past century, rapid industrial growth has followed the exploitation of oil and natural gas. Despite this, less than 20% of the population is employed in industry. One of Mexico's problems has been how to attract foreign (mainly US) investment without risking a loss of economic sovereignty.

The two main industries in Mexico are the manufacture of textiles and steel. The rise of large-scale cotton growing in northern Mexico resulted in the appearance of cotton manufacture as the main textile industry, located chiefly in the traditional centres of Mexico City and Puebla. The main steel plants at Monterrey and Monclova now produce two-thirds of the national output.

The development of industry has been highly localized. On the central plateau, Mexico City and the neighbouring towns of Puebla and Toluca form one industrial area; and, in the north, Monterrey with its iron and steel industry forms the centre of a second industrial district. This densely populated region accounts for almost two-thirds of the industrial population and is the only appreciable home market for manufactured goods.

Although coal, oil and natural gas provide a power source for the industries of Mexico, the most important development has been the swift growth in hydro-electric power. By harnessing the energy of the rivers as they descend the eastern and southern escarpments of the central plateau, a number of hydro-electric systems have been developed. There are now seven plants of over 50,000 kw capacity, providing a growing power source for industry.

Transport. A vital part of the '*marcha al mar*', the attempt to develop the lowlands has been the building of new roads and railways. Between the late 1950s and early 1970s, some 18,750 mi (30,000 km) of paved or asphalted roads were laid. The post-war growth in car-borne tourists from the United States has been one stimulus to the expansion of the road network which now extends for 125,000 mi (200,000 km) of which one-quarter comprises all-weather roads. An extensive rail network covers the central plateau.

International Trade. The main export markets for Mexican products are the United States, north-west Europe, and Japan and the ports on the Gulf Coast handle almost all of this trade. Agricultural produce, chiefly cotton, sugar, coffee and cattle, accounts for almost two-thirds of all exports. Imports included vehicles, fuel oil, machinery and fertilizers.

Dependencies. Mexico's offshore islands include Guadaloupe Island and the Revilla Gigedo Islands in the Pacific Ocean.

GUATEMALA

Area: 42,042 sq mi (108,889 km²)
Population: 5,348,000
Population Growth Rate: 2·9%
Capital: Guatemala City, pop 769,000
Language: Spanish. Maya-Quiché Indian dialects
Religion: Roman Catholic
Currency: Quetzal = 100 centavos

Guatemala is the second largest, most populous and scenically spectacular of the Central American countries, with beautiful mountain lakes set against a dramatic skyline of volcanoes. Since independence from Spain (1821), Guatemala has endured dictatorships and political unrest – in recent years fomented by opposed left- and right-wing guerrilla groups. Political conflicts, an explosive population growth and poor health facilities cloud the outlook for broad material progress.

The Land. The great Sierra Madre mountain range crosses southern-central Guatemala. From its general level of 3,500–8,000 ft (1,070–2,440 m) towering volcanoes (some still active) rise to over 11,000 ft (3,350 m), with dormant Tajumulco in western Guatemala the country's highest peak at 13,845 ft (4,220 m).

The volcanoes have blanketed large areas of western Guatemala with thick fertile layers of ejected ash and volcanic dust. Volcanic ash dams and volcanic foundering have also created beautiful mountain lakes such as Lake Atitlán.

Earthquakes are a natural hazard of the mountains, and Guatemala City, the present capital, was almost completely destroyed in 1917–18. Locally, too, volcanic explosions, landslides and lava flows have caused deaths and considerable destruction.

The mountain belt forms a dramatic backdrop to Guatemala's narrow Pacific coastal plain in the south.

North of the mountains, the land falls away to an extensive plain. Here, in the east, Guatemala has a short frontage on the Gulf of Honduras which is entered by the Motagua River, the country's largest, which drains the north-eastern slopes of the mountains.

Climate, Vegetation and Wildlife. Guatemala's climate is mainly warm and humid. Temperatures average around 77°F (25°C) in the lowlands, falling to about 55°F (13°C) in the high mountain basins. The Pacific side has summer rainfall but much of the Caribbean side is wet throughout the year with up to an annual 150 in (3,750 mm) of rain. The Caribbean also experiences autumn hurricanes.

Tropical evergreen forest grows in wet lowland Caribbean areas but the drier Pacific lowlands support partly deciduous forest and savanna. Evergreen oaks and pines flourish in the mountains but give way to tall grasses above 10,000 ft (3,050 m).

Monkeys, peccaries, deer and the rarer jaguars and tapirs are among Guatemala's indigenous wild animals.

Indians and Ladinos. Indians make up roughly half the population. Ladinos, the other major group, are europeanized peoples mainly of mixed Spanish and Indian descent. Whites, Negroes and mulattoes form tiny minorities. Guatemalans speak Spanish and Indian tongues; and most are Roman Catholics. About 38% are illiterate and poverty is extensive. Guatemala's rapidly increasing population is leading to intense pressure on the land. Many of the tiny agricultural plots, averaging less than 5 ac (2 ha), cannot fully support a family and 250,000 Indian peasants annually migrate to help with the coffee harvest and thus earn a desperately needed income supplement.

The Pacific side of Guatemala is the most densely settled, and the northern lowland forests the most sparsely peopled. More than one-eighth of all Guatemalans live in the capital, Guatemala City (768,987) which is twice as large as its nearest rival in Central America, and has one-fifth of all the industrial establishments in Central America.

Guatemala's government features an executive president (who appoints a cabinet) and a legislative Congress with a single house.

Economic Production is largely agricultural, with coffee as the main cash crop, grown on rich volcanic mountain soils. Other commercial crops include cotton (on the Pacific lowlands) and bananas (in the lower Motagua Valley). Maize and beans are the chief subsistence crops.

Other products include hardwood timber from the forests that cover most of Guatemala, but their inaccessibility leaves the forests underexploited. Zinc, lead and nickel are the only commercially important minerals. Manufacturing (largely of consumer goods) is expanding but suffers from a lack of developed power supplies.

Transport and Trade. Guatemala's 720 mi (1,159 km) of railway link the Pacific and Caribbean coasts, and connect the country with neighbouring Mexico and El Salvador. There are about 8,050 mi (12,950 km) of mainly unpaved road, including part of the Pan-American Highway. Airlines provide domestic and international services. Puerto Barrios is the main port.

Guatemala trades mainly with the USA and West Germany. The major exports are coffee, cotton and bananas. Manufactured metal products, textiles and flour are among the leading imports. There is normally a trading deficit.

BELIZE

Area: 8,867 sq mi (22,965 km²)
Population: 124,000
Population Growth Rate: 3·5%
Capital: Belmopan, pop 40,000 approx
Language: English
Religion: Largely Christian
Currency: Belize dollar = 100 cents

Belize, formerly British Honduras and still a dependent territory, was renamed in July 1973. Internal self-government was granted by Britain in 1964, but it remains Central America's only country lacking independence. The area's smallest country, it is also the only one with no Pacific coastline.

This hot, humid land of swamp and forest is thinly peopled and has poor prospects of emerging as a thriving independent nation, especially since its large western neighbour, Guatemala, claims the country as a Guatemalan province.

Forest and Highlands. The Caribbean-facing coastal plain, low and swampy in the north, is guarded by a chain of coral islands or cays which form a remarkable barrier reef, second in length only to the Great Barrier Reef off Queensland, Australia. The low-lying north of the country is crossed by a number of rivers, of which the Belize is the most important. Alluvial soils and tropical forest, including fine stands of mahogany, cover most of the land.

The southern half is dominated by the Maya Mountains, which rise abruptly from the coastal plain. These mountains reach their highest point in Victoria Peak (3,650 ft; 1,270 m) in the Cockscomb Range. Farther inland the mountains decline and their forests give way to pine savanna growing in poor sandy and limestone soils.

Wild animals include the manatee and many reptiles.

Climate. Belize averages 79°F (26°C) and only the higher mountains ever experience frosts. Rainfall increases rapidly southwards from about 50 in (1,270 mm) on the Mexican border to 180 in (4,572 mm) in the south, with even heavier amounts in the high mountains of the interior. The dry season (February-April) is used for clearing and burning the land for agriculture.

September, October and November are the hurricane months. The coastal areas where most of the people live are particularly exposed.

The People. More than half the population are so-called 'Creoles', English-speaking Negroes, who predominate in Belize City (43,000), the largest town. The British background of the colony, and its position facing the Caribbean, make their life close in character to that in former British West Indian islands.

Another group of Negroes lives in and around Stann Creek, south of Belize City. They are Carib-speaking descendants of slaves who escaped from captivity on the island of St Vincent and absorbed local Indian culture before their deportation in the 1800s.

The colony has attracted Spanish-speaking political and other refugees from neighbouring republics, and German-speaking Mennonites from Mexico seeking religious freedom. There are also Arabs, Chinese and a growing number of Americans. The indigenous Indian population consists of Mayan peasant cultivators and Kekchis, who live mostly on reservations in the south. Most British Hondurans are Roman Catholics. Children aged 6–14 receive compulsory primary education.

The new seat of government is Belmopan, established 50 mi (80 km) inland. It became the capital in 1970 after a hurricane had largely shattered coastal Belize City (the old capital) in 1961.

The Economy. Nearly half of the land area is forested and forest products have been important since settlers began trading in logwood (an effective dye for woollen and cotton goods) in the 18th century. The mahogany trade which replaced log wood in importance has declined since the 1920s. But other woods, such as Caribbean pine and tropical cedar, have been exploited. Most of the timber comes from the north and the west.

Since the early 1950s the traditional export role of timber has been taken over by sugar, grown on plantations around Corozal in the north. Citrus fruits, chiefly grapefruit, are becoming commercially important. Fresh-frozen lobsters, flown direct to the United States, are another recent addition to the very limited export trade. Machinery, food and manufactured goods are major imports.

The terrain is very difficult and so hampers internal communications, but there are 400 mi (644 km) of all-weather roads. Belize City is the chief seaport, and has an international airport.

A sugar cane vendor in Belize, main port and largest city of the country of the same name.

HONDURAS

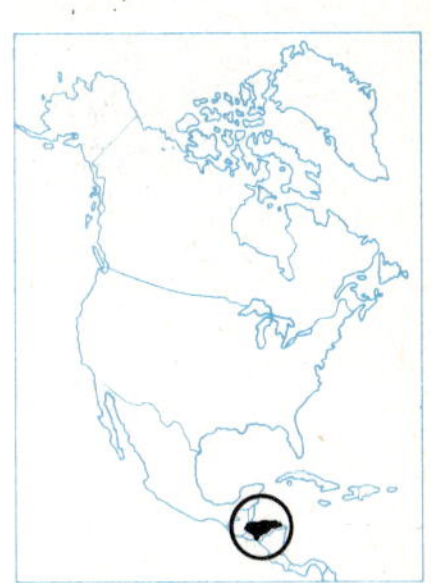

Area: 43,227 sq mi (112,088 km²)
Population: 2,582,000
Population Growth Rate: 3·4%
Capital: Tegucigalpa, pop 232,000
Language: Spanish. English spoken
Religion: Roman Catholic
Currency: Lempira = 100 centavos

Honduras (a nation a little bigger than East Germany) is the most central and the largest of the Central American republics. It is also the poorest. Since independence in 1838 it has lacked strong economic growth, despite the successful introduction of the banana.

Honduras disputes control of the Swan Islands with the USA and its southern neighbour El Salvador, with which it clashed in 1969 and 1970 over El Salvadoran 'squatters' in Honduras.

The Land and Climate. Rivers draining north-eastwards into the Gulf of Honduras have carved fertile alluvial valleys. But more than half of the country is mountainous and difficult of access, featuring sharply etched peaks that reach heights of over 9,000 ft (2,700 m).

The tropical climate is strongly modified by altitude. The lowlands of the north coast are warm and wet, with temperatures around 80°F (27°C) and up to 100 in (2,540 mm) of rain a year. But inland mountain areas at 7,000 ft (2,130 m) average 58°F (14°C), and less than 40 in (1,000 mm) of rain normally falls in the centre of the country. In the dry season (December-April) this area is brown and parched. Open savanna used for grazing is the characteristic vegetation of the interior and occurs, with tropical forest, along the short Pacific coast. In the east, pines cover large tracts of land. Jaguars and crocodiles are among the larger predatory animals found in forest areas, and the country generally abounds in reptiles, birds and insects.

The People. Most Hondurans are mestizos. But English-speaking Negroes, descended from labourers imported to work the banana plantations, form a substantial proportion of the population of northern Honduras. The population lives mainly in small towns or villages. The capital, Tegucigalpa (232,000), has few modern buildings. Education is theoretically free and compulsory but the illiteracy rate ran at over 40% in 1969.

The Economy. Minerals have always played a role in the economy of Honduras. It was the search for precious metals that first drew the Spaniards to conquer the territory. But though some gold and silver mining continues, the old silver mines have been almost exhausted. Lead and zinc are worked, but deposits of antimony, coal, copper, iron and mercury remain largely unexploited.

It is as a source of bananas that Honduras is now famous: no other Central American country grows so many. This is a 20th century development made possible by the investments of big US companies, notably the United Fruit Company and the Standard Fruit Company. The Caribbean coast continues to be dominated by the large American companies but their once considerable influence over the country's government had dwindled.

Coffee is another important cash crop, but maize (grown for subsistence) takes up over half of all crop land. Cattle-raising is important in the interior and retains such traditional trappings as the cowboy, the cattle drive to fresh pastures, the self-sufficient ranch. Honduras' extensive forests provide a source of mahogany and pine, but disease hit the forests hard in the 1960s.

Bananas, coffee, coconuts, cattle, timber and metals are among the leading exports. Honduras has little manufacturing industry of its own, and so imports include a wide variety of manufactured products for the domestic market. The United States is Honduras' main trading partner.

Transport systems remain poor. There are only 250 mi (400 km) of paved roads and most of the 900 mi (1,450 km) of railway serve the banana plantations. Most towns are only linked by dirt track largely travelled by mules and horses. Tegucigalpa must be one of few capitals in the world where visitors may still ride in on horseback and tie up in the main square.

Hopes that Honduras' membership of the Central American Common Market would help economic expansion receded in 1971 when Honduras withdrew from the agreements on free trade and taxes.

Tegucigalpa, capital of Honduras. Honduras depends on American owned banana plantations.

EL SALVADOR

Area: 8,236 sq mi (21,393 km²)
Population: 3,541,000
Population Growth Rate: 3·8%
Capital: San Salvador, pop 349,000
Language: Spanish. Nahuiti by some Indians
Religion: Roman Catholic
Currency: Salvadoran colon = 100 centavos

El Salvador is the most densely populated of the Central American republics, and the smallest (with an area less than that of Massachusetts); it is also the only one with no Caribbean coastline. El Salvador has some of the world's richest coffee growing land and ranks as the world's third largest coffee exporter.

El Salvador became a republic in 1856: Although its constitution is one of the most liberal in Central America, power has been largely concentrated in military hands. However, economic progress has been considerable since the creation of the Central American Common Market (1961).

Land and Climate. Behind a narrow coastal plain facing the Pacific Ocean, two chains of volcanoes form mountain ranges between which is a central plateau, 2,000 ft (600 m) high and 320 mi (512 km) wide. Some volcanoes, including Izalco (6,250 ft; 1,890 m), show occasional signs of activity, and Lake Ilopango near the capital San Salvador, has mysterious turbulences and a fluctuating level.

Plateau temperatures average 71–76°F (22-24°C) and annual rainfall is 45–60 in (1,143–1,524 mm). The lowlands are hotter and wetter, the mountains cooler and wetter.

The People. The population are 75% mestizo (mixed white and Indian). Of the remainder, 20% are Indians and 5% are Whites. Few Indians live a traditional Indian way of life. The population is concentrated on the central plateau and is growing fast (at 3·8% a year). Population pressures have caused international friction with Honduras where Salvadoran emigrants seek work. Only a quarter of the population live in cities, and most Salvadorans (62%) work in agriculture,

The rich volcanic soils of El Salvador's highlands provide an excellent coffee crop.

mainly on the large plantations. Illiteracy is high at 50% although education is free and compulsory between the ages 7 and 13.

The Economy. Coffee is grown on the plateaux and lower mountain slopes. The soil is excellent and labour cheap. Cotton is grown on the coastal plain. Together cotton and coffee account for 90% of the export revenue. Cattle rearing is widespread and 35% of the land is under permanent pasture.

El Salvador greatly benefited from the creation of the Central American Common Market (1961) and there has been growth in light industrial manufacturing and processing (especially cotton). El Salvador is now the most industrialized country in Central America. This growth owes much to a system of railways and roads (including the Pan-American Highway) that run the length of the land.

The pervasive poverty of Central America, however, limits the country's potential growth, and a general increase in the wealth of its own people and wider outlets abroad are necessary to ensure economic stability.

NICARAGUA

Area: 57,143 sq mi (148,000 km²)
Population: 1,912,000
Population Growth Rate: 3·7%
Capital: Managua, pop 300,000
Language: Spanish
Religion: Roman Catholic
Currency: Cordoba = 100 centavos

Nicaragua is a republic a little smaller than Greece, located between Honduras and Costa Rica. One of Central America's poorest and most turbulent countries, it is also one of that area's most strategically situated. Long Caribbean and Pacific coastlines and an extensive lake and river system made Nicaragua a likely alternative to Panama as a site for a canal linking the Atlantic and Pacific Oceans. Since the Panama Canal's construction, speculation has cited Nicaragua as the likely site of a second such canal. The nation's strategic position has been one reason why the US has repeatedly intervened in Nicaraguan internal affairs as when US marines occupied the country from 1912 to 1933 in part of a bid to stabilize conditions. There followed decades of largely dictatorial rule by the Somoza family which remained in power in the early 1970s, with little profit to the Nicaraguans in general.

Nicaragua is a potentially rich country, being naturally fertile. Yet the 67% of the active population engaged in farming are miserably poor. One in five farm labourers is paid in kind not in money and half the cultivable land is unclaimed. Medicine and sanitation are still primitive and the infant mortality rate is very high. In 1972 conditions locally worsened after an earthquake devastated the capital, Managua causing the deaths of hundreds of people.

Land and Climate. Nicaragua can be divided into three regions: a flat, forested Pacific lowland area; a mountainous central block which bisects the country except in the extreme south; and a wide belt of eastern lowlands, much of it marshy and dissected by river valleys.

Average temperatures vary from 60–70°F (16–21°C) in the mountains to 80°F (27°C) in the lowlands. Rainfall is high, especially along the Caribbean coast which intercepts the north-east trade winds from May to October.

The People. The majority of the population are mestizos (of mixed Spanish and Indian blood). There are also minority groups of Negroes, Indians and Europeans. Most of the population is concentrated in the west: between the Pacific Ocean and a line formed by a row of volcanoes in the north-west and lakes Managua and Nicaragua. Here were built the capital city, Managua (300,000), Granada, León and other centres. The highlands are only sparsely populated.

The population is chiefly Spanish speaking and Roman Catholic and nearly two-thirds of the people are illiterate.

The Economy. It is estimated that nearly 30% of the land area is cultivable, but less than 5% is cultivated. The Nicaraguan system of land tenure effectively renders some 90% of the rural population landless. The major crops are cotton, coffee, sugar, bananas, cocoa, maize, beans and rice. Livestock rearing is being encouraged.

Mining contributes relatively little to economic growth and is mainly controlled by foreign capital.

Manufacturing industry (traditionally weak) should profit from Nicaragua's entry into the Central American Common Market. It includes petroleum refining and the production of textiles, shoes, soap and cement. Attempts have also been made to develop the tourist industry.

The main imports are manufactured goods, machinery, vehicles, chemicals and foodstuffs; chief exports are cotton, coffee, sugar, meat and timber. There is usually an unfavourable balance of trade.

Nicaragua's economy suffers from a poor transport system comprising a short railway in the west, only a few paved roads, and limited air services.

COSTA RICA

Area: 19,653 sq mi (50,900 km²)
Population: 1,786,000
Population Growth Rate: 3·8%
Capital: San José, pop 211,000
Language: Spanish
Religion: Roman Catholic
Currency: Costa Rica colon = 100 centimos

Costa Rica ('Rich Coast') may be the second smallest republic in Central America but it has that area's highest standard of living. This politically stable and scenically attractive land lies between Nicaragua on the north and Panama on the south-east and has coastlines on the Pacific and the Caribbean. Politics cannot be influenced by the army, for that was abolished in 1948.

Volcanic Landscape. A mountain backbone (The Continental Divide) dominates the country as three ranges (from north-west to south-east the cordilleras Guanacaste, Central and Talamanca). Both the Guanacaste and Central cordilleras contain active volcanoes. The eruption of Irazú (11,500 ft; 3,505 m) in 1963 caused widespread damage and cast heavy showers of ash over a wide area of central Costa Rica.

Peaks of the Cordillera Central flank the Meseta Central, two small basins with floors at 3,000–5,000 ft (910–1,500 m). This region is the geographical, economic, political and demographic heart of the country. Farther south is the Talamanca Range, rising in many places to more than 9,000 ft (2,740 m).

The descent from the mountainous interior

Puerto Limon, on the Caribbean coast is the country's main banana-exporting port.

to the coastal plains is steep. The Caribbean coastline (131 mi; 211 km) is shorter and less indented than the Pacific (631 mi; 1,015 km). Its swampy, forested lowlands constitute about a third of Costa Rica's total land area.

Climate and Vegetation. There are sharp contrasts between the hot, wet climate of the coastal lowlands and the more temperate climate of the Meseta Central. Temperatures of the lowlands range between 77°F and 100°F (25–38°C), while San José on the Meseta averages 67°F (19°C). Rainfall in the southern Pacific lowlands and the Caribbean coastal plain is heavy, averaging 250 in (6,350 mm).

The southern Pacific lowlands, like the northern parts of the Caribbean coast, are covered by tropical rain forest, except where cleared for bananas. In the northern part of the Pacific coast, the natural open deciduous forest has been cleared by fire to promote tropical grasslands. In the cooler uplands of the country, grass becomes more common and forest less dense. Wild animals include tapirs, porcupines and crocodiles.

The People. Unlike other Central American peoples most Costa Ricans are of European (Spanish) descent. There are also mestizos, people of mixed European and Indian descent, and a significant English-speaking Negro minority on the Caribbean coast. In 1970 the population was an estimated 1,786,000; its annual growth rate (3·8%) was one of the highest in the world.

More than a third of the population live in the towns and cities, of which the largest is the capital, San José (211,000), on the Meseta Central – the home of over half the population. Population pressure is now forcing people out into remoter regions.

Most Costa Ricans are Spanish speaking, Roman Catholic and literate (although few complete even primary education).

A Plantation-Based Economy. Agriculture employs over half the working population. The Meseta Central provides most of Costa Rica's coffee, dairy products and sugar as well as a substantial part of other food crops, such as beans, maize and potatoes.

Costa Rica's famous coffee thrives on the deep, friable, volcanic soils of the basin, and on the surrounding mountain slopes up to about 4,200 ft (1,280 m). It is grown mainly on small *fincas* (estates) of less than 25 ac (10 ha), many of which are worked entirely by the farmer and his family.

The cooler slopes above the coffee *fincas* are devoted to dairying and mixed farming. Sugar-cane is grown on the western section of the Meseta. The patchwork landscape of small carefully tended fields and white-walled, red-tiled farm buildings recalls Mediterranean Europe.

The Caribbean lowlands are entirely different. Much of the region is still virgin rain forest, but banana plantations flank the railway and indeed largely finance it. Bananas are also grown commercially around the Golfo Dulce on the Pacific Coast. Higher up the Pacific coast, in Guanacaste province, the large cattle ranch is the rule, and social relationships are more those of the big *haciendas* found elsewhere in Latin America than of the small-farmer democracy of the rest of Costa Rica.

Industry is mostly small-scale and includes sugar refining and the manufacture of fertilizers, farm machinery and consumer goods.

Transport systems include railways linking both coasts, and 2,020 mi (3,250 km) of all-weather roads. Puerto Limón, Puntarenas and Golfito are the main ports.

Costa Rica trades (largely in coffee and bananas) mainly with the United States, West Germany and fellow members of the Central American Common Market. The balance of payments situation is unstable, but Costa Rica's economy has advanced by an average of 6·5% since 1960, a rate that many Central American countries must envy.

PANAMA

Area: 29,201 sq mi (75,650 km²)
Population: 1,478,000
Population Growth Rate: 3·3%
Capital: Panama, pop 418,000
Language: Spanish
Religion: Roman Catholic (93%) Protestant (6%)
Currency: Balboa = 100 centesimos

Panama stands at one of the world's great geographical cross-roads. It occupies a land bridge 480 mi (770 km) long and 30–120 mi (50–190 km) wide joining North and South America, and thus forms a narrow barrier between the Atlantic and Pacific oceans. This barrier is breached by ships using the Panama Canal whose US-controlled Canal Zone bisects the country. Panama's total area of 29,200 sq mi (75,650 km²) excluding the Canal Zone, makes it a little larger than Ireland.

The country's high birth-rate of 38 per 1,000 and low death-rate of 7 per 1,000 pose a threat to the economy, but economic growth has so far just outpaced the increase in population. Moreover, the soil is rich and agriculture can be developed to absorb some of the otherwise surplus labour.

The Face of the Land. Much of the republic is mountainous and heavily dissected by streams on both the Caribbean and the Pacific sides. The mountain ranges trend generally north-west to south-east, and include volcanic cones. The highest, at the

Costa Rican border, is 11,000 ft (3,350 m). Between the ranges are fertile plains, and there are generally narrow coastal plains.

Panama's tropical location ensures a mean lowland temperature around 80°F (27°C), but high mountain areas average as low as 45°F (7°C). Because of the onshore trade winds the Caribbean coast receives more rainfall than the Pacific coast.

The high temperature and high rainfall support areas of dense tropical vegetation and evergreen forest covers nearly two-thirds of the country. Savanna occurs on the relatively dry Pacific coast. Panama abounds in colourful insects, birds and reptiles, and its mammals include porcupines and tapirs.

The Miraflores Locks on the Panama Canal raise passing ships almost 30 ft.

The People. Panama has fewer people than any other central American republic, and three-quarters of the land is almost uninhabited. Of the total population of 1·4 million (1968), a third live in the metropolitan areas of Panama City and Colón, at opposite ends of the Canal. In addition there were some 48,000 persons, largely United States citizens, in the Canal Zone. There is a substantial rural population west of the Canal on the Pacific side. The interracially mixed (Spanish, Indian and Negro) population is mainly Spanish speaking and Roman Catholic.

The Economy is based partly on agriculture (rice, sugar, bananas, cocoa, coconuts, coffee, tobacco and hemp); partly on industry (including oil and sugar refining); partly on shipping registration fees; and partly on citizens' earnings from the Canal Zone.

Panama's exports include bananas, shrimps and petroleum products. Imports include raw materials, manufactured products and food.

Transportation. The republic is served by a short state-owned railway, and the Canal Zone by two railways owned by the United States government. There are over 4,160 mi (6,700 km) of roads, mostly unpaved, including one across the isthmus between Colón and Panama City. In the early 1970s engineers working in difficult jungle terrain were extending the Pan American Highway into Colombia to complete the last section of this great intercontinental route. The Compañía Panameña de Aviación and other airlines provide Panama City with international air links. Panamanian-registered merchant ships had the world's 11th-highest tonnage in 1971 but most were foreign-owned.

PANAMA CANAL ZONE

The Panama Canal Zone is a narrow strip of land across the Isthmus of Panama, flanking the Panama Canal through which up to 14,000 ships a year pass between the Pacific and Atlantic oceans. The Zone extends for about 51 mi (82 km) from Balboa on the Pacific to Cristobal on the Caribbean, and is 10 mi (16 km) wide. Within the Zone 60,000 civilians and 6,000 troops (most of them US citizens) live under American law and with American comforts which contrast sharply with conditions in the flanking Republic of Panama. Almost all the inhabitants work for the Panama Canal Company or for the Canal Zone government (two US government agencies). Canal Company employees maintain the 51 mi (82 km) artificial waterway and its 12 locks by which ships climb then descend 85 ft (26 m) as they cross the isthmus.

The Control of the Canal Zone. The Canal Zone came under the jurisdiction of the United States as a result of a treaty signed between that state and the Republic of Panama in 1903. The treaty gave the US the right to construct and maintain the canal; and the perpetual right to use, occupy and control a 5 mi (8 km) zone on each side of the entire length of the canal. Further, the treaty stipulated that the cities of Panama and Colón would remain under the control of Panama but the US should have certain jurisdiction pertaining to public health and sanitation in both cities, and the use of certain islands in Panama Bay. In return for these concessions the US paid Panama $10 million in cash, and $250,000 annually.

Panama's demands for a larger share in the economic benefits flowing from the canal have led to long and sour negotiations between both countries. In 1955, the annuity was increased to $1·93 million and in 1964 Panama was allowed to fly its flag alongside that of the US over the Zone. A year later, the US government agreed to discuss new treaties recognizing Panama's sovereignty over the Zone and integrating the Zone politically, socially and economically with Panama.

CUBA

Area: 44,206 sq mi (114,524 km²)
Population: 8,657,000
Population Growth Rate: 2·1%
Capital: Havana, pop 1,755,000
Language: Spanish
Religion: Roman Catholic
Currency: Cuban peso = 100 centavos

Cuba, the largest of the Caribbean islands, was described by Christopher Columbus as 'heaven itself' when he discovered it in 1492. Spanish colonists who followed him saw Cuba as the key to the New World, for it commands the Straits of Florida, the Yucatan Channel and the Windward Passage, major routes into the Caribbean Sea and the Gulf of Mexico.

From 1933 the island was dominated by the dictator Batista, who ruled through puppet presidents and was at times himself president. In 1953, Fidel Castro Ruz, a former student leader, launched a prolonged guerilla war and in 1959 seized power. Cuba became the first communist state in the Americas and embarked on radical social and economic reforms that alienated Cuba's former main trading partner, the USA. Cuba's hostility to the USA was emphasized by the missile crisis of 1962, when Russian-supplied atomic missiles based in Cuba were dismantled only after United States intervention and the threat of a global nuclear disaster.

The loss of the United States' market for Cuban sugar and tobacco, the fall in the world price of sugar and the collapse of the lucrative tourist industry all added to Cuba's economic problems. By 1968 all industrial production, trade and commerce had been brought under government control, and most of the cultivated land had been reorganized as state farms or collectives, but economic difficulties expressed in food rationing and other austerities persisted. Moreover the vital sugar crops that made Cuba the world's second largest sugar producer have been extremely uneven. Nevertheless, Castro had ended Cuba's economic dependence on the USA and begun some major social advances, improving standards of education and health, eliminating unemployment, introducing rent control and equalizing incomes.

Island Landscape. Cuba is the westernmost of the Caribbean islands. It is about 760 mi (1,220 km) long but only some 50 mi (80 km) across. This long, narrow island a little larger than East Germany is guarded by archipelagos of low coral cays, and the large Isle of Pines off the south-west coast – partly forested with mahogany and West Indian cedar, and the only commercially important offshore possession. Cuba's extensive plains and areas of low hills are broken by three substantial mountain systems. The most impressive is the eastern, which has two main ranges: the Sierra Cristal and Sierra Sagua-Baracca in the north, and the Sierra Maestra running along the south-east coast and containing Turquino Peak (6,576 ft; 2,005 m), Cuba's highest mountain. The central mountains rise between Cienfuegos, Trinidad and Sancti Spíritus, while the western system consists of the Sierra del Rosario and Sierra de los Órganos in Pinar del Río province.

Nearly all Cuba's rivers are small, flow only seasonally, and run from north to south or south to north. Many of the southward-flowing rivers end in coastal swamps, the largest of which are in the Zapata Peninsula.

Climate, Vegetation and Wildlife. Cuba straddles the Tropic of Cancer, and its mean temperatures generally range between 70 and 81°F (21 and 27°C), although mountain zones are rather cooler.

The island lies in the track of the northwest trade winds and annually averages 54 in (1,372 mm) of rain, although much heavier rainfall occurs in the mountains.

The November-April period is essentially a dry season, but cold winds called 'Northers' sometimes blow from North America, prolonging the rains into December and January. The wet season partly coincides with the hurricane season. Hurricanes usually strike the eastern provinces of Camagüey and Oriente. In 1963 Hurricane Flora devastated the sugar crop.

The natural vegetation includes tropical hardwood and pine forests, with savanna grassland on porous soils and mangrove swamps along the coast. In the relatively dry south-east, drought-resisting plants like cacti are common.

Man's activities, however, have disrupted this simple pattern, largely replacing forest trees with crops like sugar-cane, cacao and bananas. However, the cultivated plains still carry numbers of royal palms, the national tree, that provides thatch for the bohios (mud-walled rural homesteads).

Cuba lacks large land mammals, and has relatively few reptiles, but is rich in insect species.

The sandy tobacco-growing region of Pinar del Rio in western Cuba.

The People. Most Cubans are whites, the descendants of Spanish colonists and immigrants, but there are many mulattoes and Negroes. Negro slaves were brought in to work the sugar plantations and although freed in 1886, remained an impoverished group subject to various kinds of discrimination until the revolution of 1959. Like the mulattoes, they have made notable contributions to Cuban art, literature and music.

With more than half the population living in the cities and towns, Cuba is the most urbanized of Caribbean countries. In the countryside the larger settlements are usually near the sugar *centrales* (factories); elsewhere there are only scattered homesteads, except where the government has built *pequeñas ciudades* (literally 'small towns', but in fact modest rural housing projects) of substantial concrete houses with all basic services provided.

The destruction of the old social order and the emigration of thousands opposed to the Castro regime have radically altered ways of life. Nowhere is this more apparent than in Havana, the capital, where the commercial centre now houses government offices, the stock exchange is closed, retailing is confined to state stores, and housing vacated by refugees has been either reallocated or converted into schools.

Despite communist discouragement of religion, the Roman Catholic Church still has many followers, and there are also Protestant groups. Castro promised to maintain freedom of worship. It is claimed that illiteracy has been eliminated.

Farms, Forests and Fisheries. Cuba is still dependent upon one crop, sugar, and strenuous efforts have been made to increase production. Although part of the 1969 crop was held back, and voluntary labour was massively applied, the output in 1970 fell short of the 10-million-ton target by 1·5 million tons.

Tobacco, the second most important crop, is grown mainly in western Cuba. Cuba's cigars are esteemed throughout the world. Coffee is grown in the Oriente province. Cuba hopes soon to be self-sufficient in rice, a major food crop whose production is highly mechanized. Other crops include maize, cotton, potatoes and pangola, a fodder crop. Pineapples, citrus fruits, pimentos and tomatoes are exported. Henequen and kenaf provide fibre for making sacking and ropes. Several million head of cattle are raised on the central and eastern grasslands. Pigs, sheep and goats are also reared.

Large pine forests remain in the largely inaccessible mountain areas, and new forests, mostly of pine and eucalyptus, are being planted to improve timber stocks and help soil conservation.

Cuban waters are rich in tropical fish. The fishermen are now organized in co-operatives and a deep-sea fleet has been built up for operations in the Gulf of Mexico and off Argentina and Greenland.

Minerals and Manufacturing. All mineral resources were nationalized in 1960. They include large deposits of iron ore, nickel, cobalt, copper and manganese, mostly in the mountains of Pinar del Río and Oriente provinces. The small petroleum industry is based on the north coast and in the heart of the island; there are four refineries.

Such manufacturing as existed before the revolution was on a small scale and was financed by American capital. In 1960 the government seized some $200 million worth of assets, nationalizing the public service companies, the fish-processing plants and leather factories. An industrialization programme was launched, but failed because the available resources were spread over too many projects, and because many of the new factories depended on imported raw materials which Cuba had difficulty in buying. Since 1963 industrialization has proceeded at a much reduced pace.

Transport. The Carretera Central (Central Highway) traverses the island from Pinar del Río to Santiago de Cuba. Other roads, many surfaced, thread across the plains and into the mountains. Much of the rail network is narrow-gauge track for transporting sugarcane to the *centrales*, but Havana has rail links with the provincial capitals. Havana is also the chief seaport and airport.

Trade. Before the revolution Cuba traded mainly with the United States. Today the island depends upon trade with the Soviet Union and other communist countries. China has participated, but has proved a rather capricious trading partner. Commercial relations with Spain, Britain and other countries have meanwhile been improved.

CAYMAN ISLANDS

Cayman Islands is a self-governing British colony, including Grand Cayman, Little Cayman and Cayman Brac, lying about 200 mi (320 km) west-north-west of Jamaica. The coral islands are low lying and edged by reefs. They have no rivers.

Havana, largest city in the West Indies, was a major tourist centre before the revolution.

About half the population of 12,000 are of mixed Negro and white descent, a third are European, and the rest are Negroes. Four-fifths of the population live on Grand Cayman and most islanders live by fishing. Turtle and shark products provide valuable revenue, and the income from tourism is increasing. Education is free and compulsory.

JAMAICA

Area: 4,411 sq mi (11,424 km²)
Population: 1,897,000
Population Growth Rate: 3·0%
Capital: Kingston, pop 506,000
Language: English
Religion: Protestant (75%)
Roman Catholic (5%)
Currency: Jamaican dollar = 100 cents

Jamaica, which became an independent sovereign state in 1962, is the largest member of the Commonwealth of Nations in the Caribbean Sea.

But the island is only about 150 mi (240 km) long by 50 mi (80 km) wide and its area of only 4,411 sq mi (11,424 km²), makes it smaller than Connecticut (which ranks 48th in size among the US states). This picturesquely mountainous and wooded tropical island south of Cuba traditionally yields such agricultural produce as bananas, sugar, spices and rum. But Jamaica is turning increasingly to tourism and industry as aids to economic expansion.

The Land. Jamaica is very mountainous. The highest peaks are in the eastern Blue Mountains (which rise to 7,400 ft; 2,220 m). East and west of this range are limestone plateaux, most extensive in the west where their pockmarked surface occupies two-thirds of the island. Dozens of streams cascade off the plateaux and carve ravines in the abrupt seaward-facing slopes. Below these slopes there are broad coastal plains in the southern and western parts of the island; those in the north are narrower. The fertile plains support agriculture largely of the plantation type.

Climate, Vegetation and Wildlife. While the coastal areas experience temperatures of 70–90°F (21–32°C), areas at 5,000 ft (1,640 m) are cool at 40–45°F (4–7°C). Seasonal temperature variation is slight. Rainfall ranges from 40 in (1,016 mm) in the south to over 100 in (2,540 mm) in the north which receives the north-east trade winds. Jamaica lies in the path of hurricanes which periodically sweep across the Caribbean and ravage exposed areas.

Natural vegetation includes coastal mangrove swamps, remnants of tropical rain forest, savanna areas, and (in the dry south) cacti and other drought-resistant plants.

There are no large native land mammals. Colourful birds include humming-birds and parrots; there are also exotic butterflies, a wealth of land molluscs and over 50 reptile and amphibian species.

The People. A history of Spanish and then English colonization, the latter period seeing the import of Negro slaves and indentured labour from India and China, has produced a generally harmonious multi-racial society with a negroid majority and considerable racial intermixing. However, there is a class

Harvesting sugar cane. Sugar, molasses and rum have long been Jamaica's chief products.

system, subtly associated with wealth and skin colour. The lower income group is predominantly dark-skinned, the better off are largely of British stock. Long experience of British culture and institutions has resulted in a strongly British-influenced society. Parallel with this, however, has emerged a Jamaican or West Indian culture which seeks to express itself in colourful music, dress and dancing.

Over 90% of the population are Negroes or mulattoes (of mixed white and Negro ancestry), with small minorities of whites, Asian Indians and Chinese. In 1970 the population was estimated at 2 million and the growth rate at around 3%. Thus the pressure of the population on the land is very heavy and the overall density of over 450 persons per sq mi (173 per km²) is higher than that of El Salvador, the most crowded of the Central American republics. Most Jamaicans live and work in rural areas as farm labourers. But nearly one-fifth of the population lives in the Kingston metropolitan area. Other centres include the far smaller Montego Bay and Spanish Town.

Agriculture. There is a wide variety of crops. The leading cash crops are sugar-cane and bananas, followed by cocoa, coffee, citrus fruits, pimento and ginger. Domestic food crops include rice, maize, and root crops. An estimated 37% of the total cultivable land (much of it owned by absentee landlords) lies idle and thus adds to the island's problems of over-population and unemployment. Legislation means that failure to develop land can result in government acquisition and redistribution, but this law is not fully enforced. Stagnant agriculture has accordingly encouraged migration from the land to the slums of Kingston and smaller towns.

Minerals and Industry. Largely to help Jamaica's population and employment problem, the government has set out to develop the island's mineral and manufacturing potentials. Jamaica is claimed to have the world's largest reserves of commercial bauxite, and bauxite revenues have increased from $1 million in 1957 to much above $20 million in recent years. The bauxite is exported in the raw state and in the processed form of alumina. A variety of manufactures has been established since Jamaica's independence under three incentive laws. These aim to help Jamaican and foreign investors to establish manufacturing plants in Jamaica. The manufactures involve local and imported raw materials and include textiles, footwear, paints, building materials, television sets, electronic and telecommunications equipment, gramophone records and tyres. The tourist industry has greatly expanded, but is largely financed by foreign capital and the profits by no means all remain in Jamaica. The main tourist centres are along the north coast and include Montego Bay, Ocho Rios and Port Antonio.

Bauxite, alumina and tropical produce are Jamaica's main exports. Leading imports include manufactured goods, machinery, vehicles and foods. Most trade is with the USA, Britain and Canada. International airports serve Kingston and Montego Bay.

HAITI

Area: 10,700 sq mi (27,750 km²)
Population: 4,969,000
Population Growth Rate: 2·0%
Capital: Port-au-Prince, pop 340,000
Language: French. Haitian Creole more widely spoken
Religion: Roman Catholic, Voodoo widely practised
Currency: Gourde = 100 centimes

Haiti, the western third of the West Indian island of Hispaniola, is historically famous as the world's first Negro republic and the second nation in the Americas to win independence. Haiti nonetheless retains the official use of French – the language of its former colonial rulers. Around Port-au-Prince, the capital (340,000), luxury hotels and villas adorn the hillsides, suggesting a lucrative international tourist industry, but elsewhere villages are made of wooden, straw and mud huts. One of the most densely populated countries in the Western Hemisphere, Haiti is also one of the poorest, with a notorious record of corrupt and dictatorial government. The country is also one of the strongholds of voodoo, a powerful mixture of Christian and African beliefs which persists beside Roman Catholicism, the official religion.

The Land. Haiti is shaped like a giant horseshoe flanked by the Atlantic and the Caribbean and enclosing a great bay – the Gulf of Gonave containing the large Gonave Island. Over three-quarters of the land is mountainous, the principal massifs being westward extensions of those in the Dominican Republic. The highest point is Pic la Selle (8,793 ft; 2,680 m) in the extreme south-east.

Between the mountains there are plains and valleys (one of them containing the Artibonite, Haiti's only sizable river). Lowland areas also fringe the Gulf of Gonave and the north coast.

Haiti has a tropical climate with regional variations. Annual precipitation ranges from over 100 in (2,540 mm) in the Montagnes Noires in northern-central Haiti, to less than 20 in (508 mm) on some parts of the Gonave coastline. April-May and September-November are the wettest months. Destructive hurricanes, occur with some frequency; in 1966 Hurricane Ines took over 500 lives and caused extensive damage. Summer temperatures vary from 80°F (27°C) to 96°F (36°C) and winter temperatures from 55°F (13°C) to 80°F (27°C), the mountains being cooler than the lowlands.

The natural forest cover of Haiti's wetter areas has been largely stripped off by peasant farmers. Moreover prolonged overcultivation has led to progressive soil erosion. Humid forests now account for less than 10% of the uncultivated two-thirds of Haiti. The country's forest trees include mahogany and cedar, and royal palms are widespread. Mangroves fringe swampy coastal areas. On 40% of the total area there are only dry scrub, semi-arid savanna and drought-resistant desert plants.

Haiti has a wealth of insects and birds but no large native mammals.

The People. The vast majority of the people are Negroes descended from African slaves brought to the country by the French in the 17th and 18th centuries. The upper and middle classes of Haiti are mainly mulattoes (mixed Negro and white). Many of these have been educated in France and there is traditional tension between the Negroes and mulattoes. Comparatively few people live in the towns which mainly lie strung out around the coast.

Haiti is one of the poorest and most densely peopled nations in the Western Hemisphere. Life expectancy, at around 40 years, is desperately low. The vast Negro majority work as peasant farmers or labourers, and live at bare subsistence level in flimsy huts with little furniture and no piped water. Almost all are illiterate, speak a Creole patois, and practice voodoo. Many suffer from malnutrition, tuberculosis and malaria. It is difficult to see how the general poverty of the people can be even slightly alleviated without massive foreign assistance. US aid ceased during the ruthless dictatorship of President Francois Duvalier (Papa Doc) who ruled from 1956 to 1971 with the terrorist help of his secret police. Since he died there have been signs of renewed US aid. But despite the liberalizing tendencies of Jean-Claude Duvalier, Papa Doc's son and presidential successor, Haiti's future is extremely uncertain.

Economic Production. Agriculture forms the basis of the weak Haitian economy. With the exception of sugar and sisal production, mainly organized in large-scale commercial units, production for both domestic and export markets is based on small farms. There were approximately 550,000 such units in 1970, and the average farm size was estimated at a mere 1·5 ac (0·6 ha). The principal products for the domestic market are maize, millet and yams, supplemented by rice, beans, pigeon-peas and a wide variety of tropical fruits.

Real agricultural income has fallen in the recent past, and in the absence of far-reaching changes it is unlikely to improve much in the near future.

Hardwood production provides the raw material base of the domestic furniture and handicraft industries, although forest reserves are now much diminished.

Manufacturing is virtually undeveloped except on a handcraft basis. The major exceptions to this are several Port-au-Prince factories. The country is almost entirely dependent on imports for manufactured goods.

The only significant mining activities are concerned with the extraction of bauxite and copper ore.

Transport and Trade. Economic development is hindered by weak internal communications. The road network badly lacks all-weather surfaces and there are no functioning passenger railways. Port-au-Prince has an international airport and is the main seaport. The telephone system suffers from deficient maintenance.

The USA is Haiti's major trading partner. Coffee, sugar, sisal, bauxite and copper are among the main visible exports. Tourism has fluctuated in value as a source of invisible exports. In the early 1970s it was picking up after a decline influenced by unfavourable publicity about the Duvalier regime. Haiti has to import considerable amounts of basic foodstuffs (fats, oils, wheat, flour and dairy produce), also machinery, textiles, fuel oils and industrial raw materials.

Horses crossing a marsh near Port-au-Prince. Haiti is a mountainous country with sections of wet low-lying coastal plain. Its economy is one of the poorest in the western hemisphere.

DOMINICAN REPUBLIC

Area: 18,700 sq mi (48,442 km²)
Population: 4,188,000
Population Growth Rate: 3·6%
Capital: Santo Domingo, pop 671,000
Language: Spanish
Religion: Roman Catholic (95%)
Currency: D.R. Peso = 100 centavos

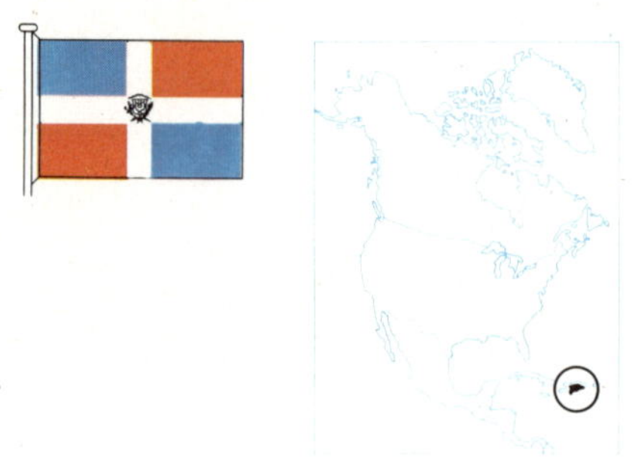

The Dominican Republic occupies the eastern two-thirds of Hispaniola, the largest island in the Caribbean after Cuba, the western third forms the Republic of Haiti. The island, known to its Pre-Columbian Carib and Arawak population as Quisqueya, was discovered by Christopher Columbus in 1492.

A Dominican attempt to win independence in 1821 failed. From 1822 until 1844, the whole of Hispaniola was under Haitian rule. Following a short period of Spanish re-colonization (1861–63), a more lasting independence was achieved. Its presidential system of government has often been abused, notably by the dictator Rafael Leonidas Trujillo Molina, whose oppressive control was ended only by his assassination.

Mountains and Fertile Plains. The Dominican Republic has a lush landscape of great beauty. The west and centre are mountainous, with four ranges running roughly parallel from north-west to south-east, the greatest elevation occurring in the Cordillera Central, where Pico Duarte rises to 10,417 ft (3,175 m). This massif is drained westwards by the Artibonite and is separated from the Cordillera del Norte by the rivers Yaque del Norte and Yuna.

To the south, the San Juan Valley separates the Cordillera Central from the Sierra de Neiba, which is separated in turn from the south coastal range of the Sierra de Bahoruco by a lowland area containing Lake Enriquillo, 150 ft (48 m) below sea level and the lowest point in the West Indies. The eastern part of the country, apart from the Cordillera Oriental flanking the southern shore of Samana Bay, is lowland.

Climate and Vegetation. The Dominican climate is mainly tropical, with average temperatures ranging from 65°F (18°C) to 82°F (28°C), although sub-tropical and temperate conditions prevail at higher elevations. Precipitation varies from more than 78 in (2,000 mm) to less than 18 in (500 mm). The country lies within the hurricane belt, the south and south-west being particularly vulnerable. Most of Santo Domingo, the capital, was destroyed by hurricane in 1930.

Vegetation ranges from the thorn bush and other drought-resistant plants of the driest mountain areas to the extensive pine forests of the more sandy uplands and the tropical rain forests of the lowlands. About 12% of the total surface area is forested.

Insects, birds and molluscs abound, but there are no large native mammals.

The People. About 70% of the population is mulatto; about 15% of pure European (chiefly Spanish) descent; and a further 15% of pure Negro origin.

Traditionally the Dominicans were a country people living, as many still do, in such favoured areas as the Yuna, Yaque del Norte and Yaque del Sur valleys. By 1970, however, about half of the total population was concentrated in the urban areas, movement into the towns being focused largely on Santo Domingo (671,000) the capital and chief port at the mouth of the Ozama River. The second largest city in the republic is Santiago de los Caballeros (125,000), the traditional centre of the Cibao, standing on the Yaque del Norte River.

Unlike the neighbouring Haitians, who have been shaped by strong African and French influences, the Dominicans have derived their culture, language and religion from Spain. In general the standard of living is much higher in the Dominican Republic than in Haiti.

Government. Since 1961 there have been several periods of corporate rule by civil or military juntas. The civil war of 1965, in which US forces intervened, was eventually ended by mediation by the Organization of American States. Government was by decree until 1966, when Dr Joaquín Balaguer was elected president for a 4-year term. He was re-elected in 1970. The constitution of 1966 provides for an elected National Congress of two houses: a 27-member Senate and a 74-member Chamber of Deputies.

Agriculture is the most important activity and provides about 24% of the Gross National Product. About 65% of the working population are farmers. Sugar is the major export crop, the largest plantations being in the south and south-east. Some are owned by American interests; others, formerly owned by the Trujillo family, now belong to the state. Subsistence farming is the livelihood of perhaps 25% of the population. Although shifting cultivation is still common in the highlands, subsistence farming centres mainly on smallholdings of 2·5 ac (1·0 ha) or even smaller areas, where farming methods are backward and yields low. In some areas, notably the eastern Cibao, conditions are better, and coffee and tobacco are grown as well as rice. Both coffee and leaf tobacco are exported, but cocoa is the second principal crop. Other crops include maize, peanuts and bananas.

Stock-raising on a commercial scale is concentrated in the north and on the eastern lowlands where beef cattle are reared mainly for domestic consumption. Dairying occurs mostly in the south and on the north coast at Sosua, where a Jewish immigrant colony was established in 1935.

Minerals. For many years the most important mineral has been bauxite, mined on the Barahona peninsula. Most of the output is shipped to the USA for processing. Rock salt and small amounts of copper are also mined, and silver and platinum have been found. A new nickel mine and primary processing plant are being opened near Bonao.

Industry. Manufacturing is largely a matter of processing sugar and other primary agricultural products. The scarcity of capital and skilled labour, high costs of production, severe competition in overseas markets and the limited size of the domestic market, make it unlikely that manufacturing will expand rapidly in the near future. The Dominican Republic will probably therefore remain dependent on imports of both capital and consumer goods for some time to come.

Transport. The republic has some 3,000 mi (4,828 km) of surfaced roads. Many routes, especially in the mountains, cannot be used in bad weather. There are no railways other than those operated by the sugar companies, and few internal air services. Santo Domingo is the chief seaport, and Puerto Plata the chief Atlantic port.

Trade. The chief exports are sugar, coffee, cocoa, chocolate, tobacco, molasses and bauxite. Imports include machinery, vehicles, fuel and vegetable oils, wheat, electrical and rubber products, paper and pharmaceuticals. The republic's major trading partner is the USA, which buys most of the sugar exports. The republic has little trade with other Caribbean countries and virtually none at all with its neighbour, Haiti.

PUERTO RICO

Area: 3,435 sq mi (8,891 km²)
Population: 2,757,000
Population Growth Rate: 1·7%
Capital: San Juan, pop 445,000
Language: English and Spanish
Religion: Roman Catholic
Currency: United States currency

Puerto Rico is one of the most densely populated islands in the world, lying about 60 mi (96 km) east of Hispaniola and 1,000 mi (2,400 km) south-east of Florida. Puerto Rico forms one of a line of island 'boundary stones' between the Caribbean Sea to the south and the Atlantic Ocean to the north. It is a fertile and attractive island much favoured by American tourists: a million came in 1968. The island is politically linked with the United States, and Puerto Ricans are US citizens. Since World War II intense efforts have been made to develop industry. The island has a higher per capita income than any other Latin American country, although far behind that of the United States.

The Land. Three-quarters of Puerto Rico consists of mountains and hills, and the island's backbone is a central range 3,000–4,000 ft (900–1,200 m) high made up of volcanic and metamorphic rocks. These central hills are flanked by limestone foothills. There are narrow coastal plains to the north and south, where sugar-cane is grown. These are the site of the two large cities and much industry.

Climate and Vegetation. The prevailing winds are north-easterly trades. The north-east facing hills and mountain slopes have over 200 in (5,080 mm) of rain a year while the south-western part of the island receives under 30 in (760 mm) a year. Most of the southern part of the island is too dry to be cultivated without irrigation. The extreme south-west is so dry that salt is produced commercially by evaporation from salt water lagoons. As with other Caribbean islands the seasonal temperature variation is small, ranging from 75°F (24°C) in January to 80°F (27°C) in August. Ninety hurricanes have been recorded in the past 400 years.

Pockets of forest and woodland still cover the low hills, particularly to the north and west. However, the development of densely populated rural areas has led to much forest clearance and four-fifths of the island is under crops and pasture.

The People. Puerto Rico was a Spanish colony from 1509 to 1898 when it was ceded to the United States, and although American influence has been very great since then, the language and culture of the people remains largely Spanish.

The population doubled from 935,000 in 1899 to 2 million in 1940 and is now more than 2,700,000. Birth control and emigration have begun to slow down the growth rate but population density remains among the highest in the world with around 770 persons per sq mi (300 per km²). The humid northern plain is the most densely peopled area and its cities are the main centres of trade and industry. The major city is the port of San Juan with a metropolitan population of over 400,000.

Ninety per cent of the people can read and write and nearly a third of Puerto Rico's budget goes on education.

Agriculture. In the 18th and 19th centuries, cattle-rearing, tobacco and ginger were the island's main exports. The ceding of the island by Spain to the United States in 1898 led to great changes. Sugar output increased rapidly and today this crop occupies half of the cultivated land, particularly on the coastal plains and interior valleys. The sugar is bought in bulk by the United States. Tobacco is second to sugar as an export and the other major crops are coffee, bananas, maize and pineapples.

Since World War II the Puerto Rican government has carried out land reform. Large estates were bought from big American corporations and government-owned profit-sharing farms were established. Some land was sold to small farmers in 25 ac (10 ha) plots. This has provided many peasants with land and improved living standards.

Industry. Since World War II more than 1,000 manufacturing plants have been established, many of them American, with the incentives of ready-made sites and roads,

Puerto Rico is a mountainous island with a rich alluvial coastal plain.

also loans and tax exemptions designed to attract industry. Manufacturing industries now earn twice as much as agriculture despite a lack of minerals. The absence of a tariff barrier between the island and the United States is another important asset. Despite this new industrial growth the processing of sugar-cane is still the dominant manufacturing industry.

The most recent growth industry has been tourism which is third in importance after manufacturing and agriculture and earns around $160 million each year. Puerto Rico's exports include sugar, tobacco, rum, electrical appliances and plastics. Food and raw materials are the main imports. The USA accounts for around 90% of overseas trade.

VIRGIN ISLANDS

The Virgin Islands of the United States, comprise two-thirds of the Virgin Islands group in the West Indies and lie between the Anegada Passage in the east and Puerto Rico in the west. The largest islands are St Croix (82 sq mi; 212 km²), St Thomas (28 sq mi; 73 km²) and St John (20 sq mi; 52 km²). The islands were bought by the United States from Denmark in 1917 for $25 million in order to control the Anegada Passage to the northern Caribbean.

The Land. Largely of volcanic origin, the group consists of the exposed peaks of a mainly submarine range that also forms Puerto Rico. Steep slopes rise sharply from sea level to over 1,000 ft (305 m).

Average annual rainfall is 50 in (1,270 mm), but as temperatures are never below 65°F (18°C) the rate of evaporation is high. The steep slopes also result in water loss by run off. Thus lack of water is a major problem, and water is largely imported from Puerto Rico.

The People. The total population was 63,000 in 1970; 70% of it was Negro. St Croix and St Thomas had the most people. The capital is Charlotte Amalie (12,370) a former supply base and coaling station on St Thomas.

Industry and Trade. The island's main products are sugar, livestock, rum and alumina. Exports total only one-third of imports and the deficit is made up by revenue from tourism, now the main economic base of the US Virgin Islands. By the 1970s over a million tourists were annually visiting the islands and spending around $100 million.

The British Virgin Islands comprise about one-third of the 100 West Indian islands in the Virgin Islands group. The islands are located 50–100 mi (80–160 km) east of Puerto Rico. The main British possessions are Tortola (24 sq mi; 67 km²); Anegada (13 sq mi; 33 km²); Virgin Gorda (9 sq mi; 23 sq km²); and Jost Van Dyke (4 sq mi; 10 km²).

The largest islands are mountainous with harsh relief. Tortola rises to 1,780 ft (538 m) and Virgin Gorda to 1,370 ft (418 m).

The soils are poor and subsistence farming is the main economic base. The land is largely rough pasture or scrubland supporting cattle-rearing. Fish form the second main export (after cattle).

In 1968 the British Virgin Islands had an estimated population of 9,000. Individual islands in the 1960 census showed the following returns – Tortola 6,200, Anegada 287, and Virgin Gorda 564. The islands are administered as a crown colony by the Overseas Development Ministry in London.

ST KITTS, NEVIS AND ANGUILLA

St Christopher (St Kitts)-Nevis is a British associated state in the Leeward Islands, comprising three islands: St Christopher, also called St Kitts (68 sq mi; 176 km²), Nevis (50 sq mi; 129 km²) and Sombrero (2 sq mi; 5 km²). (A fourth island, Anguilla, returned to direct British rule in 1971). British claims to the islands date from the 17th century. The group has a total population of 45,500, mostly of mixed Negro-European descent. English and a local patois are the common languages. The capital is Basseterre (14,000), on St Christopher.

The islands are mainly volcanic in origin and have fertile soils. St Christopher and Nevis are dominated by peaks rising to over 3,000 ft (900 m). The climate is tropical and humid. The temperature range is 62–92°F (16–33°C) and average annual rainfall is 55 in (1,397 mm).

St Christopher produces sugar-cane and cotton and its main exports are sugar and molasses (shipped from Basseterre). The chief product and export of Nevis is cotton. As in other parts of the Caribbean, tourism is of growing importance to the island group. Anguilla is a British dependency in the Leeward Islands. It returned to direct British rule in 1971 after revolting against union with St Christopher-Nevis (1967–69). Anguilla is a tiny low-lying tropical island of 35 sq mi (91 km²), with about 6,000 inhabitants of mainly Negro descent engaged in collecting salt, fishing and stock raising.

ANTIGUA

Antigua (108 sq mi; 280 km²) in the Leeward Islands is a British associated state. Two other islands are linked with it as dependencies: Barbuda (62 sq mi; 160 km²) and the uninhabited Redonda (1 sq mi; 2·59 km²). The population, largely of mixed Negro-European origin, totals 63,000 – the majority living on Antigua. The people speak English and a local patois. The capital is St John's.

Antigua and Barbuda are coral limestone islands with a flat, rolling terrain. The climate is tropical but, owing to the lack of rivers and scarcity of springs, Antigua often suffers from drought.

The economy is an agricultural one. About half the labour force is engaged in producing the two chief exports, sugar and sea-island cotton. Tourism is playing an increasingly important economic role.

MONTSERRAT

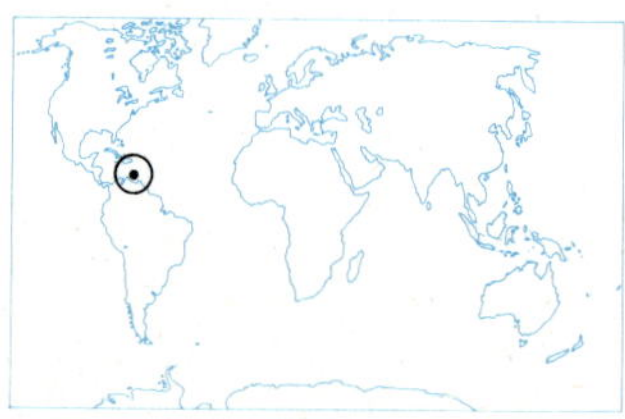

Montserrat is a British colony in the Leeward Islands, the island of Montserrat is 27 mi (43 km) south-west of Antigua and has an area of 33 sq mi (85 km²). The population is largely of mixed Negro and European descent and numbers 15,000. The common languages are English and a local patois. The island's capital is Plymouth (4,000).

Montserrat has three principal groups of volcanic mountains. The Soufrière Hills in the south are the highest group, reaching a height of slightly over 3,000 ft (914 m). The climate is humid and tropical, with an average temperature of 81°F (27°C) and average annual rainfall of 94 in (2,387 mm).

Apart from tourism, agriculture provides the mainstay of the island's economy. Sea-island cotton and tomatoes and other vegetables are exported.

GUADELOUPE AND MARTINIQUE

Guadeloupe and Martinique, 130 mi (209 km) to the south, are islands in the Windward group of the Lesser Antilles, separated by the island of Dominica in the British West Indies.

With Guadeloupe's five dependencies they form the French West Indies, having been under French control since 1635, except for temporary periods of occupation by the British. They remained French colonies until 1946 when they were raised to the status of overseas departments of France.

Guadeloupe consists of two main islands, Basse-Terre and Grande Terre, and five small islands (Marie Galante, Désirade, Les Saintes, St Barthélemy and the northern part of St Martin), covering a total of 687 sq mi (1,779 km^2). Basse-Terre and Grande Terre are divided by the Rivière Salée, a tidal channel which is spanned by a drawbridge. Most of the inhabitants of these islands are of mixed Negro and white extraction, descendants of the original French settlers and African slaves.

Martinique (431 sq mi; 1,110 km^2) was discovered by Christopher Columbus in 1502, and is well-known for its volcanic mountains, in particular Mt Pelée, which erupted with disastrous results in 1902. The island was also the birthplace of the Empress Joséphine, first wife of Napoleon I. Both islands are now enjoying a growing tourist industry, though their economies are still primarily based on agriculture.

Land and Climate. Geologically many of the islands are volcanic, rising to peaks of nearly 5,000 feet with spurs extending seawards. Mt La Grande Soufrière (4,869 ft; 1,480 m) on Basse-Terre in Guadeloupe is the highest mountain in the Lesser Antilles. The limestone island of Grande Terre, by contrast, is level and low-lying at under 450 ft (137 m). Martinique, volcanic and mountainous, is dominated by Mt Pelée (4,554 ft; 1,384 m). The devastating eruption of this active volcano in May 1902 destroyed the town of St Pierre and killed about 30,000 people. The only large expanse of level lowland on Martinique is the Plain of Lamentin, a region of sugar-cane plantations near the Bay of Fort-de-France.

Numerous streams and several large rivers drain the mountainous islands; Grande Terre, however, depends on reservoirs and cisterns for its water supply.

Vegetation in the higher, wetter areas of the islands is that of dense tropical rain forest, especially in the mountains of Basse-Terre in Guadeloupe. Elsewhere much of the forest has been cleared for agriculture, and in the southern, drier areas, scrub woodland predominates.

The climate of all the islands is basically tropical, though Guadeloupe experiences much greater variations of temperature and rainfall than does Martinique, where temperatures remain equable. The average temperature of the hottest month, September, is 81°F (27°C), and that of the coolest month, January, 76°F (24°C). These temperatures are moderated by the trade winds. The annual average rainfall is 80 in (2,032 mm). Hurricanes frequently occur during the rainy season between July and November. Guadeloupe was hard hit by such a storm in 1928, when high winds and a massive tidal wave did immeasurable damage.

The People. In 1971 the population of Guadeloupe was estimated to be 337,900; that of Martinique, 345,000. The pressure of population on land is very acute, especially in Martinique where the population density is nearly 800 persons per sq mi (309 per km^2). Negroes and mulattoes make up the greater part of the population, except in Les Saintes in Guadeloupe, which has a small all-white population, descendants of the original French immigrants.

The main settlement on Basse-Terre is the town of Basse-Terre (15,833), the capital of Guadeloupe. The other large town and chief commercial centre is Pointe-à-Pitre (29,538) on Grande Terre, the second island. The capital and commercial centre of Martinique is Fort-de-France (100,000).

Economy. Agriculture forms the basis of the economy of the islands, with sugar as the chief crop. Most of the agricultural produce is exported to France, from which come imports including textiles, foodstuffs and petroleum products.

Bananas are an important export crop in Guadeloupe, where about 100,000 tons are produced annually. Other minor export crops include coffee, pineapples, cocoa, sisal and citrus fruits.

There are no commercial minerals to support mineral-based industries on a large scale, though small amounts of salt and sulphur are produced. Because of the inaccessibility of the forests of Basse-Terre in Guadeloupe, timber is largely imported. However, the government is encouraging the development of new, light industries by offering tax exemptions. Existing manufactures are based largely on sugar and its by-products: molasses, rum and alcohol. There are sugar refineries at Pointe-à-Pitre, Basse-Terre and Grande Anse in Guadeloupe. Martinique, afflicted with a perennial problem of high unemployment, still depends heavily on both direct and indirect subsidies from the French government, and a substantial proportion of the population has emigrated to France in search of work.

Income from tourism, however, has become increasingly important to the islands in recent years.

DOMINICA

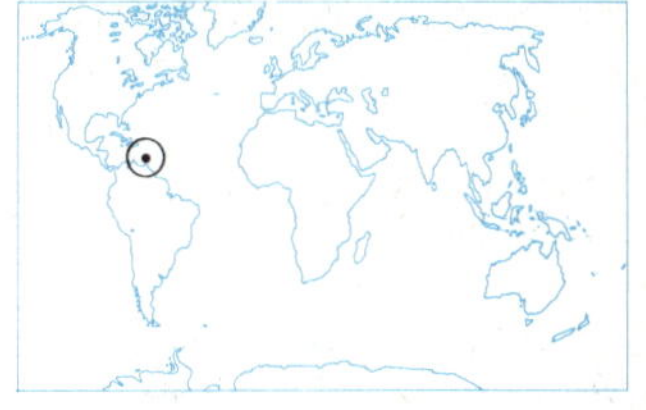

Dominica is a self-governing West Indian island associated with Great Britain, which controls defence and external affairs. It occupies some 290 sq mi (750 km^2) and lies in the Lesser Antilles, between Guadeloupe and Martinique. Dominica became self governing in 1967 as one of the West Indies Associated States.

The island is volcanic in origin and largely mountainous, with Morne Diablotin (4,747 ft; 1,447 m) as the highest point. The temperature range is 78–90°F (26–32°C) and rainfall is high, averaging over 250 in (5,200 mm) in the mountains (which are largely forested).

The People and Their Work. The population of 70,302 includes Negroes, mulattoes, Europeans and individuals of partly Carib Indian blood. English is officially spoken but peasants use a French patois. Most people are Roman Catholics. Primary education is mandatory.

Many Dominicans work on the land, raising limes, bananas, cacao, coconuts and other food crops; or in factories processing such products. Bananas, cocoa, copra and fruit juice figure among the main exports; food, machinery and textiles are among the chief imports. International trade is largely routed through the ports of Roseau and Portsmouth. Jet airliners call at Melville Hall Airport.

ST LUCIA

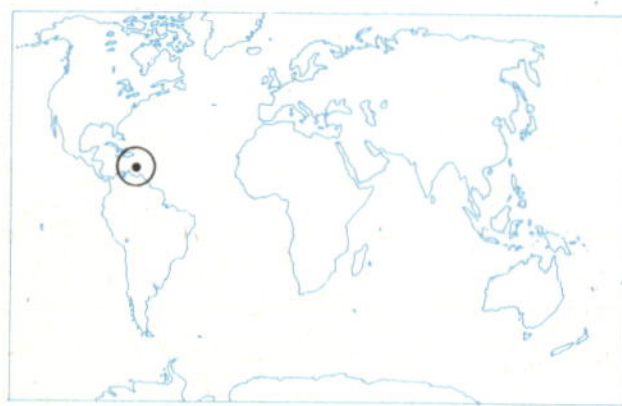

St Lucia is a British associated state in the Windward Islands with an area of 238 sq mi (616 km²). Most of the 101,000 inhabitants are of mixed European, Negro and Carib Indian descent. The two common languages are English and a French-African patois. The island's capital is Castries (45,000).

St Lucia is volcanic in origin. Like many of the other Windward Islands, it has a central mountain range running north to south. The highest point is Canaries Mountain (3,145 ft; 959 m). The island has a tropical maritime climate, with average temperatures of 69–78°F (21–25°C). Rainfall is heavy in the summer, with annual averages of 51–117 in (1,295–2,971 mm).

The island is thickly covered with tropical vegetation. Its economy is based on agriculture and its chief crops and exports are bananas, copra, cocoa and coconut oil. About half the labour force is engaged in agriculture.

ST VINCENT

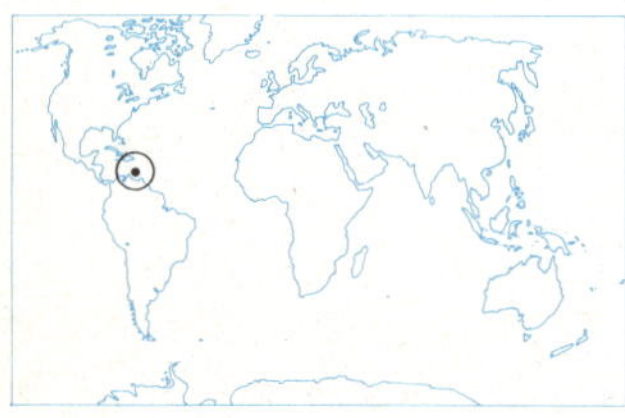

St Vincent is situated between the islands of Grenada and St Lucia. The island became a British associated state in 1969 – two years after the other British possessions in the Lesser Antilles. Together with the northern Grenadine Islands which it administers, the state has an area of 150 sq mi (389 km²). The population of 100,000 is of mixed descent: European, Negro and Carib Indian. Islanders speak both English and the French-African patois common in the area. The capital is Kingstown (21,000).

The island is dominated by heavily forested mountains of volcanic origin, stretching from north to south. The volcano Soufrière (4,018 ft; 1,224 m) is the highest peak. The climate is tropical, with temperatures averaging 67–89°F (19–32°C) and annual rainfall varying from 50 to 150 in (1,270–3,810 mm).

Most of the labour force is engaged in farming, and the island's agricultural economy is based on the export of bananas, arrowroot, sea-island cotton, copra and spices. The island grows almost all the vegetables it needs. As elsewhere in the area, tourism is of increasing importance to the economy.

BARBADOS

Area: 166 sq mi (430 km²)
Population: 239,000
Population Growth Rate: 1·1%
Capital: Bridgetown, pop 18,789
Language: English
Religion: Anglican (70%) Roman Catholic, Methodist, Moravian
Currency: East Caribbean dollar = 100 cents

Barbados has a Caribbean setting, a Portuguese name, an English culture and a predominantly Negro population. When the Portuguese briefly visited this easternmost Caribbean island in 1536 they noted trees with gracefully drooping branches and called them *barbudos* (bearded fig trees). The British came in 1627, and remained. Their three centuries of rule gave the island its distinctive Englishness; even today Barbados is sometimes called 'Little England' in the Caribbean. Barbados was one of the 10 British colonies which formed the short-lived Federation of the West Indies (1958–62), and became an independent sovereign state within the Commonwealth of Nations in 1966.

Land and People. The island is formed of coral limestone rising gently in flat terraces to Mt Hillaby (1,102 ft; 336 m) in the centre. Average temperatures are around 80°F (27°C), but the north-east trade winds have a cooling effect. During the rainy season (June–November), the short downpours are heaviest in the north. Hurricane occasionally occur. Barbados is known for its many beautiful flowering trees and shrubs.

More than 70% of the population is Negro; about 10% is European (mainly British). Most Barbadians belong to the Anglican Church, and their Englishness extends from their parliamentary system of government to a love of cricket. Educational standards are high and illiteracy is negligible. Bridgetown, the capital (18,789) and port on Carlisle Bay, has a collegiate branch of the University of the West Indies.

The Economy. Most of the island is intensively cultivated and sugar-cane is still the chief crop. Sugar, rum and molasses are leading exports, along with shellfish (mainly shrimps). The island has no minerals except limestone and a little natural gas. Industrial products include plastics and electronic components.

Sugar accounts for more than 53% by value of Barbadian exports. The island trades mainly with the UK, the USA and Canada, but also handles West Indies transit trade.

GRENADA

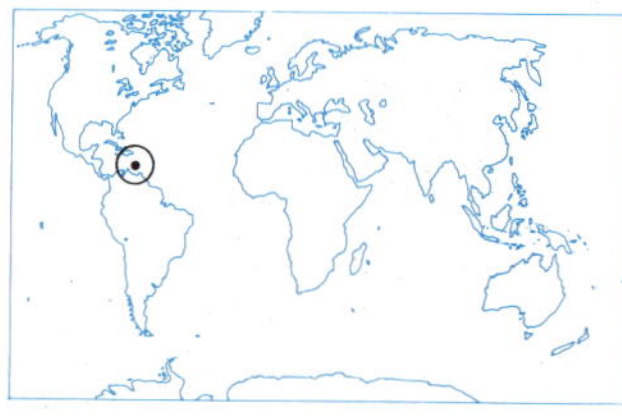

Grenada is a British associated state comprising Grenada, the southernmost of the British Windward Islands, and the southern Grenadines, the largest of which is Carriacou. The total area is 133 sq mi (344 km²) and the population – of mixed European, Negro and Carib Indian descent – numbers 104,000. The common language is English, but the people also use a French-African patois. The capital is St George's (22,893).

The island is heavily wooded, with mountains of volcanic origin ranging across it from north to south. The highest point is Mt St Catherine (2,756 ft; 840 m). The crater of an extinct volcano in the centre of these mountains is occupied by a large lake, the Grand Etang. The climate is tropical, with heavy summer rainfall.

Though tourism is of growing importance, Grenada's economy is agricultural. The chief export crops are cocoa, nutmegs and bananas. Other crops include coconuts, citrus fruit, sugar-cane, cotton and spices.

SOUTH AMERICA

South America ranks fourth in area among the continents but only fifth in size of population. Steaming tropical forests, vast rolling grasslands and barren deserts cover much of the surface and climates range from hot, wet equatorial in the Amazon basin to alpine in the high Andes Mountains. Ranches, plantations and mines flourish on and near the populous coastal fringe. But much of the interior is undeveloped and almost uninhabited. Millions of South Americans live in traditional peasant poverty, which is deepened locally by political violence and rampant inflation. Yet South America has vast untapped mineral and agricultural resources.

Conquest and Colonization. When Christopher Columbus sighted it in 1498, South America was thinly populated by Indians who ranged from primitive hunter-farmers to the civilized Incas of the Andes Mountains. Inca gold attracted Spanish military explorers in the early 1500s, followed by Spanish colonists.

Meanwhile the Portuguese were building up a considerable empire in what is now Brazil. Wealthy colonists became increasingly frustrated with controls imposed by the colonial powers that ruled them.

The New Nations. From the struggles for independence from Spanish rule in the early 19th century, eventually emerged 10 new republics: Argentina, Bolivia, Brazil, Chile, Colombia, Ecuador, Paraguay, Peru, Uruguay and Venezuela. New democratic constitutions were usually disrupted by violent coups, to the detriment of economic progress. Boundary disputes also led to disastrous wars between republics.

By 1966 the entire continent except for French Guiana and (Dutch) Surinam was politically decolonized. Political instability still dogs economic progress, but land reforms, power and highway projects and Brazil's founding of a new capital in the remote interior suggest a more hopeful future.

The Face of the Continent

The Land. South America covers some 6,889,000 sq mi (18,398,000 km^2). From its tropical north on the Caribbean Sea, it stretches southwards for about 4,750 mi (7,650 km) to Cape Horn, less than 500 mi (800 km) from Antarctica. Its greatest width is 3,200 mi (5,150 km).

The continent includes four major highland areas – the Andes Mountains, the Guiana Highlands, the Brazilian Highlands and Plateau, and the Patagonian Plateau. The Andes are the longest continuous mountain system in the world. Running close to the Caribbean and Pacific coasts, these mountains effectively isolate the relatively narrow Pacific coastal region from the rest of the continent.

The Andes are considerably narrower than the western cordilleras of North America, although in Bolivia they widen to about 400 mi (644 km). They are, however, generally higher and contain breath-taking peaks soaring above 18,000 ft. Aconcagua reaches 22,834 ft (6,960 m) and is the highest mountain in the Western Hemisphere. There are numerous volcanic cones, many of them active. Seismic shocks occur frequently in many areas, especially in Chile and Peru.

The high plateau country of the Andes, the *altiplano*, extends from southern Peru to northern Argentina, reaching its greatest extent and elevation in Bolivia. It includes Lake Titicaca, the world's highest navigable sheet of water (12,505 ft; 3,812 m).

South America east of the Andes is dominated by three upland blocks, the Guiana, Brazilian and Patagonian highlands. From the ledges of the often precipitous south-eastern Guiana Highlands drop some of the world's highest waterfalls on little-known tributaries of the Orinoci, Essequibo and Mazaruni rivers. At the southern edge of the Brazilian Highlands, in southern Brazil and eastern Paraguay, rivers such as the Paraná and Iguaçu plunge in spectacular staircases of falls down the edge of the Paraná Plateau.

Between these upland blocks and the

Over: The high and arid peaks of the Cordillera Réal in the Andes near La Paz. *Below left:* Mt Roraima lies in the rain forest of south Guyana. *Below right:* Peruvian Indians in a market in Cuzco, the ancient capital of the Inca Empire. The Inca language Quechan is still spoken.

SOUTH AMERICA
vegetation
tundras and high mountain flora
mixed forest of temperate and subtropical zones
irrigated areas
tropical rain forest
savannah, monsoon forest and thorn scrub
steppe- and mountain grassland
desert and semi-desert
cultivated areas
swamp
scale 1: 35,000,000
0 500 1000 st. miles
ATLANTIC OCEAN
PACIFIC OCEAN
CARIBBEAN SEA
GREATER ANTILLES
LESSER ANTILLES
LEEWARD ISLANDS
WINDWARD ISLANDS
GALÁPAGOS ISLANDS
FALKLAND ISLANDS
SOUTH GEORGIA
SOUTH SHETLAND ISLANDS
SOUTH ORKNEY ISLANDS
Drake Passage
Equator
Tropic of Capricorn
Caracas
Trinidad
Bogotá
Medellín
Cartagena
Quito
Guayaquil
Lima
Cuzco
La Paz
Arica
Antofagasta
Valparaiso
Santiago
Mendoza
Córdoba
Rosario
Buenos Aires
Montevideo
Porto Alegre
São Paulo
Rio de Janeiro
Brasilia
Salvador
Recife
Belém
Manaus
Iquitos
Cuiabá
Asunción
Paramaribo
Puerto Montt
Punta Arenas
Amazon
Orinoco
Rio Negro
Madeira
Paraná
Paraguay
Rio de la Plata
Pico Bolívar 16411
Roraima 9219
Angel Fall
Cotopaxi 19344
Chimborazo 20577
Huila 18856
Tolima 17110
Huascarán 22205
Coropuna 21720
Illampu 21490
Illimani 21151
Tocorpuri 22162
Incahuasi 21720
Llullaillaco 22146
Ojos del Salado 22590
Aconcagua 22834
Tupungato 17356
Tronador 11660
San Valentín 13314
Murallón 11811
Sarmiento 7546
Itatiaia 9255
Itambé 6670
Cape Horn
Tierra del Fuego
Strait of Magellan
Juan Fernández Islands
Fernando de Noronha

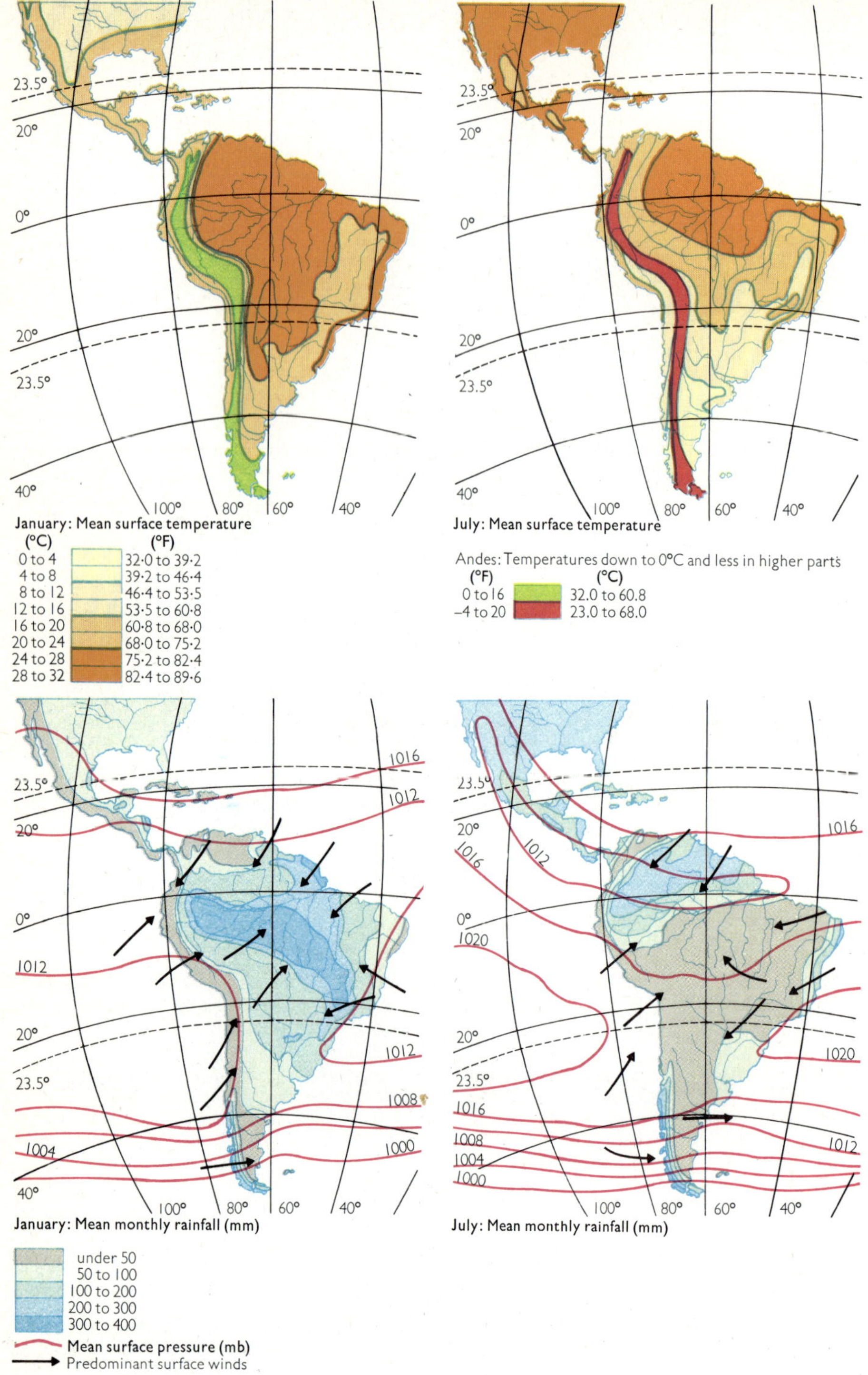

January: Mean surface temperature

July: Mean surface temperature

January: Mean monthly rainfall (mm)

July: Mean monthly rainfall (mm)

Andean cordilleras in the west is the huge sinuous curve of the central plains of South America which extend south to Argentina's Patagonian Plateau, a rocky, upland area extending east from the Andes almost to the Atlantic coast.

South America has three major river systems: Orinoco, Amazon, and Paraná-Paraguay (La Plata). Each is navigable by seagoing ships of some size for a considerable part of its length, and all empty into the Atlantic Ocean; no major river flows to the Pacific. The Amazon is slightly shorter than the Nile, the world's longest river, but it is the world's greatest in volume. Roughly 4,000 mi (6,400 km) long, it has about 15,000 tributaries and drains an area of nearly 3 million sq mi (7,770,000 km²).

Climate. The great extremes in elevation and a latitudinal range of 68° give South America a striking variety of climates including equatorial, tropical, warm temperate and cool temperate. More than two-thirds of the continent, however, lies within the tropics. In the lowlands near the equator, high even temperatures continue throughout the year and it is very humid.

The valley of the Amazon is the world's most extensive area of high temperature and high rainfall. At Manaus (Brazil) temperatures remain around 80°F (27°C), and the vast equatorial and tropical rain forest has an annual rainfall of about 80 in (2,032 mm). Other areas of high rainfall include the north-east and north-west coastlands and south-western Chile.

As South America tapers southward, so the climate is increasingly tempered by marine influences. Winter (June–September) and summer (December–March) are more clearly differentiated and are without extremes of hot or cold. In the extreme south, the moderating marine influences keep the temperatures of Tierra del Fuego's coldest month above freezing point.

The climate of west and east coast South America is affected by ocean currents, the best known being the cold Peruvian Current of the south-west coast which so cools the lower air that northern Chile and coastal areas of Peru have very little rain.

Vegetation and Wildlife. The Amazon Basin is covered by the *selvas* – immense tropical rain forests. The trees present a monotonous appearance from above, and from ground level the forests seem to consist of little but creepers and lush undergrowth beneath a thick, green canopy through which the sky is rarely glimpsed. But in fact the forests contain an immense variety of plant forms. These include commercially valuable trees – the rubber tree, hardwoods, and nut-bearing trees. The forest persists to a high level on the eastern Andean slopes, where tall ferns and cinchona (the source of quinine) abound. North and south of the Amazonian forests are the *llanos* or tropical grasslands of the Orinoco Basin, and the Brazilian Plateau.

The forests of the sub-tropical region to the south of this area provide pine and maté tea. To their south are the great pampas, the temperate grasslands of Argentina; and south of them drought resistant scrub and steppe vegetation covers Patagonia.

West of the Andes, southern Chile has dense forests of pine and deciduous trees, but northern Chile is arid and barren. The high country of Bolivia and Peru is semi-arid and supports only coarse grasses and shrubs.

Wildlife is rich and varied. Mammals include jaguars, pumas, ocelots, monkeys, sloths, guanacos, tapirs, armadillos and vampire bats. Reptiles include poisonous and constricting snakes, caymans and lizards. There are many species of beautiful birds and butterflies.

The People

South America is the home of about 195 million people. They include small communities of Caribs and Arawaks, living in the

SOUTH AMERICA
political
CITY population more than 1,000,000
CITY population more than 500,000
CITY population more than 100,000
City population less than 100,000
railways
roads
airport
scale 1: 35,000,000
0 250 500 750 st. miles
INDEPENDENT AFTER 1945
former British territory
ATLANTIC
OCEAN
PACIFIC
OCEAN
CARIBBEAN SEA
Equator
Tropic of Capricorn
BRAZIL
ARGENTINA
CHILE
PERU
BOLIVIA
COLOMBIA
VENEZUELA
ECUADOR
PARAGUAY
URUGUAY
GUYANA
SURINAM (NETH. GUIANA)
FR. GUIANA
GALÁPAGOS ISLANDS (ECUADOR)
FALKLAND ISLANDS (BR)
SOUTH GEORGIA (BR)
SOUTH SANDWICH ISLANDS (BR)
SOUTH ORKNEY ISLANDS (BR)
SOUTH SHETLAND ISLANDS (BR)
Juan Fernandez (Chile)
Drake Passage
Strait of Magellan
Cape Horn
Tierra del Fuego
Port Stanley
SANTIAGO DE CUBA
KINGSTON
PORT-AU-PRINCE
SANTO DOMINGO
SAN JUAN
Belize
GUATEMALA
TEGUCIGALPA
SAN SALVADOR
MANAGUA
SAN JOSÉ
PANAMA
Tobago
PORT-OF-SPAIN
TRINIDAD
CARACAS
MARACAIBO
BARRANQUILLA
CARTAGENA
MEDELLÍN
BOGOTÁ
CALI
QUITO
GUAYAQUIL
LIMA
CALLAO
CUZCO
AREQUIPA
LA PAZ
COCHABAMBA
Sucre
Potosí
ASUNCIÓN
SANTIAGO
VALPARAÍSO
CONCEPCIÓN
BUENOS AIRES
LA PLATA
ROSARIO
CÓRDOBA
MENDOZA
TUCUMÁN
MONTEVIDEO
PÔRTO ALEGRE
SÃO PAULO
SANTOS
RIO DE JANEIRO
BELO HORIZONTE
BRASÍLIA
SALVADOR
RECIFE
FORTALEZA
BELÉM
MANAUS
GEORGETOWN
PARAMARIBO
Cayenne
Macapá
Santarém
Pôrto Velho
Rio Branco
Iquitos
Punta Arenas
Río Gallegos
Ushuaia
Comodoro Rivadavia
Bahía Blanca
MAR DEL PLATA
Antofagasta
Iquique
Arica
Amazon
Negro
Madeira
São Francisco
Paraná
Paraguay
Titicaca
Rio de la Plata

northern coastal regions; other aboriginal Indians, living largely in Columbia and the highlands of Peru, Bolivia and Ecuador, and in Brazil's Amazonian Forests where some tribal groups still pursue a virtually stone age way of life almost untouched by Western civilization.

Much more numerous than South America's aboriginal Indians are the *mestizos* (people of mixed European and Indian stock) and *criollos* (people of European stock born in South America). The criollo element predominates in Argentina and Uruguay, and is important elsewhere, especially in the socially developed areas of Brazil and Chile. Colombia, Venezuela, Paraguay and Chile, however, are predominately mestizo, the Indian element in the last two being Garaní and Araucanian respectively. Ecuador, Bolivia and Peru have Indian majorities dominated by white and mestizo elements.

Brazil has a particularly varied population including the descendants of Negro slaves, mulattoes (peoples of mixed Negro and European stock), American Indians, mestizos, Mongolians and whites of Portuguese, Italian, German, Swiss and other ancestry.

Urbanization has occurred in South America largely ahead of industrial development, and present-day South American culture is largely urban.

Shanty towns around the cities and peasant hovels in the countryside beyond are two indications of widespread poverty and malnutrition in South America. Illiteracy is another major problem. Until recent years, the conservative forces of landowners and the Roman Catholic Church have tended to perpetuate poverty by bolstering governments that retained the social status quo. In countries like Chile and Argentina, social discontent has found an increasingly powerful voice in the growing numbers of industrial workers. In general, however, social reforms are slow to take effect.

The Economy

Economic Development. In the colonial period the Spaniards were mainly interested in the production of precious metals. In Brazil, the Portuguese concentrated on growing sugarcane. Argentina's great grasslands were used for rearing cattle and sheep, but meat exports to Europe expanded only after the 19th century invention of refrigeration. Spain and Portugal had a commercial monopoly in South America that remained unbroken until their colonies began to hive off in the early 1800s. Thus the economic organization of South America as a whole remained static until the second half of the 19th century. By this time Chile had become an important producer of copper, Brazil had begun large-scale production of coffee, and Argentina was expanding its pastoral activities.

Before the century was out, European money was financing expansion. Railway building to link mines and farms with seaports heralded a phase of relatively rapid development. Chile began exporting nitrates; Argentina shipped out large quantities of cereals, as well as mutton and beef; and beef became Uruguay's main export. For all three countries, and Peru, wool provided an important source of foreign currency. Bolivian minerals, including tin, commanded a ready market, while the oil of Venezuela, now a leading world producer, has enabled that country to pay off its national debt. Today petroleum deposits are also exploited in Argentina, Brazil, Ecuador, Peru and Colombia. The considerable mineral resources of Brazil, including iron ore, have yet to be fully exploited.

Some South American countries suffer from dependence on one or two commodities (for instance bananas and coffee) which are highly vulnerable to disease or price fluctuation in world markets. Chile suffered a major setback with the development of synthetic substitutes for natural nitrates, though copper has become an economic mainstay. Diversification has not proved easy, although Brazil has had some success in this and in developing manufacturing industries to reduce the imports bill. São Paulo has become the most rapidly expanding industrial centre in South America and the world's fastest-growing city.

British investment, at one time paramount in Argentina and elsewhere, has virtually ceased and American capital now largely influences South American economics.

Transport. The physical obstacles to communications are probably greater in South America than in any other continent. The Andes and the Amazonian Forest are the biggest barriers. In the Andes, many of the old Indian trails are still used, and in some areas pack animals are the only form of transport. In the Amazonian Forest, rivers are still virtually the only highways.

Road and rail development occurred most rapidly in countries with expanding economies, notably Argentina, Brazil and Chile. The Pan-American Highway now runs from Colombia to Chile, and a trans-Amazonian road is being built. High costs will for many years prohibit the general extension of transport systems in undeveloped areas. Meanwhile this vast continent has only some 87,000 mi (140,000 km) of metalled roads and 61,000 mi (98,000 km) of railways.

Air transport was early established locally in South America, and there is now a widespread network of air services.

International Trade is based mainly on South America's role as a supplier of food and raw materials to the more industrially advanced continents. The USA is the chief trading partner of almost all the South American republics. Changes in the commercial pattern since World War II are indicated by the increasing standing of Japan and West Germany. Spain, once dominant in South American commerce, now plays a minor part.

COLOMBIA

Area: 439,737 sq mi (1,138,919 km²)
Population: 22,490,500
Population Growth Rate: 3·2%
Capital: Bogotà, pop 2,818,300
Language: Spanish
Religion: Roman Catholic
Currency: Colombian peso = 100 centavos

Colombia, the fourth largest country in South America, occupies the north-west corner of the continent and has some tiny offshore islands including San Andrés in the Caribbean Sea. It has coasts on both the Caribbean Sea to the north and the Pacific Ocean to the west. Its name commemorates Christopher Columbus, who neared its Caribbean coast in 1502.

Since 1830 Colombia has suffered from numerous uprisings and civil wars. Some 200,000 perished in *La Violencia* (1948–1953), a bewildering period of anarchy and bloodshed. The outcome of this bloody feud was an arrangement whereby the two major parties took turns to appoint the president for a four-year term of office. Meanwhile sporadic violence and unrest continued, fed by massive unemployment and other serious social problems.

Land of the Northern Andes. Entering Colombia from Ecuador, the Andes Mountains divide into three chains or cordilleras – western, central and eastern – which run north-eastwards almost parallel to one another. The (western) Cordillera Occidental has several high peaks, including the volcano of Cumbal (16,049 ft; 4,892 m). Peaks in the Cordillera Central are higher and snow-capped while the (eastern) Cordillera Oriental reaches 18,375 ft (5,591 m) into Alto Ritacova in the Sierra Nevada de Cocuy. Separating the cordilleras

SOUTH AMERICA
Flora and fauna
Gila monster
Heloderma
pineapple
Ananassa
Giant cactus
Cereus
macaw
Ara
mate
Ilex
Two toed sloth
Choloepus
lungfish
Lepidosiren
Para rubber
Hevea
Side neck turtle
Chelys
tapir
Tapirus
Darwin's finches
Geospiza
Vampire bat
Desmodus
caimen
Caimen
llama
Lama
quillaja
Quillaja
cocoa
Theobroma
peccary
Tayassu
Hairy armadillo
Dasypus
rhea
Rhea
Surinam toad
Pipa
Geology
Cenozoic
Mesozoic
Upper and Lower Paleozoic
Precambrian
Extrusive
Intrusive
Geological structure
Regions of Tertiary mountain folding
Direction of movement of continental plates
Earthquake centers
Volcanoes
Mountain belts
0 300 600 900 1200 mi
0 400 800 1200 1600 2000 km
Alpine and tundra
Evergreen forest
Grassland
Coniferous forest
Deciduous forest
Steppe
Desert

are the valleys of two major rivers, the Cauca and the Magdalena. The latter, 969 mi (1,550 km) long, is Colombia's longest river and chief commercial waterway.

In the north, the western and central cordilleras decline as they approach the Caribbean lowlands, but the eastern cordillera divides and thrusts one branch into Venezuela and the other towards the Guajira Peninsula. In the isolated mountain mass of the Sierra Nevada de Santa Marta are Colombia's highest peaks.

Nearly two-thirds of Colombia lies east of Andes. Here a vast lowland plain slopes gently eastwards, drained by tributaries of the Orinoco and Amazon. Much less extensive are the lowlands north and west of the cordilleras which drain towards the Caribbean and Pacific respectively.

Climate, Vegetation and Wildlife. Colombia lies near the equator and has a mainly tropical climate, greatly modified however by altitude, and with wide differences from region to region. Rainfall varies a great deal. Places west of the Cordillera Occidental receive up to 400 in (10,160 mm) annually (making some areas the wettest in South America), while dry coastal lowlands along the Caribbean regularly receive less than 10 in (254 mm).

There is great variety in the vegetation, which includes the tropical jungle of the Amazonian region, moist grasslands to its north, the mangrove swamps of the Pacific coast, the desert scrub of La Guajira, tropical scrub forest and grasslands along parts of the Cauca and Magdalena rivers, and alpine plants on the high mountains. Wildlife is rich, reflecting the varied habitats, and includes such large mammals as jaguars and spectacled bears; over 1,500 species and subspecies of birds; and a wealth of reptiles, amphibians and insects.

The People. Racially the Colombian population is a mixture of three elements: the original Indians, the whites who came from Spain, and the Negroes who were shipped to Colombia from the time of the Spanish conquest. These elements are now much intermingled. The majority of the people are either mestizos (mixed white/Indian) or mulattoes (mixed white/Negro). Indians form only some 5% of the population.

Colombia's population is increasing alarmingly, by more than 600,000 a year, thereby making it difficult to reduce unemployment and raise the standard of living. Family-planning campaigns have had only limited success, partly due to the traditionally conservative attitude of the powerful Roman Catholic Church.

Most of the people live in the mainly mountainous western third of the country. Rural poverty encourages a considerable movement of population from the country into the towns. Unfortunately the towns cannot provide enough work for the newcomers, many of whom are forced to live in slums and shanty-towns without piped water or sanitation.

The major cities are widely dispersed. Bogotá, the capital, stands at 8,660 ft (2,640 m) on the Sábana de Bogotá, a plateau in the Cordillera Oriental. Medellín, the second city and the chief commercial centre, stands northwest of Bogotá in the Cordillera Central at an altitude of 5,046 ft (1,535 m); while Cali, the third largest city and an important manufacturing centre, stands in a rich agricultural area of the Cauca valley. The main port is Buenaventura on the Pacific Coast. Cartagena founded by the Spaniards in 1553 lies on the Caribbean coast.

Primary education is free yet not compulsory, and mountainous terrain makes school attendance difficult in remote areas. In the late 1960s over 2·7 million pupils attended primary and secondary schools. There are more than 20 universities, with over 60,000 students. Over 20% of Colombians are illiterate (compared with 90% in 1900).

Government. Colombia has been described as a 'presidential democracy'. Legislative power is vested in a Congress of two houses, the Senate and the House of Representatives, whose members are elected for a 4-year term. The president is also elected for a 4-year term by direct vote, but cannot be re-elected until four years later. All Colombians over 21 have the vote.

A Coffee-Based Economy. Although only some 4% of the land is cultivated, agriculture is a major industry in Colombia and employs nearly half the working population. By far the most important of the many crops is high quality coffee, which in 1969 provided nearly 70% of the republic's export earnings. Second only to Brazil in world production of coffee, Colombia is also the leading producer of mild coffee.

Other important crops include bananas, sugar-cane, rice, maize and potatoes. Colombian farmers also grow wheat, barley, cotton, tobacco and fruits. Cattle rearing is important in some areas, especially in the Caribbean lowlands and on the plain at the eastern foot of the Andes, and dairying is common around the large cities.

Agriculture, however, generates only 29% of the gross national product. Its relatively low productivity has several causes, not least the poor health of the rural population and the lack of educational facilities. The average farmer also suffers from a lack of working capital, machinery and fertilizers.

More significant still is the effect on production of the inequitable distribution of land. Outside the coffee-producing areas, where the average plantation rarely has more than 3,000 or so trees, most land is concentrated into large estates owned by a very small number of families. Most farmers live and work on the *minifundia*, tiny farms of rarely more than 5 ac (2 ha). There are also many rural families who have no land at all.

Minerals. Colombia has considerable mineral wealth. Petroleum, iron ore, coal, limestone, gold, emeralds and salt are all produced, mainly for domestic use, and new deposits and other minerals are being found in hitherto unexplored areas.

Colombia is a leading South American oil producer. The main fields are at Barrancabermeja, on the east bank of the Magdalena River; north of Cúcuta, whence oil is piped over the Andes to the Caribbean coast; and in the Putumayo district in the far south of the country.

Iron ore and coal, mined in the Cundina-

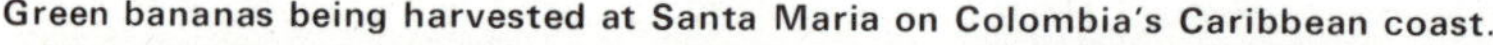

Green bananas being harvested at Santa Maria on Colombia's Caribbean coast.

marca and Boyacá departments, are used mainly in the iron and steel works at Paz de Rio. There are also important coal mines at Cerrejón and La Jagua, in La Guajira and César departments respectively, and in Valle de Cauca department (where the coal is used chiefly to generate electricity). Colombia has the largest coal reserves and is the leading coal producer in Latin America. She ranks sixth in world production.

Emeralds are mined mainly in Boyacá department. Gold and silver are mined or dredged in Antioquia department; platinum in Chocó. The great salt mine at Zipaquirá has been worked for centuries, but is unlikely to be exhausted in the next hundred years or so. Its showpiece is an underground cathedral carved out of the rock salt.

Industry. Manufacturing is mainly concentrated in the three largest cities: Bogotá, Medellín and Cali. The main boost to industrialization came during World War II when the policy of import substitution was adopted. As a result Colombia has become largely self-sufficient in food, beverages, textiles and rubber products, for example, but remains heavily dependent on imported machinery and other capital equipment. The future expansion of the local car, paper, electrical, chemical and steel industries depends on foreign expertise and capital, although efforts are being made to reduce this dependence. The government's development plans aim at much more selective use of foreign finance, and at the development of agriculture and secondary industry. Public expenditure is focused on improving power and water supplies, education and transport. More than two-thirds of Colombia's power comes from hydro-electric stations, and increasing use is made of natural gas.

Transport. The great mountain barriers across the country, the large areas of tropical forests, and immense problems of cost have severely limited road construction. Most of the larger towns and centres are now interconnected but there are still important exceptions. Medellín and Bucaramanga, for example, the second and sixth largest cities of the republic, have no direct road link.

There are only some 2,164 mi (3,483 km) of narrow-gauge railways, all state-owned. In many cases a journey by bus is quicker than one by train, although rail services have been greatly speeded in recent years. Despite seasonal drought, the Magdalena River remains a vital waterway for passenger and freight traffic. Some 900 mi (1,450 km) of its course are navigable, steamers ascending the river as far as La Dorada. Colombia is justly proud of its airline system. It is claimed that the world's first commercial airline operated in the republic (1919) and *Avianca*, the leading airline, is known for its efficient domestic and international services.

Trade. Colombia depends to a great extent upon one major export – coffee, of which more than $343 million worth was exported in 1969. This leaves the economy vulnerable to changes in world prices. Other major exports include petroleum, bananas, fuel oil and sugar. The republic constantly faces a balance of payments problem.

VENEZUELA

Area: 352,143 sq mi (912,050 km²)
Population: 10,721,522
Population Growth Rate: 3.6%
Capital: Caracas, pop 2,183,935
Language: Spanish
Religion: Roman Catholic
Currency: Bolivar = 100 centimos

Venezuela lies on the north coast of South America, with a 2,000 mi (3,000 km) shoreline on the Caribbean Sea. It is bordered on the west by Colombia, on the south by Brazil, and on the east by Guyana. In terms of size, It is in the middle rank of South American states.

Venezuela's Spanish name (meaning 'Little Venice') originated in the early 16th century when Spanish explorers found Indian fishermen's dwellings raised above the shallow waters of Lake Maracaibo in a manner reminiscent of Venetian buildings. Over 300 years later Venezuela became independent from Spain. Lake Maracaibo was found to be a fabulously rich oil field. The oil gave Venezuela one of the fastest growing economies in the world (1935–60), and about one-seventh of the world's petroleum requirements were being met by Venezuela by 1960. The resulting industrial growth has helped to produce a mainly urban society centred on the capital Caracas, and to launch big new urban and power projects. By the 1970s Venezuela's great need was to reduce its reliance on oil revenue, for known supplies were expected to run dry by 1990.

The Four Regions. Venezuela can be divided into four well-contrasted major regions: first, the Northern Highlands extending north-eastward from Colombia to form a mountainous belt along much of the Caribbean coast; second, the Maracaibo lowland embayment in the north-west; third, the Orinoco River plains towards the centre; and finally the Guiana Highlands in the south and south-east.

The Northern Highlands is a complex of ranges, rift valleys and intermontane basins, and forms part of the Andean system which divides in Colombia and Venezuela to form separate chains splaying out towards the Caribbean. To the north-west, the Sierra de Perija, the Segovia Highlands and the Cordillera de Mérida (over 16,000 ft; 4,900 m) enclose the Maracaibo Lowlands. Eastwards, steep, abruptly rising ranges curve along the Caribbean coast to form what are known as the Central Highlands, and eventually trail out to the Paria Peninsula which points a sharp finger at the island of Trinidad. Much of the region lies at 2,500–4,500 ft (760–1,370 m), particularly a number of important intermontane basins in the Central Highlands such as those of Caracas, Valencia and Maracay.

The large shallow Lake Maracaibo, fills the central floor of the Maracaibo Lowlands to leave a fringing plain approximately 20–60 mi (30–45 km) wide, between the lake shore and the steep encircling mountain slopes. The southern part of the lake is fresh water which become brackish towards the very constricted northern exit. This narrow strait opens out over a sand bar to a second great embayment known as the Gulf of Venezuela.

Behind the great wall of mountains swinging in an arc around northern Venezuela lie the low, almost featureless plains of the River Orinoco, the heart of one of the great drainage networks of South America. The river rises in the Guiana Highlands and hugs the edge of this massive block for nearly 1,500 mi (2,410 km) before ending in a huge delta. The whole 120,000 sq mi (300,000 km²) region is criss-crossed by water-courses whose banks are seasonally flooded. Extensive swamps lie to the south.

The ancient block of the Guiana Highlands capped with hard sandstones, reaches plateau summit levels of just over 9,000 ft (2,740 m). The massif takes up about half the total national area, but the Guiana Highlands is the least known and least explored region of Venezuela. On the periphery, spectacular waterfalls drop their long white columns over the sides of the plateau; Angel Falls, 3,212 ft (979 m) high, is the highest uninterrupted waterfall in the world.

Climate, Vegetation and Wildlife. Venezuela's climate and plant life are influenced by the easterly rain-bearing trade winds, but there are strong local contrasts. In the Andean zone, the *tierra caliente* (hot land) is found between sea level and approximately 3,000 ft (900 m). Here average annual temperatures of more than 80°F (26°C) show little seasonal variation, but vegetation ranges from tropical forest on wet slopes to scrub on slopes sheltered from the rain-bearing winds. The cooler *tierra templada* (temperate, or warm land) characterizes the most densely populated zone between 3,000 ft and 6,000 ft (900–2,000 m), with average annual temperatures of 65–75°F (13–18°C) and plants including tree-ferns and orchids. Above these distinctive zones the high pastures of the *tierra fría* extend to about 16,000 ft (4,900 m) where meadows end at the snow line.

Farther south, seasonal drought, alternating with widespread flooding, conditions vegetation on the Orinoco plains. Heavy summer rains from June to October convert the plains into thousands of square miles of shallow open water. During the winter dry season (December-March) the floods gradually recede leaving chains of swamps among the coarse savanna grasses and stands of palm and scrub woodland. Only on the 'islands' of slightly higher ground are the better soils and more nutritious pastures found.

In south-east Venezuela, the Guiana Highlands are mantled for the most part in dense evergreen and deciduous forests, with rolling savanna areas of relatively high quality pasture grasses.

Venezuela's large mammals include tapirs, jaguars and two bear species. Among reptiles there are caymans, turtles and many kinds of snake including the anaconda, among the largest snakes on earth. Birds include ducks, cranes and herons and the cave-dwelling oil-bird prized for the rich oil content of its body.

The People. Some 70% of the population of 10,400,000 is mestizo (mixed Amerindian and European) and 20% is of European stock. There are smaller Negro and mulatto (8%) and Amerindian (2%) minorities.

Over 98% of the population is concentrated in the northern mountain and coastal belt; the greatest urban and rural densities are found in the Caracas valley. Caracas itself, Venezuela's capital city, contains a population of over 2,000,000. Oil revenues are responsible for the impressive high-rise blocks, landscaped parks and fast expressways which make Caracas one of the urban showpieces of Latin America. But its status as one of Latin America's fastest growing cities creates grave housing problems. Newly constructed residential areas have failed to keep pace with the steady stream of migrants from the rural areas flocking in to find employment.

Depressing shanty-towns accordingly disfigure the perimeter although progress is being made in improving housing, schools and social services.

Other notable centres within the Andean plateaux, valleys and basins are Barquisimeto, Valencia and Maracay. Coastal centres include the chief port, La Guaira, near Caracas, and a number of tourist resorts. Maracaibo, Puerto Cabello and Puerto La Cruz are the other major port cities.

Deep in the interior wilderness lie two virtually isolated cities. Ciudad Bolívar (70,000) on the Orinoco functions as a river port, assembly point and trading centre between the Orinoco plains and the Guiana Highlands. Downstream, the new city of Santo Tomé de Guayana had attracted over 100,000 inhabitants by the early 1970s and was planned to hold 1,000,000.

Most Venezuelans are Roman Catholics; some 95% speak Spanish; and about 80% are literate. Improved medical facilities helped to cut the death rate to 6·8 per 1,000 by 1966 (among the lowest in the world).

Agriculture. Less than 3% of the land is cultivated, and most of this is in the north. Bananas, cacao, sugar-cane, rice, cotton and tobacco are all produced in the *tierra caliente*. The *tierra templada* is dominated by coffee production – although maize and citrus fruits are also important. In the cooler *tierra fría*, small fields of wheat, barley, beans, potatoes, peaches, cherries and maize chequer the mountain sides.

Behind the mountains lie the Orinoco plains where, despite flood-waters, flies, disease, and indifferent stock, the cattlemen (*llaneros*) reign supreme. More profitable ranches are found along the fringe of the area, especially along the rolling country of the Andean piedmont where fenced, irrigated, sown pastures escape flood and drought, and provide the leading areas of meat production. Though forests cover about half the country, inaccessibility and a limited market leave much of the country's reserves untouched.

Minerals and Manufacturing. The richness of Venezuela's petroleum resources has created one of South America's wealthiest states from what would otherwise have remained one of its poorest. In 1970 Venezuela's output was exceeded only by the USA and the USSR. It rose swiftly after 1920s exploration and investment programmes had confirmed Maracaibo as one of the greatest oil fields in the world, particularly for heavy fuel oils. The lowland, the waters of the lake and the surrounding foothills bristle with derricks, pumps and storage tanks, and important fields extend eastward towards Barcelona and the Orinoco delta to make this Caribbean belt an unbelievably rich source of 'black gold' for Venezuela. Deposits of copper and coal also occur here.

The Guiana Highlands are endowed with a variety of minerals, among them iron, manganese, bauxite, nickel, chromium, gold and diamonds, but inaccessibility has delayed their exploitation.

Venezuela must diversify its economy, for known oil reserves will soon dry up. The northern cities of Caracas, Maracay and Valencia already contain manufacturing plants involved in textiles, engineering, car assembly, tyres, paper, cement, food processing and an increasing range of domestic consumer goods. The petro-chemical industry is expanding rapidly, and so is aluminium and tin plate production. Power is mostly derived from natural gas piped in from the Maracaibo field.

In south-east Venezuela, the new steel town of Santo Tomé de Guayana (or Ciudad Guayana) founded in 1961, uses abundant local supplies of iron ore and hydro-electric power. By 1990, the Caroni River power system and the varied heavy industries grouped around Ciudad Bolívar will probably have combined to make this area one of South America's most important manufacturing centres.

Trade and Transport. Oil revenues provided 90% of Venezuela's export earnings in the early 1970s. Iron ore is exported to the USA by North American companies working the deposits at Cerro Bolívar and El Pau close to the River Orinoco. Small quantities of gold, diamonds, and other minerals, as well as coffee, sugar, hardwoods and cacao are exported. Foodstuffs and manufactured goods figure prominently among the imports, about half of which enter from the USA.

Limited railway construction has linked the Caribbean ports with a few of the valley and mountain basin centres, but there are only 291 mi (475 km) of track. There is more interest in road construction, typified by the new Angostura Bridge over the Orinoco which carries the Pan-American Highway to Ciudad Bolívar. It is the longest suspension bridge in Latin America and ties this interior growth zone to Venezuela's main highway system. Internal air services are important since much of Venezuela remains otherwise inaccessible except by very rough trails.

The llanos, or grassland plains of central Venezuela provide abundant pasture for stock.

Caracas from the top of the 27-storey Bolivar centre. Since the discovery of oil at Lake Maracaibo after World War II, Caracas has prospered. It now has a population of over 2,000,000.

NETHERLANDS ANTILLES

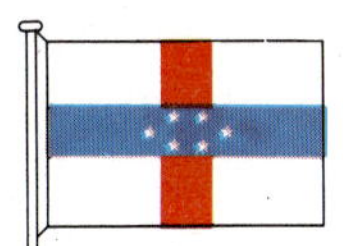

The Netherlands Antilles (Dutch West Indies) comprise two groups of tropical islands. The three main islands of Aruba, Bonaire and Curacao lie just north of western Venezuela; and the islands of St Eustatius and Saba, with the southern part of St Martin (*Sint Maarten*), are in the Lesser Antilles. With Surinam on the South American mainland, they form a part of the Kingdom of the Netherlands. Willemstad in Curaçao is the capital.

The Islands and Their People. The five and a half islands have a total area of 390 sq mi (1,010 km^2) of which 360 sq mi (932 km^2) comprise the three main islands. These are largely low lying, reef fringed and sparsely covered with cacti. They tend to be dry and hot (83–86°F; 29–30°C). The more northerly islands are made of steep-sided volcanic rocks, receive more rain than the southern group and are subject to hurricanes.

With a total population of 228,246 (1971) the islands are crowded. The population is heterogeneous, though predominantly negroid, and speaks Dutch (the official language), English and Spanish. The local dialects are Papiamento and 'Creole' English.

Since 1954 the Netherlands Antilles have had political autonomy within the Kingdom of the Netherlands.

The Economy. Aruba, Bonaire and Curaçao have a precarious economy. There is little land for agriculture and virtually no exploitable resources except calcium phosphates in Curaçao. Droughts and the enforced importation of almost all the islands' needs are further handicaps. The economy rests on the refining of oil imported from Venezuela and Aruba has one of the world's biggest refineries. Such installations require relatively few workers and unemployment is acute. Fishing and tourism could be expanded but they promise only a measure of employment. The northern islands produce sugar-cane and cotton, and St Eustatius conducts an entrepôt trade. Despite their associate membership in the European Economic Community, these Dutch islands face an uncertain economic future.

TRINIDAD AND TOBAGO

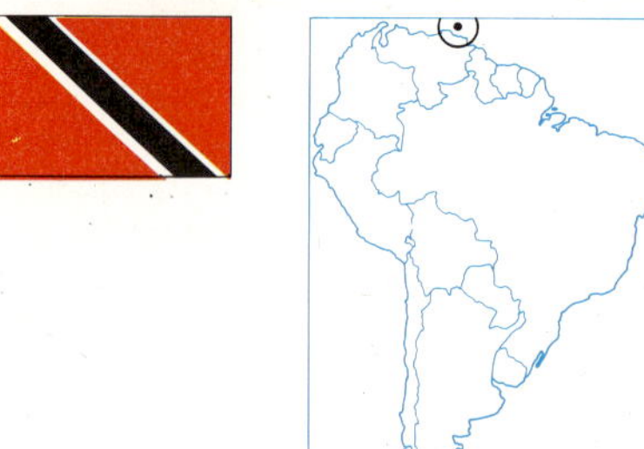

Area: 1,980 sq mi (5,128 km²)
Population: 1,026,750
Population Growth Rate: 2·2%
Capital: Port-of-Spain, pop 68,000
Language: English
Religion: Christian (71%) Hindu (23%) Moslem (6%)
Currency: T and T dollar = 100 cents

Trinidad and Tobago is an independent state in the Commonwealth of Nations. It comprises two small tropical islands and various islets in the Atlantic Ocean off South America. Trinidad, lies 7 mi (11 km) from the mouth of the Orinoco River in Venezuela. Tobago is 20 mi (32 km) north of Trinidad.

Both islands became independent from Britain in 1962. Trinidad's colonial history, associated with the slave trade and the subsequent importation of Indians and Chinese labourers from South and South-East Asia, has influenced its cultural development. A distinctive Trinidadian culture has emerged, characterized by steel band music, the limbo dance, calypso ballads, poetry and folklore. Trinidad has been the scene of racial unrest, in part caused by a high rate of unemployment.

Land and Climate. Trinidad is almost rectangular in shape and has three mountain ranges, separated by synclinal basins drained by three main rivers: the Caroni, Ortoire and Oropuche. Intermittent mud volcanoes occur in the south-west. Tobago consists of a mountainous northern area of volcanic rocks and a southern platform of coral limestone which extends far out to sea and provides magnificent beaches and opportunities for excellent underwater photography and scuba diving.

Lying close to the equator, Trinidad and Tobago has a high average temperature of 80°F (27°C) with a range of 65–90°F (18–32°C). The rainfall varies from 50 in (1,270 mm) in the west to 100 in (2,540 mm) in the north-east. Tropical Trinidad has thickly wooded areas on the east and west coasts, and areas of swamps which are being drained. The three mountain ranges are heavily forested and are a source of much-needed timber.

A steel-drum band in Port-of-Spain Trinidad, just off the north coast of Venezuela.

The People. The population of 1·1 million is racially mixed. The majority are Negroes, descendants of African slaves. Other groups include East Indians and Chinese, descendants of indentured labourers imported to work on the plantations after slavery had been abolished. In addition, there are minority groups of British, French, Portuguese, Lebanese, Syrians and Spanish. The largest settlements are Port-of-Spain, the capital, San Fernando and Arima, all on Trinidad. Most people are Christians, Hindus or Moslems. There is an extensive education system that provides schooling for over 250,000 children.

Agriculture. Subsistence crops are subordinate to cash crops, such as sugar-cane, cocoa, coconuts, citrus fruits, bananas, tobacco and coffee. This emphasis on cash crops for export, however, necessitates the importation of costly foods, including flour, dairy products, meat, fish and rice.

Minerals and Industry. Apart from Jamaica, Trinidad is the only West Indian Commonwealth island with reserves of commercial minerals. Petroleum has been exploited from the turn of the century, and the recent development of offshore wells in the Gulf of Paria increased production so that by 1968 production totalled 67 million barrels. Crude oil is refined locally at Point-a-Pierre and San Fernando. In addition to oil, Trinidad has reserves of natural asphalt. Its famous pitch lake is the world's largest single source of asphalt.

Since independence, the nation has provided incentives to encourage new industries, including the manufacture of foodstuffs, fertilizers, cement, chemicals, tyres and clothing. The islands have a growing tourist industry.

Transport and Trade. Trinidad is well served by main and secondary roads; British, European, American and two inter-island vessels; and a number of airlines. Tobago's facilities are narrower than Trinidad's but include roads and an airfield.

External trade is confined to the export of crude oil, petroleum products, sugar, fruits, vegetables and coffee. Most of this trade is with the USA. Imports include food and various manufactured goods, most of which come from the United Kingdom.

GUYANA

Area: 83,000 sq mi (214,970 km²)
Population: 721,098
Population Growth Rate: 3·0%
Capital: Georgetown, pop 195,000
Language: English. Indian, Chinese and Portuguese also spoken
Religion: Christian (57%) Hindu (34%) Moslem (9%)
Currency: Guyana dollar = 100 cents

Guyana (formerly British Guiana) is the most westerly of the three small countries (Guyana, Surinam and French Guiana) which lie to the east of Venezuela. The British controlled the country from 1874 until 1966 and Guyana became a republic in 1970.

Coast, Forest and Savanna. Guyana has three distinct areas: coast, forest and savanna. The populous coastal lowlands are barely 10 mi (16 km) wide except along the banks of the main rivers where lowland extends inland for 40 mi (64 km). This area is 4–5 ft (1·2–1·5 m) below sea level and is protected by an elaborate system of dams. Inland the land rises gently at first to an area of dense rain forests and mountains. The south-west, the Rupununi area, is characterized by open savanna. The whole country is drained by large rivers: Demerara, Berbice, Essequibo, Courentyne.

Climate. The climate is tropical with an average temperature about 80°F (27°C). Coastal areas are influenced by sea breezes. The annual rainfall varies between 60 and 120 in (1,500 and 3,050 mm). Most of it falls in two marked seasons, April to August and November to January.

The People. The population is very varied, comprising mainly East Indians (377,256) and Negroes (227,091). There are also smaller numbers of Amerindians, Portuguese, Chinese and Europeans. Nearly 25% of these peoples live in Georgetown, the capital (195,250). Guyana's Negroes are descended from slaves shipped in from Africa. After slavery had been abolished, hired East Indians and Chinese were imported to work the sugar plantations. There is considerable rivalry and tension between the present Negro and East Indian communities.

The Economy. Agriculture and mining dominate economic output. Farming occupies only about 1% of the total area and is confined largely to the coastal plain, but it employs and supports much of the population. The two principal crops are sugar-cane and rice; the first a plantation crop, the second a peasant crop. In recent years new lands have been cleared, drained or irrigated.

Guyana is endowed with deposits of bauxite (the second main export product after sugar), manganese, gold and diamonds. Guyana is among the world's top five bauxite producing nations. A seven-year development programme, launched in 1966 was aimed at increasing economic growth to 5–6% per annum and diversifying the economy. Surveys are being made into the development of hydro-electricity and aluminium smelting.

The town of Tumatumari on the Essequibo river. Much of Guyana is still unexplored.

SURINAM

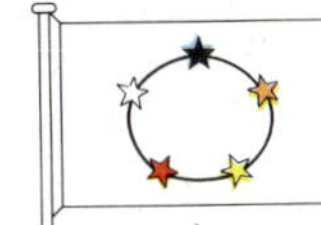

Surinam, an overseas territory of the Netherlands, has frontiers with French Guiana to the east, Brazil to the south and the Republic of Guyana to the west. The Caribbean Sea borders it on the north. First colonized by the British in 1630 it was held periodically by the British and Dutch till 1815 when it passed into enduring Dutch control.

The Land. There are three distinct regions: a coastal lowland area some 350 mi (564 km) long, featuring rivers, canals, sandy ridges and swamps; a savanna area of sandy soils, roughly 30–60 mi (48–64 km) wide, covered with grass and shrubs; and, further inland, the interior highlands, exceeding 4,000 ft (1,200 m), dissected by streams and supporting valuable stands of timber.

The climate is sub-tropical and is modified by the north-east trade winds which are felt throughout the year. The mean annual rainfall is 76–90 in (1,900–2,300 mm). The two rainy seasons are from April to August and from November to February.

The People. The population (about 400,000 in 1970) includes Negroes, the descendants of Africans who were forced into slavery by English and Dutch planters; Chinese East Indians, Indonesians and Japanese, the descendents of indentured labourers brought from South and South-East Asia to work on the estates after the abolition of slavery in 1863; mulattoes (with East Indians comprising the majority of the population); Amerindians; and Europeans. Nearly 40% of the population lives in the capital of Paramaribo (150,000).

The constitutional head of the government is a governor, supported by a Council of Ministers. The legislative council, the Staten, is a representative body which is elected by adult suffrage every four years. In 1954 the country gained autonomy in internal affairs and since then has been moving towards full sovereignty.

The Economy. Plantation agriculture (sugar-cane, coffee and cotton) was introduced in the 17th century but the abolition of the slave trade dealt it a severe blow. Now rice growing occupies nearly 80% of the cultivated land, which is in small holdings and dependent on hand labour and irrigation during the dry season.

However, the basis of the cash economy is bauxite. Surinam's bauxite production is second in world production after Jamaica's. Although much of the bauxite is exported elsewhere for aluminium smelting, the policy is increasingly to do the smelting in Surinam, which possesses the hydro-electric potential for producing the large quantities of cheap electricity that the process demands.

Altogether bauxite, alumina and aluminium contribute nearly 90% of the total exports; the remaining 10% comes largely from timber and wood products. Imports include manufactured goods, foodstuffs, fuels and lubricants. Attempts by the government to encourage industrialization are frustrated by the lack of capital and a limited domestic market.

All-weather roads link Paramaribo with the bauxite areas and the international airport at Zandery, but the road network is thin and there is only one single-line railway. Most inland transport is by water.

The Gran Rio is dammed at Brokopondo to form Lake Blommestein near the former town of Affobakka.

FRENCH GUIANA

French Guiana is bordered by Brazil on the east and south, Surinam on the west, and the Atlantic on the north. It is the smallest of the three Guianas with an area of 35,135 sq mi (90,000 km²), one-sixth that of France. The country depends very heavily on France for economic assistance. In 1946 it became an overseas department of France, having the same laws, regulations and administration as Metropolitan France.

The Land and Climate. French Guiana's relief is akin to that of Surinam and the Republic of Guyana. The land slopes gradually upward from a coastal plain to a narrow zone of savannas and thence to the interior mountains which rise to about 2,600 ft (780 m). The three zones are deeply incised by streams which make east-west movement difficult and north-south penetration is only possible by barges, dugout canoes and other shallow craft. Nine-tenths of the land is blanketed by dense tropical rain forest inhabited by animals, including jaguars, tapirs, sloths, deer and brilliantly coloured forest birds such as macaws and cocks of the rock. Caymans, anacondas and manatees occur in the rivers.

Located only a few degrees north of the equator, French Guiana has an average temperature of 80°F (27°C) and rainfall varying from about 20 in (500 mm) along the coast to over 80 in (2,000 mm) in the highlands.

The People. The entire population totals only about 51,000. Nearly all the people are Negroes, descended from slaves. There are small numbers of Amerindians, Chinese, Indians, Syrians and Europeans. Most of the population is concentrated on the coast. Illiteracy runs at the rate of about 25%.

The Economy. French Guiana remains very largely undeveloped. The coastal waters along the 200 mi (320 km) sea front are stocked with fish but there is little fishing except for shrimps – an activity monopolized by American companies for the American market. By the 1970s, agriculture occupied under 12 sq mi (31 km²) although drainage of the swampy coast lands could open up potentially rich soils for farming. Sugar-cane is the only cash crop of importance; there are subsistence crops of rice, sweet potatoes, manioc and bananas. Farmers also raise pigs, cattle and sheep.

Though some hardwood is exported and there is a limited production of rosewood extract, the vast timber reserves, the largest potential source of wealth, are barely touched. The same is true of mineral resources.

The sea and rivers are the main lines of communication. Surfaced roads total less than 200 mi (320 km). Cayenne is the main port and has an airport nearby.

Timber and shrimps are the main exports. Most trade is with France. There is a persistent trading deficit.

BRAZIL

Area: 3,286,488 sq mi (8,512,004 km²)
Population: 95,408,000
Population Growth Rate: 3·2%
Capital: Brasilia, pop 517,000
Language: Portuguese, Immigrant and Indian languages
Religion: Roman Catholic (over 90%)
Currency: Cruzeiro = 100 centavos

Brazil, dominating eastern South America, is a giant in Latin American affairs. First it is easily the biggest Latin American nation and ranks fifth in the world. Roughly as large as the United States, Brazil is four times bigger than Mexico (its nearest Latin American rival) and occupies about half of all South America, where its borders touch those of 10 out of that continent's other 12 states, and its eastern edge comprises 4,600 mi (7,400 km) of coastline facing the Atlantic Ocean. About half of all South Americans live in Brazil. Favoured parts of Brazil's vast hinterland yield more coffee than any nation on earth and its rocks contain some of the world's richest deposits of iron and manganese. The giant natural wealth of this former Portuguese colony (Brazil gained independence in 1822) has supported the mushroom growth of such cities as São Paulo and Rio de Janeiro. Yet 90% of the people live in only one-third of the land – and few civilized Brazilians venture into the dense tropical forests that clothe huge tracts of the interior where native Amerindian tribes still practise primitive hunting and farming. Wealth as well as population is unevenly spread, and millions of Brazilian peasants subsist at poverty level. Such social problems have stirred up political turbulence that led to authoritarian military rule, beginning in 1964.

With South America's richest potential resources, Brazil seems set for massive economic advance – but only provided its peoples can open up the undeveloped interior and solve their pressing socio-political problems.

The Land. Despite its immense size, Brazil lacks the extremes of altitude which characterize most states of Andean South America to the west. There are four main physical regions.

In the extreme north is the southern edge of the Guiana Highlands.

South of this region, low-lying plains occupy much of the north and are drained by the intricate pattern of rivers which thread their way eastwards through the dense forests of Amazonia.

The plateaux and mountains of the centre and east comprise the country's third main region – and the one with most of the population. This includes the great 'shoulder' of Brazil, whose Cape São Roque juts into the ocean to within 1,800 mi (2,900 km) of West Africa. Here there is a narrow coast plain which broadens to about 60 mi (100 km) along the valley of the lower São Francisco River but narrows considerably farther south, and locally disappears altogether. Steep coastal escarpments characterize this portion of Brazil, rising abruptly from the sea to heights of about 3,000 ft (915 m). In parts of the south-east, slopes continue climbing steeply to reach an altitude of over 9,000 ft (2,740 m) in the Serra da Mantiqueira, only 60 mi (100 km) behind the coast at Rio de Janeiro. This is the highest range in the whole of Brazil, but summits lie below 10,000 ft (3,050 m) and are thus low compared with those of the Andes. East of this mountain rim, a great, gently-tilted plateau block, averaging between 1,200 and 3,000 ft (370–900 m), slopes gradually westward towards the centre of the continent. This old, hard, crystalline shield-block is mantled locally with sandstones, limestones and sheets of basalt. These rocks, together with the more resistant outcrops of crystalline rock, form a series of ranges known collectively as the Brazilian Highlands, which rise above the general plateau level to reach altitudes of 5,000–6,000 ft (1,525–1,820 m). But the ancient basement block's generally westward dip remains the dominant factor and results in its eventual

Rio de Janeiro by night. Rio is the most beautiful and most famous city of South America. Its site was discovered by Vespucci on 1 January, 1502.

burial beneath the enormous silt spreads of the Amazon and Paraguay rivers in the continental interior.

Brazil's fourth region comprises the basins in southern Brazil through which flow the Paraná and Paraguay rivers and their tributaries.

The course or alignment of many Brazilian rivers has made them unsuitable as waterways to be reached from the sea. Swift, unnavigable streams descend the eastern escarpments. Others, rising on the western slopes of the eastern mountains, flow towards the interior, pursuing long, circuitous journeys of several thousand miles before reaching the sea. The Paraguay and Paraná rivers, the latter interrupted by several spectacular falls, leave Brazil to enter Paraguay and Argentina. The Amazon, however, is essentially a Brazilian waterway and remains navigable throughout its almost 2,000 mi (3,200 km) in Brazil. There, this mighty river receives tributaries from the Andes, the Guiana Highlands and the Brazilian Plateau (for instance the rivers Negro, Madeira, Tapajós, Xingu and Tocantins). Their combined waters drain more land than any river system in the world and no river on earth transports as much water as the Amazon does.

Climate, Vegetation and Wildlife. Contrasts in both climate and vegetation are more frequently the results of variation in amounts and seasonal distribution of rainfall than of any striking inequalities in temperature. Although average annual temperatures increase steadily from the sub-tropical and tropical south towards the equatorial north, central Amazonia (lying almost astride the equator) experiences mean annual temperatures of only 80–82°F (27–28°C), with so little variation that the region has one of the most monotonous climates in the world. Much of the east coast maintains somewhat similar temperatures: 75–80°F (24–27°C) averages are found as far south as Rio de Janeiro and Santos. Still farther south, towards the boundary with Uruguay, slightly cooler conditions yield pleasant annual means of 62–66°F (17–19°C), still without any very pronounced seasonal extremes. Lower temperatures also prevail in the Brazilian Highlands, where average annual temperatures of 64–70°F (18–21°C) are found. Frost can occur in the south.

Temperatures become more extreme wherever rainfall, cloudiness and humidity decrease. More than 80 in (2,032 mm) of rain falls annually in lower Amazonia, on parts of the east coast, and in the Brazilian Highlands. Temperatures tend to soar, however, in regions experiencing a pronounced dry season, thus often aggravating drought conditions.

Brazil's notorious 'drought pocket' lies in the country's extreme eastern shoulder, especially in the states of Ceará and Paraíba where the dry season is most pronounced in winter (May-October), but which also suffers long, severe and unpredictable summer droughts. Annual rainfall here is as little as 20–25 in (510–640 mm). A winter dry season belt extends right across central Brazil.

Botanists have identified several major vegetation zones in Brazil. Selvas or dense tropical rain forests with a bewildering variety of evergreen broadleaf trees cover the rainy Amazon basin and wet coastal areas. Forests of mixed deciduous and evergreen trees line the south-east coast in areas with seasonal rainfall. Deciduous thorn trees and cacti are typical of the dry north-east. Woodland savanna covers much of the Brazilian interior south of the Amazonian selvas. Savanna with palm trees occurs in the seasonally flooded plain of the River Paraguay. Open prairies and pine-dominated forests are found in the extreme, frost-prone, south.

Apart from jaguars and tapirs, Brazil is poor in large mammals. But it is rich in colourful birds such as macaws, toucans and humming-birds; in reptiles, including anacondas, iguanas, caymans and turtles; and in insects of which big beautiful butterflies are the most spectacular.

The People. The mainly Roman Catholic and Portuguese-speaking population of over 93,200,000 is made up of chiefly European stock. Peoples of Portuguese ancestry predominate but there are large German, Italian and Spanish communities, and significant Japanese and Lebanese minorities. Mulattoes (people of mixed Negro and European stock) and a small number of mestizos (people of mixed Amerindian and European stock) account for 25%, and Negroes for 15%. Slightly over half the population live in towns and cities. There are large numbers of migrants or *cabocles* who provide casual labour in rural areas.

The Brazilians have generally settled in clusters along the eastern seaboard, forming dense rural and urban concentrations in the south and south-east, particularly in the states of São Paulo, Minas Gerais, Paraná and Rio Grande do Sul. Over 90% of the total population live in only one-third of the total national territory, and it is the attempts to open up and develop the interior (the *sertão*) which have prompted the government's current highway and regional development programmes.

The practice of establishing new cities on the edge of the Brazilian wilderness has in fact long been part of the country's development strategy. But the decision to build a new federal capital (Brasília) was the most significant single attempt to draw population away from the overcrowded coastal fringe into the underdeveloped central plateau. The new capital, replacing Rio de Janeiro and inaugurated in April 1960, has become a world-famous example of modern architecture and town planning. Whether Brasília becomes the centre of a new agricultural or ranching frontier area remains to be seen; but signs are encouraging and the population of the Federal District numbered over 500,000 by the early 1970s.

Brazil's living standards are locally variable. In the south-east and south, they are better than those of the poorest European countries, but elsewhere they are worse. Even wealthy Rio de Janeiro is backed by squalid hillside slums. In rural areas malnutrition and disease keep life expectancies low, and in spite of educational improvements less than half of all school-age children were being taught in the late 1960s. But Brazil has more than 20 universities and one of the best records of cultural achievement in Latin America.

Agriculture in an Evolving Economy. Brazil is still predominantly an agricultural country, despite the rapid advances made in manufacturing production. Slightly over half the population remains directly dependent upon the land, and many have remained subsistence farmers and graziers. Significantly, nearly 80% of Brazilian exports are at present still derived from the agricultural sector, so that this remains a vital aspect of the national economy. The principal crops are coffee, cacao, sugar-cane, tobacco, maize, rice and tropical fruits, and in all of these commodities Brazil ranks as a major world producer.

The present agricultural pattern can be partly traced back to the 16th century when Portugal established coastal Brazil as a plantation colony. Timber cutting and sugar planting sustained a number of early settlements strung along the east coast, particularly around Recife (Pernambuco), Salvador (Bahia), and São Vicente, near the present site of Santos. Planters began importing Negro slaves from West Africa as early as 1538, while overseas markets for sugar rapidly expanded. As the hot, humid coastal belt of north-east Brazil became the world's major sugar-cane producing area in the 17th and early 18th centuries, it became also the core of colonial Brazil, with its prosperous land-owning planters. Sugar-cane remains a major product and today the country is the world's second-largest producer of sugar-cane.

In the 19th century another agricultural commodity – coffee – assumed great importance in the Brazilian economy. Coffee plantations (*fazendas*) began to spread rapidly behind Rio de Janeiro and São Paulo after 1830, as thousands of square miles were planted out on the rolling, frost-free plateau country in the states of São Paulo and Paraná. As world demand for coffee soared in the later 1800s, the great Brazilian coffee boom gained momentum. The railway network spread quickly across the plateau through the coffee-growing areas and the city of São Paulo became the world's coffee capital.

In terms of quantity Brazil has remained unrivalled as a world producer of coffee, and despite increasing competition Brazil still supplies a quarter of total world production. The government, however, through the Brazilian Coffee Institute, is encouraging low-yield, disease-prone areas to be taken out of production and replanted with more profitable crops such as cotton, maize, peanuts, citrus fruits and vegetables.

A second great economic boom based upon a plant-derived commodity proved relatively short-lived. It occurred in the remote equatorial rain forests of Amazonia which were found to be the world's only important source of wild rubber during the 19th century. But after 1913, plantation rubber from South-East Asia came into commercial production from trees grown from rubber seeds that had been smuggled out of Brazil. Crushed by Asian competition, the Brazilian wild rubber boom collapsed. Today, Brazil produces less than 1% of the world's rubber.

Cacao and cotton production have both shown substantial increase during the present century. Cacao requires hot, wet conditions, and 90% of Brazil's cacao comes from the narrow, heavily-forested coastal strip in the southern part of the state of Bahia. Brazil is now the world's third-largest cacao producer.

The hot, moist conditions along this coast have also encouraged large-scale banana planting. Brazil is the world's largest producer of bananas, and second only to the USA in the output of oranges. Cotton production is heavily concentrated farther south in São Paulo state, although cotton is also important in the north-east (both for its fibres and for cotton-seed cattle-cake production).

In addition, Brazil remains one of the world's greatest ranching countries (ranking as the third-largest producer of beef), with almost twice as many cattle as Argentina. Cattle are raised in large numbers on the open ranges of the interior, and many are brought closer to the urban markets for fattening on improved pastures before slaughter. Loans from the World Bank are at present assisting Brazil in its attempts to develop and improve beef, mutton and wool production.

Forestry. Over half of Brazil is forested and Amazonia has the world's largest concentration of tropical hardwoods. These include many species of mahogany, rosewood, walnut, acacia and laurel, and an immense number of unique hardwoods. Labour shortages and the difficulty of achieving profitable access to major markets have always limited the development of any significant lumber industry in the northern forests.

The dense, eastern coastal hardwood forests are cut more heavily for the domestic market. In terms of the export market, however, at present only the mixed southern temperate forests of Paraná pine (*Araucaria*) are systematically worked.

Fisheries. Despite an extensive coastline Brazil is not among the major fishing countries of the world. Sovereignty over fishing rights was extended to 200 mi in 1971, however, as part of a new federal programme designed to strengthen the fishing industry.

Minerals. Intensive exploitation has long been hampered by inadequate transport, limited ground survey, the lack of cheap power, and distance from the manufacturing centres. Nevertheless, it is now clear that Brazil possesses one of the world's greatest storehouses of minerals. Iron ore is found in the state of Minas Gerais, notably at Itabira where estimated reserves of 35,000 million tons are among the largest known in the world. There are rich deposits of iron and manganese ores at Urucúm near Corumbá in Mato Grosso; of manganese in Amapá; and of copper in Bahia. Chrome, tungsten, silver, lead, mica, zirconium, titanium, beryllium, bauxite (for aluminium) and asbestos (Brazil ranks third in world output) are all produced. Coal occurs widely but quality is poor and, as in the case of petroleum, Brazil has had to rely heavily upon imports. Gold is found in nearly every state, although Minas Gerais is the major producer. In addition, Brazil is important for diamonds, especially industrial diamonds (in Minas Gerais, Goiás and Mato Grosso). Semi-precious gem stones of excellent quality, such as aquamarine, topaz and amethyst, also abound, giving rise to a richly varied jewellery industry.

Manufacturing. In recent years Brazil has been rapidly consolidating and expanding its role as Latin America's major industrial power. Two-thirds of Brazil's manufacturing production lies within the small south-eastern triangle formed by Belo Horizonte, São Paulo and Rio de Janeiro. Steel production began in 1946 at a new plant built with financial and technical aid from the USA. Sited in the Paraíba valley, within relatively easy reach of Rio de Janeiro and São Paulo, the state-owned Gilherme Guinle Plant, the largest in Latin America, makes the greatest single contribution to Brazil's annual steel production figure of 6·2 million tons (1970). An annual figure of 20 million tons was forecast by 1980. Coal imports supplement the rather meagre domestic supplies, while charcoal (now mostly from planted eucalyptus) is used in place of coke in some of the smaller iron and steel plants. The manufacture of textiles, especially cotton cloth, vehicles, tyres, cement, paper, domestic hardware, chemicals and fertilizers is also important within the general framework of the economy. Petro-chemicals are a particularly significant growth industry.

Future industrial expansion depends largely on the impressive hydro-electric power project currently underway on the Paraná river system. This waterway has enormous power potential, and Brazil has ambitious plans to construct what amounts to a 700 mi (1,100 km) chain of dams and reservoirs to provide power in the upper valley for mining, manufacturing, and new town development in the states of São Paulo, Minas Gerais, Paraná Grosso.

The harnessing of the Urubupungá Falls is centred on two key plants – Jupiá (final phase 1,400 mw) which began production in 1969, and Ilha Solteira (final phase 3,200 mw), scheduled for first-phase operation in 1973. The major left-bank tributaries considerably extend the range and power potential of the Paraná, particularly the Grande and Tietê rivers, which are significantly located in relation to Brazil's major manufacturing and residential centres.

International Trade. Coffee accounts for about 40% of the export earnings while cotton, cocoa, sugar, pine-wood, tobacco, fibres, nuts, vegetable oils, together with iron and manganese ores are all significant. Recently, the export of Brazilian manufactured goods has begun to increase. The USA, West Germany, Argentina, Italy and the Netherlands are the principal markets for agricultural goods. Imports of machinery and components, fuels and lubricants, chemical products, heavy land transport, and wheat come mainly from the USA, West Germany, Argentina, Venezuela and Japan.

Transport. Dense forests, steep gradients close to the eastern seaboard, ravines, river barriers and the lack of population centres in the Brazilian interior have all limited the development of a nationally integrated system of land transport. Over large areas, air services make the only regular contacts between the principal towns and the more remote, outlying centres.

But in the north, the Amazon forms the world's largest system of inland navigation in what is one of the world's most thinly

A coffee worker returning from hoeing. The rich red soil of the Parana valley is ideal for the crop. Coffee is Brazil's major export.

populated regions. Iquitos in Peru, 2,300 mi (3,700 km) from the Atlantic, lies at the western head of navigation, while ocean-going vessels can reach Manaus, 1,000 mi (1,600 km) upstream.

Brazilian railways, totalling some 15,850 mi (25,000 km) and operating over tracks of five different gauges, only link restricted sections of the hinterland with the coast. Large sums have recently been allocated to nationalization and modernization of the railways.

Highways are now penetrating the central and northern areas to effect the country's first primary road network. The capital city, Brasília, is linked to Belém and Santarém on the Amazon, while the Trans-Amazonian Highway thrust through almost uninhabited and unexplored portions of northern-central Brazil to the Peruvian frontier, promises to establish the first trans-continental highway link between Lima and the major Brazilian cities and Atlantic ports.

What the results of this bold and expensive venture will be in terms of economic and demographic development remains to be seen. Long-distance express bus services, however, are already playing a significant part in linking north-central Brazil to the more densely populated areas farther south and east.

Dependencies. Brazil owns several small South Atlantic islands: Fernando de Noronha (comprising a territory), Rocas, and remote volcanic St Peter and St Paul Rocks, Trinidad Island and the Martin Vaz Islands.

ECUADOR

Area: 109,483 sq mi (283,561 km²)
Population: 6,508,000
Population Growth Rate: 3·4%
Capital: Quito, pop 551,163
Language: Spanish
Religion: Roman Catholic
Currency: Sucre = 100 centavos

Ecuador (Spanish for 'equator') takes its name from its position astride the equator on South America's Pacific coast. The mighty Andes Mountains dominate this third-smallest among South American republics (which is nonetheless rather larger than West Germany).

Long divided against itself by geographical and human barriers, Ecuador has never realized its full economic potential. To some extent its backwardness is the legacy of Spanish colonists, progenitors of a small European ruling class that despised the Indian and mestizo majority most of whom live at poverty level. Also to blame is rivalry between the state and the powerful Roman Catholic Church; and between Ecuador's two great cities, Quito and Guayaquil. Such sectional divisions are one reason why Ecuador had the second-lowest per capita income among all South American republics during the 1960s. Ecuador's eastern region, the *Oriente*, vast and largely unexplored, may hold the keys to a brighter economic future, for encouraging oil discoveries have been reported.

The Three Regions. The Andes, running for 400 mi (640 km) north to south, forms the country's mountainous backbone (the *Sierra*). Its two parallel chains, the *Occidental* (Western) and *Oriental* (Eastern) cordilleras, contain over 30 volcanoes (some still active), many of them among the highest and most beautiful cones in the world. Lava flows, ash falls or volcanic dust coat much of the mountainsides.

Between the Western and Eastern cordilleras runs an irregularly-collapsed trough, 25–40 mi (16–25 km) wide. The edges of this trough are joined at intervals by a series of cross-ranges, like the rungs of a ladder, and between each 'rung' lies an intermontane basin.

A second region, the coastal lowland, 20–100 mi (32–160 km) wide and with hills rising to 2,500 ft (760 m), lies between the Andes and the Pacific. The main coastal embayment is the 100 mi (160 km) wide Gulf of Guayaquil, at the head of which lies the great delta of the Guayas River. The Guayas lowland, extending north from this delta, is the most fertile part of the coastal region.

About one-third of Ecuador, the *Oriente*, lies east of the Andes. Here the land falls away from the Andes, at first steeply, then gently as it merges into the plains of the Amazon basin.

Climate, Vegetation and Wildlife. Sharp contrasts in climate and vegetation are found with the Andes. Temperatures fall from a uniform 75°F (24°C) or so at 3,000 ft (910 m) to below freezing on the highest snow-clad peaks. Perpetually spring-like conditions prevail at Quito, almost on the equator, but at a height of 9,400 ft (2,860 m). There is a single rainy season (November-May), with an average rainfall of about 60 in (1,520 mm). Mountain forest covers many of the wetter slopes to 10,000 ft (3,050 m). Above this altitude grasses and shrubs extend to the snow line.

The relatively short hot coastland of Ecuador incorporates the most dramatic changes in climate and vegetation found anywhere in Pacific South America. In the north, two wet seasons yield nearly 80 in (2,030 mm) of rainfall which supports dense tropical rain forest and mangrove swamp. Farther south rainfall diminishes, supporting only thorn forest and savanna grassland. The extreme southern coast is an almost rainless desert.

The *Oriente* is always hot and humid with temperatures around 80°F (27°C) and 80 in (2,030 mm) of annual rainfall. Tropical rain forest covers the area.

Ecuador's larger animal species include the jaguars and tapirs of the *Oriente* region. There are some 1,500 bird species including winter migrants from North America, and a host of insects including giant and phosphorescent beetles.

The People. The percentage of Amerindians in the population may be as high as 60%. Some are directly descended from the Incas who once ruled the region. There is also a large percentage (perhaps 40%) of mestizos. The white elite, accounting for 1 in 10, are chiefly descended from Spanish families and live in the Andean region. Negroes and mulattoes, concentrated on the coastal plain, also account for 1 in 10. The population in general is somewhat denser in the nine main intermontane basins of the Andes than on the coastal plain.

The *Oriente* contains less than 2% of Ecuador's population. Semi-nomadic Indian tribes including the feuding, head-hunting Jivaros inhabit the forests.

The two chief cities are highland Quito and coastal Guayaquil. The capital, Quito (551,163) was the northern capital of the Inca Empire conquered by the Spaniards in the 16th century, and part still retains the charm of an old colonial town, with its steep, narrow, lantern-lit streets, stone or adobe houses, huge wooden doors, barred windows and richly-carved balconies. Hot, humid and prosperous Guayaquil (835,812), the chief seaport and largest city in Ecuador, is composed mainly of European and mestizo elements and displays the more bustling, outward looking commercial side of Ecuadorean life.

Despite Guayaquil's relative prosperity, the national standard of living is low, with many Andean Indians still living in painted stone or mud huts. Infant mortality and malnutrition are rife and illiteracy remains widespread in spite of theoretically compulsory education.

Government. Ecuador is a presidential republic with a Congress comprising a Senate and a Chamber of Deputies; but army juntas have repeatedly seized power. The country is divided into 19 provinces and the National Territory of the Galápagos Islands.

An Agricultural Economy. In the Andean intermontane basins, subsistence crops of potatoes are grown by the Indians up to nearly 12,000 ft (3,660 m). At lower levels farmers produce wheat, barley, maize, vegetables, and fruits, and raise livestock.

The coastal region is dominated by commercial banana production, but there is enormous variety in what is one of the richest agricultural areas in the humid tropics. Cacao, coffee, rice, sugar-cane, manioc, cotton, pineapples, citrus fruits and livestock augment banana production. Banana cultivation by intensive methods was developed in the 1930s and increased rapidly after World War II. Cacao production revived in the 1960s and coffee is successful on the Andean flanks and coastal hills.

An agrarian reform and colonization law was passed in 1964 in an attempt to redistribute land, but progress has been slow.

Forestry and Fishing. Nearly three-quarters of Ecuador is forested, but the superb hardwood resources of the *Oriente* suffer heavily from problems of transport, labour shortage and inaccessibility. In the coastal ranges, shredded fibre from a local tree in Manabí province is used to manufacture the famous but misnamed Panama hats. Lightweight balsa wood, kapok, and castor oil seeds are among other forest products.

With Peru and Chile, Ecuador shares the rich fishing grounds of the south-east Pacific.

Minerals and Manufacturing. Oil now holds the key to future development of the *Oriente* – over a dozen new oil fields have been discovered there since 1964. A trans-Ecuadorian pipeline (318 mi: 508 km) will feed much of the oil to the Pacific port of Esmeraldas, and petroleum may well become Ecuador's leading export. The output of other minerals is generally small.

Manufactures are limited, but include cotton and woollen textiles, hats, shoes, folk-art objects, ceramics, tyres and cement. There is also oil-refining, pyrethrum extraction, sugar-refining, and the processing of other foods.

Transport and Trade. Transport systems are largely restricted by the Andes. There are around 5,000 mi (8,050 km) of all-weather roads, and 800 mi (1,290 km) of railway. River transport is important east and west of the Andes. Guayaquil handles over three-quarters of Ecuador's international trade.

Bananas, coffee and cacao dominate the export economy, and Ecuador is the world's largest exporter of bananas, producing nearly 25% of the total supply. Other exports include rice, balsa wood, pyrethrum and pharmaceutical products, straw hats and fish products.

Dependencies. The Galápagos Islands, about 600 mi (960 km) west of Ecuador, are a territory of some 15 large and many small islands totalling 3,075 sq mi (7,964 km²). They are volcanic in origin with arid shores but moister, forested peaks. Only five are inhabited and the total population is less than 4,000.

Their giant tortoises, marine iguanas, Darwin's finches and other species form a unique wildlife assemblage that helped to inspire Charles Darwin's theory of evolution when he visited the islands in 1835.

PERU

Area: 496,225 sq mi (1,285,223 km²)
Population: 14,015,000
Population Growth Rate: 3·1%
Capital: Lima, pop 2,415,700
Language: Spanish. Quecha and Aymara also spoken
Religion: Roman Catholic
Currency: Sol = 100 centavos

Peru is an Andean state bordered by Ecuador and Columbia to the north, Brazil and Bolivia to the east, and Chile to the south. It is the third-largest country in South America, exceeded in size only by Brazil and Argentina.

Peru was once the centre of the Inca Empire which stretched from northern Ecuador south into Chile and Argentina. Peru's Spanish colonists, coming in the early 16th century, gained independence from Spain in the 1820s. But damaging wars, political turmoil and resistance to social reform by the rich landed minority hampered economic development. Even now that Peru's great mineral and fishery resources are being strongly exploited, the low national expectation of life reflects widespread poor living conditions.

A Land of Three Regions. From west to east Peru features three main natural regions: the *Costa* (the narrow coast zone), the *Sierra* (the Andes which dominate most of Western Peru), and the *Montaña* (the eastern mountain valleys and the great interior plains of Amazonia).

The *Costa* is a narrow ribbon of desert that faces the Pacific along the entire length of Peru, a total of about 1,400 mi (2,230 km). In the extreme north, it forms a dune-covered plain some 50 mi (80 km) wide, backed by the Andes. Farther south this lowland narrows considerably. From Trujillo, southwards, the coast plain is largely absent and the harsh, barren slopes of the Andes, or a fronting coastal range, rise steeply out of the ocean in a series of ledges and coastal plateaux. In the central section, particularly, the mountains are slashed by valleys at whose mouths large alluvial fans of silt and rubble provide economically valuable strips of lowland along the coast.

The *Sierra* forms about 28% of Peru. This part of the Andes consists of a series of high parallel mountain ranges divided by deep valleys, except in the south where outer Andean ranges enclose a broad plateau containing Lake Titicaca (South America's largest lake) which extends south-eastwards into Bolivia. This plateau lies at an altitude of over 12,000 ft (3,650 m) but there are seven Andean peaks that exceed 19,000 ft (5,790 m). Mt Huascarán in west-central Peru has the nation's highest summit at 22,205 ft (6,768 m). Some of the Andean peaks are volcanic cones, among them El Misti, a beautifully symmetrical snow-capped peak at 19,240 ft (5,835 m) overlooking the city of Arequipa in southern Peru.

The *Montaña* (eastern Andean valleys and the eastern lowlands) comprises nearly two-thirds of Peru. The eastern slope of the Andes has been deeply carved into a complex series of gorges by some of the principal headwaters of the Amazon, including the Marañón, Huallaga, and Ucayali-Urubamba rivers. In northern and central Peru, these forested eastern border valleys form a broad, inaccessible region. In north-eastern Peru, the valleys give way to a great lowland plain ribboned by meandering rivers of the Amazon system.

Climate, Vegetation and Wildlife. The equator almost touches northern Peru but much of the country has a far from equatorial climate. The Peruvian coast is largely cool, cloudy but rainless – a result of the cold north-westward flowing Peruvian or Humboldt Current offshore. Moist air coming in off the sea is cooled by this current and the atmospheric moisture condenses as heavy mist or fog (*garúa*), or as low cloud which yields no rain. Occasionally the cold Peruvian Current is displaced by the warm southward flowing *El Niño* Current which produces unstable air conditions resulting in heavy rain.

Temperature falls with increasing altitude in the Andes, Cuzco, at 11,000 ft (3,350 m) averages 51°F (11°C) and snow lies permanently above about 16,500 ft (5,030 m) where average temperatures are below freezing. Most precipitation occurs between October and April.

The *Montaña* area has a warm humid climate. Temperatures average around 78°F (26°C) and rainfall may exceed 80 in (2,030 mm) a year.

Peru's vegetation patterns follow the climatic zones. The dry Pacific coast is barren. Drought-resistant shrubs and grasses grow on the relatively dry slopes of the western Andes below the snow line. Tropical rain forest covers the lower *Montaña* area and the relatively warm wet eastern Andean slopes support mountain rain forest to an altitude of over 10,000 ft (3,050 m).

Peru's barren coast provides nesting sites for pelicans, cormorants and other seabirds preying on the immense fish shoals associated with the Peru Current. Condors and guanacos range the high mountain slopes. The *Montaña* is rich in tropical wildlife including jaguars and macaws.

The People. The racial composition of the country is dominated by Amerindians (46%) and mestizos (42%) with a small but important white section (12%). About 60% of the population live in the *Sierra*. This includes the majority of the rural Indian communities distributed in scattered clusters over the high plateau and in the more sheltered basins and valleys, as well as those around the principal mining centres. Arequipa (250,000) is the largest Andean city. To the east, about 13% of the population may be found within the deep valleys of the Andean slope. The Amazon river port of Iquitos, the only major city in the eastern lowlands, now contains about 100,000 people.

The coastal desert oases contain fewer people than the high *Sierra* (about 27% of the total) but nevertheless represent the economic and commercial heart of Peru. Three of the country's four largest cities are found there: Lima, the capital (2,400,000), its port of Callao, and Trujillo.

Almost all Peruvians are at least nominal Roman Catholics. Education is theoretically compulsory between the ages of 7 and 16 but there are too few schools and most Peruvians remain illiterate.

Fisheries. Peru has recently emerged as the world's leading fishing nation by exploiting the immensely rich fish resources of the offshore Peruvian Current. Tuna, bonito and shrimp fishing were established before the 1950s. Since then there has been an immense increase in fishing for anchovies, used in the manufacture of fish meal (a protein-rich poultry and livestock feed concentrate in demand by North American and European markets). Peru now produces nearly half the world's fish meal, and its annual fish catch of 8–12 million tons is roughly one-seventh of the world total. To protect its fisheries, Peru claims territorial waters extending 200 mi (320 km) offshore.

Agriculture and Forest Resources. The 'gardens' of Peru lie, paradoxically, within the coastal desert zone. Only 10 permanent streams cross the desert to reach the Pacific; over 40 other rivers peter out on the way and dry up completely between July or August and October. Irrigation, however, transforms these otherwise barren tracts into dazzling green ribbons of intensive agricultural production, especially in northern Peru. Cotton, sugar-cane, rice, vines, olives, maize, alfalfa, avocados, bananas, citrus fruits and vegetables are staple products of these coastal oases.

High above, in the *Sierra*, potatoes are grown by Indian communities at altitudes around 14,000 ft (4,270 m); beans and cereals, mostly maize, wheat, barley and millets, thrive in more favourable zones a few thousand feet below. The sheltered Cuzco basin at 11,000 ft (3,350 m), heart of the old Inca Empire, and the Lake Titicaca region are two areas of dense agricultural settlement. The grazing of sheep, llamas, and alpacas is widespread, both on the high rough pastures below the snow line, and on any other steep, stony or very dry slopes which are unsuitable for cultivation.

Most of the land has for centuries been divided into the large estates (*haciendas* or *comunidades*) owned by a small land-owning elite. In June 1969, however, Peru introduced a new agrarian reform law involving proposals for widespread land expropriation and redistribution. Real progress is likely to be slow; nearly 60% of the population work on the land although only 3% of the total area is cultivated, and food imports are a major burden on the economy. Colonization of the eastern valleys has been encouraged at intervals over a long period, but development has always remained limited.

More than 60% of the country is forested. North-eastern Peru is densely covered with a variety of tropical hardwoods but exploitation is difficult.

Minerals. Peru is the world's third most important silver producer. The centre of Peru's mining area is at Cerro de Pasco, over 14,000 ft (4,270 m) up in the Andes, 110 mi (177 km) north-east of Lima. In addition to silver, copper, zinc, lead, gold, bismuth, vanadium and coal are all recovered by mining or open-cast working in this extraordinarily mineral-rich zone.

There are large reserves of low-grade copper ore in the coastal desert area of southern Peru, and iron ore is mined at Marcona, about 300 mi (480 km) south of Lima. The (nationalized) petroleum industry lies in the extreme north, along the coast north and south of Talara. It is hoped that Peru may also share fully in the promising future development of oil from the east Andean slope and piedmont zone. Oil has been discovered in the Ucayali valley, and also close to Lake Titicaca. Tungsten, phosphate and potash deposits add to Peru's richly varied mineral resources. Peru annually recovers some 54,000 tons (55,000 metre tons) of guano (seabird excrement used as fertilizer) from its offshore islands.

Industry. Manufacturing industries are concentrated overwhelmingly in the Lima-Callao area. Plants produce steel, automobiles, tyres, petro-chemicals, paint, cement, fertilizers, cotton and woollen textiles, shoes, and a wide range of consumer goods. The assembly, processing and re-packaging of imported bulk manufactured material is an important activity. Mineral smelters, petroleum and sugar refineries, and the dramatically expanded fish meal factories along the coast form part of the manufacturing sector. The fishing industry has also stimulated a new interest in nylon net making and in shipbuilding at Chimbote.

Problems of Transport. Economic progress is hampered by difficulties of transport. Steep gradients, earthquake damage, landslides, washouts, areas of almost unbroken forest, and the virtually uninhabited tracts separating the main centres of population all combine to present a daunting task to the engineer, and a serious barrier to national cohesion.

There are only 2,200 mi (3,540 km) of railway in Peru. Short tracks link several of the irrigated desert oases to their nearest ports. Two lines (both built with guano revenue) penetrate the interior more deeply to link the coast to the *Sierra*.

One of these is the famous Central Railway which climbs to nearly 16,000 ft (4,870 m) to link Lima with towns in the central Peruvian Andes.

The other links the south coast port of Mollendo with Lake Titicaca and Cuzco.

The main feature of the (mainly unpaved) road system is the Peruvian stretch of the Pan-American Highway (2,121 mi; 3,413 km) which runs through the coastal desert centres, and has Andean road links with Arequipa, Puno and Bolivia. The trans-Andean part of the transcontinental Rio de Janeiro-Lima highway is scheduled to be completed in the 1970s.

Peru's remote Amazonian lowlands depend for transport largely upon the River Amazon – an outlet leading into Brazil. But Iquitos, at the head of Amazonian navigation for ocean freighters, lies 2,300 mi (3,700 km) from the Atlantic Ocean. The nation's main oceanic port is Callao on the Pacific coast.

Airways perform an indispensable role in maintaining rapid contact between otherwise isolated regions. There are also international air services from Lima to the rest of South America.

The Inca stronghold of Machu Picchu was not discovered until 1911 and then only as a result of a casual remark by a local Indian.

International Trade. Peru's principal exports (mainly to the USA, Japan and West Germany) are copper, fish and fish meal, silver, iron ore, sugar, cotton, lead, zinc and coffee. Imports include machinery, foodstuffs (mainly cereals), fats and oils, vehicles and transport equipment, chemicals, clothing and metal products.

BOLIVIA

Area: 424,165 sq mi (1,098,587 km²)
Population: 5,062,500
Population Growth Rate: 2·6%
Capital: La Paz, pop 562,500
Language: Spanish and Indian (Quechua, Aymara) dialects spoken
Religion: Roman Catholic
Currency: Peso Bolivian = 100 centavos

Bolivia, the fifth largest republic in South America, is a landlocked country bordered by Chile and Peru on the west, Brazil on the north and east, and Paraguay and Argentina on the south. It is named after Simón Bolívar (1783–1830), 'The Liberator', hero of South America's struggle for independence from Spain: Bolivia's own independence was proclaimed in 1825. With a long history of wars and recurring military coups, Bolivia is one of the poorest countries in South America.

Highlands and Lowlands. Bolivia, a geologically young country of the Andes, has two contrasting regions: the highlands (the Andean ranges and high plateau), and the lowlands stretching eastwards from the Andean foothills.

The Andes are at their widest (about 400 mi; 650 km) and most complex in Bolivia, and cover about a third of the country. The Western Cordillera, separating Bolivia from Chile, has mountains of 19,000–22,000 ft (5,800–6,700 m) and a number of active volcanoes. The snow-capped peaks of the Eastern Cordillera (Cordillera Real), about 100 mi (160 km) to the west, include Sajama (21,391 ft; 6,520 m), Illimani (21,201 ft; 6,462 m) and others of more than 20,000 ft (6,000 m). Between the two cordilleras is the *Altiplano*, much of it 13,000 ft (4,000 m) above sea level. At the northern end, shared with Peru, is Lake Titicaca, drained southwards by the Desaguadero River into shallow Lake Poopó. Titicaca is South America's largest lake and the world's highest stretch of navigable water. Farther south the *Altiplano* becomes a dry and barren tableland glistening with huge salt flats called *salares*.

Beyond the Eastern Cordillera, in central Bolivia, the Andes are a jumble of ranges enclosing fertile valleys and basins. In the north, the Eastern Cordillera falls sharply to the plains, its slopes carved by narrow valleys and gorges called *Yungas*.

From Cochabamba southwards, the Eastern Cordillera rises above the *Altiplano* to the eastward-sloping *Puna*, a bleak and almost treeless area 13,000–15,000 ft (4,000–4,600 m) above sea level. From the Andean foothills, the rivers descend rapidly and then wind sluggishly and silt-laden across the lowland two-thirds of Bolivia, either northwards to the Amazon or southwards to the Paraguay river systems.

Climate, Vegetation and Wildlife. Climatic regions range from cold windswept mountains to steaming hot tropical jungles and swamps. The rainless belt of coastal Peru and northern Chile extends south-eastwards across the Western Cordillera and the south Bolivian *Altiplano*. Rainfall, about 25 in (635 mm) yearly around Lake Titicaca, decreases southwards to give desert conditions. Daily temperature ranges of 70°F (38·5°C) are not uncommon, and nights are bitterly cold. In parts of the *Puna*, the searing cold winter wind, bringer of lung diseases, is called 'the Harvest of Death'.

The moister parts of the *Altiplano* and the adjacent *sierras* are covered by coarse tufted grass (*paja brava*) which provides grazing for llamas. The dense stands of *totora* reeds fringing Lake Titicaca are used for rafts and thatching. Trees are rare, but eucalyptus has been introduced successfully in some sheltered areas. The warm and humid *Yungas*, thickly covered with mountain jungle, receive about 30 in (762 mm) of rain annually. Farther south, the valleys are more open and rainfall declines to 12 in (305 mm) or less. Scrub covers much of this part of central Bolivia.

Most of the eastern lowlands have their heaviest rain between December and March, the extensive flooding alternating with severe drought (July-October). The northern lowlands have dense rain forest (*selva*) associated with the upper Amazon basin; in the south, the forest is broken by open swamp and savanna. The Chaco verges in south-east Bolivia suffer seven-month droughts which permit only thicket and scrub vegetation.

Highland Bolivia has a distinctive wildlife which includes the guanaco, chinchilla and the now rare vicuña. Domesticated animals include the llama and the alpaca. The tropical lowlands are the home of jaguars, pumas, monkeys and other animals. Alligators and freshwater turtles abound in the rivers, and snakes and beautifully plumaged birds are numerous.

The People. Bolivia is one of the most sparsely-populated countries in Latin America, and its population of some 5 million is very unevenly distributed. About 75% of the people are concentrated in Andean Bolivia where they live in well separated clusters around the mining centres of the northern *Altiplano* and in the larger, warmer valleys and basins of the Andean interior.

The largest city is La Paz, the seat of government and actual capital (562,500). Lying in the La Paz river valley, 12,795 ft (3,900 m) above sea level, it is the world's highest national centre of government, although Sucre (84,900) is the original and still the legal capital. Cochabamba (157,000) is the second largest city, while Santa Cruz (108,720) is the only important city in eastern Bolivia.

Pure Amerindians – Quechuas, Aymaras and Guaranís – make up about 70% of the population, and mestizos (peoples of mixed Amerindian and European stock) about 25%. Some 5% is of pure European stock.

Spanish, the official language, is spoken by less than half of the people, the Amerindians mostly speaking either Quechua, the language of the Incas, or Aymara. Education, free and compulsory from 7 to 14, is retarded by a shortage of schools and teachers, and many children have to work to bring much-needed money into their families. The illiteracy rate is high, about 68%.

Town buildings reflect Spanish colonial styles. In rural areas, mud-walled huts with thatched or corrugated iron roofs are common. Basic foods include maize, *quinoa* (a kind of millet) and *chuño* (dehydrated potatoes).

Minerals, especially tin, dominate Bolivia's economy. Potosí, with the neighbouring mountain of Cerro Rico, became one of the world's richest sources of silver after its discovery by the Spaniards (1534). Silver, gold and copper, all mined during the colonial period, were overshadowed in the present century by the tin, wolfram, antimony, lead, tungsten, bismuth and zinc of the Eastern Cordillera.

Bolivia is the second largest tin producer in the world, and the only significant producer of this metal in the Americas. The mines, nationalized in 1952, provide about 60% of exports by value.

Production of oil and natural gas along the Andean foothills, especially around the south of Santa Cruz, has been steadily developed. Important new strikes were made in the 1950s and 1960s.

Large deposits of iron and manganese, as yet unexploited, have been found at Mutún near the Brazilian border.

Industry. Due to lack of iron, coal and hydro-electric power, little manufacturing has been developed. Bolivia still relies heavily on imported manufactured goods. Small factories, mostly in La Paz and Cochabamba, produce textiles, cement, processed foods, beer, pharmaceuticals, shoes, tyres and sawn timber. A petro-chemical plant has been planned for Santa Cruz.

Agriculture. About 80% of the people are *campesinos* (peasant farmers) but Bolivia is not yet self-sufficient in food, although the potential is there. The redistribution since 1953 of some 12 million ac (4·86 million ha) of large estates to nearly 160,000 families has resulted in a revolutionary change in the national way of life.

The densest concentration of rural population on the *Altiplano* is found around Lake Titicaca where, at altitudes of 12,000–14,000 ft (3,650–4,300 m), the chief crops are potatoes, barley, *quinoa* and beans. Sheep are the most important livestock, followed by llamas and alpacas.

Farther east, the warmer intermontane basins and valleys, often called 'the gardens of Bolivia', support large farming communities raising maize, wheat, barley, tobacco, dairy cattle, fruit and vegetables. Beyond the Andes, the city of Santa Cruz is the regional centre of an important ranching and colonization development area whose products include sugar-cane, rice, maize, cotton, citrus fruits, bananas and oilseeds.

Citrus fruits, cocoa, coffee, bananas, papaya and many other crops are grown in the steep valleys of the *Yungas*. In the adjoining central foothill region of Chaparé, where rainfall increases, cacao, yuca (manioc) and rice are additional crops. Transport and marketing problems which occur throughout this region have meant that its potential is as yet largely undeveloped.

Inaccessibility and limited market outlets are more serious problems in the extreme north of the country. Only small quantities of wild rubber and Brazil nuts are won from the tropical rain forest lying north of the Beni ranching country.

Forests and Fisheries. About 40% of Bolivia is forested. Extensive belts of evergreen and deciduous tropical hardwoods of excellent quality and variety lie in the north and along the hot, wet Andean foothills of central Bolivia, but transport difficulties have so far prohibited the development of any significant lumber industry.

Lake Titicaca contains three or four varieties of edible freshwater fish and the introduction of salmon-trout is proving a commercial success.

The Problem of Transport. Bolivia has only some 11,358 mi (18, 312 km) of roads, mostly unsurfaced, and 2,190 mi (3,524 km) of railways. The country's poverty, the physical difficulties of its terrain, and the relative isolation of its main centres have all combined to make lack of adequate transport a perennial problem and one which has aggravated other obstacles to progress.

Most of Bolivia's trade has to pass through three foreign Pacific ports: Antofagasta and Arica in Chile, and Matarani in Peru (by way of Lake Titicaca, which breaks the otherwise

Left: At 12,001 ft La Paz is the highest major city in the world. It is Bolivia's largest city and the seat of its government. *Right:* Two-thirds of Bolivia's population farm but the country must still import food.

all-rail route to Matarani). Other connections, by rail, road or navigable river through Argentina, Brazil or Paraguay, are relatively insignificant, although new highways under construction in Brazil may provide faster outlets eastwards.

The chief towns of highland Bolivia were linked by rail following the development of tin mining and the need for mineral export to the coast. The skeletal road network is slowly being lengthened. The 320 mi (515 km) Cochabamba-Santa Cruz paved highway was completed in 1954.

Road links are being improved between the Andean cities and the small head-of-navigation ports on rivers draining the eastern plains, especially on the headstream tributaries of the Mamoré River. Apart from dry-season trails, surface transport in lowland Bolivia depends almost entirely on navigable rivers.

Regular international flights link Bolivia with other South American countries and with the USA. Internal services by Lloyd Aereo Boliviano, the national airline, link key urban centres and help to identify the remote eastern plains as Bolivian and counteract separatist tendencies there.

International Trade. Mineral concentrates form more than 90% of the country's exports. Petroleum is piped into Argentina from the Andean foothill oil fields, and from Caranda (Santa Cruz) to Arica (Chile), whence it is shipped to southern California for refining. A natural gas pipeline from Santa Cruz to the Argentine border is under construction. Exports of oil (and presently natural gas) help to pay the cost of railway (and some highway) construction by Argentina in south and central Bolivia.

Other exports include small quantities of wild rubber, Brazil nuts, skins (alligator, lizard, jaguar, llama, alpaca, vicuña), wool, coca and citrus fruits. Manufactured goods and foodstuffs are imported. Bolivia trades mainly with the United States, West Germany, Argentina, and Japan.

PARAGUAY

Area: 157,047 sq mi (406,752 km²)
Population: 2,395,614
Population Growth Rate: 3·2%
Capital: Asunción, pop 437,000
Language: Spanish (official) Guarani spoken
Religion: Roman Catholic
Currency: Guarani = 100 centimos

Paraguay is a landlocked republic bordered by Bolivia to the north, Brazil to the east, and Argentina to the south and west. With a total area of 157,047 sq mi (406, 752 km^2) Paraguay is the fourth-smallest country in South America. Its name derives from a local Indian word meaning 'the decorated river', probably a reference to the huge rafts of water hyacinth adorning the River Paraguay.

After independence from Spain came in 1811, Paraguay's economic progress was slow, hampered by savage wars that decimated the population, by political instability and by lack of a sea outlet. These and other factors make Paraguay probably the poorest nation in South America today.

The Land. The River Paraguay, flowing from north to south through the country, separates two distinct landscapes. East of the river lies over one-third of Paraguay, an attractively varied countryside which extends from the river's flood plain to a gently rolling hill belt, 1,000–2,000 ft (300–600 m) high, before falling again towards the River Paraná (the upper or Alto Paraná) which forms the south-eastern boundary of the state. Part of the upper Paraná flows across the edge of the Paraná Plateau which dominates the neighbouring region of southern Brazil. The plateau ends in a strikingly dramatic series of scarps and cliffs, from which the upper Paraná pours in huge staircases of rapids and falls. As far downstream as Encarnacíon, the upper Paraná flows in a deep valley cut through the dark walls of old lava flows. Below this point the river heads westward, carrying Paraguay's boundary with Argentina across the low flood plains of the continental interior.

West of the River Paraguay, the great central 'stem' of the country, lies the Gran Chaco. This immense plain, built from the fine silts deposited by rivers draining from the Andes into the Paraguay, comprises over one-half of the country. No permanent streams cross this monotonous tract; but swamps, abandoned channels, brackish pools and complex meandering patterns all bear witness to the erratic movements of rivers and flood waters.

Climate, Vegetation and Wildlife. Most of Paraguay has a congenial sub-tropical climate. Summer temperatures average more than 80°F (27°C), winter temperatures over 60°F (16°C), although a cool south wind called the *surazo* occasionally lowers temperatures to as little as 35–45°F (2–7°C).

Annual rainfall is highest in the extreme east, on the Paraná Plateau (about 80 in; 2,030 mm), and tends to decrease towards the west. Dense forests of tropical evergreen and deciduous trees clothe the wettest slopes of the plateau and hill belt. There are also areas of more open woodland and scrub.

Central Paraguay, including the region around Asunción, receives an annual rainfall of about 50 in (1,270 mm) with a pronounced summer maximum between November and March. The Paraguay river plains are prone to heavy flooding. Land liable to prolonged flooding usually supports only grass or mixed stands of grass and palm savanna. Economically important hardwood trees extend along the western side of the River Paraguay.

The length of the dry season increases farther west in the scrub and thorn thicket of the Chaco. There, the (April-September) winter season is one of searing drought in which swamps dwindle and waterholes dry up.

Paraguay's wild animals include jaguars, deer, armadillos, parrots and toucans.

The People. Mestizos (peoples of mixed Amerindian and European stock) account for 98% of the population. Amerindians account for most of the rest and there are a few thousand Mennonite and Japanese colonists. Spanish is the official language, but outside the capital Asunción most people speak the native Tupi-Guaraní language and indeed most Paraguayans retain the easy-going, fun-loving disposition of the Guaraní Indians who figure largely among their ancestors. More than two-thirds of all Paraguayans live in rural areas where most exist by subsistence farming and inhabit one-roomed mud huts. Living standards are low and the illiteracy rate is high.

The capital city and river port of Asunción (440,000) is the only large town. Choosing a low hill beside the east bank of the River Paraguay, the Spaniards established Asunción among the gentle, amenable, Guaraní Indians as Spain's colonial centre for eastern South America. After the rise of Buenos Aires in the 17th century, Asunción declined in importance. Virtually no colonial architecture survives, and the port and administrative sections of Asunción are small and undistinguished in appearance. Paraguay's other main centres are Concepción, Coronel Oviedo, Encarnación and Villarrica – all east of the River Paraguay.

The Chaco, the area west of the river is still virtually uninhabited except for some 15,000 Mennonites, mostly from Canada, the USSR and Germany.

The very slow population growth, poverty, and meagre development in Paraguay are not solely the result of the country's isolation and landlocked continental location. Paraguay has twice suffered a catastrophic loss of manpower during wars with neighbouring states: one in the 1860s against Argentina, Brazil and Uruguay, and a second in the 1930s against Bolivia. As much as 89% of the male population reportedly died in the Paraguayan War of 1864–70.

The Economy. The country gave promise of becoming one of South America's most important agricultural regions after 1537 when the Spanish *Conquistadores* established Asunción as their main centre for the exploration of southern South America. In the early 1600s, Jesuits began collecting formerly scattered groups of Indians into 30 missions in eastern Paraguay. There, the Jesuits introduced cattle ranching and developed the cultivation of a wide range of crops. These included indigenous plants such as maize, tobacco and *yerba maté* (Paraguayan tea); also plants imported from Spain, such as grape vines, cotton, sugar-cane, orange trees and rice. Thus the basis of Paraguay's pastoral and farming economy was laid. But when the Jesuits were eventually expelled in 1767, their Indian protégés reverted to the jungle or were virtually enslaved by less progressive secular landowners.

Ranching and cultivation survived, however, despite the perennial problems of drought in the west, and seasonal floods in the riverine plains which combine to limit good cultivable land. Over half the total working population is today engaged in agriculture, although only 4% of the country is farmed and this mainly on a subsistence basis with peasants wresting a bare living from the land.

Sugar-cane, cotton, maize, tobacco, rice, manioc, *yerba maté*, oil seeds, coffee, and a variety of vegetables and fruits (especially oranges, grapefruit, pineapples and bananas)

Left: **The modern centre of the capital, Asunción. Paraguay is impoverished with poor agriculture and no exploitable minerals.** ***Above:*** **A backyard tannery. Paraguay produces quantities of Quebracho bark, used to make tannin.**

form the chief field and forest crops. Cattle ranching, however, remains the most important single aspect of Paraguayan agriculture; herds total about 6 millions.

Just over half of Paraguay is covered with forest, but so far only one-third of this potential has been exploited, largely because of isolation and inadequate communications. Quebracho forests on the fringes of the Chaco yield important supplies of hardwood timber and tannin which is extracted at mills sited along the Paraguay.

Mineral resources are generally poor, and Chaco oil prospecting has proved unrewarding.

Manufacturing is limited and largely consists of the processing of the country's agricultural and forest products. These include meat and meat products, *yerba maté*, vegetable oils (palm nut, castor, coconut, cottonseed, tung) and petitgrain oil, which is distilled from the leaves of bitter orange trees, and exported for use in perfume manufacture. Expansion is planned, particularly of small and medium-sized industries. New hydro-electric schemes on the Acaray and Monday rivers in eastern Paraguay are now supplying power to Asunción (previously dependent on thermo-electric plants fuelled by oil and wood).

Meat products and hides provide about one-third of Paraguay's export earnings and, with investment loans concentrated on the livestock industry, it is hoped substantially to increase exports by the mid 1980s. Agriculture, livestock and forestry together account for 95% of the export earnings (mostly coming from Argentina, Uruguay, the USA and West European countries). Paraguay's imports (mainly from the same countries) include wheat, fuel oils and manufactured products including vehicles and machinery.

Transport. There are only some 300 mi (480 km) of railway but more than 10 times that mileage of road. New highway construction involves Paraguay in trans-continental programmes linking Brazil's Atlantic port of Paranaguá with Pacific coast ports in Chile and Peru. The highway was completed westward as far as Asunción in 1969. A dirt road runs out across the Chaco from Asunción, through a group of Mennonite colonies in the central Chaco, to reach the Bolivian border.

The River Paraguay has been the country's traditional transport artery and trade outlet. Regional development projects for improved communications between the Plate basin countries include provision for the reliable maintenance of a minimum 13 ft (4 m) depth in the River Paraguay below Asunción. Paraguay has free transit through Buenos Aires, and free port facilities in Uruguay and Brazil. Both Argentina and Paraguay are jointly engaged on plans to improve navigation of the River Paraná. Paraguay relies heavily on air transport for fast communication.

CHILE

Area: 286,395 sq mi (741,763 km²)
Population: 9,780,000
Population Growth Rate: 2·4%
Capital: Santiago, pop 3,362,000
Language: Spanish
Religion: Roman Catholic
Currency: Chilean escudo = 100 condores

Chile occupies a narrow ribbon of land on the Pacific coast of South America. Bordered in the north by Peru, in the north-east by Bolivia and in the east by Argentina, it stretches southwards for more than 3,000 mi (4,828 km) to Horn Island within about 500 mi (800 km) of Antarctica.

Until recently much of the wealth in land and commerce was held by a comparatively few families. The traditional conservatism of the *fundo* owners has always been a feature of Chilean politics, and it is only since 1964 that serious land reforms have been introduced. Large-scale expropriation of land was the outstanding feature of the Popular Unity regime of South America's first and democratically elected Marxist president, Dr Salvador Allende.

Until the coup of September 1973 in which a military junta overthrew Allende, Chile's record of democratic government, with free elections and secret ballots, was remarkable, particularly in comparison with other Latin American countries.

A Long, Narrow Land. Chile's dominant feature is a great mountain range comprising the Chilean section of the Andes which forms a natural barrier between Chile and neighbouring Bolivia and Argentina. This mountain spine runs the entire length of eastern Chile, reaching heights of over 18,000 ft (5,490 m) in the north and centre and including Ojos del Salado (22,539 ft; 6,870 m), the Western Hemisphere's second highest peak. The mountains become fragmented in the south which has beautiful mountain lakes, glaciers and fjords and myriads of offshore islands.

Cross sections through northern, central and southern Chile reveal a structural pattern that holds good almost throughout the length of the land. The closely-parallel ridges of the Andes form the eastern boundary. West of the Andes lies the well-defined Pacific trench, the main lowland area of Chile, extending from the northern border to Puerto Montt. South of Puerto Montt the sea has invaded this lowland area. West of the trench is a coastal range, linked in places with the Andes by transverse spurs, but in the far south the range has subsided and appears only as an archipelago of islands. Throughout Chile short rivers flow steeply down to the sea from the Andes.

Located in a region where the earth's crust is unstable, Chile is prone to earthquakes: more than 100 have occurred in the last four centuries.

Climate, Vegetation and Wildlife. The country's great length spans five distinct climatic regions.

The desert North (Norte Grande) extends from the Peruvian border at Arica to Copiapó. It includes the 750 mi (1,200 km) long Atacama Desert, one of the driest regions in the world, where years pass with no measurable rain. Northern Chile lies within 18° of the equator, but the cold offshore Peruvian Current moderates temperatures (which average only about 75°F in summer at Arica). The extreme north is virtually barren. But farther south intermittent streams, dew and fog locally support valley oases and such plants as cacti and drought-resistant shrubs. A transitional semi-desert region (Norte Chico) extends southwards for about 450 mi (724 km) from Copiapó to Illapel. In the valleys of the westward-flowing rivers such as the Huasco, de los Choros, Limari and Aconcagua, crops are grown under irrigation. Both the climate and the vegetation change imperceptibly until the true Mediterranean heartland of Chile is reached.

The Mediterranean heartland (Zona Central) extends from Illapel, in the Choapa river basin, to Concepción on the Bió-Bió River, a distance of some 425 mi (684 km). It is the home of 70% of the population and has one of the most congenial climates in the world. Temperatures range from about 80°F (27°C) in January to around 50°F (10°C) in July. Annual rainfall (concentrated in the winter months) is about 30 in (762 mm) with local variations. The cold Peruvian Current, flowing northwards from the Antarctic, cools the coastal areas whose summer temperatures may be 15°F (8°C) lower than those of the Central Valley. In general the area's vegetation is adapted to a winter rainfall regime.

Forest Chile (Zona Sur) stretches 400 mi

Shanty towns in Santiago de Chile are sometimes called 'poblaciones callampas' – 'toadstool towns'.

(644 km) from Concepción to Puerto Montt. This is a region of equable temperatures and increasing rainfall. Frosts are rare in winter, and temperatures seldom rise above 70°F (21°C) in summer. The coastal range, which has a rainfall of about 100 in (2,540 mm), gives some protection to the inland areas, which receive about 55 in (1,283 mm) yearly. Although winds are mainly from the sea, occasional stormy winds known as *puelches* sweep through the Andes to the east. The dense forests on the mountain slopes are mainly of false beech, but include conifers.

Archipelagic Chile, the fifth climatic region, extends from Puerto Montt to Cape Horn, a distance of more than 1,200 mi (1,931 km). This area (which includes the Strait of Magellan and western Tierra del Fuego) is one of the stormiest regions in the world. Annual rainfall may be as much as 200 in (5,080 mm), and snows are heavy in winter, especially on the Patagonian steppe. The annual temperature range is 24–59°F (−4–15°C). Less than 2% of Chile's total population lives in this cool, wet and windy region. Much of this southernmost third of Chile is densely forested with trees including false beech and conifers.

In each region altitude modifies climate and there are distinctive Andean plant species including the Chilean pine (or monkey puzzle tree), a variety of flowering plants and low-growing cushion plants which occur just below the snow line.

Chilean wildlife includes the puma, guanaco, chinchilla and the pudu (the smallest deer). But the Andes have isolated Chile from the wealth of wildlife found in central and eastern South America.

The People. Although the early European settlers were Spaniards, Chile now has a markedly cosmopolitan population. During the 1700s there was an influx of Basques who soon featured among the leading families. Land grants attracted other Europeans, especially the British who rapidly became the keystone of the business community. Other immigrants entered Chile at the end of both world wars.

The long period of intermarriage between Spaniards and native Indians has produced a large percentage of mestizos, but disease and warfare decimated Chile's pure Araucanian and other Indian groups. However, scores of thousands of Araucanians persist in southern-central Chile, speaking both Spanish, the official language of Chile, and Araucanian. But the primitive Fuegians of the extreme south are virtually extinct.

Unlike that of some other South American republics, Chile's population includes a large and influential middle class and politically powerful industrial class – factors influencing social reforms that were under way by the 1970s. Most Chileans are Roman Catholics. Education is free and compulsory from the age of 7 to 15, and the illiteracy rate of 12% is low for Latin America. The University of Chile, founded in 1842, is at Santiago, and there are other universities at Concepción, Valparaíso, Valdivia and Antofagasta.

The Pattern of Population. About 68% of the population now lives in urban areas. The capital, Santiago, beautifully situated on a wide plain backed by the Andes, is the home of about 3,362,000 people – more than a third of the total population of the country. Santiago is not only the seat of government but the centre of economic and cultural life.

Valparaíso, the second city, has a population of 296,000. Built on the shores of a broad bay, and ringed by hills, it is the chief port of Chile and the chief commercial city on the Pacific coast of South America. Concepción (178,000) Chile's third city, has been rebuilt five times after earthquakes, and now stands some 3 mi (5 km) from the original site, and some 9 mi (14 km) from its port of Talcahuano (which is also Chile's chief naval base) and the nearby Huachipato iron and steel works. Through its proximity to the coal fields, its abundant hydro-electric power and good communications, Concepción has become a leading industrial centre.

The Economy: Minerals. Copper dominates the Chilean economy and its control has been a major factor in Chilean politics. Under President Allende all mineral deposits were nationalized (1971), compensation being negotiated with the American companies in the copper industry.

Chile is estimated to have 30% of the world's reserves of copper. She is the world's fourth largest producer and the second largest exporter (after Zambia). Chuquicamata, in the Atacama Desert, is the world's largest open-cast copper mine; and El Teniente is the world's largest underground copper mine. The main smelters are at Potrerillos and Quintero, although most copper is exported in its raw state.

The nitrate industry, Chile's first great source of wealth, declined in importance with the advent of artificial fertilizers, but survives in a reduced form. It is based on nitrate deposits formed in the beds of now dried up lakes which occur in the northern deserts. Water piped from the Andes supplies the otherwise waterless nitrate and other mining communities, whose produce is shipped out from ports built on small low-lying shelves among the precipitous cliffs of Chile's northern coast. Many abandoned workings and ghost towns recall the lost hey day of nitrate extraction.

Coal is mined on a relatively small scale. Extensive iron ore deposits occur mainly in the semi-desert region with Copiapó, the capital of Atacama province, as a leading centre and the port of Caldera as the main sea outlet. Oil is found in southern Chile, in the Magallanes oil field which straddles the Strait of Magellan. The chief wells, in northern Tierra del Fuego, are linked by pipelines to tanker ports on the strait. The State Oil Company (ENAP) now produces almost enough oil to meet domestic needs.

Agriculture. Some 24·7 million ac (10 million ha) are under crops and permanent pasture, and about 29% of the working population is employed in farming of one kind or another. In the north, crops are grown in the semi-desert region with the help of water tapped from the rivers that cross it. But extensive farming is concentrated in the Mediterranean and temperate forest zones (Zona Central and Zona Sur) further south, especially in the fertile and well irrigated Central Valley. This beautiful countryside of fruitful valleys, large farms and vineyards is not intensively cultivated and there is scope to improve crop yields and quality by scientific management.

Maize is grown throughout the Mediterranean heartland; and fruit, especially peaches and apples, in the centre and south. Large areas around Santiago are devoted to market gardening, and there are extensive rice fields around Talca. Since some 70% of the total population lives in this Zona Central, the farms are not far from their markets. But the region's agricultural prosperity is limited by under-developed communications and variable production.

Grapes are among the leading crops of the central Mediterranean region. More than 247,105 ac (100,000 ha) are devoted to viticulture, and the total output is about half that of Portugal. Some Chilean wines are of very high quality, but the Chileans themselves are prodigious wine-drinkers and comparatively little wine is available for export.

Middle Chile provides most of the barley harvest and has more than a third of the country's cattle. Forest Chile (Zona Sur) has the rest of the cattle and most of the pigs, and grows all the oats and most of the wheat. But large imports of wheat and other basic foodstuffs are necessary. In the south of the country more than 6 million sheep graze on the Patagonian steppe and on Tierra del Fuego.

As part of the attempt to redistribute land once owned by a few big landowners, the government has set up state farms called centres for agrarian reform. Problems of land reorganization have, however, contributed to a decline in farm production. Thus in 1972 about 53% of the national requirement of wheat had to be imported, compared with 19% in 1969.

Forests and Fisheries. Chile's main forests are in the Zona Sur between Concepción and Puerto Montt. This area's large forests have been little touched commercially. Development plans have been prepared involving reafforestation that will probably feature the Monterrey pine, which matures in as little as 20 years. There are pulp and paper plants at Constitución, Concepción, Valdivia and other centres.

Fish is not a major item in the Chilean diet. But fish abound off the southern shores, and the government has recently been trying to promote fish consumption and the development of the fishing industry. Shellfish of all kinds are particularly abundant.

The Chiquicamata copper mine on Chile's Atacama desert is the most productive in the world. Chile is the world's third largest copper producer.

Industry and Tourism. Chile is one of the most highly industrialized countries in Latin America. In addition to the integrated iron and steel works at Huachipato near Concepción, the country has engineering, car assembly and chemical plants, paper and newsprint mills, textile, glass, cement and other industries. Santiago and Valparaíso are the main manufacturing centres. Valparaíso has metallurgical, wood working, leather and other industries. Talca, capital of Talca province, is another big manufacturing centre but less important than Concepción which produces woollen textiles, paper, glass, cellulose, newsprint, and cement. Arica and Punta Arenas in the extreme north and south have been made free ports to attract foreign capital and industry. But the cost of importing materials to support Chile's industrial base remains a problem, and Chilean industry can operate only under heavy protection.

With fine mountain scenery, a beautiful lake district and a fjordland southern archipelago, Chile has considerable potential as a tourist country. There were nearly 200,000 foreign visitors in 1968.

Transport. Ships have played an important part in the development of this long maritime country. Coastal services are still a major means of transport; the southern provinces are accessible only by sea or air. But good harbours are few.

About half of Chile's 34,635 mi (55,416 km) of roads are all-weather roads. The Chilean section of the Pan-American Highway has now been completely paved from Arica to south of Puerto Montt, but its lateral roads are mostly poor. Chile has five rail links with her neighbours. Bolivia is reached by the Arica-La Paz and Antofagasta-Ollagüe lines; Peru, by the Arica-Tacna line; and Argentina, by the Antofagasta-Salta line and the Transandean Railway.

There are internal and international air services.

International Trade. Although less than 6% of the working population is employed in mining, minerals account for almost 80% of the total value of exports. Copper is far and away the most important, and the world price of copper dominates the Chilean economy. Other exports include iron ore, chemical products, paper and pulp, liquid gas and small quantities of wine and dried fruit. Though essentially a farming country, Chile exports little agricultural produce (less than 6% of the total value of exports). Throughout the years the adverse balance of trade has been largely due to the cost of imported foodstuffs and manufactured products including machinery and electrical equipment. The United States, Britain, West Germany, and Japan are among Chile's main trading partners.

Dependencies. In the eastern South Pacific, Chile owns the tiny Juan Fernández and Desventurados island groups and Easter Island and Sala y Golnez Island. Chilean Antarctic Territory (Antarctica between 53 and 90°W) overlaps Argentinian and British claims.

URUGUAY

Area: 75,172 sq mi (186,925 km²)
Population: 2,921,000
Population Growth Rate: 1·2%
Capital: Montevideo, pop 1,350,000
Language: Spanish
Religion: Roman Catholic
Currency: Uruguayan peso = 100 centesimos

Uruguay is the smallest independent state in South America: a republic a little larger than the state of Washington in the USA with a population little more than that of Washington DC. By the mid 20th century this small nation had built a reputation as one of the most prosperous in Latin America – a reputation resting largely upon the ranching profits from Uruguay's extensive grasslands.

The region's grazing possibilities attracted Spanish colonists, who eventually gained independence in 1828. Uruguay's subsequent economic progress was remarkable, but it has suffered serious setbacks since World War II. An ambitious social welfare programme proved a heavy financial burden; economic competition increased from neighbouring Brazil and Argentina; and inflation accelerated at an astronomic rate – factors that have threatened the very democratic fabric of the state. In the 1970s, Urban guerrillas (the *Tupamaros*), striking successfully at public figures throughout the country, were a violent manifestation of social discontent.

Plains and Rolling Hills. The land bordering the Uruguay and Plate rivers and the Atlantic coast forms a low flood plain, which includes lagoons, belts of sand dunes and beautiful chains of sandy beaches on the seaward side. But most of Uruguay is gently rolling hill country, nowhere higher than 1,650 ft (500m). The north-western uplands, formed by the southern fringes of the Brazilian Plateau, feature old lava flows with somewhat steeper slopes and tabular hills.

Lying close to the sea, Uruguay has a temperate climate. The weather is mild, damp and windy for much of the year, with summer temperatures of 70–75°F (21–24°C) and winter temperatures of 50–60°F (10–16°C). There is no pronounced dry season; much of the average annual rainfall of about 40–50 in (1,014–1,270 mm) comes in the autumn and winter, particularly in April and May, while thunderstorms are very common during the summer months.

Rich prairie grass covers most of the land. Scattered trees are found along water-courses and in the south-eastern region. Little more than 3% of the country is forested. Uruguay's native animals include capybaras, caymans and parakeets.

The People. The population of 2,921,000 is 90% of European stock. Most Uruguayans are descended from 18th- and 19th-century immigrants from Spain and Italy. There is a small mestizo element (10%) and there are a few thousand negroes and mulattoes.

The capital city, Montevideo (1,350,000), claims nearly half the country's entire population and is the fourth largest city in South America. The city region is a national centre for the transport network, 90% of all port activity, manufacturing, and the political, social and cultural life of Uruguay. To the east, a number of modern coast resorts have developed along beautiful beaches of sparkling white sand, a series of small bays set among wooded hills, rocky promontories and shallow lagoons. The most famous resort is Punta del Este, which has magnificent beaches backed by mimosa, pine and eucalyptus trees.

Other towns are sited along the east bank of the River Uruguay, notably Salto, Paysandú and Fray Bentos. Mercedes, on the River Negro, an old colonial port 30 mi (48 km) above the confluence with the River Uruguay, is often described as the country's prettiest town.

Colonia, founded on the north bank of the River Plate by Brazilian Portuguese settlers in 1680, remains the best preserved colonial town in Uruguay. Small regional centres in the interior, such as Minas, Florida, Durazno, Tacuarembó, Rivera and Artigas each contain between about 15,000 and 25,000 people.

All told, over 80% of Uruguayans live in cities or towns, mainly in new apartments or small brick or concrete houses with tiled roofs. In the country, ranch hands often occupy one-roomed homes built of mud brick with thatched roofs. The densest rural population is found in the south and west.

Uruguayans are Spanish speaking and predominantly Roman Catholic. There are extensive welfare and education systems and the literacy level is very high at over 90%.

Ranching: the Heart of the Economy. Grasses are Uruguay's greatest natural resource and the country's agriculture is dominated by pastoral activities. The story of the development of the great grazing industry in Uruguay from the 17th century, when cowboys (*gauchos*) casually worked the cattle on the open plains and rolling hill country, has much in common with that of its neighbour Argentina. In Uruguay, however, the scale of the enterprise was smaller. There are now between 7 and 8 million cattle and about 20 million sheep.

Other types of agriculture are limited, partly because of the traditional prestige attached to the pastoral way of life. Commercial cultivation, as in Argentina, was largely developed by 19th-century European immigrants in the south-west of the country, notably along the River Uruguay south of Paysandú, and the River Plate plains as far as Montevideo. Irrigation has subsequently also been brought to the north-east coast. Wheat, maize, barley, flax, rice, sugar-cane and oil seeds (especially peanuts and sunflower seeds) are the principal crops, and efforts have been made, particularly since World War II, to increase their acreages in order to reduce imports and diversify the economy.

Citrus fruits, grapes, olives and dairy produce are gaining in importance in the list of agricultural products. In all, only about 12% of the country is cultivated.

Industry. Development is severely hampered by the lack of power, minerals and raw materials, and the very small internal consumer market. Emphasis remains on the primary processing of the meat, wool, hides and other animal products, on oil-seed crushing, winemaking, and sugar refining. Manufactures include textiles, plastic and rubber goods, cement, electrical apparatus, pharmaceuticals, paper, glass and domestic hardware. Most of the manufacturing industries are concentrated in Montevideo; other smaller centres are Artigas, Paysandú, Fray Bentos and Colonia. Plans to increase hydro-electric power for industrial and residential purposes are focused upon El Palmar on the River Negro, and on Salto Grande, a major waterfall on the River Uruguay in the international section shared with Argentina. A joint development programme for power and improved navigation forms part of the current scheme for regional economic co-operation among the five countries of the River Plate basin.

Transport Systems. Uruguay's small size and gentle gradients present few difficulties to the engineer, and the country is adequately served by surface transport systems. There are 1,874 mi (3,000 km) of railway all of which like the main highway system, converge upon Montevideo; the railways (begun in 1868) were built by British interests but have passed into state ownership since 1948. Inland waterways like the Plate and Uruguay rivers are an important means of transport. Fray Bentos, the head of navigation for ocean-going shipping, is also the site of a new international road bridge over the River Uruguay to Puerto Unzue in Argentina. Ferry-boats and hydrofoils frequently ply across the River Plate between Montevideo and Colonia, and Buenos Aires. An internal airline links Montevideo with all the important Uruguayan towns. International services link Montevideo with

Watering crops with a hose on Uruguay's rich pastureland. 90% of the 40 million acres devoted to agriculture are accounted for by livestock. Crops are grown mainly in the south.

The Gauchos who established Uruguay's cattle industry are regarded as national heroes.

North America, Europe and with other parts of the world.

International Trade. Agricultural commodities, consisting mainly of wool, meat, meat products and hides, account for over four-fifths of Uruguay's export earnings which come mainly from Western Europe and the USA. The country's agricultural exports have suffered from its high-cost production methods and from fluctuations in world prices. Imports of fuel oils, raw materials, machinery, manufactured goods and timber come mainly from the USA, Argentina, Brazil and West Germany. In most recent years, import controls have failed to prevent an overall trading deficit. This is somewhat reduced by income from tourism. An average of more than 300,000 people normally visit the country every year.

ARGENTINA

Area: 1,072,163 sq mi (2,776,902 km²)
Population: 23,392,000
Population Growth Rate: 1·5%
Capital: Buenos Aires, pop 9,000,000
Language: Spanish. Immigrant European languages.
Religion: Roman Catholic (95%) Protestant
Currency: Argentine peso = 100 centavos

The Republic of Argentina which dominates the South American land mass south of the Tropic of Capricorn, is the continent's second largest country. Only Brazil is bigger. Argentina is bordered on the west by Chile; on the north by Bolivia and Paraguay; on the east by Brazil and Uruguay, and has an Atlantic coastline of some 1,600 mi (2,575 km).

Part of the land is barren desert or mountain, and Argentina is poor in many mineral resources. But much of the country consists of natural pastures which support more than two cattle and nearly two sheep for every human inhabitant. Argentina is the world's largest exporter of raw meat, and its high export earnings have encouraged industrial growth and urban expansion. By 1970 Uruguay was the only South American nation with a higher percentage of non-agricultural workers, and Buenos Aires was the Southern Hemisphere's biggest city. By Latin American standards the people were prosperous: there were more cars and doctors per 1,000 inhabitants than in any other South American state. Yet Argentina has suffered guerrilla violence provoked by sectional poverty and repressive regimes, and inflation largely erodes the benefits of fast industrial growth.

Spanish explorers discovered the area in 1516, and it was later misnamed Argentina ('Land of Silver') from the silver ornaments worn by its Indians, who had in fact obtained the silver in Bolivia. Argentina remained a Spanish possession for nearly three centuries. Independence was proclaimed by the colonists in 1816. Thereafter the political history of Argentina followed the familiar South American pattern of coups, revolutions and revised constitutions. Its most famous recent ruler was Juan Domingo Perón, who won the loyalty of the working classes with sweeping social and economic reforms (1945–55).

The Face of the Land. Argentina is a land of lofty mountains and vast lowlands. The Andes extend 2,400 mi (3,900 km) along its western flank, forming a bold but relatively narrow frontier zone through which runs Argentina's boundary with Chile.

Argentina is not primarily an Andean country, but it contains the highest peak in the Andes – Aconcagua (22,834 ft; 6,960 m), a majestic extinct volcano which is also the highest mountain in the Western Hemisphere. Within this same group lie Mercedario (21,878 ft; 6,668 m), Tupungato (21,484 ft; 6,548 m), and Amarillo (20,631 ft; 6,288 m).

The northern and central sections of the Argentine Andes are broadened by the high, windswept plateau of the Puna de Atacama, a southward extension of the Bolivian *altiplano* 11,000–13,000 ft (3,400–4,000 m) above sea level, and by a wide piedmont zone at altitudes between 3,000 and 8,000 ft (900–2,500 m). In the piedmont (foothill plateau) valleys and basins lie some of the oldest and most beautifully-sited cities in Argentina.

The Northern Plains consist of the Chaco country and north-eastern Mesopotamia. The Chaco, a vast semi-arid tropical lowland, is shared by Argentina with Paraguay and Bolivia. The Argentine Chaco is a gently tilted alluvial tract crossed in wide sweeping courses by the Pilcomayo, Bermejo and Salado rivers.

Mesopotamia lies between two rivers, the Paraná and Uruguay in north-eastern Argentina. In the north-east it projects between Brazil and Paraguay like a curled finger, rising onto the rolling Paraná Plateau, where rivers cut deeply through the lava surfaces, forming spectacular canyons and waterfalls.

The most famous of these, and one of the most magnificent natural wonders in South America, are the Iguaçu Falls on the Iguaçu River, which Argentina shares with Brazil. Set in luxuriant virgin forest, where orchids and begonias, palm and ferns, and brilliantly coloured birds and butterflies abound, the Iguaçu Falls are higher and wider than Niagara. The highest in the series plunges 230 ft (72 m) over a 2,700 yd (2,500 m) frontage, throwing up a rainbow cloud of mist 100 ft (30 m) high.

The heart of Argentina is the pampas, a virtually flat, featureless series of plains averaging 200–300 ft (60–90 m) above sea level and extending southward from the Chaco between the Andean foothills and the Atlantic – in all '250,000 sq mi (647,500 km²) of almost unrelieved monotony'. The pampas (Spanish for 'plains') are drained chiefly by the rivers Paraná and Uruguay, and have as their natural focus the wide estuary of the Río de la Plata (River Plate).

South of the Colorado River, the featureless level expanses are left behind, and as the continental land mass narrows rapidly, it is marked by the gaunt, windswept plateaux and wide, shallow valleys of Patagonia. This bleak, rugged, lake-strewn landscape is one of the most remote and most desolate areas in the world.

Climate. Extending over some 33° of latitude, Argentina has a variety of climates. Important factors include the moist south-east trade winds north of about latitude 35°S; the wet westerlies south of that latitude; the Andean barrier creating a dry rain-shadow area east of the crest line; and the moderating effects of the sea on temperatures except in the inland tropical north (where South America's highest temperatures occur).

Most of the north has dry winters with temperatures averaging 55°F (13°C), and hot, wet summers, with temperatures averaging 77°F (25°C). The Chaco receives 20–40 in (508–1,016 mm) of rain during the summer.

The pampas have mild winters and hot summers. The Argentines distinguish the Humid Pampa, the wetter region east of Córdoba and Bahía Blanca where many areas receive 35 in (889 mm) of rain annually, from the western Dry Pampa, a dusty expanse where the lower and less reliable rainfall decreases westward.

In semi-arid Patagonia, winter temperatures average about 35°F (2°C), and summer temperatures up to 70°F (21°C). Severe winters are rare but Patagonia remains for most of the year a bleak and rather cheerless land, its coasts often shrouded in fog or buffeted by wind.

Vegetation and Wildlife. Within the Andean zone, the Puna de Atacama forms part of the high cold desert characterized by *salares* (salt flats), scattered coarse tussock grasses and low, drought-resistant shrubs. The eastern face of the Puna and many of the lower ranges are forested, mainly with beech and occasional Chilean pine. Contrasts are provided by the rain-shadowed valleys and basins where steppe vegetation prevails. In the drier foothill areas south of Tucumán, salt flats and saline depressions again become common.

The Chaco has patches of deciduous woodland; thorn thickets, savanna and swamp. In the luxuriant sub-tropical forests of the north-east, the *yerba maté* grows profusely. The leaves of this evergreen provide maté tea popular in Argentina and other South American countries. The pampas have few trees and are open grasslands of varying quality. Patagonia is mostly desert or semi-desert steppe.

Wildlife includes the jaguars, pumas, tapirs, giant anteaters, alligators, parrots and water birds of the northern forests; the condor and camel-like guanaco of the Andes; and the rhea (South American ostrich) and burrowing viscacha of the plains.

The People. The Argentines are a predominantly white people, descended for the most part from early Spanish colonists or from later Spanish or Italian migrants. Official sources estimate 97% of the population of over 20,000,000 to be of unmixed European descent, but this ignores the considerable number of *mestizos* (mixed Amerindian-European stock) along the Chilean, Bolivian and Paraguayan borders. The dwindling number of pure-bred Indians are found mainly in the north-west highlands, the Chaco and southern Patagonia.

As in many other spheres, the Humid Pampa dominates the pattern of population. This region covering 200,000 sq mi (518,000 km^2) – about 22% of the total land area – is the home of some 15,000,000 people (about 63% of the total population). Despite its importance as the chief ranching and grain-producing area, the Humid Pampa has a comparatively low rural population; its people are mostly urban dwelling.

The cities of Rosario (810,840), Córdoba (798,663), Sante Fe (270,000), and Bahía Blanca (160,000) are notable regional centres. But the capital city Buenos Aires completely dominates the population pattern.

Greater Buenos Aires now contains 9,000,000 people and is the home of more than 1 in 3 Argentines. It is the focal point of all transport systems, port activities, manufacturing and processing, the consumer market and practically every aspect of national life. Not only is Buenos Aires the largest city in Argentina; it is the largest urban centre in the entire Southern Hemisphere.

There are relatively few regional centres along the Andean foothills. They include Tucumán (San Miguel de Tucumán) with more than 300,000 people; Mendoza, with some 112,000 (and a centre for some 250,000 people, many of Italian descent); and Salta with about 125,000. There are other, smaller, centres in the north and north-west.

Patagonia still has less than one person per sq mi (0·39 per km^2). Only one town, Comodoro Rivadavia, has more than 50,000 inhabitants. Taken as a whole, Argentina remains sparsely peopled.

Creed and Culture. Most Argentines belong to the Roman Catholic Church, which is state supported, but there is complete religious freedom. Argentina has one of the highest literacy rates (91%) in Latin America. Education is free and compulsory from 6 to 14 years of age. The oldest of the eight major national universities is Córdoba, founded in 1613; the largest, Buenos Aires. Spanish is the official language, but immigration has been so great that other languages, such as Italian, German and French may be heard.

In the popular imagination, Argentina is symbolized by the *gaucho*, the romantic hard-riding herdsman of the pampas, who has given Argentina many folk songs and dances.

Urban life however, is very different. Many small towns are characterized by colonial type architecture. Many large cities have fine Spanish colonial buildings, but also the most modern functional structures.

The Argentines are prodigious meat-eaters, with perhaps the highest consumption in the world. Sport is popular, especially association football.

Ranching and Farming. The dramatic 19th century transformation of the pampas from a grassy wilderness into one of the world's greatest meat and grain producing regions dominates both the geography and the history of Argentina. Swift and spectacular changes began in the 1860s, the introduction of strong smooth-wire and barbed-wire fencing; deeper wells, wind pumps, and reservoirs to stabilize water supply in the drier areas; improved grasses and alfalfa pastures; and the selective breeding of high-quality stock.

Refrigeration enabled perishable products to be carried safely through the tropics and to expanding overseas markets. Familiar British breeds of beef cattle suitable for European tastes began to replace the existing herds of longhorns which had excellent hides but were too lean. New attention was given to grain production, both to supplement stock feed, and for export.

The development of commercial agriculture in the 1880s brought the higher rainfall zone of the pampas into focus as the major economic region of Argentina. Here, large-scale maize, wheat, flax and alfalfa production was introduced. A stream of immigrants entered as tenant farmers on the great *estancias*, responsible for cultivation, rather than for grazing operations which remained the interest of the landowners and the *gauchos*.

The 425 ft wide Avenida 9 de Julio, reputedly the widest street in the world, commemorates Argentina's independence in 1816.

Meat-chilling and packing plants multiplied rapidly in the major river ports at the end of the 19th century, and Argentina joined the ranks of world beef, wool, maize, wheat and linseed suppliers.

The Coming of the Railways. Within a radius of approximately 300 mi (483 km) behind Buenos Aires, delineated in a great arc by the cities of Bahía Blanca, Mercedes, Córdoba and Santa Fe, developed one of the world's densest railway networks. The first railway began operating in 1857.

In the 1870s and 1880s, British capital and British engineers covered the Humid Pampa with more than 18,000 mi (28,968 km) of railway through the ranching and grain lands, and extended some 8,000 mi (12,874 km) of tracks over other parts of Argentina. These steam railways in turn provided a major market for British coal. British managers, accountants, bankers and merchants arrived, thereby increasing the British interests already represented by such enterprises as the Bovril Estates and the Liebig Meat Extract Company.

Beyond the Humid Pampa. Beyond the extraordinary focus of pastoral and agricultural activity in the Humid Pampa lie the relatively few outer regional centres, such as Mendoza, Tucumán and Salta to the west and north-west, and, in the north-east, Formosa, Corrientes and Posadas, which are smaller riverine settlements serving the pastoral, quebracho forest, and irrigated cotton centres on the drier Chaco fringe, or the *yerba maté* and other tea growers in the wetter lands of Misiones province. Farther south, Mendoza is the largest of Argentina's vineyard oases. In 1967, Argentina became the world's third largest producer of wine.

South again lies the cool, dry sheep country of Patagonia. Wool is the region's staple product, although irrigated pastures and fruits are found along the valley floors of the Colorado, Negro and lower Chubut.

Mining. With the exception of petroleum (of which Argentina is South America's second-largest supplier) the country is not rich in minerals. Small quantities of asbestos, lead, zinc, silver, copper, tungsten, mica and lignite are found, chiefly in the north-west. Patagonia contains coal and probably substantial reserves of iron ore.

Most oil and natural gas comes from Patagonia; the Comodoro Rivadavia district is responsible for 60% of the total production. Like Neuquén on the River Negro, it pipes supplies to Buenos Aires. Oil and gas have also been found along the Strait of Magellan and in Tierra del Fuego. Other oil fields lie around Mendoza and in the extreme north in Salta and Jujuy provinces, all linked by pipeline to Buenos Aires and other important centres on the pampas.

Manufactures. Argentina still has a primarily agricultural export economy, and much of its industry is concerned with meat processing, meat packing and a wide range of animal by-products. However, there has been rapid growth in the plastics, steel, textile, motor vehicle, engineering, cement and petro-chemical industries under state sponsorship.

The country's major steel plant is at San Nicolas on the bank of the River Paraná below Rosario. Nearly one half of the country's steel consumption must at present be imported, but Argentina aimed to become self-sufficient in iron and steel in the early 1970s.

San Lorenzo, 15 mi (24 km) upstream from Rosario, is an important oil and gas pipeline junction and refining centre which has become the nucleus of a huge complex of petrochemical industries including plastics, fertilizers and synthetic rubber.

International Trade. Agriculture is responsible for about 90% of the export earnings. Meat (principally beef), grains (wheat, maize, oats), vegetable oils and oil-seeds, wool, hides and other animal products dominate the export trade.

European countries take about 90% of the total exports of meat and also provide the main market for grain.

Imports, chiefly from the United States, Brazil and West Germany, consist mainly of iron and steel, non-ferrous metals, machinery, vehicles and timber.

Transport. The railways passed to Argentine state ownership after World War II. There are nearly 25,000 mi (43,450 km) of track, more than in any other Latin American state. International rail communication extends into Bolivia, Paraguay, Brazil and Uruguay, while the most beautiful and spectacular lines link Argentina with Chile. The famous 900 mi (1,448 km) trans-Andean railway links Buenos Aires with Santiago/Valparaíso via the Uspallata Pass (12,650 ft; 3,856 m) where a famous bronze statue of Christ the Redeemer marks the boundary of Argentina and Chile. Farther north a second trans-Andean line, opened in 1948, links Salta with Antofagasta.

Main railways are paralleled by branches of the Pan-American Highway, and current projects include improvement of the primary highway network to promote fuller regional integration of the Plate Basin lands. Riverboats navigate the long Plate, Paraná, Paraguay and Uruguay system. There is a nightly river-boat service across the Plate estuary between Buenos Aires and Montevideo, the capital of Uruguay. The main seaports are Buenos Aires and Bahía Blanca. Internal air services are good and international flights connect Buenos Aires with all major capitals in Europe and the Americas.

Dependencies. Argentina claims a 475,000 sq mi (1,230,000 km²) sector of Antarctica that overlaps Chilean and British claims and remote southern islands comprising the British-held Falkland Islands (Islas Malvinas), South Orkneys (Orcadas), South Sandwich Islands and South Georgia.

FALKLAND ISLANDS

Area: 4,700 sq mi (12,220 km²)
Population: 2,045
Population Growth rate: 0·5%
Capital: Stanley, pop 1,100
Language: English
Religion: Christian
Currency: Falkland pound = 100 pence

The Falkland Islands are a British colony comprising an archipelago in the South Atlantic, discovered in 1592 by an English sea captain. The Falklands lie about 320 mi (510 km) off the Argentine coast and the entrance to the Strait of Magellan, and 480 mi (610 km) north-east of Cape Horn. Argentina has disputed ownership with Britain since 1832. Although the group comprises over 200 islands totalling nearly 4,700 sq mi (12,200 km²), life is centred on the two largest islands – East Falkland (2,580 sq mi; 6,495 km²) and West Falkland (2,038 sq mi; 5,880 km²), separated by Falkland Sound.

A Rugged, Windswept Landscape. The Falkland Islands form a rugged, isolated group. In West Falkland the land rises to 2,315 ft (705 m) at Mt Adam and in East Falkland the Wickham Heights reach 2,315 ft (705 m) at Mt Usborne. From the coast (which features deep fjords, rocky promontories and sheltered bays) the stony terrain rises inland to form generally rolling hill and plateau country covered in rough pasture, peat and moorland. There are very few trees.

High winds are an inescapable feature of life in the Falkland Islands. Winds reach gale force on an average of one day in five. The weather is characteristically raw, cloudy, bracing and very changeable. Mean monthly temperatures range from 36°F (2°C) in July to 49°F (9°C) in January. Annual rainfall, however, is only 26–28 in (660–711 mm). Fish-rich seas support the islands' penguin rookeries, petrel and albatross colonies, sea lions and elephant seals.

How the People Live. The total population is about 2,000, roughly half of it concentrated in Stanley which was established as the capital in 1844 on a sheltered port site on East Falkland. The inhabitants are almost all of British descent.

The islands are almost entirely given over to sheep farming; there are approximately 640,000 sheep whose wool clip, mostly for the London market, forms the only export of any significance. Some sheepskins and hides are also sent out. Only small quantities of oats and potatoes are grown.

Rough overland tracks, coastal launches and an island seaplane service link the scattered settlements. Outside connection is made by the monthly shipping service to Montevideo, about 1,000 mi (1,600 km) to the north, whence sea or air link is made to London.

Falkland Islands Dependencies. These comprise South Georgia Island (1,450 sq mi; 3,755 km²) and the South Sandwich Islands (130 sq mi; 367 km²). South Georgia is about 1,500 mi (2,400 km) south-east of the Falklands, and South Sandwich (uninhabited) a further 470 mi (750 km) south-east again. There are about 150–200 people living permanently on South Georgia. Whale meat, whale oil and seal oil are exported, mostly to Japan.

OCEANIA

AUSTRALIA

Area: 2,967,894 sq mi (7,686,845 km²)
Population: 12,728,461
Population Growth Rate: 2·0%
Capital: Canberra, pop 158,594
Language: English
Religion: Anglican (35%) Roman Catholic (25%) Methodist (10%) Others (30%)
Currency: Australian dollar = 100 cents

The Commonwealth of Australia alone among nations occupies an entire continent. It is the smallest and least populated of the continents, but the nation is nonetheless one of the world's wealthiest and most dynamic.

Progress has been particularly spectacular since World War II. From being a predominantly pastoral country providing wool, meat and dairy products, Australia has become a leading supplier of lead, zinc, iron ore and other minerals, and a major exporter of manufactured goods. At the same time the old imperial ties with Britain have been loosened. World War II showed that Britain could no longer play a protective role, and in military and economic matters Australia now looks more to the United States and Japan. Australia has accepted defence responsibilities in South-East Asia and the South-West Pacific. Japan has supplanted Britain as Australia's chief overseas customer, and co-operation between the two countries in Pacific and South-East Asian affairs seems likely to develop. While both Britain and Japan have provided capital for the development of Australia's great mineral wealth, the USA is now the chief overseas investor.

Australia in History

The continent is geologically very ancient; the country historically young. Its late discovery and considerably later settlement are explained mainly by its comparative isolation from the rest of the world, located entirely in the Southern Hemisphere and encompassed by the Indian and Pacific Oceans. Australia did not initially attract European attention because it was on no route to the spice lands of the East.

James Cook, the great English navigator, explored the eastern coast and formally claimed it for Britain (1770). For many years Australia served Britain as a conveniently distant penal colony, although free settlers also came.

Climatic and overland transport problems were important factors in determining the pattern of settlement which emerged around the coasts. The country's economy evolved gradually with the introduction of merino sheep, the development of wheat farms and cattle stations, and the discovery of gold and other minerals. Serious efforts to unite the separate Australian colonies began in the 1880s, and on the first day of 1901 the Commonwealth of Australia came into being.

Today the Commonwealth of Australia is a federation of six states (New South Wales, Victoria, Queensland, South Australia, Western Australia and Tasmania) and two territories – the Northern Territory and the Australian Capital Territory.

The Face of Australia

The Flattest Continent. Australia is the flattest continent, and flatness characterizes the Western Plateau or Australian Shield which extends from the north-west coast across about half the continent, sloping gently eastwards and reaching its lowest point in South Australia around Lake Eyre, 43 ft (13 m) below sea level. Only the most resistant rocks have withstood the erosion of ages to remain above the general level of the terrain in such features as the rugged Macdonnell and Musgrave ranges in the 'Red Centre' of the continent, where the landscape is reddish-brown. South-west of Alice Springs are two of the world's largest single rocks: Ayers Rock at almost the exact geographical centre of Australia, and Mount Olga.

Plains extend, almost without interruption, from the Gulf of Carpentaria in the north to Spencer Gulf in the south. In the north, a number of rivers flow into the Gulf of Carpentaria, some descending from the Barkly Tableland. Much of the south is drained by the Murray River and its tributaries, the Darling and the Murrumbidgee; the Murray, 1,600 mi (2,575 km) long, is Australia's major river. Many of the rivers marked boldly on maps come to life only during the rains. This is especially true of streams feeding Lake Eyre, which itself is usually no more than a dry expanse of salt encrusted mud.

In the east and south-east the plains rise to form the western slopes of the Great Dividing Range extending from northern Queensland to Tasmania. These eastern highlands are a series of broken, but interlocking, plateaux and tablelands 300 mi (480 km) wide in places which culminate, in south-east New South Wales and north-east Victoria, in the Australian Alps, the continent's highest mountains, where Mt Kosciusko rises to 7,328 ft (2,234 m).

Along the Queensland coast is the Great Barrier Reef, the world's largest coral reef, stretching to the mouth of the Fly River in western Papua and covering with its islets, cays and submerged coral about 80,000 sq mi (207,200 km²).

Patterns of Climate. Australia lies within latitudes 10–40°S and for the most part has a warm, dry climate with an average of more than eight hours of sunshine a day. In the south, the hottest months are January and February; in the north, November and December. Except in the high country of the south-east and in the extreme south-west, summer can be extremely hot with maximum shade temperatures well above 100°F (38°C). Winter temperatures range from 50°F (10°C) in the south to 75°F (24°C) in the north.

Climatic patterns are dominated largely by the seasonal developments and shifts of pressure cells over the continent. In summer low pressure cells and their associated hot, humid and unstable airmasses (the north-west monsoon) dominate northern Australia bringing heavy but unreliable rainfall. 'The Wet', as this season is called, is marked by occasional violent cyclones with winds over 100 mph (160 kph); the torrential rains cause widespread flooding and damage to property.

The southern summer in both the south-west and south-east is characterized by successive heat-waves associated with fine anticyclonic weather. Welcome relief comes from the cooler airmasses associated with the low pressure cells which move south of the continent in a west-east direction. Along the east coast, the prevailing onshore trade winds at times give way to a hot, dry, offshore wind whose sweltering heat is presently mitigated by a cooler south wind. In the Sydney area the cycle occurs with remarkable regularity.

In winter, fine anticyclonic weather prevails in the north: but the south has cloud, thunder and rain as the depression belt south of Australia moves north. Thus Australia can be basically divided into a northern 'monsoon' zone of wet summers and dry winters, and a southern 'Mediterranean' zone of summer drought and winter rains, with a transitional zone between the two.

Rainfalls of more than 177 in (4,492 mm) per annum are recorded in parts of eastern Queensland and of over 145 in (3,680 mm) in western Tasmania. But rainfall decreases rapidly away from the coast towards the arid centre of the continent where some places receive under 5 in (127 mm) of rain a year. Only a small part of Australia enjoys plentiful rainfall. In most areas the high rate of evaporation means that much of the rainfall is lost before it can be used, and experiments in methods of water conservation, such as spreading a chemical film over the surface of

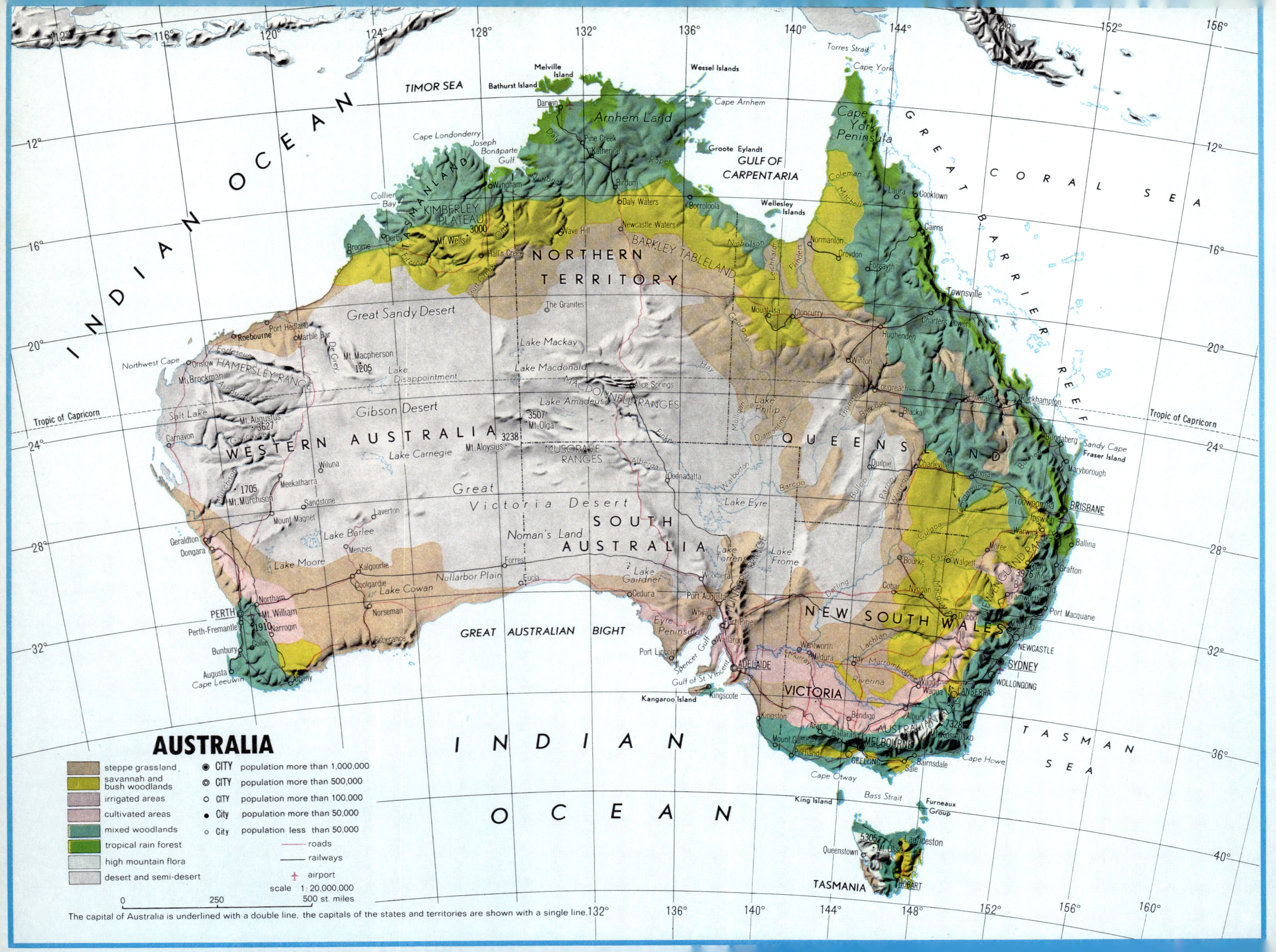
AUSTRALIA
steppe grassland
savannah and bush woodlands
irrigated areas
cultivated areas
mixed woodlands
tropical rain forest
high mountain flora
desert and semi-desert
CITY population more than 1,000,000
CITY population more than 500,000
CITY population more than 100,000
City population more than 50,000
City population less than 50,000
roads
railways
airport
scale 1: 20,000,000
0 250 500 st. miles
The capital of Australia is underlined with a double line, the capitals of the states and territories are shown with a single line.
INDIAN OCEAN
TIMOR SEA
CORAL SEA
GREAT BARRIER REEF
TASMAN SEA
GREAT AUSTRALIAN BIGHT
GULF OF CARPENTARIA
Tropic of Capricorn
WESTERN AUSTRALIA
NORTHERN TERRITORY
SOUTH AUSTRALIA
QUEENSLAND
NEW SOUTH WALES
VICTORIA
TASMANIA
Great Sandy Desert
Gibson Desert
Great Victoria Desert
Noman's Land
Nullarbor Plain
Arnhem Land
Cape York Peninsula
KIMBERLEY PLATEAU
BARKLEY TABLELAND
HAMERSLEY RANGE
MACDONNELL RANGES
MUSGRAVE RANGES
NEW ENGLAND RANGE
PERTH
ADELAIDE
MELBOURNE
SYDNEY
BRISBANE
CANBERRA
HOBART
Darwin
NEWCASTLE
WOLLONGONG
GEELONG
Alice Springs
Kalgoorlie
Townsville
Rockhampton
Cairns
Mount Isa
Broome
Carnarvon
Geraldton
Esperance
Launceston
Lake Eyre
Lake Torrens
Lake Frome
Lake Amadeus
Lake Mackay
Lake Disappointment
Lake Carnegie
Kangaroo Island
Melville Island
Bathurst Island
Groote Eylandt
King Island
Fraser Island
Bass Strait
Torres Strait
Cape York
Cape Arnhem
Cape Londonderry
Northwest Cape
Cape Leeuwin
Cape Howe
Cape Otway

Mt Olga and Ayers Rock (background), sandstone monoliths some 230 million years old, stand near the exact geographical centre of the continent.

reservoirs, have been tried. Periods of below-average rainfall may mean the virtual failure of a season's crops over huge areas.

Vegetation. Most of the varied plant life is purely Australian and adapted to very dry conditions. Eucalyptus trees form a widespread genus represented by several hundred species ranging in height from a mere 2 ft (0·6 m) to more than 320 ft (98 m). Usually called gum trees in Australia, eucalyptus trees include the familiar blue gum which has narrow leaves and peeling bark. Eucalyptus leaves exude an oil whose sweet smell contributes to the haunting aroma of the Australian bush. There are about 600 species of acacia, known in Australia as wattles. This sturdy shrub is typical of the tableland bush, and its golden flower is often used as a national emblem. Among the many beautiful flowering plants are the kangaroo paw, one of the best known western Australian wild flowers, and the dark red waratah of east coast, the floral emblem of New South Wales.

In the coastal south-east, where the growing season exceeds nine months, live plant species adapted to humid and semi-humid conditions. Queensland and northern New South Wales have coastal areas of evergreen tropical forest, while western Tasmania has temperate rain forest, mainly of southern beech. Other parts of the south-east are covered by leathery-leaved bush.

A second zone includes those areas with a five to nine month growing season: the northern coast, the eastern highlands in Queensland, south-east South Australia and south-west Western Australia. In areas of summer rain, leathery-leaved shrubs and open woodlands are common; where there is summer drought the main vegetation is mallee, a stunted eucalypt scrub.

Yet another vegetation zone has only one to five months of plant growth, and is characterized by extremely drought-resistant plants. In Queensland and New South Wales low forest gives way in the south to grassland plains and savanna woodlands, or to mallee and mulga (dwarf acacia scrub). Bordering the deserts in the east is a shrub steppe with salt-bush, blue-bush and mulga. In the arid 'Red Centre', grass hummocks mark the ridges of the sandy deserts and harsh spinifex grass grows in the lower parts. The Australian Alps and the mountains of Tasmania have small areas of alpine vegetation.

The vegetation pattern has been greatly modified by forest-clearing and the introduction of exotic plants, in some cases with economically disastrous results. The American prickly pear cactus, introduced in Queensland, soon spread at the alarming rate of one million ac (404,686 ha) a year. This 'Green Octopus', as it was called, was eventually controlled biologically by the introduction of the *Cactoblastis* moth.

The Unique Wildlife. Australia has been called the land of living fossils, and its unique fauna supports the view that the continent was cut off from the rest of the world during the Mesozoic era between 220 million and 70 million years ago. It includes the most primitive surviving mammals, the platypus and the spiny anteater, and many kinds of marsupials, including the teddy-bear-like koala, the kangaroo, the wallaby, wombat and the Australian opossum. Placental mammals comprise bats, rodents and dingo, a lone and cunning nocturnal sheep-killer which is as great an enemy of farmers as the rabbit. There are many kinds of birds including the unique emu and cassowary, and the kookaburra, an expert snake-killer.

Reptiles include two species of crocodile, fresh-water tortoises and lizards in abundance, and over 100 kinds of snake, including the deadly tiger snake. There are deadly spiders and Australia is notorious for its wealth of poisonous animal species.

The People

Culture and Customs. About 86% of all Australians are of British origin, a fact that contributes to a remarkable homogeneity of customs and outlook through the nation.

The British and other predominantly European origins of Australia's population are reflected strongly in religion. Anglicans account for about a third of the people; Roman Catholics come next, followed by Methodist, Presbyterian and other Christian groups. In education (compulsory to the age of 15), British influence is seen in the exclusive private schools of England, but most children in Australia attend the government schools. Both types together had nearly 2,770,000 pupils in 1970. There were also about 117,000 students attending Australia's 15 universities.

Australia has one of the highest per capita incomes in the world and little unemployment and most families are house- and car-owners. Australia is an egalitarian country with smaller differences between wealthier and poorer residential districts than most Western countries. Bungalow-style homes built of timber or brick predominate, some raised on stilts for coolness. Skyscrapers, urban congestion and the rapid pace of living in Australian cities contrasts with the low verandah-fronted buildings, and wide, quiet streets, of small country towns.

The generally warm dry climate has helped such activities as swimming and surfing to become national pastimes: sport is indeed almost a fetish.

The sheer size of the continent has done much to shape national patterns of living. The Flying Doctor service and radio lessons are features of the lonely Australian Outback.

Australians Old and New. Since World War II the Australian government has stepped up immigration with the twofold aim of exploiting the country's increasingly apparent natural wealth and building a nation numerically powerful enough to deter any would-be aggressor. More than two million immigrants have entered since 1945. Today, nearly 1 in every 5 of the population was born outside Australia, and 1 in 13 in the British Isles. Large numbers of immigrants have come from Italy, Yugoslavia, Greece, Germany, The Netherlands and the United States, but the British element predominates.

One in every six immigrants is a craftsman or factory worker; one in every three, a child or student – a fact that helps to explain why as much as 46% of the population is under 25 years old. The New Australians have tended to go to the large cities – industrial centres like Geelong or Wollongong – or to major construction projects. The Restrictive Immigration Policy formalized in 1901 prohibited permanent settlement by non-whites, but did not bar the entry of students, businessmen and other visitors. In 1970, however, a number of professional Asians were allowed to become residents.

The Aborigines. In 1788 Australia was inhabited by about 300,000 Aborigines – rugged featured dark-skinned individuals who went nearly naked and followed a primitive hunting and food-gathering way of life. Today there are about 140,000 including more than 80,000 full-blooded Aborigines. The Aboriginal population is highest in the remoter parts of the Northern Territory, Queensland and Western Australia. Most are detribalized rural workers or live on government pensions on supervised reserves or in shanties on the outskirts of inland country

Canberra, which began as a squatter settlement, is now Australia's federal capital. It was designed by an American architect Walter B. Griffin.

CHINA
Yangtze tjiang
AMOY
CANTON
HONG KONG (BR.)
VICTORIA
MACAO (P.)
TAIPEI
FORMOSA (T'AI-WAN)
RYUKYU ISLANDS
Okinawa
AMAMI ISLANDS (JAP.)
DAITO ISLANDS (JAP.)
BONIN ISLANDS (JAP.)
Chichi
VULCANO ISLANDS (JAP.)
Marcus (Jap.)
Tropic of Cancer
BATAN ISLANDS
Parece Vela (Jap.)
Asuncion
Pagan
Saipan
Rota
Guam (U.S.)
MARIANA ISLANDS (U.S.)
Wake (U.S.)
HAINAN
SOUTH CHINA SEA
LUZON
MANILA
QUEZON CITY
PHILIPPINES
Mindoro
Samar
Panay
CEBU
Negros
Palawan
SULU SEA
DAVAO
MINDANAO
SOUTH VIETNAM
MICRONESIA
Eniwetok
Bikini
MARSHALL ISLANDS (U.S.)
Jaluit
Yap
Truk
Ponape
Elato
CAROLINE ISLANDS (U.S.)
PALAU ISLANDS (U.S.)
TALAUD ISLANDS
BRUNEI
Brunei
MALAYSIA
CELEBES SEA
Morotai
Halmahera
MANADO
PACIFIC
Nauru
Ocean (Br.)
MELANESIA
BISMARCK ARCHIPELAGO (AUSTR.)
Manus
ADMIRALTY ISLANDS (AUSTR.)
New Ireland
Rabaul
New Britain
SOLOMON ISLANDS
Bougainville (Austr.)
Choiseul (Br.)
Equator
KALIMANTAN
PONTIANAK
BALIKPAPAN
SULAWESI CELEBES
Manokwari
Djajapura
NEW WEST IRIAN
NEW GUINEA
N.E. NEW GUINEA
Ceram
Amboina
PAPUA (TERR.)
BANDA SEA
ARU ISLANDS
TROBRIAND ISLANDS (AUSTR.)
Nw. Georgia (Br.)
D'ENTRECASTEAUX (AUSTR.)
Malaita (Br.)
SANTA CRUZ ISLANDS
Guadalcanal (Br.)
San Christobal (Br.)
INDONESIA
BANDJARMASIN
MAKASAR
TANIMBAR ISLANDS
Merauke
Port Moresby
Torres Strait
Cape York
LOUISIADE ARCH. (AUSTR.)
SURABAJA
Sumbawa
Flores
PORT. TIMOR
Timor
ARAFURA SEA
JAVA
Sumba
TIMOR SEA
C. Arnhem
Darwin
CORAL SEA
Espiritu Santo
NEW HEBRIDES (BR.FR.)
Efate
LOYALTY ISLANDS (FR.)
Huon (Fr.)
Belep (Fr.)
CHESTERFIELD ISLANDS (FR.)
Townsville
NEW CALEDONIA (FR.)
Noumea
Î. des Pines
Walpole
Cloncurry
AUSTRALIA
Alice Springs
BRISBANE
Toowoomba
Onslow
Norfolk (Austr.)
LORD HOWE ISLANDS (AUSTR.)
Darling
NEWCASTLE
SYDNEY
WOLLONGONG
CANBERRA
TASMAN SEA
Tropic of Capricorn
Kalgoorlie
ADELAIDE
GREAT AUSTRALIAN BIGHT
PERTH
Ballarat
MELBOURNE
GEELONG
Cape Howe
Albany
C. Leeuwin
Bass Strait
Launceston
HOBART
TASMANIA
INDIAN OCEAN
Invercargill
30°
20°
10°
0°
10°
20°
30°
40°
100°
110°
120°
130°
140°
150°
160°

OCEANIA
MIDWAY ISLANDS
HAWAIIAN ISLANDS
Lisianski
Layson
La Pérouse Pinnacle
Kauai
Oahu
HONOLULU
Pearl Harbor
Maui
Hawaii
Hilo
(U.S.)
Johnston (U.S.)
Tropic of Cancer
International Date Line
Sunday
Monday
PACIFIC OCEAN
POLYNESIA
Kingman Reef (U.S.)
Palmyra (U.S.)
CENTRAL POLYNESIAN SPORADES (BR. U.S.)
Washington (Br.)
Fanning (Br.)
Christmas (Br.)
Jarvis (U.S.)
LINE ISLANDS (BR.)
Malden (Br.)
Starbuck (Br.)
GILBERT ISLANDS (BR.)
Tarawa
Howland (U.S.)
Baker (U.S.)
PHOENIX
Canton
Enderbury
ISLANDS (BR.)
Equator
ELLICE ISLANDS (BR.)
Funafuti
TOKELAU ISLANDS (Nw.Z.)
(UNION ISLANDS)
Swains
Rotuma
Wallis (Fr.)
Futuna (Fr.)
WESTERN SAMOA
Savai'i
Upolu
Apia
AM. SAMOA
Tutuila
Pago Pago
MANUA ISLANDS
Tongareva (Penrhyn) (Nw.Z.)
MANIHIKI ISLANDS (Nw.Z.)
Vostok (Br.)
Caroline (Br.)
Nuku-Hiva
MARQUESAS ISLANDS (FR.)
Hiva-Oa
Suvorov (Nw.Z.)
Flint (Br.)
FIDJI ISLANDS
Vanua Levu
Viti Levu
Suva
TONGA ISLANDS
Vava'u
Ha'apai
Nuku'alofa
Tongatapu
Niue (Nw.Z.)
COOK ISLANDS (Nw.Z.)
Aitutaki
Rarotonga
TUAMOTU ARCHIPELAGO (FR.)
Rangiroa
Pukapuka
Borabora
Raiatea
Moorea
Papeete
Tahiti
SOCIETY ISLANDS
Anaa
Hao
Rurutu
Tubuai
TUBUAI ISLANDS (AUSTRAL ISLANDS)
Rapa
Raoul (Sunday)
KERMADEC ISLANDS (Nw.Z.)
GAMBIER ISLANDS (FR.)
Tropic of Capricorn
Oeno
Henderson
Ducie
Pitcairn
PITCAIRN ISLANDS (BR.)
THREE KINGS ISLANDS
North Cape
NORTH ISLAND
AUCKLAND
Hamilton
Napier
Cook Strait
HUTT
WELLINGTON
NEW ZEALAND
CHRISTCHURCH
SOUTH ISLAND
CHATHAM ISLANDS (Nw.Z.)
BOUNTY ISLANDS (Nw.Z.)
ANTIPODES ISLANDS (Nw.Z.)
180°
170°
160°
150°
140°
130°
30°
20°
10°
0°
CITY population more than 1,000,000
CITY population more than 500,000
CITY population more than 100,000
City population more than 50,000
City population less than 50,000
roads
railways
airport
INDEPENDENT AFTER 1945
former British territory
former Dutch territory
scale 1: 33,400,000
0
500
1000 st. miles

towns. Their economic influence is negligible. Culturally they have provided apposite names on the Australian map such as *Woomera*, the rocket range so named from the wooden stick with which an Aboriginal warrior launches his spears. Aboriginal words such as *billabong* (a cut-off river loop) and *boomerang* have become part of Australian English. Many artists and designers in Australia now use Aboriginal motifs.

There have been protests and demonstrations by militant groups to draw attention to the Aborigines' generally depressed conditions and to such vital issues as land rights. Official policy is to give all Aborigines 'a similar manner and standard of living to those of other Australians'.

Where People Live. Population has always been concentrated around coastal centres. Today most Australians live in towns or cities; few have first-hand experience of the Outback. Moreover more than two-thirds of the population live in the three cool and relatively humid south-eastern states: New South Wales, Victoria and Tasmania. Sydney and Melbourne alone have nearly 40% of the population, and in the early 1970s 9 out of Australia's 10 cities with over 100,000 inhabitants were located in the south-east or east (the exception being Perth in the south-west).

Sydney, with 2,800,000 people, is Australia's largest and oldest city. It stands on a beautiful harbour spanned by one of the world's longest steel arch bridges. Melbourne, with 2,400,000 people, is several decades younger; it houses the headquarters of more than half of the country's large corporations and is the financial hub of Australia.

In order of population size, the other top 8 cities are Brisbane and Adelaide (both with over 800,000), Perth, Newcastle, Wollongong, Hobart, Canberra and Geelong. Canberra (160,000), in south-eastern Australia was purpose-built as the federal capital. Most of the big cities combine the functions of industrial centre and seaport. Six of the top 10 cities are also state capitals.

The rural population has declined steadily since the early 1930s. Density varies according to the local economy, being highest where there is dairying, sugar-cane, intensive irrigated cultivation or market gardening on the outskirts of towns. The smallest land holdings, averaging 544 ac (220 ha), are in Victoria. Holdings in the Northern Territory average more than 570,000 ac (230,850 ha). The typical Outback cattle station consists of a main homestead and several out-stations. Many are operated by a few permanent hands, with contract workers carrying out jobs like fencing and dam-building. Settlement here still has a pioneering flavour.

The North was long held to be too barren for human habitation. Since World War II, however, it has been strongly argued that it must be populated and developed for defence reasons. The agricultural potential, where water is available for irrigation, has been demonstrated by experimental cultivation of sorghum, other tropical grasses, rice and cotton. Important mineral discoveries have led to the formation of new communities with good social conditions and transport facilities.

The Economy

A Leading Agricultural and Pastoral Country. Nearly two-thirds of Australia's total area is devoted to farming, although much of this land has a very low production capacity. Individuals or families own most of the 250,000 rural land holdings, although there are some company enterprises, especially in the sheep and cattle industries. Farms are highly mechanized and involve a large capital outlay. Specialized transporters and other bulk-handling facilities have been developed, and large sums are spent on crop research, soil improvement and marketing studies.

The old saying that Australia rides on the sheep's back is no longer true. But Australia still provides about 30% of the world's wool, mainly from merino sheep, a fine-woolled breed first introduced in 1797. New South Wales has always had the most sheep. Mutton and lamb come mostly from this state and from Victoria, more than two-thirds of the production being marketed within the country.

Nearly 25% of Australia's cattle are dairy breeds. Dairying flourishes in the eastern and south-eastern coastal areas and on the inland Darling Downs of Queensland. Victoria leads with nearly half the cattle and about 60% of the production. About 60% of milk is converted into butter, and 10% into cheese.

Beef cattle are more widespread. The Hereford strain dominates in eastern Australia, shorthorns in tropical Queensland, the Northern Territory and the Kimberleys. The United States and Japan are the chief export markets.

Aridity or difficult terrain limits Australia's potentially cultivable land to around 8% of the total area. But much of this is grazed by cattle or sheep. On the mere 2% that yields crops, wheat is the major product. Two-thirds of the crop is exported, mainly (as unmilled grain) to China, India, Britain and Japan. In 1970 Australia ranked as the world's ninth-largest producer of wheat. Other cereals grown include oats, barley, maize, sorghum and rice. So-called industrial crops (including cotton, flax, peanuts and sugar-cane) form an important group. Other agricultural products include fruits, which range from paw-

Mount Goldsworthy is one of the most productive iron mines in Western Australia. Ore deposits are estimated at 15,000 million tons.

paws and mangoes in the tropics to currants, grapes and strawberries in the south-east.

Australia's primary producers still depend heavily upon overseas markets, and are thus highly vulnerable to world gluts and price slumps. For example sugar, the second most valuable export crop, faced low world prices and restrictive international quotas in the early 1970s. Government aid, especially at federal level, has become increasingly necessary. Dairy farmers are subsidized; wool, wheat and tobacco prices are stabilized; the sugar industry is government controlled.

Water Conservation and Irrigation. Although Australia as a whole is an exceedingly dry continent, it has perhaps the world's largest artesian water supplies. The Great Artesian Basin, stretching from the Gulf of Carpentaria to South Australia, covers 676,250 sq mi (1,753,000 km^2) and is tapped by about 18,000 bores.

Coober Pedy, west of Lake Eyre, obtains fresh water by solar desalination, and atomic desalination plants to irrigate the deserts are under consideration. Irrigation from rivers, however, is most important.

Since 1915 the Murray River Commission has been responsible for dams and other works along the Murray and its tributaries and for allocating water to Victoria, New South Wales and South Australia. The most important reservoirs are the Hume on the Murray, the Eildon on the Goulburn tributary and the Menindee Lakes on the Darling. In New South Wales more than half of the 1·4 million ac (567,000 ha) under irrigation are used for pasture. Cereals, rice, fruit, vegetables and fodder crops are grown intensively in the Murrumbidgee Irrigation Area, centred on Griffith and Leeton. Most of Victoria's irrigated land is improved pasture, but there are irrigated stone-fruit orchards around Shepparton, and citrus orchards and vineyards in the Mildura-Swan Hill section of the Murray valley. Lesser areas are irrigated in South Australia for fruit including grapes, and in Queensland mainly for sugar-cane.

Western Australia's irrigated area will be more than doubled by the Ord River project in the Kimberleys, which will also serve part of the Northern Territory.

The most dramatic increase in water supplies has been provided by the Snowy Mountains Scheme, completed in October 1972, after nearly 23 years' work and an expenditure of $A800 million. This project, centred on the Snowy and Eucumbene rivers in the Australian Alps, has 16 major dams, 7 power stations and many miles of tunnels and aqueducts. Its waters will not only irrigate some 1,000 sq mi (2,590 km^2) but produce about $A60 million worth of power annually.

Forests and Fisheries. Commercial forests occupy a relatively small area and timber has to be imported. Hardwoods include the karri of south-west Western Australia, the mountain ash (Victoria and Tasmania) and river red gum; these and other eucalypts provide 75% of the sawn timber. Other hardwoods, including walnut, oak and other cabinet woods, are found in high rainfall areas; softwoods are represented mainly by the white cypress and other pines. In the south, serious damage is often caused by bush fires, despite strict government regulations and summer supervision. Man-made forests producing Radiata pine have been established in the wetter south-east area by both government and private enterprise.

Commercial fishing takes place along most coasts, nearly half the catch by value coming from West Australian waters. Mullet are caught in the estuaries, tuna and snoek on the continental shelf, whiting and other fish in deeper waters. Grayling, prawns, oysters, and scallops have become important with the growth of the American market. Production of cultured pearls has expanded in tropical waters.

Minerals: The Age of Abundance. The 1960s ushered Australia into a mineral age far more profitable than that of the 1850s. The country is now self-sufficient in almost all minerals and able to export many in large quantities. The government encourages the new boom with subsidies for exploration, tax concessions for mining companies and other inducements that have helped to accelerate mineral prospecting and to attract foreign investment.

The old saying that Australia rode on the sheep's back is prompted by the traditional pre-eminence of wool as an export. There are 15 sheep per head of population in Australia.

Important minerals have been found in nearly all parts of the continent. Among the older centres of world significance are Kalgoorlie (Western Australia), still a major gold producer; Broken Hill in New South Wales, known for its lead; Mount Lyell (Tasmania), where copper has been mined since the 1890s; and Rum Jungle, where uranium was discovered after World War II.

Iron ore was first mined on a large scale near Whyalla in the Middleback Ranges of South Australia. Today the chief deposits are in the Hamersley and Ophthalmia ranges in north-west Western Australia. Some mines here feed a pelletizing plant at the port of Dampier, others ship their output through Port Hedland. There are more than six other commercially-important deposits including Yampi Sound, the birthplace of the Western Australian mining boom; Koolyanobbing, which supplies the iron and steel complex at Kwinana near Perth; Mount Bundy and Frances Creek in the Northern Territory; and the Savage River district of north-west Tasmania which feeds a pelletizing plant at Port Latta. A huge iron and steel plant, with new port facilities, is being built at Westernport in Victoria.

Australia has been a major lead producer since the development of Broken Hill whose ore is refined at Port Pirie in South Australia. Mount Isa remains the chief copper deposit, and has copper and lead smelters. Copper is also mined at Cobar for refineries at Port Kembla, and at Mount Morgan, Queensland.

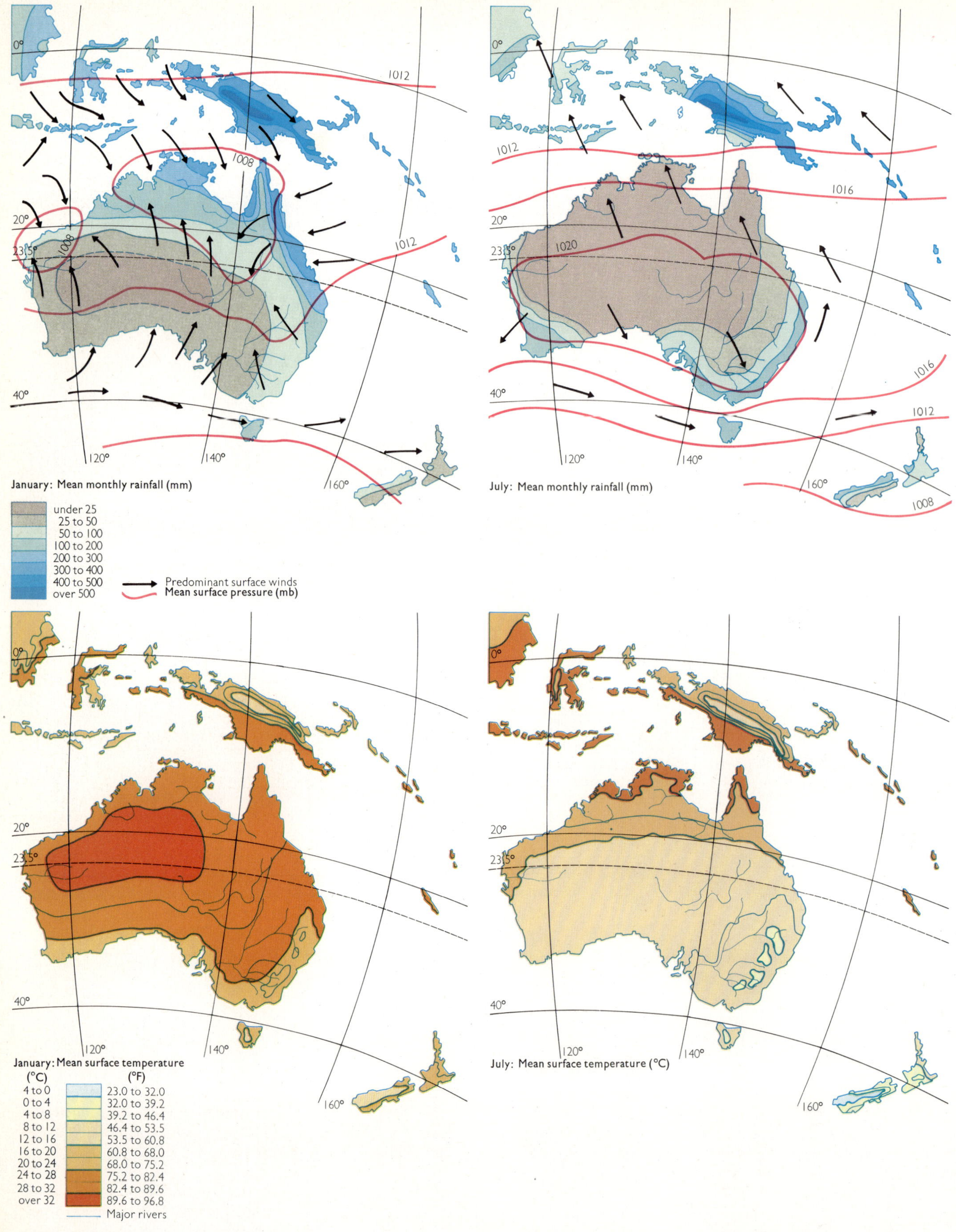

0°
20°
23.5°
40°
120°
140°
160°
1012
1008
1016
1020
January: Mean monthly rainfall (mm)
July: Mean monthly rainfall (mm)
under 25
25 to 50
50 to 100
100 to 200
200 to 300
300 to 400
400 to 500
over 500
Predominant surface winds
Mean surface pressure (mb)
January: Mean surface temperature
July: Mean surface temperature (°C)
(°C)
(°F)
4 to 0
0 to 4
4 to 8
8 to 12
12 to 16
16 to 20
20 to 24
24 to 28
28 to 32
over 32
23.0 to 32.0
32.0 to 39.2
39.2 to 46.4
46.4 to 53.5
53.5 to 60.8
60.8 to 68.0
68.0 to 75.2
75.2 to 82.4
82.4 to 89.6
89.6 to 96.8
Major rivers

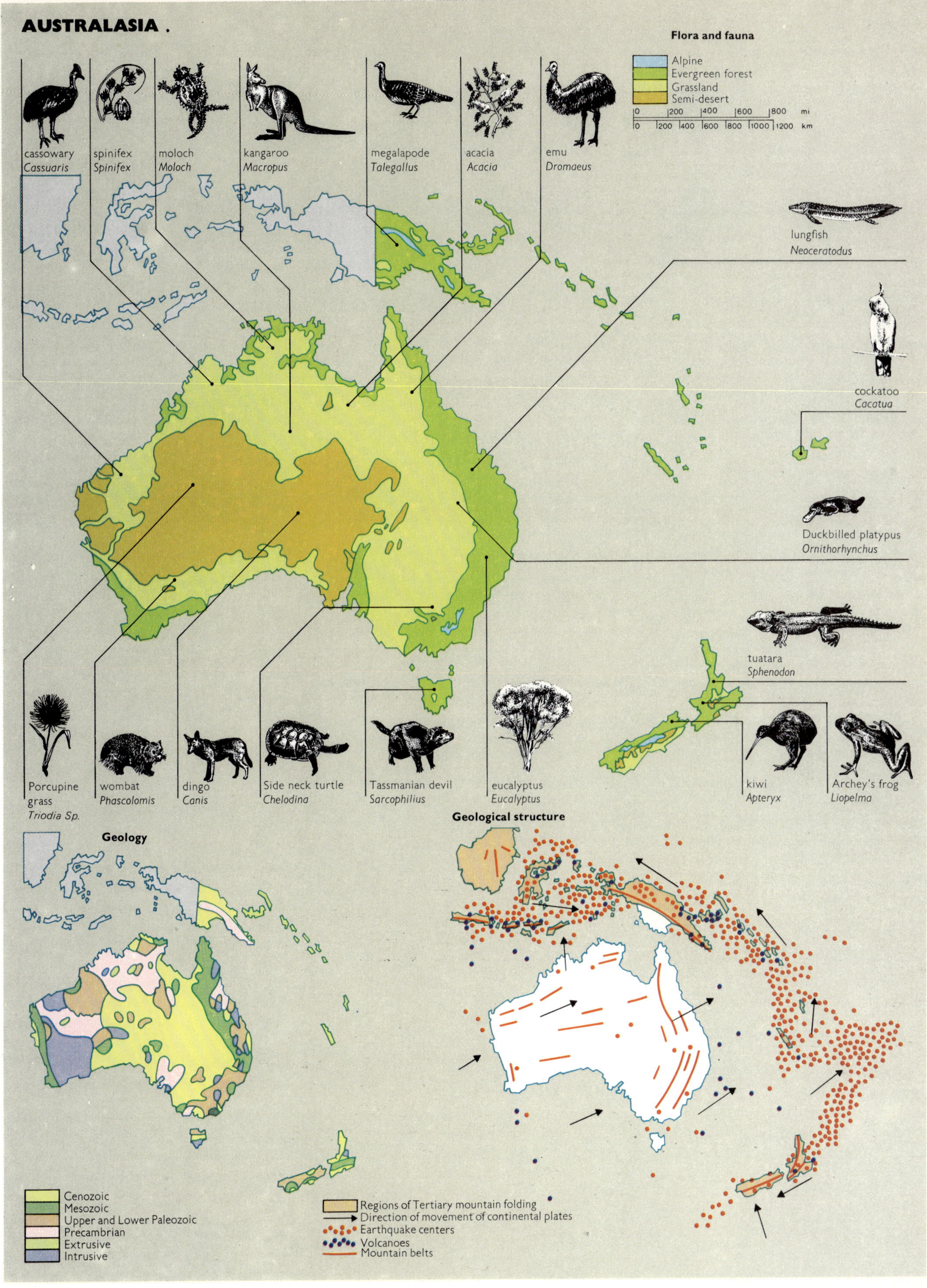
AUSTRALASIA
Flora and fauna
Alpine
Evergreen forest
Grassland
Semi-desert
0 200 400 600 800 mi
0 200 400 600 800 1000 1200 km
cassowary Cassuaris
spinifex Spinifex
moloch Moloch
kangaroo Macropus
megalapode Talegallus
acacia Acacia
emu Dromaeus
lungfish Neoceratodus
cockatoo Cacatua
Duckbilled platypus Ornithorhynchus
tuatara Sphenodon
Porcupine grass Triodia Sp.
wombat Phascolomis
dingo Canis
Side neck turtle Chelodina
Tassmanian devil Sarcophilius
eucalyptus Eucalyptus
kiwi Apteryx
Archey's frog Liopelma
Geology
Cenozoic
Mesozoic
Upper and Lower Paleozoic
Precambrian
Extrusive
Intrusive
Geological structure
Regions of Tertiary mountain folding
Direction of movement of continental plates
Earthquake centers
Volcanoes
Mountain belts

Manganese is worked mainly in the Pilbara-Peak Hill area, but there are much larger reserves in the Gulf of Carpentaria.

Nickel, the prestige mineral of the early 1970s, is mined chiefly at Kambalda in south-central Western Australia and refined at Kwinana. The discovery in 1966 of large phosphate deposits at Lady Anne and Duchess in north-west Queensland is of great importance to Australian farmers, for supplies from Ocean Island and Nauru will cease within a few decades. The Herberton tin mines south-west of Cairns in Queensland, together with earlier finds in other parts of the continent, make Australia self-sufficient in this metal.

The chief bauxite deposits are at Weipa on the Cape York Peninsula, at Gove in Arnhem Land, and in the Darling Ranges of Western Australia. There are aluminium smelters at Bell Bay (Tasmania), Point Henry (Victoria) and Kurri Kurri (New South Wales), and alumina refineries at Gladstone (Queensland).

Black coal is produced in every state except Victoria, but the bulk comes from New South Wales and Queensland. Coal is exported to Japan and Europe, and to New Caledonia where it is used by the nickel refining industry. Victoria has the world's largest known deposits of brown coal in the Latrobe valley. Exploited in open cuts, this coal provides cheap electricity for the state.

Natural gas and oil have been found in many states. Commercial quantities of oil were first struck in Queensland's Moonie field in 1961. Natural gas from the same sedimentary basin at Roma and Rolleston is piped to Brisbane. In Western Australia the Barrow Island field began oil production in 1967, while Perth and Kwinana are to receive natural gas from the Dongara field near Geraldton. A 480 mi (770 km) pipeline takes natural gas to Adelaide from the Gidgealpa and Moomba fields in northern South Australia. The Bass Strait reserves, tapped off-shore and processed at Dutson, are the most important, however. Melbourne, Geelong and Westernport already receive gas from Bass Strait, and Sydney may be linked to this field during the 1970s.

Industry. Manufacturing received a great boost during World War II, when imports were limited and large-scale production of armaments and war supplies involving a high degree of mechanization became essential. Today Australia has a mature and balanced industrial structure and 27% of the labour force are engaged in manufacturing.

Some types of manufacturing, such as the food and drink industries, are widespread. Generally, however, the concentration is heaviest in the south-east, especially in New South Wales and Victoria, where nearly 75% of the factory workers live. Sydney dominates the engineering industry, while Melbourne leads in textiles and leather processing.

The problem of the 1970s became the search for overseas markets. Early on the automobile industry took a strong lead, exporting to New Zealand, the Pacific islands and Africa, and fixing its sights on the Japanese and South-East Asian markets. By contrast, some industries, such as printing and electrical equipment, are shielded from foreign competition by protective tariffs.

Power. Energy demands were expected to double in the 1970s. Petroleum will be increasingly important as coal declines. By 1980 natural gas will be providing about a tenth of Australia's fuel needs. Hydro-electric power is vital only in Tasmania, but its mainland importance has been increased by the Snowy Mountains and Kiewa (north-east Victoria) projects. Australia's first nuclear power station, at Jervis Bay will start generating electricity for New South Wales by 1975.

Transport. How to provide efficient transport over Australia's huge thinly peopled areas is a difficult problem. Roads are of prime importance, and, all told, Australia has about 550,000 mi (885,000 km) of them, nearly half sealed or surfaced. Road haulage between capital cities is increasing and has long dominated short distance movement. Railways still face the problem of different gauges of track between states, although standard gauge routes now link Sydney with Melbourne and Perth. This route spans 2,460 mi (3,960 km) and there are lines serving much of the east, south-east and south-west. Shipping has recently prospered. Australian cargo ships now sail to Japan, Britain and North America, thus freeing Australian exporters from their previously complete dependence on foreign shipping lines. All passenger services, however, still belong to overseas companies.

Australia has a network of internal air services, and Qantas, a major airline with international routes.

International Trade. Except for the early 1940s and mid-1960s, Australia has generally had a continuously favourable balance of trade. Agricultural and pastoral products, once pre-eminent, now account for only some 40% of the value of exports, being overtaken by minerals and metal goods. Nearly half the imports are materials for use in manufacturing, and a further quarter represent capital equipment. Australia trades with many countries, but its chief overseas customers are now Japan and the United States.

Dependencies. In the early 1970s Australia administered the eastern half of the island of New Guinea and certain nearby Pacific islands. The territory attained self government on 1 December 1973. Australia controls Norfolk Island, Lord Howe Island and Macquarie Island in the Pacific; Christmas Island, the Cocos (Keeling) Islands, Heard Island and the McDonald Islands in the Indian Ocean; and the Australian Antarctic Territory.

Trains are essential for transporting goods across the western plains. Practically all of the western plateau is desert or arid scrub land, traversed only in a few places by rail and road.

PAPUA NEW GUINEA

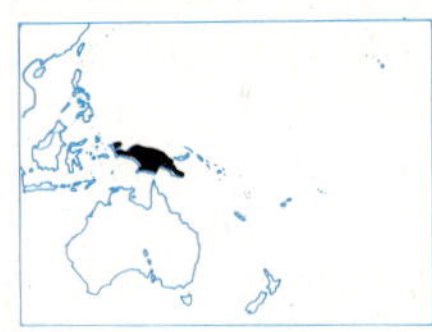

Papua New Guinea, which on 1 December 1973 emerged as a new nation, comprises the eastern half of the island of New Guinea and its adjacent islands. Lying just south of the equator, this island region of lofty mountains and swampy plains – largely jungle-clad and still unexplored – became dimly known to the West only after Portugal's Antonio de Abrea sighted New Guinea in 1511. New Guinea was later divided between the Dutch, Germans and British.

Papua New Guinea eventually emerged as a union of the former British protectorate of Papua, transferred to Australia in 1906, and the UN trust territory of New Guinea, the former German New Guinea. Papua, the south-eastern section (90,540 sq mi; 234,400 km^2), includes south-eastern New Guinea and the D'Entrecasteaux, Trobriand, Woodlark and Louisiade groups of islands. The trust territory (93,000 sq mi; 240,900 km^2) comprises north-eastern New Guinea together with Bougainville and other northern Solomon Islands, and the Bismarck Archipelago.

The Face of the Islands. The major land area is the eastern half of New Guinea, the world's second largest island which is altogether about 1,500 mi (2,400 km) long; some 480 mi (770 km) wide in the centre; and covers about 317,000 sq mi (788,000 km^2).

Running the length of the island from south-east to north-west is the lofty central cordillera including the Owen Stanley and Bismarck ranges in Papua New Guinea, whose highest point is Mt Wilhelm (14,762 ft; 4,500 m) compared with 16,535 ft (5,040 m) for Mt Djaja in West Irian, the island's highest peak. North of the cordillera, beyond the intermontane trough, lie other ranges and the northern coastal lowlands. South of the central cordillera in Papua is a foothill zone containing Mt Murray and other volcanoes which separates the high ranges from the southern lowlands. Here are the swampy plains of the Fly and other rivers.

The smaller islands are largely of volcanic or coral origin. Bougainville in the Solomons, and New Britain in the Bismarck Archipelago (the largest islands after New Guinea), are exposed parts of largely submerged mountain chains.

Climate, Vegetation and Wildlife. Lying close to the equator, the islands forming Papua New Guinea are mainly hot and humid throughout the year, with lowland temperatures averaging 80°F (27°C) and rainfall averaging 50 in (1,270 mm) in the lowlands and over 200 in (5,080 mm) in the mountains of the island of New Guinea.

Vegetation on New Guinea itself ranges from the mangroves, nipa palms and tropical rain forests of the coast and lowlands to the coniferous forests and grasslands of the higher mountain slopes whose highest summits are perennially snow covered. Dense forests cover much of the surface of the small islands.

The People. Papua New Guinea has a population of about 2,467,000. The largest town is Port Moresby (66,244), the administrative capital.

The indigenous peoples of the island of New Guinea fall mainly within two major groups: the dark skinned Melanesians, who live mostly along the coasts, and the woolly haired Papuans, who mostly live inland. A third group consists of the small and primitive Negrito peoples of the remote mountain areas. Melanesians predominate in the Solomon Islands and the Bismarck Archipelago. Papua New Guinea also has Polynesian, Asian and European minorities.

Fewer than 80,000 pupils attended government and mission schools by 1970, but a university had been opened in 1966. Even head-hunting and cannibalism have not entirely disappeared. In recent years, however, much progress has been made in bringing forward the indigenous peoples so that they can increasingly manage their own affairs.

Economic Production. Primitive Stone Age hunting persists in the remote New Guinea interior, but the economic mainstay is agriculture. In New Guinea, subsistence food crops like sago, yams, taro, manioc and sweet potatoes are mostly produced under a system of shifting cultivation ('bush fallowing'). The only domesticated food animals kept by most aboriginal peoples are pigs and poultry. European colonists introduced plantation farming in certain areas, notably in the Bismarck Archipelago and in south-eastern New Guinea. Major cash crops now include coconuts, cacao, coffee and rubber. Other crops include peanuts, passion fruit, rice, pineapples and pyrethrum.

Prawns, crayfish and cultured pearls figure among commercial sea products.

Gold is mined locally, and newly-discovered copper deposits are being exploited. More than $A100 million has been spent on oil exploration in Papua New Guinea, and while commercially worthwhile deposits have still to be found, significant flows of natural gas have been discovered in the Gulf District of Papua.

Industry is mainly limited to processing farm and forest products. Activities include producing palm oil and copra, processing cacao, sawmilling and ship repairing. Palm-oil and livestock industries are being developed and timber production is expanding.

Transport and Trade. Because of its dense vegetation and mountainous character, New Guinea has few roads. Ships and aircraft thus assume great importance. The island's chief ports are Port Moresby, Samarai, Lae and Madang. Other ports include Rabaul, Kavieng and Lorengau in the Bismarck Archipelago.

Cash crops (notably coconuts, cacao, coffee and rubber) provide about 85% of the country's exports. Imports include machinery, vehicles, fuels and foodstuffs. Australia, Japan and Britain are the main trading partners.

A highland village near Goroka. Much of New Guinea's interior has not been visited by Europeans.

NEW ZEALAND

Area: 103,736 sq mi
(268,676 km²)
Population: 2,961,869
Population Growth Rate: 1·5%
Capital: Wellington, pop 328,800
Language: English. Maori also spoken
Religion: Protestant (65%) Roman Catholic (15%)
Currency: New Zealand dollar = 100 cents

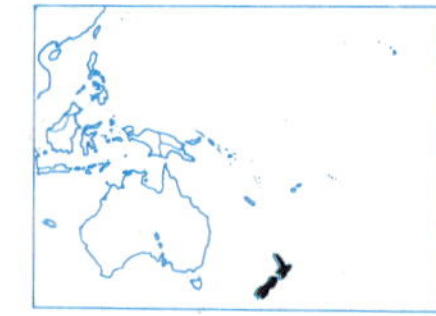

New Zealand is a little bigger than the United Kingdom (from whose peoples most New Zealanders are descended), yet it has only about one-twentieth as large a population. Endowed with a mild South Pacific climate and fertile plains, the islanders have built a wealthy economy renowned for its progressive social services.

Archaeological evidence helps to confirm that Polynesian ancestors of the Maoris arrived from the north about 1,000 years ago. British sovereignty was recognized by the Maori chiefs in the Treaty of Waitangi (1840) which cleared the way for systematic colonization. The young colony became self-governing in all but a few respects in 1852; dominion status was granted in 1907. Today New Zealand has one of the world's most comprehensive welfare systems, little unemployment and an enviably high standard of living. Prosperity literally has grass roots, for it has long stemmed from the lush island pastures and the large numbers of sheep and dairy cattle which they sustain.

In international affairs New Zealand has strengthened its interests in the Pacific area since World War II. It has drawn closer to Australia and the United States and is actively promoting the advance of indigenous Pacific peoples. Western Samoa, which achieved independence in 1962, was formerly administered by New Zealand, which continues to act for it in foreign relations. Wellington, the New Zealand capital, was the scene of the first meeting (1971) of the South Pacific Forum, at which common problems were discussed by Fiji, Tonga, Western Samoa, Nauru and the Cook Islands.

A Nation of Islands. New Zealand's geographical boundaries contain two major islands (the North Island and the South Island); and the smaller Stewart Island; Chatham Islands; Kermadec Islands; Campbell Island; and the uninhabited Three Kings; Snares; Antipodes; Solander; Bounty and Auckland Islands.

New Zealand as a whole is geologically young, structurally a part of 'the girdle of fire' – the long volcano-dominated, earthquake-prone zone that encircles the Pacific Ocean. Much of it is hilly or mountainous and geologically recent faulting accounts for abrupt changes from one type of landscape to another. There are also striking differences between the two main islands.

The North Island has an almost continuous succession of hills and ranges such as the Raukumura Range, the Kaimanawa Mountains and Tararua Range extending from East Cape south-westwards to Cook Strait. Hills and mountains make up some 63% of the island. The central feature is the Volcanic Plateau formed of ash-covered ignimbrite (a distinctive volcanic rock) and dominated by the volcanic peaks of Ruapehu (9,175 ft; 2,797 m), smoking Ngauruhoe (7,515 ft; 2,291 m) and Tongariro (6,517 ft; 1,986 m). Away to the west, in Taranaki, is the remarkably symmetrical cone of Mt Egmont (8,260 ft; 2,518 m). The plateau has numerous boiling mud pools, hot springs and solfataras, the best-known being at Whakarewarewa near Rotorua. Such activity may be the dying phases of volcanicity, but the explosion in 1886 which tore a 9 mi (15 km) rift across Mt Tarawera and spread ash over some 6,000 sq mi (15,540 km²) suggests that further violent activity is not impossible.

The volcanic district, which includes Taupo and other lakes, extends north-east to the Bay of Plenty, where White Island has its steaming fumaroles. The Coromandel Peninsula and the Waitakere hills near Auckland are composed of old volcanic rocks.

The North Island has more modest plains than the South Island. They include the Manawatu Plain, west of the Tararua Range; the lower valley of the Waikato, New Zealand's longest river; the Hauraki Plains around the head of the Firth of Thames; and the Tauranga Harbour lowlands. Among the features of the North Island's coasts are the white sand beaches flanking the Tasman Sea in Northland, the towering cliffs of the Coromandel Peninsula, and the fine natural harbours of Waitemata (an inlet of Hauraki Gulf) and Port Nicholson which serve Auckland and Wellington respectively.

The South Island, separated from the North Island by Cook Strait, has mountains and hill country on a far grander scale. Here are such strongly glaciated mountains as the Kaikoura, Seaward Kaikoura and St Arnaud ranges, the Spenser Mountains, and the Southern Alps containing Mt Cook (12,349 ft; 3,764 m), New Zealand's highest peak, and numerous glaciers and ice fields. In the extreme south-west is Fiordland, an intensely glaciated plateau area with steep-sided valleys and deep, narrow fjords, here called sounds.

East of the Southern Alps is the Canterbury Plain, the largest and best-known in Newland, which slopes steadily eastwards from about 1,000 ft (300 m) to sea level. This slope reflects its origin, for it has been formed from material deposited by such rivers as the Rakaia and Waimakariri. The narrow Westland Plain and the Southland Plain are the only other noteworthy lowland areas. Of the South Island's many lakes, the largest is Te Anau east of Fiordland.

Climate. New Zealand has a temperate equable climate. Extremes of precipitation and temperature are rare, although the latitudinal extent and mountainous nature of the country ensure that important differences do occur. In the South Island, for example, the west-facing slopes of Westland and Fiordland may receive more than 100 in (2,540 mm) precipitation annually, while in sheltered localities of central Otago, the other side of the mountains, as little as 13 in (330 mm) may fall. In general, rainfall is distributed uniformly throughout the year, and large areas receive 40 in (1,016 mm) or more. Temperatures decrease from a mean annual figure of 59°F (15°C) in the north to 49°F (9°C) in the south. Except in the mountains, summers are warm and winters mild.

Vegetation. In the North Auckland peninsula there are still stretches of the once extensive kauri pine forests. Other native trees such as rimu, matai and totara occur in the hills and mountains of the central North Island. The South Island has extensive southern beech forests, seen at their finest in the north-west and in Fiordland.

The New Zealand 'bush' has many interesting trees and shrubs such as the scarlet-flowered pohutukawa (Christmas tree), subtropical nikau palms and giant tree ferns.

Wildlife. New Zealand has no native land mammals except bats. The colonists introduced deer, now regarded as a pest, and other animals. There is, however, the unique, lizard-like tuatara, often described as a living fossil, which now survives only on certain offshore islands. Birds include the kiwi (New Zealand's national bird), the weka (wood hen) and kakapo (owl-parrot).

The People. The population is primarily of British stock. The Maoris declined rapidly during the first few years of contact with the Europeans, but since 1896 have increased their numbers to some 233,000. Maoris and *pakehas* (whites) are well integrated. The

Maoris, still proud of their culture, have provided outstanding figures in many fields. There are small numbers of Yugoslavs, Chinese, Indians, Dutch, and Pacific Islanders.

A feature of the population is its great mobility. Immigration has always been important, and the present annual intake is about 40,000 persons, mainly from Britain, but also from Australia and The Netherlands. Within the country, people tend to move from the South Island to the North Island, and from the country areas to the towns. The result has been to give the North Island over 70% of the total population and to make New Zealand one of the world's most highly urbanized countries. The drift of Maoris from the outlying areas of Northland and Eastland has been particularly significant. Auckland now has the largest concentration of Polynesians in the world.

Four metropolitan centres are pre-eminent – Auckland, Wellington-Hutt, Christchurch and Dunedin – and they are also the main hubs of industry and commerce. Most of the smaller towns serve their local farming community, but a few have special interests. Kawerau and Tokoroa, for example, are both forestry towns.

Beliefs and Culture. Religion was an influential factor in the settlement of New Zealand. Canterbury, for example, was settled by Anglicans, while Otago was founded by members of the Free Church of Scotland. Today Anglicans are most numerous, followed by Presbyterians, Roman Catholics, Methodists and Baptists.

New Zealand has an impressive educational record and illiteracy is negligible. Education is free and compulsory from 6 to 15, and there is also a government-aided free kindergarten organization for the very young. There are six universities.

Agriculture is the life-blood of New Zealand. The area under cultivation is relatively small, for this is a land where vivid emerald pastures dominate the landscape, except in the High Country of the South Island were tawny natural tussock grass is still widespread.

Sheep take pride of place, more than half the flock of 59 million being reared in the North Island. Farms in Eastland, the Manawatu and South Auckland (and also in Southland, in the South Island) raise sheep primarily for wool, but also for fat lambs. The Canterbury Plain has mixed crop and livestock farms where sheep (and some cattle) graze on pastures rotated with wheat, barley, oats, green fodder and root crops. The traditional large sheep 'runs' (farms) occupying several thousand acres are now found only in the High Country.

New Zealand is no less famous for its dairy farming, which rose to prominence after the advent of refrigeration in 1882. The most important dairying areas are the Hauraki-Piako lowlands and the Waikato (especially for butter). Dairying is also important along the Bay of Plenty and in Taranaki, noted for its cheese. Beef cattle are reared mainly in Eastland.

Commercial orchards are the fourth major element. Large quantities of apples and pears are grown near Nelson and Hawke Bay; stone fruits (apricots, cherries and peaches), near Roxburgh in central Otago; citrus fruits and vines near Auckland and Gisborne, and especially at Kerikeri in Northland.

Forestry. The indigenous forests still provide valuable timber, but the forestry industry is dominated by its conifer plantations, chiefly on the Volcanic Plateau, which feed the pulp and paper plants of Kawerau, Tokoroa, Rotorua, Whakatane and other centres. Forest products represent about 41% of all manufactured exports.

A cable car ascending the hill rising from Wellington harbour towards the suburb of Kelburn.

The hot springs at Wairakei on the North Island are used to generate electricity. This area of geysers and hot springs is a recent volcanic formation and is of great interest to geologists.

Minerals. Coal is mined in more than 25 localities scattered through both main islands, sub-bituminous coal coming mainly from the Waikato field in the North Island, and bituminous coal from the Buller field in the South Island, whose Ohai and Mataura fields have recently increased production. The black iron-sands of the west coast beaches, especially those at Waikato North Head, are being exploited in connection with the new iron and steel mill at Glenbrook near Auckland. Limestone is quarried extensively for agricultural use and for cement. Lead-zinc ores are mined near Te Aroha. Small quantities of oil are produced, and natural gas from Kapuni in Taranaki is now piped to Auckland, Wellington and other centres in the North Island.

Industry. Manufacturing now employs about 25% of the labour force, and while it is concentrated mainly in the four metropolitan centres, it is assuming new importance at lesser towns such as Hamilton, Whangarei and Tauranga in the North Island. Regional specializations include the South Island's woollen towns (Mosgiel, Kaiapoi and Milton), the pulp and paper industry of the Bay of Plenty area, and the food-processing industry which is dispersed near its various raw materials. Most industry, including the recently developed light aircraft industry, is on a small scale. Larger concerns include the meat freezing works (Southdown and Westfield in Auckland, Mataura and other centres), the oil refinery at Whangarei, the pulp and paper industry, Glenbrook with its iron and steel, and Bluff with its new aluminium smelter.

Power. Hydro-electric power is produced by plants on the Waikato and other major rivers and at Lake Waikare Moana. The North Island taps Benmore and other sources in the South Island through a cable across Cook Strait. Geothermal power, tapped at Wairakei near Taupo, is used to generate electricity and nuclear power stations may be built.

Tourism is fast developing. New Zealand has not only a pleasant climate and fine scenery, but also winter sports, hunting and some of the world's best fishing.

Transport. New Zealand has good road and rail networks, inter-island shipping services, and domestic and international air services. The chief seaports are Auckland, Wellington, Lyttelton and Dunedin. Of growing importance are Whangarei, the main port for oil imports, and Tauranga, a timber port.

International Trade. Dairy produce, meat and wool overshadow all other exports, with timber coming up fast. Manufactured goods, machinery, transport equipment, petroleum, fertilizers and chemicals are the chief imports. Britain was still a major trading partner in the early 1970s, although its share in New Zealand's trade had decreased significantly since 1965. Japan has acquired fresh prominence in New Zealand's trade, which also embraces many other countries including the USA. The Australian market is opening to New Zealand goods as a result of the 1965 free trade agreement.

Dependencies. Overseas territories comprise Niue Island (one of the Cook Islands); the Tokelau Islands and Ross Dependency in Antarctica. The Cook Islands (except Niue) are a self-governing overseas territory.

NORFOLK ISLAND

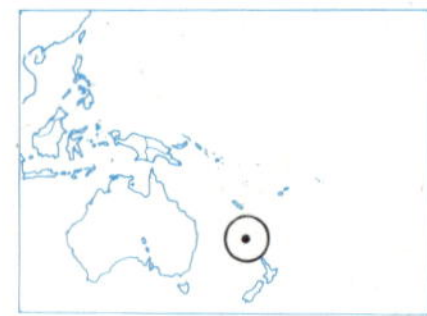

Norfolk Island, a beautiful, hilly, wooded island of just over 13 sq mi (34 km²) lies about 1,000 mi (1,609 km) north-east of Sydney, New South Wales. This Australian territory was once a penal colony, the first free settlers being descendants of the *Bounty* mutineers transferred from Pitcairn Island in 1856. The present population (1,240) is mainly engaged in the flourishing tourist industry.

COOK ISLANDS

Cook Islands in the South Pacific are a scattered group named after Capt James Cook, who discovered many of them (1773). They have been a New Zealand territory since 1901, and internally self-governing since 1965.

The archipelago consists of two groups: the Northern (Nassau, Palmerston, Penrhyn, Manahiki, Rakahanga, Pukapuka and Suwarrow): and the Southern or Lower group (Rarotonga, Mangaia, Atiu, Aitutaki, Mauke, Mitiaro, Manuae and Te au-o-tu). The total area is some 93 sq mi (240 km²). The northern islands are low atolls; the southern, high and volcanic with fringing reefs.

People and Economy. The population of 21,227 consists mainly of Cook Island Maoris, a Polynesian people closely related to the Maoris of New Zealand. Most of the people live in the southern islands, but pressure on land and lagoon resources in both groups is now so heavy that many islanders are migrating to New Zealand. Avarua (11,398), the capital and chief port, is on Rarotonga, the largest and most beautiful island.

The islanders live largely by subsistence farming and fishing. The chief cash crops are copra, citrus and other fruits. Nearly all external trade is with New Zealand.

NIUE

Niue Island is a coral island in the Pacific Ocean. Geographically it is one of the Cook Islands, but has been separately administered since 1903, and since 1949 has been an overseas territory of New Zealand. The island covers 100 sq mi (259 km²) and has a mainly Polynesian population of 4,988, mostly engaged in producing copra and bananas.

FIJI

Fiji comprises more than 320 islands in the south-west Pacific, of which only some 106 are inhabited. After tribal wars and disease had ravaged the islands, the Fijian chiefs asked Britain for protective supervision and in 1874 Fiji became a British colony. Indians were first brought to the islands in 1879 as indentured labour for the sugar plantations and outnumbered the indigenous Fijians by 1970 when Fiji became an independent dominion within the Commonwealth of Nations. Racial friction is one of Fiji's problems, especially in matters of land ownership. Most of the land is owned by the indigenous Fijians, while the Indians are commercially powerful.

Modern commerce and the jet age are transforming Fiji. There has been considerable investment by Australian, British, Japanese and other interests, and the main islands are increasingly feeling the impact of tourism. Such influences make it all the more difficult for the Fijians to continue their traditional patterns of life.

The Islands. Viti Levu and Vanua Levu are the two largest islands and constitute about 87% of the total land area of 6,852 sq mi (18,272 km^2). Volcanic in origin, they are mountainous, well watered, and protected by a maze of coral reefs. The other islands, some volcanic, others merely sand on coral, include the Lau and Moala groups, the Lomaiviti islands scattered about the Koro Sea, and the Kandavu and Yasawa groups.

From November until April the climate is oppressively hot and humid with severe thunderstorms. For the rest of the year the south-east trade winds prevail, bringing heavy rainfall to the south-east coasts, which are consequently covered with tropical rain forest, while the drier leeward slopes have only lighter forest, casuarinas, reeds and grasses.

The People. Indians descended from plantation workers form more than half the population. The indigenous Fijians, primarily Melanesian but with some Polynesian blood, take second place, numbering some 220,000 against the Indian 263,000. There are also about 7,000 Europeans, 10,000 Euronesians, 13,000 Rotumans and other islanders, and some 5,000 Chinese. Nearly half the population is Christian (mainly Methodist); most of the rest is Hindu. Suva, the capital and chief port (60,000) is the only city, and about 75% of its inhabitants are Indians.

The Economy. Commercial agriculture is dominated by the sugar crop, which provides more than half of Fiji's exports by value. It is grown mainly on Vitu Levu and Vanua Levu on small farms worked by Indian tenant farmers. By contrast, the Fijians grow little to sell. Their chief food crops are taro, yams, tapioca, coconuts and tropical fruits. Surpluses of copra and bananas are marketed. The forests provide mahogany and pine, and forest industry is expected to increase as a result of recent planting programmes.

Gold ranks high among Fiji's exports, although the output is now limited to a single mine at Vatukoula on Viti Levu. There are small deposits of silver and manganese, and copper has been found recently in northern Vanua Levu. Suva has a wide range of light industrial activities such as copra-processing, making aluminium products and boat-building. Industries outside the capital include cane-crushing, pineapple-canning and gold-refining.

Fiji is a major place of call on trans-Pacific air and sea routes. Its chief trading partners are Britain, Australia, Japan and the United States. Britain buys 50% of Fiji's sugar and much of her copra and coconut oil.

TOKELAU ISLANDS

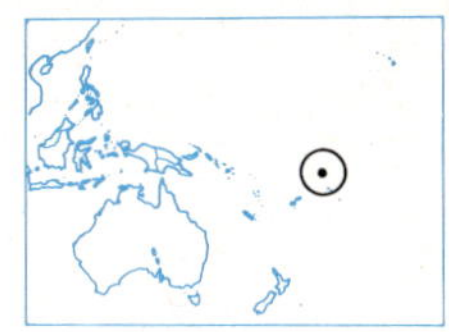

The Tokelau (formerly Union) Islands, a small group about 270 mi (435 km) north of Western Samoa, form an overseas territory of New Zealand. Covering a mere 4 sq mi (10 km^2), it consists of three atolls (Atafu, Nukunonu and Fakaofo) and has a Polynesian population of 1,687 engaged in subsistence farming and copra production. Because of the unfavourable economic outlook, plans were made in 1965 to transfer the islanders to New Zealand, but by 1971 only a few hundred had been resettled.

NAURU

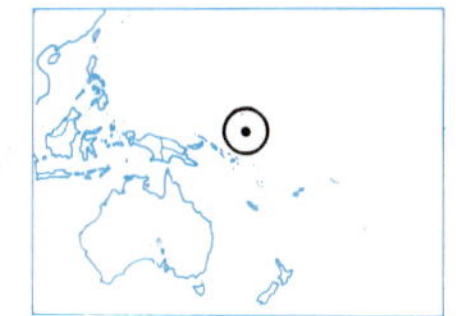

Nauru is a tiny coral-reefed island in the Western Pacific. It was a UN trusteeship territory administered by Australia until 1968 when it became an independent republic with special membership of the Commonwealth of Nations. Covering only 8·5 sq mi (22 km^2), it has a narrow fertile coastal strip and an inland, phosphate-rich plateau which provides the 6,768 islanders (including about 3,400 Micronesian Nauruans) with their only export and (in 1969) the highest income per head in the world. Some of the profits from the phosphates are being set aside against the time when the deposits run out, probably by the early 1990s.

TONGA

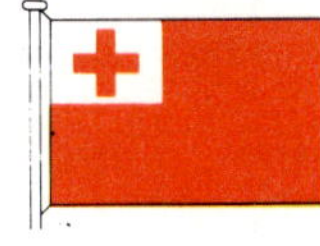

The Kingdom of Tonga (formerly the Friendly Islands) in the South Pacific became independent within the Commonwealth of Nations in 1970 after some 70 years as a British protected state. Lying east of Fiji, Tonga consists of more than 150 islands, the main groups being Tongatapu, Haapai and Vavau.

The westernmost islands are rugged and volcanic, a continuation of the volcanic chain that stretches from New Zealand through the Kermadec Islands to Savaii in Western Samoa. In 1946 a violent eruption caused Niuafoo to be evacuated, and isolated volcanic activity still occurs. Most of the eastern islands are coral limestone; some lie almost at sea level, but many have been uplifted to over 500 ft (152 m).

The climate is cooler and drier than in neighbouring Fiji and Samoa, temperatures averaging 75°F (24°C) and rainfall varying between 45 in (1,143 mm) and 80 in (2,032 mm). Hurricanes may occur from November to March, but the prevailing winds are the south-east trades.

The People. Apart from about 900 Europeans and part-Europeans, the rapidly increasing population (87,000) is almost entirely Tongan, a Polynesian people. Education and medical services are free.

The Economy. Tonga is an interesting example of how a traditional agricultural economy can survive successfully in modern conditions. European or American plantations do not exist. Since 1862 the basic land unit has been the 8·5 ac (3·4 ha) granted to each male Tongan at a nominal rent when he attains the age of 18. A variety of crops is grown on these holdings, including coconuts, bananas, oranges, yams and taro. Copra and bananas are exported. Oil is being sought, and tourism may offer hope of further diversification.

SOLOMON ISLANDS

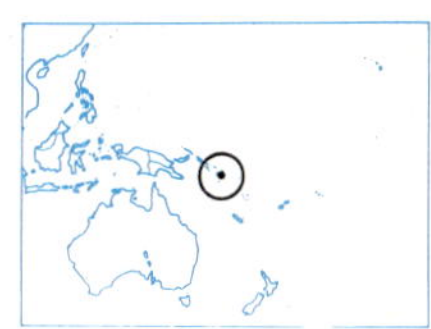

The British Solomon Islands, a British protectorate, are located in the south-west Pacific between latitudes 5° and 12° south and longitudes 155° and 170° east. There are six large high islands: San Cristóbal, Guadalcanal, Malaita, Santa Isabel, New Georgia, and Choiseul. And there are also the smaller Rennell Island and the Ontong Java and Santa Cruz groups. These run in two parallel lines, are basically continental in origin, and are surrounded by many smaller volcanic islands and both low and raised atolls. High humidity and malaria make the islands uncomfortable for European settlers. The rainfall is high. Dense rain forest is the typical vegetation, with mangrove swamps on some coasts.

How the People Live. The population (173,500) of the main islands is predominantly Melanesian, while Polynesians predominate on some peripheral ones such as Tikopia. There are also considerable Micronesian and European minorities. Malaita has half of the territory's population, but the main concentration is at the capital, Honiara (Guadalcanal), where an estimated 8,000 people lived in 1968. The Solomon Islands are being educated towards self-government. A unique experiment began in 1970 to give executive responsibility to a series of committees, as political parties had not yet emerged.

Copra is the only significant agricultural product. A little over half is grown by the islanders on small farms. The small-holdings are, however, more severely affected by seasonal fluctuations due to cyclones and dry spells than are the plantations, owned by large overseas companies.

Rice is now the most important subsistence crop in the Solomons. Protein in the diet is provided by fishing. Since 1965, the timber industry has expanded rapidly, mainly for the Japanese market. The kauri pine forests, especially on Vanikoro (Santa Cruz Group), are the most valuable commercial resource.

With so many scattered, outlying islands, transport is an immense problem. Air links are vital, but small trading vessels and canoes are more commonly used. Imports are primarily foodstuffs, manufactured goods and petroleum products. Copra and timber are exported to Britain, Japan and Australia.

WESTERN SAMOA

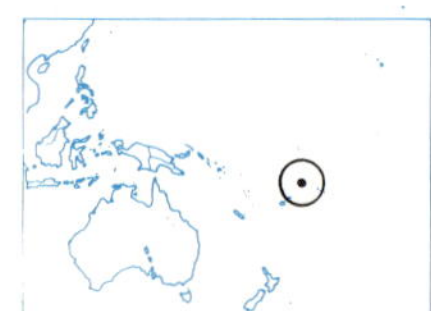

Western Samoa was the first Polynesian country to become fully independent (1962). Previously it had been administered by New Zealand under a UN trusteeship agreement, which still manages its foreign relations.

The country consists of two large islands of volcanic origin, the rugged and mountainous Savai'i and Upolu; two small islands, Manono and Apolima; and numerous uninhabited coastal islets. All are fringed by coral reefs. The highest point is Mt Silisili on Savai'i (6,095 ft; 1,858 m).

The tropical climate is moderated by the south-east trade winds, and temperatures are fairly constant around 77°F (25°C). Rainfall is heavy – at least 100 in (2,540 mm) annually in most areas, and more than 250 in (6,350 mm) in the highlands. Most of the islands are covered by dense tropical vegetation.

Because of the mountainous interior most of the 141,000 people live along the coasts, especially near Apia, the capital and chief port on the north coast of Upolu.

The chief export crops are copra, bananas and cacao, while subsistence crops include coconuts, taro and breadfruit. There are no known mineral resources, and the conservatively-inclined Samoans have restricted industry to small timber, soap and soft drinks establishments. New Zealand, Australia, Britain and West Germany are her chief trading partners.

AMERICAN SAMOA

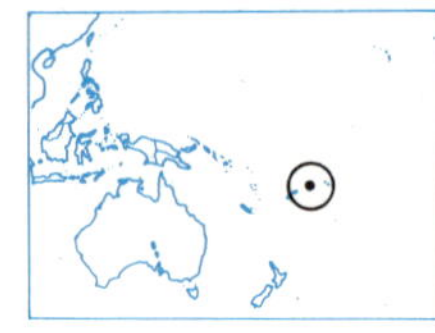

American Samoa, comprising the eastern islands of the Samoan group, has been under American control since 1899 and constitutes an 'unorganized unincorporated territory'. Covering some 76 sq mi (197 km²), it includes Tutuila Island (the largest) with the fine natural harbour of Pago Pago; the Manua group, and Swains and Rose islands. The population of 27,769 is almost entirely Polynesian, and most people are US nationals.

GILBERT AND ELLICE ISLANDS

The Gilbert and Ellice Islands, a British crown colony, are spread over a vast area of the western Pacific Ocean. The colony comprises Ocean Island; the Gilbert Islands; the Ellice Islands; Christmas, Fanning, Washington, Starbuck, Malden, Flint, Caroline and Vostock Islands in the Line Islands; and the Phoenix Islands (except Canton and Enderbury Islands, administered jointly by Britain and the United States). All, except Ocean Island, are low atolls with coconut palms, pandanus and breadfruit among the most common vegetation. The approximate total area is 396 sq mi (1,026 km²). The climate is warm (70–90°F; 21–32°C) and rainfall is often heavy, but can vary from year to year. The population of some 53,517 includes about 44,200 Gilbert Islanders. They speak Gilbertese and, like the Ocean Islanders, are Micronesians. The people of the other islands are Polynesians.

Subsistence agriculture and fishing are the main activities, with some commercial copra production. Ocean Island, an uplifted atoll, has high-grade phosphate deposits which will probably be exhausted by 1980. Following UN advice, many Ocean Islanders are now seeking a future as merchant seamen. The colony exports phosphates to Australia and New Zealand, and copra to Britain.

HAWAII

Hawaii, the only non-mainland state of the United States, consists of over 100 mid-Pacific islands. The Hawaiian Islands were first settled by Polynesians, probably from Tahiti, about the 5th century AD. In 1900 the islands became a US territory. The growing demand for statehood after World War II was recognized in 1959 when Hawaii became the 50th state of the Union.

A Chain of Many Islands. Hawaii consists of 132 islands stretched out in a chain in the north central Pacific with the principal islands – Hawaii, Maui, Oahu, Kauai, Molokai, Lanai, Niihau and uninhabited Kahoolawe – at the eastern end. These islands make up all but 3 sq mi (8 km²) of the state's total area of 6,450 sq mi (16,705 km²).

The islands are of volcanic origin and generally consist of an elevated central mass surrounded by a narrow coastal strip. Some of the larger islands have coral reefs. Those in the west are sand and coral atolls or mere shoals. The island of Hawaii is the largest and has the state's highest mountains, Mauna Kea (13,796 ft; 4,205 m), and Mauna Loa (13,680 ft; 4,170 m), an active volcano.

The most populous island is Oahu, on whose south coast plain stand Honolulu, the state capital and chief port, and the naval base of Pearl Harbor.

The lowlands are consistently warm or hot, with temperatures rarely below 60°F (16°C) in winter and rising to around 80°F (27°C) in summer. Rainfall, mainly in winter, varies from over 300 in (7,620 mm) on the north-eastern mountain slopes to less than 15 in (381 mm) on leeward south-western slopes. Thick forest with dense undergrowth covers the lower slopes of the larger islands, and there are many beautiful palms, ferns, orchids and hibiscus.

The People. About 82% of the population of 796,913 live on the island of Oahu. Intermarriage between the main groups (Polynesian, Chinese, Japanese and North American) has produced a people of mixed blood, but ethnic differences are still apparent. Honolulu (324,871), the state capital, is the largest city, and as a major port of call on trans-Pacific sea routes has attracted shipbuilding and oil refining industries. It also has sugar refineries, canneries and steel plants.

The Economy. As a major US base in the Pacific, Hawaii benefits from the earnings of service personnel and civilian staff – the state's major source of income. Little less important is tourism based on Waikiki, south-east of Honolulu, and newer resort complexes.

Agriculture also makes a major contribution. Sugar dominates and is grown on large plantations on the islands of Hawaii, Kauai, Maui and Oahu. Pineapples, grown on all the larger islands except Hawaii, are exported canned (as fruit or juice). Other crops include sisal, coffee, fruit and vegetables. Beef cattle are reared on the island of Hawaii. Apart from bauxite on Kauai, the state has few commercially attractive minerals.

Waikiki beach, most famous of Hawaii's beaches. In the background is Diamond Head, an extinct volcano recently saved from the plans of developers.

NEW CALEDONIA

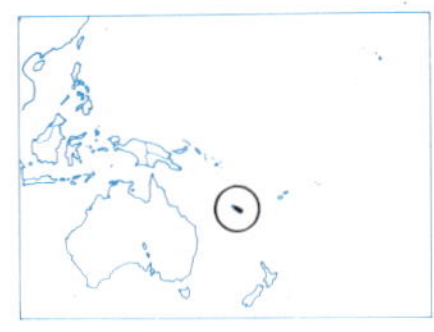

New Caledonia is the largest French possession in the Pacific and probably the only French overseas territory which can support itself. It lies east of north-eastern Australia and consists of the large island of New Caledonia, about 250 mi (400 km) long and 25 mi (40 km) wide, and many smaller islands – Île des Pins, the Loyalty, Huon and Chesterfield groups and others – the whole territory occupying some 7,374 sq mi (19,099 km²).

The island of New Caledonia is girdled by a coral barrier reef and has a mountainous interior rising to more than 5,000 ft (1,524 m) in Mt Panié. On the east coast the mountains rise sheer from the sea; in the west they descend to a broad alluvial plain. Rain forests blanket the eastern slopes, while araucarian or kauri pine, and the rather drab *niaouli*, which resembles the scrub eucalyptus of Australia, are common west of the mountains. Much forest has been cleared for pasture. The east coast is very humid, while the west coast lies in a partial rain-shadow. Annual rainfall averages 42 in (1,070 mm) at Nouméa on the south-west coast, but more than 130 in (3,300 mm) on the east coast. Temperatures at Nouméa range from 57°F (14°C) in July to 91°F (33°C) in November.

The People. Nearly half the population of over 100,000 is Melanesian; nearly one-third are European (mainly French). There are also some 6,000 Vietnamese and Indonesians, brought in to work on the plantations and in the mines and towns, and over 9,000 Tahitians and Wallis Islanders, who are temporary immigrant workers. Most of the Europeans and Asians live in the capital, Nouméa (about 58,000), which is also important as the headquarters of the South Pacific Commission. Most of the Melanesians live in tribal reserves, subsisting on yams and taro.

The Economy. Minerals are the principal natural resource, the most important being nickel, which is processed at Doniambo near Nouméa. Chrome, iron ore and manganese are also mined. Hydro-electric power is provided by the Yaté Falls on the east coast.

The main export crops are coffee, grown in the valley bottoms, and copra, produced on the east coast and in the Loyalty Islands. Crops of rice, maize and vegetables are insufficient to meet local needs owing to a shortage of farm labour.

Because of its nickel exports, New Caledonia enjoys a favourable balance of trade. New Caledonia's chief trading partners are France, Australia, the USA and Japan.

NEW HEBRIDES

The New Hebrides, a double chain of 13 large and 80 small islands in the south-west

Pacific, is a condominium ruled jointly by Britain and France since 1906. In the 18th century, the New Hebrides were the haunt of European pirates, and wanted men on the run. The indigenous Melanesian population, despite having been decimated during these years, makes up the overwhelming majority.

Joint European control is responsible for many of the condominium's problems, for neither administration has been eager to initiate reforms, and at times conflicting policies have been pursued. But the 1970 five-year development plan held promise of much-needed progress.

The islands have a total area of 5,700 sq mi (14,760 km²), the largest being Espíritu Santo. They are mountainous and densely-forested, with raised coral terraces on all eastern slopes. There are three active volcanoes. Earth tremors are frequent. The annual average temperature is 75°F (24°C), and rainfall at Vila, on Efate, averages 81 in (2,055 mm). The islands are malarial.

The People. The population of some 78,000 includes more than 72,000 Melanesians, nearly 4,000 French and about 1,600 British. Vila – the capital, has about 3,000 inhabitants. English and French are both official languages and dialects, and Pidgin serves as the *lingua franca*. Missionaries have been active since 1848, and there are Presbyterian, Anglican and Roman Catholic communities, although primitive cults persist in the interior.

The Economy. The major cash crops – copra, cocoa and coffee – are produced on European-owned plantations on the coastal plains of Efate and Espíritu Santo. Subsistence crops include yams, taro, manioc and bananas. Pigs are very important in the native economy, and sheep are reared in the savanna areas of Eromanga. This island and Aneityum also provide kauri pine timber. Fishing is mainly a subsistence activity, but Espíritu Santo has a tuna freezing works supplying the Japanese market. Manganese is mined at Forari on Efate. Trade is mostly with Australia and France.

FRENCH POLYNESIA

French Polynesia is formed of islands spread across a wide area of the eastern Pacific. Annexed in 1887, and since 1946 an overseas territory of France, they form five groups: the Marquesas, Tubuai (Austral), Gambier and Society Islands and the Tuamotu Archipelago. Their total area is estimated at 1,544 sq mi (4,000 km²), of which a quarter makes up the island of Tahiti in the Society group.

Except for the Tuamotu Archipelago, which consists of coral atolls, all are high volcanic islands with jagged crests, deep gorge-like valleys and swiftly-flowing streams. Beautiful Tahiti, where Gauguin lived and painted, has twin peaks rising to 7,377 ft (2,238 m). Most of the islands have coral barrier reefs and lagoons. The climate is relatively hot and humid, and annual rainfall averages 73 in (1,852 mm). Temperatures vary because of the north-south spread of the islands; the Gambier group is too cool for coconut palms to grow. Dense rain forest is the usual vegetation, except on the porous atolls, covered with coconut groves.

The Islanders. The population, nearing 120,000, is largely made up of tall, light-skinned Polynesians, most of whom live in small villages. More than 60% of the people live on Tahiti, whose main town, Papeete (79,500), is the territorial capital. Here, too, live most of the Europeans (mainly French) and Asians (mainly Chinese). French is the official language, but Polynesian languages are also spoken. There are about 70,000 Protestants and 35,000 Roman Catholics.

The Economy. Taro, yams, sweet potatoes and breadfruit are grown for food. The main commercial crops are vanilla, coffee and copra; pepper, sugar-cane and citrus fruits have been introduced. The lagoons and seas abound in fish, which form the basic diet of the Polynesians, who are skilled sailors and fishermen. Pearling and the gathering of mother-of-pearl shell are commercially important in the Tuamotu and Gambier groups. The high-grade phosphates mined on Makatéa in the Tuamotu Archipelago were exhausted in 1966. Tourism is of growing importance, especially on Tahiti, Bora-Bora and Mooréa.

The atoll of Mururoa in the Tuamotu group is the site of French atomic tests.

PITCAIRN ISLAND

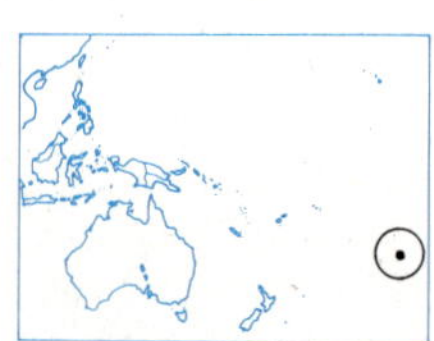

Pitcairn Island is a tiny British colony in the South Pacific comprising inhabited Pitcairn Island and uninhabited Henderson, Ducie and Oeno islands. In 1790 Pitcairn Island was occupied by 10 mutineers from Captain Bligh's ship *Bounty*, who brought with them 18 Tahitians.

Their English-speaking descendants, numbering about 82, still live on this cliff-girt, fertile tropical island of only 1·75 sq mi (4·6 km²), Adamstown being their settlement. They live by subsistence fishing and farming and sell fruit, vegetables and wood carvings to passing ships.

WALLIS AND FUTUNA ISLANDS

Wallis and Futuna, a French overseas territory north-east of Fiji, comprises the 23 reef-encircled Wallis Islands (106 sq mi; 275 km²), mostly hilly and wooded; and the Futuna, or Hoorn, Islands (56 sq mi; 145 km²) consisting of the densely-wooded islands of Futuna and Alofi. The population of both groups (about 9,900) is mainly Polynesian. Copra is the chief product.

GUAM

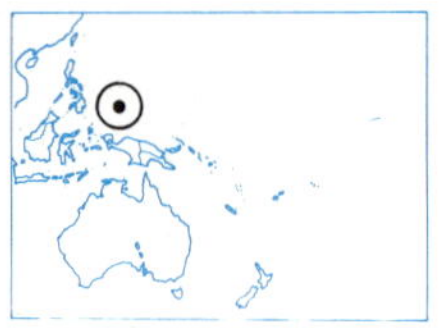

Guam, a strategically important island in the west Pacific and the largest and southernmost of the Marianas, is an 'unincorporated territory' of the United States and a major air and naval base, which saw fierce fighting during World War II.

Land and people. Covering 212 sq mi (549 km²) and fringed by coral reefs, the island is volcanic in origin. The north is a coral limestone plateau, and the south hilly. The climate is tropical, with temperatures around 81°F (27°C) and an annual rainfall of about 70 in (1,778 mm). Typhoons sometimes cause widespread damage, and the island is also earthquake-prone. Coconut groves cover much of the land.

The mixed population of 86,926 includes about 30,000 mainly Indonesian-Spanish Chamorros and more than 20,000 transient residents on military or government service. Agana (2,100) is the capital, and Apra Harbor the chief port. Nearly all Guamanians are Roman Catholics. The official language is English.

Economy. Crops of many kinds are grown, and livestock, including pigs and cattle, reared. Fishing is also of local importance, and a tourist industry is being developed. Guam is the only 'free trade' US territory, only tobacco, liquid fuel and spirits being subject to excise duties. By far the most important factor in the economy, however, is the military base.

Other American Outposts in the Pacific. Midway comprises two small islands forming part of an atoll some 1,200 mi (1,930 km) north-west of Honolulu. It has a population of 2,356 and, like Guam, is an air and naval base. In 1942 it was the scene of a major air/sea battle in which United States forces decisively defeated a Japanese invasion.

Wake Island, in the west-central Pacific, is an atoll comprising three islets (Wake, Wilkes and Peale) and a population of 1,978. Like Guam it was occupied by the Japanese in World War II, and it is now an important civil aviation base.

Johnston Island lies 700 mi (1,127 km) south-east of Hawaii and is mainly a dumping ground for old munitions and, since 1970, a storage centre for nerve gas. Privately-owned Palmyra Island, like Howland, Baker and Jarvis Islands, and Kingman Reef, is uninhabited.

TRUST TERRITORY OF THE PACIFIC ISLANDS

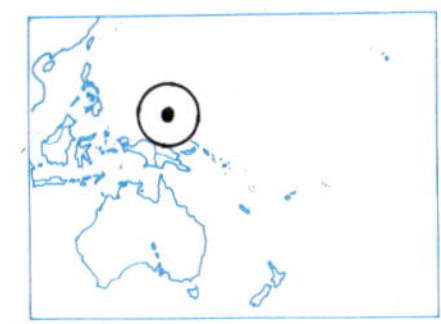

The Trust Territory of the Pacific Islands, administered for the United Nations by the United States, comprises the Caroline, Marshall and Mariana island groups (excluding Guam). These total 2,141 islands and atolls scattered across 3,000,000 sq mi (8,000,000 km²) of the Pacific Ocean. Only 96 of the islands are inhabited, and the population of over 100,000 is Micronesian. Saipan, in the Marianas, is the administrative centre.

ANTARCTICA

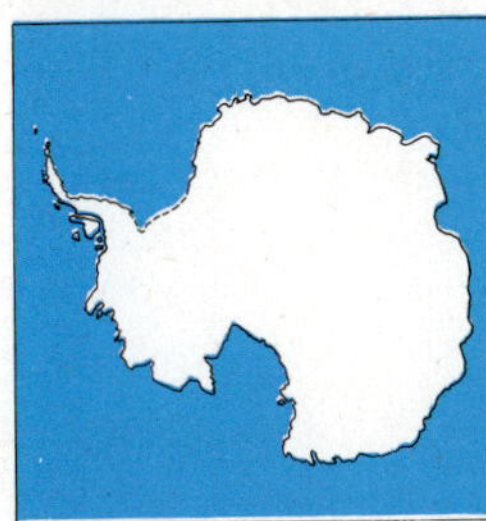

Antarctica, centred asymmetrically on the South geographical Pole, is the fifth largest continent. It is also the world's coldest and most desolate, and the only continent with no permanent human inhabitants. Snow and ice averaging nearly 1 mi (1·6 km) thick, cover most of the land. Antarctica is surrounded by the partly ice-covered Southern or Antarctic Ocean (formed by the southern parts of the Atlantic, Indian and Pacific Oceans) whose northern boundary approximates to a line joining the southern tips of the surrounding continents. The emptiness of this ocean waste is relieved only by a few scattered islands, again largely ice-covered. To the north of the ice edge island groups with somewhat less rigorous climates lie in a region known as the Sub-Antarctic.

There are a number of boundaries which geographers use to demarcate the Antarctic proper. First there are political boundaries at latitude 60°S; then there is the Antarctic Circle at 66°33′S, not a very meaningful geographical concept as there are many truly Antarctic characteristics well to the north of it. The boundary which is generally accepted on scientific grounds is an oceanographic one – the Antarctic Convergence, a belt of water 20–30 mi (30–50 km) wide, which girdles the Southern Ocean between latitudes 50°S and 62°S. The Antarctic regions may be defined as the lands and seas south of the Antarctic Convergence, namely: the continent Antarctica, the southern half of the Southern Ocean, the South Shetland Islands, the South Sandwich Islands, the South Orkney Islands, South Georgia, Bouvet Island, Heard Island, the Balleny Islands, Scott Island and Peter Island. Islands to the north of this boundary are classed as Sub-Antarctic, though some have many Antarctic characteristics.

The continent of Antarctica covers about 5 million sq mi (14·3 million km²). It has a coastline of some 18,500 mi (30,000 km). Antarctica is the most remote of the continents, being some 600 mi (960 km) from the southern tip of South America at its nearest point, the Antarctic Peninsula, and 2,500 mi (4,000 km) distant from both South Africa and New Zealand. Roughly circular in shape, Antarctica is almost entirely within the Antarctic Circle, its regularity broken only by the two great indentations of the Ross and the Weddell seas. Two minor seas are the Amundsen and the Bellinghausen.

Floating Ice. Surrounding Antarctica is the Southern Ocean which in the winter months is frozen to depths exceeding 9 ft (3 m). The maximum extent of the pack-ice occurs in August or September. During this period, the area of the ice-continent is virtually doubled. In summer, Antarctica is surrounded by a broken fringe of pack-ice which may be separated from the coast by a belt of water a few miles wide. In the Ross Sea the ice may disappear almost entirely, and even parts of the notoriously ice-ridden Weddell Sea may be virtually open during this season. The pack-ice is never entirely still, winds and currents keeping it constantly on the move, usually from east to west.

The passage of ships to and from Antarctica is not a major problem since sufficient is known about these seasonal variations in ice distribution to ensure that vessels reach their destination without being ice bound.

Land Ice. Very little of the land of Antarctica is visible on the surface. Over many thousands of years snow has slowly piled up, compacting into a mass of ice that now covers nearly 95% of the land beneath. Averaging 6,000 ft (2,000 m) in thickness and exceeding 14,000 ft (4,700 m) in some places, the Antarctic ice-sheet represents 90% of all the ice in the world. Were it to melt, world sea levels would rise by some 200 ft (66 m), inundating ports and low lying regions.

The ice sheet is by no means inert. Its viscous and static properties enable it to flow and bend under pressure. The enormous weight of the ice causes it to flow slowly outwards and downwards towards the coast as a result of acceleration due to gravity. The rate of flow is only a few yards a year. As the ice nears the coast its velocity increases, its

Icebreakers are essential in the Antarctic but even these powerful ships can be overwhelmed by the ice. The USS Atka (behind) is here shown before returning to New Zealand for repairs to a damaged propeller.

thickness decreases, and eventually it reaches the sea either as an ice front or in the form of glaciers, ice-streams or ice-shelves.

Where the inland ice-sheet begins to thin out near the coast, isolated peaks, called *nunataks*, or even whole mountain ranges, may protrude through the ice channelling it into relatively fast-flowing glaciers. Some of the most spectacular of these are in Victoria Land and include Beardmore Glacier, 100 mi (160 km) long and averaging 12 mi (19 km) wide, up which explorers Shackleton and Scott toiled en route to the South Pole. On the other side of the continent, in MacRobertson Land, is the Lambert Glacier, probably the world's largest valley glacier. In some places rivers of ice debouch into the sea.

Over 10% of the Antarctic glacier sheet does not terminate at the coast but spreads out over the sea as floating ice-shelves. These are partly fed by moving glaciers descending from the inland ice-sheet, but their principal source of nourishment is the accumulation of snow on their surface. Sometimes larger areas of the seaward edge of these ice-shelves break away and float out to sea as giant flat-topped or tabular icebergs. One of the largest ice-shelves is the Ross Ice Shelf discovered by Captain James Clark Ross in 1841. It covers an area of some 310,000 sq mi (806,000 km^2) and varies in thickness from 600 ft (200 m) at its seaward edge to 2,000 ft (670 m) at its junction with the inland ice-sheet.

The surface of the Antarctic ice-sheet is far from being a smooth, featureless dome: great terraced steps or wavelike undulations reflect most probably the outline of the landscape buried thousands of feet below the surface. Other features, such as crevasses, are often not obvious until viewed from the air. Some of the most heavily crevassed areas are in regions where an ice-shelf meets the land ice or in fast-flowing valley glaciers. On the inland ice-sheet, a crevassed area may be many miles in extent with individual crevasses extending 100 ft (30 m) in depth and 50 ft (15 m) or more across. Smaller crevasses, often invisible at ground level because of their snow cover, can present a deadly hazard to the traveller.

Other small-scale ice surface features caused by the action of wind on snow include the wave-like dunes called *sastrugi*. Often reaching 6 ft (2 m) in height, they are formidable obstacles to land transport.

The Land. Though ice is the predominant feature of the landscape, there are small areas of exposed rock, which provide the only direct evidence of the structural history of the continent underlying the ice. The two main regions of Antarctica where rock outcrops are most abundant are Victoria Land and the Antarctic Peninsula.

In Victoria Land, the Royal Society Range provide a magnificent backcloth to the historic McMurdo Sound region. Flanked by the Admiralty Range and the Queen Alexandra and Queen Maud ranges, they form a section of one of the world's greatest mountain chains, the Transantarctic Mountains, which stretch from Victoria Land to Coates Land, a distance of some 2,000 mi (3,200 km). Ringing the continent are numerous other mountain ranges. In the Australian Antarctic Territory lie the high peaks of Enderby Land which include the Thula Range, the Prince Charles Range of MacRobertson Land and the Vestfold Hills of Princess Elizabeth Land. The Antarctic Peninsula, in particular the west coast and the offshore islands, provides some of the most magnificent scenery in the whole continent with fjords and mountains similar to those of northern Norway. In Marie Byrd Land lie the Ellsworth Mountains whose Sentinel Range contains the highest peaks in Antarctica – Mt Tyree at 16,290 ft (4,968 m) and the Vinson Massif (16,860 ft; 5,142 m).

During summer, a few areas in the coastal regions of Antarctica are partly or wholly ice-free. They are known as oases, though the term is a relative one as they have exceedingly desert-like climates with low precipitation and low summer temperatures. Oases are valuable places in which to study processes such as erosion of rocks and are of special interest for their unique saline lakes. The largest is the McMurdo Oasis in Victoria Land, 93 mi (150 km) long and between 9 and 15 mi wide (15–25 km).

Antarctica is divided into two areas by the Transantarctic Mountain chain. The larger of these two areas is termed Greater or East Antarctica. Here the rocks are among the most ancient found on earth, having crystallized some 600 million years ago, and forming a shield comparable with that of northern Canada.

The rock structure of the smaller area, called Lesser or West Antarctica, differs in many ways from Greater Antarctica. Here the mountains tend to be more recent in origin and there is much evidence of volcanic activity. On Ross Island, in McMurdo Sound, Mt Erebus is a dormant volcano – as is Mt Melbourne on the coast of Victoria Land. In 1967 an unexpected volcanic eruption on Deception Island, one of the South Shetland group off the Antarctic Peninsula, caused the speedy evacuation of several long-established scientific stations.

Geologically the peninsula resembles the Andes of South America, 600 mi (970 km) to the north-west, of which it is probably a continuation. If its ice cover were removed, Lesser Antarctica would appear as a region of low-lying plains and numerous archipelagos and fjords. Evidence for this hypothetical picture of the sub-ice continent is derived from deep drilling to the bed-rock and by seismic and radar sounding.

Climate. Observations from Antarctic land stations date only from the beginning of the present century and these, for the first 50 years or so, were limited to coastal regions. Not until the International Geophysical Year (IGY) of 1957–58 was a fully coordinated network of weather stations established around the coast, as well as inland. Their reports, analysed at the International Antarctic Meteorological Research Centre in Melbourne, Australia, are incorporated in daily synoptic charts covering the whole of the Southern Hemisphere. As well as providing day-to-day information on Antarctic weather, these charts are the basis of long-term climatic studies.

Antarctica is colder than the Arctic because of its great average height – over 6,000 ft

Left: The arrival of a supply ship in Antarctica is an eagerly awaited event. *Right:* Drifted snow around Williams Field (US) on Ross Island.

Bourgeois Fjord is cut into the Antarctic mainland. The mountains in the background are on Adelaide Island and rise to a height of 7,000 ft.

(2,000 m) above sea level. This fact alone would make Antarctica 22°F (12°C) colder than the mainly sea-level Arctic. In winter, the freezing of the surrounding ocean effectively doubles the area of the continent, removing it even further from the warming influence of open water. By contrast, the Arctic regions consist of a large ocean whose underlying currents of relatively warm water have a moderating effect on the coastal zones of the surrounding continents.

Antarctica ranks as one of the world's great deserts. Above the high inland ice-sheet, the air is extremely dry and precipitation consists entirely of fine granular snow whose annual average water equivalent has been calculated at only about 2 in (55 mm).

The world's record low temperatures all originate in Antarctica. At the Soviet station Vostok, over 10,000 ft (3,040 m) above sea level, a record low temperature of −126°F (−88°C) was recorded in 1960.

Atmospheric Phenomena. The very dry and dust-free atmosphere of Antarctica is almost completely without the haze which usually shrouds the view in lower latitudes. Mountains have been sighted at a distance of 300 mi (480 km) and distance judging is difficult. Mirages are not uncommon. Light reflected and refracted from millions of tiny ice-crystals in the air can cause luminous spots and arcs in the air known as parhelia or mock suns and mock moons (parselenae). While fog is not a major problem in Antarctica, blowing snow or blizzards can obscure everything for days on end. Another optical hazard is 'whiteout', when the normal contrast due to light and shade is absent and it is impossible to judge the rise or fall in the landscape. In such conditions it is easy to stumble into a crevasse or fall over an ice cliff.

One of the most beautiful and spectacular phenomena is the aurora – glowing patterns in the night sky very occasionally seen in middle latitudes but more often occurring in the Antarctic and Arctic.

Scientific stations study auroral phenomena. Another related study is the observation of cosmic rays, especially those originated from the sun. Because of the shielding effect of the earth's magnetic field, it is only in the Antarctic and Arctic that these lower energy particles can reach the earth's surface.

Wildlife. The low temperatures and geologically long separation of Antarctica from other continents explain why no land mammals, reptiles or amphibians live there today. The largest land animal on the continent is a wingless fly found on the west coast of the Antarctic Peninsula. But bordering the continent is a sea exceedingly rich in chemical nutrients – nitrates, carbonates, phosphates and silica. These minerals, together with sunlight, nourish billions of tiny plants collectively named phytoplankton, which in their turn feed shoals of tiny animals or zooplankton. Typical of the zooplankton is the red shrimp-like crustacean known as krill which is the principal food of the numerous fish, penguins and other sea birds, seals and baleen whales which inhabit the Southern Ocean.

Living near the ocean floor is an abundance of sea life collectively called benthos. Common creatures in this grouping include sponges, sea-urchins, sea anemones, starfish and various marine worms. There are also molluscs including mussels, limpets, octopuses and squids.

Antarctic fish are few in species and most belong to the group of Nototheniform fishes and include the Antarctic dragon-fish, plunder fish, Antarctic cod and ice-fishes.

Most obvious of the animals round the Antarctic coastline are the birds. Against the 120 or so species breeding in the Arctic, Antarctica (the area south of the Antarctic Convergence) can claim fewer than 50, and only 16 on the continent itself. But this bird

A number of King penguins on Grace Glacier, South Georgia.

life makes up in sheer numbers what it lacks in variety. The marine Antarctic birds belong to five main groups: penguins; petrels, skuas, gulls and terns; shags; and sheathbills. Almost all migrate in winter to the edge of the expanding sea-ice that would otherwise cut off their supply of food. The Emperor Penguin is unique in not only coming ashore for the Antarctic winter but breeding in it: laying a solitary egg on shelf ice and incubating it upon its feet beneath an abdominal skin fold.

Inhabiting the pack-ice off the coast are four species of seal, all belonging to the group known as true seals. Most southerly is the Weddell seal, perhaps the best studied of all Antarctic seals. It is known to be able to dive to depths approaching 2,000 ft (660 m) and to remain under water for up to 40 minutes at a time. Breeding on the islands of the Antarctic north of the pack-ice is the largest member of the true seals, the southern elephant seal; it can weigh up to 2–3 tons.

Those seals were at one time commercially important as a source of oil. Also of former commercial importance were the fur seals, members of the second family of seals – the eared seals.

The other group of marine mammals associated with the Antarctic are the whales which are migratory in their habits, leaving the Antarctic for more northerly waters during the winter. The largest Antarctic species is the blue whale – which can measure over 90 ft (30 m) in length and weigh up to 150 tons (152 metric tons). Other whales in the Southern Ocean belong to the family of toothed whales which have teeth in place of baleen 'strainers'.

Plants. Antarctic land plants are small in size and restricted in species. The high plateau is a cold desert where only hardy lichen finds a hold on exposed rock surfaces. At lower altitudes in coastal regions, the brief summer growing season permits the growth of a few mosses and liverworts. Farther north, in the northern half of the Antarctic Peninsula, two small flowering plants, a grass and a pearlwort have been recorded. The Sub-Antarctic islands still farther north, have a more favourable climate and can boast 30 or so flowering grasses and small flowering plants and ferns.

Man in the Antarctic. Exploration of the Antarctic proper can be said to date from Captain James Cook's second circumnavigation of the world (1772–75), when the Antarctic Circle was crossed for the first time and a farthest southern latitude of 71°S achieved. Later landmarks in discovery included Charles Wilkes' 1,500 mi (2,400 km) voyage along the coast in 1838 which established Antarctica as a continent; discovery of the South Pole by Roald Amundsen in 1911; and the first transantarctic crossing (by air) by Lincoln Ellsworth in 1935.

Up to World War II Antarctic exploration was essentially national in character and tended to lack co-ordination and continuity. The first real international attempt to complete the exploration of Antarctica and coordinate scientific programmes dates from the International Geophysical Year (IGY) of 1957–58 when 12 nations set up stations round the coast of Antarctica and at the South Pole itself.

At the present time Antarctica has a population of 800 or so scientists in winter and more in summer, when tourists help to bring the total population to around 2,000. But there are no settled communities on Antarctica. Stanley, capital of the Falkland Islands, has around 1,000 settlers and South Georgia a mere handful of scientists and administrators.

Of the 40-odd stations in the Antarctic only the United States' McMurdo Station on Ross Island in Victoria Land can rank as a township. Uniquely sophisticated for Antarctica, the station includes church, shop, living quarters, workshops and scientific laboratories powered by a central nuclear reactor.

Since the early 1950s, Antaractic transport has been revolutionized by transport aircraft equipped with skis and capable of flying-in bulk fuel and all the equipment needed to establish a complete scientific station. Though there are virtually no roads in Antarctica, a wide range of tracked vehicles is available to pull loaded sledges or caravans for hundreds of miles over the ice-sheets. Dog-teams still have their uses in some areas.

Economic Resources. Of the two staple Antarctic industries based on the riches of the sea, sealing is not currently profitable and whaling is in rapid decline. The Antarctic fur seal industry began in the late 18th century, but by the 1830s the stocks of fur seals were exhausted and the sealers switched to the elephant seals, valued largely for their oil. These in their turn were hunted almost to extinction. Both species are currently reviving as a result of careful conservation.

The expansion of the Antarctic pelagic whaling industry since the 1930s, based on floating factory ships attended by highly mobile catchers, increased the need for rigidly enforced international control. A future world shortage of edible protein might be combated by using the Southern Ocean's vast stocks of plankton.

The shield area of Greater Antarctica includes many potentially valuable metals: chromium, gold, copper, lead, manganese, molybdenum and zinc. In Lesser Antarctica coal-bearing strata are exposed. The cost of mining ore from the frozen ground and shipping it to distant parts of the world through ice-infested seas, makes commercial production appear unlikely, although Antarctic minerals could provide a reserve against future shortages.

Who Owns Antarctica? Claims, based on priority of discovery and continuity of occupation and administration, show on the map as segments with their apexes at the South Pole. The present claimants are Australia (Australian Antarctic Territory, 45–160°E excluding Adélie Land); France (Adélie Land, 136–142°E, part of the French Southern and Antarctic Territories which include St Paul Island, Amsterdam Island, and the Kerguelen and Crozet Islands – all in the southern Indian Ocean); New Zealand (Ross Dependency, 160°E–150°W); Chile (Chilean Antarctic Territory, 90–53°W); the United Kingdom (British Antarctic Territory, 80–20°W, including the South Shetland and South Orkney Islands); Argentina (Argentine Antarctic Territory, 74–25°W); Norway (Queen Maud Land, 20°W–45°E). British, Chilean and Argentine claims overlap.

The USA has long been active in unclaimed Marie Byrd Land but has made no claims to it, nor does she recognize other political claims. The Soviet Union likewise recognizes none of these claims but maintains numerous stations in Antarctica. Before the IGY in 1957–58, scientific co-operation between the various nations had often been frustrated by these problems of sovereignty.

In 1959, a meeting of all the interested powers was convened in Washington DC and an Antarctic treaty was signed by 12 interested powers which froze all political claims for 30 years and guaranteed the use of the continent for peaceful scientific purposes only.

INDEX

The page numbers in *italics* refer to illustrations.

ACKNOWLEDGMENTS

Aspects of Geography, the introduction was written by John David Yule, M.A. (Cantab.). The main text is an abridged section of *The World and Man* an encyclopedia of geography five times larger in size to which 200 specialist authors contributed under the chief editorship of Professor Emrys Jones.

The population figures are based on census figures or the latest reliable estimate derived from official sources.

Diagrams including thematic maps were made by Diagram Ltd based on drafts by Peter Hutchinson.

Maps (topographical) were produced by the photographic department of Elsevier Nederland B.V., Amsterdam.

Photographs, agents and photographers (with their agents where applicable) are listed alphabetically:

J Allan Cash: 98, 181, 182, 264(bottom), 279, 371
Ambassador College, St Albans, Herts., 32
Amsterdam Harbour Authorities: 48
M Andrews: 328, 348, 367
Association Nederland, USSR: 127, 128, 133
Australian News and Information Bureau, London: 31, 376, 377, 380, 381, 384
Barnaby: 152, 172, 190, 194(bottom), 224, 226, 304, 327, 366
Bergmans: 50, 57, 81, 82, 84, 87, 100, 104, 105, 119, 161, 194(top), 218(top right), 246, 263(top), 264(top), 319, 334, 351
Berkey K & L New York: 232
M & E Bernheim (S Griggs): 251, 254, 259
Botswana High Commission: 282(top)
J Burton: 33
Camera Press: 6(left) Iguaçu falls, Brazil
Colorific: 131, 241, 282(bottom), 311(bottom), 311(top right), 357, 391
Colour Library International: 46, 144, 207, 303, 307, 309, 335
Danish Tourist Board: 61, 62
D Davies: 65, 68
P T Denwood: 185
J Drysdale: 228
M Edwards: 146, 177
Elsevier, Nederland B.V.: 6(right), 8(centre), 9, 13, 14, 15, 16(3), 17(2), 28, 34(9), 42, 44, 51, 56, 59, 69, 70, 72, 73, 75, 81, 92, 107, 108, 109, 114(2), 115(2), 116, 117, 118, 125(2), 150, 153, 158, 179, 188(2), 189, 191, 192, 193, 200, 201(2), 209, 222, 224, 234, 250, 253, 273, 274, 284, 292, 293, 298, 299, 305, 308, 315, 321, 322, 331, 333, 350, 353(right), 363(2), 364, 393, 394(2), jacket front flap
V Engelbert (S Griggs): 7(centre), 135, 218(top left), 252
Fotofass: 180, 202(bottom)
Fiat: 39, 109
Professor C A Fisher: 220
Florida Development Commission: 292
Foto Leimbach: 361
Fotolink: 23, 66, 166, 217, 258, 260, 278(bottom), 288, 301
D J Fox: 330
M Freeberne: 197, 198, 204, 387(bottom)
French Travel: 95
German Institute, London: 78
German Travel, Amsterdam: 77, 79
G Hall (S Griggs): 320
S Halliday: 151
Dr. F E I Hamilton: 83
R Harding: 2–3 (Deneki Lakes, McKiney Park, Alaska), 4–5 (salt extraction; photo Jon Gardey), 102(bottom), 148, 154, 168, 176, 215(bottom), 244, 263(bottom), 271(bottom), 293(top), 302, 310(bottom), 313, 385, 387(top), back of jacket (Mt. Russell, Alaska range, Alaska), back flap of jacket (fishing Besar village, Malaysia; photo Robert Cundy)
Drs. A A M Van Der Heyden: 8(left), 108(right), 157(2), 242, 243, 245, 287, 311(top left), 316, 324, 325
R Hill: 205
D Hilling: 257, 266, 267
Idaho Department of Commerce: 317
Israel Government Tourist Office: 156(right)
Jewish National Fund, London: 156(left)
Kees Scherer: 55, 237, 337
Keystone: 197, 203, 249, 290, 318(top)
Dr. R King: 160
Las Vegas News Bureau: 312(top)
M A Lowenthal: 7(left), 41
Manitoba Department of Industry and Commerce: 300
J Massey Stewart: 129, 130
L van der Meer: 120
D Moore (Colorific): 8(right), 373
NASA (Colorific): front jacket
Natural Science Photographs: 162
New Brunswick Travel Bureau: 305(top)
Dr. R C Y Ng: 213, 214
Ontario House: 366
Oregon State Highway Department: 310(top)
J Pickerell (Colorific): 210
Picturepoint: 5(above) (Manhattan from East River), 53, 80, 85, 113, 175, 212, 248, 255, 283, 318(bottom), 328(top), 352, endpapers (Antarctica)
Popperfoto: 86, 96, 97, 167, 211, 219, 269
Reflejo (S Griggs): 101, 102(top), 364(bottom right), 369, 370
C Rentmeester: 221, 223
J Rosen (P Phipp): 58
Scan-Globe, Copenhagen: front of jacket(right)
Servicio Editoriale Fotografico: 89, 112, 230, 262, 276, 278(top), 281, 314
Spectrum: 110, 121, 123, 134, 164, 171, 184, 202(top), 277
Professor R. W. Steel: 271(top)
Dr. K Sutton: 7(right), 227, 238
Swiss National Tourist Office: 94
Dr. C Swithinbank: 395
N. Taylor: 195, 332
Thai Information Service: 215(top)
Varig Brazilian Airlines: 355
R W Vaughan: 396
R Waller (Associated Freelance Artists): 178
A Warren (Ardea): 342, 353(left)
Dr. A Warren: 239, 240
C Waterson (Foto Leimbach): 341
R Wood: 169, 186, 187
A Woolfitt (S Griggs): 91, 312(bottom)
I Yeomans (S Griggs): 63